Frontiers in Entomopathogenic Fungi

Sunil Kumar Deshmukh
Kandikere Ramaiah Sridhar
Gurmeet Kour Bali • Belle Damodara Shenoy
Editors

Frontiers in Entomopathogenic Fungi

Diversity, Ecology and Bioprospects

Springer

Editors
Sunil Kumar Deshmukh
Research and Development Division
Greenvention Biotech Pvt. Ltd.
Uruli-Kanchan, Maharashtra, India

Kandikere Ramaiah Sridhar
Central Research Laboratory, NITTE (Deemed to be University)
KS Hegde Medical Academy (KSHEMA)
Mangalore, Karnataka, India

Gurmeet Kour Bali
Microbial Type Culture Collections
CSIR-Institute of Microbial Technology
Chandigarh, Punjab, India

Belle Damodara Shenoy
Regional Centre Visakhapatnam
CSIR-National Institute of Oceanography
Visakhapatnam, Andhra Pradesh, India

ISBN 978-981-95-6528-3 ISBN 978-981-95-6529-0 (eBook)
https://doi.org/10.1007/978-981-95-6529-0

This Springer imprint is published by the registered company Springer Nature Singapore Pte Ltd.
The registered company address is: 152 Beach Road, #21-01/04 Gateway East, Singapore 189721, Singapore

Preface

Fungi are a relatively understudied group, although they have great biotechnological potential. They are useful in agriculture, pharmaceutical, food, textile, leather, paper, and beverage industries. They play a promising role in environmentally friendly approaches such as biological control and bioremediation. Insects being the most populous group among eukaryotes, their diversity has a profound influence on many ecosystem functions. Research has focused on their negative impact on human society, especially serving as vectors for many diseases and destroying agricultural crops and stored products.

Entomopathogenic fungi (EPF) are globally known but less studied pathogens, which are capable of infecting and controlling insect populations. EPF are phylogenetically diverse heterotrophic eukaryotes perpetuated by sexual and/or asexual phases. There are about 750–1000 species of EPF classified in about 100 genera. They belong to six fungal taxa: Ascomycota, Basidiomycota, Blastocladiomycota, Chytridiomycota, Entomophthoromycota, and Microsporidia. Some of the prominent genera of EPF are *Beauveria*, *Isaria*, *Lecanicillium*, *Metarhizium*, *Paecilomyces*, and *Trichoderma*. The lifestyle of EPF depends on several environmental factors, like temperature, oxygen, humidity, ultraviolet radiation, and minerals. The dual lifestyle of EPF as endophytes, capable of colonizing live tissues of plants, enhances plant growth and protects against herbivores, which has become a new area of crop protection. *Beauveria*, *Metarhizium*, and *Lecanicillium* are among the most common endophytic fungi that colonize plants, thus protecting them from insect herbivores. There are about 170 fungal species designated as mycopesticides, as they are generally regarded as safe (GRAS) for commercial applications. They are useful as biopesticides in agriculture as harmless alternatives to toxic insecticides and pesticides. To understand the global distribution, ecological importance, and host specificity, an open-access database, EntomoFun 1.0, was launched in 2024 (https://doi.org/10.5061/dryad.1zcrjdg17). This database facilitates the identification of gaps in our knowledge and directs future research on EPF.

The book *Frontiers in Entomopathogenic Fungi* deals with diversity, distribution, ecology, phylogeny, bioactive metabolites, agricultural applications as biopesticides, pest management, biosafety, and futurology in five parts: (1) The first part, in five chapters, records global distribution, their taxonomy, molecular phylogeny, coevolution, and ecological roles; (2) in the second part, bioactive compounds, enzymes, mycotoxins, and pharmaceutical applications are dealt with in four chapters; (3) three chapters in the third part cover agricultural applications as biocontrol agents, protection of *Agave* spp. in Mexico, and the role of endophytic EPF in soil fertility and pesticide degradation; (4) the fourth part deals with the compatibility of biocontrol agents, endophytic EPF, and the management of insect pests; and (5) the final part encompasses biosafety, biopesticide formulations, and future prospects. The expert contributors are in various fields of EPF and hail from ten countries: Australia, Brazil, the Czech Republic, Egypt, India, Iran, Japan, Mexico, the United Kingdom, and Vietnam.

This book is intended to cater to the needs of researchers and students from agriculture, biochemistry, biotechnology, botany, ecology, entomology, environmental biology, field biology, forestry, microbiology, mycology, and zoology. The editors appreciate the excellent gestures of the contributors and reviewers for delivering chapters, evaluations, and revisions of chapters on the scheduled time. We are grateful to Springer Nature for their persistence and keen interest in publishing this book within the stipulated time frame.

Uruli-Kanchan, Maharashtra, India — Sunil Kumar Deshmukh
Mangalore, Karnataka, India — Kandikere Ramaiah Sridhar
Chandigarh, Punjab, India — Gurmeet Kour Bali
Visakhapatnam, Andhra Pradesh, India — Belle Damodara Shenoy

Contents

Editors and Contributors

About the Editors

Sunil Kumar Deshmukh is a Scientific Advisor to Greenvention Biotech, Uruli Kanchan, Pune, India, and AGPharmBioinnovations LLP, Patiala, Punjab, India. He is a veteran industrial mycologist who spent a substantial part of his career in drug discovery at Hoechst Marion Roussel Limited [now Sanofi India Ltd.], Mumbai, and Piramal Enterprises Limited, Mumbai. He has also served at the TERI-Deakin Nanobiotechnology Centre, TERI, New Delhi, and as an Adjunct Associate Professor at Deakin University, Australia. He has to his credit 8 patents, 156 publications, and 26 books on various aspects of fungi and natural products of microbial origin. He is the president of the Association of Fungal Biologists (AFB) and a past president of the Mycological Society of India (MSI). He is a fellow of the Mycological Society of India, the Association of Biotechnology and Pharmacy, the Society for Applied Biotechnology, and the Maharashtra Academy of Sciences. He is Associate Editor of the Frontiers in Microbiology series, editor of Progress in Mycological Research published by CRC Press, and Managing Editor of *Mycological Spectrum*, the official journal of the Association of Fungal Biologists. He serves as a referee for more than 20 national and international journals. He has approximately four decades of research experience in getting bioactives from fungi and various aspects of keratinophilic fungi.

Kandikere Ramaiah Sridhar is an Adjunct Professor in the Department of Biosciences, Mangalore University, and Visiting Professor, NITTE (Deemed to be University), Mangalore, India. His main areas of research are: Diversity and Ecology of Fungi of the Western Ghats and Maritime Habitats. He was an NSERC postdoctoral fellow and visiting professor at Mount Allison University, Canada; the Helmholtz Centre for Environmental Research at UFZ and Martin Luther University, Germany; and the Centre of Biology, University of Minho, Portugal. He was a recipient of the Dr. Shome Memorial Award (2004), Vice President (2013), President (2018), and Lifetime Achievement Awardee (2019) of the Mycological Society of India. He was a recipient of the Fellowship of the Indian Mycological Society, Kolkata (2014), Distinguished Asian Mycologist (2015), and Outstanding Leader in Education and Research, Association of Agricultural Technology of Southeast Asia (2016). He was awarded a UGC-BSR Faculty Fellowship (2014–2017). He is one of the top 2% of scientists in the field of mycology (2020–2024). He has published over 500 research articles, reviewed over 200 research articles, 25 project proposals, 12 book proposals, and 25 PhD theses (Australia, India, Mauritius, and Thailand). He is on the editorial board of several national and international journals and editor-in-chief of the journal *MycoIndia*. He has edited 20 books published by Apple Academic Press, CRC Press, Fungal Diversity Press, IK International Publishing House, and Springer International. He has guided 25 PhD students in Environmental Science, Food Science, Microbiology, and Mycology.

Gurmeet Kour Bali is presently working as a senior project associate in the Department of Microbial Type Culture Collection (MTCC), CSIR-IMTECH, Chandigarh, India. She also held the position of Assistant Professor in the Department of Zoology, DAV University, Jalandhar, -India. She graduated with a Master of Philosophy in Zoology from Guru Nanak Dev University, Amritsar, and her PhD from the Department of Zoology, Savitribai Phule Pune University, Pune, India. Her expertise is in the field of entomopathogenic fungal diversity and its efficacy against insect pests, and its application as a biocontrol agent. In her research, she isolated indigenous strains of entomopathogenic fungi from the different soil samples and characterized them with advanced molecular techniques. She was able to identify different sets of proteins elevated during insect-fungal interaction, and her work was published in the reputed journal "Scientific Data." She also presented her paper at the international conference "SOL2023 International Solanaceae Genome Conference," conducted by the Department of Plant Sciences, McGill University, Canada. She has attended many ICAR and DST workshops in the fields of mycology and entomology, giving an edge to her research work. She has seven national and international research papers to her credit.

Belle Damodara Shenoy is a distinguished fungal taxonomist and the Principal Scientist at CSIR-National Institute of Oceanography, Regional Centre Visakhapatnam, with a PhD (2007, University of Hong Kong) and postdoctoral training in Canada. His research spans marine mycology, environmental biotechnology, and fungal ecology, most notably microbial degradation of marine pollutants, trace-metal impacts on fungal diversity, and discovery of bioactive marine fungi. With over 50 peer-reviewed publications, he is recognized internationally for advancing fungal taxonomy and microbial ecology. As Editor-in-Chief of *MycoAsia*, Managing Editor of *MycoIndia*, and Associate Editor for multiple journals, he plays a leading role in shaping scholarship in fungal and marine microbiology. His unique combination of expertise in rigorous taxonomy and applied environmental science, his editorial leadership, and his dedicated regional focus on the Global South contribute significantly to advancing scientific knowledge and fostering collaboration in these critical areas.

Contributors

Yamini Agrawal Department of Botany, University of Delhi, New Delhi, Delhi, India

Rania F. Ahmed Chemistry of Natural Compounds Department, National Research Centre, Dokki, Giza, Egypt

T. K. Arun Kumar Department of Botany, The Zamorin's Guruvayurappan College (Affiliated to the University of Calicut), Kozhikode, Kerala, India

Fathalla Ayoob Chemistry of Tanning Materials and Leather Technology Department, National Research Centre, Giza, Egypt

Prakash Balu Department of Biotechnology, School of Life Sciences, Vels Institute of Science, Technology and Advanced Studies (VISTAS), Pallavaram, Chennai, Tamil Nadu, India

Sanjan Kumar Bharti ICAR-National Research Centre on Litchi, Muzaffarpur, Bihar, India

Liza Handique Bhattacharyya Department of Botany, Jagannath Barooah University, Jorhat, Assam, India

Pranaba Nanda Bhattacharyya Department of Botany, Nanda Nath Saikia College, Titabar, Jorhat, Assam, India
Institutional Biotech Hub (IBT-HUB), Nanda Nath Saikia College, Titabar, Jorhat, Assam, India
Asia/Pacific Network, Common Microbial Biotechnology Platform (CMBP) for Sustainable Agriculture in Asia-Pacific (www.cmbp-network.org), Hanoi, Vietnam

Tanmaya Kumar Bhoi Forest Protection Division, ICFRE - Arid Forest Research Institute (ICFRE-AFRI), Jodhpur, Rajasthan, India

Dhruba Jyoti Borgohain C. S. Central Library, Nanda Nath Saikia College, Titabar, Assam, India

Sashidhar Burla ATGC Biotech Pvt. Ltd., Hyderabad, Telangana, India

Ravinder Singh Chandel Department of Entomology, CSK Himachal Pradesh Krishi Vishvavidyalaya, Palampur, Himachal Pradesh, India

Shalini Chandel Department of Botany, Sri Sai University, Kangra, Himachal Pradesh, India

Praneet Chauhan Department of Plant Pathology, Dr Khem Singh Gill Akal College of Agriculture, Eternal University, BaruSahib, Sirmaur, India

Hemraj Chhipa College of Agriculture, Kota (Agriculture University, Kota), Kota, Rajasthan, India

Shouvik Chowdhury Symbiotic Sciences Pvt. Ltd., Gurgaon, Haryana, India

Ha Huy Cong Faculty of Biology, Hanoi National University of Education, Hanoi, Vietnam

Carlos I. Cortes Martínez Instituto Politécnico Nacional, CIIDIR Oaxaca, Santa Cruz Xoxocotlán, Oaxaca, Mexico

Filippe Elias de Freitas Soares Department of Chemistry, Universidade Federal de Lavras, Minas Gerais, Brazil

Sunil Kumar Deshmukh Research and Development Division, Greenvention Biotech Pvt. Ltd., Uruli-Kanchan, Maharashtra, India

Vanshdeep Dhanda Department of Entomology, Chaudhary Sarwan Kumar Himachal Pradesh Krishi Vishvavidyalaya, Palampur, India

Amanda do Carmo Alves Department of Chemistry, Universidade Federal de Lavras, Minas Gerais, Brazil

Saurabh Dubey Krishi Vigyan Kendra, Madhopur, West Champaran, Dr. Rajendra Prasad Central Agricultural University, Pusa, Samastipur, Bihar, India

Mai M. Elghonemy Chemistry of Natural Compounds Department, National Research Centre, Dokki, Giza, Egypt

Asmaa S. Abd Elkarim Chemistry of Tanning Materials and Leather Technology Department, National Research Centre, Dokki, Giza, Egypt

Abdelsamed I. Elshamy Chemistry of Natural Compounds Department, National Research Centre, Dokki, Giza, Egypt
Faculty of Pharmaceutical Sciences, Tokushima Bunri University, Yamashiro-cho, Tokushima, Japan

Maunata Ghorui Symbiotic Sciences Pvt Ltd, Gurgaon, Haryana, India

Subrata Giri Department of Botany, Bajkul Milani Mahavidyalaya, Kismat Bajkul, Purba Medinipur, West Bengal, India

Rajadurai Gothandaraman Department of Plant Biotechnology, Centre for Plant Molecular Biology and Biotechnology, Tamil Nadu Agricultural University, Coimbatore, India

Carlos Granados-Echegoyen SECIHTI- Instituto Politécnico Nacional, CIIDIR Oaxaca, Santa Cruz Xoxocotlán, Oaxaca, Mexico

Emran Habibi Centre for Natural Products Discovery, School of Pharmacy and Biomolecular Sciences, Liverpool John Moores University, Liverpool, UK
Medicinal Plants Research Centre, Institute of Herbal Medicines and Metabolic Disorders, Mazandaran University of Medical Sciences, Sari, Iran

Jafrin Farha Hussain Department of Zoology, Nanda Nath Saikia College, Titabar, Jorhat, Assam, India

Hiroshi Imagawa Faculty of Pharmaceutical Sciences, Tokushima Bunri University, Yamashiro-cho, Tokushima, Japan

Imtinungsang Jamir Krishi Vigyan Kendra, Lada, Dr. Rajendra Prasad Central Agricultural University, Pusa, Samastipur, Bihar, India

Pooja Kapoor Flavorite Hydroponics Tomatoes, Shepparton, VIC, Australia

Yusuke Kasai Faculty of Pharmaceutical Sciences, Tokushima Bunri University, Yamashiro-cho, Tokushima, Japan

Anurag Kashyap Department of Plant Pathology, Assam Agricultural University, Jorhat, Assam, India

Abhishek Katoch Flavorite Hydroponics Tomatoes, Shepparton, VIC, Australia

Shabnam Katoch Department of Plant Pathology, Chaudhary Sarwan Kumar Himachal Pradesh Krishi Vishvavidyalaya, Palampur, India

J. Komal Central Silk Board-Basic Tasar Silkworm Seed Organization, Bilaspur, Chattisgarh, India

Mansi Kothari Fungal Biotechnology and Invertebrate Pathology Laboratory, Department of Biological Sciences, Rani Durgavati University, Jabalpur, Madhya Pradesh, India

Jithu Unni Krishnan Division of Forest Health, Kerala Forest Research Institute, Thrissur, India

Amit Kumar ICAR-National Research Centre on Litchi, Muzaffarpur, Bihar, India

Duong Minh Lam Faculty of Biology, Hanoi National University of Education, Hanoi, Vietnam
Institute of Natural Sciences, Hanoi National University of Education, Hanoi, Vietnam

Esperanza Loera-Alvarado SECIHTI-Universidad Autónoma Chapingo (CRUCO), Lomas del Valle, Morelia, Michoacán, Mexico

Rosa Elvira López-López Instituto Politécnico Nacional, CIIDIR Oaxaca, Santa Cruz Xoxocotlán, Oaxaca, Mexico

Chetna Mahajan Department of Plant Pathology, Chaudhary Sarwan Kumar Himachal Pradesh Krishi Vishvavidyalaya, Palampur, India

Rishi Mahajan Microbial Ecology and Applied Microbial Sciences Research Group, Department of Microbiology, College of Basic Sciences, Chaudhary Sarwan Kumar Himachal Pradesh Agricultural University, Palampur, India

Deepak Kumar Mahanta Forest Entomology Discipline, Forest Protection Division, Indian Council of Forestry Research and Education (ICFRE)-Forest Research Institute (ICFRE-FRI), Dehradun, Uttarakhand, India

Bidisha Mukherjee Shaheed Rajguru College of Applied Sciences for Women, University of Delhi, Delhi, India

Lutfun Nahar Centre for Natural Products Discovery, School of Pharmacy and Biomolecular Sciences, Liverpool John Moores University, Liverpool, UK

C. V. Nandeesha Department of Plant Pathology, Junagadh Agricultural University, Junagadh, Gujarat, India

Bharat Chandra Nath Department of Plant Pathology, Assam Agricultural University, Jorhat, Assam, India

Ondrej Novak Laboratory of Growth Regulators, Palacký University and Institute of Experimental Botany, The Czech Academy of Sciences, Olomouc, Czech Republic

Joginder Pal Department of Plant Pathology, Chaudhary Sarwan Kumar Himachal Pradesh Krishi Vishvavidyalaya, Palampur, India

Prakash Pradhan West Bengal Biodiversity Board, Kolkata, West Bengal, India

Prakriti Department of Plant Pathology, Chaudhary Sarwan Kumar Himachal Pradesh Krishi Vishvavidyalaya, Palampur, India

Rakhi Kunnath Radhakrishnan Division of Forest Health, Kerala Forest Research Institute, Thrissur, India

Ansh Raj ICAR-National Research Centre on Litchi, Muzaffarpur, Bihar, India

Vaishnavi Rajwar Department of Botany, University of Delhi, New Delhi, Delhi, India

Karthik Ramappa Department of Entomology, CSK Himachal Pradesh Krishi Vishvavidyalaya, Palampur, Himachal Pradesh, India

Abhishek Rana Krishi Vigyan Kendra, Gopalganj, Dr. Rajendra Prasad Central Agricultural University, Pusa, Samastipur, Bihar, India

Murali Revanasiddappa Division of Forest Health, Kerala Forest Research Institute, Thrissur, India

Ahmed Sabt Chemistry of Natural Compounds Department, National Research Centre, Giza, Egypt

Bimal Kumar Sahoo Tamil Nadu Agricultural University, Coimbatore, India

Ipsita Samal ICAR-National Research Centre on Litchi, Muzaffarpur, Bihar, India

Sardul Singh Sandhu Bio-Design Innovation Centre, Rani Durgavati University, Jabalpur, Madhya Pradesh, India

Suman Sanjta Department of Entomology, Chaudhary Sarwan Kumar Himachal Pradesh Krishi Vishvavidyalaya, Palampur, India

Indrani Sarkar Amity Institute of Biotechnology, Amity University Noida, Noida, Uttar Pradesh, India

Satyajit D. Sarker Centre for Natural Products Discovery, School of Pharmacy and Biomolecular Sciences, Liverpool John Moores University, Liverpool, UK

Bhaskar Sarma Department of Botany, Dhemaji College, Dhemaji, Assam, India

Parismita Sarma Department of Botany, Rangia College, Rangia, Assam, India

Parthasarathy Seethapathy Department of Plant Pathology, Amrita School of Agricultural Sciences, Amrita Vishwa Vidyapeetham, Coimbatore, India

Shikha Sharma Department of Plant Pathology, Chaudhary Sarwan Kumar Himachal Pradesh Krishi Vishvavidyalaya, Palampur, India

Atham Bhavana Sruthi Department of Agricultural Entomology, College of Agriculture, Kerala Agricultural University, Thrissur, India

Anju Mathiyil Subramanian Division of Forest Health, Kerala Forest Research Institute, Thrissur, India

Sakshi Suman Department of Botany, University of Delhi, New Delhi, Delhi, India

Dusmanta Kumar Sut Department of Mathematics, Nanda Nath Saikia College, Jorhat, Assam, India

Rehab F. Taher Chemistry of Natural Compounds Department, National Research Centre, Giza, Egypt

Eduardo Alfonso Treviño-López Centro de Investigación en Química Aplicada, San José de los Cerritos, Saltillo, Coahuila, Mexico

Le Thi Tuoi Faculty of Biology, Hanoi National University of Education, Hanoi, Vietnam
Institute of Natural Sciences, Hanoi National University of Education, Hanoi, Vietnam

Alfonso Vásquez-López Instituto Politécnico Nacional, CIIDIR Oaxaca, Santa Cruz Xoxocotlán, Oaxaca, Mexico

Jeer Vinayaka Krishi Vigyan Kendra, Saran, Dr. Rajendra Prasad Central Agricultural University, Pusa, Samastipur, Bihar, India

Part I

Diversity, Taxonomy, and Ecology

Global Ecological Niches of *Cordyceps militaris*, *Cordyceps tenuipes*, and *Ophiocordyceps nutans*

1

Subrata Giri and Prakash Pradhan

Abstract

This chapter investigates the global potential distribution of three entomopathogenic fungi, *Cordyceps militaris*, *Cordyceps tenuipes*, and *Ophiocordyceps nutans*, using ecological niche modeling via MaxEnt. Habitat suitability was predicted under current environmental conditions based on occurrence data from global repositories, literature, and expert records, integrated with a suite of biotic, climatic, and environmental variables. The resulting habitat suitability maps revealed distinct biogeographic patterns, strongly influenced by specific covariates. Notably, host suitability emerged as the primary driver of the distribution of *C. militaris* and *O. nutans*, accounting for the highest variable contributions (74.2% and 41.2%, respectively). Among bioclimatic variables, precipitation of the driest month (bio_14) consistently influenced all three fungal species. *C. militaris* exhibited high suitability in temperate regions, closely emulating its host distribution, while *C. tenuipes* displayed a broader potential range across humid environments. *O. nutans* showed a highly restricted potential distribution concentrated in East Asia, despite a wider potential range for its hosts. Five climatic covariates (bio_14, vapr_02, srad_10, vapr_08, and srad_07) consistently emerged as key determinants across the fungal and host models, emphasizing the importance of dry-season moisture availability and seasonal insolation. These findings highlight the critical role of host availability and specific climatic factors in shaping the global distribution of these

S. Giri
Department of Botany, Bajkul Milani Mahavidyalaya, Kismat Bajkul, Purba Medinipur, West Bengal, India

P. Pradhan (✉)
West Bengal Biodiversity Board, Prani Sampad Bhawan, Kolkata, West Bengal, India

S. K. Deshmukh et al. (eds.), *Frontiers in Entomopathogenic Fungi*,
https://doi.org/10.1007/978-981-95-6529-0_1

entomopathogenic fungi, providing valuable insights into their ecological niches and potential responses to environmental change.

Keywords

AUC · Ecological niche modeling · Entomopathogenic fungi · MaxEnt · TSS · Worldclim

1 Introduction

Entomopathogenic fungi, known for their parasitic relationships with insect larvae and pupae, are a diverse and ecologically important group in the Kingdom *Fungi*. These organisms have evolved specialized life cycles, often manipulating host physiology and behavior to enhance their transmission and survival (Evans 1982). In recent decades, entomopathogenic fungi have attracted considerable scientific interest for their ecological roles and potential in biological pest control (Mantzoukas et al. 2022). Their capacity to naturally regulate insect populations makes them attractive alternatives to chemical pesticides, aligning with global efforts for sustainable agriculture and integrated pest management.

Cordyceps sensu lato and *Ophiocordyceps* sensu stricto are of particular scientific and economic interest within the entomopathogenic fungi. These fungi have broad host ranges, complex ecological interactions, and the ability to synthesize bioactive compounds with significant pharmacological potential. These features make them valuable as natural biocontrol agents and potential sources of new therapeutic agents (Sung et al. 2007; Olatunji et al. 2018; Tapingkae et al. 2024).

A thorough understanding of the ecological niches and biogeographic distributions of entomopathogenic fungi is essential for identifying their potential habitats, discovering new metabolites, and developing strategies for their sustainable use. The spatial distribution of these fungi is closely linked to the ecology, behavior, and distribution of their insect hosts. Analyzing host-fungus associations spatially is crucial for understanding habitat suitability, coevolutionary dynamics, and ecological community structure, and for elucidating their role in regulating insect populations (Hajek and St Leger 1994).

Global assessments of the distribution of entomopathogenic fungi contribute to the field of fungal biogeography, revealing the historical, ecological, and evolutionary processes that have shaped their current distributions and dispersal capacities (Evans 1982; Lomolino et al. 2016). Detailed, species-specific knowledge of their environmental tolerances and geographic ranges is crucial for selecting fungal strains for pest control in specific agroecological contexts. Ecological niche modeling (ENM) is an indispensable tool for conservation planning, helping identify regions of high fungal endemism and predicting the effects of climate change on their future distributions (Walther et al. 2002).

Among *Cordyceps* species, *C. militaris* (L.) Fr., known as Yong Chong Cao in China, is a well-known and extensively studied member. It belongs to

Cordycipitaceae, Ascomycota, and is recognized by its bright orange to red stromata that typically emerge from Lepidoptera pupae (Kobayashi 1941). While its primary hosts are Lepidoptera, *C. militaris* can also parasitize other insect orders, including Coleoptera, Diptera, and Hymenoptera (Shrestha et al. 2012; Chou et al. 2024), with a recorded host range of at least 48 species. The fungus has a wide geographic distribution, favoring humid temperate and subtropical regions across North America, Europe, and East and South Asia (Winker 2010). However, it typically thrives in moist microhabitats within forest ecosystems, exhibits some host specificity, and requires particular environmental conditions for optimal growth and reproduction (Lee et al. 2019). Its rarity in nature and the difficulties in wild harvesting make it expensive and challenging to commercialize through natural collection (Lou et al. 2019; Rathod et al. 2023). This has led to large-scale cultivation efforts, driven by the demand for its bioactive compounds like polysaccharides, D-mannitol, cordycepin, adenosine, ergosterol, and superoxide dismutase (SOD) (Pradhan et al. 2023; Rathod et al. 2023). Cordycepin, in particular, is a potent bioactive molecule with anticancer, antiviral, antifungal, antidiabetic, antioxidant, and antihyperlipidemic activities and potential therapeutic applications against COVID-19 (Dubhashi et al. 2023; Rathod et al. 2023; Chou et al. 2024). The recognition of *C. militaris* as a functional food and its historical use in traditional Chinese medicine highlight its importance in the pharmaceutical and wellness industries (Fig. 1.1).

Cordyceps tenuipes (Peck) Kepler, B. Shrestha & Spatafora is another notable entomopathogenic fungus, though it is less studied than *C. militaris*. Unlike *C. militaris*, *C. tenuipes* has a more restricted and patchy geographic distribution, mainly concentrated in eastern North America, with scattered populations in parts of Central and South America, Europe, and Asia. Similar to *C. militaris*, its primary hosts are mostly Lepidoptera, although host specificity may vary geographically. In eastern North America, *C. tenuipes* typically inhabits deciduous and mixed forests, indicating its adaptation to specific microclimatic and edaphic conditions (Shrestha et al. 2017).

Fig. 1.1 Fruit bodies of *Cordyceps militaris* (left), *C. tenuipes* (center), and *Ophiocordyceps nutans* (right) observed in the Eastern Himalayan landscape. Photographs by PP: *C. militaris* from Khoppi Dara, near Singhalila National Park, Darjeeling (lat 27.0007282, long 88.1218609); *C. tenuipes* from Neora Valley National Park (lat 27.059139, long 88.770083); and *O. nutans* from Ghoom Bhanjyang, Darjeeling (lat 27.0128776, long 88.2342589)

Ophiocordyceps nutans (Pat.) G.H. Sung, J.M. Sung, Hywel-Jones & Spatafora is a third species of interest, classified under *Ophiocordyceps* after taxonomic revisions (Sung et al. 2007). It has a highly restricted geographic range, predominantly found in East Asia—especially Japan, with rare and localized occurrences elsewhere. Suggesting adaptation to specific local environmental conditions or specialized ecological requirements, the narrow East Asian distribution of *O. nutans* also raises questions about its evolutionary history and biogeography, potentially indicating a relictual species limited by dispersal or expansion.

The study of ecological niche models for *Cordyceps* and *Ophiocordyceps* on a global scale is increasingly recognized as a critical research priority (Shrestha and Bawa 2014; Zhang et al. 2022; Pradhan 2024; Chen and He 2025; Peng et al. 2025). ENM offers a robust framework for analyzing biodiversity patterns, informing conservation strategies, optimizing bioprospecting efforts, and understanding ecological roles in changing environmental contexts. Global-scale analyses help identify broad distributional patterns, centers of diversity, and areas of high endemism that localized studies may miss (Kier et al. 2009). Given the taxonomic changes within these genera, it is important to integrate phylogenetic information with spatial distribution data for understanding their evolutionary trajectories and biogeographic histories (Sung et al. 2007). Global perspectives are essential for conservation planning, especially among the increasing threats from habitat loss, deforestation, and climate change. ENM can help in identifying vulnerable species, predicting habitat shifts, and prioritizing conservation actions (Huang et al. 2021). Furthermore, the potential for bioprospecting novel bioactive compounds from these fungi also drives global ENM initiatives, as targeted exploration of high-diversity regions can enhance the discovery of new compounds (Wu et al. 2021). Understanding the environmental factors influencing metabolite production can also refine biotechnological applications (Calvo et al. 2002). From a macroecological standpoint, global ENM is vital for understanding how *Cordyceps*/*Ophiocordyceps*-host interactions shape insect communities and how these interactions may change with environmental shifts, which is important for predicting the ecosystem-level impacts of biodiversity loss (Araújo et al. 2019).

While the regional occurrences of *C. militaris*, *C. tenuipes*, and *O. nutans* are moderately documented, their potential global distributions remain largely unexplored. This chapter aims to present a comprehensive ENM study to: (1) characterize the ecological niches of studied fungal species and the hosts of *C. militaris* and *O. nutans*; (2) identify the key biotic and environmental drivers influencing their distributions; (3) generate predictive global distribution maps showing areas of high, medium, and low habitat suitability; (4) compare and contrast distribution patterns among the three species and their corresponding host insects; and (5) explore the biogeographical implications of the observed patterns.

The outcomes of this study are expected to enhance our understanding of the ecological biogeography of these important fungi and to support informed decisions on spatial conservation planning, sustainable management, and the potential utilization of these organisms.

2 Methodology

The global potential distributions of *C. militaris*, *C. tenuipes*, and *O. nutans* were investigated using ENM, a technique that utilizes species occurrence data and environmental variables to predict the distribution of species across geographical landscapes (Peterson et al. 2011). ENM, also known as species distribution modeling (SDM), is based on the concept of the ecological niche, which describes the biotic and environmental conditions under which a species can survive and reproduce (Hutchinson 1957). By examining environmental conditions at known locations of occurrence, ENM algorithms extrapolate the ecological requirements of species and project these onto global landscapes, allowing for predictions of suitable habitats in regions not previously surveyed.

The occurrence records consist of geographic coordinates with marked/documented locations where each species was observed or encountered, and they form an indispensable part of the ENM process. At the very beginning, research-grade occurrence records of the species were compiled from major databases, including the Global Biodiversity Information Facility (GBIF) via the *rgbif* package (Chamberlain et al. 2025) and iNaturalist through the *rinat* package (Barve and Hart 2022) in R, as well as from GlobalFungi (2025) and MyCoPortal (2025). Additionally, relevant occurrence data from published literature, field surveys, and personal observations of the authors in the Eastern Himalaya were incorporated (Kamble and Agre 2012; Karun and Sridhar 2013a, b; Pradhan et al. 2016; Nguyen et al. 2023; Rathod et al. 2023).

To address host species and biotic interactions, habitat suitability layers for *C. militaris* and *O. nutans* hosts were incorporated into the respective fungal ENMs. For *C. militaris*, host suitability was based on cumulatively modeling 48 insect species from Coleoptera (Scarabaeidae, Scolytidae, Tenebrionidae), Diptera (Tipulidae), Hymenoptera (Cimbicidae), and Lepidoptera (Arctiidae, Bombycidae, Drepanidae, Geometridae, Hepialidae, Lasiocampidae, Lymantriidae, Noctuidae, Notodontidae, Saturniidae, Sphingidae, and Thyatiridae) (Shrestha et al. 2012). The *O. nutans* host suitability layer was derived from cumulatively modeling 32 insect species within the order Hemiptera, suborder Heteroptera (Acanthosomatidae, Coreidae, Dinidoridae, Pentatomidae, Plataspidae, Pyrrhocoridae, Reduviidae, Tessaratomidae, and Urostylididae) (Karun and Sridhar 2013a, b; Shrestha et al. 2017; Nguyen et al. 2023). Due to a lack of detailed host information, a suitability layer for *C. tenuipes* could not be included.

The occurrence dataset underwent a thorough data cleaning process to remove duplicate records and spatial outliers. To mitigate sampling bias and spatial autocorrelation arising from clustered occurrence records, the spatial thinning technique was applied using the *spthin* package in R (Aiello-Lammens et al. 2015), which led to the retention of a single occurrence point per 2.5 arc-minute raster grid resolution, resulting in a more spatially even dataset across the study area.

The rasters of environmental variables (bioclimatic variables, solar radiation, vapor pressure, and wind speed) necessary for modeling species distributions, along with elevation data for analyzing fungal and host distribution across elevational gradients, were acquired from the WorldClim dataset (Fick and Hijmans 2017), representing near-current environmental conditions. A spatial resolution of 2.5 arc-minutes was selected for these environmental rasters, a scale deemed appropriate for capturing broad biogeographic patterns relevant to the global extent of this study, while also being less computationally demanding than finer resolutions (e.g., 0.5 m, approximately 30 arc-seconds). To address potential issues of multicollinearity among the 55 initial candidate environmental variables, a variance inflation factor (VIF) analysis was conducted using *usdm* package in R, resulting in the retention of 16 variables with VIF values below the threshold of 10 (Naimi et al. 2014; Pradhan 2016; Pradhan and Setyawan 2021). The final set of uncorrelated predictor variables comprised nine bioclimatic variables (bio_2: Mean Diurnal Range, bio_3: Isothermality, bio_8: Mean Temperature of Wettest Quarter, bio_9: Mean Temperature of Driest Quarter, bio_13: Precipitation of Wettest Month, bio_14: Precipitation of Driest Month, bio_15: Precipitation Seasonality, bio_18: Precipitation of Warmest Quarter, bio_19: Precipitation of Coldest Quarter); three monthly solar radiation values (srad_03 [March], srad_07 [July], srad_10 [October], in kJ m^{-2} day^{-1}); two monthly water vapor pressure values (vapr_02 [February], vapr_08 [August], in kPa); and two monthly wind speed values (wind_01 [January], wind_06 [June], in m s^{-1}).

MaxEnt (version 3.4.4), which employs the maximum entropy algorithm (Phillips et al. 2006), was used to construct ecological niche models due to its robustness in handling presence-only data and its ability to model complex relationships between species occurrences and environmental variables, even at global spatial extent. The modeling was conducted on an 11th Gen Intel(R) Core (TM) i5-11300H 3.11 GHz computer with 16 GB RAM, of which 8.192 GB of RAM was made available to the MaxEnt program by editing the *maxent.bat* file. The modeling process for individual host/fungus involved 10 replicate runs, each with randomized initial seeds, to account for stochasticity inherent in the machine learning algorithm and to ensure reproducibility (Yan et al. 2024).

The model training process used 70% of the occurrence data, while 30% of the data was reserved for model testing. The model training process was constrained by setting a regularization multiplier of 1, which controls model complexity and prevents overfitting. In the absence of true absence data, 100,000 random pseudo-absence points were generated by selecting locations with environmental dissimilarity to known occurrence locations (Phillips and Dudík 2008). The maximum number of iterations was set to 10,000, with a convergence threshold of 0.00001, which ensured model optimization at the desired resolution. To optimize model performance, samples were added to the background for inclusion of species occurrence locations as part of the background sampling across the global study domain. The 10-percentile training presence threshold rule was applied to generate binary presence–absence predictions (Radosavljevic and Anderson 2014).

The complementary log-log (cloglog) transformation was selected as the MaxEnt output format, as it is well-suited for modeling species with low prevalence or spatially skewed distributions. Its theoretical basis aligns with ecological processes in which multiple limiting factors can act multiplicatively on species occurrence

(Phillips and Dudík 2008). This method proved useful for modeling the three fungi and two hosts in this study, reflecting their inherently low prevalence in the landscape, which is a characteristic feature of their parasitic life cycles. Model performance was evaluated using the Area Under the Receiver Operating Characteristic Curve (AUC) and the True Skill Statistic (TSS) across all replicate runs to assess the model's predictive accuracy and consistency (Allouche et al. 2006). The mean output of 10 replicated (randomly seeded) runs was selected for studying the global potential distributions of *C. militaris*, *C. tenuipes*, and *O. nutans* and the hosts of *C. militaris* and *O. nutans*. Habitat suitability values ranging from 0 to 0.2 were considered unsuitable, 0.2 to 0.45 low, 0.45 to 0.7 medium, and 0.7 to 1 high (Zhang et al. 2022).

Characterization of the ecological niches and the latitudinal and elevational ranges of the studied fungal species and their hosts was performed using habitat suitability (ENM) and elevation (DEM) rasters. Niche breadth was quantified as ecological generalization using Levins' *B* index (Levins 1968), calculated from continuous suitability values. Pairwise niche overlap—a measure of similarity in geographic suitability distributions (0 = no overlap, 1 = complete overlap)—was estimated using a raster-based approximation of Schoener's *D* index (Schoener 1986). All niche analyses and range characterizations were conducted in R (R Core Team 2025) using the *terra* package (Hijmans 2024) and custom functions.

3 Observations and Discussion

MaxEnt modeling was employed to predict the potential global habitat suitability for *C. militaris* hosts, *C. militaris*, *C. tenuipes*, *O. nutans* hosts, and *O. nutans* based on environmental variables, primarily bioclimatic factors, and for the fungi, the potential distribution of their respective hosts. The resulting suitability maps were analyzed using predefined thresholds to categorize areas as unsuitable (0–0.2), low (0.2–0.45), medium (0.45–0.7), and high (0.7–1) suitability. The spatial distribution of these suitability categories varied among the studied organisms, reflecting differences in their ecological niches and requirements across the globe. The analysis revealed broad regional patterns and suitability zones across continents and major subregions, including countries and areas that were visually apparent at the 2.5 arc-minute raster resolution.

The MaxEnt projections indicated that *C. militaris* occupied a total suitable area of approximately 9.44×10^6 km^2, of which 17.3% fell into the high (0.7–1.0), 27.8% into the medium (0.45–0.7), and 54.9% into the low-suitability class (0.2–0.45). The combined distribution of its insect hosts was far broader (3.19×10^7 km^2), yet only 3.2% of that area was classified as highly suitable for fungal development. Nearly half of the host range (49.7%) lay within the medium suitability bracket, while 47.2% was in the low suitability bracket (Table 1.1).

O. nutans exhibited the smallest projected fungal range at 1.12×10^6 km^2, but 38.1% of that area was rated as high suitability—a substantially greater proportion than observed for the other allied *Cordyceps* species. Medium and low suitability

Table 1.1 Habitat suitability areas (km^2) for *Cordyceps militaris*, *Cordyceps tenuipes*, *Ophiocordyceps nutans*, and their hosts

	High (0.7–1)	Medium (0.45–0.7)	Low (0.2–0.45)
Cordyceps militaris	16,36,149.96	26,26,320.37	51,79,090.73
Hosts of *C. militaris*	10,09,335.00	1,58,18,202.76	1,50,23,000.96
Ophiocordyceps nutans	4,26,508.56	1,65,278.50	5,28,591.08
Hosts of *O. nutans*	27,41,972.44	85,72,530.56	89,29,454.95
C. tenuipes	7,79,102.69	21,20,859.70	55,27,847.66

areas comprised 14.8% and 47.2%, respectively. Its hosts, however, spanned 2.02×10^7 km^2, with only 13.5% of that extent deemed as highly suitable, 42.3% as medium, and 44.1% as low. These results indicate that each host-dependent entomopathogenic fungus occupies a substantially narrower and more specialized climatic niche than its hosts (Table 1.1).

While *C. tenuipes* was predicted to occur across 8.43×10^6 km^2, with 9.2% of the area classified as high suitability, 25.2% as medium, and 65.6% as low. The much smaller high-suitability zones, particularly for *O. nutans* and *C. tenuipes*, highlight target areas for field validation and conservation prioritization within the broader host-insect distributions (Table 1.1).

Figure 1.2 presents the percent contribution (PC) and permutation importance (PI) of environmental variables used in MaxEnt models for the three focal fungal species. These metrics quantify the relative influence of each predictor in determining species distribution. Notably, for *C. militaris* and *O. nutans*, host-specific variables accounted for the highest contributions (PC = 74.2% and 41.2%, respectively) and permutation importance (PI = 69.8% and 19.8%, respectively). This emphasizes their obligate parasitic life strategy, highlighting host availability as the principal factor shaping fungal distribution across ecosystems.

Among the bioclimatic variables, *bio_14* (precipitation of the driest month) emerged as a critical environmental constraint. It contributed between 1.04% and 59.68% (mean = 30.3% ± 27.7%) and showed permutation importance ranging from 1.14% to 10.88% (mean = 5.16% ± 4.51%). In biomes characterized by seasonal aridity and distinct dry periods, water scarcity likely imposes physiological stress on both fungi (e.g., spore desiccation, reduced mycelial viability) and insect hosts (e.g., desiccation, reduced food availability). The variable influence of *bio_14* reflects the high sensitivity of host–fungus interactions to moisture deficits during phenologically critical periods. *Vapr_02* (February vapor pressure) also showed considerable variability in influence, with percent contributions ranging from 0.78% to 30.56% (mean = 16.28% ± 14.47%) and permutation importance ranging from 2.12% to 38.68% (mean = 16.7% ± 16.53%). Globally, rising humidity in late winter may correspond with the emergence or resurgence of susceptible insect populations following dormant or stressful periods, creating favorable conditions for fungal spore germination and initial host colonization. Solar radiation—especially ultraviolet-B—is known to reduce the viability of entomopathogenic fungal propagules on plant surfaces (Quesada-Moraga et al. 2024). In this context, *srad_10* (October solar radiation) showed moderate contributions (PC range = 1.62%–

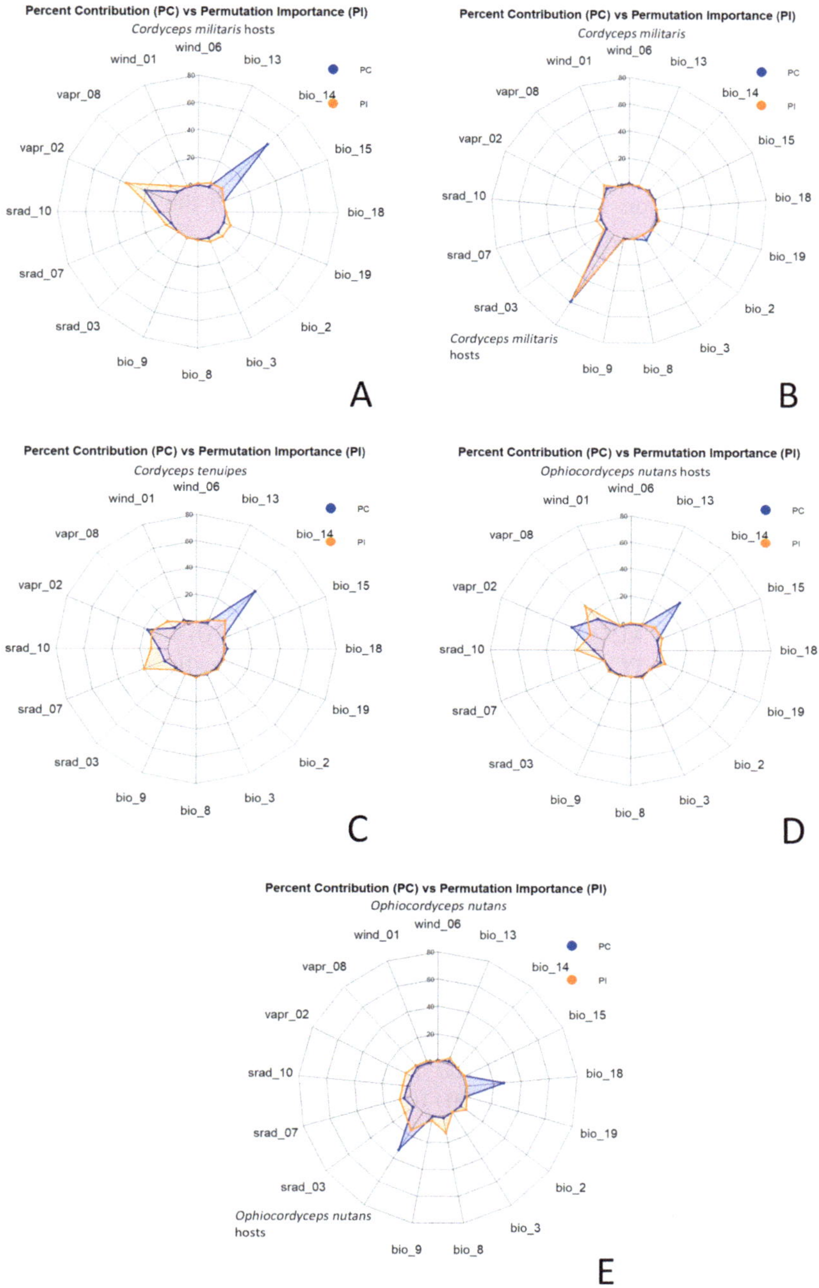

Fig. 1.2 Percent contribution (PC) and permutation importance (PI) of abiotic and biotic variables contributing to the development of mean MaxEnt output for: (**a**) *Cordyceps militaris hosts*, (**b**) *C. militaris*, (**c**) *Cordyceps tenuipes*, (**d**) *Ophiocordyceps nutans* hosts, and (**e**) *O. nutans*

8.66%; mean = 5.58% ± 3.49%) and relatively higher permutation importance (PI range = 2.12%–21.5%; mean = 11.64% ± 8.40%). In regions experiencing peak radiation concurrent with fungal reproduction or host insect aggregation, insolation may indirectly affect host–pathogen dynamics through changes in insect activity, microhabitat quality, or immune responses.

Other bioclimatic variables—such as *bio_13* (precipitation of the wettest month), *bio_15* (precipitation seasonality), *bio_18* (precipitation of the warmest quarter), *bio_19* (precipitation of the coldest quarter), *bio_2* (mean diurnal range), *bio_3* (isothermality), and *bio_8* (mean temperature of the wettest quarter)—generally showed lower average percent contributions (mean = 0.64%–1.54%) and permutation importance (mean = 0.5%–3.84%), suggesting limited direct influence. Additional environmental factors such as *srad_03*, *srad_07*, *vapr_08*, *wind_01*, and *wind_06* also exhibited low mean contributions and permutation importance, though with varying ranges. *Bio_9* (mean temperature of the driest quarter) contributed the least among all predictors.

The ENM for hosts of *C. militaris* (mean AUC = 0.799 ± 0.001, TSS = 0.9 ± 0) identified *bio_14* (PC = 62.3%, PI = 5.6%), *vapr_02* (PC = 26.0%, PI = 43.9%), and *srad_10* (PC = 8.9%, PI = 12.4%) as the most influential variables. All remaining variables, including *vapr_08*, *srad_07*, wind parameters, and secondary bioclimatic variables, each contributed less than 3% to model gain. The ENM for *C. militaris* (mean AUC = 0.969 ± 0.002, TSS = 0.9001 ± 0) was primarily driven by host suitability (PC = 74.2%, PI = 69.8%). Secondary predictors included *bio_3* (PC = 5.4%, PI = 0.1%), *vapr_08* (PC = 4.8%, PI = 8.5%), and *srad_07* (PC = 1.3%, PI = 5.5%). All other variables contributed minimally (<2%), suggesting redundancy or negligible influence. The ENM for hosts of *O. nutans* (AUC = 0.899 ± 0, TSS = 0.9 ± 0) showed *bio_14* (PC = 36.9%, PI = 5.6%), *vapr_02* (PC = 31.7%, PI = 13.9%), *vapr_08* (PC = 16.3%, PI = 32.1%), *srad_10* (PC = 7.7%, PI = 23.1%), and *bio_8* (PC = 2.3%, PI = 16.2%) as key determinants, while the remaining predictors had minimal influence (<5%). The ENM for *O. nutans* (mean AUC = 0.986 ± 0.011, TSS = 0.9048 ± 0) was driven mainly by host presence (PC = 41.2%, PI = 19.8%), followed by *bio_18* (PC = 34.5%, PI = 1.1%), *srad_07* (PC = 6.3%, PI = 10.2%), and *srad_03* (PC = 2.8%, PI = 11.6%). Other predictors had negligible influence. For *C. tenuipes*, *bio_14* emerged as the dominant predictor (PC = 49.2%, PI = 12.2%), followed by *vapr_02* (PC = 21.9%, PI = 17.8%) and *srad_10* (PC = 7.7%, PI = 15.1%). The model (mean AUC = 0.899 ± 0, TSS = 0.901109 ± 0.00003) also highlighted *srad_07* for its highest PI (25.2%) despite a lower PC (5.3%), indicating unique explanatory power. All other predictors contributed less than 4%.

Across the ENMs of *C. militaris*, *C. tenuipes*, *O. nutans*, and their respective host's five climatic covariates consistently emerged as key determinants in at least three out of five models: *bio_14*, *vapr_02*, *srad_10*, *vapr_08*, and *srad_07*. Therefore, these variables represent the core environmental envelope influencing both fungal niches and host distributions, emphasizing the roles of dry-season moisture availability and seasonal insolation gradients. In contrast, wind speed metrics (*wind_01*, *wind_06*) and minor bioclimatic predictors contributed little explanatory power.

The latitudinal and elevational profiles and quantitative analysis of niche breadth (Levins' B) and niche overlap (Schoener's D) are important for dissecting the structure and dynamics of complex biological systems (Nan et al. 2023). The study integrated these niche metrics with the concerned fungi and their hosts' geographic distributions to elucidate their interacting ecological strategies.

Niche breadth analysis revealed a clear gradient of specialization. The hosts, exemplified by the broad niche of the *C. militaris* hosts (B = 0.0556), exhibited eurytopic characteristics, indicating their capacity to thrive across a wide spectrum of environmental conditions. On the other hand, the *C. militaris* displayed greater specialization. *O. nutans* demonstrated extreme stenotopy (B = 0.0048), suggesting highly specific ecological requirements, potentially linked to host specificity, specific local conditions, or specialized ecological needs. *C. militaris* (B = 0.0189) and *C. tenuipes* (B = 0.0218) occupied intermediate positions along this specialization spectrum among the fungi (Table 1.2).

The extent of shared ecological space within the system was illuminated by pairwise niche overlap coefficients (Schoener's D). High overlap was consistently observed between the host-parasite pairs (*O. nutans* / *O. nutans* Host: D = 0.627; *C. militaris* / *C. militaris* Host: D = 0.573), highlighting the strong niche congruence characteristic of specialized parasitism, where the parasite's ecological requirements are closely aligned with those of its host (Filz and Schmitt 2015) (Table 1.2). However, the markedly low niche overlap between *C. militaris* and *O. nutans* (D = 0.140) provides compelling quantitative evidence for niche partitioning, likely driven by differentiation in host utilization, microhabitat preferences, or phenological patterns, crucial for potentially mitigating interspecific competition and facilitating their coexistence within the broader ecosystem (Chesson 2000). Overlap with non-associated hosts was generally low (*O. nutans* / *C. militaris* Host: D = 0.166), further supporting the specificity of host-parasite interactions. An intriguing exception was the relatively high overlap between *C. militaris* and the *O. nutans* host (D = 0.548), comparable to its overlap with its own host, potentially indicating a broader fundamental niche for *C. militaris* or shared environmental tolerances between the two host taxa. *C. tenuipes* displayed moderate overlap across all other entities (D = 0.354–0.489), suggesting a distinct ecological strategy involving partial niche intersection with multiple components of the system (Table 1.2).

The geographic representation of habitat suitability provided additional ecological understanding. *C. tenuipes* exhibited the broadest latitudinal span (−48.02° to 47.40°) and reached the highest elevational maximum (4242 m), suggesting a wide thermal and geographic tolerance (Table 1.3). However, given that this species was modeled without incorporating host habitat suitability, its broad range may represent an inflated fundamental niche. Conversely, *C. militaris* and *O. nutans*—modeled with host habitat suitability layers—demonstrated narrower latitudinal and elevational ranges, reflecting thermogeographic specialization potentially shaped by host distribution and ecological filtering. Notably, the integration of host suitability significantly increased the estimated niche breadth for both *C. militaris* (0.0189–0.0556) and *O. nutans* (0.0048–0.0355), highlighting the critical role of

Table 1.2 Niche Breadth (Levins' B) and Pairwise Niche Overlap (Schoener's D Approximation) for *Cordyceps militaris*, *Ophiocordyceps nutans*, *Cordyceps tenuipes*, and their associated hosts

ENM	*Cordyceps militaris*	*Cordyceps militaris* Hosts	*Ophiocordyceps nutans*	*Ophiocordyceps nutans* Hosts	*Cordyceps tenuipes*	Niche Breadth (Levins' B)
C. militaris	NA	0.572	0.140	0.548	0.428	0.018887
Host insects of *C. militaris*	0.572	NA	0.166	0.627	0.354	0.055648
O. nutans	0.140	0.166	NA	0.236	0.359	0.004838
Host insects of *O. nutans*	0.548	0.627	0.236	NA	0.489	0.035542
C. tenuipes	0.428	0.354	0.359	0.489	NA	0.021774

Niche Breadth values indicate ecological specialization, while Niche Overlap values quantify shared niche space (range 0–1, where 1 is complete overlap)

Table 1.3 Range of latitude and elevation within the predicted suitable areas (using data derived from MaxEnt outputs) for each fungus and host group, offering insight into the core and extreme geographic and altitudinal limits of their potential distributions

	Latitude						Elevation					
Species/Host	Min	p5	p95	Max	Mean	SD	Min	p5	p95	Max	Mean	SD
Cordyceps militaris	22.56	38.52	59.69	67.52	49.44	6.73	−18	7	827	4065	253.6	310.52
Cordyceps tenuipes	−48.02	−40.35	43.60	47.395	20.297	30.45	−19	12	1690	4242	481.84	553.74
Ophiocordyceps nutans	−21.85	24.69	42.69	44.31	35.12	6.38	−4	11	1521	3149	444.44	490.64
Cordyceps militaris hosts	22.56	39.44	64.56	80.48	50.86	8.37	−30	13	1096	3791	315.80	364.88
Ophiocordyceps nutans hosts	15.895	27.27	51.69	54.85	40.92	7.21	−30	10	905	3533	300.98	327.84

Lat latitude, *Elev* elevation, *p5* 5th percentile, *p95* 95th percentile, *SD* standard deviation biotic associations in defining the realized niche (Wisz et al. 2013). The low niche overlap between *C. militaris* and *O. nutans* (D = 0.140) further supports their functional ecological divergence. The restricted range of elevation and latitude values in the host-constrained models of *C. militaris* and *O. nutans* suggests ecological stability within specific bioclimatic windows, potentially indicating their sensitivity to disruptions in host availability or environmental change.

3.1 Areas of High Habitat Suitability (0.7–1.0)

Regions predicted to have high habitat suitability represented environments where the environmental conditions were most congruent with those associated with known occurrences of the organisms, based on the MaxEnt models (Figs. 1.3, 1.4, 1.5, 1.6, and 1.7). These areas likely offered optimal combinations of climatic factors and, for the fungi, potential host presence necessary for their life cycles within specific biogeographic contexts, facilitating successful infection, development, and reproduction (Yan et al. 2024).

For the hosts of *C. militaris*, areas of high suitability were notably concentrated in the temperate regions of the Northern Hemisphere, largely within the Nearctic and Palearctic realms. In the Nearctic, a significant and contiguous expanse of high suitability was evident across the eastern United States, covering states from the Northeast (e.g., Pennsylvania, New York) southward through the Mid-Atlantic (e.g., Virginia, North Carolina) and into parts of the Southeast and Midwest, extending northeastward into southeastern Canadian provinces like Ontario and Quebec. This vast region is characterized by extensive temperate deciduous and mixed forests, biomes known for supporting high insect diversity and biomass in forest floor and soil habitats where many potential *Cordyceps* hosts are known to occur (NASA

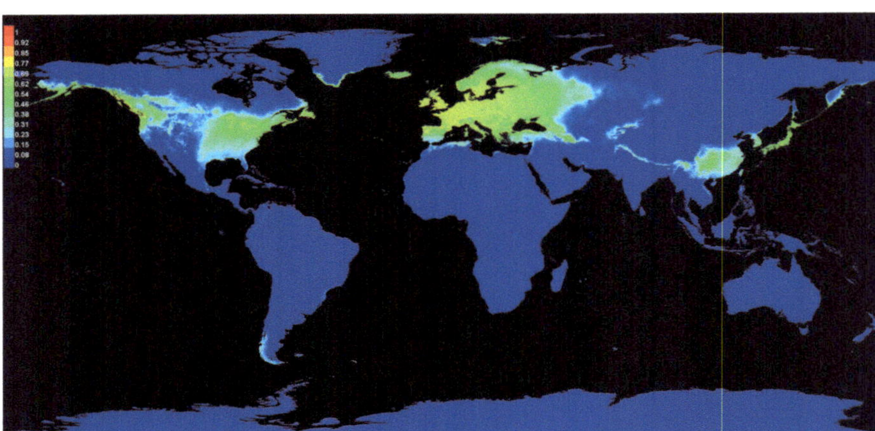

Fig. 1.3 Global potential distribution of hosts of *Cordyceps militaris*. Cooler colors indicate low habitat suitability and warmer colors indicate higher habitat suitability (on a scale of 0–1)

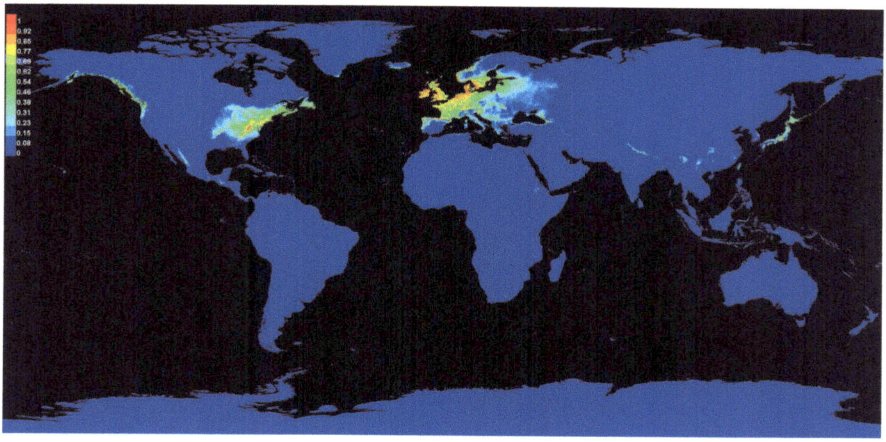

Fig. 1.4 Global potential distribution of *Cordyceps militaris*. Cooler colors indicate low habitat suitability and warmer colors indicate higher habitat suitability (on a scale of 0–1)

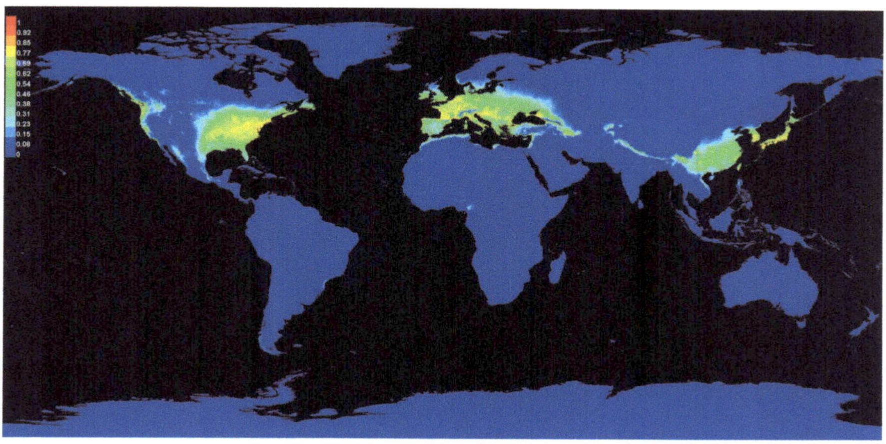

Fig. 1.5 Global potential distribution of hosts of *Ophiocordyceps nutans*. Cooler colors indicate low habitat suitability and warmer colors indicate higher habitat suitability (on a scale of 0–1)

2025). The climate experiences distinct seasons with cold winters necessary for insect diapause or overwintering life stages and warm, humid summers ideal for insect activity and fungal growth/dispersal. The moderate annual temperatures and consistent moisture throughout the year, characteristic of these productive temperate forests, appeared to support robust host populations at densities likely required for a density-dependent pathogen like *Cordyceps* (De Gouvenain and Silander 2017). In the Palearctic, a similarly extensive band of high suitability spanned a large portion of Western and Central Europe, including countries like France, Germany, the United Kingdom, and Poland, and extended into parts of Scandinavia (e.g., southern

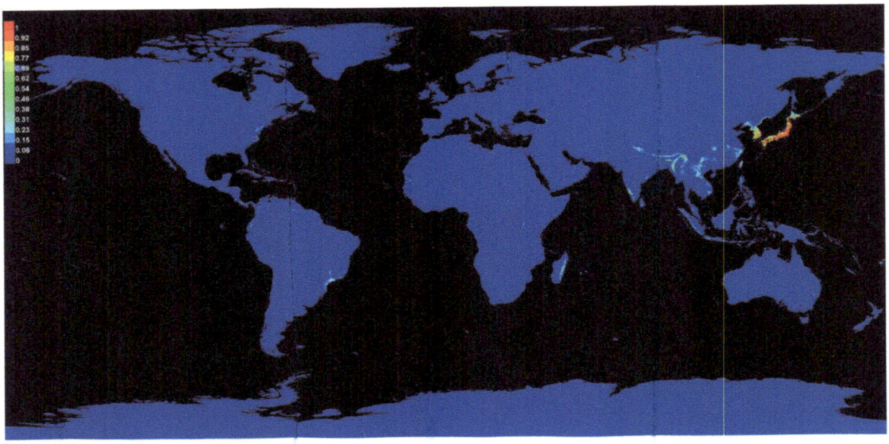

Fig. 1.6 Global potential distribution of *Ophiocordyceps nutans*. Cooler colors indicate low habitat suitability and warmer colors indicate higher habitat suitability (on a scale of 0–1)

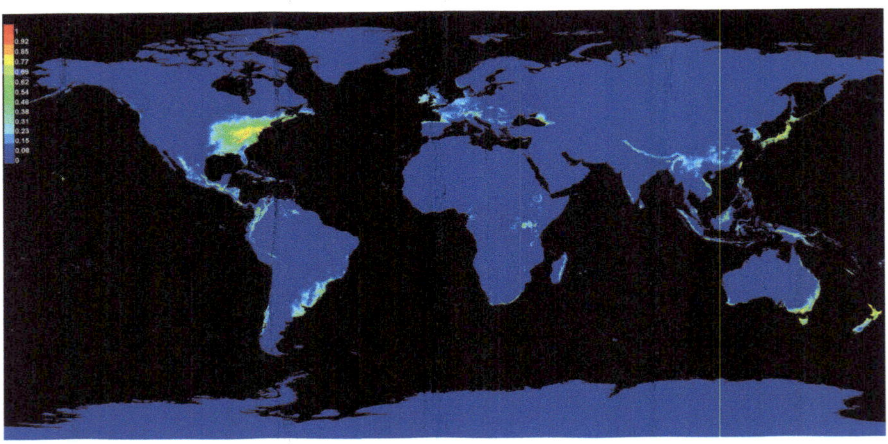

Fig. 1.7 Global potential distribution of *Cordyceps tenuipes*. Cooler colors indicate low habitat suitability and warmer colors indicate higher habitat suitability (on a scale of 0–1)

Sweden and Denmark) and Eastern Europe. This European zone also largely corresponded to temperate broadleaf and mixed forests, featuring moderate temperatures and ample precipitation distributed throughout the year. The comparable seasonal temperature and moisture regimes in European temperate forests likely provided analogous favorable conditions for the life cycle of *C. militaris* hosts as in the Nearctic, supporting dense and active insect populations susceptible to fungal parasitism within similar forest floor microhabitats. In East Asia, high suitability was predicted for significant parts of China (particularly central and eastern regions), South Korea, and Japan. These areas, dominated by temperate forest biomes like deciduous and evergreen broadleaf forests, shared climatic similarities with the European and North American high-suitability zones, exhibiting clear seasonality, including cold winters and warm, humid summers, and sufficient moisture (Hu et al. 2022).

The habitat suitability map for *C. militaris* itself showed a distribution pattern of high suitability remarkably similar to and largely overlapping with that of its hosts, indicating a strong ecological coupling and dependence on host availability within suitable environmental conditions across these temperate realms. High suitability areas for the fungus were predominantly found in the same key temperate regions of the Nearctic (e.g., eastern US states and southeastern Canadian provinces), Palearctic Europe (e.g., Western and Central European countries), and East Asia (e.g., parts of China, South Korea, and Japan) described for its hosts. This near-complete spatial congruence across these distinct but climatically similar realms highlighted the critical dependence of *C. militaris* on the broad-scale distribution and abundance of its suitable insect hosts inhabiting these temperate forest biomes. The environmental conditions favorable for the hosts, such as temperature seasonality, adequate moisture through precipitation, specific temperature ranges during insect development, and the presence of humid substrates like leaf litter, were noted to be equally important, if not more so, for the fungal pathogen's life cycle.

In contrast, *C. tenuipes* exhibited a different pattern of high suitability, suggesting a potentially distinct niche or host association across biogeographic realms compared to *C. militaris* and *O. nutans*, possibly adapted to slightly different environmental cues or host guilds. The most prominent and extensive area of high suitability for *C. tenuipes* was predicted across a substantial part of the Nearctic, covering much of the eastern United States, but importantly, extending further south into areas with warmer temperate or humid subtropical climates than the core high suitability zone for *C. militaris* hosts and *C. militaris*. This included states along the Atlantic coast and extending inland, encompassing a variety of forest types from northern mixed deciduous to more humid subtropical forests towards the transition to the Neotropical realm. This southward extension suggested that *C. tenuipes* or its specific hosts might be adapted to a slightly different, possibly warmer overall temperature regime or require higher ambient humidity levels for optimal development or host interaction within the broader eastern North American temperate zone compared to *C. militaris*. While not as extensive as in the Eastern Nearctic, smaller pockets of high suitability were also visible in scattered locations globally, including some areas in East Asia (Palearctic/Oriental transition), potentially limited areas in the Neotropical (e.g., parts of southeastern South America) and Australasian realms (e.g., parts of eastern Australia, New Zealand), suggesting a potentially wider, although fragmented and perhaps climatically more varied, global distribution for this species or its hosts compared to the more strictly temperate focus of *C. militaris* and the highly restricted *O. nutans*.

The habitat suitability for the hosts of *O. nutans* displayed a pattern of high suitability that shared significant similarities with *C. militaris* hosts but also showed distinct differences across realms, hinting at potential variations in host specificity or environmental preferences within the broader insect groups they parasitize. High suitability was predicted for the portions of the Nearctic, overlapping with the high suitability zone for *C. militaris* hosts across eastern US states and southeastern Canadian provinces. The Palearctic, particularly the Western and Central parts (e.g., France, Germany), also showed extensive areas of high host suitability, again similar to the pattern for *C. militaris* hosts in terms of broad geographic coverage within temperate Europe. However, in East Asia, situated at the confluence of the Palearctic and Oriental realms, while parts of China and South Korea showed high suitability for *C. militaris* hosts, the highest suitability for *O. nutans* hosts appeared more concentrated in specific areas, including Japan and coastal regions of China.

Strikingly, the habitat suitability map for *O. nutans* itself showed a much more restricted and highly localized distribution of high suitability compared to its predicted host distribution or the distributions of *C. militaris* and *C. tenuipes*. The primary and almost exclusive area of high suitability for *O. nutans* was predicted in Japan and immediately adjacent coastal areas of East Asia, such as parts of South Korea and northeastern China.

This very narrow predicted distribution of high suitability for the fungus, juxtaposed with a broader predicted distribution for its hosts across other temperate regions within the Nearctic and Palearctic realms, highlighted a pattern of apparent niche decoupling and strongly suggested that factors beyond the mere presence of potential hosts at a broad scale were critical in determining the actual suitable habitat for *O. nutans*. This specialization might be attributed to requirements for specific climatic seasonality, temperature/humidity regimes characteristic of its East Asian habitat, particular microhabitat conditions (e.g., soil, litter, stable microclimates), dependence on a narrower set of host species, historical biogeographic constraints such as limited dispersal, adaptation to a unique East Asian niche, or a recent evolutionary origin.

3.2 Areas of Medium Habitat Suitability (0.45–0.7)

Regions classified as having medium habitat suitability represented environments where conditions were moderately favorable, potentially supporting populations, but perhaps not at peak densities or with consistent reproductive success compared to high-suitability areas. These zones often bordered areas of high suitability or occurred in transitional climate zones and adjacent biogeographic subregions, where the balance of environmental factors was less optimal than in the core areas but still permissive for persistence, as indicated by the broader patterns on the global maps.

For the hosts of *C. militaris*, medium suitability extended around the high-suitability zones in the Nearctic, covering parts of the central US (e.g., states in the Midwest and Great Plains) and southern Canada, transitioning from temperate forests to grasslands and more continental climates. These areas might experience greater temperature extremes (hotter summers, colder winters), lower overall precipitation, or less consistent seasonality compared to the high-suitability forest zones, potentially impacting host insect populations or the timing of their life cycles, rendering conditions only moderately favourable. In the Palearctic, it encompassed areas south of the main high-suitability band, including parts of Southern Europe (e.g., Italy, Spain, and Greece) and the Balkans, representing transitions to Mediterranean and drier temperate climates with potentially hotter, drier summers or milder, wetter winters. It also included peripheral areas in Northern (e.g., central Scandinavia, southern boreal regions) and Eastern Europe, where cooler overall temperatures or shorter growing seasons might limit host activity or fungal development. In Asia, medium suitability was predicted across broader areas of China and parts of Southeast Asia (extending into the Oriental realm) and stretching westward into Central Asia, encompassing diverse transitional climates and biomes ranging from drier temperate grasslands to subtropical zones.

C. militaris also showed medium suitability in areas surrounding its high-suitability zones across these realms, reflecting the coupled nature of its distribution with its hosts and its own climatic tolerances. This included broader parts of the central Nearctic (e.g., the Midwest US), more extensive areas of the Southern and Eastern Palearctic (e.g., Balkan countries, Ukraine), and wider regions of East and Central Asia. The congruence with its host distribution at the medium suitability level was also apparent, reinforcing the link between the fungus's potential distribution and the availability of suitable hosts under moderately favorable environmental conditions in these transitional biogeographic areas.

C. tenuipes displayed medium suitability extending beyond its core high-suitability zone in the Eastern Nearctic, reaching into the central US and parts of the Caribbean and Central America (Neotropical realm), suggesting potential adaptability to slightly warmer or more humid conditions than *C. militaris*. Medium suitability was also predicted in scattered regions globally, including parts of South America (e.g., southeastern South America, Andean foothills in the Neotropical realm) and isolated areas in Europe (Palearctic), Asia (Palearctic/Oriental), and Oceania (Australasian realm). This broader scattering of medium suitability across multiple realms suggested a potentially wider, although less consistently optimal or more fragmented, potential global range for *C. tenuipes* compared to *C. militaris* or *O. nutans*. This could be due to a broader environmental tolerance allowing it to persist in a wider variety of mesic environments or dependence on host species with a more widespread but patchy distribution across these diverse biogeographic regions.

The hosts of *O. nutans* showed medium suitability in areas adjacent to their high-suitability zones in the Nearctic (e.g., central US, southern Canada), Palearctic Europe (e.g., Southern and Eastern Europe), and East Asia (e.g., broader areas in China, extending into Southeast Asia and the Oriental realm). The pattern suggested that while these areas might not offer peak conditions for *O. nutans* hosts compared to the core high-suitability temperate forest biomes, they could still support insect populations in transitional climates and biomes.

For *O. nutans* itself, areas of medium suitability were extremely limited globally. They appeared as small zones immediately surrounding the high-suitability region in Japan and adjacent East Asia, within the Palearctic/Oriental transition, changing over rapidly to low or unsuitable conditions elsewhere across all other realms. This further emphasized the apparently narrow environmental niche or strong geographic constraints on this specific fungal species. Even in areas broadly predicted as medium suitability for its hosts in other temperate regions (like Europe or North America), the specific environmental requirements for *O. nutans* itself were apparently not met consistently, highlighting a likely mismatch between the fungus's precise niche and the broader range of its host group even at moderately favorable conditions.

3.3 Areas of Low Habitat Suitability (0.2–0.45)

Areas with low habitat suitability were predicted to be marginally suitable, where environmental conditions might allow for occasional or sparse populations but were generally suboptimal for consistent presence or reproduction of the organisms. These zones often represented the periphery of the potential distribution, often bordering unsuitable or extremely different biogeographic zones, as indicated by the gradual decrease in suitability values on the maps, suggesting increasing environmental stress or resource limitation that pushes conditions towards the limits of tolerance.

Low suitability for the hosts of *C. militaris* was widespread across the globe, including parts of western North America (e.g., states in the Great Plains and Rocky Mountains, transitioning to arid and montane environments), large areas of South America (Neotropical, Andean), Africa (Afrotropical), and Australia (Australasian), as well as Southern Europe (Mediterranean fringe), the Middle East and Central Asia (drier Palearctic/Oriental regions), and large parts of Southeast Asia (Oriental tropics). These regions covered diverse climates and biomes, including drier temperate zones with insufficient moisture or high temperature stress; Mediterranean climates with pronounced dry seasons unfavorable for moisture-dependent hosts or fungal stages; parts of the subtropics; and higher elevations in some areas with shorter growing seasons or colder temperatures.

C. militaris also showed widespread low suitability across similar regions to its hosts, indicating the direct and indirect influence of environmental limitations across realms. This included western North America, much of South America, Africa, Australia, and extensive parts of Asia. The presence of low suitability for the fungus in areas with low host suitability further highlighted the intertwined nature of their distributions. In these diverse regions, spanning multiple biogeographic realms, even if temperature and moisture conditions were occasionally permissive for fungal growth or spore survival, the scarcity or absence of suitable hosts, the infrequency of favorable periods for the entire infection cycle, or high rates of mortality from environmental stress would limit the fungus's ability to establish and persist effectively, based on the broad patterns shown.

C. tenuipes displayed low suitability across vast areas of the globe, including much of the Nearctic outside the eastern high/medium suitability zone and large parts of the Neotropical (excluding core rainforest), Afrotropical, Palearctic,

Oriental, and Australasian realms. This broad distribution of low suitability suggested that while optimal conditions were restricted, *C. tenuipes* might be able to survive in a wider range of marginally suitable environments across different continents and climates than *C. militaris* or *O. nutans*, perhaps utilizing a broader range of less-preferred hosts, having a slightly wider environmental tolerance, or being less sensitive to certain environmental extremes found at the edges of its core niche. Persistence in these low suitability zones might rely on occasional favorable years, the presence of isolated, highly specific microhabitats acting as refugia, or sporadic dispersal from more suitable areas.

The hosts of *O. nutans* showed low suitability across extensive parts of the world, including most of the Nearctic outside the high or medium zones, large parts of the Neotropical, Afrotropical, Palearctic (especially northern and southern fringes), Central Asian, and Australasian realms. Similar to *C. militaris* hosts, this indicated that suitable hosts or their required environmental conditions were generally absent or scarce in these varied climatic zones and associated biomes across different biogeographic regions, limiting the potential for fungal interaction due to a lack of the essential resource.

For *O. nutans*, low suitability was predicted for a few scattered, very limited areas globally, mainly as narrow transitional zones around the medium/high suitability regions in East Asia, within the Palearctic or Oriental transition zone. Much of the global map outside the core East Asian region was predicted to be unsuitable, indicating a very narrow tolerance range or strong geographic constraints for this species even in marginally favorable climates across all other realms. Even in regions broadly predicted as having low suitability for its hosts in other temperate realms (like peripheral areas of Europe or North America), the specific, likely stringent, environmental conditions necessary for *O. nutans* development and reproduction were apparently not met, suggesting a limitation beyond just marginal host availability.

3.4 Areas of Unsuitable Habitat (0–0.2)

The unsuitable habitat represented regions where environmental conditions were highly unfavorable, unlikely to support persistent populations of the organisms or their hosts. These areas, often characterized by climatic extremes or biomes fundamentally different from occurrence locations, acted as significant biogeographic barriers and defined the outer limits of each species' potential ecological niche. Vast regions globally were predicted to be unsuitable for the hosts of both *C. militaris* and *O. nutans*. These included the Arctic and Antarctic realms, specifically mentioning areas like Greenland and Antarctica (due to extreme cold), large deserts such as the Sahara in North Africa, the Arabian Desert in the Middle East, the Australian Outback, and the Gobi Desert in Asia (due to aridity and temperature extremes). High mountain ranges like the Himalayas in Asia and parts of the Andes in South America were also unsuitable (likely due to altitude and cold). Furthermore, large portions of the tropics, including the Amazon basin in South America, the

Congo basin in Africa, and large parts of Central Africa and Southeast Asia, were unsuitable (due to continuous high temperatures/humidity lacking necessary seasonality for temperate hosts). Similarly, large areas were predicted to be unsuitable for the fungi themselves, often overlapping extensively with their hosts' unsuitable zones. For *C. militaris*, unsuitable regions closely emulated those of its hosts across polar regions (Arctic, Antarctic), major deserts (Sahara and Gobi), high mountain areas (Himalayas, Andes), and large areas of the tropics (e.g., equatorial Africa, Amazon basin). This reiterated its dependence on both host presence and compatible climate, where extreme cold, heat, prolonged dryness, or consistently high humidity without suitable cycles rendered habitats unsuitable for both fungal development and host survival.

C. tenuipes also showed vast unsuitable areas globally, including the polar regions (Arctic, Antarctic), major deserts (Sahara, Gobi), and high mountain ranges, where climatic extremes likely exceed physiological limits. However, compared to *C. militaris*, some tropical areas in the Neotropical and Oriental realms appeared to fall into low or even localized medium suitability categories for *C. tenuipes*, suggesting a potentially broader tolerance for warmer, more humid conditions. Nevertheless, large parts of the deep tropics remained unsuitable, likely due to incompatibility with specific life cycle requirements.

O. nutans exhibited the most extensive areas of predicted unsuitable habitat. Almost the entire world outside of Japan and immediate neighboring areas in East Asia was classified as unsuitable across all major biogeographic realms, including the Nearctic, most of the Palearctic, Neotropical, Afrotropical, and Australasian realms. This severe restriction highlighted that *O. nutans* likely possesses a very narrow environmental niche or is highly dependent on geographically limited factors specific to its core East Asian range.

4 Global Scale Insights

The MaxEnt modeling approach, primarily utilizing environmental variables, revealed distinct global patterns of habitat suitability for the studied *Cordyceps* and *Ophiocordyceps* species and their hosts, providing valuable insights into their potential ecological niches and how they interact with the environmental conditions and host availability across major biogeographic realms.

4.1 Main Suitability Patterns and Environmental Associations

The observed distributions of high suitability were predominantly concentrated in the temperate regions of the Northern Hemisphere (Nearctic and Palearctic), characterized by moderate temperatures, distinct seasonality, and sufficient precipitation. These conditions were generally favorable for the diverse temperate insect hosts these fungi infect and for the various stages of the fungal life cycle, such as spore germination, hyphal growth within the host, and the emergence of fruiting bodies. For *C. militaris* and its hosts and *O. nutans* hosts, major suitability hubs were identified in these key temperate regions, aligning with the distribution of temperate broadleaf and mixed forests, biomes that offer the necessary environmental structure, microclimates (e.g., stable humidity in forest litter), and host communities (Hu et al. 2022). Unsuitable areas consistently represented climatic extremes and fundamentally different biomes (polar, desert, high mountain, and core tropics), acting as significant environmental barriers.

4.2 Comparative Fungus–Host Relationships: Coupling Versus Decoupling

A critical insight gained from comparing the maps was the contrasting relationship between fungus and host predicted suitability. The strong spatial congruence between the predicted suitability of *C. militaris* and its hosts across the Nearctic and Palearctic revealed a pattern of apparent niche coupling at the macro-scale. This suggested that the environmental niche of *C. militaris* was largely contained within, or closely emulated, the broad environmental niche of its host group within these temperate biomes, indicating that the fungus's potential distribution was primarily driven by the environmental suitability for its hosts and macroclimate, which had a direct influence on the fungal life cycle itself. In such a coupled system, factors benefiting the host group under favorable temperate climates also largely benefit the pathogen (Haverkost et al. 2010). This pattern is consistent with a scenario where the pathogen has broad environmental tolerances and/or a broad host range, allowing its distribution to largely track its resource base across a wide environmental space. Conversely, the significant disparity between the broad potential distribution of *O. nutans* hosts and the highly restricted predicted range of *O. nutans* fungus presented a clear pattern of apparent niche decoupling. This finding highlighted that while host presence was necessary, it was not sufficient; *O. nutans* appeared to require a more specific or stringent combination of environmental conditions than its hosts or faced limitations not related to broad host group distribution. This suggested that environmental filtering might be acting more selectively or intensely on *O. nutans* than on its host group at a broad scale, potentially due to narrower physiological tolerances, dependence on specific fine-scale environmental factors, or reliance on a more restricted subset of host species whose distribution is also more limited than the overall group modeled. This represents a scenario where the fungus's niche is a much smaller subset of the host group's niche within the environmental landscape.

4.3 Interspecific Fungal Niche Differentiation

The comparison of the three fungal habitat suitability maps revealed interspecific niche differentiation among *C. militaris*, *C. tenuipes*, and *O. nutans*. *C. militaris* exhibited a widespread temperate niche across both the Nearctic and Palearctic, indicating adaptation to a broad range of conditions within these realms. *C. tenuipes* showed a distinct pattern with its highest suitability concentrated in the eastern Nearctic and broader potential including warmer, more humid environments and scattered areas in other realms, suggesting a somewhat different climatic envelope or adaptation that allows it to potentially thrive in regions less optimal for *C. militaris*. *O. nutans* displayed a remarkably narrow niche, primarily confined to East Asia, indicating potentially stricter environmental requirements than either *C. militaris* or *C. tenuipes*. This differentiation indicated that despite being related entomopathogens, these species likely have evolved distinct environmental tolerances or host associations that allow them to occupy different parts of the global environmental space, potentially reducing direct competition for resources (hosts) or environmental space. Analyzing the specific environmental variables driving suitability for each fungus, potentially through comparing MaxEnt response curves, would be crucial to quantify these niche differences and understand the ecological factors promoting their coexistence or spatial separation.

4.4 Host Group Niche Differences and Implications

Comparing the ENM for *C. militaris* hosts and *O. nutans* hosts also provided insight into the potential ecological preferences of the insect communities they parasitize. While both host groups were broadly associated with temperate Northern Hemisphere biomes and shared extensive areas of high suitability, subtle differences were visible in the extent or concentration of high suitability in regions, particularly East Asia and potentially at the warmer or drier edges of the temperate zones. These potential differences in the macroenvironmental preferences of the host groups themselves could partly contribute to the differing distributions of the fungi that parasitize them.

4.5 Biogeographic Discrepancies and Puzzles

The comparative analysis, especially the pronounced contrast between the widespread predicted distribution of *O. nutans* hosts and the highly restricted range of the *O. nutans* fungus, presents a biogeographic puzzle that calls for further investigation into historical processes. This pattern strongly suggested that factors beyond the potential ecological niche defined by broad climate and host group presence were limiting the realized distribution of *O. nutans*. Potential explanations, highlighted by this comparison, include the possibility that *O. nutans* possesses a very narrow environmental niche, requiring a specific combination of environmental conditions (perhaps microclimatic or related to fine-scale seasonality) that is consistently met only in its core East Asian range, making it highly sensitive to environmental filtering. Alternatively, this discrepancy could be primarily driven by historical biogeographic factors, such as a relatively recent evolutionary origin in East Asia, followed by limited dispersal capabilities preventing *O. nutans* from colonizing other seemingly suitable temperate areas after its divergence. Dispersal barriers like oceans, vast arid zones, or cold regions may have been more effective impediments for *O. nutans* compared to *C. militaris*, which appeared to have successfully colonized multiple temperate landmasses. Adaptation to unique biotic factors (e.g., highly specific endemic host lineages or unique competitor/mutualist interactions) prevalent only in East Asia could also restrict its establishment elsewhere (Sasaki et al. 2012). The contrast with the more widespread *C. militaris* suggests differential abilities to overcome historical barriers or adapt to variations within temperate biomes across continents, making the *O. nutans* case a compelling example of how biogeographic history can constrain a species' distribution within its potential ecological niche.

5 Limitations and Way Forward

While global-scale MaxEnt models provided a valuable macroecological and biogeographic framework—revealing broad habitat suitability patterns and highlighting distinct ecological niches—they also highlighted limitations necessitating further, more targeted investigation, particularly when interpreting the comparisons and the apparent niche decoupling and differentiation observed. The broad resolution of global environmental data constrained the ability to assess suitability at fine spatial scales, where microclimatic variations (e.g., humidity in litter, temperature fluctuations near the ground), soil conditions (e.g., pH, moisture retention), and local habitat structure (e.g., forest type, canopy cover) play critical roles in creating the specific microhabitats required by both fungi and ground-dwelling insects, varying significantly even within broadly suitable regions. Furthermore, the host distribution inputs were based on predicted suitability for broad host groups, not the precise distribution and density of individual host species, a limitation, especially for interpreting the *O. nutans* pattern if it is a narrow specialist on a host with a more restricted range than the broader group modeled. While the study inferred potential roles for environmental filtering and host availability, factors like dispersal limitations, biotic interactions beyond the primary host (e.g., hyperparasites, competitors, facilitators), and fine-scale habitat structure were not explicitly modeled, especially relevant for understanding the restricted range of *O. nutans*.

The comparative analysis, however, identified key areas for targeted future research to overcome these limitations and address the puzzles raised. One direction involves the targeted investigation of mismatch areas, specifically regions predicted as suitable for hosts but not the associated fungus (like *O. nutans* hosts vs. the fungus outside East Asia), which are prime locations for hypothesis-driven field research. Focused surveys in these specific mismatch areas should aim to confirm fungal absence and collect fine-scale environmental and biological data; comparing this data from mismatch areas to those from high-suitability areas (e.g., Japan for *O. nutans*) could pinpoint the precise limiting factors (environmental or biotic) not captured by the broad-scale model, explaining the observed decoupling and geographic restriction. Building upon this, high-resolution regional modeling in identified highly suitable or mismatched areas (e.g., Eastern North America, Europe, East Asia), using much higher-resolution environmental data and more localized occurrence records for the fungi and specific hosts, if possible, would provide more accurate suitability maps at sub-country and local scales, allowing for validation of predictions and a better understanding of habitat partitioning. Quantitatively identifying the key environmental variables responsible for the observed niche differentiation among the fungi and host groups is also essential; quantifying niche differences and drivers can be achieved by accessing and analyzing variable importance metrics and response curves from the MaxEnt models, providing empirical support for inferred ecological drivers. To address the biogeographic puzzle, particularly *O. nutans*'s restricted range, integrating ENM results with phylogeographic studies is vital; integrating biogeographic and genetic data through genetic analysis of *O. nutans* populations and its key host lineages across different realms can reveal patterns that support or refute hypotheses about dispersal limitations, historical isolation, or adaptation to specific endemic conditions (as suggested by studies like Sasaki et al. 2012). Finally, a fundamental aspect for strengthening ecological niche modeling, especially in data-poor regions, involves enhancing occurrence data through institutional and citizen science initiatives (Rathnayaka et al. 2024). Investing in and supporting efforts focused on documenting observations of these fungi and their hosts can significantly strengthen publicly available datasets, providing invaluable fine-scale occurrence data for more robust and validated models in the future.

6 Conclusion

Global MaxEnt modeling revealed distinct habitat suitability for *C. militaris*, *C. tenuipes*, and *O. nutans*, highlighting unique ecological niches and distributional constraints. Suitable habitats are concentrated within temperate zones of the Northern Hemisphere (Nearctic and Palearctic), offering moderate temperatures, seasonality, and precipitation ideal for both fungi and their hosts. Temperate forests were key biomes, providing stable microclimates and host communities. *C. militaris* displayed broad environmental tolerance and a wide predicted distribution overlapping with its host group. This strong niche coupling suggests that the fungus is a generalist, its range embedded within the hosts' niche and availability in temperate regions. In contrast, *O. nutans* showed a notably limited pattern, suitable almost exclusively within Japan and adjacent East Asia. This restriction, despite a broader host distribution, signifies niche decoupling and dependence on more specific environmental conditions than modeled, potentially including fine-scale factors like microhabitat humidity or narrow temperature ranges. This pattern may also point towards higher host specificity or limited dispersal, tying it to its historical East Asian origin.

Comparing the three fungi highlights clear interspecific niche differentiation: *C. militaris* occupies a widespread temperate niche, *C. tenuipes* prefers warmer and more humid environments, and *O. nutans* is confined to East Asia. These differences reflect evolutionary divergence in niche preferences driven by host associations, tolerances, or historical barriers. While host groups predict broad temperate suitability, subtle distribution differences may influence fungal ranges, especially for specialists like *O. nutans*. The disparity between the potential *O. nutans* host range and its predicted fungus range strongly suggests intriguing potential biogeographic influences beyond immediate ecological suitability. Hypotheses include a relatively recent East Asian origin with insufficient time for wider dispersal or dependence on highly specific host lineages or biotic partners.

The scope of global-scale MaxEnt modeling offers valuable macroecological perspective and opens avenues for further investigation. The current study provides a foundational understanding despite reliance on coarse data and general host groups. Future research can build on these findings: Field validation for *O. nutans* in potential areas outside East Asia helps distinguish fine-scale unsuitability vs. dispersal barriers. High-resolution regional modeling with localized data will refine predictions. Analyzing environmental response curves can identify key niche drivers. Integrating genetic and phylogeographic data holds the potential to illuminate how history and environment shape these fungi's distributions.

References

Aiello-Lammens ME, Boria RA, Radosavljevic A, Vilela B, Anderson RP (2015) spThin: an R package for spatial thinning of species occurrence records for use in ecological niche models. Ecography 38(5):541–545. https://doi.org/10.1111/ecog.01132

Allouche O, Tsoar A, Kadmon R (2006) Assessing the accuracy of species distribution models: prevalence, kappa and the true skill statistic (TSS). J Appl Ecol 43(6):1223–1232. https://doi.org/10.1111/j.1365-2664.2006.01214.x

Araújo MB, Anderson RP, Márcia Barbosa A, Beale CM, Dormann CF et al (2019) Standards for distribution models in biodiversity assessments. Sci Adv 5(1):eaat4858. https://doi.org/10.1126/sciadv.aat4858

Barve V, Hart E (2022) rinat: Access 'iNaturalist' Data Through APIs. R package version 0.1.9. Retrieved from https://CRAN.R-project.org/package=rinat. Accessed 20 April 2025

Calvo AM, Wilson RA, Bok JW, Keller NP (2002) Relationship between secondary metabolism and fungal development. Microbiol Mol Biol Rev 66(3):447–459. https://doi.org/10.1128/MMBR.66.3.447-459.2002

Chamberlain S, Barve V, Mcglinn D, Oldoni D, Desmet P, et al (2025) rgbif: Interface to the Global Biodiversity Information Facility API. R package version 3.7.7. Retrieved from https://CRAN.R-project.org/package=rgbif. Accessed 20 April 2025

Chen J, He D (2025) Potential geographical distribution of *Cordyceps cicadae* and its two hosts in China under climate change. Front Microbiol 15:1519560. https://doi.org/10.3389/fmicb.2024.1519560

Chesson P (2000) Mechanisms of maintenance of species diversity. Ann Rev Ecol Evol Syst 31: 343–366. https://doi.org/10.1146/annurev.ecolsys.31.1.343

Chou YC, Sung TH, Hou SJ, Khumsupan D, Santoso SP et al (2024) Current progress regarding *Cordyceps militaris*, its metabolite function, and its production. Appl Sci 14:4610. https://doi.org/10.3390/app14114610

De Gouvenain RC, Silander JA (2017) Temperate forests. In: Reference module in life sciences. https://doi.org/10.1016/b978-0-12-809633-8.02310-4

Dubhashi S, Sinha S, Dwivedi S, Ghanekar J, Kadam S et al (2023) Early trends to show the efficacy of *Cordyceps militaris* in mild to moderate COVID inflammation. Cureus 15(8): e43731. https://doi.org/10.7759/cureus.43731

Evans HC (1982) Entomogenous fungi in tropical forest ecosystems: an appraisal. Ecol Entomol 7(1):47–60. https://doi.org/10.1111/j.1365-2311.1982.tb00643.x

Fick SE, Hijmans RJ (2017) WorldClim 2: new 1-km spatial resolution climate surfaces for global land areas. Int J Climatol 37(12):4302–4315. https://doi.org/10.1002/joc.5086

Filz KJ, Schmitt T (2015) Niche overlap and host specificity in parasitic Maculinea butterflies (Lepidoptera: Lycaenidae) as a measure for potential extinction risks under climate change. Org Divers Evol 15:555–565. https://doi.org/10.1007/s13127-015-0210-1

Hajek AE, St Leger RJ (1994) Interactions between fungal pathogens and insect hosts. Annu Rev Entomol 39:293–322. https://doi.org/10.1146/annurev.en.39.010194.001453

Haverkost TR, Gardner SL, Peterson AT (2010) Predicting the distribution of a parasite using the ecological niche model, GARP. Rev Mex Biodiv 81(3):895–902

Hijmans R (2024) terra: Spatial Data Analysis. R package version 1.7–78. https://CRAN.R-project.org/package=terra

Hu JJ, Zhao GP, Tuo YL, Qi ZX, Yue L et al (2022) Ecological factors influencing the occurrence of macrofungi from eastern mountainous areas to the Central Plains of Jilin Province, China. J Fungi 8:871. https://doi.org/10.3390/jof8080871

Huang E, Chen Y, Fang M, Zheng Y, Yu S (2021) Environmental drivers of plant distributions at global and regional scale. Glob Ecol Biogeogr 30(3):697–709. https://doi.org/10.1111/geb.13251

Hutchinson GE (1957) Concluding remarks. Cold Spring Harb Symp Quant Biol 22:415–427. https://doi.org/10.1101/SQB.1957.022.01.039

Kamble VR, Agre DG (2012) Reinvestigation of *Cordyceps militaris* (L. ex St. Amaans) Link insect parasitic fungus from Maharashtra. Bionano Front 5(2):224–225

Karun NC, Sridhar KR (2013a) Incidence of Entomophagous medicinal fungus, *Ophiocordyceps nutans* on Stink Bug, *Halyomorpha halys* (Stål) in the Western Ghats of India. J Biol Control 27(2):139–143

Karun NC, Sridhar KR (2013b) The stink bug fungus *Ophiocordyceps nutans* – a proposal for conservation and flagship status in the Western Ghats of India. Fungal Conserv 3:43–49

Kier G, Kreft H, Lee TM, Jetz W, Ibisch PL et al (2009) A global assessment of endemism and species richness across Island and mainland regions. Proc Natl Acad Sci USA 106(23): 9322–9327. https://doi.org/10.1073/pnas.0810306106

Kobayashi Y (1941) The genus *Cordyceps* and its allies. Sci Rep Tokyo Bunrika Daigaku, Sect B 5(84):53–260

Lee SK, Lee JH, Kim HR (2019) Improved cordycepin production by *Cordyceps militaris* KYL05 using casein hydrolysate in submerged conditions. Biomolecules 9(9):461. https://doi.org/10.3390/biom9090461

Levins B (1968) Evolution in changing environments: some theoretical explorations (MPB-2). In: Monographs in population biology. Princeton University Press. https://doi.org/10.2307/j.ctvx5wbbh

Lomolino MV, Riddle BR, Whittaker RJ (2016) Biogeography, 5th edn. Sinauer Associates Inc, USA

Lou H, Lin J, Guo L, Wang X, Tian S, Liu C, Zhao Y, Zhao R (2019) Advances in research on *Cordyceps militaris* degeneration. Appl Microbiol Biotechnol 103(19):7835–7841. https://doi.org/10.1007/s00253-019-10074-z

Mantzoukas S, Kitsiou F, Natsiopoulos D, Eliopoulos PA (2022) Entomopathogenic fungi: interactions and applications. Encyclopedia 2:646–656. https://doi.org/10.3390/encyclopedia2020044

MyCoPortal (2025). Retrieved from https://www.mycoportal.org/portal/collections/map/index.php. Accessed 20 April 2025

Naimi B, Hamm NAS, Groen TA, Skidmore AK, Toxopeus AG (2014) Where is positional uncertainty a problem for species distribution modelling. Ecography 37:191–203. https://doi.org/10.1111/j.1600-0587.2013.00205.x

Nan Q, Zhang Q, Li X, Zheng D, Li Z, Zhao L (2023) Niche and Interspecific Association of the Dominant Species during the Invasion of *Alternanthera philoxeroides* in the Yangtze River Basin, China. Agriculture 13:621. https://doi.org/10.3390/agriculture13030621

NASA (2025) Temperate Deciduous Forest. Retrieved from https://earthobservatory.nasa.gov/biome/biotemperate.php. Accessed 20 April 2025

Nguyen TM, Doan PTT, Nguyen VH (2023) Observations of the host habitat of *Ophiocordyceps nutans* in the Bidoup Nui Ba National Park, Lam Dong, Vietnam. IOP Conf Ser Earth Environ Sci 1155:012010. https://doi.org/10.1088/1755-1315/1155/1/012010

Olatunji OJ, Tang J, Tola A, Auberon F, Oluwaniyi O et al (2018) The genus *Cordyceps*: an extensive review of its traditional uses, phytochemistry and pharmacology. Fitoterapia 129: 293–316. https://doi.org/10.1016/j.fitote.2018.05.010

Peng Y, Xu D, Ali H, Liu Z, Zhuo Z (2025) Quantifying potentially suitable geographical habitat changes in Chinese Caterpillar fungus with enhanced MaxEnt Model. Insects 16(3):262. https://doi.org/10.3390/insects16030262

Peterson AT, Soberón J, Pearson RG (2011) Ecological niches and geographic distributions (MPB-49). In: Levin SA, Horn HS (eds) Monographs in population biology. Princeton University Press. https://doi.org/10.23943/princeton/9780691136868.001.0001

Phillips SJ, Dudík M (2008) Modeling of species distributions with Maxent: new extensions and a comprehensive evaluation. Ecography 31:161–175. https://doi.org/10.1111/j.0906-7590.2008.5203.x

Phillips SJ, Anderson RP, Schapire RE (2006) Maximum entropy modeling of species geographic distributions. Ecol Model 190(3–4):231–259. https://doi.org/10.1016/j.ecolmodel.2005.03.026

Pradhan P (2016) Strengthening MaxEnt modelling through screening of redundant explanatory Bioclimatic Variables with Variance Inflation Factor analysis. Research 8(5):29–34. https://doi.org/10.7537/marsrsj080516.05

Pradhan P (2024) Biogeography and impacts of climate change on the distribution of *Ophiocordyceps sinensis*. In: Deshmukh SK, Sridhar KR (eds) *Cordyceps* and allied species. Springer, Singapore. https://doi.org/10.1007/978-981-97-6345-0_1

Pradhan P, Setyawan AD (2021) Filtering multi-collinear predictor variables from multi-resolution rasters of WorldClim 2.1 for Ecological Niche Modeling in Indonesian context. Asian J For 5(2):111–122. https://doi.org/10.13057/asianjfor/r050207

Pradhan P, Dutta AK, Paloi S, Roy A, Acharya K (2016) Diversity and distribution of macrofungi in Eastern Himalayan ecosystem. EurAsian J BioSci 10:1–12. https://doi.org/10.5053/ejobios.2016.10.0.1

Pradhan P, De J, Acharya K (2023) Strategies in artificial cultivation of two entomopathogenic fungi *Cordyceps militaris* and *Ophiocordyceps sinensis*. In: Singh SK, Kumar D, Shamim M, Sharma R (eds) Applied mycology for agriculture and foods industrial applications. Apple Academic Press, partnered with CRC Press (Taylor & Francis). https://doi.org/10.1201/9781003369868

Quesada-Moraga E, González-Mas N, Yousef-Yousef M, Garrido-Jurado I, Fernández-Bravo M (2024) Key role of environmental competence in successful use of entomopathogenic fungi in microbial pest control. J Pest Sci 97:1–15. https://doi.org/10.1007/s10340-023-01622-8

R Core Team (2025) R: a language and environment for statistical computing. R Foundation for Statistical Computing, Vienna, Austria. https://www.R-project.org/

Radosavljevic A, Anderson RP (2014) Making better Maxent models of species distributions: complexity, overfitting and evaluation. J Biogeogr 41:629–643. https://doi.org/10.1111/jbi.12227

Rathnayaka AR, Tennakoon DS, Jones GEB, Wanasinghe DN, Bhat DJ et al (2024) Significance of precise documentation of hosts and geospatial data of fungal collections, with an emphasis on plant-associated fungi. New Zealand J Bot 63(2–3):462–489. https://doi.org/10.1080/0028825X.2024.2381734

Rathod LB, Pasha SM, Rajan A, Pasha C (2023) Sequential strain improvement of *Cordyceps militaris*. J Adv Microbiol 23(3):13–20. https://doi.org/10.9734/JAMB/2023/v23i3711

Sasaki F, Miyamoto T, Yamamoto A, Tamai Y, Yajima T (2012) Relationship between intraspecific variations and host insects of *Ophiocordyceps nutans* collected in Japan. Mycoscience 53(2): 85–91. https://doi.org/10.1007/S10267-011-0137-0

Schoener TW (1986) Mechanistic approaches to community ecology: a new reductionism. Am Zool 26(1):81–106. https://doi.org/10.1093/icb/26.1.81

Shrestha UB, Bawa KS (2014) Impact of climate change on potential distribution of Chinese Caterpillar fungus (*Ophiocordyceps sinensis*) in Nepal Himalaya. PLoS One 9(9):e106405. https://doi.org/10.1371/journal.pone.0106405

Shrestha B, Zhang W, Zhang Y, Liu X (2012) The medicinal fungus *Cordyceps militaris*: research and development. Mycol Prog 11:599–614. https://doi.org/10.1007/s11557-012-0825-y

Shrestha B, Tanaka E, Hyun MW, Han JG, Kim CS et al (2017) Mycosphere Essay 19. *Cordyceps* species parasitizing hymenopteran and hemipteran insects. Mycosphere 8(9):1424–1442. https://doi.org/10.5943/mycosphere/8/9/8

Sung GH, Hywel-Jones NL, Sung JM, Luangsa-Ard JJ, Shrestha B et al (2007) Phylogenetic classification of *Cordyceps* and the clavicipitaceous fungi. Stud Mycol 57:5–59. https://doi.org/10.3114/sim.2007.57.01

Tapingkae T, Yachai M, Tapingkae W (2024) The medicinal fungus *Cordyceps militaris.* Academia to industry. J Nepal Biotechnol Assoc 5(1):69–71

Walther GR, Post E, Convey P, Menzel A, Parmesan C et al (2002) Ecological responses to recent climate change. Nature 416:389–395. https://doi.org/10.1038/416389a

Winker D (2010) *Cordyceps sinensis*: a precious parasitic fungus infecting Tibet. Field Mycol 11: 60–67. https://doi.org/10.1016/j.fldmyc.2010.04.009

Wisz MS, Pottier J, Kissling WD (2013) The role of biotic interactions in shaping distributions and realised assemblages of species: implications for species distribution modelling. Biol Rev 88: 15–30. https://doi.org/10.1111/j.1469-185X.2012.00235.x

Wu X, Wu T, Huang A (2021) New insights into the biosynthesis of typical bioactive components in the traditional Chinese medicinal fungus *Cordyceps militaris*. Front Bioeng Biotechnol 7(9): 801721. https://doi.org/10.3389/fbioe.2021.801721

Yan C, Hao H, Sha S (2024) Comparative assessment of habitat suitability and Niche overlap of three *Cytospora* species in China. J Fungi (Basel) 10(1):38. https://doi.org/10.3390/jof10010038

Zhang M, Sun X, Miao Y, Li M, Huang L (2022) *Cordyceps cicadae* and *Cordyceps gunnii* have closer species correlation with *Cordyceps sinensis*: from the perspective of metabonomic and MaxEnt models. Sci Rep 12:20469. https://doi.org/10.1038/s41598-022-24309-z

Taxonomy and Molecular Phylogeny of Entomopathogenic Fungi

2

Sakshi Suman, Vaishnavi Rajwar, and Yamini Agrawal

Abstract

Entomopathogenic fungi (EPF) are crucial regulators in ecosystems, influencing host populations and biodiversity. Despite their ecological significance and potential as biocontrol agents, our understanding of EPF remains limited due to gaps in knowledge related to taxonomic clarity, host specificity, and geographical distribution. However, advancements in molecular techniques, including DNA barcoding, multilocus sequencing, and whole-genome sequencing (WGS), are transforming EPF classification. These methods are overcoming the limitations of traditional morphology, revealing previously hidden diversity and intricate phylogenetic relationships across major fungal phyla. *Hypocreales* (*Ascomycota*) is the most prominent among EPF, comprising key genera such as *Metarhizium*, *Beauveria*, and *Ophiocordyceps*. Comparative studies indicate that ecological factors, rather than host specificity alone, often drive speciation. Therefore, an integrative taxonomic approach, combining morphological, ecological, and genomic data, is critical for accurate species identification, understanding host-pathogen dynamics, predicting species invasions or extinctions, and optimizing EPF for sustainable pest management and ecological conservation.

Keywords

Classification · Cryptic · Hypocreales · Polyphyletic · Systematics

Sakshi Suman and Vaishnavi Rajwar contributed equally with all other contributors.

S. Suman · V. Rajwar · Y. Agrawal (✉)
Department of Botany, University of Delhi, New Delhi, Delhi, India

S. K. Deshmukh et al. (eds.), *Frontiers in Entomopathogenic Fungi*,
https://doi.org/10.1007/978-981-95-6529-0_2

1 Introduction

Fungi represent one of the most diverse groups of organisms on Earth, second only to insects. Recent estimates suggest that there may be over 2 to 4 million fungal species globally, yet only about 2.5 million species have been estimated to be formally described to date (Hawksworth and Lücking 2017; Niskanen et al. 2023; Hyde et al. 2024). Fungi inhabit nearly every ecological niche, playing critical roles in ecosystem function, pathogenic and symbiotic relationships, nutrient cycling, and saprotrophism, making them evolutionary and ecologically significant. However, despite their importance, our understanding of fungal taxonomy, evolution, and ecology remains inadequate, with over 90% of fungal diversity still unexplored and undescribed.

Pathogens are key regulators of species populations and hold ecological significance in ecosystems (De Wint et al. 2024). It is quite evident to find their increased impact with the introduction of climate change and anthropogenic activities in due course of time (Singh et al. 2023). Host-pathogen interactions are seen to get affected by the dynamics of the host population (Liu et al. 2007; Brown 2023). Therefore, accurate identification of pathogen-infected hosts is important to enable effective disease control and ecosystem conservation strategies (Burdon and Thrall 2008; Goncalves 2025).

EPF are the notable pathogens that infect and kill insects and other arthropods from different evolutionary groups worldwide (Kaishian et al. 2024). It is recognized as one of the most effective and desirable eco-friendly alternatives to chemical insecticides, playing a critical role in regulating insect populations in an ecosystem (Sharma et al. 2020). The infection begins when the spores attach to the host and penetrate through the cuticle. Later on, it proliferates by growing mycelium inside the hemocoel. Under favorable conditions, it emerges through the host body and forms fruiting bodies that generate and produce new spores to continue the infection cycle (Mora et al. 2018). Fungi and insects often co-occur in overlapping ecological niches, suggesting their profound ecological interdependence and adaptation to similar habitats (Skrzecz et al. 2024; Koyuncu et al. 2025).

Morphological description is the foundation of every taxonomic study. However, these are often limited, as many fungal lineages lack distinctive morphological characters. Many species have pleomorphic life cycles, producing teleomorph and anamorph stages, which are difficult to resolve using traditional morphological approaches (Shenoy et al. 2007; Ban et al. 2009; Lucking et al. 2020). Some fungal lineages are very complex, containing several cryptic species that are not distinguishable only on the basis of phenotypic features (Kaishian et al. 2024; Ekanayaka et al. 2025). It demands the use of integrative taxonomic methods, merging morphology with molecular and genomic data to precisely resolve species boundaries.

A noteworthy revolution in fungal taxonomy occurred with the integration of molecular data into fungal taxonomy, initially in the 1970s (De Bertoldi et al. 1973). Since then, the use of DNA-based approaches has become crucial in fungal systematics (Wei et al. 2023; Fan et al. 2024). Recent advances highlight that the appropriate selection of the molecular markers is imperative to correctly reconstruct the phylogenetic relationships (Schoch et al. 2012; Lucking et al. 2020). The nuclear ribosomal internal transcribed spacer (ITS) region has been established as the universal DNA barcode for fungi due to the presence of conserved flanking gene regions, presenting higher resolution at the species level (Schoch et al. 2012). The use of additional informative gene regions, such as housekeeping genes and nuclear protein-coding genes, has improved the resolution at the interspecies or intergenus level in phylogenetic analysis (Meyling 2008; Raja et al. 2017). To resolve the deep phylogenetic relationships within the *Ascomycota*, taxonomists commonly use protein-coding loci such as partial sequences of the largest and the second-largest subunits of RNA polymerase II (RPB1, RPB2), translation elongation factor 1-alpha (TEF-1α), and the nuclear ribosomal large subunit (nrLSU) (Hofstetter et al. 2007; Sung et al. 2007; Rehner et al. 2011). These protein-coding genes are comparatively easy to align, and their intronic sites are under low selection pressure, which can be useful for revealing phylogenetic relationships (Hofstetter et al. 2007; Chang et al. 2024). Microsatellite loci, or single-nucleotide polymorphisms, are considered effective, reliable, and informative genetic markers for intraspecific phylogeny (Morin et al. 2004; Oliveira and Azevedo 2022). To delineate cryptic species, the genealogical concordance phylogenetic species recognition (GCPSR) criterion is now widely used. This approach applies the concordance of multiple gene genealogies, overcoming the challenges of phylogenetically less informative morpho-taxonomic characters (Taylor et al. 2000; Bickford et al. 2007; Shenoy 2025).

Biological evolution is driven by changes in DNA. Recent advancements in genome-wide sequencing have significantly increased our understanding of fungal biology (Dannemiller et al. 2014; Wei et al. 2023). High-throughput sequencing has helped in understanding the evolution of fungal genes, chromosomes, regulatory networks, metabolic pathways, and mechanisms of pathogenicity (Priest et al. 2020; Aragona et al. 2022). These genomic insights are crucial for unraveling the complexities of fungal-insect interactions, adaptation, infection processes, persistence, and reproduction strategies in natural environments (Wang et al. 2016; Zaccaron and Stergiopoulos 2025).

Technological advances in genome sequencing are essential for filling knowledge gaps in fungal phylogenetics and ecology (Lofgren and Stajich 2021). The availability of sequenced genomes will facilitate rapid identification of virulence factors, host-pathogen interaction genes, and biosynthetic gene clusters (Leger 2024; Zaccaron and Stergiopoulos 2025). These genomic resources not only enhance our understanding of the entomopathogenic lifestyle but also offer novel targets for improving the efficacy of fungal biocontrol agents (Iwanicki et al. 2022). Therefore, expanding genome sequencing and applying genomics-enabled phylogenetic tools to *Hypocreales* is a high priority for advancing fungal biology and integrated pest management.

2 Historical Accounts

The study of EPF has a long history, with early descriptions focusing on *Cordyceps* species due to their distinct and often brightly colored fruiting bodies (Lim 2024; Sheng and Gerasimova 2024). In 1694, Wang Ang illustrated *Ophiocordyceps sinensis* (formerly *Cordyceps sinensis*) in his book *Ben Cao Bei Yao* (Lu 2023). French scientist de Réaumur published the first report on *Cordyceps*, referring to them as "vegetable growths" (Sheng and Gerasimova 2024). In 1878, Saccardo formally named the species Cordyceps sinensis, a name still in use (Lim 2024; Alviti Kankanamalage et al. 2025).

The nineteenth century marked a turning point, with focused studies on EPF, particularly for managing diseases in the silkworm industry. In 1835, entomologist Agostino Bassi discovered that a fungus was causing "Pebrine disease" in silkworms, known as "white muscardine disease" (Mascarin and Jaronski 2016; Dannon et al. 2020). The pathogen was initially described as *Botrytis bassiana* but later renamed as *Beauveria bassiana* (Vega and Blackwell 2005; Kepler et al. 2017). In 1878, Elias Metschnikoff identified a fungal disease, "green muscardine," affecting the pest *Anisoplia austriaca*. He described the agent as *Entomophthora anisopliae*, later revising it to *Isaria destructor*, which is now classified as *Metarhizium anisopliae* (Zimmermann et al. 1995). Metschnikoff was also one of the first to attempt biological control using entomopathogens, successfully infecting larvae of *A. austriaca* and the sugar beet weevil (Zimmermann et al. 1995). Since the 1970s, there has been a growing interest in EPF diversity.

3 Classification

The classification of EPF has undergone multiple changes over the years based on molecular and comparative genomics studies. In earlier studies, EPF were grouped into morphological categories, but as per recent studies, they are now categorized into seven major fungal phyla: *Ascomycota, Basidiomycota, Blastocladiomycota, Chytridiomycota, Entomophthoromycota, Kickxellomycotina,* and *Microsporidia* (Hibbett et al. 2007; Humber 2012; Kaczmarek and Boguś 2021; Alviti Kankanamalage et al. 2025). Among these, the order *Hypocreales* comprises major EPF such as *Beauveria*, *Lecanicillium*, *Metarhizium*, *Nomurae*, and *Ophiocordyceps* that primarily target hemipterans and lepidopterans (Bihal et al. 2023). *Entomophthorales*, placed under *Entomophthoromycota*, also comprises a diverse range of insect pathogenic fungi, such as *Entomophthora*, *Erynia*, *Eryniopsis*, and *Entomophaga*, representing a narrow range of host specificity. This group reflects a comparatively distinct evolutionary lineage from Ascomycota and targets aphids and lepidopterans (Koiri et al. 2017). Insect pathogens such as *Septobasidium* and *Fibulorhizoctonia* (Basidiomycota) are some of the lesser-explored EPF and parasitize scale insects or termites (He et al. 2022). Members of *Microsporidia* were earlier placed with protists but are now recognized as obligate insect parasites attacking mosquitoes, copepods, and silkworms (Park and Poulin

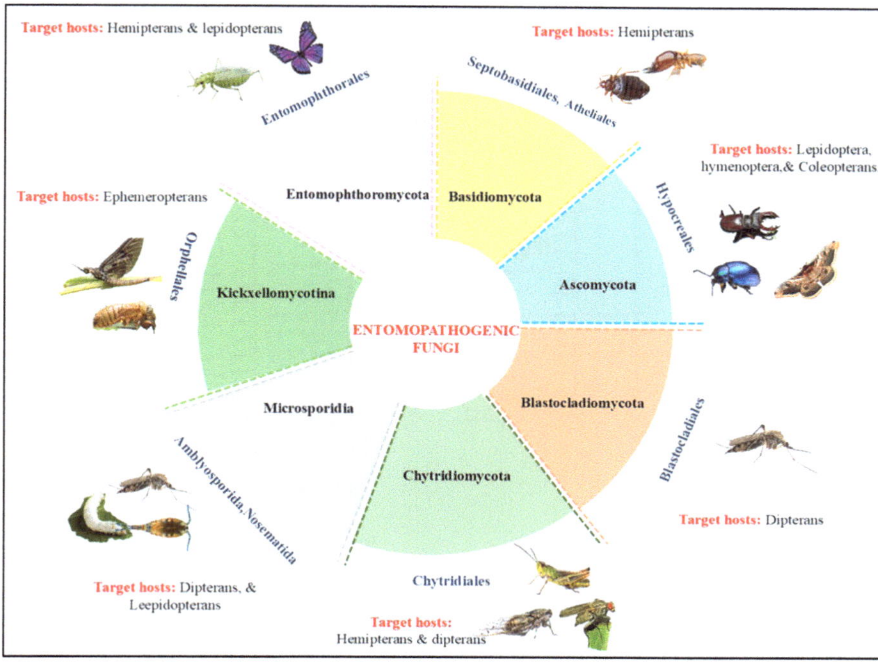

Fig. 2.1 Overview of EPF classification across different fungal lineages and their host specificity (Hibbett et al. 2007; Humber 2012; Kaczmarek and Boguś 2021; De Wint et al. 2024)

2021). Fungi adapted to the aquatic habitat are also known to act as insect pathogens, viz., *Chytridiales* and *Blastocladiales*, which are grouped under *Chytridiomycota* and infect hemipterans and dipterans (Khachatourians 2009). Both *Chytridiomycota* and *Blastocladiomycota* represent flagellated fungi, whereas *Kickxellomycotina* comprises arthropod-associated fungi, often acting as gut symbionts (Reynolds et al. 2023). Figure 2.1 provides an overview of the classification of EPF based on their hierarchical taxonomy and host range, respectively.

4 Major Taxonomic Lineages: Characteristics and Diversity

4.1 *Ascomycota*

It is one of the largest phyla in the kingdom *fungi*, including about 30 classes and 171 orders, comprising many well-known entomopathogenic genera (Wijayawardene et al. 2017; www.indexfungorum.org). Figure 2.2a represents the phylogenetic location of *Ascomycota* in the kingdom *fungi*. *Sordariomycetes* is the largest class of *Ascomycota* as presented in Fig. 2.2b, exhibiting the most highly diversified lineage among fungal-insect pathogens (Wijayawardene et al. 2022). Many of these fungi form specialized trapping structures and exploit a wide variety

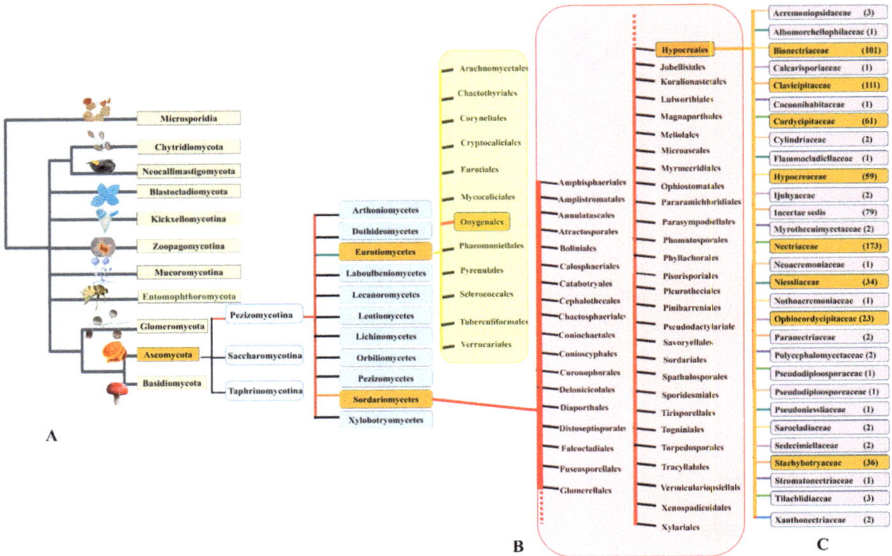

Fig. 2.2 Evolutionary diversity of EPF: (**a**) Phylogenetic distribution of phyla in the kingdom *Fungi*, highlighting phyla comprising EPF. (**b**) Phylogenetic diversity of *Sordariomycetes*, the largest class of *Ascomycota* containing major EPF. (**c**) Phylogenetic distribution of families in *Hypocreales,* the largest order dominated by EPF, indicating the number of genera known to date in parentheses. Families representing a high diversity of EPF are highlighted in orange (Kirk et al. 2008; Hyde et al. 2020; Wijayawardene et al. 2017, 2022)

of small arthropods as hosts. These fungi infect a vast range of insect orders, including *Lepidoptera, Coleoptera, Hemiptera*, and *Diptera*. They infect and proliferate within the host body and eventually kill the insect. Subsequently, they grow saprophytically on the insect cadaver and produce spores for further dissemination (Dissanayake and Liu 2025; Zhang et al. 2025).

Hypocreales Lindau (*Sordariomycetes*, *Ascomycetes*) represents the most prominent group of EPF (Hyde et al. 2020). As of now, it is classified into 29 families as shown in Fig. 2.2c (Wijayawardene et al. 2022, https://www.speciesfungorum.org/). *Bionectriaceae, Clavicipitaceae, Cordycipitaceae, Hypocreaceae, Nectriaceae, Niessliaceae, Ophiocordycipitaceae,* and *Stachybotryaceae* represent the most diverse families within this order, comprising many well-known insect pathogenic genera. These families are notable for their wide host range, ecological versatility, and significant roles in natural and applied biological control (Sharma et al. 2020).

4.2 *Basidiomycota*

Basidiomycota is one of the major fungal phyla and ranks second in terms of species richness. Despite the vast number of species within this phylum, only a few genera are known to be entomopathogenic. These include *Fibulorhizoctonia*, which infects termite eggs; *Uredinella* (e.g., *U. coccidiophaga* and *U. spinulosa*); and the largest and most representative genus, *Septobasidium,* which attacks 240 species of scale insects (Hemiptera) (Wijayawardene et al. 2024). The genus *Septobasidium* forms dense hyphal mats that cover colonies of scale insects on their plant hosts, acting as protective and parasitic structures (Aime et al. 2014). Within *Septobasidiales*, the two entomopathogenic genera, *Uredinella* and *Septobasidium,* differ significantly in their biology and host range. While *Uredinella* typically targets a single insect host and produces distinctive binucleate uredospores, *Septobasidium* can parasitize entire colonies of scale insects, sometimes infecting up to 250 individuals, but does not produce uredospores (Araújo and Hughes 2016; Choi et al. 2016).

4.3 *Blastocladiomycota*

Previously considered a monophyletic group within *Chytridiomycota* (Powell 2017), *Blastocladiomycota* has since been known as a separate phylum based on morphological and molecular evidence. For example, *Myiophagus*, originally classified under *Chytridiomycota* (Blackwell and Powell 2020), was reclassified into *Blastocladiomycota* based on its distinct morphological characteristics (James et al. 2014; Powell 2017). Within this phylum, *Coelomycetaceae* includes *Coelomomyces*, comprising around 101 species pathogenic to insects such as mosquitoes and flies (Powell 2017; https://www.speciesfungorum.org/). *Coelomomyces* species are obligate aquatic parasites, primarily known for infecting mosquito larvae and other aquatic dipterans, such as members of *Chironomidae, Psychodidae, Simuliidae,* and *Tabanidae* (Scholte et al. 2004; Shen et al. 2020). These fungi play a significant role in managing insect populations in aquatic ecosystems.

4.4 *Chytridiomycota*

Chytridiomycota is considered the earliest diverging lineage among the true fungi (James et al. 2006). This phylum comprises zoosporic fungi that are phylogenetically related to true fungi and uniquely characterized by the presence of motile cells (zoospores) at some stage of their life cycle (Kaczmarek and Boguś 2021). Although many chytrids are recognized as parasites, only a few genera include species that parasitize insects internally. One such example is *Myrmicinosporidium,* with *M. durum* identified as an obligate endoparasite of ants (Lapeva-Gjonova 2014). Another notable example is *Nephridiophaga*, which consists of single-celled parasites that inhabit the Malpighian tubules of insects, especially cockroaches and beetles, where they are typically localized in the lumen (Radek et al. 2023). These early diverging fungi represent an important and understudied component of entomopathogen diversity.

4.5 *Entomophthoromycota*

Entomophthoromycota is characterized by the presence of hyphal bodies bearing conidiophores and is taxonomically divided into three classes: *Basidiobolomycetes*, *Entomophthoromycetes*, and *Neozygitomycetes* (Humber 2012; Wijayawardene et al. 2024). Among these, Entomophthoromycetes is distinguished by its rod-shaped to variably shaped hyphal bodies and the absence of uniformly uninucleate cells (Humber 2012). Most entomopathogenic species within this phylum are found in *Entomophthorales*, which includes highly specialized, obligate, and host-specific insect pathogens (Manfrino et al. 2020). Representative examples include *Zoopthora radicans*, *Z. phalloides*, *Entomophthora muscae*, and *Pandora neoaphidis*, all of which could be utilized as effective biocontrol agents against flies and aphids (Sharma et al. 2020).

4.6 *Kickxellomycota*

Kickxellomycota includes fungi known for their roles as insect parasites and coprophilous organisms (Tretter et al. 2014). Among its members, the species of *Orphella* are notable obligate endoparasites, highly specialized to infect the hindgut of mayfly nymphs (order *Ephemeroptera*), where they complete their life cycle with remarkable host specificity (Strongman and White 2006). This phylum represents a relatively small but ecologically distinct lineage within the fungal kingdom.

4.7 *Microsporidia*

Microsporidia are a group of highly specialized, unicellular, spore-forming organisms that have undergone extreme evolutionary reduction. They have lost many fundamental cellular functions, including the presence of mitochondria, and are incapable of performing oxidative phosphorylation (Kaczmarek and Boguś 2021). These obligate intracellular parasites are highly host-specific and entirely dependent on their host for survival and reproduction, existing outside the host only as resilient, environmentally resistant spores (Vávra and Lukeš 2013). Among the notable entomopathogenic genera, *Amblyospora* is known to infect up to 79 different insect species within the order *Diptera*, particularly mosquitoes (Becnel and Andreadis 1999; Araújo and Hughes 2016). Another significant genus is *Nosema*, which targets the majority of *Lepidoptera* species (Zhu et al. 2011). *N. bombycis* and *N. ceranae* are of ecological and economic concern, as they infect honey bees and are associated with great losses in apiculture by shortening the life span of bees and weakening colonies (Higes et al. 2006).

5 Methods and Tools Used for the Characterization

5.1 Morphological Characterization

It involves the detailed observation and description of physical and structural features of fungi and remains one of the most traditional and fundamental methods for species identification and classification. The macro-morphological characteristics include colony shape and size, color at different growth stages (initial and mature), texture (e.g., cottony, velvety, powdery, slimy), and growth rate (fast, moderate, or slow growth on standard media) (Alsohaili and Bani-Hasan 2018). The micro-characteristics include hyphal features: septate or aseptate, branched or unbranched, pigmentation, and hyphal diameter; spore morphology: type (e.g., conidia, sporangiospores), shape (globose, oval, fusiform), and arrangement (solitary, chains, or clusters); reproductive structures: presence, arrangement, and structure of conidiophores, sporangiophores, sporangia, asci, or basidia; and specialized structures: presence of structures such as appressoria, columella, haustoria, phialides, rhizomorphs, or sclerotia (Ozolina et al. 2023). This method, while sometimes limited in resolving cryptic or closely related species, remains essential for preliminary identification and taxonomic placement, especially when integrated with molecular tools. Table 2.1 summarizes the colony and spore features of representative EPF genera, the major bioactive metabolites they produce, their primary insect hosts, and supporting literature references.

5.2 Molecular Characterization

It involves a series of precise steps, starting with DNA extraction, followed by gene amplification, sequencing, and phylogenetic inference to resolve taxonomic relationships and species boundaries. High-quality genomic DNA is extracted either by the cetyltrimethylammonium bromide (CTAB) method or by commercial extraction. Specific genes such as ITS, nrLSU, TEF-1α, and RPB1 are commonly targeted for PCR amplification using gene-specific primers. The amplified products are then sequenced through Sanger sequencing for individual loci or next-generation sequencing for more comprehensive genomic data. These molecular markers play a vital role in accurate species identification, strain differentiation, and population studies. Table 2.2 and Fig. 2.3 provide a detailed overview of the commonly used genetic markers and their application, along with supporting references.

6 Phylogenetic Analysis

The DNA sequences obtained after sequencing are first analyzed using Basic Local Alignment Search Tool (BLAST) (Altschul et al. 1997) to identify the closest known relatives. For precise evolutionary inference, the sequences are aligned using multiple sequence alignment tools like MAFFT (Kuraku et al. 2013), ClustalW

Table 2.1 Morphological characteristics, toxin profiles, and target insect pests of key entomopathogenic fungal species

Entomopathogenic fungus	Characteristics	Toxin	Host	References
Beauveria bassiana, *B. velata*, *B. brongniartii*, *B. amorpha*, and *B. caledonica*	A filamentous fungus, colonies are white with cottony aerial mycelium. Conidiophores are single or branched, cylindrical, or flask-shaped. Conidia are globose to oval in shape and hydrophobic.	Beauvericin, bassianolide, isarolides, and beauverolides	Pales weevils (*Hylobius pales*), black bug (*Scotinophara lurida*), zigzag leafhopper (*Recilia dorsalis*), spotted pod borer (*Maruca testulalis*), yellow hairy caterpillar, shoot borer, leaf webber, western corn rootworm, and ladybird beetle	Feng et al. (2015), Caloni et al. (2020)
Hirsutella thompsonii, *H. gigantea*, and *H. citriformis*	Colonies show fluffy mycelial growth, a brownish to grayish-green substratum color. The hyphae are smooth. Conidia are single at the tip of the branch.	Hirsutellin A	*N. lugens*, *N. virescens*, citrus rust mite (*Phyllocoptruta oleivora*), hoppers and bug pests, whiteflies, and red mites.	Pérez-González et al. (2022)
Lecanicillium lecanii, *L. attenuatum*, *L. longisporum*, *L. muscarium*, and *L. nodulosum*	The mycelium is hyaline, multinucleate, and septate. Conidiophores are formed in verticillate whorls. Hydrophilic conidia borne in slime or mucus balls vary in shape and size; they form an ascus.	VD toxins	*Aphis craccivora*, *Mylabris pustulata*, aphids, whiteflies, grasshoppers, *Pericallia ricini*, thrips, brown planthopper, hairy caterpillar, and stem borer.	Arakere et al. (2022)
Metarhizium anisopliae, *M. hiemalis*, *M. acridum*, *M. album*, *M. flavoviride*, *M. brunneum*, *M. pingshaense*, *M. robertsii*, *M. guizhouense*, *M. lepidiotae*,	The colony color is white, and later turns green. The hyphae are hyaline, septate, and branched. Conidiophores are short, erect, septate, simple, or branched. Conidia may be hyaline, lilac, brown, or green, smooth, and form chains. Shape may be cylindrical, globose, or ellipsoidal. Hydrophobic conidia are due to the presence of cysteine-rich proteins called hydrophobins in the cell wall.	Destruxins and cytochalasins	Brown plant hopper (*Nilaparvata lugens*), western corn rootworm (*Diabrotica virgifera*), green leafhopper, coconut rhinoceros beetle, diamondback moth, early shoot, top shoot, and internode borer of sugarcane, and red cotton bug	Steinwender et al. (2014)

M. majus, and *M. globosum*				
Nomuraea rileyi and *N. atypicola*	Colonies are white and later turn green. Hyphae are smooth, septate, and hyaline. Conidiophores are long, dense, compacted clusters of phialides. The conidia are pale green.	Diketopiperazine	Tobacco caterpillar, American bollworm (*H. armigera*), *Helicoverpa*, *Bombyx mori*, pod borers, cutworms, and cabbage borers.	Bharathi and Duraimurugan (2022)
Paecilomyces lilacinus, *P. marquandii*, *P. farinosus*, *P. fumosoroseus*, *P. tenuipes*, *P. javanicus* and *P. variotii*	The colonies are white, red, or yellow. Well-developed synnematous conidiophores are found in many species. Conidia are unicellular, hyaline to light colored.	Leucinostatins	*Aproaerema modicella*, *Pericallia ricini*, yellow and red mites, and whiteflies	Moreno-Gavíra et al. (2020)

Table 2.2 Some common molecular tools and techniques used in fungal phylogenetic studies

Tool/ technique	Marker/gene targeted	Application	References
DNA barcoding	ITS (ITS1–5.8S–ITS2 or ITS4–5.8S–ITS5)	Species-level identification; universal fungal barcode	White et al. (1990)
GCPSR	ITS, CAL, CHS, PR1, ACT, and NTL	Confirms species based on gene tree concordance; resolves cryptic species	Bidochka et al. (2005)
Mitochondrial DNA phylogeny	Intergenic mtDNA + conserved mt genes	Intra-/interspecific resolution; cryptic species identification	Kortsinoglou et al. (2020)
MLST	TEF1-α, RPB1, RPB2, TUB2, CAL and CHS	High-resolution species delimitation	Patwardhan et al. (2014)
Multigene phylogeny	Concatenated genes (e.g., ITS, TEF1-α, RPB1, and RPB2)	Reconstruction of evolutionary relationships; species delimitation	Raja et al. (2017)
PCR-RFLP; RAPD	ITS, 18S, 28S rDNA; Random genomic regions	Cryptic species discrimination; strain level differentiation	Oliveira and Azevedo (2022)
SSR	Microsatellite loci	Population genetics, clonal structure, and genotypic diversity	Steinwender et al. (2014)

ACT actin; *CAL* calmodulin; *CHS* chitin synthase; *ITS* internal transcribed spacer; *MLST* multilocus sequence typing; *mtDNA* mitochondrial DNA; *NTL* neutral trehalase; *PR1* subtilisin-like protease; *RPB1* largest subunit of RNA polymerase II; *RPB2* second-largest subunit of RNA polymerase II; *SSR* simple sequence repeat; *TEF1α* translation elongation factor 1α; *TUB2* β-tubulin

(Thompson et al. 2003), or MUSCLE (Edgar 2004). Manual editing and fine-tuning of the alignments, like the removal of ambiguous or misaligned regions, are performed using software like MEGA (Kumar et al. 2024) or BioEdit (Hall et al. 2011). A variety of methods can be employed to construct the phylogenetic tree depending on the dataset and desired resolution, such as Neighbor-Joining (Saitou and Nei 1987), Maximum Parsimony (Fitch 1971), or Maximum Likelihood (Felsenstein 1981). Widely used tools for tree construction include MEGA, PAUP (Swofford 1993), RAxML (Silvestro and Michalak 2012), IQ-TREE (Nguyen et al. 2015), and PhyML (Guindon and Gascuel 2003), each providing different analyses for sequence evolution and divergence.

6.1 Resolution of Cryptic Species and Ecological Complexity

There are several occurrences where morphological data are not satisfactory in resolving species boundaries and host-pathogen complexity. For example, the most common EPF, *B. bassiana* and *M. anisopliae*, largely display morphological overlap, signifying species complexes and genetic diversity (Bidochka et al. 2001; Shenoy et al. 2007; Bischoff et al. 2009). The advancement in molecular phylogenetics, including multilocus phylogenies and WGS data-driven phylogeny,

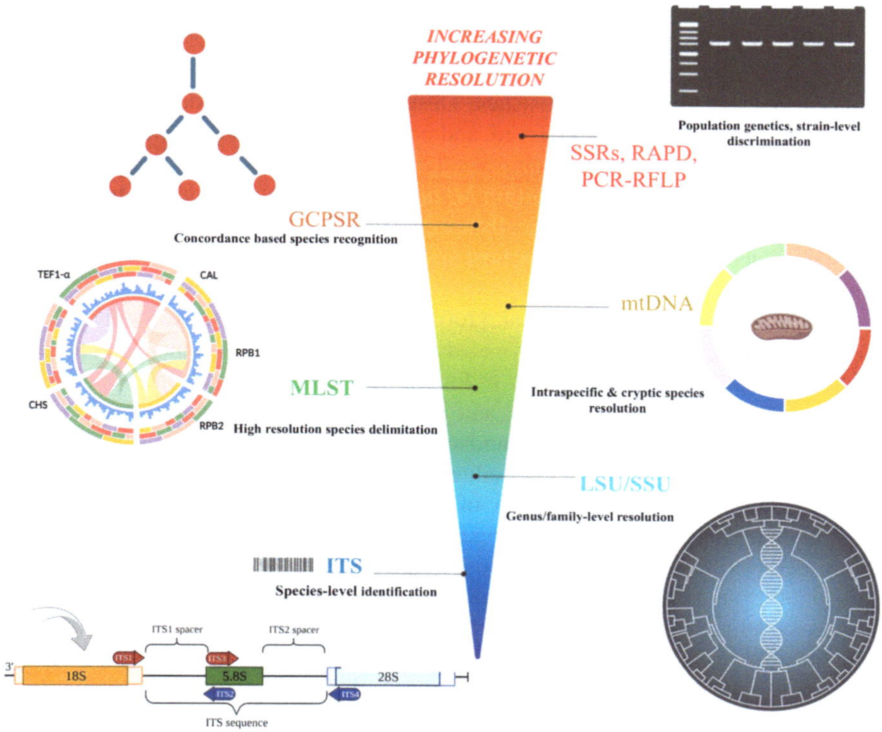

Fig. 2.3 Schematic illustration of molecular markers commonly employed in fungal phylogenetics, arranged in increasing order of resolution from broad species level to high-resolution population structure (Caramalho et al. 2019; Delavaux et al. 2024; Panchariya et al. 2024)

provided a deeper understanding of fungal taxonomy, offering new insights into evolutionary relationships, host-pathogen interactions, and ecological specialization among them (Kepler et al. 2017; Rizal et al. 2024).

6.2 *Metarhizium*

Metarhizium (*Clavicipitaceae, Hypocreales*) represents the most widely known EPF, extensively studied for its adaptations to varied environments and potential in biological pest control (Brunner-Mendoza et al. 2019). They may exhibit either a broad or narrow host range, depending on their infection ability; thereby, gaining increased attention toward understanding their ecology, diversity, evolutionary dynamics, and phylogeny (Roberts and St Leger 2004; Gao et al. 2011). One of the earliest studies involved analysis of the population structure of *M. anisopliae* isolates from two different habitats in Ontario, Canada (Bidochka et al. 2001). Two reproductively isolated groups were identified, one from a forested environment, active at low temperature, while the other one was prevalent in an agricultural habitat and was heat-tolerant and UV resistant. Despite genetic distinctions, no differences in mitochondrial DNA were observed, and no strong relationship with specific insect hosts was detected. These findings suggested the presence of cryptic species, *M. brunneum* and *M. robertsii*, within *M. anisopliae*; each was adapted to a distinct environment and was thus ecologically specialized. The study, therefore, highlighted the importance of habitat type over host specificity leading to divergence (Bidochka et al. 2001).

After a few years, Bidochka et al. (2005) again took some isolates and classified them into cold-active and heat-active ecotypes. They took into account six loci (ITS, CAL, CHS, *pr1*, ACT, and NTL) following the multigene approach and further utilized the GCPSR in order to test and justify the species boundaries. The results further revealed that the heat-active and cold-active ecotypes formed two concordant monophyletic clades, which supported recombination among them but not between them, strongly supporting the idea of sympatric cryptic species. These results brought new insights into the phylogenetic studies of *M. anisopliae,* which was earlier believed to exist as a single, host-adapted species (Bidochka et al. 2005).

In another study, a multilocus analysis of *Metarhizium* was conducted in a Danish agroecosystem (Steinwender et al. 2014). The study revealed the co-occurrence of four species: *M. brunneum* (the most abundant—78.8%), *M. robertsii* (14.6%), and the less frequent *M. majus* and *M. flavoviride* (each 3.3. %). SSR genotyping further depicted the predominance of *M. brunneum*, suggesting its possible adaptive advantage to local agricultural soil conditions. Overall, the study highlighted the ecological significance of genotype-specific adaptations within a community along with high genetic diversity in *Metarhizium* (Steinwender et al. 2014).

Further, a detailed comparative study between multilocus phenotyping approaches and WGS methods was performed to assess the effectiveness in species-level resolution of *Metarhizium* isolates (Rizal et al. 2024). A total of 43 *Metarhizium* isolates were used, collected from five distinct vegetation types and four suburban habitats in Southeast Queensland. Three agricultural insect pests were used as bait. Initial species identification using the ITS region showed limited resolution, failing to differentiate species within the genus clearly. Further, multilocus phylogenies using the TEF1α 3′, TEF1α 5′, RPB1, and RPB2 genes provided better species-level resolution, except for one unresolved clade. WGS using SNP analysis provided the highest resolution, revealing more genetic clusters, suggesting the presence of potentially unrecognized species. Interestingly, these genetic clusters showed no consistent association with distinct vegetation type or host species, implying that neither environmental nor host specificity strongly influences the genetic structuring of *Metarhizium* in this region. Overall, the study emphasizes the limitations of traditional gene-marker approaches and highlights the recent WGS-based advances in identifying hidden genetic diversity. The results suggested that *Metarhizium* contains greater genetic complexity than previously recognized and that some of the WGS-identified clusters may represent novel or cryptic species, suggesting further taxonomic investigation (Rizal et al. 2024).

6.3 *Beauveria*

Beauveria (Balsamo-Crivelli) Vuill. (*Cordyciptaceae, Hypocreales*) represents another one of the most widely known genera of necrotrophic EPF (Roy et al. 2010; Chen et al. 2018). Species of *Beauveria* have been extensively studied for their role as mycoinsecticides and as sources of secondary metabolites such as bassianolide, beauvericin, polyketides, and terpenoids (Thomas and Read 2007; Ames and Walsh 2010; Feng et al. 2015; Caloni et al. 2020). Traditionally, the distinction of *Beauveria* species is based on their conidia characteristics. Conidia are typically 1.5–5.5 μm in length, smooth-walled, and hyaline and vary from globose, ellipsoidal, cylindrical, and vermiform to reniform in shape (Kirk et al. 2008; Rehner et al. 2011; Kepler et al. 2017). In contrast to the generic delineation of *Beauveria*, defining species-level boundaries is complicated due to overlapping morphological features.

Preliminary molecular data revealed *B. bassiana* with significant intraspecific genetic variability, suggesting a role of geographic origin, rather than host association, in determining population structure (Rehner and Buckley 2005). This suggests that *B. bassiana* represents a species complex, signifying multiple cryptic lineages with their adaptation to distinct habitats.

Over the years, identification techniques and delimitation of *Beauveria* species have seen major improvements because of the integration of multigene sequence data and sophisticated phylogenetic analysis tools (Rehner et al. 2011). Deviating from the traditional morpho-taxonomic approach, the taxonomy of the genus is revised using a multigene phylogenetic approach. Molecular phylogenetics based on four molecular markers, nuclear transgenic factor (Bloc), TEF1α, RPB1, and RPB2, has recognized six new species (*B. asiatica, B. australis, B. kipukae, B. pseudobassiana, B. sungii,* and *B. varroae*) within monophyletic *Beauveria* (Rehner et al. 2011). Following this evolutionary framework, numerous other species were identified, including *B. lii* from China (Zhang et al. 2013); *B. sinensis*, *B. araneola,* and *B. majiangensis* from Guizhou, China (Chen et al. 2013, 2017, 2018). In India, a new cryptic species, *B. rudraprayagi*, was identified following the same multigene phylogeny (Agrawal et al. 2014). Similarly, *B. hoplocheli* was described from Réunion Island (Robène-Soustrade et al. 2015). A key inclusion to *Beauveria* taxonomy occurred with the addition of six new species, including *B. acridophila*, *B. blattidicola*, *B. diapheromeriphila*, *B. locustiphila*, *B. scarabaeidicola*, and *B. staphylinidicola,* based on similar multilocus sequence typing (Kepler et al. 2017). Additionally, *B. peruviensis* has been identified based on an identical molecular framework (Bustamante et al. 2019). Further, three more species, *B. polyrhachicola*, *B. songmingensis*, and *B. subscarabaeidicola*, were investigated (Wang et al. 2022). In all these studies, species shared morphological overlap and presented an absence of clear phenotypic distinctions, thereby emphasizing the crucial role of molecular data in species delimitation.

In a recent study on phylogenetic diversity within *Beauveria*, multilocus phylogenies and WGS-SNP data are employed to resolve the species complexes (Rizal et al. 2024). Initial species identification using ITS provided inadequate resolution. However, multilocus phylogeny permitted better species resolution with the identification of one more species in the *B. bassiana* species complex. Remarkably, the WGS approach provided the best resolution through the detection of additional genetic clusters (three more clades in *B. bassiana* and two clades in *B. australis*) that were not revealed using multilocus phylogenetics, suggesting further investigation in this direction (Manfrino et al. 2024; Rizal et al. 2024).

6.4 *Ophiocordyceps*

Ophiocordyceps (*Ophiocordycipitaceae, Hypocreales*) comprises a diverse genus of EPF that are known to attack a broad range of insects. Their perithecia are usually either embedded or superficial, and they infect at various stages of the host. Earlier, taxonomists placed it under *Cordyceps* as a subgenus on the basis of conventional morphological traits (Kobayasi 1941, 1982; Mains 1958). However, molecular phylogenetic analyses revolutionized the taxonomy of *Cordyceps*-like fungi (Sung et al. 2007; Sanjuan et al. 2015). Molecular data discovered *Cordyceps* as polyphyletic (Sung et al. 2007). This led to the inclusion of *Ophiocordyceps* into its new family, *Ophiocordycipitaceae,* supported by phylogenetic analyses and discrete phenotypic features such as darkly pigmented and pliant to fibrous stromata (Sung et al. 2007). Studies further continued to emphasize the importance of molecular tools in the resolution of evolutionary complexities and in the identification of cryptic species within *Ophiocordyceps* (Chen et al. 2021; Mongkolsamrit et al. 2021; Sun et al. 2022; Dai et al. 2024).

In one of the studies, taxonomists described *O. pingbianensis,* a novel species isolated from a tiger beetle larva from China (Chen et al. 2021). They employed a multigene nuclear dataset, including ITS, TEF1α, nrSSU, RPB1, RPB2, and nrLSU, as well as 14 mitochondrial protein-coding genes. Phylogenetic analysis inferred that the new species belonged to the *Hirsutella nodulosa* clade within *Ophiocordyceps.* This study highlighted how mitogenomic information combined with nuclear gene phylogeny can help determine and discover new species, thus providing better insights (Chen et al. 2021).

In another study, cryptic diversity of *Ophiocordyceps* infecting Coleopteran hosts in Thailand was resolved using molecular phylogenetic analyses (Mongkolsamrit et al. 2021). They identified six novel species (*O. albostroma*, *O. brunnea*, *O. kohchangensis*, *O. ratchaburiensis*, *O. capilliformis*, and *O. phitsanulokensis*), each positioned in well-supported clades.

In a recent study, species boundaries within the *O. sinensis* species complex were resolved utilizing a similar molecular phylogenetic approach (Dai et al. 2024). It was previously believed to be a single taxon of high medicinal and ecological value.

Phylogenetic analyses based on sequencing of six nuclear loci (nrSSU, nrLSU, TEF1α, RPB1, RPB2, and ITS) revealed that the *O. sinensis* complex comprised at least ten discrete species, including *O. megala*. Further confirmation was done using the insect cadavers and amplification of their mitochondrial *cox1* genes. In addition to it, the ITS-based strategy was also employed to strengthen the results and refine the intra-lineage relationships. This study highlighted that cryptic species within *O. sinensis* could be distinguished through high-resolution, multilocus molecular phylogenetics combined with host identity and ecological modeling (Dai et al. 2024).

A study in Southwestern China employed an integrative approach from classical to advanced genome sequencing, providing a powerful framework for fungal taxonomy (Sun et al. 2022). The study utilized both nuclear and mitochondrial genome sequencing to reveal a new species, *O. nujiangensis*. The mitogenomic data supported its uniqueness and helped resolve its evolutionary relationships, while metagenomic analysis suggests ecological interactions with other microbial communities (Sun et al. 2022).

Here again, the role of molecular data in species delimitation is emphasized, along with genomics-enabled taxonomy, marking significant advancements in the resolution of species boundaries with greater accuracy. Figure 2.4 summarizes the important molecular tools along with morphological traits in the above-mentioned genera.

7 Integrative Phylogeny

Although molecular methods have improved our understanding of phylogeny, morphological characteristics remain essential for formal species descriptions and ecological studies (Kobmoo et al. 2024). Moreover, in the initial phase of taxonomic classification, the host range and habitat preferences of fungi also provide key information at the familial or genus level, covering the ecological aspects of the study (Araújo and Hughes 2016). With the help of WGS data, fungal cryptic diversity is being discovered (Passer et al. 2019).

Following an integrative or holistic taxonomy, an integration of molecular, morphological, ecological, and physiological information is the most appropriate method for classification of fungi (Boluda et al. 2019). Morphological traits provide further background in interpreting molecular data and assist in the recognition of phenotypic characters associated with ecological roles. For instance, although species of *Metarhizium* are genetically different, their conidial morphology can still provide information about their mechanisms of spore dispersal and stress tolerance (Bich et al. 2021). In a recent study, an integrative approach was followed, combining multilocus nuclear phylogeny with mitochondrial genome analysis, and identified two new entomopathogenic fungal species, *O. albastroma* and *O. nigristroma*, isolated from the tissue of infected Hepialid larvae (Sun et al. 2024). As a result, comparative mitogenomics analysis, along with nuclear phylogeny, helped resolve differences in genome length, emphasizing their utility in fungal

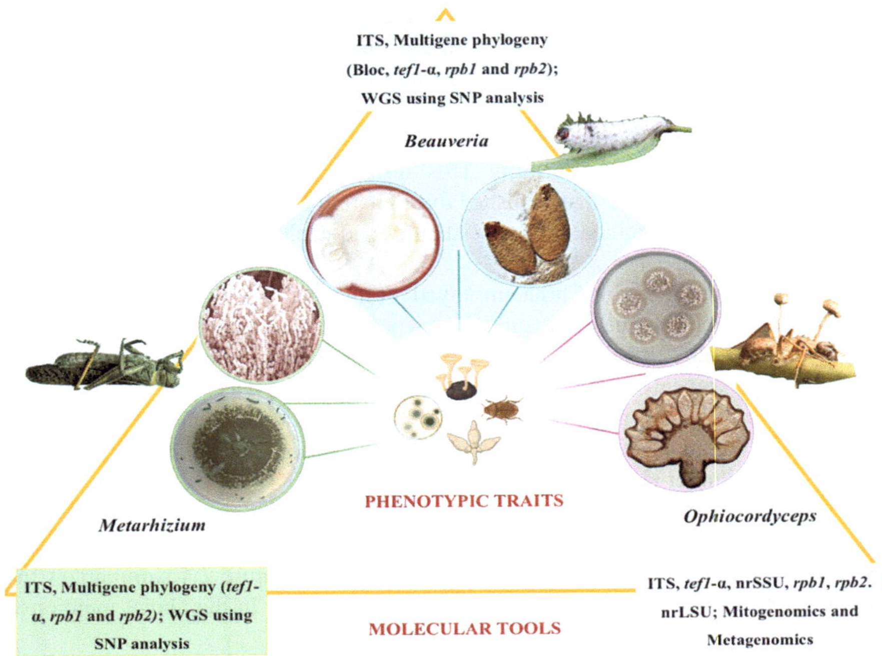

Fig. 2.4 Key morphological characteristics of genera *Metarhizium, Beauveria,* and *Ophiocordyceps*, highlighting common molecular tools used for the resolution of cryptic species and taxonomic clarification (Sung et al. 2011; Pavlov et al. 2018; Abarna and Vishnupriya 2022; Kobmoo et al. 2024)

taxonomy. The addition of molecular data to phylogenetic studies following an integrative approach not only provides the ultimate method for validation of new taxa but also establishes robust phylogenetic relationships, thereby refining taxonomy and our understanding of speciation and cryptic diversity (Fig. 2.5).

8 Conclusion

The phylogenetic distribution of EPF extends across major fungal phyla, emphasizing their remarkable evolutionary and ecological versatility. Within these, the order *Hypocreales* includes the most predominant entomopathogenic genera, known for their diverse host ranges, such as *Beauveria*, *Metarhizium*, and *Ophiocordyceps*. Advances in molecular tools and techniques, especially multilocus phylogenetics, WGS, and mitogenomic analyses, have significantly offered higher resolution of species boundaries, revealed hidden genetic diversity, and simplified complex evolutionary relationships. Moreover, integrative taxonomy has become the most robust framework for understanding EPF diversity and adaptation by combining morphological, molecular, and ecological concepts. These efforts will

Fig. 2.5 Venn diagram representing integrative phylogeny encompassing aspects of morphological, molecular, and ecological phylogeny (Boluda et al. 2019)

improve our understanding of fungal evolution and ecology but also support the development of innovative, eco-friendly, and environmentally sustainable pest management strategies.

References

Abarna VP, Vishnupriya R (2022) DNA barcoding reveals the natural occurrence of *Beauveria bassiana* (Balsamo) Vuillemin in two-spotted spider mite, *Tetranychus urticae* Koch in Bhendi [*Abelmoschus esculentus* (L.) Moench] ecosystem in Coimbatore district of Tamil Nadu. Egypt J Biol Pest Control 32(1):128. https://doi.org/10.1186/s41938-022-00623-6

Agrawal Y, Mual P, Shenoy BD (2014) Multi-gene genealogies reveal cryptic species *Beauveria rudraprayagi* sp. nov. from India. Mycosphere 5:719–736. https://doi.org/10.5943/mycosphere/5/6/3

Aime MC, Toome M, David J, McLaughlin DJ (2014) Systematics and evolution. Mycota 2(10): 271–294. https://doi.org/10.1007/978-3-642-55318-9_10

Alsohaili SA, Bani-Hasan BM (2018) Morphological and molecular identification of fungi isolated from different environmental sources in the Northern Eastern desert of Jordan. Jordan J Biol Sci 11(3):329–337

Altschul SF, Madden TL, Schäffer AA, Zhang J, Zhang Z et al (1997) Gapped BLAST and PSI-BLAST: a new generation of protein database search programs. Nucleic Acids Res 25(17):3389–3402. https://doi.org/10.1093/nar/25.17.3389

Alviti Kankanamalage HP, Yang JY, Karunarathna SC, Tibpromma S, Kumla J et al (2025) Entomopathogenic fungi: insights into recent understanding. World J Microbiol Biotechnol 41(6):179. https://doi.org/10.1007/s11274-025-04377-9

Ames BD, Walsh CT (2010) Anthranilate-activating modules from fungal nonribosomal peptide assembly lines. Biochemist 49:3351–3365. https://doi.org/10.1021/bi100198y

Aragona M, Haegi A, Valente MT, Riccioni L, Orzali L et al (2022) New-generation sequencing technology in diagnosis of fungal plant pathogens: a dream comes true? J Fungi 8(7):737. https://doi.org/10.3390/jof8070737

Arakere UC, Jagannath S, Krishnamurthy S, Chowdappa S, Konappa N (2022) Microbial bio-pesticide as sustainable solution for management of pests: achievements and prospects. In: Rakshit A, Meena VS, Abhilash PC, Sarma BK, Singh HB et al (eds) Advances in Bio-inoculant Science, Biopesticides. Woodhead Publishing, pp 183–200. https://doi.org/10.1016/B978-0-12-823355-9.00016-X

Araújo JP, Hughes DP (2016) Diversity of EPF: which groups conquered the insect body? Adv Genet 94:1–39. https://doi.org/10.1016/bs.adgen.2016.01.001

Ban S, Sakane T, Toyama K, Nakagiri A (2009) Teleomorph–anamorph relationships and reclassification of *Cordyceps cuboidea* and its allied species. Mycoscience 50(4):261–272. https://doi.org/10.47371/mycosci.MYC50261

Becnel JJ, Andreadis G (1999) Microsporidia in insects. In: Wittner M, Weiss LM (eds) The Microsporidia and Microsporidiosis. ASM Press, Washington, DC, pp 447–501

Bharathi E, Duraimurugan P (2022) Pathogenicity of EPF *Beauveria bassiana* and *Metarhizium (Nomuraea) rileyi* against castor hairy caterpillar, *Euproctis fraterna* (Lepidoptera: Lymantriidae). J Oilseeds Res 39(1):58–60. https://doi.org/10.56739/jor.v39i1.137339

Bich GA, Castrillo ML, Kramer FL, Villalba LL, Zapata PD (2021) Morphological and molecular identification of EPF from agricultural and forestry crops. Floresta Ambient 28:e20180086. https://doi.org/10.1590/2179-8087-FLORAM-2018-0086

Bickford D, Lohman DJ, Sodhi NS, Ng PK, Meier R et al (2007) Cryptic species as a window on diversity and conservation. Trends Ecol Evol 22:148–155. https://doi.org/10.1016/j.tree.2006.11.004

Bidochka MJ, Kamp AM, Lavender TM, Dekoning J, De Croos JA (2001) Habitat association in two genetic groups of the insect-pathogenic fungus Metarhizium anisopliae: uncovering cryptic species? Appl Environ Microbiol 67(3):1335–1342. https://doi.org/10.1128/AEM.67.3.1335-1342.2001

Bidochka MJ, Small CLN, Spironello M (2005) Recombination within sympatric cryptic species of the insect pathogenic fungus Metarhizium anisopliae. Environ Microbiol 7(9):1361–1368. https://doi.org/10.1111/j.1462-5822.2005.00823.x

Bihal R, Al-Khayri JM, Banu AN, Kudesia N, Ahmed FK et al (2023) Entomopathogenic fungi: an eco-friendly synthesis of sustainable nanoparticles and their nanopesticide properties. Microorganisms 11(6):1617. https://doi.org/10.3390/microorganisms11061617

Bischoff JF, Rehner SA, Humber RA (2009) A multilocus phylogeny of the *Metarhizium anisopliae* lineage. Mycologia 101(4):512–530. https://doi.org/10.3852/07-202

Blackwell WH, Powell MJ (2020) A nomenclatural and systematic note on the genus *Myiophagus* (ex Chytridiomycota). Phytologia 102(1):5–8

Boluda CG, Rico VJ, Divakar PK, Nadyeina O, Myllys L et al (2019) Evaluating methodologies for species delimitation: the mismatch between phenotypes and genotypes in lichenized fungi (*Bryoria* sect. Implexae, Parmeliaceae). Pers Mol Phylogeny Evol Fungi 42(1):75–100. https://doi.org/10.3767/persoonia.2019.42.04

Brown AJ (2023) Fungal resilience and host–pathogen interactions: future perspectives and opportunities. Parasite Immunol 45(2):e12946. https://doi.org/10.1111/pim.12946

Brunner-Mendoza C, Reyes-Montes MDR, Moonjely S, Bidochka MJ, Toriello C (2019) A review on the genus *Metarhizium* as an entomopathogenic microbial biocontrol agent with emphasis on its use and utility in Mexico. Biocontrol Sci Tech 29(1):83–102. https://doi.org/10.1080/09583157.2018.1531111

Burdon JJ, Thrall PH (2008) Pathogen evolution across the agro-ecological interface: implications for disease management. Evol Appl 1:57–65. https://doi.org/10.1111/j.1752-4571.2007.00005.x

Bustamante DE, Oliva M, Leiva S, Mendoza JE, Bobadilla L et al (2019) Phylogeny and species delimitations in the entomopathogenic genus *Beauveria* (Hypocreales, Ascomycota), including the description of *B. peruviensis* sp. nov. MycoKeys 58:47. https://doi.org/10.3897/mycokeys.58.35764

Caloni F, Fossati P, Anadón A, Bertero A (2020) Beauvericin: the beauty and the beast. Environ Toxicol Pharmacol 75. https://doi.org/10.1016/j.etap.2020.103349

Caramalho R, Madl L, Rosam K, Rambach G, Speth C et al (2019) Evaluation of a novel mitochondrial pan-Mucorales marker for the detection, identification, quantification, and growth stage determination of Mucormycetes. J Fungi 5(4):98. https://doi.org/10.3390/jof5040098

Chang X, Li X, Li Z, Hywel-Jones N, Li G et al (2024) Comparative Mitogenomics analysis revealed evolutionary divergence among *Purpureocillium* species and gene arrangement and intron dynamics of Ophiocordycipitaceae. Microorganisms 12(10):2053. https://doi.org/10.3390/microorganisms12102053

Chen M, Huang B, Li Z, Spatafora JW (2013) Morphological and genetic characterisation of Beauveria sinensis sp. Nov. from China. Mycotaxon 124(1):301–308. https://doi.org/10.5248/124.301

Chen WH, Han YF, Liang ZQ, Jin DC (2017) A new araneogenous fungus in the genus *Beauveria* from Guizhou, China. Phytotaxa 302(1):57–64. https://doi.org/10.11646/phytotaxa.302.1.5

Chen WH, Man L, Huang ZX, Yang GM, Han YF et al (2018) *Beauveria majiangensis*, a new entomopathogenic fungus from Guizhou, China. Phytotaxa 333:243–250. https://doi.org/10.11646/phytotaxa.333.2.8

Chen S, Wang Y, Zhu K, Yu H (2021) Mitogenomics, phylogeny and morphology reveal *Ophiocordyceps pingbianensis* Sp. Nov., an Entomopathogenic fungus from China. Life 11(7):686. https://doi.org/10.3390/life11070686

Choi IY, Lee WH, Lee JJ, Park MJ, Ko JA et al (2016) Characterization of a *Septobasidium* sp. associated with felt disease of *Schisandra chinensis*. Mycobiology 44(1):58-62. https://doi.org/10.5941/MYCO.2016.44.1.58

Dai Y, Chen S, Wang Y, Wang Y, Yang Z et al (2024) Molecular phylogenetics of the *Ophiocordyceps sinensis*-species complex lineage (Ascomycota, Hypocreales), with the discovery of new species and predictions of species distribution. IMA Fungus 15(1):2. https://doi.org/10.1186/s43008-023-00131-8

Dannemiller KC, Lang-Yona N, Yamamoto N, Rudich Y, Peccia J (2014) Combining real-time PCR and next-generation DNA sequencing to provide quantitative comparisons of fungal aerosol populations. Atmos Environ 84:113–121. https://doi.org/10.1016/j.atmosenv.2013.11.036

Dannon HF, Dannon AE, Douro-Kpindou OK, Zinsou AV, Houndete AT et al (2020) Toward the efficient use of *Beauveria bassiana* in integrated cotton insect pest management. J Cotton Res 3(1):24. https://doi.org/10.1186/s42397-020-00061-5

de Bertoldi M, Lepidi AA, Nuti MP (1973) Significance of DNA base composition in classification of *Humicola* and related genera. Trans Brit Mycol Soc 60:77–85. https://doi.org/10.1016/S0007-1536(73)80062-2

De Wint FC, Nicholson S, Koid QQ, Zahra S, Chestney-Claassen G et al (2024) Introducing a global database of entomopathogenic fungi and their host associations. Sci Data 11(1):1418. https://doi.org/10.1038/s41597-024-04103-4

Delavaux CS, Ramos RJ, Stürmer SL, Bever JD (2024) An updated LSU database and pipeline for environmental DNA identification of arbuscular mycorrhizal fungi. Mycorrhiza 34(4):369–373. https://doi.org/10.1007/s00572-024-01159-3
Dissanayake AJ, Liu JK (2025) Ascomycota: diversity, taxonomy and phylogeny, 2nd edition: editorial. J Fungi 11(6):419. https://doi.org/10.3390/jof11060419
Edgar RC (2004) MUSCLE: multiple sequence alignment with high accuracy and high throughput. Nucleic Acids Res 32(5):1792–1797. https://doi.org/10.1093/nar/gkh340
Ekanayaka AH, Karunarathna SC, Tibpromma S, Dutta AK, Tennakoon DS et al (2025) Species evolution: cryptic species and phenotypic noise with a particular focus on fungal systematics. Front Cell Infect Microbiol 151497085. https://doi.org/10.3389/fcimb.2025.1497085
Fan Q, Yang T, Li H, Wang XM, Liao HF et al (2024) Molecular phylogeny and morphology reveal two new entomopathogenic species of *Ophiocordyceps* (Ophiocordycipitaceae, Hypocreales) parasitic on termites from China. MycoKeys 103:1–24. https://doi.org/10.3897/mycokeys.103.116153
Felsenstein J (1981) Evolutionary trees from DNA sequences: a maximum likelihood approach. J Mol Evol 17(6):368–376. https://doi.org/10.1007/bf01734359
Feng P, Shang Y, Cen K, Wang C (2015) Fungal biosynthesis of the bibenzoquinone oosporein to evade insect immunity. Proc Natl Acad Sci USA 112:11365–11370. https://doi.org/10.1073/pnas.1503200112
Fitch WM (1971) Toward defining the course of evolution: minimum change for a specific tree topology. Syst Biol 20(4):406–416. https://doi.org/10.2307/2412116
Gao Q, Jin K, Ying SH, Zhang Y, Xiao G et al (2011) Genome sequencing and comparative transcriptomics of the model entomopathogenic fungi *Metarhizium anisopliae* and *M. acridum*. PLoS Genet 7:e1001264. https://doi.org/10.1371/journal.pgen.1001264
Goncalves MF (2025) Plant pathogenic fungi special issue: genetics and genomics. Microorganisms 13(4):925. https://doi.org/10.3390/microorganisms13040925
Guindon S, Gascuel O (2003) A simple, fast, and accurate algorithm to estimate large phylogenies by maximum likelihood. Syst Biol 52(5):696–704. https://doi.org/10.1080/10635150390235520
Hall T, Biosciences I, Carlsbad CJ (2011) BioEdit: an important software for molecular biology. GERF Bull Biosci 2(1):60–61
Hawksworth DL, Lücking R (2017) Fungal diversity revisited: 2.2 to 3.8 million species. Microbiol Spectr 5(4):10–128. https://doi.org/10.1128/microbiolspec.funk-0052-2016
He MQ, Zhao RL, Liu DM, Denchev TT, Begerow D et al (2022) Species diversity of Basidiomycota. Fungal Divers 114(1):281–325. https://doi.org/10.1007/s13225-021-00497-3
Hibbett DS, Binder M, Bischoff JF, Blackwell M, Cannon PF et al (2007) A higher-level phylogenetic classification of the Fungi. Mycol Res 111(5):509–547. https://doi.org/10.1016/j.mycres.2007.03.004
Higes M, Martín R, Meana A (2006) *Nosema ceranae*, a new microsporidian parasite in honeybees in Europe. J Invertebr Pathol 92(2):93–95. https://doi.org/10.1016/j.jip.2006.02.005
Hofstetter V, Miadlikowska J, Kauff F, Lutzoni F (2007) Phylogenetic comparison of protein-coding versus ribosomal RNA-coding sequence data: a case study of the Lecanoromycetes (Ascomycota). Mol Phylogen Evol 44:412–426. https://doi.org/10.1016/j.ympev.2006.10.016
Humber RA (2012) Entomophthoromycota: a new phylum and reclassification for entomophthoroid fungi. Mycotaxon, 120(1), 477–492. https://doi.org/10.5248/120.477
Hyde KD, Norphanphoun C, Maharachchikumbura SSN, Bhat DJ, Jones EBG et al (2020) Refined families of Sordariomycetes. Mycosphere 11(1):305–1059. https://www.nstda.or.th/openarchive/nstda-research-publications/refined-families-of-sordariomycetes/
Hyde KD, Saleh A, Aumentado HDR, Boekhout T, Bera I et al (2024) Fungal numbers: global needs for a realistic assessment. Fungal Divers 128(1):191–225. https://doi.org/10.1007/s13225-024-00545-8
Iwanicki NSA, Botelho ABRZ, Klingen I, Júnior ID, Rossmann S, Lysøe E (2022) Genomic signatures and insights into host niche adaptation of the entomopathogenic fungus *Metarhizium*

humberi. G3 Genes|Genomes|Genetics 12(2):jkab416. https://doi.org/10.1093/g3journal/jkab416

James TY, Kauff F, Schoch C, Matheny PB, Hofstetter V et al (2006) Reconstructing the early evolution of fungi using a six-gene phylogeny. Nature 443:818–822. https://doi.org/10.1038/nature05110

James TY, Porter TM, Martin WW (2014) 7 Blastocladiomycota. In: McLaughlin DJ, Spatafora JW (eds) Systematics and evolution: part A. Springer, Berlin Heidelberg, Berlin, Heidelberg, pp 177–207

Kaczmarek A, Boguś MI (2021) Fungi of entomopathogenic potential in Chytridiomycota and Blastocladiomycota, and in fungal allies of the Oomycota and Microsporidia. IMA Fungus 12(1):29. https://doi.org/10.1186/s43008-021-00074-y

Kaishian P, Lubbers M, Ben Fekih I, Groot MDD, Schilthuizen M et al (2024) Definitions of parasites and pathogens through time. Authorea. https://doi.org/10.22541/au.165712662.22738369/v2

Kepler RM, Luangsa-Ard JJ, Hywel-Jones NL, Quandt CA, Sung GH et al (2017) A phylogenetically-based nomenclature for Cordycipitaceae (Hypocreales). IMA Fungus 8(2):335–353. https://doi.org/10.5598/imafungus.2017.08.02.08

Khachatourians GG (2009) Insecticides, microbial. In: Schaechter M (ed) Encyclopedia of microbiology, 3rd edn. Elsevier, Amsterdam, pp 95–109. https://doi.org/10.1016/B978-012373944-5.00124-3

Kirk PM, Cannon PF, Minter DW, Stalpers JA (2008) Dictionary of the fungi, 10th edn. CAB International, Wallingford, UK

Kobayasi Y (1941) The genus *Cordyceps* and its allies. Sci Rep Tokyo Bunrika Daigaku, Sect B 5:53–260. https://www.science-open.com/document?vid=0fa805f8-29da-4082-b30f-f1d4708e6dff

Kobayasi Y (1982) Keys to the taxa of the genera *Cordyceps* and *Torrubiella*. Nippon Kingakkai Kaiho 23:329–364. https://www.scienceopen.com/document?vid=74f7ff8b-cd2d-4882-a902-d87963e26322

Kobmoo N, Mongkolsamrit S, Khonsanit A, Cedeño-Sanchez M, Arnamnart N et al (2024) Integrative taxonomy of *Metarhizium anisopliae* species complex, based on phylogenomics combined with morphometrics, metabolomics, and virulence data. IMA Fungus 15(1):30. https://imafungus.biomedcentral.com/articles/10.1186/s43008-024-00154-9

Koiri RK, Naik RA, Rawat D, Chhonker SK, Ahi JD (2017) Bioecological perspective of entomopathogenic fungi with respect to biological control. J Appl Microb Res 1:7–14

Kortsinoglou AM, Saud Z, Eastwood DC, Butt TM, Kouvelis VN (2020) The mitochondrial genome contribution to the phylogeny and identification of Metarhizium species and strains. Fungal Biol 124(10):845–853. https://doi.org/10.1016/j.funbio.2020.06.003

Koyuncu MÖ, Görmez V, Sevindik M, Krupodorova T, Eraslan EC (2025) Multilayered interactions between lepidoptera and fungi: Spore dispersal, mycophagy, and entomopathogenic relationships. Symbiosis. https://doi.org/10.1007/s13199-025-01068-9

Kumar S, Stecher G, Suleski M, Sanderford M, Sharma S et al (2024) MEGA12: molecular evolutionary genetic analysis version 12 for adaptive and green computing. Mol Biol Evol 41(12):msae263. https://doi.org/10.1093/molbev/msae263

Kuraku S, Zmasek CM, Nishimura O, Katoh K (2013) ALeaves facilitates on-demand exploration of metazoan gene family trees on MAFFT sequence alignment server with enhanced interactivity. Nucleic Acids Res 41(W1):W22–W28. https://doi.org/10.1093/nar/gkt389

Lapeva-Gjonova A (2014) Cataglyphis aenescens- a newly discovered ant host of the fungal parasite *Myrmicinosporidium durum*. Bulg J Agric Sci 20(Supp. 1):157–159

Leger RJS (2024) The evolution of complex *Metarhizium*-insect-plant interactions. Fungal Biol 128(8):2513–2528. https://doi.org/10.1016/j.funbio.2024.01.001

Lim LT (2024) Diversity, traditional knowledge, and historical perspectives on wild *Cordyceps*. In: Deshmukh SK, Sridhar KR (eds) *Cordyceps* and allied species. Springer Nature, pp 63–83. https://doi.org/10.1007/978-981-97-6345-0_3

Liu WC, Bonsall MB, Godfray HCJ (2007) The form of host density-dependence and the likelihood of host–pathogen cycles in forest-insect systems. Theor Popul Biol 72:86–95. https://doi.org/10.1016/j.tpb.2007.01.002

Lofgren LA, Stajich JE (2021) Fungal biodiversity and conservation mycology in light of new technology, big data, and changing attitudes. Curr Biol 31(19):R1312–R1325. https://doi.org/10.1016/j.cub.2021.06.083

Lu D (2023) The Global Circulation of Chinese Materia Medica, 1700–1949: a microhistory of the Caterpillar fungus. Springer Nature. https://doi.org/10.1007/978-3-031-24723-1

Lucking R, Aime MC, Robbertse B, Miller AN, Ariyawansa HA et al (2020) Unambiguous identification of fungi: where do we stand and how accurate and precise is fungal DNA barcoding? IMA Fungus 11(1):14. https://doi.org/10.1186/s43008-020-00033-z

Mains EB (1958) North American entomogenous species of *Cordyceps*. Mycologia 50(2):169–222. https://doi.org/10.1080/00275514.1958.12024722

Manfrino RG, Castrillo LA, López Lastra CC, Toledo AV, Ferrari W et al (2020) Morphological and molecular characterization of entomophthorales (Entomophthoromycota: Entomophthoromycotina) from Argentina. Acta Mycol 55(2):5522. https://doi.org/10.5586/am.5522

Manfrino R, Gutierrez A, Ben Gharsa H, Schuster C, López Lastra C et al (2024) Molecular taxonomic characterization and infra-specific diversity of entomopathogenic *Beauveria bassiana* fungi from Argentina. Fungal Biol 128(3):1800–1805. https://doi.org/10.1016/j.funbio.2024.04.003

Mascarin GM, Jaronski ST (2016) The production and uses of *Beauveria bassiana* as a microbial insecticide. World J Microbiol Biotechnol 32(11):177. https://doi.org/10.1007/s11274-016-2131-3

Meyling NV (2008) PCR-based characterisation of entomopathogenic fungi for ecological studies. Department of Agriculture and Ecology, University of Copenhagen, Denmark

Mongkolsamrit S, Noisripoom W, Pumiputikul S, Boonlarppradab C, Samson RA et al (2021) *Ophiocordyceps flavida* sp. nov. (Ophiocordycipitaceae), a new species from Thailand associated with *Pseudogibellula formicarum* (Cordycipitaceae), and their bioactive secondary metabolites. Mycol Prog 20:477–492. https://doi.org/10.1007/s11557-021-01683-y

Mora MAE, Castilho AMC, Fraga ME (2018) Classification and infection mechanism of entomopathogenic fungi. Arq Inst Biol 84:e055201. https://doi.org/10.1590/1808-1657000552015

Moreno-Gavíra A, Huertas V, Diánez F, Sánchez-Montesinos B, Santos M (2020) *Paecilomyces* and its importance in the biological control of agricultural pests and diseases. Plants 9(12):1746. https://doi.org/10.3390/plants9121746

Morin PA, Luikart G, Wayne RK, the SNP workshop group (2004) SNPs in ecology, evolution and conservation. Trends Ecol 19:208–216. https://doi.org/10.1016/j.tree.2004.01.009

Nguyen LT, Schmidt HA, Von Haeseler A, Minh BQ (2015) IQ-TREE: a fast and effective stochastic algorithm for estimating maximum-likelihood phylogenies. Mol Biol Evol 32(1): 268–274. https://doi.org/10.1093/molbev/msu300

Niskanen T, Lücking R, Dahlberg A, Gaya E, Suz LM et al (2023) Pushing the frontiers of biodiversity research: unveiling the global diversity, distribution, and conservation of fungi. Annu Rev Environ Resour 48:149–176. https://doi.org/10.1146/annurev-environ-112621-090937

Oliveira M, Azevedo L (2022) Molecular markers: an overview of data published for fungi over the last ten years. J Fungi 8(8):803. https://doi.org/10.3390/jof8080803

Ozolina A, Long M, Beales PA (2023) Identification of fungi based on morphological characteristics. In: Lane CR, Beales PA, Hughes KJD (eds) Fungal plant pathogens: applied techniques. CABI, pp 77–156. https://doi.org/10.1079/9781845936686.0074

Panchariya DC, Dutta P, Ananya MA, Chawade A et al (2024) Genetic marker: a genome mapping tool to decode genetic diversity of livestock animals. Front Genet 15:1463474. https://doi.org/10.3389/fgene.2024.1463474

Park E, Poulin R (2021) Revisiting the phylogeny of microsporidia. Int J Parasitol 51(10):855–864. https://doi.org/10.1016/j.ijpara.2021.02.005

Passer AR, Coelho MA, Billmyre RB, Nowrousian M, Mittelbach M et al (2019) Genetic and genomic analyses reveal boundaries between species closely related to *Cryptococcus* pathogens. MBio 10(3):e00764. https://doi.org/10.1128/mbio.00764-19

Patwardhan A, Ray S, Roy A (2014) Molecular markers in phylogenetic studies - a review. J Phylogenet Evol Biol 2(2):131. https://doi.org/10.4172/2329-9002.1000131

Pavlov IN, Litovka YA, Golubev DV, Astapenko SA, Chromogin PV (2018) New outbreak of *Dendrolimus sibiricus* tschetv. in Siberia (2012–2017): monitoring, modeling and biological control. Contemp Probl Ecol 11:406–419. https://link.springer.com/article/10.1134/S1995425518040054

Pérez-González O, Gomez-Flores R, Tamez-Guerra P (2022) Insight into biological control potential of *Hirsutella citriformis* against Asian citrus psyllid as a vector of citrus Huanglongbing disease in America. J Fungi 8(6):573. https://doi.org/10.3390/jof8060573

Powell MJ (2017) Blastocladiomycota. In: Archibald J, Simpson A, Slamovits C (eds) Handbook of the protists. Springer, Cham, pp 1497–1521. https://doi.org/10.1007/978-3-319-28149-0_17

Priest SJ, Yadav V, Heitman J (2020) Advances in understanding the evolution of fungal genome architecture. F1000Res 9:F1000 Faculty Rev-776. https://doi.org/10.12688/f1000research.25424.1

Radek R, Wurzbacher C, Strassert JF (2023) New nephridiophagid genera (Fungi, Chytridiomycota) in a mallow beetle and an earwig. MycoKeys 100:245. https://doi.org/10.3897/mycokeys.100.111298

Raja HA, Miller AN, Pearce CJ, Oberlies NH (2017) Fungal identification using molecular tools: a primer for the natural products research community. J Nat Prod 80(3):756–770. https://doi.org/10.1021/acs.jnatprod.6b01085

Rehner SA, Buckley E (2005) A *Beauveria* phylogeny inferred from nuclear ITS and EF1-α sequences: evidence for cryptic diversification and links to *Cordyceps* teleomorphs. Mycologia 97(1):84–98. https://doi.org/10.3852/mycologia.97.1.84

Rehner SA, Minnis AM, Sung GH, Luangsa-Ard JJ, Devotto L et al (2011) Phylogeny and systematics of the anamorphic, entomopathogenic genus *Beauveria*. Mycologia 103(5): 1055–1073. https://doi.org/10.3852/10-302

Reynolds NK, Stajich JE, Benny GL, Barry K, Mondo S et al (2023) Mycoparasites, gut dwellers, and saprotrophs: phylogenomic reconstructions and comparative analyses of Kickxellomycotina fungi. Genome Biol Evol 15(1):evac185. https://doi.org/10.1093/gbe/evac185

Rizal LM, Hereward JP, Brookes DR, Furlong MJ, Walter GH (2024) Hidden diversity within *Beauveria* and *Metarhizium*–comparing morphology, barcoding, multilocus phylogenies and whole-genome sequences. Fungal Ecol 67:101304. https://doi.org/10.1016/j.funeco.2023.101304

Robène-Soustrade I, Jouen E, Pastou D, Payet-Hoarau M, Goble T et al (2015) Description and phylogenetic placement of *Beauveria hoplocheli* sp. Nov. Used in the biological control of the sugarcane white grub, *Hoplochelus marginalis*, on Reunion Island. Mycologia 107(6): 1221–1232. https://doi.org/10.3852/14-344

Roberts DW, St Leger RJ (2004) *Metarhizium* spp., cosmopolitan insect-pathogenic fungi: mycological aspects. Adv Appl Microbiol 54(1):1–70. https://doi.org/10.1016/s0065-2164(04)54001-7

Roy HE, Vega FE, Chandler D, Goettel MS, Pell J, Wajnberg E (2010) The ecology of fungal entomopathogens. Springer, Dordrecht, p 198. https://doi.org/10.1007/978-90-481-3966-8

Saitou N, Nei M (1987) The neighbor-joining method: a new method for reconstructing phylogenetic trees. Mol Biol Evol 4(4):406–425. https://doi.org/10.1093/oxfordjournals.molbev.a040454

Sanjuan TI, Franco-Molano AE, Kepler RM, Spatafora JW, Tabima J et al (2015) Five new species of entomopathogenic fungi from the Amazon and evolution of neotropical *Ophiocordyceps*. Fungal Biol 119(10):901–916. https://doi.org/10.1016/j.funbio.2015.06.010

Schoch CL, Seifert KA, Huhndorf S, Robert V, Spouge JLCA et al (2012) Fungal Barcoding Consortium. Nuclear ribosomal internal transcribed spacer (ITS) region as a universal DNA barcode marker for Fungi. Proc Natl Acad Sci USA 109:6241–6246. https://doi.org/10.1073/pnas.1117018109

Scholte E-J, Knols BGJ, Samson RA, Takken W (2004) EPF for mosquito control: A review. J Insect Sci 4(1):19. https://doi.org/10.1093/jis/4.1.19

Sharma A, Srivastava A, Shukla AK, Srivastava K, Srivastava AK et al (2020) Entomopathogenic fungi: a potential source for biological control of insect pests. In: Solanki MK, Kashyap PL, Kumari B (eds) Phytobiomes: current insights and future vistas, pp 225–250. https://doi.org/10.1007/978-981-15-3151-4_9

Shen D, Wang J, Dong Y, Zhang M, Tang Z et al (2020) The glycoside hydrolase 18 family chitinases are associated with development and virulence in the mosquito pathogen *Pythium guiyangense*. Fungal Genet Biol 135:103290. https://doi.org/10.1016/j.fgb.2019.103290

Sheng J, Gerasimova K (eds) (2024) Sustainable Management of Cordyceps: supply chains and resource management policies. Taylor & Francis, Informa UK Limited, UK, p 204

Shenoy BD (2025) Do you believe in new fungal species? Mycol Spectr 1(1):47–58. https://doi.org/10.64993/MS.1.1.4

Shenoy BD, Jeewon R, Hyde KD (2007) Impact of DNA sequence-data on the taxonomy of anamorphic fungi. Fungal Diver 26:1–54

Silvestro D, Michalak I (2012) raxmlGUI: a graphical front-end for RAxML. Org Divers Evol 12(4):335–337. https://doi.org/10.1007/s13127-011-0056-0

Singh BK, Delgado-Baquerizo M, Egidi E, Guirado E, Leach JE et al (2023) Climate change impacts on plant pathogens, food security and paths forward. Nat Rev Microbiol 21:640–656. https://doi.org/10.1038/s41579-023-00900-7

Skrzecz I, Sierpińska A, Tumialis D (2024) Entomopathogens in the integrated management of forest insects: from science to practice. Pest Manag Sci 80(6):2503–2514. https://doi.org/10.1002/ps.7871

Steinwender BM, Enkerli J, Widmer F, Eilenberg J, Thorup-Kristensen K et al (2014) Molecular diversity of the entomopathogenic fungal *Metarhizium* community within an agroecosystem. J Invert Pathol 123:6–12. https://doi.org/10.1016/j.jip.2014.09.002

Strongman DB, White MM (2006) New species of Lancisporomyces, *Orphella*, and *Paramoebidium*, endosymbionts of stonefly nymphs from streams in Nova Scotia, Canada. Botany 84(9):1478–1495. https://doi.org/10.1139/b06-107

Sun T, Zou W, Dong Q, Huang O, Tang D et al (2022) Morphology, phylogeny, mitogenomics and metagenomics reveal a new entomopathogenic fungus *Ophiocordyceps nujiangensis* (Hypocreales, Ophiocordycipitaceae) from Southwestern China. MycoKeys 94:91. https://doi.org/10.3897/mycokeys.94.89425

Sun T, Chen Y, Wang D, Dai Y, Zou W et al (2024) Mitogenomics, phylogeny and morphology reveal two new entomopathogenic species of *Ophiocordyceps* (Ophiocordycipitaceae, Hypocreales) from south-western China. MycoKeys 109:49–72. https://doi.org/10.3897/mycokeys.109.124975

Sung GH, Hywel-Jones NL, Sung JM, Luangsa-ard JJ, Shrestha B et al (2007) Phylogenetic classification of *Cordyceps* and the clavicipitaceous fungi. Stud Mycol 57:5–59. https://doi.org/10.3114/sim.2007.57.01

Sung GH, Shrestha B, Han SK, Sung JM (2011) Growth and cultural characteristics of *Ophiocordyceps longissima* collected in Korea. Mycobiology 39(2):85–91. https://doi.org/10.4489/myco.2011.39.2.085

Swofford DL (1993) PAUP, phylogenetic analysis using parsimony. version # 3.1. Computer program distributed by the Illinois Natural History Survey, Illinois, USA, p 280

Taylor JW, Jacobson DJ, Kroken S, Kasuga T, Geiser DM et al (2000) Phylogenetic species recognition and species concepts in fungi. Fungal Genet Biol 31:21–32. https://doi.org/10.1006/fgbi.2000.1228

Thomas MB, Read AF (2007) Can fungal biopesticides control malaria? Nat Rev Microbiol 5:377–383. https://doi.org/10.1038/nrmicro1638

Thompson JD, Gibson Toby J, Higgins DG (2003) Multiple sequence alignment using ClustalW and ClustalX. Curr Protoc Bioinformatics 1:2–3. https://doi.org/10.1002/0471250953.bi0203s00

Tretter ED, Johnson EM, Benny GL, Lichtwardt RW, Wang Y et al (2014) An eight-gene molecular phylogeny of the *Kickxellomycotina*, including the first phylogenetic placement of Asellariales. Mycologia 106(5):912–935. https://doi.org/10.3852/13-253

Vávra J, Lukeš J (2013) Microsporidia and 'the art of living together'. Adv Parasitol 82:253–319. https://doi.org/10.1016/b978-0-12-407706-5.00004-6

Vega FE, Blackwell M (eds) (2005) Insect-fungal associations: ecology and evolution. Oxford University Press (New York, NY, 2005; online edn, Oxford Academic, 31 Oct. 2023), https://doi.org/10.1093/oso/9780195166521.001.0001

Wang JB, Leger RS, Wang C (2016) Advances in genomics of entomopathogenic fungi. Adv Gen 94:67–105. https://doi.org/10.1016/bs.adgen.2016.01.002

Wang Y, Fan Q, Wang D, Zou WQ, Tang DX et al (2022) Species diversity and virulence potential of the Beauveria bassiana complex and *Beauveria scarabaeidicola* complex. Front Microbiol 13:841604. https://doi.org/10.3389/fmicb.2022.841604

Wei D, Gentekaki E, Wanasinghe D, Tang S, Hyde K (2023) Diversity, molecular dating and ancestral characters state reconstruction of entomopathogenic fungi in Hypocreales. Mycosphere 13(1):281–351. https://doi.org/10.5943/mycosphere/si/1f/8

White TJ, Bruns T, Lee SJWT, Taylor J (1990) Amplification and direct sequencing of fungal ribosomal RNA genes for phylogenetics. In: Innis MA, Gelfand DH, Sninsky JJ, White TJ (eds) PCR protocols: a guide to methods and applications. Academic Press Inc, New York, pp 315–322

Wijayawardene NN, Hyde KD, Rajeshkumar KC, Hawksworth DL, Madrid H et al (2017) Notes for genera: Ascomycota. Fungal Divers 86:1–594. https://doi.org/10.1007/s13225-017-0386-0

Wijayawardene NN, Phillips AJ, Pereira DS, Dai DQ, Aptroot A et al (2022) Forecasting the number of species of asexually reproducing fungi (Ascomycota and Basidiomycota). Fungal Divers 114(1):463–490. https://doi.org/10.1007/s13225-022-00500-5

Wijayawardene NN, Hyde KD, Mikhailov KV, Péter G, Aptroot A et al (2024) Classes and phyla of the kingdom Fungi. Fungal Divers 128:1–165. https://doi.org/10.1007/s13225-024-00540-z

Zaccaron AZ, Stergiopoulos I (2025) The dynamics of fungal genome organization and its impact on host adaptation and antifungal resistance. J Genet Genom 52(5):628–640. https://doi.org/10.1016/j.jgg.2024.10.010

Zhang SL, He LM, Chen X, Huang B (2013) *Beauveria lii* sp. nov. isolated from *Henosepilachna vigintioctopunctata*. Mycotaxon 121(1):199–206. https://doi.org/10.5248/121.199

Zhang ZX, Shang YX, Liu QY, Li DH, Yin CZ et al (2025) Deciphering the evolutionary and taxonomic complexity of Diaporthales (Sordariomycetes, Ascomycota) through integrated phylogenomic and divergence time estimation. Fungal Divers 132(1):1–125. https://doi.org/10.1007/s13225-025-00551-4

Zhu F, Shen Z, Guo X, Xu X, Tao H et al (2011) A new isolate of *Nosema* sp. (Microsporidia, Nosematidae) from *Phyllobrotica armata* Baly (Coleoptera, Chrysomelidae) from China. J Invertebr Pathol 106(2):339–342. https://doi.org/10.1016/j.jip.2010.10.005

Zimmermann G, Papierok B, Glare T (1995) Elias Metschnikoff, Elie Metchnikoff or Ilya Ilich Mechnikov (1845-1916): a Pioneer in insect pathology, the first describer of the Entomopathogenic fungus *Metarhizium anisopliae* and how to translate a Russian name. Biocontrol Sci Tech 5(4):527–530. https://doi.org/10.1080/09583159550039701

Ecological Role and Functions of Entomopathogenic Fungi in Insect Population Dynamics

3

Jithu Unni Krishnan, Anju Mathiyil Subramanian, Rakhi Kunnath Radhakrishnan, and Murali Revanasiddappa

Abstract

Entomopathogenic fungi (EPF) play a critical ecological role in regulating insect populations and maintaining the balance of terrestrial ecosystems. These fungi, especially species of *Beauveria*, *Metarhizium*, and *Ophiocordyceps*, serve as natural predators of many insects, helping to control their populations and prevent outbreaks. This chapter explores the multifaceted interactions between EPF and insects, highlighting the behavioral, physiological, and immune defences insects employ to resist fungal invasion, including grooming, thermoregulation, cuticular defences, and immune pathway activation. It also examines how plants modulate these interactions through chemical cues and volatiles that can either suppress or enhance fungal virulence, as well as through epicuticular waxes that affect spore adhesion. Some EPF can function as endophytes within plant tissues, providing dual benefits by promoting plant health and offering indirect protection against herbivores. Environmental conditions such as temperature, humidity, UV radiation, and soil composition further influence the efficacy and persistence of these fungi in natural habitats. Notably, some EPF exhibit host-specific behavioral manipulations, such as summit disease and death-grip phenomena, which enhance fungal dispersal and transmission. Collectively, these complex biotic and abiotic interactions determine the success of EPF in insect population regulation and their ecological significance in both natural and managed ecosystems. The chapter highlights the importance of understanding these dynamics for improving the use of EPF in integrated pest management (IPM) and predicting their responses under scenarios of environmental change. By bridging insights from entomology, plant biology, and microbial ecology, the chapter aims to

J. U. Krishnan (✉) · A. M. Subramanian · R. K. Radhakrishnan · M. Revanasiddappa
Division of Forest Health, Kerala Forest Research Institute, Thrissur, Kerala, India

S. K. Deshmukh et al. (eds.), *Frontiers in Entomopathogenic Fungi*,
https://doi.org/10.1007/978-981-95-6529-0_3

provide a comprehensive understanding of the role and functions of entomopathogenic fungi in shaping insect population dynamics and contributing to ecosystem resilience.

Keywords

Hexapodes · Mycopathogens · Mycoinsecticides · Insect-immunity · Endophytes

1 Introduction

Early pest control relied on natural predators and pathogens before the advent of synthetic insecticides. Entomopathogenic microbes (fungi, bacteria, nematodes, and viruses) are widely used as part of integrated pest management (IPM) in horticulture, forestry, and agriculture (Inglis et al. 2000). Entomopathogenic fungi (EPF) play a crucial role in insect pest control and population dynamics due to their ability to target a wide range of insect groups, from highly specific to more generalist ones. Recent research has explored how EPF can be affected by and even remove various anthropogenic contaminants (e.g., pesticides, alkylphenols, organotin compounds, synthetic estrogens, and hydrocarbons) that pollute farmlands and the environment (Swathy et al. 2024). Beyond their biocontrol potential, EPF produce numerous secondary metabolites that show promise for applications in medicine and as antimicrobial agents (Jinna et al. 2025).

In 1888, Metchnikoff conducted field trials for EPF, prominently utilizing *M. anisopliae*. His methods notably included the mass production of fungal conidia for crop application. His research inspired global interest in using "friendly fungi" for pest control. Boverin, a mycoinsecticide from *Beauveria bassiana*, was developed in 1965 for pest control. Increased environmental awareness and resistance to chemicals boost EPF application (Um et al. 2018). Different environmental conditions, such as UV light, temperature, and humidity, influence the efficiency of EPF in fields. The wide range of insect orders, including Lepidoptera, Coleoptera, Hymenoptera, Orthoptera, and Diptera, are infected by EPF. Fungi within the order Hypocreales can infect a broad spectrum of insects. While those from Entomophthorales typically exhibit high specificity, targeting only a single type of insect. EPF can identify and effectively infect a wide variety of insect pests and mite species, such as aphids, lepidopterous larvae, and thrips, which are recognized as the major pests in the crops. Due to the host-specific activity, these will cause deadly diseases to the specified host (Deka et al. 2021).

The mode of infection involves a few steps: EPF are mostly soil-borne and infect hosts by attaching to the cuticle. Germination leads to the formation of specialized structures, including a germ tube that penetrates the host integument. Enzymes facilitate cuticle penetration, enabling fungal growth within the insect. Fungi disseminate through the insect's body, causing death within three to 14 days. Opportunistic pathogens cause mortality through nutritional deficiency and tissue destruction. Mycotoxins produced during infection act as poisons to the host.

After the host's death, fungi sporulate on cadavers, completing their life cycle. Obligate pathogens primarily colonize tissues without relying heavily on toxins (Um et al. 2018).

2 Mechanism of Pathogenesis

The effective infectious mechanism developed by EPF helps establish them as promising biocontrol agents against pests and develop them as cost-effective biocontrol agents. EPF primarily belong to the phyla Ascomycota and Entomophthoromycota. Notably, the order Hypocreales within Ascomycota is the prominent group involved in insect pathogenesis. Their journey begins with the production of spores, specifically conidia or blastospores. When conidia come into contact with a suitable substrate (the host insect), they begin to germinate and develop a germ tube, which then penetrates the host cuticle, leading to the systemic infection stage (Hajek and St Leger 1994; Krishnan et al. 2024). Some infective sessile spores (Capilliconidia) possess an adhesion drop for attachment (Goettel and Glare 2010). Fungal structures, such as appressoria (infection pegs), aid in germ tube penetration (Hajek 1997). The mucus layer of spores consists of proteins and glucans, which help adhere to the substratum. The hydrolytic enzymes (lipases, proteases, and chitinases) produced by the germinating spore, along with mechanical energy, help penetrate the insect as the germ tube (Xiao et al. 2012). The fungus develops within the host body, and fungal hyphae colonize the entire infection. The fungus proliferates within the host body and spreads hundreds of spores, which are then ready to infect new hosts (Deka et al. 2021).

Beyond the direct infection, EPF employ several strategies to paralyze and kill the host. In most cases, death is caused by mechanical damage resulting from mycelia development and toxins produced by the fungi. These toxins disrupt normal physiological activities, reducing feeding and mortality. The toxins produced by fungi such as *Beauveria*, *Metarhizium*, and *Tolypocladium*, including destruxin, beauvericin, and efrapeptins, play a crucial role in pathogenesis (Hajek and St Leger 1994).

While the initial stages of infection may not exhibit obvious symptoms, behavioral changes typically appear days before death, including reduced coordination, feeding, and activity (e.g., in grasshoppers and locusts infected with *M. anisopliae*). Other observed behavioral responses include "behavioral fever" (seeking warmer temperatures, e.g., locusts/grasshoppers infected with *M. anisopliae* and *B. bassiana*), increased feeding (e.g., Colorado potato beetles infected with *B. bassiana*), altered photo- or geotropism, and changes in mating behavior (Noma and Strickler 2007; Blanford and Thomas 2001). Some are manipulating the behavior of the host to enhance the spore dispersal. The zombie ant that climbs to elevate its position before death, thereby optimizing spore spread (Bekker et al. 2014).

3 Spore Production, Spore Density, Transmission, and Dispersal

While some fungi, such as *Strongwellsea castrans*, sporulate in living flies through abdominal holes, spore production most commonly occurs after the host dies, when the fungus colonizes the cadaver. Many Entomophthorales species can forcibly discharge primary spores from the cadaver, enabling them to spread and find new hosts (Luz and Fargues 1998; Eilenberg and Meadow 2003).

EPF employ various strategies to increase its chances of encountering new hosts. For groups that produce abundant spores, such as Hyphomycetes, rain, wind, and invertebrates facilitate the transmission of spores. Wind is a primary means of spore distribution, and growing hyphae from insect cadavers also contributes significantly to conidia dispersal. High humidity and moisture are crucial for the germination and sporulation of terrestrial fungi, thus impacting their transmission. The fungi can infect multiple insect life stages, facilitating the spread of the disease within insect populations; for instance, winged insects can carry and disperse spores (Dromph 2001).

Many insects contend with multiple natural enemies, including parasitoids, predators, and other pathogens, leading to competition for the same host resources. An insect can be simultaneously infected by more than one natural enemy or pathogen. Interactions can also occur where natural enemies feed on already infected hosts. Conversely, EPF can infect natural enemies or other non-target insects, which then become a source of fungal spore dispersal (Feng et al. 1990; Pell et al. 2010).

4 Host Range and Specificity

EPF exhibit considerable variation in host specificity and range, reflecting both taxonomic diversity and the ecological contexts in which they operate. Host specificity refers to the ability of an EPF strain to infect one or more insect hosts, ranging from specialists with narrow host ranges to generalists that can infect diverse insect taxa (Brodeur 2012). Among the well-studied generalist fungi are *Beauveria bassiana* and *Metarhizium anisopliae*, both capable of infecting insects from multiple orders, including Lepidoptera, Coleoptera, Diptera, Hemiptera, Orthoptera, and Hymenoptera. Despite being generalists, individual isolates of these fungi show distinct patterns of specificity and virulence. This variability often reflects the origin of the fungal strain, as strains tend to be most virulent toward the insect species from which they were originally isolated (Goettel et al. 1990). Notably, highly specific entomopathogens tend to be more virulent and often obligate in nature, while generalist strains are typically less virulent (Goettel et al. 1995).

Hajek and Butler (2000) further differentiate host range into two categories: *ecological host range*, which encompasses naturally infected hosts under field conditions, and *physiological host range*, which includes all insect species that can be infected under laboratory conditions. This distinction is critical for understanding the full potential and risks of EPF as biocontrol agents. Experimental assays highlight the variation in host specificity among EPF strains. For instance, two *B. bassiana* strains and one *B. hoplocheli* strain were tested for pathogenicity against nine insect pests from Lepidoptera, Coleoptera, and Diptera (Rohrlich et al. 2018). All three strains were pathogenic to Lepidoptera and Diptera. In Coleoptera, *B. hoplocheli* was effective against *Hoplochelus marginalis*, while *B. bassiana* strains targeted *Alphitobius diaperinus*. However, *B. hoplocheli* displayed reduced virulence in Lepidoptera and Diptera compared to *B. bassiana*, highlighting differences in strain efficacy and host range (Rohrlich et al. 2018).

Further illustrating host specificity, *Entomophaga maimaiga* was shown to be highly specific to Lepidoptera, with no infection observed in Coleoptera, Orthoptera, or Hymenoptera under experimental conditions (Hajek 1999). Similarly, *Beauveria brongniartii* strains isolated from *H. marginalis* were uniquely effective against this beetle species, whereas strains from other hosts were largely ineffective (Neuvéglise et al. 1994). These findings emphasize the importance of host origin in determining the infectivity of EPF strains. The potential for unintended impacts on non-target insects, especially when using generalist fungi, cannot be disregarded without a comprehensive evaluation of the ecological host range (Rohrlich et al. 2018). For example, although *Coelomomyces psorophorae* successfully infected seven out of twelve mosquito species under laboratory conditions, field infections were not observed, highlighting the need to validate laboratory findings in natural settings (Zebold et al. 1979).

While nearly all major insect orders are susceptible to infection by EPF, the following is a consolidated attempt to highlight representative EPF associated with key insect orders, along with a few notable species from each group.

4.1 Orthoptera

Orthoptera includes the grasshoppers and locusts predominantly and forms the dominant insect group in dry grasslands and deserts. When the number of these insects increases, they exhibit gregarious and migratory behavior, forming swarms. They have very high reproductive rates and can reproduce rapidly under favorable conditions. The brown locust and the Senegalese grasshopper, *Oedaleus senegalensis*, have two or three annual generations (Batten 1969). Different stages of insects respond differently to pathogens, which significantly affects the pathogen's establishment, development, fecundity, host resistance, and mortality in insect hosts (Mackauer 1973). Nymphal stages are more susceptible to infection than adults (Wraight et al. 1998) (Table 3.1).

4.2 Lepidoptera

The order Lepidoptera is the second-largest insect order, including moths and butterflies. This order comprises numerous insect pests that should be controlled to prevent agricultural losses in temperate, subtropical, and tropical regions (Trajković

Table 3.1 List of entomopathogenic fungi associated with different species of Orthoptera

Orthoptera species	Entomopathogenic fungus	Reference
Anabrus simplex	*Metarhizium acridum*, *M. runneum*, *M. uizhouense* and *M. robertsii*	Rangel et al. (2022)
Camnula pellucida	*Entomophaga grylli*	Kistner and Belovsky (2013)
Dociostaurus maroccanus	*Metarhizium anisopliae* and *Beauveria bassiana*	Crespo and Santiago-Alvarez (1997)
Gryllotalpa gryllotalpa	*Beauveria bassiana*, *Clonostachys rogersoniana*, *Metarhizium anisopliae*, and *Myriodontium keratinophilum*	Sönmez et al. (2016)
Hieroglyphus sp.	*Aspergillus* sp., *Beauveria* sp., and *Metarhizium flavoviride*	Sultana et al. (2013)
Locusta migratoria capito	*Sorosporella* sp.	Welling et al. (2016)
Locusta migratoria manilensis	*Beauveria bassiana*	Sangbaramou et al. (2018)
Locusta migratoria migratorideis	*Beauveria bassiana* and *Metarhizium anisopliae*	Fathy et al. (2025)
Melanoplus bivittatus	*Lecanicillium lecanii*	Johnson et al. (1988)
Melanoplus differentialis	*Entomophaga grylli*	Ramoska et al. (1988)
Melanoplus packardii	*Lecanicillium lecanii*	Johnson et al. (1988)
Melanoplus sanguinipes	*Beauveria bassiana*, *Entomophaga gryll* and *Lecanicillium lecanii*	Johnson et al. (1988), Jeff et al. (1997)
Oxya velox	*Entomophthora grylli*	Gupta et al. (2011)
Oxya vicinia	*Entomophthora grylli*	Gupta et al. (2011)
Poekilocerus pictus	*Beauveria bassiana* and *Metarhizium anisopliae*	Ramanujam et al. (2022)
Ruspolia differens	*Aspergillus flavus*, *A. tamarii*, *Clonostachys rosea*, *Fusarium equiseti*, and *Mucor fragilis*	Leonard et al. (2023)
Schistocerca gregaria	*Beauveria bassiana*, *B. brogniartii*, *Metarhizium acridum*, *M. anisopliae*, *M. robertsi*, and *Steinernema carpocapsae*	Gillespie et al. (2000), Levchenko et al. (2020), Xia et al. (2000)
Zonocerus variegatus	*Aspergillus niger*, *Beauveria bassiana*, *Metarhizium anisopliae*, *Mucor* sp. *Penicillium* sp.	Balogun and Fagade (2004)
New species of Orthoptera	*Metarhizium gryllidicola*, *Neotorrubiella chinghridicola*, and *Ophio cordyceps kobayasii*	Thanakitpipattana et al. (2020)

and Žikić 2023). Both larvae and adults are significant pests of crops, plants, and trees, and many lepidopteran pests are expanding their geographical ranges; for example, the diamondback moth, *Plutella xylostella*, has already expanded its geographical range (Karthi et al. 2024) (Table 3.2).

Table 3.2 List of EPF associated with different species of Lepidoptera

Lepidoptera species	Entomopathogenic fungus	Reference
Chaetoprocta odata	*Beauveria bassiana* and *Isaria fumosorosea*	Shazia et al. (2024)
Cydia pomonella	*Aspergillus flavus*, *Beauveria pseudo-bassiana*, *Fusarium oxysporum*, *F. polyphialidicum*, *Metarhizium robertii*, and *Purpureocillium lilacinum*	Abaajeh (2014)
Helicoverpa armigera	*Beauveria bassiana*, *Isaria fumosorosea* and *Metarhizium anisopliae*	Nguyen et al. (2007)
Lobesia botrana	*Beauveria bassiana* and *Metarhizium robertsii*	López Plantey et al. (2019)
Lymantria dispar	*Beauveria bassiana*, *Entomophaga maimaiga*, *Fusarium polyphialidicum*, *Paecilomyces farinosus* and *Verticillium lecanii*	Hajek et al. (1996), Hajek (1997)
Pieris rapae	*Beauveria bassiana*, *Fusarium* sp., *Metarhizium anisopliae*, and *Rhizopus* sp.	Embaby and Lotfy (2015)
Plodia interpunctella	*Beauveria bassiana*, *Isaria fumosorosea*, and *Metarhizium anisopliae*	Sabbour and Abd-El-Aziz (2012)
Plutella xylostella	*Beauveria bassiana*, *Isaria fumosorosea*, and *Metarhizium anisopliae*	Ibrahim and Low (1993), Shehzad et al. (2021)
Prays oleae	*Beauveria bassiana*	Oliveira (2012)
Spodoptera frugiperda	*Aspergillus* sp., *Beauveria bassiana*, *Cladosporium tenuissimum*, *Nomuraea rileyi*, and *Penicillium citrinum*	Ruiz-Nájera et al. (2013), Idrees et al. (2021)
Spodoptera littoralis	*Beauveria bassiana*, *Clonostachys rogersoniana*, *Metarhizium anisopliae*, and *M. brunneum*	Resquín-Romero et al. (2016), Şahin and Yanar 2021
Spoladea recurvalis	*Beauveria bassiana*, *Isaria fumosorosea*, and *Metarhizium anisopliae*	Opisa et al. (2018)
Thaumatotibia leucotreta	*Beauveria bassiana* and *Metarhizium anisopliae*	Coombes (2012)
Thaumetopoea pityocampa	*Beauveria bassiana*, *B. pseudo-bassiana*, *B. varroae*, *Isaria farinosa*, *I. fumosorosea*, *Lecanicillium lecanii* and *Metarhizium anisopliae*	Paparatti and Fabozzi (1988), Vargas-Osuna et al. (1994), Er et al. (2007), Sevim et al. (2010), Tarasco et al. (2015), Akinci et al. (2017), Barta et al. (2019)

4.3 Hemiptera

Hemiptera is considered one of the most prevalent insect orders, which comprises the majority of horticulture and forestry pests. They are found in terrestrial and aquatic ecosystems (Kaur and Banu 2019). They comprise a diverse group, including scale insects, aphids, psyllids, and whiteflies (Sternorrhyncha), true bugs (Heteroptera), and the Auchenorrhyncha (leafhoppers, planthoppers, treehoppers, spittlebugs, and cicadas). Hemiptera exhibit a range of feeding habits, from phytophagy to predation, including ectoparasitism and hematophagy (Table 3.3).

Table 3.3 List of EPF associated with different species of Hemiptera

Hemiptera species	Entomopathogenic fungus	References
Acyrthosiphon pisum	*Aspergillus* spp.	Seye et al. (2014)
Aphis gossypii	*Lecanicillium attenuatum*, *L. lecani*	Hall (1982), Kim and Roberts (2012)
Aelia rostrata	*Beauveria bassiana* and *Isaria farinosa*	Mustu et al. (2011)
Bemisia tabaci	*Paecilomyces fumosoroseus*	Eslamizadeh et al. (2014)
Corythucha ciliata	*Beauveria bassiana*, *B. pseudo-bassiana*, *Isaria fumosorosea*, and *Metarhizium anisopliae*	Sevim et al. (2013)
Diaphorina citri	*Lecanicillium attenuatum* and *L. psalliotae*	Lu et al. (2015)
Dysdercus cingulatus	*Metarhizium anisopliae*	Sahayaraj and Francis Borgio (2009)
Dysdercus peruvianus	*Metarhizium anisopliae*	Santi et al. (2011)
Elasmolomus pallens	*Aspergillus flavus*, *A. tamarii*, *F. proliferatum*, *Metarhizium Anisopliae*, and *Trichoderma atroviride*	Umaru and Simarani (2020, 2022)
Eurygaster integriceps	*Beauveria bassiana*	Abdulhai et al. (2010)
Lygus lineolaris	*Beauveria bassiana* and *Metarhizium anisopliae*	Liu et al. (2002)
Oebalus mexicana	*Beauveria bassiana* and *Fusarium moniliforme*	Salazar (2002), Martínez et al. (2005), Cortez-Madrigal et al. (2022)
Plautia stali	*Beauveria bassiana* and *Metarhizium anisopliae*	Ihara et al. (2001)
Trialeurodes vaporariorum	*Lecanicillium lecani*	Chandler et al. (1993)

4.4 Coleoptera

Coleoptera is the most diverse order in the animal kingdom. Its members inhabit a wide variety of habitats. Beetles are rich in diversity, size, form, and ecological strategies. They impact agriculture and forestry and serve as beneficial model organisms for various areas of science. They have helped maintain ecosystem health and monitoring, as well as providing ecosystem services such as decomposition, dung removal, pollination, and biological control (Bouchard et al. 2017) (Table 3.4).

4.5 Hymenoptera

The order Hymenoptera includes hornets, bees, ants, sawflies, and parasitic wasps. Ecologically and economically, they are one of the important groups of insects. Bees are recognized for their vital role in pollinating ecosystems in both natural and managed environments (Gallai et al. 2009). Parasitic Hymenoptera control the

Table 3.4 List of EPF associated with different species of Coleoptera

Coleoptera species	Entomopathogenic fungus	Reference
Anoplophora glabripennis	*Beauveria bassiana* and *B. brongniartii*	Dubois et al. (2024)
Aplosonyx chalybaeus	*Heterorhabditis indica*, *Steinernema thermophilum*, and *S. glaseri*	Yadav and Lalramliana (2012)
Holotrichia oblita, *Anomala corpulenta*	*Beauveria bassiana*, *B. brongniartii*, and *Metarhizium anisopliae*	Nong et al. (2011)
Hypothenemus hampei	*Beauveria bassiana*	Posada-Flórez (2008)
Ips stebbingi	*Beauveria bassiana*, *Lecanicillium lecanii*, and *Metarhizium anisopliae*	Khanday et al. (2018)
Oryzaephilus surinamensis and *Rhyzopertha dominica*	*Beauveria bassiana*	Lord (2001)
Osphranteria coerulescens	*Beauveria bassiana and Metarhizium anisopliae*	Mohammadyani et al. (2016)
Pantorhytes plutus	*Beauveria bassiana*	Prior et al. (1988)
Polyphylla fullo	*Beauveria bassiana* and *Metarhizium anisopliae*	Erler and Ates (2015)
Popillia japonica	*Metarhizium brunneum* and *Beauveria brongniartii*	Graf et al. (2023)
Scolytus amygdali	*Beauveria bassiana*	Batta (2007)
Sinoxylon anale	*Beauveria bassiana*	Kumar and Attri (2016)
Xyleborus glabratus	*Beauveria bassiana* and *Isaria fumosorosea*	Carrillo et al. (2015)
Xylosandrus crassiusculus	*Beauveria bassiana* and *Metarhizium brunneum*	Carrillo et al. (2013)

Table 3.5 List of entomopathogenic fungi associated with various Hymenoptera species

Hymenoptera species	Entomopathogenic fungus	Reference
Aphelinus asychis	*Paecilomyces fumosoroseus*	Lacey et al. (1997), Mesquita and Lacey (2001)
Apis mellifera	*Ascosphaera apis*, *Beauveria bassiana*, and *Metarhizium anisopliae*	Gilliam and Lorenz (1993), Potrich et al. (2018)
Camponotus pennsylvanicus	*Beauveria bassiana* and *Metarhizium anisopliae*	Kelley-Tunis et al. (1995)
Chelostoma florisomne	*Ascosphaera callicarpa*	Wynns et al. (2013)
Dolichoderus thoracicus	*Aspergillus nomius*	Lin et al. (2021)
Encarsia formosa and *Trialeurodes vaporariorum*	*Beauveria bassiana*	Oreste et al. (2016)
Formica selys	*Beauveria bassiana*, *Metarhizium brunneum*, and *Purpureocillium lilacinus*	Reber and Chapuisat (2012)
Megachile relativa	*Ascosphaera aggregata*	Goerzen et al. (1990)
Megachile rotundata	*Ascosphaera subglobosa*, *A. proliperda*	Youssef et al. (1984), Wynns et al. (2012)
Melipona beecheii, *Scaptotrigona mexicana* and *Tetragonisca angustula*	*Beauveria bassiana*, *Isaria fumosorosea* and *Metarhizium anisopliae*	Toledo-Hernández et al. (2016)
Solenopsis invicta	*Metarhizium anisopliae*	Hassan et al. (2024)
Vespula vulgaris	*Aspergilus niger*, *A. niveus*, *A. restricus*, *Beauveria bassiana* and *Metarhizium anisopliae*	Gambino and Thomas (1988), Harris et al. (2000)
Xylocopa augusti	*Ascosphaera apis*	Reynaldi et al. (2015)
Xylocopa californica arizonensis	*Ascosphaera apis*	Gilliam and Lorenz (1993)

populations of phytophagous insects (Tscharntke et al. 1998) and are effective agents for controlling insect pests (Johnson et al. 2013). Some phytophagous Hymenoptera are closely associated with their host plants (Nyman et al. 2006) and function as parasitoids (Table 3.5).

4.6 Diptera

The order Diptera consists of true flies or two-winged flies. They are among the most diverse groups of insects, including mosquitoes, black flies, horse flies, and midges. The larval stages are found in aquatic, semi-aquatic, and moist terrestrial environments. They feed on decaying organic materials, plant sap, and animal exudates. They are classified as parasites, predators, or parasitoids depending on their feeding mode. Some flies are vectors for several diseases. (Fusari et al. 2018) (Table 3.6).

Table 3.6 List of EPF associated with different species of Diptera

Diptera species	Entomopathogenic fungus	Reference
Aedes aegypti and *A. albopictus*	*Isaria farinosa*, *I. flavovirescens*, *I. fumosorosea*, *Lecanicillium* spp. and *Metarhizium anisopliae*	Scholte (2007), Darbro et al. (2011)
Anopheles gambiae	*Beauveria bassiana* and *Metarhizium anisopliae*	Mnyone et al. (2011)
Bactrocera cucurbitae and *B. Zonata*	*Beauveria bassiana*, *Isaria fumosorosea* and *Metarhizium anisopliae*	Sookar et al. (2008)
Bactrocera zonata	*Metarhizium anisopliae*, *Beauveria bassiana* and *Verticillium lecanii*	Murtaza et al. (2022)
Ceratitis capitata	*Beauveria bassiana*, *Isaria fumosorosea* and *Metarhizium anisopliae*	Beris et al. (2013)
Ceratitis capitata	*Isaria fumosorosea* and *Metarhizium anisopliae*	Castillo et al. (2002)
Ceratitis capitata, *C. rosa* var. *fasciventris*	*Isaria fumosorosea* and *Metarhizium anisopliae*	Ekesi et al. (2002)
Drosophila melanogaster	*Metarhizium brunneum*	Liao et al. (2024)
Drosophila suzukii	*Entomophthora muscae*	Becher et al. (2018)
Hematobia irritans	*Beauveria bassiana*, *Isaria fumosorosea*, and *Metarhizium anisopliae*	Angel-Sahagún et al. (2005)
Hermetia illucens	*Beauveria bassiana*	Opare et al. (2024)
Musca domestica	*Entomophthora muscae*	Zurek et al. (2002)
Odagmia sp.	*Coelomycidium simulii*	Levchenko et al. (1974)
Stomoxys calcitrans	*Metarhizium anisopliae*	Baleba et al. (2021)

5 Compatibility with the Host

The success of EPF infection largely depends on host-pathogen compatibility. The fungal spores, or conidia, are responsible for the infection and are dispersed, landing on the insect host. The conidia attach and germinate if the host is compatible, initiating a chain reaction in which they recognize and activate enzymes (Samson et al. 1988). Once the fungus has penetrated the cuticle of the external insect skeleton, it invades the insect's body and circulatory system, known as hemolymph. Structures and processes for the invasion of insect tissues are similar to those of plant pathogens, including the formation of germ tubes, appressoria, and penetration pegs (Samson et al. 1988). The population of *Delia platura* is only infected by the pathogen *Entomophthora muscae*. When the mixed population of *Delia platura* and *Delia brassicae* occurred, *Delia platura* was infected only with *Entomopthora muscae*, while *Delia brassicae* was only infected with *Strongwellsea castrans* (Nair and McEwen 1973). Depending on the species and their developmental stages, insects have different susceptibilities to fungal pathogens (Clarkson and Charnley 1996). Differences in virulence may be influenced by the insects' developmental stage, cuticle biochemical composition, physiological characteristics, and/or inoculation method (Tang et al. 1999).

6 Spore Density and Dispersal

Spore efficacy requires a critical concentration for infection, and high spore density increases the chances of disease transmission by providing adequate infective propagules. Dispersal mechanisms also play a crucial role in reaching new hosts. The spore germination rate was used to measure viability (Soetopo 2004). The capacity of a fungal pathogen to consume host nutrients is considered a virulence factor. Many different strains of fungi exhibit considerable variation in their virulence levels and host ranges. Spore viability might be one factor that influences virulence. Innate virulence and conidial viability are more prominent determinants of an isolate's effectiveness in causing mortality against aphid nymphs of *Aphis gossypii* (Herlinda 2010). The capacity of the fungal pathogen to consume the host nutrients is considered one of the virulence factors (Meshrif et al. 2010).

There are four categories of entomopathogens:

Accidental pathogens: They cannot penetrate the insect cuticular barrier themselves. Instead, they must be introduced into the body by making wounds or scars.

Occasional pathogens are widespread saprophytes that can only enter the host under certain natural or experimentally induced conditions.

Facultative pathogens: These pathogens possess the ability to develop in specific natural environments without their hosts. They can also overcome competition from antagonistic organisms in their environment.

Obligate pathogens: These fungi can grow only in their hosts or, in some cases, in axenic cultures on simple or complex media. These ecologically obligate pathogens often respond to competition from their antagonists by developing resistant structures.

7 Insect Host Factors

7.1 Behavioral Responses

Parasites use two mechanisms to alter host behavior: direct and indirect. Parasites directly influence the host by interacting with the host's nervous system or muscles. However, an indirect mechanism, which infects tissues other than neurons and muscles, mediates changes in behavior (Thomas et al. 2005). Behaviorally manipulated hosts often occupy distinct niches from healthy hosts to increase parasite transmission (Moore 2002).

The two-spotted spider mite males prefer females infected with *Neozygites floridana* to healthy females (Trandem et al. 2015). *Ophiocordyceps unilateralis* modifies the behavior of its ant hosts, *Camponotus leonardi.* This behavioral modification is a fine-tuned fungal adaptation and an adaptive extended phenotype. The first modification the fungus makes to manipulate the ants is to bite the vegetation before death, which is called a death grip. It appears to be a strategy to optimize conditions for fungal development and reproduction (Andersen et al. 2009). The fly *Musca domestica* infected with *Entomopthora muscae* tends to move to a high place before death, a phenomenon known as the summit behavior, which facilitates spore transfer (Krasnoff et al. 1995). There is a chance that the insect will be caught by predators or its natural enemies, making it challenging to determine whether the summit behavior benefits the fungus (Masoudi et al. 2024).

The self-grooming behavior of *Aedes aegypti* removes the lower amounts of fungal conidia of *Metarhizium anisopliae*, which is the mosquito's behavioral response to avoid fungal infections (Mehmood et al. 2022). Nymphs of the migratory locust *Locusta migratoria manilensis* increase their thoracic temperature when infected with *Beauveria bassiana* during basking. The increased temperature limits the conidial germination and hyphal development of *B. bassiana*. Behavioral thermoregulation can adversely affect the fungal growth in *Locusta migratoria manilensis* (Sangbaramou et al. 2018).

The infection of *Metarhizium anisopliae* on the fire ant *Solenopsis invicta* changed their physiology and behavior. The infected ants showed disturbed locomotion and reduced foraging behavior. There was a significant decrease in the distance traveled and velocity; at the physiological level, the catalase and antifungal activities increased (Hassan et al. 2024). *Metarhizium* anisopliae cannot complete its lifecycle in the colony of the subterranean *Coptotermes formosanus*. This is mainly due to the other colony members burying the infected individual or adopting cannibalism to prevent the spread of the disease in the colony (Chouvenc et al. 2008). In the laboratory assay of *Anthocoris nemorum*, they avoid the *Beauveria bassiana*-infected leaf surface to reduce the mortality risk (Meyling and Pell, 2006). Mole crickets can detect and avert *Beauveria bassiana* in the soil but cannot recognize the conidia at distances of 8 cm. This suggests a chemosensory detection mechanism, which may involve volatile cues, contact chemoreception, or postingestional mechanisms (Thompson et al. 2014). *Entomophthora muscae* infected *Musca domestica* alerts healthy individuals by exhibiting body vibrations, which reduces the chance of infection in other flies (Kalsbeek et al. 2001).

An important example of hygienic behavior involves worker honeybees (*Apis mellifera*) being able to detect and remove larvae infected with the fungus *Ascosphaera apis*, also known as chalkbrood, from their nest (Swanson et al. 2009). Some ants groom their metapleural glands when pathogens are present and use the secretion to avoid the pathogenicity (Fernández-Marín et al. 2006). The locusts Schistocerca increase their movements when infected with *M. anisopliae*, making them easier targets for predators and thereby avoiding further spread within the population (Arthur and Thomas 2001).

7.2 Insect Cuticle: A Physical Barrier

The insect integument serves as a crucial line of defense against pathogenic infections. The insect cuticle is the first line of defense against pathogenic infections and water evaporation (Reynolds and Samuels 1996). The insect cuticle comprises two layers: the outermost epicuticle and the inner procuticle. The procuticle, which is composed of an exocuticle and endocuticle, has numerous layers of interwoven fibrous chitin and proteins (Reynolds and Samuels 1996). The insect cuticle is the primary and most important barrier to EPF infection, composed of chitin and tanned proteins that are resistant to degradation. The cuticle is also rich in short-chain fatty acids, which provide an antimicrobial character (Leger et al. 1991). Apart from other microorganisms, EPF can directly penetrate the insect cuticle. The resident microbial community and gland secretion on the cuticle make the penetration more difficult. *Delphacodes kuscheli* and *Dalbulus maidis* are two hemipterans that have *Bacillus pumilus* on their cuticle, making them resistant to *B. bassiana* (Toledo et al. 2011). The fatty amides of the cuticle of *Liposcelis bostrychophila* prevent dry conidial adhesion due to the reduced hydrophobicity and static charge of the entomopathogenic fungus on the insect (Howard and Lord 2003). Short-chain fatty acids on the surface of the silkworm (*Bombyx mori*) and the fall webworm (*Hyphantria cunea*) inhibited the conidial germination and hyphal growth of *B. bassiana* and *Paecilomyces fumosoroseus* (Saito and Aoki 1983). Caprylic acid on the surface of corn earworm, *Heliothis zea*, and fall armyworm, *Spodoptera frugiperda* larvae inhibits the germination of *B. bassiana* (Smith and Grula 1982). The aldehyde (E)-2-decenal in the cuticle of the stink bug has the mycostatic potential against *M. anisopliae* (Sosa-Gomez et al. 1997). The metapleural glands of ants produce secretions that deter their fungal pathogens. When pathogens are present, they groom their metapleural glands, using the secretion to mitigate pathogenicity (Fernández-Marín et al. 2006).

The mortality rate may be reduced in cases where the insect host molts before the fungus invades the hemocoel; the insect effectively discards much of the fungal inoculum and the discarded exuvium. This phenomenon occurs more frequently in the early instars of small insects, as they molt at short intervals, and spore germination may be reduced (James et al. 2003). The early nymphal stage of the cotton aphid *Aphis gossypii* is less affected by *Lecanicillium attenuatum* due to the low number of conidia and conidial germination, and rapid ecdysis prevents the germ tube penetration inside the host (Kim and Roberts 2012). The protein Bombyx mori single domain von Willebrand factor type c (BmSVWC) was isolated from the molting fluid of *Bombyx mori*. During molting, BmSVWC is released into the molting fluid. The fungal infection reduced considerably when the BmSVWC increased (Han et al. 2017).

7.3 Immune Response of Insects

Insects defend against pathogens through integrated cellular and humoral immune responses (Lemaitre and Hoffmann 2007). Cellular defence is mediated by hemocytes such as prohemocytes, plasmatocytes, granulocytes, and oenocytes, which vary with the insect developmental stage and infection status (Li et al. 2022). Infections by *Beauveria bassiana* in *Eurygaster integriceps* and *Conogethes punctiferalis* elicit hemocyte-mediated phagocytosis and nodulation. In *Galleria mellonella*, fungal attack activates hemocyte responses; however, infections progress when hemocyte numbers decline or cytoskeletal functions are compromised (Wang et al. 2023). Humoral immunity involves antimicrobial peptides (AMPs)—small, cationic, heat-stable molecules under 10 kDa—produced mainly in the fat body and secreted into hemolymph (Wang and Lai 2010; Hancock et al. 2016). Their expression is regulated via Toll, IMD, Jak/STAT, and RNAi pathways (Kaisho and Akira 2002; Geng et al. 2020). For example, *Metarhizium rileyi* infection in *Helicoverpa armigera* promotes translocation of gut bacteria, exploiting the host's immune regulation to outcompete other microbes (Wang et al. 2024). Blood-fed *Aedes aegypti* resist fungal attack through early Toll/IMD pathway activation (Cabral et al. 2020). In *Bombyx mori*, apolipophorin-III contributes antifungal activity and modulates immune gene expression (Wu et al. 2021).

7.4 Pathogenesis-Related Genes

The evolutionary expansion of gene families, such as chitinase, lipase, and protease, in *Beauveria bassiana* (Xiao et al. 2012), *Cordyceps militaris* (Zheng et al. 2012), and *Metarhizium anisopliae* (Gao et al. 2011), is linked to their broad host range and pathogenicity. Virulence-related genes, including *PHI* and a subgroup of chitinase genes, have been identified in *Hirsutella thompsonii*, *B. bassiana*, and *Ophiocordyceps sinensis* (Agrawal et al. 2015). Genes such as *mad1* and mad2 (Wang and St Leger 2007), as well as Hyd1A-F and Hyd2A-C (Feng et al. 2025), mediate adhesion and colonization. The *BbATG11* gene supports autophagy during stress (Ding et al. 2018), while *bbmcm1* aids in growth and conidial viability (Zhao et al. 2019). Appressorium formation is governed by *Mpl1*, *MrSkn7*, *MaSte12*, and *MapacC* (Wang and St Leger 2007; Shang et al. 2015; Wei et al. 2017; Zhang et al. 2020). Genes like *PY450*, *CYP52*, *CYP53*, *Pr1*, *Pr2*, and *chit1–3* (Bogo et al. 1998; Zhang et al. 2012; Staats et al. 2013; Sánchez-Pérez et al. 2014; Huarte-Bonnet et al. 2018) produce enzymes for host cuticle penetration. Blastospore formation is regulated by *Fkh2*, *Bbsnf1*, *ATG7*, *MAD1*, and *ATM1* (Wang et al. 2014; Jin et al. 2015; Wang et al. 2015; Lin et al. 2019), while conidial production within hosts involves *Mrmep1*, *Mrmep*, and *MrStuA* (Yang et al. 2018; Zhou et al. 2018).

8 Plant-Mediated Effects

Plants play a crucial role in modulating the interactions between entomopathogenic fungi and insect hosts through various direct and indirect mechanisms. One major pathway involves herbivore-induced plant volatiles (HIPVs), which can attract or repel entomopathogens or their insect hosts. While certain plant volatiles can suppress fungal activity, others enhance it. For instance, volatiles released from plants suppress *Neozygites tanajoa* conidiation in green mites, whereas herbivore-induced volatiles promote *N. tanajoae* conidiation (Hountondji et al. 2005). Similarly, the volatiles menthol and methyl salicylate, produced in Arabidopsis thaliana following infestation by *Lipaphis erysimi*, increased the sporulation, germination, and pathogenicity of *Lecanicillium lecanii* (Lin et al. 2019). However, *Beauveria bassiana*-colonized plants emit a unique volatile blend that repels natural enemies and exhibits antibacterial activity (González-Mas et al. 2021). Volatile emissions from tobacco plants under aphid attack can even inhibit conidial germination, suggesting a plant-induced delay in fungal infectivity (Brown et al. 1995).

The physical and chemical properties of plant surfaces also significantly affect fungal efficiency. Epicuticular waxes contain both stimulatory and inhibitory compounds; their removal from Chinese cabbage enhanced *the conidial germination of Metarhizium anisopliae* on the mustard beetle *Phaedon cochleariae*, suggesting fungistatic roles for these waxes (Inyang et al. 1999). Similarly, reduced wax content in peas led to higher mortality in pea aphids due to increased adhesion and germination of fungal spores (Duetting et al. 2003). Differences in host plant species can alter susceptibility to fungal pathogens, as seen in *Bemisia argentifolii* nymphs, which were less affected by entomopathogens on cotton than on melon (Poprawski and Jones 2001).

Plants can potentially manipulate the presence of fungi in their tissues to enhance their defence against herbivores. Root-associated fungi, such as *Metarhizium brunneum*, can alter plant traits that influence herbivore interactions, thereby increasing the occurrence and infection rates of fungi (Cotes et al. 2020). In olive plants, plant organ-specific responses control the abundance and functioning of *B. bassiana* (Oliveira et al. 2012). Such mutualistic relationships support the concept of entomopathogens serving as microbial "bodyguards" for plants against herbivores (Elliot et al. 2000). These complex tritrophic interactions are influenced by numerous biotic and abiotic factors, including allelochemicals such as tomatine and camptothecin, which may also inhibit entomopathogens (Lacey et al. 1997).

9 Abiotic Factors and Their Role in Insect Population Dynamics

EPF are an integral biological control agent that regulates insect populations, with its effectiveness strongly influenced by various abiotic factors. Understanding the interplay of these environmental variables is essential for optimizing the use of EPF in integrated pest management (IPM) programs and for comprehending how they impact insect population dynamics through infection processes.

9.1 Temperature

Temperature is a critical determinant of EPF activity, affecting germination, growth, sporulation, and host infection rates (Benz 1987). Generally, lower temperatures reduce the speed of fungal germination and growth, thereby slowing host mortality, although cumulative mortality may remain substantial (Yeo et al. 2003). For example, different strains of *Metarhizium anisopliae* exhibited varied growth and conidia production across temperatures of 8, 15, and 22 °C, with cold-active isolates capable of growing even at 8 °C (Amritha De Croos and Bidochka 1999). In contrast, *Beauveria bassiana* shows significant sensitivity to high temperatures and low water availability, which severely limit its growth under stressful tropical field conditions (Uma Devi et al. 2008). Some fungi, such as *Isaria* spp., tolerate elevated temperatures, surviving up to 35 °C and growing optimally at 30 °C in vitro (Cabanillas and Jones 2009). The combined stress of heat, water scarcity, and elevated CO_2 levels negatively impacts growth and sporulation in several fungal strains, including *B. bassiana*, *M. anisopliae*, and *Isaria farinosa*, when tested against *Schistocerca gregaria* (Borisade and Magan 2015). The virulence of *B. bassiana* against malaria vector mosquitoes was notably higher at moderate temperatures of 21 °C and 25 °C, regardless of the insecticide resistance of the mosquitoes (Kikankie et al. 2010). Additionally, fluctuating temperature regimes reduce fungal virulence compared to constant temperature conditions (Meissle et al. 2024). Thermal optima for mycelial growth among isolates of *Cordyceps* and *Paecilomyces* lie between 15 and 25 °C, while *Nomuraea rileyi* grows best at 25 °C with no growth above 35 °C (Ignoffo et al. 1976).

9.2 Solar Radiation and UV Light

Solar radiation and UV light constitute another significant abiotic influence, as conidia of EPF are vulnerable to ultraviolet (UV) light damage. Solar radiation acts through photic effects (UV and visible light) and thermal effects (near-infrared radiation), which increase ambient temperature (Rougier et al. 1993). Rapid exposure to UV-B radiation drastically reduces conidial viability of *B. bassiana* (Inglis et al. 1995). This UV sensitivity partly explains why fungal infections in the field are less prevalent under warm, sunny conditions compared to shaded greenhouse environments, where grasshoppers showed higher susceptibility to fungal infection (Inglis et al. 1997). The survival of conidia decreases proportionally with increasing exposure duration to simulated sunlight, highlighting the importance of UV protection for effective field applications (Fargues et al. 1996). Strains of *M. anisopliae* differ in their tolerance to UV radiation, which affects germination and conidial survival (Braga et al. 2001). Incorporating UV protectants enhances persistence of *B. bassiana* conidia under UV stress (Inglis et al. 1995). Studies on *Beauveria* and *Metarhizium* isolates found no detrimental effects on infection potential when exposed to growth light and UV light, with some isolates even showing improved survival and pathogenicity (Rossouw et al. 2023). Light conditions also influence sporulation; *Nomuraea rileyi* requires illumination for efficient sporulation on host cadavers, with reduced sporulation observed in darkness (Tang 2001).

9.3 Humidity

Humidity plays a pivotal role in EPF efficacy, with high relative humidity generally promoting fungal infection and mortality of insect hosts (Hajek et al. 1990). For instance, *B. bassiana* caused higher mortality in *Rhodnius prolixus* at humidity levels above 96% (Luz and Fargues 1999). Interestingly, in some cases, lower humidity increased mortality in hide beetles exposed to certain EPF strains (Lord 2011). The interaction between temperature and humidity influences pathogen virulence, as demonstrated in houseflies infected with *B. bassiana*, which exhibited 100% mortality at 30 °C and relative humidity levels of 90%–100% (Mishra et al. 2015). Similarly, the virulence of *Verticillium lecanii* against aphids (*Rhopalosiphum padi*) increased with higher humidity, resulting in faster mortality, extensive fungal growth, and sporulation. Conversely, low humidity delayed sporulation and facilitated the growth of competing saprophytes such as *Mucor* species (Hsiao et al. 1992). In natural ecosystems, temporal and spatial variations in humidity influence the occurrence and transmission dynamics of fungal species. Behavioral manipulation by *Ophiocordyceps* species, such as infected carpenter ants climbing vegetation before death, optimizes transmission in relation to humidity gradients. Higher air moisture is correlated with a tenfold increase in the abundance of infected *Camponotus bispinosus* (Neto et al. 2019).

9.4 Rainfall and Wind

Rainfall and wind contribute to the dispersal of fungal spores, thereby influencing EPF population dynamics and infection rates. Rainfall acts as a natural dispersal mechanism, particularly where insect vectors are absent (Schuh 1986). However, rain can also reduce the efficacy of fungal spores by washing them off host plants (Inyang et al. 2000). The dispersal of *B. bassiana* spores from soil to plants is facilitated by rainfall (Bruck and Lewis 2002). Splash dispersal of spores and mycelial fragments has been documented for fungi like *Hirustella cryptosclerotium*, where droplets maintain size but carry fewer spores with increasing distance from the source (Fernández-Garcıa and Fitt 1993).

9.5 Soil Properties

Soil properties significantly affect the distribution, survival, and effectiveness of EPF in soil environments. Fungal presence correlates positively with soil texture, pH, and organic matter content, with sandy, alkaline soils showing lower fungal abundance (Qayyum et al. 2021). The occurrence of specific fungal species depends on habitat and soil parameters, including altitude and geography (Medo and Cagáň 2011). Soil type influences the control of insect pests, such as Mediterranean fruit fly puparia, by EPF, with *B. bassiana* spores adhering more to clay particles. In contrast, *M. anisopliae* spores predominate in sandy soils (Garrido-Jurado et al. 2011). Higher soil pH and calcium ion levels promote *the growth of B. bassiana but inhibit that of Metarhizium robertsi* (Sharma and Sharma 2021).

The efficacy of entomopathogenic fungi in regulating insect populations depends strongly on environmental factors such as temperature, solar radiation, humidity, rainfall, and soil characteristics. The complex interactions of these abiotic factors influence fungal germination, survival, dispersal, and virulence, thereby shaping infection dynamics and their role in maintaining ecological balance by naturally suppressing insect pest populations.

10 Insect Population Regulation—Classic Examples

The European gypsy moth (*Lymantria dispar dispar*) is one of the most notorious forest pests in North America. Accidentally introduced near Boston, Massachusetts, in 1869, this species has slowly expanded its range across the continent (Belme 1995). Despite its over 150 years of presence, it occupies only about one-third of its suitable habitat. A slow spread is attributed in part to the female's flightlessness and in part to management efforts that limit dispersal. Currently, the gypsy moth is established in approximately 20 eastern and midwestern U.S. states, with a typical delay of 5–10 years from establishment in a state to population outbreaks causing significant defoliation (Blackburn and Hajek 2018). The historical expansion in the eastern United States is documented through the designation of quarantine areas by the U.S. Department of Agriculture (Dwyer and Elkinton 1993). Over the past 25 years, populations have expanded westward into Illinois, Indiana, Michigan, Ohio, and Wisconsin, and southward into North Carolina, Virginia, and West Virginia (Carnegie and Bashford 2012).

In both its native Eurasian range and its invaded North American range, gypsy moth populations exhibit periodic eruptions, causing widespread defoliation of key hosts, such as oaks and poplars. The Asian fungal pathogen *Entomophaga maimaiga* plays a crucial role in suppressing gypsy moth outbreaks. First detected in North America in 1989, *E. maimaiga* now regularly causes epizootics across multiple locations (Hajek et al. 1995). Unlike the gypsy moth nuclear polyhedrosis virus (LdMNPV), which typically induces mortality only during high-density outbreaks, *E. maimaiga* can cause significant mortality even in low-density host populations (Blackburn and Hajek 2018). The fungus produces two spore types: asexual conidia, which are ejected from the cadaver and infect early instar larvae but have a short lifespan, and resting azygospores produced in later instars, which persist in the environment during adverse conditions such as winter or host scarcity, surviving for at least 9 months and germinating regardless of host presence (Hajek and Roberts 1991). Surveys reveal widespread *E. maimaiga* presence in gypsy moth larvae, even in control sites without fungal release, though infection levels are lower in these locations. The emergence of this pathogen correlates with a decline in the intensity and frequency of regional gypsy moth outbreaks, illustrating its role in natural population regulation (Blackburn and Hajek 2018).

Another example is the spotted alfalfa aphid (*Therioaphis trifolii*), introduced into Australia in 1977, which became a serious pest of pasture legumes (Milner and Lutton 1983). The native Israeli fungus *Zoophthora radicans* can cause severe epizootics in these aphid populations (Kenneth and Olmert 1975). Due to climatic similarities, an isolate of *Z. radicans* was imported from Israel to Australia and released experimentally at four sites using various methods: introduction of live, lab-infected aphids onto lucerne; addition of dead, infected aphids; and placement of sporulated Petri dishes on plants under high humidity conditions (Milner and Lutton 1986). After 5 weeks, the fungus was detected at release sites and up to 100 m away. Its resting spores allow it to persist across seasons, ensuring long-term population control. The fungus works synergistically with the parasitoid wasp *Trioxys complanatus*, which also reduces aphid populations and pest damage (Pell et al. 2001).

In Seychelles, the Indian strain of *Lecanicillium lecanii* was introduced from the Imperial Mycological Institute for biological control of scale insects infesting coconuts, where it has provided effective control (Squibbs 1935). The fungus is also applied successfully against *Coccus viridis* on coffee crops, demonstrating its utility across diverse pest-host systems.

11 Coevolutionary Arms Race Between Insects Versus Fungi

The interaction between insects and their fungal pathogens exemplifies a classic coevolutionary arms race, defined as reciprocal adaptive genetic changes in interacting populations. In host-pathogen dynamics, both parties exert selective pressures on each other, resulting in the evolution of defense and counter-defense mechanisms (Joop and Vilcinskas 2016).

An illustrative model is the interaction between the tenebrionid beetle *Tribolium castaneum* and *B. bassiana*. The beetle secretes benzoquinones antimicrobial compounds such as methyl-1,4-benzoquinone (MBQ) and ethyl-1,4-benzoquinone (EBQ)) from its stink glands, which serve as a primary defense against fungal infection. In response, *B. bassiana* produces a benzoquinone reductase enzyme (BqrA) that detoxifies these compounds, facilitating infection. Laboratory experiments knocking out the BqrA gene significantly reduce fungal virulence, while overexpression increases it, underscoring the molecular arms race between host defensive chemistry and pathogen detoxification (Pedrini et al. 2015).

At the molecular level, this arms race involves sophisticated signaling and counter-signaling. Fungal cell wall molecular patterns are recognized by insect receptors, triggering production of proteinase inhibitors and antifungal peptides such as metchnikowin. In turn, the fungus produces chymotrypsin and metalloproteases that degrade these antifungal molecules. Insects can detect these fungal enzymes, which induce further synthesis of antimicrobial peptides. These reciprocal molecular responses are governed by transcriptional reprogramming, mediated through histone acetylation, deacetylation, and DNA methylation (Vilcinskas 2019).

Experimental coevolution studies with *B. bassiana* and *T. castaneum* reveal the rapid dynamics of this race. Over relatively short evolutionary periods, *B. bassiana* adapts by increasing its resistance to benzoquinones through the upregulated expression of the benzoquinone oxidoreductase (BbbqrA) gene, thereby enhancing fungal virulence. Conversely, the beetle can evolve increased secretion of defensive compounds. The balance of resistance and susceptibility fluctuates with the initial conditions of the host population, demonstrating the rapid and dynamic nature of host-pathogen coevolution (Joop and Vilcinskas 2016).

12 Conclusion

Entomopathogenic fungi play a pivotal ecological role in shaping the dynamics of insect populations across both natural and managed ecosystems. Acting as natural regulators, these fungi play a significant role in suppressing herbivorous insect populations, often preventing outbreaks and promoting ecological stability. Their ability to infect a broad spectrum of insect hosts, sometimes even at low population densities, positions them as key components of density-independent and density-dependent regulatory mechanisms.

Beyond direct mortality, EPF influence insect populations through sublethal effects, behavioral alterations, and interactions with other natural enemies, further reinforcing their functional role in trophic networks. The persistence of fungal propagules in the environment, their adaptability to diverse ecological niches, and their involvement in coevolutionary interactions with hosts all underscore their evolutionary significance in host-pathogen dynamics. Collectively, these functions underline the importance of EPF not merely as biocontrol agents but as integral elements of ecosystem processes, deserving deeper ecological investigation and thoughtful integration into sustainable pest management and biodiversity conservation frameworks. The present study tries to compose the available understanding of the ecological role and functions of entomopathogenic fungi in insect population dynamics.

References

Abaajeh AR (2014) Evaluation of entomopathogenic fungi (Ascomycota) for the control of *Cydia pomonella* (Lepidoptera: Tortricidae). Doctoral dissertation, Cape Peninsula University of Technology

Abdulhai M, El-Bouhssini M, Jamal M, Trissi AN, Sayyadi Z et al (2010) *Beauveria bassiana* characterization and efficacy vs. Sunn Pest *Eurygaster integriceps* Puton (Hemiptera: Scutelleridae). Pak J Biol Sci 13:1052–1056. https://doi.org/10.3923/pjbs.2010.1052.1056

Agrawal Y, Khatri I, Subramanian S, Shenoy BD (2015) Genome sequence, comparative analysis, and evolutionary insights into chitinases of entomopathogenic fungus *Hirsutella thompsonii*. Genome Biol 7(3):916–930. https://doi.org/10.1093/gbe/evv037

Akinci HA, Ozman-Sullivan SK, Diler H, Celik N, Sullivan GT et al (2017) Entomopathogenic fungi isolated from *Thaumetopoea pityocampa* and their efficacies against its larvae. Fresenius Environ Bull 26:5251–5257

Amritha De Croos JN, Bidochka MJ (1999) Effects of low temperature on growth parameters in the entomopathogenic fungus *Metarhizium anisopliae*. Can J Microbiol 45(12):1055–1061. https://doi.org/10.1139/w99-098

Andersen SB, Gerritsma S, Yusah KM, Mayntz D, Hywel-Jones NL et al (2009) The life of a dead ant: the expression of an adaptive extended phenotype. Am Nat 174:424–433. https://doi.org/10.1086/603640

Angel-Sahagún CA, Lezama-Gutiérrez R, Molina-Ochoa J, Galindo-Velasco E, López-Edwards M et al (2005) Susceptibility of biological stages of the horn fly, *Haematobia irritans*, to entomopathogenic fungi (Hyphomycetes). J Insect Sci 5(1):50. https://doi.org/10.1093/jis/5.1.50

Arthur S, Thomas M (2001) Effects of temperature and relative humidity on sporulation of *M. Anisopliae var. acridum* in mycosed cadavers of *Schistocerca gregaria*. J Invertebr Pathol 78:59–65. https://doi.org/10.1006/jipa.2001.5050

Baleba SB, Agbessenou A, Getahun MN, Akutse KS, Subramanian S et al (2021) Infection of the stable fly, *Stomoxys calcitrans*, L. 1758 (Diptera: Muscidae) by the entomopathogenic fungi *Metarhizium anisopliae* (Hypocreales: Clavicipitaceae) negatively affects its survival, feeding propensity, fecundity, fertility, and fitness parameters. Front Fungal Biol 2:637817. https://doi.org/10.3389/ffunb.2021.637817

Balogun SA, Fagade OE (2004) Entomopathogenic fungi in population of *Zonocerus variegatus* (l) in Ibadan, south west, Nigeria. Afr J Biotechnol 3(8):382–386. https://doi.org/10.5897/AJB2004.000-2074

Barta M, Horáková MK, Georgieva M, Mirchev P, Zaemdzhikova G et al (2019) Entomopathogenic fungi (Ascomycota: Hypocreales) as natural antagonists of the pine processionary moth *Thaumetopoea pityocampa* (Denis & Schiffermüller, 1775) (Lepidoptera: Notodontidae) in Bulgaria. Acta Zool Bulg 15:89–96

Batta YA (2007) Biocontrol of almond bark beetle (*Scolytus amygdali* Geurin-Meneville, Coleoptera: Scolytidae) using *Beauveria bassiana* (Bals.) Vuill. (Deuteromycotina: Hyphomycetes). J Appl Microbiol 103(5):1406–1414. https://doi.org/10.1111/j.1365-2672.2007.03369.x

Batten A (1969) The Senegalese grasshopper, *Oedaleus senegalensis* Krauss. J Appl Ecol 6:27–45. https://doi.org/10.2307/2401299

Becher PG, Jensen RE, Natsopoulou ME, Verschut V, De Fine Licht HH (2018) Infection of *Drosophila suzukii* with the obligate insect-pathogenic fungus *Entomophthora muscae*. J Pest Sci 91:781–787. https://doi.org/10.1007/s10340-017-0915-3

Bekker CD, Quevillon LE, Smith PB, Fleming KR, Ghosh D et al (2014) Species-specific ant brain manipulation by a specialized fungal parasite. BMC Evol Biol 14(1):166. https://doi.org/10.1186/s12862-014-0166-3

Belme DM (1995) Gypsy moths (*Lymantria dispar*) in the Niagara Region: population density variation, introduction of an entomopathogenic fungus (*Entomophaga maimaiga*) and occurrence of nuclear polyhedrosis virus. Dissertation, Brock University

Benz S (1987) Environment. In: Fuxa JR, Tanada Y (eds) Epizootiology of insect diseases. Wiley, New York, pp 177–214. https://doi.org/10.1016/B978-012078147-8/50060-8

Beris EI, Papachristos DP, Fytrou A, Antonatos SA, Kontodimas DC (2013) Pathogenicity of three entomopathogenic fungi on pupae and adults of the Mediterranean fruit fly, *Ceratitis capitata* (Diptera: Tephritidae). J Pest Sci 86:275–284. https://doi.org/10.1007/s10340-012-0468-4

Blackburn LM, Hajek AE (2018) Gypsy moth larval necropsy guide, vol 179. US Department of Agriculture, Forest Service, Northern Research Station, Newtown Square, pp 1–30. https://doi.org/10.2737/NRS-GTR-179

Blanford S, Thomas MB (2001) Adult survival, maturation, and reproduction of the desert locust *Schistocerca gregaria* infected with the fungus *Metarhizium anisopliae* var *acridum*. J Invertebr Pathol 78(1):1–8. https://doi.org/10.1006/jipa.2001.5031

Bogo MR, Rota CA, Pinto JH, Ocampos M, Correa CT et al (1998) A chitinase encoding gene (chit1 gene) from the entomopathogen *Metarhizium anisopliae*: isolation and characterization of genomic and full-length cDNA. Curr Microbiol 37:221–225. https://doi.org/10.1007/s002849900368

Borisade OA, Magan N (2015) Resilience and relative virulence of strains of entomopathogenic fungi under interactions of abiotic stress. Afr J Microbiol Res 9(14):988–1000

Bouchard P, Smith AB, Douglas H, Gimmel ML, Brunke AJ et al (2017) Biodiversity of Coleoptera. In: Foottit RG, Adler PH (eds) Insect biodiversity: science and society. Wiley, Hoboken, pp 337–417

Braga GU, Flint SD, Messias CL, Anderson AJ, Roberts DW (2001) Effect of UV-B on conidia and germlings of the entomopathogenic hyphomycete *Metarhizium anisopliae*. Mycol Res 105(7): 874–882. https://doi.org/10.1017/S0953756201004270

Brodeur J (2012) Host specificity in biological control: insights from opportunistic pathogens. Evol Appl 5(5):470–480. https://doi.org/10.1111/j.1752-4571.2012.00273.x

Brown GC, Prochaska GL, Hildebrand DF, Nordin GL, Jackson DM (1995) Green leaf volatiles inhibit conidial germination of the entomopathogen *Pandora neoaphidis* (Entomopthorales: Entomophthoraceae). Environ Entomol 24(6):1637–1643. https://doi.org/10.1093/ee/24.6.1637

Bruck DJ, Lewis LC (2002) Rainfall and crop residue effects on soil dispersion and *Beauveria bassiana* spread to corn. Appl Soil Ecol 20(3):183–190. https://doi.org/10.1016/S0929-1393(02)00022-7

Cabanillas HE, Jones WA (2009) Effects of temperature and culture media on vegetative growth of an entomopathogenic fungus *Isaria* sp. (Hypocreales: Clavicipitaceae) naturally affecting the whitefly, *Bemisia tabaci* in Texas. Mycopathologia 167:263–271. https://doi.org/10.1007/s11046-008-9176-2

Cabral S, De Paula A, Samuels R, Da Fonseca R, Gomes S et al (2020) *Aedes aegypti* (Diptera: Culicidae) immune responses with different feeding regimes following infection by the entomopathogenic fungus *Metarhizium anisopliae*. Insects 11(2):95. https://doi.org/10.3390/insects11020095

Carnegie AJ, Bashford R (2012) Sirex woodwasp in Australia: current management strategies, research and emerging issues. In: Slippers B, de Groot P, Wingfield M (eds) The Sirex Woodwasp and its fungal symbiont. Springer, Dordrechet, pp 175–201. https://doi.org/10.1007/978-94-007-1960-6_14

Carrillo D, Crane JH, Peña JE (2013) Potential of contact insecticides to control *Xyleborus glabratus* (Coleoptera: Curculionidae), a vector of laurel wilt disease in avocados. J Econ Entomol 106(6):2286–2295. https://doi.org/10.1603/EC13205

Carrillo D, Dunlap CA, Avery PB, Navarrete J, Duncan RE et al (2015) Entomopathogenic fungi as biological control agents for the vector of the laurel wilt disease, the redbay ambrosia beetle, *Xyleborus glabratus* (Coleoptera: Curculionidae). BioControl 81:44–50. https://doi.org/10.1016/j.biocontrol.2014.10.009

Castillo MA, Moya P, Hernández E, Primo-Yufera E (2002) Susceptibility of *Ceratitis capitata* Wiedemann (Diptera: Tephritidae) to entomopathogenic fungi and their extracts. BioControl 19(3):274–282. https://doi.org/10.1006/bcon.2000.0867

Chandler D, Heale JB, Gillespie AT (1993) Germination of the entomopathogenic fungus *Verticillium lecanii* on scales of the glasshouse whitefly *Trialeurodes vaporariorum*. Biocontrol Sci Technol 3(2):161–164. https://doi.org/10.1080/09583159309355272

Chouvenc T, Su NY, Elliott ML (2008) Interaction between the subterranean termite *Reticulitermes flavipes* (Isoptera: Rhinotermitidae) and the entomopathogenic fungus *Metarhizium anisopliae* in foraging arenas. J Econ Entomol 101(3):885–893. https://doi.org/10.1093/jee/101.3.885

Clarkson JM, Charnley AK (1996) New insights into the mechanisms of fungal pathogenesis in insects. Trends Microbiol 4(5):197–203

Coombes CA (2012) Entomopathogenic fungi for control of soil-borne life stages of false codling moth, *Thaumatotibia leucotreta* (Meyrick) (1912) (Lepidoptera: Tortricidae). Doctoral dissertation, Rhodes University

Cortez-Madrigal H, Montores-Ramírez J, Cárdenas-Ochoa C, Ochoa-Echegollen M (2022) Epizootics of entomopathogenic fungi at overwintering sites of *Oebalus mexicana* Sailer

(Hemiptera: Pentatomidae) from Western Mexico. Egypt J Biol Pest Control 32(1):11. https://doi.org/10.1186/s41938-022-00509-7
Cotes B, Thöming G, Amaya-Gómez CV, Novák O, Nansen C (2020) Root-associated entomopathogenic fungi manipulate host plants to attract herbivorous insects. Sci Rep 10(1): 22424. https://doi.org/10.1038/s41598-020-80123-5
Crespo PH, Santiago-Alvarez C (1997) Entomopathogenic fungi associated with natural populations of the Moroccan Locust *Dociostaurus maroccanus* (Thunberg) (Orthoptera: Gomphocerinae) and other Acridoidea in Spain. Biocontrol Sci Technol 7(3):357–364. https://doi.org/10.1080/09583159730767
Darbro JM, Graham RI, Kay BH, Ryan PA, Thomas MB (2011) Evaluation of entomopathogenic fungi as potential biological control agents of the dengue mosquito, *Aedes aegypti* (Diptera: Culicidae). Biocontrol Sci Technol 21(9):1027–1047. https://doi.org/10.1080/09583157.2011.597913
Deka B, Baruah C, Babu A (2021) Entomopathogenic microorganisms: their role in insect pest management. Egypt J Biol Pest Control 31(1):121. https://doi.org/10.1186/s41938-021-00466-7
Ding JL, Peng YJ, Chu XL, Feng MG, Ying SH (2018) Autophagy-related gene BbATG11 is indispensable for pexophagy and mitophagy, and contributes to stress response, conidiation and virulence in the insect mycopathogen *Beauveria bassiana*. Environ Microbiol 20(9): 3309–3324. https://doi.org/10.1111/1462-2920.14329
Dromph K (2001) Dispersal of entomopathogenic fungi by collembolans. Soil Biol Biochem 33: 2047–2051. https://doi.org/10.1016/S0038-0717(01)00130-4
Dubois T, Hajek AE, Jiafu H, Li Z (2024) Evaluating the efficiency of entomopathogenic fungi against the Asian longhorned beetle, *Anoplophora glabripennis* (Coleoptera: Cerambycidae), by using cages in the field. Environ Entomol 33(1):62–74. https://doi.org/10.1603/0046-225X-33.1.62
Duetting PS, Ding H, Neufeld J, Eigenbrode SD (2003) Plant waxy bloom on peas affects infection of pea aphids by *Pandora neoaphidis*. J Invertebr Pathol 84(3):149–158. https://doi.org/10.1016/j.jip.2003.10.001
Dwyer G, Elkinton JS (1993) Using simple models to predict virus epizootics in gypsy moth populations. J Anim Ecol 1:1–11. https://doi.org/10.2307/5477
Eilenberg J, Meadow R (2003) Fungi for biological control of Brassica root flies *Delia radicum* and *Delia floralis*. In: Upadhyay RK (ed) Advances in microbial control of insect pests. Kluwer Law International, Alphen aan den Rijn, pp 181–189
Ekesi S, Maniania NK, Lux SA (2002) Mortality in three African tephritid fruit fly puparia and adults caused by the entomopathogenic fungi, *Metarhizium anisopliae* and *Beauveria bassiana*. Biocontrol Sci Technol 12(1):7–17. https://doi.org/10.1080/09583150120093077
Elliot SL, Sabelis MW, Janssen A, van der Geest LPS, Beerling EAM, Fransen J (2000) Can plants use entomopathogens as bodyguards? Ecology Letters 3(3):228-235. https://doi.org/10.1046/j.1461-0248.2000.00137.x
Embaby ME, Lotfy DE (2015) Control of the cabbage white butterfly (*Pieris rapae* L.) (Lepidoptera: Pieridae) using entomopathogenic fungi. Egypt J Biol Pest Control 25(3):569
Er MK, Tunaz H, Gökçe A (2007) Pathogenicity of entomopathogenic fungi to *Thaumetopoea pityocampa* (Schiff.) (Lepidoptera: Thaumatopoeidae) larvae in laboratory conditions. J Pest Sci 80:235–239. https://doi.org/10.1007/s10340-007-0177-6
Erler F, Ates AO (2015) Potential of two entomopathogenic fungi, *Beauveria bassiana* and *Metarhizium anisopliae* (Coleoptera: Scarabaeidae), as biological control agents against the June beetle. J Insect Sci 15(1):44. https://doi.org/10.1093/jisesa/iev029
Eslamizadeh R, Sajap ASB, Omar DB, Binti ADam NA (2014) Evaluation of different isolates of entomopathogenic fungus, *Paecilomyces fumosoroseus* (Deuteromycotina: Hyphomycetes) against *Bemisia tabaci* (Hemiptera: Aleyrodidae). Biocontrol 2(2):82–91. https://doi.org/10.22092/bcpp.2015.101230

Fargues J, Goettel MS, Smits N, Ouedraogo A, Vidal A et al (1996) Variability in susceptibility to simulated sunlight of conidia among isolates of entomopathogenic Hyphomycetes. Mycopathologia 135:171–181. https://doi.org/10.1007/BF00632339

Fathy Z, Muhammad J, Ali SS (2025) The pathology of native entomopathogenic fungi, *Beauveria bassiana* and *Metarhizium anisopliae* (Hypocreales: Clavicipitaceae), on biological processes and ovarian structure of the African migratory locust, *Locusta migratoria* migratorioides (Orthoptera: Acrididae). Egypt J Biol Pest Control 35(1):1–9. https://doi.org/10.1186/s41938-025-00842-7

Feng MG, Johnson JB, Kish LP (1990) Survey of entomopathogenic fungi naturally infecting cereal aphids (Homoptera: Aphididae) of irrigated grain crops in southwestern Idaho. Environ Entomol 19:1534–1542. https://doi.org/10.1093/ee/19.5.1534

Feng JR, Li M, Ying SH, Feng MG (2025) Deciphering roles of nine hydrophobins (Hyd1A–F and Hyd2A–C) in the asexual and insect-pathogenic lifecycles of *Beauveria bassiana*. Microbiol Res 292:128029. https://doi.org/10.1016/j.micres.2024.128029

Fernández-Garcıa E, Fitt BD (1993) Dispersal of the entomopathogen *Hirsutella cryptosclerotium* by simulated rain. J Invertebr Pathol 61(1):39–43. https://doi.org/10.1006/jipa.1993.1007

Fernández-Marín H, Zimmerman JK, Rehner SA, Wcislo WT (2006) Active use of the metapleural glands by ants in controlling fungal infection. Proc Biol Sci 273(1594):1689–1695. https://doi.org/10.1098/rspb.2006.3492

Fusari LM, Dantas GP, Pinho LC (2018) Order Diptera. In: Hamada N, Thorp JH, Rogers H (eds) Thorp and Covich's freshwater invertebrates. Academic Press, Boston, pp 607–623. https://doi.org/10.1016/B978-0-12-804223-6.00029-9

Gallai N, Salles JM, Settele J, Vaissiere BE (2009) Economic valuation of the vulnerability of world agriculture confronted with pollinator decline. Ecol Econ 68:810–821. https://doi.org/10.1016/j.ecolecon.2008.06.014

Gambino PA, Thomas GM (1988) Fungi associated with two *Vespula* (hymenoptera: Vespidae) species. Pan Pac Entomol 64:107–113

Gao Q, Jin K, Ying SH, Zhang Y, Xiao G et al (2011) Genome sequencing and comparative transcriptomics of the model entomopathogenic fungi *Metarhizium anisopliae* and *M. Acridum*. PLoS Genet 7(1):e1001264. https://doi.org/10.1371/journal.pgen.1001264

Garrido-Jurado I, Torrent J, Barrón V, Corpas A, Quesada-Moraga E (2011) Soil properties affect the availability, movement, and virulence of entomopathogenic fungi conidia against puparia of *Ceratitis capitata* (Diptera: Tephritidae). Biocotrol 58:277–285. https://doi.org/10.1016/j.biocontrol.2011.05.017

Geng T, Lu F, Wu H, Lou D, Tu N et al (2020) Target antifungal peptides of immune signalling pathways in silkworm, *Bombyx mori*, against *Beauveria bassiana*. Insect Mol Biol 30(1): 102–112. https://doi.org/10.1111/imb.12681

Gillespie JP, Burnett C, Charnley AK (2000) The immune response of the desert locust *Schistocerca gregaria* during mycosis of the entomopathogenic fungus, *Metarhizium anisopliae* var *acridum*. J Insect Physiol 46(4):429–437. https://doi.org/10.1016/S0022-1910(99)00128-6

Gilliam M, Lorenz BJ (1993) Enzymatic activity of strains of *Ascosphaera apis*, an entomopathogenic fungus of the honey bee, *Apis mellifera*. Apidologie 24(1):19–23. https://doi.org/10.1051/apido:19930102

Goerzen DW, Erlandson MA, Moore KC (1990) Effect of two insect viruses and two entomopathogenic fungi on larval and pupal development in the alfalfa leaf cutting bee, *Megachile rotundata* (fab.) (hymenoptera: Megachilidae). Can Entomol 122(5):1039–1040. https://doi.org/10.4039/Ent1221039-9

Goettel MS, Glare T (2010) Entomopathogenic fungi and their role in regulation of insect populations. Wallingford. CAB International, pp. 387–419

Goettel MS, Leger RJ, Bhairi S, Jung MK, Oakley BR et al (1990) Pathogenicity and growth of *Metarhizium anisopliae* stably transformed to benomyl resistance. Curr Genet 17:129–132. https://doi.org/10.1007/BF00312857

Goettel MS, Johnson DL, Inglis GD (1995) The role of fungi in the control of grasshoppers. Can J Bot 73(S1):71–75. https://doi.org/10.1139/b95-227

González-Mas N, Gutiérrez-Sánchez F, Sánchez-Ortiz A, Grandi L, Turlings TC et al (2021) Endophytic colonization by the entomopathogenic fungus *Beauveria bassiana* affects plant volatile emissions in the presence or absence of chewing and sap-sucking insects. Front Plant Sci 12:660460. https://doi.org/10.3389/fpls.2021.660460

Graf T, Scheibler F, Niklaus PA, Grabenweger G (2023) From lab to field: biological control of the Japanese beetle with entomopathogenic fungi. Front Insect Sci 3:1138427. https://doi.org/10.3389/finsc.2023.1138427

Gupta RK, Jaronski ST, Srivastava K, Bali K (2011) First record on epizootics of *Entomophthora grylli* on grasshopper in Indian subcontinent: pathogen city and bio control potential on *Oxya velox*. Arch Phytopathol Plant Protect 44(5):475–483. https://doi.org/10.1080/03235400903145244

Hajek AE (1997) Fungal and viral epizootics in gypsy moth (Lepidoptera: Lymantriidae) populations in Central New York. BioControl 10(1):58–68. https://doi.org/10.1006/bcon.1997.0541

Hajek AE (1999) Pathology and epizootiology of *Entomophaga maimaiga* infections in forest Lepidoptera. Microbiol Mol Biol Rev 63(4):814–835. https://doi.org/10.1128/mmbr.63.4.814-835.1999

Hajek AE, Butler L (2000) Predicting the host range of entomopathogenic fungi. In: Follett PA, Duan JJ (eds) Nontarget effects of biological control. Springer, New York, pp 262–276

Hajek AE, Roberts DW (1991) Pathogen reservoirs as a biological control resource: introduction of *Entomophaga maimaiga* to north American gypsy moth, *Lymantria dispar*, populations. BioControl 1(1):29–34. https://doi.org/10.1016/1049-9644(91)90098-K

Hajek AE, St Leger RJ (1994) Interactions between fungal pathogens and insect hosts. Annu Rev Entomol 39(1):293–322

Hajek AE, Humber RA, Elkinton JS, May B, Walsh SR et al (1990) Allozyme and restriction fragment length polymorphism analyses confirm *Entomophaga maimaiga* responsible for 1989 epizootics in north American gypsy moth populations. Proc Natl Acad Sci USA 87(18):6979–6982. https://doi.org/10.1073/pnas.87.18.6979

Hajek AE, Humber RA, Elkinton JS (1995) Mysterious origin of *Entomphaga maimaiga* in North America. Am Entomol 41(1):31–43. https://doi.org/10.1093/ae/41.1.31

Hajek AE, Butler L, Walsh SR, Silver JC, Hain FP et al (1996) Host range of the gypsy moth (Lepidoptera: Lymantriidae) pathogen *Entomophaga maimaiga* (Zygomycetes: Entomophthorales) in the field versus laboratory. Environ Entomol 25(4):709–721. https://doi.org/10.1093/ee/25.4.709

Hall RA (1982) Control of whitefly, *Trialeurodes vaporariorum* and cotton aphid, *Aphis gossypii* in glasshouses by two isolates of the fungus, *Verticillium lecanii*. Ann Appl Biol 101(1):1–11. https://doi.org/10.1111/j.1744-7348.1982.tb00794.x

Han F, Lu A, Yuan Y, Huang W, Beerntsen BT et al (2017) Characterization of an entomopathogenic fungi target integument protein, *Bombyx mori* single domain von Willebrand factor type C, in the silkworm, *Bombyx mori*. Insect Mol Biol 26(3):308–316. https://doi.org/10.1111/imb.12293

Hancock RE, Haney EF, Gill EE (2016) The immunology of host defence peptides: beyond antimicrobial activity. Nat Rev Immunol 16(5):321–334. https://doi.org/10.1038/nri.2016.29

Harris RJ, Harcourt SJ, Glare TR, Rose EA, Nelson TJ (2000) Susceptibility of *Vespula vulgaris* (hymenoptera: Vespidae) to generalist entomopathogenic fungi and their potential for wasp control. J Invertebr Pathol 75(4):251–258. https://doi.org/10.1006/jipa.2000.4928

Hassan A, Kang L, Zhang K, Wang L, Qin X et al (2024) Effect of entomopathogenic fungi on behavior and physiology of *Solenopsis invicta* (hymenoptera, Formicidae). J Econ Entomol 117(3):825–833. https://doi.org/10.1093/jee/toae068

Herlinda S (2010) Spore density and viability of entomopathogenic fungal isolates from Indonesia, and their virulence against *Aphis gossypii* glover (Homoptera: Aphididae). Trop Life Sci Res 21(1):11

Hountondji FC, Sabelis MW, Hanna R, Janssen A (2005) Herbivore-induced plant volatiles trigger sporulation in entomopathogenic fungi: the case of *Neozygites tanajoae* infecting the cassava green mite. J Chem Ecol 31:1003–1021. https://doi.org/10.1007/s10886-005-4244-2

Howard R, Lord J (2003) Cuticular lipids of the booklouse, liposcelis bostrychophila: hydrocarbons,aldehydes, fatty acids, and fatty acid amides. J Chem Ecol 29:615-27. https://doi.org/10.1023/A:1022806922246

Hsiao WF, Bidochka MJ, Khachatourians GG (1992) Effects of diphenols on the growth of three entomopathogenic fungi. Can J Microbiol 38(9):1000–1003. https://doi.org/10.1139/m92-163

Huarte-Bonnet C, Kumar S, Saparrat MC, Girotti JR, Santana M et al (2018) Insights into hydrocarbon assimilation by eurotialean and hypocrealean fungi: roles for CYP52 and CYP53 clans of cytochrome P450 genes. Appl Biochem Biotechnol 184:1047–1060. https://doi.org/10.1007/s12010-017-2608-z

Ibrahim YB, Low W (1993) Potential of mass-production and field efficacy of isolates of the entomopathogenic fungi *Beauveria bassiana* and *Paecilomyces fumosoroseus* against *Plutella xylostella*. Int J Pest Manag 39(3):288–292. https://doi.org/10.1080/09670879309371807

Idrees A, Qadir ZA, Akutse KS, Afzal A, Hussain M et al (2021) Effectiveness of entomopathogenic fungi on immature stages and feeding performance of fall armyworm, *Spodoptera frugiperda* (Lepidoptera: Noctuidae) larvae. Insects 12(11):1044. https://doi.org/10.3390/insects12111044

Ignoffo CM, Garcia C, Hostetter DL (1976) Effects of temperature on growth and sporulation of the entomopathogenic fungus *Nomuraea rileyi*. Environ Entomol 5(5):935–936. https://doi.org/10.1093/ee/5.5.935

Ihara F, Yaginuma K, Kobayashi N, Mishiro K, Sato T (2001) Screening of entomopathogenic fungi against the brown-winged green bug, *Plautia stali* Scott (Hemiptera: Pentatomidae). Appl Entomol Zool 36(4):495–500. https://doi.org/10.1303/aez.2001.495

Inglis GD, Goettel MS, Johnson DL (1995) Influence of ultraviolet light protectants on persistence of the entomopathogenic fungus, *Beauveria bassiana*. Biocontrol 5(4):581–590. https://doi.org/10.1006/bcon.1995.1069

Inglis GD, Johnson DL, Goettel MS (1997) Effects of temperature and sunlight on mycosis (*Beauveria bassiana*) (Hyphomycetes: Sympodulosporae) of grasshoppers under field conditions. Environ Entomol 26(2):400–409. https://doi.org/10.1093/ee/26.2.400

Inglis PW, Aragao FJ, Frazao H, Magalhães BP, Valadares-Inglis MC (2000) Biolistic co-transformation of *Metarhizium anisopliae* var. *acridum* strain CG423 with green fluorescent protein and resistance to glufosinate ammonium. FEMS Microbiol Lett 191(2):249–254. https://doi.org/10.1111/j.1574-6968.2000.tb09347.x

Inyang EN, Butt TM, Doughty KJ, Todd AD, Archer SA (1999) The effects of isothiocyanates on the growth of the entomopathogenic fungus *Metarhizium anisopliae* and its infection of the mustard beetle. Mycol Res 103(8):974–980. https://doi.org/10.1017/S0953756298007989

Inyang EN, McCartney HA, Oyejola B, Ibrahim L, Archer SA (2000) Effect of formulation, application and rain on the persistence of the entomogenous fungus *Metarhizium anisopliae* on oilseed rape. Mycol Res 104(6):653–661. https://doi.org/10.1017/S0953756200002641

James RR, Buckner JS, Freeman TP (2003) Cuticular lipids and silverleaf whitefly stage affect conidial germination of *Beauveria bassiana* and *Paecilomyces fumosoroseus*. J Invertebr Pathol 84(2):67–74. https://doi.org/10.1016/j.jip.2003.08.006

Jeff LB, Feng MG, Falkowsky JE, Khachatourians GG (1997) Infection of the migratory grasshopper (Orthoptera: Acrididae) by ingestion of the entomopathogenic fungus *Beauveria bassiana*. J Econ Entomol 90(2):383–390. https://doi.org/10.1093/jee/90.2.383

Jin K, Peng G, Liu Y, Xia Y (2015) The acid trehalase, ATM1, contributes to the in vivo growth and virulence of the entomopathogenic fungus, *Metarhizium acridum*. Fungal Genet Biol 77:61–67. https://doi.org/10.1016/j.fgb.2015.03.013

Jinna Z, Khan T, Kanwal N, Zhenji W (2025) Diversity and biological functions of fungal secondary metabolites: biocontrol agents for sustainable agriculture. A review. Cell Mol Biol 71(5):43–48. https://doi.org/10.14715/cmb/2025.71.5.7

Johnson DL, Huang HC, Harper AM (1988) Mortality of grasshoppers (Orthoptera: Acrididae) inoculated with a Canadian isolate of the fungus *Verticillium lecanii*. J Invertebr Pathol 52:335–342. https://doi.org/10.1016/0022-2011(88)90143-7

Johnson BR, Borowiec ML, Chiu JC, Lee EK, Atallah J et al (2013) Phylogenomics resolves evolutionary relationships among ants, bees, and wasps. Curr Biol 23(20):2058–2062

Joop G, Vilcinskas A (2016) Coevolution of parasitic fungi and insect hosts. Fortschr Zool 119: 350–358. https://doi.org/10.1016/j.zool.2016.06.005

Kaisho T, Akira S (2002) Toll-like receptors as adjuvant receptors. Biochim Biophys Acta Mol Cell Res 1589(1):1–3. https://doi.org/10.1016/S0167-4889(01)00182-3

Kalsbeek V, Mullens BA, Jespersen JB (2001) Field studies of Entomophthora (Zygomycetes: Entomophthorales)—induced behavioral fever in *Musca domestica* (Diptera: Muscidae) in Denmark. Biol Control 21(3):264–273. https://doi.org/10.1006/jipa.2001.5012

Karthi S, Vasantha-Srinivasan P, Senthil-Nathan S, Han YS, Shivakumar MS et al (2024) Entomopathogenic fungi promising biocontrol agents for managing lepidopteran pests: review of current knowledge. Biocatal Agric Biotechnol 58:103146. https://doi.org/10.1016/j.bcab.2024.103146

Kaur R, Banu N (2019) Review on management strategies of mealy bugs. Int J Emerg Technol Innov Res 6(1):362–367. https://ssrn.com/abstract=3691236

Kelley-Tunis KK, Reid BL, Andis M (1995) Activity of entomopathogenic fungi in free-foraging workers of *Camponotus pennsylvanicus* (hymenoptera: Formicidae). J Econ Entomol 88(4): 937–943. https://doi.org/10.1093/jee/88.4.937

Kenneth R, Olmert I (1975) Entomopathogenic fungi and their insect hosts in Israel: additions. Isr J Entomol 10:105–112

Khanday AL, Buhroo AA, Ranjith AP, Kerchev IA (2018) Evaluating the efficacy of entomopathogenic fungi against the bark beetle, Ips stebbingi Strohmeyer (Coleoptera, Curculionidae: Scolytinae) in India. Entomol Rev 98:985–994. https://doi.org/10.1134/S0013873818080055

Kikankie CK, Brooke BD, Knols BG, Koekemoer LL, Farenhorst M et al (2010) The infectivity of the entomopathogenic fungus *Beauveria bassiana* to insecticide-resistant and susceptible *anopheles arabiensis* mosquitoes at two different temperatures. Malar J 9:1–9. https://doi.org/10.1186/1475-2875-9-71

Kim JJ, Roberts DW (2012) The relationship between conidial dose, moulting and insect developmental stage on the susceptibility of cotton aphid, Aphis gossypii, to conidia of *Lecanicillium attenuatum*, an entomopathogenic fungus. Biocontrol Sci Technol 22(3):319–331. https://doi.org/10.1080/09583157.2012.656580

Kistner EJ, Belovsky GE (2013) Susceptibility to disease across developmental stages: examining the effects of an entomopathogen on a grasshopper (Orthoptera: Acrididae) pest. J Orthoptera Res 22(2):73–77. https://doi.org/10.1665/034.022.0206

Krasnoff S, Watson D, Gibson D, Kwan E (1995) Behavioral effects of the entomopathogenic fungus, *Entomophthora muscae* on its host *Musca domestica*: postural changes in dying hosts and gated pattern of mortality. J Insect Physiol 41(10):895–903. https://doi.org/10.1016/0022-1910(95)00026-Q

Krishnan JU, Raju DL, Subramanian AM, Revanasiddappa M, Kumar S et al (2024) Diversity of Cordycipitaceae in the Sahyadri mountain ranges of the Western Ghats. In: Sridhar KR, Deshmukh SK, Fung S, Mahadevakumar S (eds) Advances in cordyceps research. CRC Press, Boca Raton, pp 94–113

Kumar A, Attri AK (2016) Characterization of fungal spores in ambient particulate matter: a study from the Himalayan region. Atmos Environ 142:182–193. https://doi.org/10.1016/j.atmosenv.2016.07.049

Lacey LA, Mesquita AL, Mercadier G, Debire R, Kazmer DJ et al (1997) Acute and sublethal activity of the entomopathogenic fungus *Paecilomyces fumosoroseus* (Deuteromycotina: Hyphomycetes) on adult *Aphelinus asychis* (hymenoptera: Aphelinidae). Environ Entomol 26(6):1452–1460. https://doi.org/10.1093/ee/26.6.1452

Lacey LA, Frutos R, Kaya HK, Vail P (2001) Insect pathogens as biological control agents: do they have a future? Biol Control 21:230–248. https://doi.org/10.1006/bcon.2001.0938

Leger RS, Goettel M, Roberts DW, Staples RC (1991) Prepenetration events during infection of host cuticle by *Metarhizium anisopliae*. J Invertebr Pathol 58(2):168–179. https://doi.org/10.1016/0022-2011(91)90061-T

Lemaitre B, Hoffmann J (2007) The host defense of *Drosophila melanogaster*. Annu Rev Immunol 25(1):697–743. https://doi.org/10.1146/annurev.immunol.25.022106.141615

Leonard A, Egonyu JP, Khamis FM, Tanga CM, Ekesi S et al (2023) Identification and virulence screening of fungal and bacterial entomopathogens of the edible long-horned grasshopper *Ruspolia differens* (Orthoptera: Tettigoniidae) from Uganda. Afr Entomol 31:1–7. https://doi.org/10.17159/2254-8854/2023/a11581

Levchenko NG, Dubitskii AM, Vakker VG (1974) The entomopathogenic fungus *Coelomomycidium simulii* (Phycomycetes, Chitridiales) in larvae of Simuliids of the genus *Odagmia* (Diptera, Simuliidae) in Kazakhstan. Med Parazitol 43(1):110–112

Levchenko MV, Kononchuk AG, Gerus AV, Lednev GR (2020) Differential susceptibility of *Locusta migratoria* and *Schistocerca gregaria* (Orthoptera: Acrididae) to infection with entomopathogenic fungi. Plant Prot News 103(2):150–152. https://doi.org/10.31993/2308-6459-2020-103-2-13364

Li S, Liu F, Kang Z, Li X, Lu Y et al (2022) Cellular immune responses of the yellow peach moth, *Conogethes punctiferalis* (Lepidoptera: Crambidae), to the entomopathogenic fungus, *Beauveria bassiana* (Hypocreales: Cordycipitaceae). J Invertebr Pathol 194:107826. https://doi.org/10.1016/j.jip.2022.107826

Liao A, Cavigliasso F, Savary L, Kawecki TJ (2024) Effects of an entomopathogenic fungus on the reproductive potential of drosophila males. Ecol Evol 14(4):e11242. https://doi.org/10.1002/ece3.11242

Lin HY, Wang JJ, Feng MG, Ying SH (2019) Autophagy-related gene ATG7 participates in the asexual development, stress response and virulence of filamentous insect pathogenic fungus *Beauveria bassiana*. Curr Genet 65:1015–1024. https://doi.org/10.1007/s00294-019-00955-1

Lin WJ, Chiu MC, Lin CC, Chung YK, Chou JY (2021) Efficacy of Entomopathogenic fungus *aspergillus nomius* against *Dolichoderus thoracicus*. BioControl 66(4):463–473. https://doi.org/10.1007/s10526-021-10086-7

Liu H, Skinner M, Parker BL, Brownbridge M (2002) Pathogenicity of Beauveria bassiana, *Metarhizium anisopliae* (Deuteromycotina: Hyphomycetes), and other entomopathogenic fungi against *Lygus lineolaris* (Hemiptera: Miridae). J Econ Entomol 95(4):675–681. https://doi.org/10.1603/0022-0493-95.4.675

López Plantey R, Papura D, Couture C, Thiéry D, Pizzuolo PH et al (2019) Characterization of entomopathogenic fungi from vineyards in Argentina with potential as biological control agents against the European grapevine moth *Lobesia botrana*. BioControl 64(5):501–511. https://doi.org/10.1007/s10526-019-09955-z

Lord JC (2001) Desiccant dusts synergize the effect of *Beauveria bassiana* (Hyphomycetes: Moniliales) on stored-grain beetles. J Econ Entomol 94(2):367–372. https://doi.org/10.1603/0022-0493-94.2.367

Lord JC (2011) Influence of substrate and relative humidity on the efficacy of three entomopathogenic fungi for the hide beetle, *Dermestes maculatus* (Coleoptera, Dermestidae). Biocontrol Sci Technol 21(4):475–483. https://doi.org/10.1603/0022-0493-94.2.367

Lu L, Cheng B, Du D, Hu X, Peng A et al (2015) Morphological, molecular and virulence characterization of three *Lencanicillium* species infecting Asian citrus psyllids in *Huangyan citrus* groves. J Invertebr Pathol 125:45–55. https://doi.org/10.1016/j.jip.2015.01.002

Luz C, Fargues J (1998) Factors affecting conidial production of *Beauveria bassiana* from fungus-killed cadavers of *Rhodnius prolixus*. J Invertebr Pathol 72:97–103. https://doi.org/10.1006/jipa.1998.4774

Luz C, Fargues J (1999) Dependence of the entomopathogenic fungus, *Beauveria bassiana*, on high humidity for infection of *Rhodnius prolixus*. Mycopathologia 146(1):33–41. https://doi.org/10.1023/a:1007019402490

Mackauer M (1973) The population growth of the pea aphid biotype R1 broad bean and pea (Homoptera: Aphididae). Z Angew Entomol 74(1–4):343–351. https://doi.org/10.1111/j.1439-0418.1973.tb01815.x

Martínez JL, Pescador A, Lezama R, Rebolledo O, Molina J et al (2005) Asociación de la chinche apestosa Oebalus mexicanus con el hongo Fusarium moniliforme causante del tizón de la panoja del sorgo, en la Ciénega de Chapala, Jalisco, México. Av Investig Agropecu 9:1–7. https://www.redalyc.org/articulo.oa?id=83709104

Masoudi A, Joseph RA, Keyhani NO (2024) Viral-and fungal-mediated behavioral manipulation of hosts: summit disease. Appl Microbiol Biotechnol 108(1):492. https://doi.org/10.1007/s00253-024-13332-x

Medo J, Cagáň Ľ (2011) Factors affecting the occurrence of entomopathogenic fungi in soils of Slovakia as revealed using two methods. BioControl 59(2):200–208. https://doi.org/10.1016/j.biocontrol.2011.07.020

Mehmood N, Hassan A, Zhou W, Usman HM, Huang Q (2022) Behavioural alterations in female *Aedes aegypti* mosquito in response to entomopathogenic fungal infections. Pest Manag Sci 78(5):2065–2073. https://doi.org/10.1002/ps.6832

Meissle M, Grabenweger G, Romeis J (2024) No interaction of fluctuating or constant temperature and virulence of entomopathogenic fungi in two noctuid species. J Pest Sci 97(2):809–823. https://doi.org/10.1007/s10340-023-01673-x

Meshrif WS, Rohlfs M, Hegazi MA, Shehata MG, Barakat EM et al (2010) Humoral response and plasma changes of *Spodoptera littoralis* (Lepidoptera: Noctuidae) following injection with the entomopathogenic fungi *Beauveria bassiana* and *Nomuraea rileyi*. In: Proc 6th Int Con Biol Sci (Zool), vol 6, pp 96–102

Mesquita AL, Lacey LA (2001) Interactions among the entomopathogenic fungus, *Paecilomyces fumosoroseus* (Deuteromycotina: Hyphomycetes), the parasitoid, *Aphelinus asychis* (hymenoptera: Aphelinidae), and their aphid host. BioControl 22(1):51–59. https://doi.org/10.1006/bcon.2001.0950

Meyling NV, Pell JK (2006) Detection and avoidance of an entomopathogenic fungus by a generalist insect predator. Ecol Entomol 31(2):162–171. https://doi.org/10.1111/j.0307-6946.2006.00781.x

Milner RJ, Lutton GG (1983) Effect of temperature on *Zoophthora radicans* (Brefeld) batko: an introduced microbial control agent of the spotted alfalfa aphid, *Therioaphis trifolii* (monell) f. *maculata*. Aust J Entomol 22:167–173. https://doi.org/10.1111/j.1440-6055.1983.tb01865.x

Milner RJ, Lutton GG (1986) Dependence of *Verticillium lecanii* (fungi: Hyphomycetes) on high humidities for infection and sporulation using *Myzus persicae* (Homoptera: Aphididae) as host. Environ Entomol 15:380–382. https://doi.org/10.1093/ee/15.2.380

Mishra S, Kumar P, Malik A (2015) Morpho-molecular characterization and virulence determination of entomopathogenic fungal isolates native to Indian subcontinent. Med Mycol 1(1):3. https://doi.org/10.21767/2471-8521.100003

Mnyone LL, Kirby MJ, Mpingwa MW, Lwetoijera DW, Knols BG et al (2011) Infection of *Anopheles gambiae* mosquitoes with entomopathogenic fungi: effect of host age and blood-feeding status. Parasitol Res 108(2):317–322. https://doi.org/10.1007/s00436-010-2064-y

Mohammadyani M, Karimi J, Taheri P, Sadeghi H, Zare R (2016) Entomopathogenic fungi as promising biocontrol agents for the rosaceous longhorn beetle, *Osphranteria coerulescens*. BioControl 61:579–590. https://doi.org/10.1007/s10526-016-9745-0

Moore J (2002) Parasites and the behavior of animals. Oxford University Press. https://doi.org/10.2307/3285495

Murtaza G, Naeem M, Manzoor S, Khan HA, Eed EM et al (2022) Biological control potential of entomopathogenic fungal strains against peach fruit fly, *Bactrocera zonata* (Saunders) (Diptera: Tephritidae). Peer J 10:e13316. https://doi.org/10.7717/peerj.13316

Mustu M, Demirci F, Kocak ER (2011) Mortality effects of *Isaria farinosa* (Holm.) and *Beauveria bassiana* (Balsamo) Vuillemin (Sordariomycetes: Hypocreales) on *Aelia rostrata* Boh. (Hemiptera: Pentatomidae). Turk J Entomol 35(4):559–568

Nair KS, McEwen FL (1973) *Strongwellsea castrans* (Phycomycetes: Entomophthoraceae), a fungal parasite of the adult cabbage maggot, Hylemya brassicae (Diptera: Anthomyiidae). J Invertebr Pathol 22(3):442–449. https://doi.org/10.1016/0022-2011(73)90175-4

Neto JA, Leal LC, Baccaro FB (2019) Temporal and spatial gradients of humidity shape the occurrence and the behavioral manipulation of ants infected by entomopathogenic fungi in Central Amazon. Fungal Ecol 42:100871. https://doi.org/10.1016/j.funeco.2019.100871

Neuvéglise C, Brygoo Y, Vercambre B, Riba G (1994) Comparative analysis of molecular and biological characteristics of strains of *Beauveria brongniartii* isolated from insects. Mycol Res 98(3):322–328. https://doi.org/10.1016/S0953-7562(09)80460-7

Nguyen NT, Borgemeister C, Poehling HM, Zimmermann G (2007) Laboratory investigations on the potential of entomopathogenic fungi for biocontrol of *Helicoverpa armigera* (Lepidoptera: Noctuidae) larvae and pupae. Biocontrol Sci Technol 17(8):853–864. https://doi.org/10.1080/09583150701546375

Noma T, Strickler K (2007) Effects of *Beauveria bassiana* on Lygus hesperus (Hemiptera: Miridae) feeding and oviposition. Environ Entomol 29(2):394–402. https://doi.org/10.1093/ee/29.2.394

Nong X, Liu C, Lu X, Wang Q, Wang G et al (2011) Laboratory evaluation of entomopathogenic fungi against the white grubs, *Holotrichia oblita* and *Anomala corpulenta* (Coleoptera: Scarabaeidae) from the field of peanut, *Arachis hypogaea*. Biocontrol Sci Technol 21(5): 593–603

Nyman T, Farrell BD, Zinovjev AG, Vikberg V (2006) Larval habits, host-plant associations, and speciation in nematine sawflies (hymenoptera: Tenthredinidae). Evol 60(8):1622–1637. https://doi.org/10.1111/j.0014-3820.2006.tb00507.x

Oliveira I, Pereira JA, Lino-Neto T, Bento A, Baptista P (2012) Fungal diversity associated to the olive moth, prays oleae Bernard: a survey for potential entomopathogenic fungi. Microb Ecol 63:964–974. https://doi.org/10.1007/s00248-011-9955-z

Opare LO, Jensen AB, Lecocq A, Holm S, Esperk T (2024) Exposure to entomopathogenic fungus and high larval density induce a strong immune response and life-history costs in black soldier fly, a commercially important insect. Entomol Exp Appl 172(8):710–719. https://doi.org/10.1111/eea.13449

Opisa S, Du Plessis H, Akutse KS, Fiaboe KK, Ekesi S (2018) Effects of Entomopathogenic fungi and *bacillus thuringiensis*-based biopesticides on *Spoladea recurvalis* (Lepidoptera: Crambidae). J Appl Entomol 142(6):617–626. https://doi.org/10.1111/jen.12512

Oreste M, Bubici G, Poliseno M, Tarasco E (2016) Effects of entomopathogenic fungi on *Encarsia Formosa* Gahan. (hymenoptera: Aphelinidae) activity and behavior. BioControl 100:46–53. https://doi.org/10.1016/j.biocontrol.2016.05.011

Paparatti B, Fabozzi R (1988) A new pathogen of the pine processionary caterpillar (*Thaumetopoea pityocampa* Den. Et Schif.), Lepidoptera: Thaumetopoeidae. Inf Fitopatol 38(9):45–48

Pedrini N, Ortiz-Urquiza A, Huarte-Bonnet C, Fan Y, Juárez MP et al (2015) Tenebrionid secretions and a fungal benzoquinone oxidoreductase form competing components of an arms race between a host and pathogen. Proc Natl Acad Sci 112(28):E3651–E3660. https://doi.org/10.1073/pnas.1504552112

Pell JK, Eilenberg J, Hajek AE, Steinkraus DC (2001) Biology, ecology and pest management potential of Entomophthorales. In: Butt TM, Jackson C, Magan N (eds) Fungi as biocontrol agents: progress, problems and potential. CABI Publishing, Wallingford, pp 71–153. https://doi.org/10.1079/9780851993560.0071

Pell JK, Hannam JJ, Steinkraus JJ (2010) Conservation biological control using fungal entomopathogens. BioControl 55:187–198. https://doi.org/10.1007/s10526-009-9245-6

Poprawski TJ, Jones WJ (2001) Host plant effects on activity of the mitosporic fungi *Beauveria bassiana* and *Paecilomyces fumosoroseus* against two populations of Bemisia whiteflies

(Homoptera: Aleyrodidae). Mycopathologia 151:11–20. https://doi.org/10.1023/a:1010835224472

Posada-Flórez FJ (2008) Production of *Beauveria bassiana* fungal spores on rice to control the coffee berry borer, *Hypothenemus hampei*, in Colombia. J Insect Sci 8(1):41. https://doi.org/10.1673/031.008.4101

Potrich M, da Silva RT, Maia F, Lozano ER, Rossi RM et al (2018) Effect of entomopathogens on africanized *Apis mellifera* L. (Hymenoptera: Apidae). Rev Bras Entomol 62:23–28. https://doi.org/10.19113/sdufenbed.477889

Prior C, Jollands P, Le Patourel G (1988) Infectivity of oil and water formulations of *Beauveria bassiana* (Deuteromycotina: Hyphomycetes) to the cocoa weevil pest *Pantorhytes plutus* (Coleoptera: Curculionidae). J Invertebr Pathol 52(1):66–72. https://doi.org/10.1016/0022-2011(88)90103-6

Qayyum MA, Saeed S, Wakil W, Nawaz A, Iqbal N et al (2021) Diversity and correlation of entomopathogenic and associated fungi with soil factors. J King Saud Univ Sci 33(6):101520. https://doi.org/10.1016/j.jksus.2021.101520

Ramanujam B, Kandan A, Poornesha B, Shylesha AN, Gracy G et al (2022) Pathogenicity of *Beauveria bassiana* and *Metarhizium anisopliae* on Aak grasshopper, Poekilocerus pictus Fabr. (Orthoptera: Acridoidea). Int J Trop Insect Sci 42:2023–2026. https://doi.org/10.1007/s42690-021-00686-2

Ramoska WA, Hajek AE, Ramos ME, Soper RS (1988) Infection of grasshoppers (Orthoptera: Acrididae) by members of the *Entomophaga grylli* species complex (Zygomycetes: Entomophthorales). J Invertebr Pathol 52(2):309–313. https://doi.org/10.1016/0022-2011(88)90140-1

Rangel DE, Bignayan HG, Golez HG, Keyser CA, Evans EW et al (2022) Virulence of the insect-pathogenic fungi *Metarhizium* spp. to Mormon crickets, Anabrus simplex (Orthoptera: Tettigoniidae). Bull Entomol Res 112(2):179–186. https://doi.org/10.1017/S0007485321000663

Reber A, Chapuisat M (2012) Diversity, prevalence and virulence of fungal entomopathogens in colonies of the ant *Formica selysi*. Insect Soc 59:231–239. https://doi.org/10.1007/s00040-011-0209-3

Resquín-Romero G, Garrido-Jurado I, Quesada-Moraga E (2016) Combined use of entomopathogenic fungi and their extracts for the control of *Spodoptera littoralis* (Boisduval) (Lepidoptera: Noctuidae). BioControl 92:101–110. https://doi.org/10.1016/j.biocontrol.2015.10.007

Reynaldi FJ, Lucia M, Garcia ML (2015) Ascosphaera apis, the entomopathogenic fungus affecting larvae of native bees (*Xylocopa augusti*): first report in South America. Rev Iberoam Micol 32(4):261–264. https://doi.org/10.1016/j.riam.2015.01.001

Reynolds SE, Samuels RI (1996) Physiology and biochemistry of insect moulting fluid. In: Evans PD (ed) Advances in insect physiology. Academic Press, London, pp 157–232. https://doi.org/10.1016/S0065-2806(08)60031-4

Rohrlich C, Merle I, Mze Hassani I, Verger M, Zuin M et al (2018) Variation in physiological host range in three strains of two species of the entomopathogenic fungus *Beauveria*. PLoS One 13(7):e0199199. https://doi.org/10.1371/journal.pone.0199199

Rossouw S, Mathulwe LL, Stokwe NF, Malan AP (2023) Effect of visible light and ultraviolet light on the pathogenicity of entomopathogenic fungi to false codling moth, *Thaumatotibia leucotreta* (Lepidoptera: Tortricidae) larvae. Afr Entomol 31:1–8. https://doi.org/10.17159/2254-8854/2023/a13141

Rougier M, Fargues J, Goujet R, Smits N (1993) Photic effect of solar U.V. Radiations on the survival of entomopathogenic hyphomycete. In: Varlet Grancher C, Bonhomme R, Sinoquet H (eds) Crop structure and light microclimate: characterisation and applications. INRA, Paris, pp 453–458

Ruiz-Nájera RE, Ruiz-Estudillo RA, Sánchez-Yáñez JM, Molina-Ochoa J, Skoda SR et al (2013) Occurrence of entomopathogenic fungi and parasitic nematodes on *Spodoptera frugiperda*

(Lepidoptera: Noctuidae) larvae collected in Central Chiapas, México. Fla Entomol 96(2): 498–503

Sabbour M, Abd-El-Aziz S (2012) Efficacy of three entomopathogenic fungi alone or in combination with diatomaceous earth modifications for the control of three pyralid moths in stored grains. J Plant Prot Res 52(3):359–363. https://doi.org/10.2478/v10045-012-7

Sahayaraj K, Francis Borgio J (2009) Distribution of *Metarhizium anisopliae* (Metsch.) Sorokin (Deuteromycotina: Hyphomycetes) in Tamil Nadu, India, its biocontrol potential on *Dysdercus cingulatus* (fab.) (Hemiptera: Pyrrhocoridae). Arch Phytopathol Plant Protect 42(5):424–435. https://ssrn.com/abstract=1591785

Şahin F, Yanar Y (2021) Pathogenicity of some local entomopathogenic fungus isolates on the cotton leafworm larvae, *Spodoptera littoralis* (Boisd.) (Lepidoptera: Noctuidae). Egypt J Biol Pest Control 31:1–6. https://doi.org/10.1186/s41938-021-00494-3

Saito T, Aoki J (1983) Toxicity of free fatty acids on the larval surfaces of two lepidopterous insects towards *Beauveria bassiana* (BALS.) VUILL. and *Paecilomyces fumoso-roseus* (WIZE) BROWN et SMITH (Deuteromycetes: Moniliales). Appl Entomol Zool 18(2):225–233

Salazar E, Garciducieñas R, Salas MD (2002) Utilización del hongo entomopatógeno *Beauveria bassiana* para el control de la chinche del sorgo Oebalus mexicana en sus sitios de invernación. In: Rev Protég Vég No. 2132

Samson RA, Evans HC, Latge JP (1988) Atlas of entomopathogenic fungi. Springer Verlag, Berlin

Sánchez-Pérez L, Barranco-Florido J, Rodríguez-Navarro S, Cervantes-Mayagoitia J, Ramos-López M (2014) Enzymes of entomopathogenic fungi, advances and insights. Adv Enzyme Res 2:65–76. https://doi.org/10.4236/aer.2014.22007

Sangbaramou R, Camara I, Huang XZ, Shen J, Tan SQ et al (2018) Behavioral thermoregulation in *Locusta migratoria manilensis* (Orthoptera: Acrididae) in response to the entomopathogenic fungus, *Beauveria bassiana*. PLoS One 13(11):e0206816. https://doi.org/10.1371/journal.pone.0206816

Santi L, e Silva LA, da Silva WO, Corrêa AP, Rangel DE et al (2011) Virulence of the entomopathogenic fungus *Metarhizium anisopliae* using soybean oil formulation for control of the cotton stainer bug, *Dysdercus peruvianus*. World J Microb Biotechnol 27:2297–2303. https://doi.org/10.1007/s11274-011-0695-5

Scholte EJ, Takken W, Knols BG (2007) Infection of adult *Aedes aegypti* and *ae. Albopictus* mosquitoes with the entomopathogenic fungus *Metarhizium anisopliae*. Acta Trop 102(3): 151–158. https://doi.org/10.1016/j.actatropica.2007.04.011

Schuh W (1986) Epidemiological aspects of *Sorghum* downy mildew, caused by *Peronosclerospora sorghi* (plant protection, diseases, pests). Dissertation, Texas A&M University

Sevim A, Demir I, Demirbağ Z (2010) Molecular characterization and virulence of Beauveria spp. from the pine processionary moth, *Thaumetopoea pityocampa* (Lepidoptera: Thaumetopoeidae). Mycopathologia 170:269–277. https://doi.org/10.1007/s11046-010-9321-6

Sevim A, Demir I, Sönmez E, Kocacevik S, Demirbağ Z (2013) Evaluation of entomopathogenic fungi against the sycamore lace bug, *Corythucha ciliata* (say) (Hemiptera: Tingidae). Turk J Agric 37(5):595–603. https://doi.org/10.3906/tar-1208-55

Seye F, Bawin T, Boukraa S, Zimmer JY, Ndiaye M et al (2014) Effect of entomopathogenic aspergillus strains against the pea aphid, *Acyrthosiphon pisum* (Hemiptera: Aphididae). Appl Entomol Zool 49:453–458. https://doi.org/10.1007/s13355-014-0273-z

Shang Y, Feng P, Wang C (2015) Fungi that infect insects: altering host behavior and beyond. PLoS Pathog 11(8):e1005037. https://doi.org/10.1371/journal.ppat.1005037

Sharma R, Sharma P (2021) Fungal entomopathogens: a systematic review. Egypt J Biol Pest Control 31:57. https://doi.org/10.1186/s41938-021-00404-7

Shazia GU, Rasool A, Ahmad T (2024) Efficacy of entomopathogenic fungi for control of walnut blue butterfly (*Chaetoprocta odata* [Hewitson, 1865]) in walnut (*Juglans regia* L.) under laboratory conditions. Acta Agric Slov 120(1):1–11. https://doi.org/10.14720/aas.2024.120.1.13554

Shehzad M, Tariq M, Mukhtar T, Gulzar A (2021) On the virulence of the entomopathogenic fungi, *Beauveria bassiana* and *Metarhizium anisopliae* (Ascomycota: Hypocreales), against the diamondback moth, *Plutella xylostella* (L.) (Lepidoptera: Plutellidae). Egypt J Biol Pest Control 31(1):1–7. https://doi.org/10.1186/s41938-021-00428-z

Smith RJ, Grula EA (1982) Toxic components on the larval surface of the corn earworm (*Heliothis zea*) and their effects on germination and growth of *Beauveria bassiana*. J Invertebr Pathol 39(1):15–22. https://doi.org/10.1016/0022-2011(82)90153-7

Soetopo D (2004) Efficacy of selected *Beauveria bassiana* (Bals.) Vuill. Isolates in combination with a resistant cotton variety (PSB-Ct 9) against the cotton bollworm, *Helicoverpa armigera* (Hübner) (Lepidoptera: Noctuidae). Dissertation, University of The Philippines

Sönmez E, Sevim A, Demirbağ Z, Demir I (2016) Isolation, characterization, and virulence of entomopathogenic fungi from *Gryllotalpa gryllotalpa* (Orthoptera: Gryllotalpidae). Appl Entomol Zool 51:213–226. https://doi.org/10.1007/s13355-015-0390-3

Sookar P, Bhagwant S, Awuor Ouna E (2008) Isolation of entomopathogenic fungi from the soil and their pathogenicity to two fruit fly species (Diptera: Tephritidae). J Appl Entomol 132(9–10):778–788. https://doi.org/10.1111/j.1439-0418.2008.01348.x

Sosa-Gomez DR, Boucias DG, Nation JL (1997) Attachment of *Metarhizium anisopliae* to the southern green stink bug *Nezara viridula* cuticle and Fungistatic effect of Cuticular lipids and aldehydes. J Invertebr Pathol 69(1):31–39. https://doi.org/10.1006/jipa.1996.4619

Squibbs FL (1935) Work connected with insect pests and fungus diseases. Rev Appl Entomol 23: 241

Staats CC, Kmetzsch L, Lubeck I, Junges A, Vainstein MH et al (2013) *Metarhizium anisopliae* chitinase CHIT30 is involved in heat-shock stress and contributes to virulence against *Dysdercus peruvianus*. Fungal Biol 117(2):137–144. https://doi.org/10.1016/j.funbio.2012.12.006

Sultana R, Wagan YS, Naeem M, Wagan MS, Khatri I (2013) Susceptibility of three *Hieroglyphus* species (Hemiacridinae: Acrididae: Orthoptera) to some strains of the entomopathogenic fungi from Pakistan. Can J Pure Appl Sci 7:2325–2332

Swanson JA, Torto B, Kells SA, Mesce KA, Tumlinson JH et al (2009) Odorants that induce hygienic behavior in honeybees: identification of volatile compounds in chalkbrood-infected honeybee larvae. J Chem Ecol 35:1108–1116. https://doi.org/10.1007/s10886-009-9683-8

Swathy K, Perumal Vivekanandhan P, Yuvaraj A, Sarayut P (2024) Biodegradation of pesticide in agricultural soil employing entomopathogenic fungi: current state of the art and future perspectives. Heliyon 10:1–17. https://doi.org/10.1016/j.heliyon.2023.e23406

Tang H (2001) Effects of environmental factors on virulence of the entomopathogenic fungus, *Nomuraea rileyi*, against the corn earworm, *Helicoverpa armigera* (Lep., Noctuidae). J Appl Entomol 125(5):243–248. https://doi.org/10.1046/j.1439-0418.2001.00544.x

Tang LC, Cheng DJ, Hou RF (1999) Virulence of the entomopathogenic fungus, *Nomuraea rileyi*, to various larval stages of the corn earworm, *Helicoverpa armigera* (Lepidoptera: Noctuidae). Appl Entomol Zool 34(3):399–403. https://doi.org/10.1303/aez.34.399

Tarasco E, Triggiani O, Zamoum M, Oreste M (2015) Natural enemies emerged from *Thaumetopoea pityocampa* (Denis & Sciffermüller) (Lepidoptera Notodontidae) pupae in southern Italy. Redia 98(1):103–108

Thanakitpipattana D, Tasanathai K, Mongkolsamrit S, Khonsanit A, Lamlertthon S et al (2020) Fungal pathogens occurring on Orthopterida in Thailand. Persoonia 44(1):140–160. https://doi.org/10.3767/persoonia.2020.44.06

Thomas F, Adamo SS, Moore J, Adamo S (2005) Parasitic manipulation: where are we and where should we go? Behav Process 68:185–199. https://doi.org/10.1016/j.beproc.2004.06.010

Thompson SR, Brandenburg RL, Roberson GT (2014) Entomopathogenic fungi detection and avoidance by mole crickets (Orthoptera: Gryllotalpidae). Environ Entomol 36(1):165–172. https://doi.org/10.1603/0046-225x(2007)36[165:efdaab]2.0.co;2

Toledo AV, Alippi AM, de Remes Lenicov AM (2011) Growth inhibition of *Beauveria bassiana* by bacteria isolated from the cuticular surface of the corn leafhopper, *Dalbulus maidis* and the planthopper, Delphacodes kuscheli, two important vectors of maize pathogens. J Insect Sci 11(1):29. https://doi.org/10.1673/031.011.0129

Toledo-Hernández RA, Ruíz-Toledo J, Toledo J, Sánchez D (2016) Effect of three entomopathogenic fungi on three species of stingless bees (Hymenoptera: Apidae) under laboratory conditions. J Econ Entomol 109(3):1015–1019. https://doi.org/10.1093/jee/tow064

Trajković A, Žikić V (2023) Stuck in the caterpillars' web: a half-century of biocontrol research and application on gregarious lepidopteran pests in Europe. Sustainability 15(4):2881. https://doi.org/10.3390/su15042881

Trandem N, Bhattarai UR, Westrum K, Knudsen GK, Klingen I (2015) Fatal attraction: male spider mites prefer females killed by the mite-pathogenic fungus *Neozygites floridana*. J Invertebr Pathol 128:6–13. https://doi.org/10.1016/j.jip.2015.04.002

Tscharntke T, Gathmann A, Steffan-Dewenter I (1998) Bioindication using trap-nesting bees and wasps and their natural enemies: community structure and interactions. J Appl Ecol 35(5): 708–719. https://doi.org/10.1046/j.1365-2664.1998.355343.x

Um M, Zakaria D, Galadima IB, Gambo FM, Zakaria DJ (2018) A review on the use of entomopathogenic fungi in the management of insect pests of field crops. J Entomol Zool Stud 6(1):27–32

Uma Devi K, Padmavathi J, Uma Maheswara Rao C, Khan AA, Mohan MC (2008) A study of host specificity in the entomopathogenic fungus *Beauveria bassiana* (Hypocreales, Clavicipitaceae). Biocontrol Sci Technol 18(10):975–989. https://doi.org/10.1080/09583150802450451

Umaru FF, Simarani K (2020) Evaluation of the potential of fungal biopesticides for the biological control of the seed bug, *Elasmolomus pallens* (Dallas) (Hemiptera: Rhyparochromidae). Insects 11(5):277. https://doi.org/10.3390/insects11050277

Umaru FF, Simarani K (2022) Efficacy of entomopathogenic fungal formulations against *Elasmolomus pallens* (Dallas) (Hemiptera: Rhyparochromidae) and their extracellular enzymatic activities. Toxins 14(9):584. https://doi.org/10.3390/toxins14090584

Vargas-Osuna E, Munoz Ledesma J, Aldebis HK, Santiago-Alvarez C (1994) Pathogens and parasitoids for the control of *Thaumetopoea pityocampa* (D. Y Schiff) (Lep. Notodontidae). Bol Sanid Veg Plagas 20:511–515. https://doi.org/10.1007/s10340-007-0177-6

Vilcinskas A (2019) Evolutionary ecology of parasitic fungi and their host insects. Fungal Ecol 38: 12–20. https://doi.org/10.1016/j.funeco.2018.04.007

Wang YP, Lai R (2010) Insect antimicrobial peptides: structures, properties and gene regulation. Zool Res 31(1):27–34. https://doi.org/10.3724/sp.j.1141.2010.01027

Wang C, St Leger RJ (2007) The MAD1 adhesin of Metarhizium anisopliae links adhesion with blastospore production and virulence to insects, and the MAD2 adhesin enables attachment to plants. Eukaryot Cell 6(5):808–816. https://doi.org/10.1128/EC.00409-06

Wang XX, He PH, Feng MG, Ying SH (2014) BbSNF1 contributes to cell differentiation, extracellular acidification, and virulence in *Beauveria bassiana*, a filamentous entomopathogenic fungus. Appl Microbiol Biotechnol 98:8657–8673. https://doi.org/10.1007/s00253-014-5907-0

Wang JJ, Qiu L, Cai Q, Ying SH, Feng MG (2015) Transcriptional control of fungal cell cycle and cellular events by Fkh2, a forkhead transcription factor in an insect pathogen. Sci Rep 5(1): 10108. https://doi.org/10.1038/srep10108

Wang JL, Sun J, Song YJ, Zheng HH, Wang GJ et al (2023) An entomopathogenic fungus exploits its host humoral antibacterial immunity to minimize bacterial competition in the hemolymph. Microbiome 11(1):116. https://doi.org/10.1186/s40168-023-01538-6

Wang G, Zhang X, Chen B, Peng Y (2024) Construction of an efficient *agrobacterium tumefaciens*-based transformation system in the Entomopathogenic fungus *Metarhizium rileyi*. J Agric Food Chem 72(39):21711–21719. https://doi.org/10.1021/acs.jafc.4c05535

Wei Q, Du Y, Jin K, Xia Y (2017) The Ste 12-like transcription factor MaSte12 is involved in pathogenicity by regulating the appressorium formation in the entomopathogenic fungus, *Metarhizium acridum*. Appl Microbiol Biotechnol 101:8571–8584. https://doi.org/10.1007/s00253-017-8569-x

Welling MT, Liu L, Rose TJ, Waters DL, Benkendorff K (2016) Arbuscular mycorrhizal fungi: effects on plant terpenoid accumulation. Plant Biol 18(4):552–562. https://doi.org/10.1111/plb.12408

Wraight SP, Carruthers R, Bradley CA, Jaronski ST, Lacey LA et al (1998) Pathogenicity of entomopathogenic fungi *Paecilomyces* spp. and *Beauveria bassiana* against the silverleaf whitefly, *Bemisia argentifolii*. J Invertebr Pathol 71(3):217–226. https://doi.org/10.1006/jipa.1997.4734

Wu W, Lin S, Zhao Z, Su Y, Li R, Zhang Z, Guo X (2021) *Bombyx mori* Apolipophorin-III inhibits *Beauveria bassiana* directly and through regulating expression of genes relevant to immune signaling pathways. J Invertebr Pathol 184:107647. https://doi.org/10.1016/j.jip.2021.107647

Wynns AA, Jensen AB, Eilenberg J, James R (2012) *Ascosphaera subglobosa*, a new spore cyst fungus from North America associated with the solitary bee *Megachile rotundata*. Mycologia 104(1):108–114. https://doi.org/10.3852/10-047

Wynns AA, Jensen AB, Eilenberg J (2013) *Ascosphaera callicarpa*, a new species of bee-loving fungus, with a key to the genus for Europe. PLoS One 8(9):e73419. https://doi.org/10.1371/journal.pone.0073419

Xia Y, Dean P, Judge AJ, Gillespie JP, Clarkson JM et al (2000) Acid phosphatases in the haemolymph of the desert locust, *Schistocerca gregaria*, infected with the entomopathogenic fungus *Metarhizium anisopliae*. J Insect Physiol 46(9):1249–1257. https://doi.org/10.1016/S0022-1910(00)00045-7

Xiao G, Ying SH, Zheng P, Wang ZL, Zhang S et al (2012) Genomic perspectives on the evolution of fungal entomopathogenicity in *Beauveria bassiana*. Sci Rep 2(1):483. https://doi.org/10.1038/srep00483

Yadav AK, Lalramliana (2012) Efficacy of indigenous entomopathogenic nematodes from Meghalaya, India against the larvae of taro leaf beetle, *Aplosonyx chalybaeus* (Hope). J Parasit Dis 36:149–154. https://doi.org/10.1007/s12639-012-0139-7

Yang W, Wu H, Wang Z, Sun Q, Qiao L et al (2018) The APSES gene MrStuA regulates sporulation in *Metarhizium robertsii*. Front Microbiol 9:1208. https://doi.org/10.3389/fmicb.2018.01208

Yeo H, Pell JK, Alderson PG, Clark SJ, Pye BJ (2003) Laboratory evaluation of temperature effects on the germination and growth of entomopathogenic fungi and on their pathogenicity to two aphid species. Pest Manag Sci 59(2):156–165. https://doi.org/10.1002/ps.622

Youssef NN, Roush CF, WR MM (1984) In vivo development and pathogenicity of *Ascosphaera proliperda* (Ascosphaeraceae) to the alfalfa leafcutting bee, Megachile rotundata. J Invertebr Pathol 43(1):11–20. https://doi.org/10.1016/0022-2011(84)90059-4

Zebold SL, Whisler HC, Shemanchuk JA, Travland LB (1979) Host specificity and penetration in the mosquito pathogen *Coelomomyces psorophorae*. Can J Bot 57(24):2766–2770. https://doi.org/10.1139/b79-327

Zhang S, Widemann E, Bernard G, Lesot A, Pinot F et al (2012) CYP52X1, representing new cytochrome P450 subfamily, displays fatty acid hydroxylase activity and contributes to virulence and growth on insect cuticular substrates in entomopathogenic fungus *Beauveria bassiana*. J Biol Chem 287(16):13477–13486. https://doi.org/10.1074/jbc.M111.338947

Zhang M, Wei Q, Xia Y, Jin K (2020) MaPacC, a pH-responsive transcription factor, negatively regulates thermotolerance and contributes to conidiation and virulence in *Metarhizium acridum*. Curr Genet 66:397–408. https://doi.org/10.1007/s00294-019-01032-3

Zhao X, Yang X, Lu Z, Wang H, He Z et al (2019) MADS-box transcription factor Mcm1 controls cell cycle, fungal development, cell integrity and virulence in the filamentous insect pathogenic fungus *Beauveria bassiana*. Environ Microbiol 21(9):3392–3416. https://doi.org/10.1111/1462-2920.14629

Zheng P, Xia Y, Xiao G, Xiong C, Hu X et al (2012) Genome sequence of the insect pathogenic fungus *Cordyceps militaris*, a valued traditional Chinese medicine. Genome Biol 12:1–22. https://doi.org/10.1186/gb-2011-12-11-r116

Zhou R, Zhou X, Fan A, Wang Z, Huang B (2018) Differential functions of two metalloproteases, Mrmep1 and Mrmep2, in growth, sporulation, cell wall integrity, and virulence in the filamentous fungus *Metarhizium robertsii*. Front Microbiol 9:1528. https://doi.org/10.3389/fmicb.2018.01528

Zurek L, Watson DW, Krasnoff SB, Schal C (2002) Effect of the entomopathogenic fungus, *Entomophthora muscae* (Zygomycetes: Entomophthoraceae), on sex pheromone and other cuticular hydrocarbons of the house fly, *Musca domestica*. J Invertebr Pathol 80(3):171–176. https://doi.org/10.1016/S0022-2011(02)00109-X

Insect Parasites in the Basidiomycota

4

T. K. Arun Kumar

Abstract

Insect-parasitic fungi are of significance in agriculture and industry. Several fungal parasites of insects have been documented, collected, studied, and utilized that belong to all the major fungal lineages. However, entomoparasitic fungi of the phylum Basidiomycota are an underrepresented group. A brief review of the entomoparasitic fungi belonging to the Basidiomycota is presented in this chapter.

Keywords

Atheliales · Coccid · Pucciniomycetes · *Reticulitermes* · Septobasidiales

1 Introduction

Fungi adopt diverse nutritional modes. Among these modes, parasitism, pathogenicity, and predation have been preferred by many fungal lineages. Evolution has shaped unique patterns of parasitic/pathogenic dependencies, which have resulted in diverse specificities regarding hosts and mechanisms, with multiple origins. Host preferences across the major kingdoms of life based on nutritional requirements have been established by many fungi. These can be host-specific or generalist types of interactions. Fungal parasitism/pathogenesis on animal and plant systems has been well studied, and exclusive groups of fungal pathogens have been recognized. Entomopathogens are a group of insect-specific fungi that are of ecological and

T. K. Arun Kumar (✉)
Department of Botany, The Zamorin's Guruvayurappan College (Affiliated to the University of Calicut), Kozhikode, Kerala, India

S. K. Deshmukh et al. (eds.), *Frontiers in Entomopathogenic Fungi*,
https://doi.org/10.1007/978-981-95-6529-0_4

economic importance. Because of their importance as pathogens of insects directly beneficial to humans and due to their potential for use as biocontrol agents of insect pests, studies on insect-attacking fungi have gained special importance.

Insect pathogenic fungi are present in all the major fungal phyla and are known to attack a wide range of insect hosts (Araújo and Hughes 2016). The majority of entomopathogenic fungi (EPF) belong to Ascomycota, Entomophthoromycota, and Microsporidia. Basidiomycota is one of the least represented phyla (Araújo and Hughes 2016; Hughes et al. 2016), with six genera that are reported to exhibit insect pathogenesis so far. The genera are *Athelia* (order Atheliales, subphylum Agaricomycotina), *Auriculoscypha*, *Coccidiodictyon*, *Ordonia*, *Septobasidium*, and *Uredinella* (order Septobasidiales, subphylum Pucciniomycotina).

Atheliales is the only order of Agaricomycetes with a known insect-parasitic taxon. Atheliales is composed of about 20 genera, many of which are polyphyletic (Binder et al. 2005). The order currently includes five phylogenetically supported families: Atheliaceae, Byssocorticiaceae, Lobuliciaceae, Pilodermataceae, and Tylosporaceae (Sulistyo et al. 2021). The resupinate and loose basidiocarps that athelioid species produce show affinities with corticioid and polyporoid basidiocarps. The members are generally saprobic and cause white rot of various substrates like wood and leaf litter (Larsson 2004). Many Atheliales species form lichen associations with cyanobacteria and algae and ectomycorrhizal relationships with Pinaceae and Fagaceae trees (Hibbett et al. 2014). An exceptional case of association with termites has been reported in one species, *Athelia termitophila*.

Septobasidiales is an order of Pucciniomycetes and is the largest basidiomycetous group that contains insect-parasitic fungi (Jiang et al. 2024). Currently, the order hosts the only family, Septobasidiaceae, that includes seven genera: *Auriculoscypha*, *Coccidiodictyon*, *Glenospora*, *Johncouchia* (anamorph of *Septobasidium* species), *Ordonia*, *Septobasidium*, and *Uredinella*. Septobasidiales species are wholly or partially parasitic on scale insects that in turn parasitize certain plant families (Couch 1938), except for *Glenospora*, which is not an insect parasite. The tripartite nutritional dependency of Septobasidiales members is obligatory. The members grow as dense, perennial, organized fungal mats by partially parasitizing insect colonies on tree barks and produce basidia throughout the mycelial mass. A yeast stage is known in the life cycle of species of Septobasidiales (Aime et al. 2014). Phylogenetic relationships of members of Septobasidiales have been resolved by studies (Henk and Vilgalys 2007; Kumar et al. 2007; Zhao et al. 2017) that were based on molecular methods. Molecular analyses by Henk and Vilgalys (2016) suggest a single origin of insect symbiosis in the Pucciniomycetes but reject the monophyly of the Septobasidiales.

2 Insect Associates

2.1 *Athelia*

Athelia (Atheliaceae) produces pellicular, corticoid basidiomata that are easily separable from the substrate on which it grows. It is characterized by interwoven hyphae, clavate basidia, and inamyloid basidiospores. More than 30 species have been described in the genus so far (Maekawa et al. 2020). Atheliaceae forms a monophyletic clade and is related to Agaricales and Boletales (Hibbett et al. 2007). Some species of *Athelia* are facultative plant pathogens.

Athelia termitophila is the only species of Atheliales that has been reported to have a symbiotic relationship with insects. This fungus, which is primarily a wood-decaying saprobe, is known to parasitize termite eggs by producing "termite balls" that mimic its size, shape, and color. Brown fungal balls (fungal sclerotia) were frequently observed among egg piles in the nests of the termites of the *Reticulitermes and Captotermes* species in Japan, the United States, and China (Maekawa et al. 2020; Matsuura et al. 2000). *Fibulorhizoctonia*, the anamorphic state of *A. termitophila*, is also frequently encountered in the termite nests along with the teleomorphic state.

Athelia termitophila exhibits a rare case of mimicry and mutualistic parasitism in fungi, where the protective environment of the termite nests and the nutrition of the termite eggs are exploited. Matsuura et al. (2000) observed that the fungus mimicked the termite eggs both morphologically and chemically and manipulated the worker insects to tend the fungal sclerotia similar to termite eggs. This may appear as a mutualistic association (Komagata et al. 2022; Costa-Leonardo et al. 2024). In the absence of termite workers, the fungal sclerotia germinated and later parasitically grew by exploiting the termite eggs. Also, studies indicate that fungus has adapted to exploit ecological niches with reduced competition from other wood-rotting fungi, especially in winter (Komagata et al. 2022).

A recent study by Hu et al. (2025) elucidated the chemical characteristics of *A. termitophila* and has shown its notable production of bioactive triterpenoid compounds that have significant pharmacological properties. Moreover, chemical camouflage employed by *A. termitophila* is an adaptation of ecological importance. Characterization of the precise chemical compounds responsible for this mimicry will provide insights into mechanisms involved in remarkable ecological interactions.

2.2 *Auriculoscypha*

Auriculoscypha is a monotypic genus erected by Reid and Manimohan (1985) to accommodate a fungus species found associated with a scale insect. Although it has a stipitate-cupulate basidiomata, which distinguishes the genus from other members of Septobasidiales, its placement in the order is well supported (Lalitha et al. 1994; Kumar et al. 2007; Manimohan 2019). A close relationship of *Auriculoscypha* with

Fig. 4.1 (**a**) Basidiomata of (**a**), *Auriculoscypha anacardiicola*, and (**b**) *Septobasidium* species on living branches of *Anacardium occidentale* and *Mangifera indica*, respectively. (Photo credit: T. K. Arun Kumar)

Septobasidium has been observed. A comprehensive phylogenetic study with septal ultrastructure and rDNA sequence analyses conducted by Kumar et al. (2007) reveals the similarities of the two genera. A simple septal pore plugged on both sides by a pulley-wheel-shaped structure is characteristic of the Pucciniomycetes. This has also been observed in *Auriculoscypha*. *Auriculoscypha* has woody, perennial, centrally stipitate basidiomata with brown fibrous context, thick-walled brown septate hyphae devoid of clamp-connections, clavate or circinate transversely septate basidia with two fertile segments, and septate basidiospores. Basidiomata are fertile only during rainy seasons and are mostly found growing pendant from the bark of trees of the family Anacardiaceae. Basidiomata arise from small tubercles that are embedded in the tree bark.

Auriculoscypha is known to have a very limited distribution and has so far been reported only from the Western Ghats of India (Manimohan 2019) (Fig. 4.1a). The first report of the genus was from Kerala (Reid and Manimohan 1985), and subsequent reports were from Karnataka and Goa (Lalitha et al. 1994; Kumar et al. 2007) and from the northern Western Ghats of Maharashtra (Rajeshkumar et al. 2024). Interestingly, the fungus is restricted in occurrence to the lower sides of branches of living anacardiaceous trees like *Anacardium occidentale*, *Mangifera indica*, and *Holigarna arnottiana*. *H. arnottiana* is a tree species endemic to the Western Ghats of India. The limited distribution range may also be corresponding with the geographical range of the insect associate *Neogreenia zeylanica* (Hemiptera, Coccomorpha, Quinococcidae). *N. zeylanica* is recorded only from Sri Lanka (Wu and Nan 2012).

The obligate association of *Auriculoscypha* with a coccid, which in turn feeds on the sap of a tree, is an instance of a tripartite relationship that has evolved similarly to that in *Septobasidium* (Fig. 4.1b). Basidiomata of *Auriculoscypha* become fertile during monsoon seasons and start producing unicellular and uninucleate basidiospores. These basidiospores become multinucleate and septate and start releasing uninucleate blastospores. The blastospores undergo continuous division by budding to form a yeast (unicellular) stage. Uninucleate cells of these yeast stages eventually grow germ tubes, which later fuse with compatible cells to form binucleate cells. These fusion products develop into dikaryotic hyphae that develop as haustorial coils that establish nutritional connections with the coccid. A dikaryon is identified as the prominent phase in the life cycle of Pucciniomycetes (Aime et al. 2014) and is considered essential to pathogenesis/parasitic life modes in both plant and insect systems. Similar to the identification of the dikaryotic condition in the mycelium of *Septobasidium* in infected insects by Henk and Vilgalys (2016), all hyphae isolated from the coccids infected by *Auriculoscypha* were demonstrated by Thomas and Manimohan (2020) to be dikaryotic. It has been suggested by Thomas and Manimohan (2020) that the yeast stage is the infectious stage in the life cycle of the fungus, where contact with the insect is established. Dikaryotic fungal haustoria inside the coccid feed on the haemolymph. Oberwinkler (2017) had observed that the fusion of compatible yeast cells and subsequent formation of dikaryotic haustoria and mycelial colonies take place inside the insects. Published scientific evidence proves the insect-parasitic capability of *Auriculoscypha*.

Neogreenia zeylanica feeds on the phloem sap of anacardiaceous trees using their sucking apparatus and has been reported as a minor pest (Kumar et al. 2007, 2011). The yeast cells of *Auriculoscypha* that have entered the body of these insects gradually develop into haustoria and hyphae and start colonizing. Mycelial mats grow out through the dermal pores of the individual *Neogreenia* that are stationed with their stylets penetrated into the plant host tissue and start developing tubercles around the insects. These tubercles that may be partially or fully immersed in the plant bark immobilize the insect host (Manimohan 2019). Basidiomata develop from these tubercles, deriving nutrition from the insects that may still be alive and actively feeding on plant sap. The association in this condition may be considered partially parasitic. The extent to which individual coccids and other members of the colony benefit or suffer from this association with the fungus is yet unknown (Kumar et al. 2007). Unlike some *Septobasidium* species, *Auriculoscypha* does not produce extensive basidiomata that house or protect uninfected members of the coccid population. There is no evidence to prove the evolution of altruism and kin selection character in the *Auriculoscypha-Neogreenia* association (Kumar et al. 2007).

Kumar et al. (2011) had conducted culture-based studies and established pure cultures from spore deposits and growing tissue from the fruit bodies of *Auriculoscypha*. Slow-growing mycelial cultures developed from yeast stages. Single-spore isolates could not be established under any culture conditions. They found malt yeast peptone agar medium suitable for short-term maintenance of the fungus. However, long-term maintenance of the pure cultures and production of basidiomata were not possible in artificial media, which may be due to the obligatory nature of the fungus. Further culture-based studies are required to establish, conserve, and commercially utilize the insect parasite as a potential biocontrol agent. Puthussery et al. (2010) studied the antioxidant and anti-inflammatory properties of *Auriculoscypha*. They reported that the extracts had more free radical scavenging potential and anti-inflammatory activity than the standard reference drug diclofenac, administered at 10 mg/kg body weight, for both acute and chronic inflammation in mice.

2.3 *Coccidiodictyon*, *Ordonia*, and *Uredinella*

Coccidiodictyon and *Ordonia* are little-known genera in Septobasidiales that are currently accepted with a single species each (*C. inconspicuum*, *O. orthobasidion*). *Coccidiodictyon* was erected by Oberwinkler (1989) as a fungus that differs from other Septobasidiales members by an inconspicuous colony. *Coccidiodictyon* parasitizes scale insects. *Ordonia* typically produces crust-like basidiomata and is a parasite of scale insects. *Ordonia* is closely related to *Coccidiodictyon*, and the phylogenetic distinctiveness between the two genera is unclear (Henk and Vilgalys 2007). The ecology of the two genera is also poorly known. *Uredinella*, another scale insect-parasitic genus of Septobasidiales, was described by Couch (1937) as a fungus intermediate between rusts and the genus *Septobasidium*. Two species, *U. coccidiophaga* and *U. spinulosa*, have been described so far. The species are known to parasitize only a single insect at a time and are characterized by basidiospores formed from probasidia. *Uredinella* has characters of both rust fungi (Pucciniales) and Septobasidiales, suggesting a possible evolutionary link of transition from plant parasitic to insect-parasitic mode of nutrition (Webster and Weber 2007). *Coccidiodictyon*, *Ordonia*, and *Uredinella* have rarely been encountered and studied. Details of the morphology and ecology of the taxa are lacking.

2.4 *Septobasidium*

Septobasidium is one of the most studied genera of Septobasidiales (Pucciniomycetes) because of its ecological peculiarities. *Septobasidium* was erected by Patouillard (1892) with *S. velutinum* as the type. The genus is composed of more than 170 described species (*Index Fungorum*), and all are known to be insect parasites (Henk and Vilgalys 2016). Some *Septobasidium* species are also considered pathogens of minor diseases like felt and velvet blight of crop plants like *Camellia*, *Citrus*, and *Coffea*. The species produces resupinate and basidiomata of varying thickness on the underside of branches of trees belonging to different plant families. Basidiomata are perennial, though noticeable when fertile and actively growing during the rainy seasons. Some species form several basidiomatal layers and elaborate chambers, while the basidiomata of some species may be thin and simple. The species with complex basidiomatal chambers are known to form mutualistic associations with the scale insects (Hemiptera, Coccomorpha, Coccoidea) they parasitize. The monographic work on *Septobasidium* by Couch (1938) opines that the complex basidiomata that grow as extended structures and fungal tissue tunnels protect the uninfected members of the insect colony from predators and other abiotic stresses. About 13 genera of scale insects and 98 plant hosts have been identified as being associated with *Septobasidium* species (Kavale 2023). The basidia may be curved or straight, cylindrical to clavate, and one to three septate. Macromorphology of the basidiomata and structure of the basidium, probasidium, and haustoria are important taxonomic characters within the genus. Molecular phylogenetic analyses (Henk and Vilgalys 2016; Jiang et al. 2024) did not support natural groups and morphology-based delimitations of the genus.

Septobasidium has a cosmopolitan distribution and is reported from Africa, Asia, Australia, Europe, North America, and South America. It has both temperate and tropical occurrences and is found on the barks of a wide range of both wild and cultivated woody plants, including citrus, apple, tea, and rubber (Couch 1938; Henk and Vilgalys 2007). Dispersal of *Septobasidium* species is closely related to the occurrence of associated host insects and plants. Many new species are being discovered and described from around the world. The genus is also well documented from Korea, Costa Rica, and the southeastern United States (Choi et al. 2016). Couch (1938) had described 165 species in his world monograph, and nearly half of the species that have been collected and originally described to date are from Southeast Asia (Henk and Vilgalys 2016). More than 60 species have been collected and reported from China so far (Jiang et al. 2024). Five species of *Septobasidium* (*S. citricola*, *S. bogoriense*, *S. pilosum*, *S. sterilis*, and *S. theae*) have been reported from India (Rao 1969/70; Jamaluddin et al. 2004; Kavale 2023). *Septobasidium citricola* has been reported as causing velvet blight in citrus, and *S. bogoriense*, *S. pilosum*, and *S. theae* have been associated with tea and coffee plants. *Septobasidium sterilis* was a species recently described from Maharashtra, India, by Kavale (2023) from the leaves of *Cinnamomum zeylanicum* and *Atlantia racemosa* in association with the scale insect *Aspidiotus hartii*. Since *Septobasidium* species are mostly reported as associated with velvet blight of tea, coffee, and cinnamon trees, the states of India, particularly Assam, West Bengal, Kerala, and Tamil Nadu, with significant cultivation of these crops, will be harboring these fungi.

Septobasidium species form tripartite (fungus-insect-plant) relationships with scale insects and enter into a complex symbiosis. The insects are parasitized by the fungus but are not killed. Instead, the immobilized insects are allowed to feed on the host plant associates. Uninfected insects are protected by the fungi, and the rest of the insect colony is allowed to develop normally. This is a case of partial parasitism and mutualism between the fungus and the insect and may indicate the evolution of kin selection in coccids. A specificity with regard to the insect associates exists, and the basidiospores of the fungus infect the first instar stage of the insects that settle on the surface of the host plant bark (Couch 1938). The fungus colonizes the insect, derives nutrition using haustoria, and starts growing out to form thick mycelial mats that then fuse with adjacent fungal colonies, covering both infected and uninfected insects (Henk and Vilgalys 2016). Seasonal production of basidiospores continues the infection process to new generations. This kind of mutualistic approach differs from other insect symbiotic genera of Septobasidiales in that the parasitism is partial and is not extended to all insects of a population. The mutualistic-parasitic approach allows both populations to coexist, although one of them is considerably weakened. The fusion of hyphae of individual adjacent and independently derived fungal colonies may give rise to genetically distinct hyphae within a larger colony. This mutualism provides a model for studies on host manipulation and co-evolution. Henk and Vilgalys (2016) identified the determination of the mating system and genetic makeup of fungal colonies as imperative to understanding coevolutionary dynamics in fungus-insect symbiosis. Results of their study revealed that *Septobasidium* colonies were composed of a single heterozygous dikaryotic mycelium and that the individual species colonies outcrossed with other colonies while maintaining their genetic identity.

Septobasidium species are ecologically and economically important. *Septobasidium* species have symbiotic associations with scale insects that are pests of a large number of wild and cultivated plants belonging to diverse plant families. Insect control mechanisms using microorganisms can be explored by observing the ecology of the symbionts. Many *Septobasidium* species indirectly cause minor diseases like felt and velvet blight in many crop plants that are infested by coccids. The protection of the fungal mycelial mat given to the coccids from direct exposure to pesticides requires mechanical removal of tree branches with fungal mats and incurs additional disease control costs. *Septobasidium* species are interesting models for studying nutrient transfer mechanisms across kingdoms of life in symbiosis research. Species of *Septobasidium* are obligate symbionts and are unculturable in laboratory media, except in the case of *S. carestianum*, which forms yellowish yeast colonies (Benjamin et al. 2004). This indicates the ability of *S. carsetianum* to switch to a saprobic mode in the absence of insect hosts. *Septobasidium* species have not been explored for their biotechnological and industrial applications.

3 Conclusion

Insect-parasitic fungi have been widely studied because of their ecological and economic importance. Their potential to naturally infect insect pests of crop and forest plants has been explored, and efficient biological control agents have been developed. The ability of entomoparasitic fungi has been utilized in isolating industrially and clinically useful secondary metabolites. Entomoparasitic capability has been derived multiple times during fungal evolution and can be observed across kingdom fungi. Although a large phylum, Basidiomycota is one of the least represented with entomopathogenic fungal lineages. Septobasidiales is the only entomoparasitic order of the phylum that shows an obligate association with scale insects. The only other report from the phylum is that of an *Athelia* species of Atheliales. It is possible that many of the insect associations of the group remain unknown because of the paucity of ecological studies. Exploratory studies from different parts of the world are required to know more about the interesting insect-fungus associations that might have evolved over time in Basidiomycota.

References

Aime MC, Toome M, McLaughlin DJ (2014) 10 Pucciniomycotina. In: McLaughlin D, Spatafora J (eds) Systematics and evolution. The Mycota, vol 7A. Springer, Berlin, pp 271–294. https://doi.org/10.1007/978-3-642-55318-9_10

Araújo JPM, Hughes DP (2016) Diversity of entomopathogenic fungi: which groups conquered the insect body? Adv Genet 94:1–39. https://doi.org/10.1016/bs.adgen.2016.01.001

Benjamin RK, Blackwell ME, Chapela IH, Humber RA, Jones KG et al (2004) Insect- and other arthropod-associated fungi. In: Mueller GM, Bills GF, Foster MS (eds) Biodiversity of fungi: inventory and monitoring methods. Elsevier Academic Press, Amsterdam, pp 395–433

Binder M, Hibbett DS, Larsson KH, Larsson E, Langer E et al (2005) The phylogenetic distribution of resupinate forms across the major clades of mushroom-forming fungi (Homobasidiomycetes). Syst Biodivers 3:113–157. https://doi.org/10.1017/S1477200005001623

Choi IY, Lee WH, Lee JJ, Park MJ, Ko JA et al (2016) Characterization of a *Septobasidium* sp. associated with felt disease of *Schisandra chinensis*. Mycobiology 44:58–62. https://doi.org/10.5941/myco.2016.44.1.58

Costa-Leonardo AM, Janei V, da Silva IB (2024) New reports on the association between eggs and sclerotium-forming fungi in Neotropical termites with insights into this mutualistic interaction. Biol J Linn Soc 143:1–12. https://doi.org/10.1093/biolinnean/blae010

Couch JN (1937) A new fungus intermediate between the rusts and *Septobasidium*. Mycologia 29: 665–673. https://doi.org/10.1080/00275514.1937.12017236

Couch JN (1938) The genus *Septobasidium*. University of North Carolina Press, Chapel Hill, p 480

Henk DA, Vilgalys R (2007) Molecular phylogeny suggests a single origin of insect symbiosis in the Pucciniomycetes with support for some relationships within the genus *Septobasidium*. Am J Bot 94:1515–1526. https://doi.org/10.3732/ajb.94.9.1515

Henk DA, Vilgalys R (2016) Genetic structure within *Septobasidium* colonies suggests outcrossing and frequent non-self-fusion. Fungal Genet Biol 6(2):1000143. https://doi.org/10.4172/2165-8056.1000143

Hibbett DS, Binder M, Bischoff JF, Blackwell M, Cannon PF et al (2007) A higher-level phylogenetic classification of the fungi. Mycol Res 111(5):509–547. https://doi.org/10.1016/j.mycres.2007.03.004

Hibbett DS, Bauer R, Binder M, Giachini AJ, Hosaka K et al (2014) Agaricomycetes. In: McLaughlin DJ, Spatafora JW (eds) Systematics and evolution: the Mycota: a comprehensive treatise on fungi as experimental systems for basic and applied research, vol 7A. Springer Verlag GmbH, pp 373–429. https://doi.org/10.1007/978-3-642-55318-9

Hu F, Fang Y, Khan Z, Xing L (2025) Unveiling the synergistic effect of salicylic acid on triterpenoid biosynthesis in *Athelia termitophila*: Elucidating the molecular underpinnings. Int J Mol Sci 26:996. https://doi.org/10.3390/ijms2603096

Hughes DP, Araújo JPM, Loreto RG, Quevillon L, De Bekker C et al (2016) From so simple a beginning: the evolution of behavioral manipulation by fungi. Adv Genet 94:437–469. https://doi.org/10.1016/bs.adgen.2016.01.004

Jamaluddin S, Goswami MG, Ojha BM (2004) Fungi of India (1989–2001). Scientific Publisher, Jodhpur, p 326

Jiang Q, Kang Z, Wang X, Zhao C (2024) Molecular phylogeny and morphology reveal three new plant pathogenic fungi species (Septobasidiales, Basidiomycota) from China. MycoKeys 111: 229–248. https://doi.org/10.3897/mycokeys.111.125933

Kavale TR (2023) *Septobasidium sterilis* sp. nov. from India. J Mycopathol Res 61:267–270. https://doi.org/10.57023/JMycR.61.2.2023.267

Komagata Y, Fukasawa Y, Matsuura K (2022) Low temperature enhances the ability of the termite-egg-mimicking fungus *Athelia termitophila* to compete against wood-decaying fungi. Fungal Ecol 60:101178. https://doi.org/10.2139/ssrn.4009359

Kumar TKA, Celio GJ, Matheny PB, McLaughlin DJ, Hibbett DS et al (2007) Phylogenetic relationships of *Auriculoscypha* based on ultrastructural and molecular studies. Mycol Res 111:268–274. https://doi.org/10.1016/j.mycres.2006.12.003

Kumar TKA, Jisha KC, Jisha ES, Manimohan P (2011) Isolation and in vitro cultivation of *Auriculoscypha anacardiicola* D.A. Reid et Manim., an insect-associated and potentially medicinal fungus from India. Int J Med Mushrooms 13:273–280. https://doi.org/10.1615/intjmedmushr.v13.i3.70

Lalitha CR, Leelavathy KM, Manimohan P (1994) Patterns of basidiospore germination in *Auriculoscypha anacardiicola*. Mycol Res 98:64–66. https://doi.org/10.1016/S0953-7562(09)80338-9

Larsson KH (2004) High phylogenetic diversity among corticioid homobasidiomycetes. Mycol Res 108:983–1002. https://doi.org/10.1017/S0953756204000851

Maekawa N, Yokoi H, Sotome K, Matsuura K, Tanaka C et al (2020) *Athelia termitophila* sp. nov. is the teleomorph of the termite ball fungus *Fibularhizoctonia* sp. Mycoscience 61:323–330. https://doi.org/10.47371/mycosci.2022.10.001

Manimohan P (2019) *Auriculoscypha*: a fascinating instance of fungus-insect-plant interactions. In: Sridhar KR, Deshmukh SK (eds) Advances in macrofungi: diversity, ecology and biotechnology. CRC Press/Taylor & Francis, Boca Raton, pp 34–39

Matsuura K, Tanaka C, Nishida T (2000) Symbiosis of a termite and a sclerotium-forming fungus: Sclerotia mimic termite eggs. Ecol Res 15:405–414. https://doi.org/10.1046/j.1440-1703.2000.00361.x

Oberwinkler F (1989) *Coccidiodictyon*, new genus and *Ordonia*, two genera in the Septobasidiales. Opera Bot 100:185–192

Oberwinkler F (2017) Yeasts in Pucciniomycotina. Mycol Prog 16:831–856. https://doi.org/10.1007/s11557-017-1327-8

Patouillard NT (1892) *Septobasidium*, nouveau genre d'Hyménomycètes hétérobasidiés. J Bot 6: 61–64

Puthusseri B, Smina TP, Janardhanan KK, Manimohan P (2010) Antioxidant and anti-inflammatory properties of new medicinal fungus, Auriculoscypha anacardiicola DA Reid et Manim. (Agaricomycetideae), from India. Int J Med Mushrooms 12:391–400. https://doi.org/10.1615/IntJMedMushr.v12.i4; https://doi.org/10.1615/IntJMedMushr.v12.i4.60

Rajeshkumar KC, Desai S, Ansil PA, Rane AA, Harikrishnan K et al (2024) Phylogenetic insights and first record of *Auriculoscypha anacardiicola* on *Holigarna arnottiana* from the Northern Western Ghats, India. Kavaka 60:125–132. https://doi.org/10.36460/Kavaka/60/sp/2024/125-132

Rao VG (1969/1970) Fungi on citrus from India. Sydowia 23:215–224

Reid DA, Manimohan P (1985) *Auriculoscypha*, a new genus of Auriculariales (basidiomycetes) from India. Trans Br Mycol Soc 85:532–535. https://doi.org/10.1016/S0007-1536(85)80055-3

Sulistyo BP, Larsson KH, Haelewaters D, Ryberg M (2021) Multigene phylogeny and taxonomic revision of Atheliales s.l.: reinstatement of three families and one new family, Lobuliciaceae fam nov. Fungal Biol 125(3):239–255. https://doi.org/10.1016/j.funbio.2020.11.007

Thomas A, Manimohan P (2020) Dikaryotization seems essential for hypha formation and infection of coccid in the life-cycle of *Auriculoscypha anacardiicola*. Stud Fungi 5:66–72. https://doi.org/10.5943/sif/5/1/6

Webster J, Weber R (2007) Urediniomycetes: Uredinales (rust fungi). In: Introduction to fungi. Cambridge University Press, Cambridge, pp 609–635

Wu S, Nan N (2012) *Neogreenia lonicera* sp. nov., a new species of Margarodidae *sensu lato* (Hemiptera: Coccoidea) from China, with a key to species of *Neogreenia* MacGillivray and placement of the genus in the family Kuwaniidae. Zootaxa 3274:43–54. https://doi.org/10.11646/zootaxa.3274.1.5

Zhao R, Li GJ, Sánchez-Ramírez S, Stata M, Yang ZL et al (2017) A six-gene phylogenetic overview of Basidiomycota and allied phyla with estimated divergence times of higher taxa and a phyloproteomics perspective. Fungal Divers 84:3–74. https://doi.org/10.1007/s13225-017-0381-5

Interaction Between Entomopathogenic Fungi and Insect Immune Systems: A Coevolutionary Perspective

5

Vaishnavi Rajwar, Sakshi Suman, and Yamini Agrawal

Abstract

Entomopathogenic fungi (EPF) are the pathogenic microorganisms that attack and kill insects and other arthropods. Despite being a significant part of the ecosystem, various insect species pose a threat to economically important crops, leading to substantial agricultural losses annually. As a result, sustainable pest management has drawn significant attention over the years, and biocontrol agents play a major role in contributing to it. Application of EPF as biocontrol agents provides an effective and ecofriendly alternative to harmful pesticides and insecticides. The intricate interplay between EPF and insects is marked by a constant coevolutionary arms race occurring over thousands of years. In this consistent tug-of-war, the insect hosts utilize a diverse range of defense mechanisms and immune responses. This chapter underscores the various types of immune responses by the host, such as behavioral, physical, cellular, and humoral responses, further highlighting the various strategies employed by EPF as a countermeasure against each kind of immunity, ensuring efficient and successful host invasion. Additionally, this chapter explores the different stages of fungal infection, shedding light on the mechanisms behind pathogen invasion and the dual role of miRNA in insect immunity and how it acts as both an asset and a liability. Insights into this interplay between host and pathogen help us gain a deeper understanding of the coevolutionary arms race and can help enhance pest management strategies via genetic engineering. Consequently, the biocontrol potential of EPF can be enhanced in different aspects by overcoming the

Vaishnavi Rajwar and Sakshi Suman contributed equally with all other contributors.

V. Rajwar · S. Suman · Y. Agrawal (✉)
Department of Botany, University of Delhi, New Delhi, Delhi, India

S. K. Deshmukh et al. (eds.), *Frontiers in Entomopathogenic Fungi*,
https://doi.org/10.1007/978-981-95-6529-0_5

limitations and improving the overall virulence and pathogenicity, thereby ensuring efficient pest management.

Keywords

Countermeasures · Defenses · Hemocyte · Immunity · Management

1 Introduction

Insects constitute a large proportion (~75%) of the world's recognized fauna and are present in a wide variety of environments. They play crucial roles in several environmental processes like pollination, nutrient cycling, and food webs. However, more than half of the described insect species are pests of the crop plants and indirectly contribute to crop loss (Jankielsohn 2018; Tudi et al. 2021). Entomopathogenic fungi (EPF) are insect-pathogenic fungi and play an important role in controlling insect populations in nature. They have emerged as one of the most promising alternatives for pest control, with over 60% of naturally occurring insect diseases known to be caused by EPF (Charnley and Collins 2007; Mantzoukas et al. 2022). This makes them interesting from a biological perspective and extremely promising in sustainable pest management.

Moreover, EPF are also important from the ecological and economic points of view. In natural ecosystems, EPF are important biological control agents that keep insect populations under control. This contributes to ecological stability. Biopesticides such as EPF are a viable alternative to synthetic chemical insecticides, which have several drawbacks, such as environmental degradation, non-target effects, and development of pesticide resistance (Zhang et al. 2025). Commercial formulations of genera including *Beauveria*, *Metarhizium*, and *Hirsutella* are commonly used against insect pests in agricultural settings (Qu and Wang 2018). The interaction between insects and EPF involves several stages that may be broadly categorized as early infection, adhesion, penetration, and hemolymph growth (Fig. 5.1), and a successful infection might eventually lead to insect death. Insects exhibit both pre-contact and post-contact avoidance behaviors to evade EPF. Volatile organic compounds (VOCs) emitted by fungi can either repel or attract insects. Termites and ants, for example, use olfactory cues to detect infected individuals (Yanagawa et al. 2015). EPF conidia use hydrophobins and adhesin proteins to adhere to insect cuticles. Meanwhile, insects employ cuticular compounds and rapid molting to inhibit fungal adhesion (Zhang et al. 2011). Fungal penetration involves mechanical and enzymatic actions through appressoria. Enzymes such as proteases and chitinases help fungi breach the cuticle (Shang et al. 2016). Inside the hemocoel, EPF adopt yeast-like hyphal bodies to evade immunity (Lewis et al. 2009). They secrete metabolites like destruxins and oosporein, suppressing immune responses. EPF miRNAs like bba-milR1 also target host immune genes (Cui et al. 2019). In the

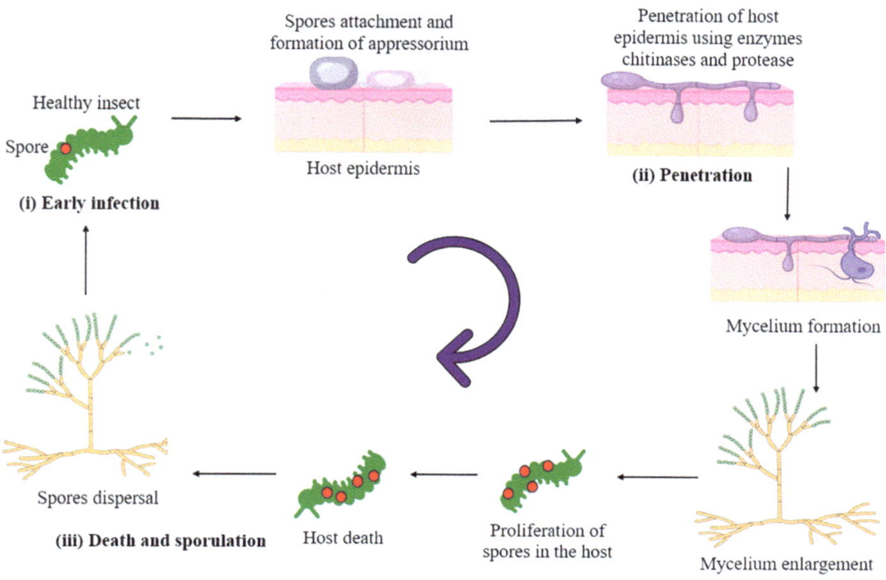

Fig. 5.1 Stages of fungal infection showing early infection, penetration, death, and sporulation (Zhang et al. 2025)

final stage, fungi sporulate from the host cadaver. Infected insects often isolate themselves to reduce the risk to the colony (Ortiz-Urquiza and Keyhani 2013). EPF secrete antimicrobial compounds to suppress microbial competitors in the carcass (Bulmer et al. 2009).

Insects, however, are not passive victims in this interaction. Over the years, they have developed complex, multilayered defense mechanisms to combat this fungal invasion (Fig. 5.2). The insect cuticle serves as the primary line of defense and acts as a physical barrier for the invading fungi. The insect's innate immune system gets activated as the fungus breaches the physical barrier and enters the cuticle. The immune responses can be broadly classified into humoral and cellular. Circulating hemolymphocytes mediate cellular responses such as phagocytosis, nodulation, encapsulation, and melanization, whereas reactive oxygen species (ROS), lysozymes, and antimicrobial peptides (AMPs) are produced during humoral responses (Qu and Wang 2018; Mahanta et al. 2023).

The humoral immune responses in insects are under the control of signaling pathways like the Toll, Imd, and JAK-STAT pathways. The most important role in fungal infection of insects is played by the Toll signaling pathway, as it regulates the production of AMPs such as drosomycin and metchnikowin (Lu and St Leger 2016). This pathway directly gets activated in response to fungal pathogen-associated molecular patterns (PAMPs) like β-1,3 glucan, and mannans. AMPs circulate in the hemolymph and slow down the growth of pathogens. In response to these immune mechanisms, EPF have also developed numerous counter-defense

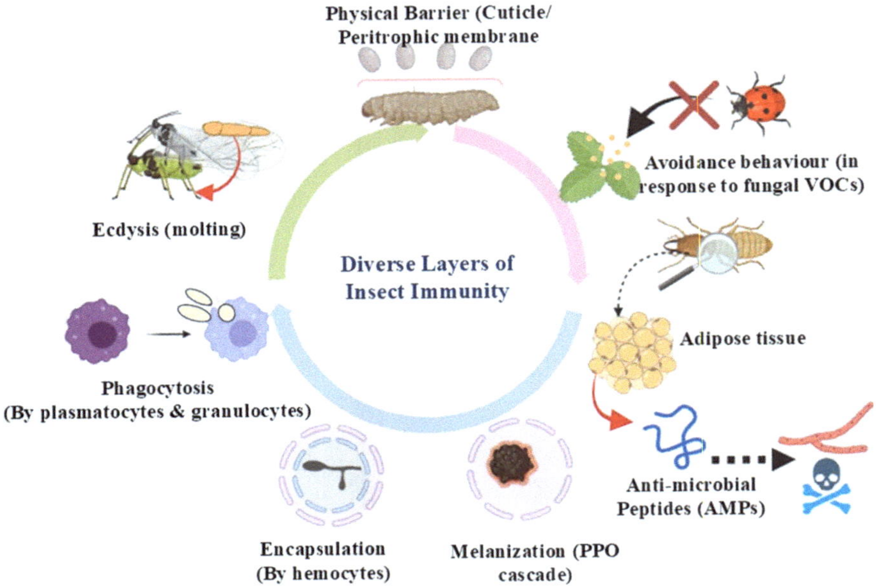

Fig. 5.2 A visual summary of insect immune defenses—highlighting behavioral (detection of fungal VOCs), physical (barrier by cuticle, integument, and peritrophic membrane), cellular (phagocytosis, encapsulation, melanization), and humoral (production of AMPs and lysozymes) immune responses (Cheung et al. 2022). (Created using https://BioRender.com)

strategies. While some release immunosuppressive substances that disrupt host signaling pathways, others produce protease inhibitors that destroy the host enzymes (Qu and Wang 2018).

Hence, there is a constant coevolutionary battle between EPF and their insect hosts, where both of them have developed strategies for survival. The most pathotypes are contained within a limited number of genotypic classes. There are costs and benefits to specialization. Specialization depends on host availability, but it reduces interspecific competition. Specialization for the selection of evasion mechanisms from immune systems that are adapted to a large range of host species must be costly, whereas selection of narrow specificity is possible only if the host species has stable populations through time (Lu and St Leger 2016). Genetic studies have revealed the presence of variable immune gene regulators in insects that are determined by their ecological niche and pathogen exposure (Lazzaro and Little 2009). Similarly, there has been an expansion in fungal gene families related to pathogenicity, such as chitinases, proteases, and secondary metabolites (Liu et al. 2023). In this chapter, we aim to emphasize the interplay between insects and EPF, focusing on different infection stages, the cross-talk mechanisms, and the development of various counter-defense strategies to obtain a better understanding of the various underlying molecular mechanisms.

2 Interaction Between Insect Immunity, Behavior, and Fungal Virulence

As insects lack adaptive immunity like other organisms, they depend solely on their innate immunity for defense. Their defense includes majorly behavioral responses, physical barriers, humoral immunity, and cellular immunity. Prior to any contact, insects can initiate avoidance behaviors by the detection of certain volatile VOCs released by fungi (Hong et al. 2023). Among the various chemically diverse fungal metabolites, such as esters, alcohols, ketones, aldehydes, acids, and terpenes, VOCs like geosmin, 1-octen-3-ol, and 2-phenylethanol help insects avoid contact with any of the entomopathogenic fungi. Avoidance behavior has been reported in *Drosophila melanogaster* in response to geosmin (Störtkuhl et al. 2005), while ladybird beetles (*Coccinella septempunctata*) avoid any contact on leaves that are colonized by *Beauveria bassiana* (Ormond et al. 2011; Thompson and Brandenburg 2005). A similar kind of avoidance behavior was also reported in *Popillia japonica* and termites (Villani et al. 1994; Staples and Milner 2000).

Additionally, certain metabolites are also known to play a dual role, such as 1-octen-3-ol (Fäldt et al. 1999; El Jaddaoui et al. 2023), which functions as a neurotoxin at relatively lower concentrations, while at higher concentrations it inhibits fungal spore germination (Chitarra et al. 2004; Hummadi et al. 2022). In addition, insects adapt to different behavioral patterns, such as grooming, habitat avoidance, and thermoregulation, as part of behavioral immunity. For example, to prevent the spread of pathogens, termites and ants use communal hygiene behaviors to identify and eliminate contaminated individuals from the colony, which helps limit further pathogen spread. Table 5.1 summarizes molecular and behavioral strategies used by fungi and insects at different infection stages.

2.1 Physical Barriers

Among the physical barriers, the cuticle forms an outer protective covering against external invaders; it is rich in chitin and proteins and encases the integument. The peritrophic membrane provides another barrier; it is present in the midgut and acts like a molecular sieve, filtering coarse food particles, thus preventing pathogen entry. The insect cuticle also plays a significant role in molting (ecdysis), which enables the insects to physically shed pathogens attached to the cuticle, thereby making a successful escape (Vandenberg et al. 1998). For example, in insects like aphids and the diamondback moth (*Plutella xylostella*), rapid molting reduces the chances of fungal adherence, thus preventing further infection (Kim and Roberts 2012). However, some EPF have evolved mechanisms to disrupt the host molting process by oxidative inactivation of host ecdysteroids, which further delays ecdysis and increases the chance of successful infection. Interestingly, insects can also modify their hydrocarbon profile to reduce fungal attachment to the cuticle.

One such genus is *Metarhizium*, which struggles to adhere to the cuticle of *Aedes aegypti* larvae due to the absence of long-chain hydrocarbons, which are likely

Table 5.1 Molecular defense mechanisms adopted by entomopathogenic fungi (EPF) and insects at different infection stages

Infection stage	Fungus/insect species	Defense mechanism or genes involved	Reference
Pre-contact	*Anopheles stephensi*	Attraction to EPF volatile organic compounds (possibly self-vaccination)	George et al. (2013)
	Ants (*Formica selysi*)	Self-grooming increases in the presence of EPF	Reber et al. (2011)
	Japanese beetle larvae	Avoid soil with high EPF concentration	Villani et al. (1994)
	Termites (*Reticulitermes flavipes*)	Alarm and aggregation in response to EPF volatile organic compounds	Myles (2002)
Adhesion	*Beauveria bassiana*	Hydrophobins (Hyd1, Hyd2) and adhesins (Adh1–3)	Zhang et al. (2011)
	Metarhizium anisopliae	Adhesins Mad1, Mad2 and CWP10 protein	Wang and St Leger (2007a, b)
	Liposcelis bostrychophila	Fatty amides decrease hydrophobicity	Lord and Howard (2004)
	Bedbugs	Cuticular aldehydes inhibit EPF germination	Ulrich et al. (2015)
Penetration	*Beauveria bassiana*	Insect cuticle melanization and PPO activation cascade	Yassine et al. (2012)
	Drosophila	Microbial surface recognition via PRRs	Govind (2008)
	Helicoverpa armigera	Reduced immune response to *Metarhizium rileyi* hyphal bodies	Li et al. (2021)
	Metarhizium anisopliae	Enzymes: Pr1, Pr2 and chitinases	Sánchez-Pérez et al. (2014)
Hemolymph colonization	*Beauveria bassiana*	Production of oosporein, tenellin	Fan et al. (2017)
	Insect fat body	The toll and Imd pathways are activated for AMP production	Li et al. (2023)
	Metarhizium anisopliae	Destruxins inhibit AMPs and PPOs	Han et al. (2022)
	Mosquito (*Aedes aegypti*)	bba-milR1 downregulates Spatzle	Cui et al. (2019)
Systemic infection	*Drosophila*	AMPs: Drosomycin, metchnikowin	Huang et al. (2023)
	Termites	Termicin, GNBP2 with antifungal activity	Da Silva et al. (2003)
Death and sorulation	Entomopathogenic fungi	Secretion of antimicrobial compounds in cadaver	Zhang et al. (2010)
	Infected insects	Behavioral isolation to prevent colony transmission	Hassan et al. (2022)

essential for EPF development on the host surface (Kiuchi et al. 2003). Additionally, some insects secrete formic acid (e.g., *Lasius neglectus*), benzoquinones (*Tribolium castaneum*), or aldehydes (*Cimex lectularius*) to inhibit conidial germination (Cho et al. 2007). To breach the first line of insect defense, fungi like *B. bassiana* and *Metarhizium anisopliae* use hydrophobins (Hyd1, Hyd2) and adhesins (Adh1–3, Mad1) to attach to host surfaces (Wang and St Leger 2007a, b) and secrete cuticle-degrading enzymes: *Pr1* proteases, *Bbchit1* chitinases, and lipases to break down the insect integument (Fang et al. 2005). The ectomicrobiomes, bacterial communities on insect surfaces, inhibit fungal growth by producing antifungals (Wanchoo et al. 2009). EPF can also breach these protective microbiotas by secreting antimicrobial peptides like BbAMP1 (Ma et al. 2024). Upon breach of physical barriers by a pathogen, the cellular and humoral responses are activated. The former gets triggered almost immediately, while the latter takes many hours to begin.

2.2 Cellular Immunity

Upon successful entry of EPF into the insect hemocoel the pathogen forms blastopores, in response to which various cellular immune responses are activated (Fig. 5.3). These include coagulation, hemocyte activation, phagocytosis, nodulation, encapsulation, melanization, and the generation of reactive oxygen species (ROS) (Li and Xia 2022). These responses are primarily mediated by hemocytes that exist in different forms and possess adhesion and phagocytosis capabilities (Prisnyi and Grebtsova 2022).

Phagocytosis is generally the initial cellular response, in which plasmocytes and granulocytes engulf small fungal propagules like conidia. However, after the fungus develops into filamentous hyphae or large hyphal bodies, the structures become too big, and the granulocytes and plasmocytes fail to phagocytose them. At this step, the host immune system transitions into the encapsulation phase, where the hemocytes form aggregates or clusters around the pathogen, forming a multilayered sheath. Encapsulation is often followed by melanization, which involves the enzymatic conversion of phenols to melanin pigments via the prophenoloxidase (PPO) cascade (Wang et al. 2012). It generates toxic intermediates like quinones and ROS and also reinforces the physical barrier surrounding the fungal element (Ashida and Brey 1995). In previous studies, PPO-deficient mutants of *Drosophila* and mosquitoes had significantly higher mortality after exposure to *Metarhizium* or *Beauveria*, further highlighting the role of the pathway in antifungal defense (Yassine et al. 2012).

To counterattack the cellular immunity, EPF, in turn, has evolved various mechanisms. For example, *M. rileyi* alters the cell wall compositions of its hyphal bodies by reducing the β-glucan content, thus making them less recognizable to host immune receptors (Li et al. 2021). In addition, some EPF secrete LysM domain proteins that are bound to fungal chitin and protect it from host recognition (Cen et al. 2017). They also produce collagenous coatings or extracellular laccases that break down immune effectors such as PPOs and inhibit ROS accumulation.

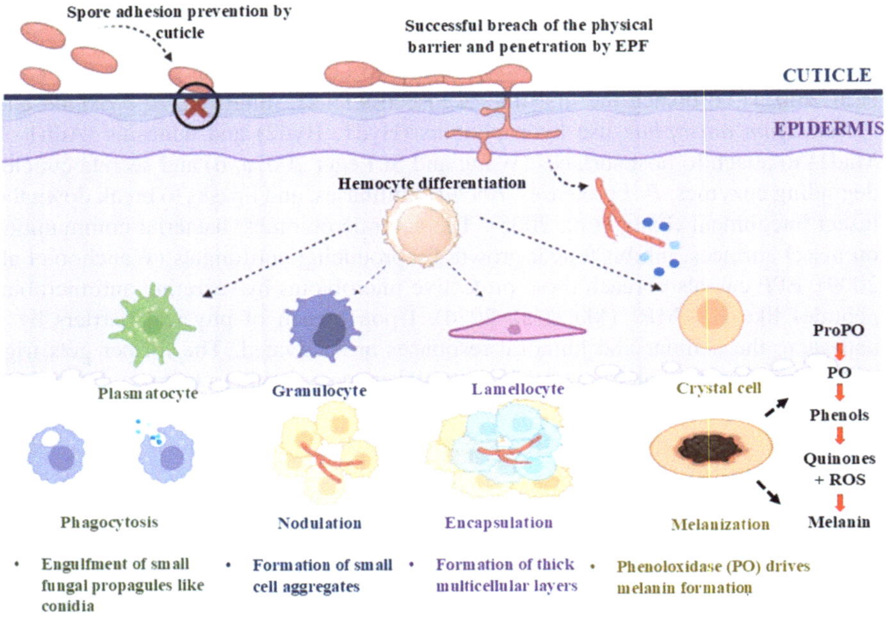

Fig. 5.3 Schematic diagram of insect cellular immunity highlighting hemocyte differentiation (granulocytes, plasmatocytes, lamellocytes, crystal cells) and their roles in phagocytosis, nodulation, encapsulation, and melanization in response to EPF invasion (Rosales and Vonnie 2017). (Created using https://BioRender.com)

Moreover, EPF deploy secondary metabolites that specifically impair immune cells. One notable example is the secretion of destruxins by *Metarhizium* species, which interfere with calcium signaling, inhibit PPO activation, and induce apoptosis in hemocytes (Wang et al. 2012, Agrawal et al. 2016). These compounds can also prevent nodule formation and suppress phagocytic efficiency.

2.3 Humoral Immunity

Parallel to cellular responses, humoral immune responses primarily involve AMPs and lysozyme, the former synthesized in the fat body and secreted into the hemocoel (Fig. 5.4). Upon recognition of fungal microbe-associated molecular patterns (MAMPs) by pattern recognition receptors (PRRs), this kind of response is triggered, which in turn activates the Toll, JAK-STAT, and IMD (immune deficiency) signaling pathways (Khaleel et al. 2013). Additionally, some studies have depicted certain mutations in these pathways, like Toll pathway components, that increased insect susceptibility to fungal infection, further highlighting the crucial role of humoral immunity in overall insect defense (Lu et al. 2015).

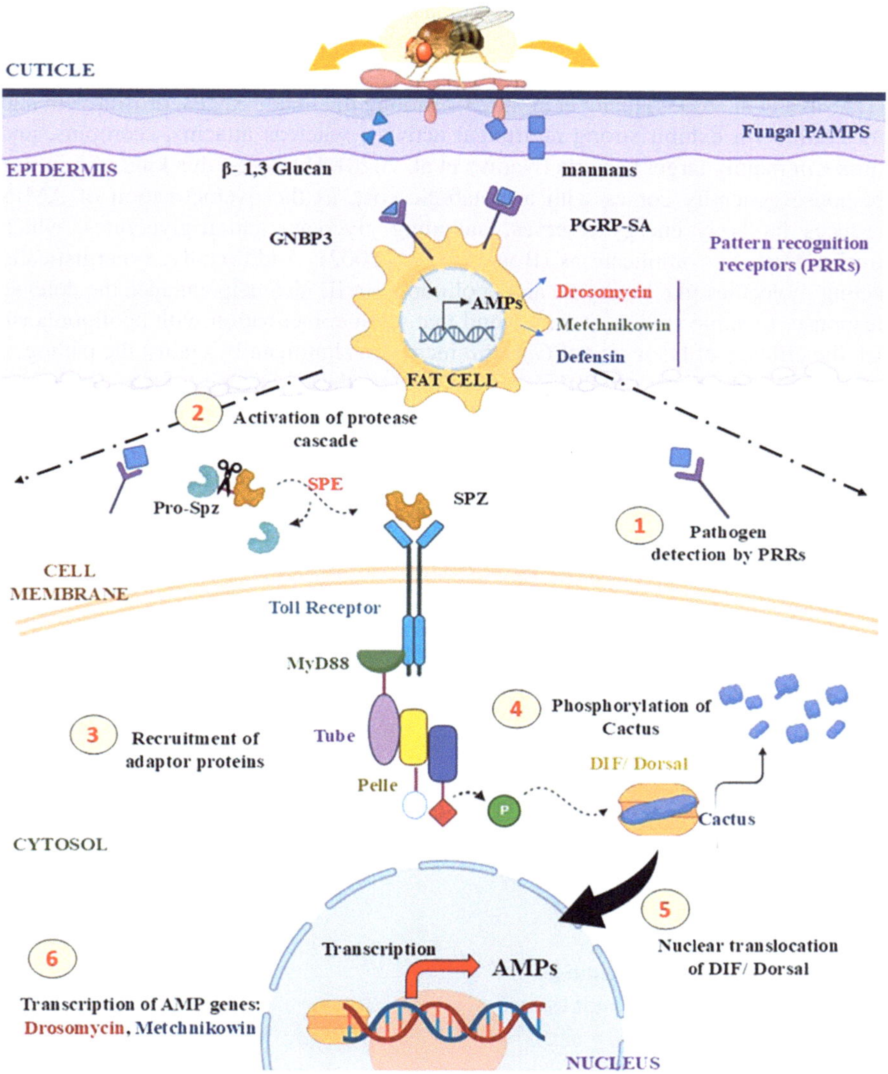

Fig. 5.4 A schematic diagram of insect humoral immunity representing the TOLL pathway and the major signaling steps involved: (1) fungal PAMPs (β-1,3-glucan, mannans) bind to GNBP3 (Gram-negative bacteria binding protein 3) and other PRRs; (2) pathogen recognition activates a serine protease cascade; Spatzle Processing Enzyme (SPE) cleaves pro-Spatzle and forms active Spatzle; (3) activated Spatzle binds to Toll; activation of the Toll receptor recruits adaptor proteins dMyD88 (myeloid differentiation primary response 88), Tube, and Pelle; (4) cactus is bound to DIF/Dorsal (NF-κB transcription factors); Pelle phosphorylates Cactus, leading to its degradation; (5) released Dorsal/DIF (Dorsal-related Immunity Factor) moves to the nucleus; (6) nuclear Dorsal/DIF upregulates AMP genes, such as Drosomycin expression, to combat infection (Stokes et al. 2015). (Created using https://BioRender.com)

These regulatory pathways upregulate the genes encoding AMPs like drosomycin, metchnikowin, baramicin A, and thanatin in *Drosophila*; termicin in termites; gallerimycin in *Galleria mellonella*; and heliomycin in *Heliothis virescens* (Da Silva et al. 2003; Huang et al. 2023). Among the major AMPs, drosomycin and metchnikowin exhibit strong antifungal activity, whereas attacins, cecropins, and drosocin mainly target bacteria (Narava et al. 2020). However, this kind of immune response generally comes with a metabolic cost, as the overactivation of AMPs reduces fat body energy reserves, including glycogen and triglycerides, which further leads to complications (Boutros et al. 2002). Additionally, synergistically acting molecules like lysozyme and apolipophorin III also help enhance the defense response. In some studies, it was found that upon coincubation with apolipophorin III, the efficacy of lysozyme of *Galleria* increased significantly against the pathogen (Zdybicka-Barabas et al. 2013). To overcome these defenses, fungi, including *M. anisopliae*, produce destruxins and microRNA-like effectors that further suppress AMP expression by targeting Toll signaling components such as Spatzle 4, thereby silencing the systemic immunity against EPF by the host (Chen et al. 2014).

3 Fungal Strategies to Overcome Host Defenses

Fungi produce immune-evasive proteins, secondary metabolites, and miRNAs that suppress insect immune pathways (Fig. 5.5). Comparative genomics reveals extensive protease/chitinase gene families that facilitate infection.

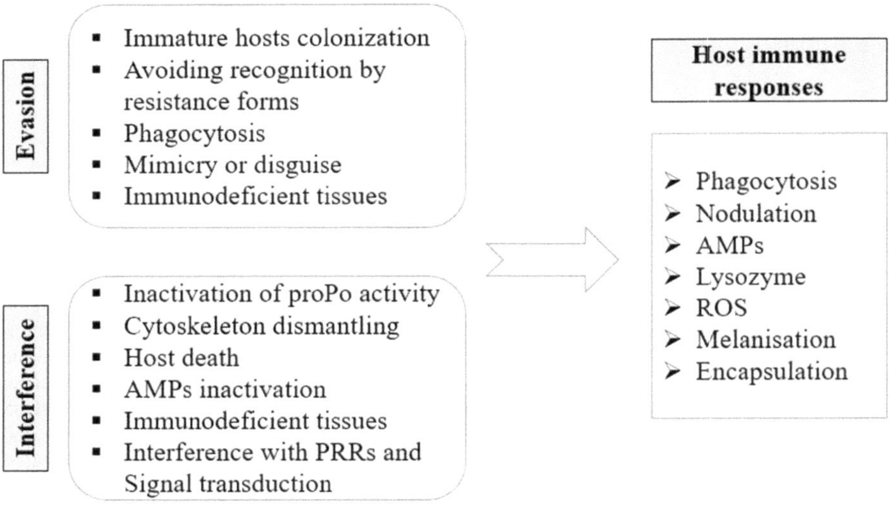

Fig. 5.5 Flowchart summarizing pathogen immune evasion, interference approaches, and host immune responses

3.1 miRNA in Insect Immunity

Apart from the behavioral, cellular, and humoral immunity, insects also rely on non-coding RNAs (ncRNAs) to defend against entomopathogenic fungi (Fig. 5.6). miRNAs, in particular, play a major role in regulating post-transcriptional gene expression. According to recent studies, these small regulators (19–21 nucleotides) have been found to participate in cross-kingdom interactions with invading pathogens, especially EPF (Lu and Chtarbanova 2022). Cross-kingdom RNA interference (ck-RNAi) allows insects to export their miRNAs into fungal cells. For example, *Anopheles stephensi* and *A. aegypti* introduce *let-7* and *miR-100* into *B. bassiana*, where they suppress fungal pathogenicity genes such as *sec2p* and *C6TF*, further affecting the fungal pathogenicity (Wang et al. 2021). In a recent study, infection by *Metarhizium anisopliae* is shown to induce expression modification of miRNA in *Solenopsis invicta*. Within the host machinery, *miR-71* and *miR-7* were seen inhibiting antifungal immune responses by directly inhibiting ModSP1-Relish and Lysozyme2-Serine protease7, which are the major constituents of the Toll, IMD, and melanization pathways (Xu et al. 2025).

The immunosuppressive role of these miRNAs was further evident due to increased susceptibility to fungal infection as a result of overexpression. In *Apis mellifera* and other honeybees, similar observations were made, where infection-induced miRNAs modified activation of Toll and IMD pathways (Wang et al. 2021; Cui et al. 2022). Interestingly, fungi such as *B. bassiana* and *M. anisopliae* have adapted to take advantage of these miRNAs. Genetically modified *B. bassiana* carrying mosquito-derived *miR-8* and *miR-375* effectively repressed Toll pathway

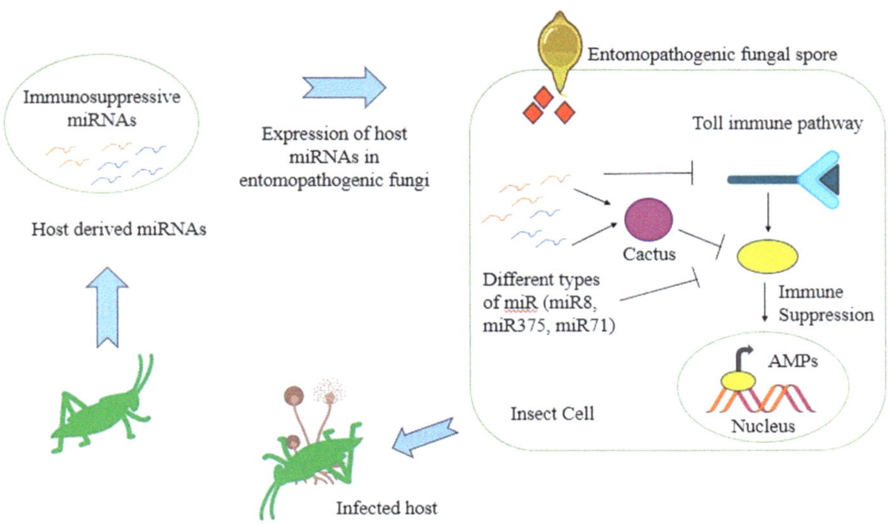

Fig. 5.6 A visual representation of the expression of insect miRNAs in entomopathogenic fungi induces pathogen-mediated host RNA interference (Cui et al. 2022)

effectors such as *RelA* and activated inhibitors such as Cactus in *A. aegypti*, increasing fungal virulence (Cui et al. 2022). Such manipulation by the pathogen further depicts an adaptive strategy in the coevolutionary race, where pathogens use the host regulatory machinery to their advantage. The dual role of host miRNAs, as both defenders and liabilities, demonstrates their strategic role in regulating host-pathogen interactions.

3.2 Fungal Strategies to Counter Host Immunity

When EPF enter inside the insect, the fungi experience a strong immunity response by the host. The pathogenic fungi interact with the innate immune system of their host. The pathogens have a wide range of immune evasion strategies. *Metarhizium* spp. function as specialized entomopathogens, producing a cyclo-hexadepsipeptidic toxin called destruxins, which play a specific role in suppressing AMP gene expression (Pal et al. 2007; Wang et al. 2023) and block phagocytosis, which inhibits V-ATPase (Chen et al. 2014). *Metarhizium* blastospores can evade hemocytes by secreting a hydrophilic collagen (Mcl1) coat as a camouflage strategy (Wang and St Leger 2007a, b; Mascarin et al. 2024). *B. bassiana* produces a quinone derivative known as oosporein, which inhibits ProPO activity and downregulates expression of an antifungal peptide, gallerimycin, in larvae of the *Galleria* sp. (wax moth) (Feng et al. 2015).

Conidia can be grown within insect phagocytic cells and avoid immune reactions (Kurtti and Keyhani 2008; Lovett and St Leger 2017). Many entomopathogens release decoy components that bind to antibodies to divert the immune system (Chai et al. 2009). A mutant screen showed that the outcome of disrupting various *Drosophila* genes affects cellular immunity, which increases susceptibility to *Metarhizium* (Lu et al. 2015), consistent with a dynamic evolutionary arms race between pathogen and host.

4 Conclusion and Future Perspectives

The study comprises biological interactions and how EPF and their insect hosts were shaped by millions of years of coevolution. The interaction confines molecular complexity, ecological consequences, and behavioral strategies, which offer insights into fundamental aspects of immunity and pathogenesis for practical applications in sustainable agriculture. Fungal pathogens have evolved in such a complex way to combat every situation, and virulence is an important factor that includes adhesion, penetration, immune suppression, and proliferation within host tissues. In parallel, insects have developed advanced defense strategies, which include a robust innate immune system, behavioral adaptations, and physical barriers. Understanding these responses and counter-responses across infection stages can become essential for improving the effectiveness of EPF in pest management. The future research should focus on how fungal effectors interact with insect immune receptors, various mechanisms, and genetic and molecular understanding of host-pathogen interaction. CRISPR/Cas9, the genomic editing tool, presents optimistic opportunities to enhance EPF virulence by targeting major immune evasion genes. Integrating EPF along with other biocontrol agents and agricultural practices can yield mutual pest management systems. As research continues, it deciphers the relationship of host-pathogen coevolution in natural systems.

References

Agrawal Y, Narwani T, Subramanian S (2016) Genome sequence and comparative analysis of clavicipitaceous insect-pathogenic fungus *Aschersonia badia* with *Metarhizium* spp. BMC Genomics 17:367–382. https://doi.org/10.1186/s12864-016-2710-6

Ashida M, Brey PT (1995) Role of the integument in insect defense: pro-phenol oxidase cascade in the cuticular matrix. Proc Natl Acad Sci USA 92(23):10698–10702. https://doi.org/10.1073/pnas.92.23.10698

Boutros M, Agaisse H, Perrimon N (2002) Sequential activation of signaling pathways during innate immune responses in *Drosophila*. Dev Cell 3(5):711–722. https://doi.org/10.1016/S1534-5807(02)00325-8

Bulmer MS, Bachelet I, Raman R, Rosengaus RB, Sasisekharan R (2009) Targeting an antimicrobial effector function in insect immunity as a pest control strategy. Proc Natl Acad Sci USA 106(31):12652–12657. https://doi.org/10.1073/pnas.0904063106

Cen K, Li B, Lu Y, Zhang S, Wang C (2017) Divergent LysM effectors contribute to the virulence of *Beauveria bassiana* by evasion of insect immune defenses. PLoS Pathog 13(9):e1006604. https://doi.org/10.1371/journal.ppat.1006604

Chai LY, Netea MG, Vonk AG, Kullberg BJ (2009) Fungal strategies for overcoming host innate immune response. Med Mycol 47(3):227–236. https://doi.org/10.1080/13693780802209082

Charnley AK, Collins SA (2007) Entomopathogenic fungi and their role in pest control. In: Kubicek C, Druzhinina I (eds) Environmental and microbial relationships, The Mycota, vol 4. Springer, Berlin/Heidelberg, pp 159–187. https://doi.org/10.1007/978-3-540-71840-6_10

Chen XR, Hu QB, Yu XQ, Ren SX (2014) Effects of destruxins on free calcium and hydrogen ions in insect hemocytes. Insect Sci 21(1):31–38. https://doi.org/10.1111/1744-7917.12028

Cheung YP, Park S, Pagtalunan J, Maringer K (2022) The antiviral role of NF-κB-mediated immune responses and their antagonism by viruses in insects. J Gen Virol 103(5):001741. https://doi.org/10.1099/jgv.0.001741

Chitarra GS, Abee T, Rombouts FM, Posthumus MA, Dijksterhuis J (2004) Germination of *Penicillium paneum* conidia is regulated by 1-octen-3-ol, a volatile self-inhibitor. Appl Environ Microbiol 70(5):2823–2829. https://doi.org/10.1128/aem.70.5.2823-2829.2004

Cho EM, Kirkland BH, Holder DJ, Keyhani NO (2007) Phage display cDNA cloning and expression analysis of hydrophobins from the entomopathogenic fungus *Beauveria* (*Cordyceps*) *bassiana*. Microbiology 153(10):3438–3447. https://doi.org/10.1099/mic.0.2007/008532-0

Cui C, Wang Y, Liu J, Zhao J, Sun P et al (2019) A fungal pathogen deploys a small silencing RNA that attenuates mosquito immunity and facilitates infection. Nat Commun 10(1):4298. https://doi.org/10.1038/s41467-019-12323-1

Cui C, Wang Y, Li Y, Sun P, Jiang J et al (2022) Expression of mosquito miRNAs in entomopathogenic fungus induces pathogen-mediated host RNA interference and increases fungal efficacy. Cell Rep 41(4):111527. https://doi.org/10.1016/j.celrep.2022.111527

Da Silva P, Jouvensal L, Lamberty M, Bulet P, Caille A et al (2003) Solution structure of termicin, an antimicrobial peptide from the termite *Pseudacanthotermes spiniger*. Protein Sci 12(3): 438–446. https://doi.org/10.1110/ps.0228303

El Jaddaoui I, Rangel DE, Bennett JW (2023) Fungal volatiles have physiological properties. Fungal Biol 127(7–8):1231–1240. https://doi.org/10.1016/j.funbio.2023.03.005

Fäldt J, Jonsell M, Nordlander G, Borg-Karlson AK (1999) Volatiles of bracket fungi *Fomitopsis pinicola* and *Fomes fomentarius* and their functions as insect attractants. J Chem Ecol 25:567–590. https://doi.org/10.1023/A:1020958005023

Fan Y, Liu X, Keyhani NO, Tang G, Pei Y et al (2017) Regulatory cascade and biological activity of *Beauveria bassiana* oosporein that limits bacterial growth after host death. Proc Natl Acad Sci USA 114(9):E1578–E1586. https://doi.org/10.1073/pnas.1616543114

Fang W, Leng B, Xiao Y, Jin K, Ma J et al (2005) Cloning of *Beauveria bassiana* chitinase gene Bbchit1 and its application to improve fungal strain virulence. Appl Environ Microbiol 71(1): 363–370. https://doi.org/10.1128/aem.71.1.36370.2005

Feng P, Shang Y, Cen K, Wang CS (2015) Fungal biosynthesis of the bibenzoquinone oosporein to evade insect immunity. Proc Natl Acad Sci USA 112(36):11365–11370. https://doi.org/10.1073/pnas.1503200112

George J, Jenkins NE, Blanford S, Thomas MB, Baker TC (2013) Malaria mosquitoes attracted by fatal fungus. PLoS One 8(5):e62632. https://doi.org/10.1371/journal.pone.0062632

Govind S (2008) Innate immunity in *Drosophila*: pathogens and pathways. Insect Sci 15(1):29–43. https://doi.org/10.1111/j.1744-7917.2008.00185.x

Han P, Gong Q, Fan J, Abbas M, Chen D et al (2022) Destruxin A inhibits scavenger receptor B mediated melanization in *Aphis citricola*. Pest Manag Sci 78(5):1915–1924. https://doi.org/10.1002/ps.6809

Hassan A, Mehmood N, Xu H, Muhammad Usman H, Wu J et al (2022) Phosphofructokinase downregulation disturbed termite social behaviors and immunity against fungal infections. Entomol Gen 42(5):799–808. https://doi.org/10.1127/entomologia/2022/1444

Hong S, Sun Y, Chen H, Zhao P, Wang C (2023) Fungus–insect interactions beyond bilateral regimes: the importance and strategy to outcompete host ectomicrobiomes by fungal parasites. Curr Opin Microbiol 74:102336. https://doi.org/10.1016/j.mib.2023.102336

Huang J, Lou Y, Liu J, Bulet P, Cai C et al (2023) A toll pathway effector protects *Drosophila* specifically from distinct toxins secreted by a fungus or a bacterium. Proc Natl Acad Sci USA 120(12):e2205140120. https://doi.org/10.1073/pnas.2205140120

Hummadi EH, Cetin Y, Demirbek M, Kardar NM, Khan S et al (2022) Antimicrobial volatiles of the insect pathogen *Metarhizium brunneum*. J Fungi 8(4):326. https://doi.org/10.3390/jof8040326

Jankielsohn A (2018) The importance of insects in agricultural ecosystems. Adv Entomol 6(2): 62–73. https://doi.org/10.4236/ae.2018.62006

Khaleel A, Nawaz M, Hindawi B (2013) Sol-gel derived Cr (III) and Cu (II)/γ-Al2O3 doped solids: effect of the dopant precursor nature on the structural, textural and morphological properties. Mater Res Bull 48(4):1709–1715. https://doi.org/10.1016/j.micromeso.2009.02.008

Kim JJ, Roberts DW (2012) The relationship between conidial dose, moulting and insect developmental stage on the susceptibility of cotton aphid, *Aphis gossypii*, to conidia of *Lecanicillium attenuatum*, an entomopathogenic fungus. Biocontrol Sci Technol 22(3):319–331. https://doi.org/10.1080/09583157.2012.656580

Kiuchi M, Yasui H, Hayasaka S, Kamimura M (2003) Entomogenous fungus *Nomuraea rileyi* inhibits host insect molting by C22-oxidizing inactivation of hemolymph ecdysteroids. Arch Insect Biochem Physiol 52(1):35–44. https://doi.org/10.1002/arch.10060

Kurtti TJ, Keyhani NO (2008) Intracellular infection of tick cell lines by the entomopathogenic fungus *Metarhizium anisopliae*. Microbiology 154:1700–1709. https://doi.org/10.1099/mic.0.2008/016667-0

Lazzaro BP, Little TJ (2009) Immunity in a variable world. Philos Trans R Soc Lond B Biol Sci 364(1513):15–26. https://doi.org/10.1098/rstb.2008.0141

Lewis MW, Robalino IV, Keyhani NO (2009) Uptake of the fluorescent probe FM4-64 by hyphae and haemolymph-derived in vivo hyphal bodies of the entomopathogenic fungus *Beauveria bassiana*. Microbiology 155(9):3110–3120. https://doi.org/10.1099/mic.0.029165-0

Li J, Xia Y (2022) Host–pathogen interactions between *Metarhizium* spp. and locusts. J Fungi 8(6): 602. https://doi.org/10.3390/jof8060602

Li L, Zhong K, Wang JL, Liu XS (2021) Mechanism of *Metarhizium rileyi* evading cellular immune responses in *Helicoverpa armigera*. Arch Insect Biochem Physiol 106(3):e21769. https://doi.org/10.1002/arch.21769

Li S, Wang J, Tian X, Toufeeq S, Huang W (2023) Immunometabolic regulation during the presence of microorganisms and parasitoids in insects. Front Immunol 14:905467. https://doi.org/10.3389/fimmu.2023.905467

Liu D, Smagghe G, Liu TX (2023) Interactions between entomopathogenic fungi and insects and prospects with glycans. J Fungi 9(5):575. https://doi.org/10.3390/jof9050575

Lord JC, Howard RW (2004) A proposed role for the cuticular fatty amides of *Liposcelis bostrychophila* (Psocoptera: Liposcelidae) in preventing adhesion of entomopathogenic fungi with dry-conidia. Mycopathologia 158:211–217. https://doi.org/10.1023/b:myco.0000041837.29478.78

Lovett B, St Leger RJ (2017) The insect pathogens. Microbiol Spectr 5(2):FUNK-0001-2016. https://doi.org/10.1128/microbiolspec.funk-0001-2016

Lu MY, Chtarbanova S (2022) The role of micro RNAs (miRNAs) in the regulation of *Drosophila melanogaster's* innate immunity. Fly 16(1):382–396. https://doi.org/10.1080/19336934.2022.2149204

Lu HL, St Leger RJ (2016) Insect immunity to entomopathogenic fungi. Adv Genet 94:251–285. https://doi.org/10.1016/bs.adgen.2015.11.002

Lu HL, Wang JB, Brown MA, Euerle C, St Leger RJ (2015) Identification of *Drosophila* mutants affecting defense to an entomopathogenic fungus. Sci Rep 5:12350. https://doi.org/10.1038/srep12350

Ma M, Luo J, Li C, Eleftherianos I, Zhang W et al (2024) A life-and-death struggle: interaction of insects with entomopathogenic fungi across various infection stages. Front Immunol 14:1329843. https://doi.org/10.3389/fimmu.2023.1329843

Mahanta DK, Bhoi TK, Komal J, Samal I, Nikhil RM et al (2023) Insect-pathogen crosstalk and the cellular-molecular mechanisms of insect immunity: uncovering the underlying signaling pathways and immune regulatory function of non-coding RNAs. Front Immunol 14:1169152. https://doi.org/10.3389/fimmu.2023.1169152

Mantzoukas S, Kitsiou F, Natsiopoulos D, Eliopoulos PA (2022) Entomopathogenic fungi: interactions and applications. Encyclopedia 2(2):646–656. https://doi.org/10.3390/encyclopedia2020044

Mascarin GM, Golo PS, de Souza Ribeiro-Silva C, Muniz ER, de Oliveira Franco A et al (2024) Advances in submerged liquid fermentation and formulation of entomopathogenic fungi. Appl Microbiol Biotechnol 108(1):451. https://doi.org/10.1007/s00253-024-13287-z

Myles TG (2002) Alarm, aggregation, and defense by *Reticulitermes flavipes* in response to a naturally occurring isolate of *Metarhizium anisopliae*. Sociobiology 40(2):243–256

Narava R, Jaba J, DV, SRK (2020) Cellular and molecular mechanisms of insect immunity. In: Ghoneim K (ed) Advances in agricultural entomology. AkiNik Publications, New Delhi, pp 89–118

Ormond EL, Thomas AP, Pell JK, Freeman SN, Roy HE (2011) Avoidance of a generalist entomopathogenic fungus by the ladybird, *Coccinella septempunctata*. FEMS Microbiol Ecol 77(2):229–237. https://doi.org/10.1111/j.1574-6941.2011.01100.x

Ortiz-Urquiza A, Keyhani NO (2013) Action on the surface: entomopathogenic fungi versus the insect cuticle. Insects 4(3):357–374. https://doi.org/10.3390/insects4030357

Pal S, St Leger RJ, Wu LP (2007) Fungal peptide Destruxin A plays a specific role in suppressing the innate immune response in *Drosophila melanogaster*. J Biol Chem 282:8969–8977. https://doi.org/10.1074/jbc.m605927200

Prisnyi AA, Grebtsova EA (2022) Morphological and functional status of hemocytes in species of the order Blattodea. AIP Conf Proc 2467:070015. https://doi.org/10.1063/5.0092439

Qu S, Wang S (2018) Interaction of entomopathogenic fungi with the host immune system. Dev Comp Immunol 83:96–103. https://doi.org/10.1016/j.dci.2018.01.010

Reber A, Purcell J, Buechel SD, Buri P, Chapuisat M (2011) The expression and impact of antifungal grooming in ants. J Evol Biol 24(5):954–964. https://doi.org/10.1111/j.1420-9101.2011.02230.x

Rosales C, Vonnie S (2017) Cellular and molecular mechanisms of insect immunity. Insect Physiol Ecol 10(3):182–190. https://doi.org/10.5772/67107

Sánchez-Pérez L, Barranco-Florido J, Rodríguez-Navarro S, Cervantes-Mayagoitia J, Ramos-López M (2014) Enzymes of entomopathogenic fungi, advances and insights. Adv Enzyme Res 2:65–76. https://doi.org/10.4236/aer.2014.22007

Shang Y, Xiao G, Zheng P, Cen K, Zhan S et al (2016) Divergent and convergent evolution of fungal pathogenicity. Genome Biol Evol 8(5):1374–1387. https://doi.org/10.1093/gbe/evw082

Staples JA, Milner RJ (2000) A laboratory evaluation of the repellency of *Metarhizium anisopliae* conidia to *Coptotermes lacteus* (Isoptera: Rhinotermitidae). Sociobiology 36(1):133–148

Stokes BA, Yadav S, Shokal U, Smith LC, Eleftherianos I (2015) Bacterial and fungal pattern recognition receptors in homologous innate signaling pathways of insects and mammals. Front Microbiol 6:19. https://doi.org/10.3389/fmicb.2015.00019

Störtkuhl KF, Kettler R, Fischer S, Hovemann BT (2005) An increased receptive field of olfactory receptor Or43a in the antennal lobe of *Drosophila reduces* benzaldehyde driven avoidance behavior. Chem Senses 30(1):81–87. https://doi.org/10.1093/chemse/bji003

Thompson SR, Brandenburg RL (2005) Tunneling responses of mole crickets (Orthoptera: Gryllotalpidae) to the entomopathogenic fungus, *Beauveria bassiana*. Environ Entomol 34(1): 140–147. https://doi.org/10.1603/0046-225X-34.1.140

Tudi M, Daniel Ruan H, Wang L, Lyu J, Sadler R et al (2021) Agriculture development, pesticide application and its impact on the environment. Int J Environ Res Public Health 18(3):1112. https://doi.org/10.3390/ijerph18031112

Ulrich KR, Feldlaufer MF, Kramer M, St Leger RJ (2015) Inhibition of the entomopathogenic fungus *Metarhizium anisopliae* sensu lato in vitro by the bed bug defensive secretions (E)-2-hexenal and (E)-2-octenal. BioControl 60:517–526. https://doi.org/10.1007/s10526-015-9667-2

Vandenberg JD, Ramos M, Altre JA (1998) Dose–response and age-and temperature-related susceptibility of the diamondback moth (Lepidoptera: Plutellidae) to two isolates of *Beauveria bassiana* (Hyphomycetes: Moniliaceae). Environ Entomol 27(4):1017–1021

Villani MG, Krueger SR, Schroeder PC, Consolie F, Consolie NH et al (1994) Soil application effects of *Metarhizium anisopliae* on Japanese beetle (Coleoptera: Scarabaeidae) behavior and survival in turfgrass microcosms. Environ Entomol 23(2):502–513. https://doi.org/10.1093/ee/23.2.502

Wanchoo A, Lewis MW, Keyhani NO (2009) Lectin mapping reveals stage-specific display of surface carbohydrates in in vitro and haemolymph-derived cells of the entomopathogenic fungus *Beauveria bassiana*. Microbiology 155(9):3121–3133. https://doi.org/10.1099/mic.0.029157-0

Wang C, St Leger RJ (2007a) The MAD1 adhesin of *Metarhizium anisopliae* links adhesion with blastospore production and virulence to insects, and the MAD2 adhesin enables attachment to plants. Eukaryot Cell 6(5):808–816. https://doi.org/10.1128/ec.00409-06

Wang C, St Leger RJ (2007b) The *Metarhizium anisopliae* perilipin homolog MPL1 regulates lipid metabolism, appressorial turgor pressure, and virulence. J Biol Chem 282(29):21110–21115. https://doi.org/10.1074/jbc.m609592200

Wang B, Kang Q, Lu Y, Bai L, Wang C (2012) Unveiling the biosynthetic puzzle of destruxins in *Metarhizium* species. Proc Natl Acad Sci USA 109(4):1287–1292. https://doi.org/10.1073/pnas.1115983109

Wang Y, Cui C, Wang G, Li Y, Wang S (2021) Insects defend against fungal infection by employing microRNAs to silence virulence-related genes. Proc Natl Acad Sci USA 118(19): e2023802118. https://doi.org/10.1073/pnas.2023802118

Wang J, Weng Q, Zhang K, Hu Q (2023) Binding proteins of destruxin a from *Metarhizium* against insect cell. BMC Microbiol 23(1):96. https://doi.org/10.1186/s12866-023-02843-8

Xu Y, Zafar J, Lin L, Wu H, Kang Z et al (2025) MicroRNA-mediated modulation of immune genes facilitates *Metarhizium anisopliae* infection in the red imported fire ant, *Solenopsis invicta*. J Integr Agric 25(1):192–206. https://doi.org/10.1016/j.jia.2025.02.009

Yanagawa A, Imai T, Akino T, Toh Y, Yoshimura T (2015) Olfactory cues from pathogenic fungus affect the direction of motion of termites, *Coptotermes formosanus*. J Chem Ecol 41(12): 1118–1126. https://doi.org/10.1007/s10886-015-0649-8

Yassine H, Kamareddine L, Osta MA (2012) The mosquito melanization response is implicated in defense against the entomopathogenic fungus *Beauveria bassiana*. PLoS Pathog 8(11): e1003029. https://doi.org/10.1371/journal.ppat.1003029

Zdybicka-Barabas A, Stączek S, Mak P, Skrzypiec K, Mendyk E, Cytryńska M (2013) Synergistic action of Galleria mellonella apolipophorin III and lysozyme against Gram- negative bacteria. Biochimica et Biophysica Acta (BBA)- Biomembranes 1828(6):1449–1456. https://doi.org/10.1016/j.bbamem.2013.02.004

Zhang Y, Zhang J, Jiang X, Wang G, Luo Z et al (2010) Requirement of a mitogen-activated protein kinase for appressorium formation and penetration of insect cuticle by the entomopathogenic fungus *Beauveria bassiana*. Appl Environ Microbiol 76(7):2262–2270. https://doi.org/10.1128/aem.02246-09

Zhang S, Xia YX, Kim B, Keyhani NO (2011) Two hydrophobins are involved in fungal spore coat rodlet layer assembly and each play distinct roles in surface interactions, development and pathogenesis in the entomopathogenic fungus, *Beauveria bassiana*. Mol Microbiol 80(3): 811–826. https://doi.org/10.1111/j.1365-2958.2011.07613.x

Zhang D, Qi H, Zhang F (2025) Parasitism by entomopathogenic fungi and insect host defense strategies. Microorganisms 13(2):283. https://doi.org/10.3390/microorganisms13020283

Part II

Bioactive Metabolites

Entomopathogenic Fungi and Their Bioactive Compounds

6

Lutfun Nahar, Emran Habibi, Ondrej Novak, and Satyajit D. Sarker

Abstract

Entomopathogenic fungi (EPF), including species such as *Beauveria bassiana* and *Metarhizium anisopliae*, are effective biological control agents against a wide range of insect pests. This chapter explores the infection mechanisms of EPF and the bioactive compounds they produce, such as destruxins and beauvericin, which exhibit insecticidal and antimicrobial properties. EPF also enhance plant nutrient uptake and stress tolerance, contributing to sustainable agriculture. Recent advances in formulation technologies, field applications, and plant-microbe interactions are discussed. The integration of EPF into pest management strategies supports biodiversity and food security, highlighting their role in environmentally friendly crop protection.

L. Nahar (✉) · S. D. Sarker
Centre for Natural Products Discovery, School of Pharmacy and Biomolecular Sciences, Liverpool John Moores University, Liverpool, UK
e-mail: L.Nahar@ljmu.ac.uk; profnahar@outllook.com

E. Habibi
Centre for Natural Products Discovery, School of Pharmacy and Biomolecular Sciences, Liverpool John Moores University, Liverpool, UK

Medicinal Plants Research Centre, Institute of Herbal Medicines and Metabolic Disorders, Mazandaran University of Medical Sciences, Sari, Iran

O. Novak
Laboratory of Growth Regulators, Palacký University and Institute of Experimental Botany, The Czech Academy of Sciences, Olomouc, Czech Republic

S. K. Deshmukh et al. (eds.), *Frontiers in Entomopathogenic Fungi*,
https://doi.org/10.1007/978-981-95-6529-0_6

Keywords

Agriculture · Antimicrobials · Biodiversity · Compounds · Fungi · Management · Metabolites · Pest · Sustainability · Toxins

1 Introduction

The increasing demand for sustainable pest control arises from the growing awareness of the risks associated with synthetic pesticides. These risks include adverse effects on human health, loss of biodiversity, and the development of pesticide-resistant pest populations (Sharma and Sharma 2021; Zhou et al. 2024). The search for effective alternatives has led to the exploration of entomopathogenic fungi (EPF) as promising biological control agents. Species such as *Beauveria*, *Metarhizium*, *Entomophaga*, and *Isaria* have demonstrated strong potential in replacing chemical pesticides (Bamisile et al. 2021a; Sharma et al. 2023). These fungi exhibit insecticidal activity and interact with the environment in ways that support agroecosystem health. Insect pests are responsible for an estimated 18%–26% loss in global crop production, which highlights the need for reliable and eco-friendly pest management solutions (Deka et al. 2021). The use of EPF offers both pest suppression and plant health benefits (Vega et al. 2009; Qin et al. 2023), although their full potential, particularly about their bioactive compounds, remains underexplored.

The production of bioactive compounds plays a central role in the effectiveness of EPF. These compounds include nonribosomal peptides, polyketides, terpenoids, and other chemical classes that contribute to insecticidal activity and plant-beneficial effects (Zhang et al. 2020; Raghunandan et al. 2024). Research has traditionally focused on insecticidal properties, but recent findings suggest broader roles in enhancing plant resilience. Many EPF colonize plant tissues as endophytes, where they improve tolerance to drought and salinity and suppress plant pathogens (Mantzoukas et al. 2022; Sharma et al. 2023). The discovery of novel compounds continues to expand the range of potential applications (San Juan-Maldonado et al. 2024), although many biosynthetic pathways remain poorly understood.

The roles of individual compounds have been clarified through gene knockout studies. Some compounds are essential for virulence, while others contribute in minor or non-essential ways (Butt et al. 2016; Qin et al. 2023). Destruxins in *Metarhizium anisopliae* and beauvericin in *Beauveria bassiana* enhance insecticidal activity but are not always required for full virulence. In contrast, bassianolide in *B. bassiana* functions as a key virulence factor (Wang and Xu 2012; Patocka 2016). These compounds display a range of bioactivities, including insecticidal, antimicrobial, and cytotoxic effects (Isaka et al. 2005; Zhang et al. 2020; Sharma et al. 2023). Beauvericin exhibits strong insecticidal and antimicrobial properties (Wang and Xu 2012), while oosporein, isolated from *B. bassiana*, influences insect immunity and provides antimicrobial protection (Weng et al. 2019; Rodríguez-Romero et al. 2023).

Other species, such as *Isaria fumosorosea* and *I. farinosa*, produce oxalic acid, which contributes to virulence by acidifying insect tissues (Patocka 2016; Raghunandan et al. 2024).

This chapter provides an overview of the biodiversity and functional potential of EPF bioactive compounds. The discussion focuses on their mechanisms of action, their contributions to pest control and plant health, and their role in advancing sustainable agricultural practices (Bamisile et al. 2021a; Chen et al. 2021). The challenges related to production, formulation, and biosafety must be addressed to fully realize the potential of these biocontrol agents (Mascarin et al. 2024). Further research in biotechnology and genomics is essential to unlocking the transformative potential of these fungi.

2 Biological Characteristics and Mechanisms

EPF serve as an effective biological control agent, playing an essential role in sustainable agriculture. These fungi exhibit specialized adaptations that enable them to infect and kill a wide range of insect pests successfully (Deka et al. 2021; Sharma and Sharma 2021).

2.1 Biological Characteristics

EPF possess several specialized adaptations for the effective infection of insect pests. The fungi primarily infect insects through direct contact (Sharma et al. 2023). They develop specialized structures known as appressoria, which facilitate secure attachment to the insect cuticle. After attachment, fungal spores germinate and penetrate the cuticle. This process allows the fungi to enter the hemocoel of the insect (Qin et al. 2023). Following penetration, rapid fungal growth occurs. This growth leads to tissue degradation and ultimately causes insect mortality. The EPF infection process is complex, involving multiple stages crucial for successful infection (Ma et al. 2024; Vivekanandhan et al. 2024). The key infection process of EPF is a multi-stage attack, illustrated in Table 6.1 and Fig. 6.1. A thorough understanding of each stage is vital for developing effective biocontrol strategies.

2.1.1 Spore Germination

The infection process begins with either the ingestion or attachment of fungal spores to the insect cuticle. Various environmental factors, such as humidity and temperature, influence spore germination (Dillon and Charnley 1990; Kurniawan et al. 2022). After attachment, spores absorb moisture and nutrients, leading to germination and initiating the infection cycle. Spores can remain viable for extended periods before germinating (Dillon and Charnley 1990; Afifah and Saputro 2020). The capacity to remain dormant yet viable in the environment is key to the effectiveness of these fungi as biocontrol agents (Afifah and Saputro 2020; Kurniawan et al. 2022).

Table 6.1 Key infection process of entomopathogenic fungi

Feature	Description	Reference
Spore germination	Spores attach or are ingested; germination is triggered by moisture and temperature	Dillon and Charnley (1990), Afifah and Saputro (2020), Kurniawan et al. (2022)
Cuticle penetration	Appressoria and enzymes (chitinases, proteases) break the insect cuticle	Charnley and St Leger (1991), Mondal et al. (2016), Valero-Jiménez et al. (2017), Ma et al. (2024)
Mycelial growth	Hyphae spread in the hemocoel; toxins disrupt host physiology and immunity	Mondal et al. (2016), Valero-Jiménez et al. (2017), Ma et al. (2024)
Adaptability	Spores survive harsh conditions; infection resumes when the environment improves	Bamisile et al. (2021b), Kryukov and Glupov (2023), Ahsan et al. (2024), Farhan et al. (2024), Gielen et al. (2024)

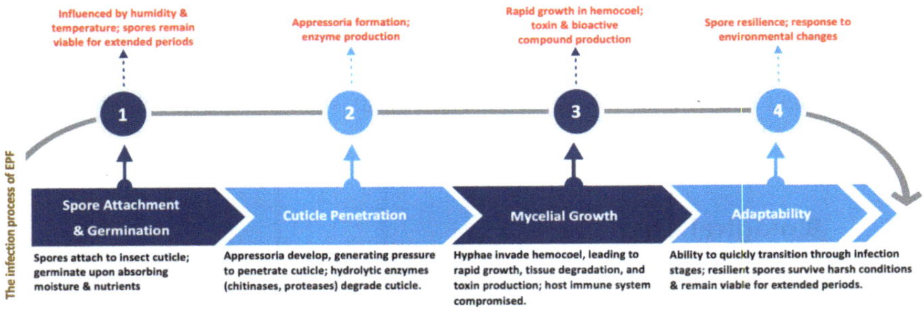

Fig. 6.1 The infection process of entomopathogenic fungi, from spore adhesion to host mortality

2.1.2 Cuticle Penetration

EPF develop specialized structures called appressoria, which generate significant turgor pressure to aid in penetrating the cuticle of the insect (Charnley and St Leger 1991; Ma et al. 2024). The penetration process is facilitated by the production of hydrolytic enzymes, such as chitinases and proteases, which degrade chitin and proteins in the insect exoskeleton. This degradation weakens the cuticle and creates a pathway for fungal hyphae to invade (Charnley and St Leger 1991; Mondal et al. 2016; Valero-Jiménez et al. 2017).

2.1.3 Mycelial Growth

Once the cuticle is breached, fungal hyphae invade the hemocoel, the insect circulatory fluid, providing access to nutrients and fostering rapid growth of the fungus (Ma et al. 2024). As the hyphae proliferate, vital physiological processes within the host become disrupted, leading to tissue degradation. The production of toxins and bioactive compounds during this growth phase compromises the immune system of the insect, ultimately resulting in insect mortality. The combination of mycelial growth and the effects of bioactive compounds exemplifies the powerful role of EPF in agricultural pest management (Mondal et al. 2016; Valero-Jiménez et al. 2017; Ma et al. 2024).

2.1.4 Adaptability

The ability of EPF to transition rapidly through various infection stages is crucial for its effectiveness as a biocontrol agent. This characteristic allows the fungi to respond quickly to changing environmental conditions (Kryukov and Glupov 2023). The production of resilient spores further enhances adaptability, enabling them to withstand harsh conditions, including extreme temperatures and limited moisture availability. Such resilience is vital for survival in diverse environments.

EPF spores often remain viable for extended periods, enduring unfavorable conditions until more suitable environments for infection occur (Ahsan et al. 2024; Farhan et al. 2024). When optimal conditions arise, the spores can germinate and begin the infection process. This ecological adaptability is essential for the success of EPF in sustainable pest management. The ability to remain viable while waiting for favorable conditions increases the chances of successful infections in target insect populations (Bamisile et al. 2021b; Gielen et al. 2024). This trait makes EPF an important resource for controlling pest populations in agricultural systems (Ahsan et al. 2024; Farhan et al. 2024).

2.2 Mechanisms of Action

The mechanisms of action of entomopathogenic fungi (EPF) involve several coordinated steps. The process begins with spore adhesion to the insect cuticle, followed by germination and penetration. Once inside the host, the fungi proliferate and produce enzymes and toxins that degrade host tissues (Mantzoukas et al. 2022). These bioactive compounds, including destruxins, beauvericin, and bassianolide, contribute significantly to fungal pathogenicity (Gibson et al. 2014; Sharma et al. 2023). For example, destruxins in *M. anisopliae* and beauvericin in *B. bassiana* enhance insecticidal activity, although they are not always essential for full virulence. In contrast, bassianolide in *B. bassiana* functions as a key virulence factor (Wang and Xu 2012; Gibson et al. 2014; Patocka 2016).

2.2.1 Multifaceted Infection Mechanisms of Action

EPF employ a complex infection process to ensure effective colonization and host mortality (Islam et al. 2021; Sharma et al. 2023). This process includes several critical stages: attachment, cuticle penetration, systemic infection, and toxin production. The multifaceted infection process of EPF is summarized in Table 6.2 and Fig. 6.2. Each stage is intricately linked and plays a crucial role in the success of fungal pathogens in overcoming host defenses and establishing infection.

Attachment Fungal spores adhere to the insect cuticle using specialized proteins and hydrophobic adhesive agents. This attachment serves as the foundation for subsequent infection steps (Ortiz-Urquiza and Keyhani 2013).

Cuticle Penetration After attachment, the spores germinate and form specialized structures called appressoria, which generate significant turgor pressure to penetrate

Table 6.2 Summary of infection mechanisms of entomopathogenic fungi

Stage	Description	Reference
Attachment	Spores adhere to the insect cuticle via proteins and adhesives	Ortiz-Urquiza and Keyhani (2013)
Penetration	Appressoria and enzymes breach the cuticle	Charnley and St Leger (1991), Mondal et al. (2016), Ma et al. (2024)
Systemic infection	Hyphae spread in the hemocoel, disrupting metabolism	Ma et al. (2024)
Mortality	The host dies; fungi sporulate and release conidia	Sharma et al. (2023)
Enzymes	Chitinases, proteases, and lipases degrade tissues	Charnley and St Leger (1991), Mondal et al. (2016)
Toxins	Destruxins and beauvericin disrupt cell function	Islam et al. (2021)

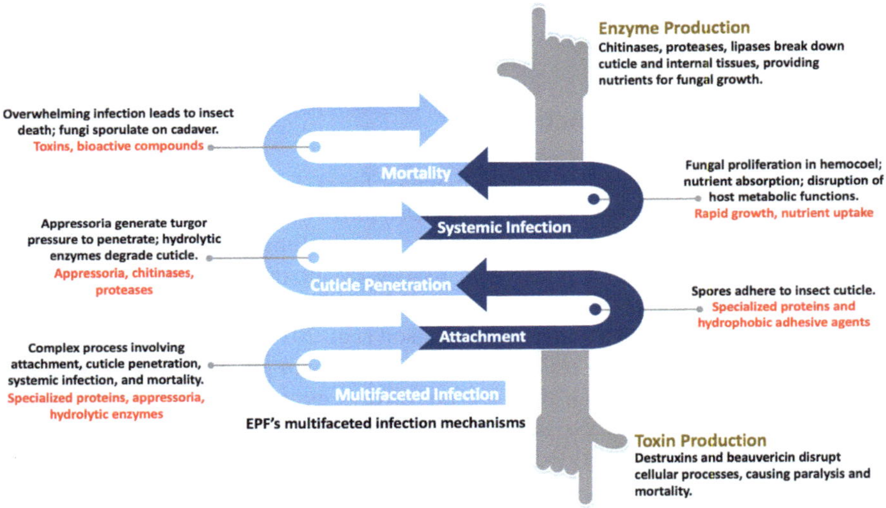

Fig. 6.2 The multifaceted infection mechanisms of action of entomopathogenic fungi

the insect cuticle (Charnley and St Leger 1991; Ma et al. 2024). Hydrolytic enzymes degrade structural components of the insect exoskeleton (Mondal et al. 2016).

Systemic Infection Once the cuticle is breached, EPF proliferate in the hemocoel, allowing nutrient absorption. This process disrupts crucial metabolic functions, causing tissue damage and systemic infection (Ma et al. 2024).

Mortality The overwhelming impact of the fungal infection results in mortality. Following host death, fungi sporulate on the cadaver, releasing conidia to initiate new infections (Sharma et al. 2023).

Role of Enzymes The pathogenicity of EPF largely relies on various enzymes, toxins, and bioactive compounds. These elements facilitate successful infection and enhance virulence (Charnley and St Leger 1991; Mondal et al. 2016). Enzymes such as chitinases, proteases, and lipases are critical for breaking down the insect cuticle. Once inside, EPF can produce additional enzymes that degrade internal tissues. This degradation provides essential nutrients for continued fungal growth (Charnley and St Leger 1991; Mondal et al. 2016).

Toxin Production The synthesis of toxic compounds by EPF negatively affects the physiological functions of insect hosts. Bioactive compounds such as destruxins (Table 6.1 and Fig. 6.5) and beauvericin (Table 6.1 and Fig. 6.5) disrupt critical cellular processes. This disruption leads to paralysis and eventual mortality of targeted pests (Islam et al. 2021).

2.3 Prominent Genera and Species

EPF utilize complex mechanisms of action that are essential for their effectiveness as biocontrol agents in agriculture (Bamisile et al. 2021a; Chen et al. 2021). The unique mechanisms through which EPF interact with insect hosts and their environments contribute significantly to their reputation as promising solutions for sustainable agricultural methods. By developing sophisticated infection systems that specifically target insect hosts, these fungi can effectively deploy bioactive compounds. These compounds not only enhance pest control efforts but also support ecosystem health (Ahsan et al. 2024; Vivekanandhan et al. 2024).

A summary of notable species, their biological control mechanisms, bioactive properties, and target insect pests is presented in Table 6.3 and Fig. 6.3. A detailed understanding of these biological characteristics and modes of interaction provides valuable insights for developing innovative pest management strategies. These strategies utilize the natural capabilities of EPF. The ongoing exploration of entomopathogenic fungi, particularly concerning their bioactive compounds and ecological interactions, remains critical for enhancing agricultural productivity and promoting environmental health (Bamisile et al. 2021b; Chen et al. 2021).

2.4 Mechanisms of Insecticidal Action

The insecticidal properties of EPF are characterized by several key mechanisms, including neurotoxicity, metabolic disruption, and immune suppression (Mantzoukas et al. 2022; Ma et al. 2024). The understanding of these mechanisms is crucial for developing effective biocontrol strategies that utilize the full potential

Table 6.3 Summary of infection mechanisms of entomopathogenic fungi

Aspect	Description	Reference
Prominent genera	*Beauveria*, *Metarhizium*, *Isaria*, *Entomophaga* used in pest control	Bamisile et al. (2021a), Chen et al. (2021)
Bioactive role	Compounds enhance pest control and ecosystem health	Ahsan et al. (2024), Vivekanandhan et al. (2024)
Neurotoxicity	Toxins disrupt nerve signals, causing paralysis or death	Gibson et al. (2014), Han et al. (2022), Ma et al. (2024)
Metabolic disruption	Toxins impair energy production, leading to mortality	Mannino et al. (2021), Sun et al. (2023)
Immune suppression	Compounds suppress insect immunity, aiding colonization	Lu and Leger (2016), Qu and Wang (2018)

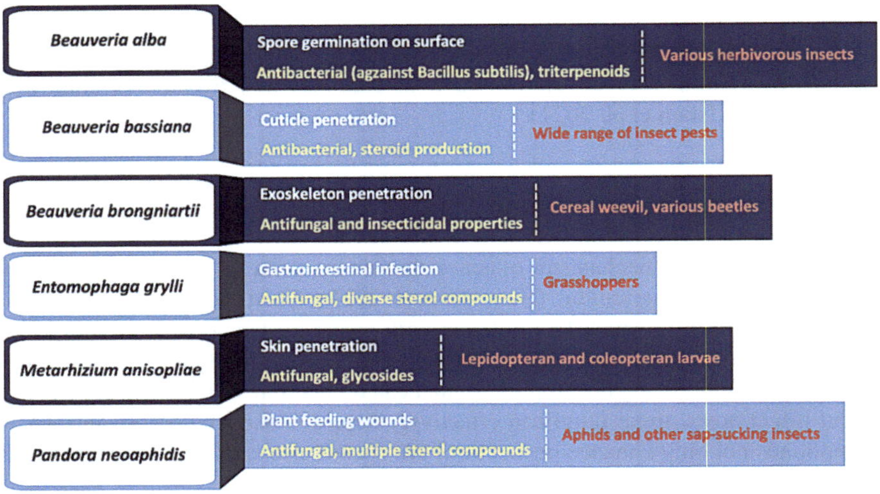

Fig. 6.3 Notable species, biological control mechanisms, bioactive properties, and target insect pests of entomopathogenic fungi

of EPF in pest management. An overview of these mechanisms is summarized in Table 6.4 and Fig. 6.4.

2.4.1 Neurotoxic Action

Certain bioactive compounds produced by EPF specifically target the nervous systems of insects. These compounds disrupt signal transmission, resulting in paralysis or death. Research indicates that destruxins (Table 6.5 and Fig. 6.5) interfere with neurotransmitter signaling, causing cytotoxic effects in target insects (Gibson et al. 2014; Han et al. 2022; Ma et al. 2024).

Table 6.4 Summary of insecticidal mechanisms of entomopathogenic fungi

Mechanism	Description	Reference
Neurotoxicity	Disrupts neural signaling; causes paralysis or death	Gibson et al. (2014), Han et al. (2022), Ma et al. (2024)
Metabolic disruption	Interferes with energy production; depletes reserves	Mannino et al. (2021), Sun et al. (2023)
Immune suppression	Inhibits host immunity; enhances colonization and virulence	Lu and Leger (2016), Qu and Wang (2018)

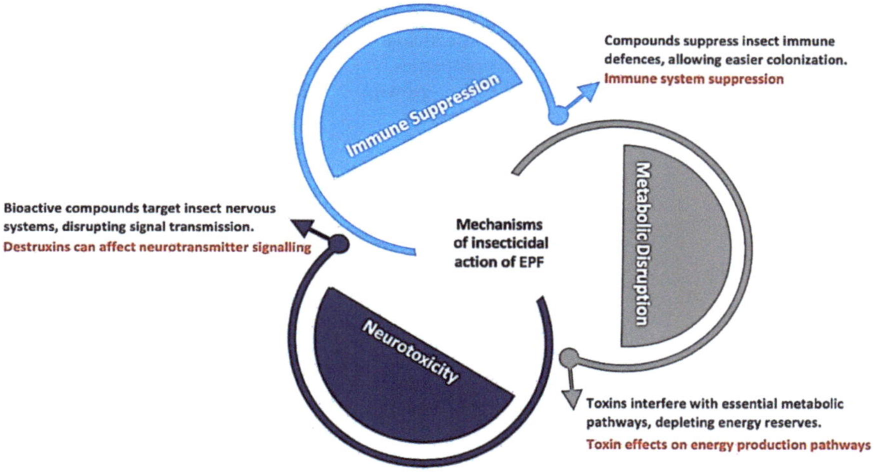

Fig. 6.4 Key insecticidal mechanisms of entomopathogenic fungi

2.4.2 Metabolic Disruption

Toxins produced by EPF can interfere with essential metabolic pathways within insects. The disruption of energy production processes can lead to the depletion of energy reserves, ultimately resulting in mortality (Mannino et al. 2021; Sun et al. 2023).

2.4.3 Immune Suppression

EPF produce specific compounds that suppress the immune defenses of insect hosts. This suppression facilitates colonization and nutrient extraction by the fungi. It enhances fungal virulence and prolongs the duration of infection within the host (Lu and Leger 2016; Qu and Wang 2018).

Table 6.5 Key bioactive compounds of entomopathogenic fungi

Bioactive compound	Fungal source	Function	Pest control	Reference
Aflatoxin	Various EPF species	Insecticidal properties, cytotoxic and genotoxic effects	Improves pest management and plant resilience	Zhang et al. (2020)
Avenacoside B	Various EPF species	Surface-active agent	Improves adhesion and distribution of fungal spores	Morcia et al. (2022)
Azadirachtin	Various EPF species	Insecticidal properties, plant growth regulator	Effective against various pests; IC_{50} of 60 μL/L air against *Rhyzopertha dominica*	Hernández et al. (2012), Otieno et al. (2016)
Bassianolide	*B. bassiana*	Interacts with ion channels at neuromuscular junctions	Increases insecticidal effectiveness	Patocka (2016), Zhang et al. (2020)
Beauvericin	*B. bassiana*, *Paecilomyces* spp.	Ionophore, disrupts ion gradients; IC_{50} of 1.9 μM for HIV-1 integrase, and ~10 μM for CHO-K1 cells	Causes rapid paralysis and death in insect larvae	Wang and Xu (2012), Gibson et al. (2014), Patocka (2016), Pedrini (2018), Weng et al. (2019), Zhang et al. (2020), Hong et al. (2024)
Caffeine	Various EPF species	Neurotoxic effects	Alter insect feeding behavior	Wink (2018), Zhang et al. (2020)
Capsaicin	Various EPF species	Alter the feeding behavior of herbivorous insects	Acts as a feeding deterrent	Patocka (2016), Raghunandan et al. (2024)
Destruxins (e.g., Destruxin A, B, E)	*M. anisopliae*	Inhibits DNA, RNA, and protein synthesis in insect cells; IC_{50} of 6.3 μg/larva in *Tenebrio molitor* and *Galleria mellonella*	Increases mortality among target insects	Fredenhagen et al. (2006), Gibson et al. (2014), Rodríguez-Romero et al. (2023), San Juan-Maldonado et al. (2024)

(continued)

Table 6.5 (continued)

Bioactive compound	Fungal source	Function	Pest control	Reference
Efrapeptin F	Various EPF species	Antibiotic peptides with insecticidal properties	Inhibits harmful microorganisms	Zhang et al. (2020), Rodríguez-Romero et al. (2023)
Farnesol	Various EPF species	Antimicrobial properties; MIC of 0.29–75.0 μM against *Cryptococcus* spp.	Inhibits certain plant pathogens	Patocka (2016), Raghunandan et al. (2024)
Griseofulvin	Various EPF species	Antifungal properties	Halts the growth of pathogenic fungi	Gibson and Krasnoff (1999), Isaka et al. (2005)
Helvolic acid	Various EPF species	Antimicrobial properties: MIC of 8.0 μg/mL for *H. pylori*, 15–30 μg/mL for *S. aureus*, and *C. albicans*	Contributes to the overall effectiveness of the fungus against pests	Zhang et al. (2020), Hong et al. (2024)
4-Hydroxy-methylazoxybenzoic acid	*E. virulenta*	Insecticidal effects	Targets specific pest populations	Farhan et al. (2024), Raghunandan et al. (2024)
Limonene	Various EPF species	Antimicrobial properties promote plant growth	Important in pest management	Farhan et al. (2024), San Juan-Maldonado et al. (2024)
Nicotine	Various EPF species	Neurotoxic effects	Functions as a natural insecticide	Wink (2018), Zhang et al. (2020), Farhan et al. (2024)
Nystatin	Various EPF species	Antifungal properties: MIC_{50} of 0.03–0.06 μg/mL for *C. albicans*, MIC_{90} of 0.06–0.23 μg/mL for *C. albicans*	Protection against plant infections	Patocka (2016), Paschapur et al. (2021), Raghunandan et al. (2024)

(continued)

Table 6.5 (continued)

Bioactive compound	Fungal source	Function	Pest control	Reference
Oosporein	*B. bassiana*	Exhibits insecticidal, antiviral, and antibacterial properties	Provides antagonistic effects	Pedrini (2018), Zhang et al. (2020), Paschapur et al. (2021), Hong et al. (2024), San Juan-Maldonado et al. (2024)
Oxalic acid	*Beauveria* spp., *M. anisopliae*	Compromises insect cuticle integrity	Promotes dehydration	Weng et al. (2019), Zhang et al. (2020), Rodríguez-Romero et al. (2023), San Juan-Maldonado et al. (2024)
2,6-Pyridinedicarboxylic acid	*Isaria* spp., *M. anisopliae*	Insecticidal properties	Enhances plant defenses	Patocka (2016), Rodríguez-Romero et al. (2023), Raghunandan et al. (2024), San Juan-Maldonado et al. (2024)
Zearalenone	Various EPF species	Hormonal effects	Affects reproduction and health	Isaka et al. (2005), San Juan-Maldonado et al. (2024)

3 Bioactive Compounds

EPF produce a diverse range of bioactive compounds crucial for their roles as biological control agents. These compounds not only kill insects and combat pathogens but also enhance plant health, highlighting their importance for sustainable agriculture (Vega et al. 2009; Zhang et al. 2020). By producing bioactive compounds, EPF gain competitive advantages in their natural environments, allowing them to interact effectively with pests and other environmental factors (Raghunandan et al. 2024). The ability to create beneficial compounds is integral to IPM strategies (Patocka 2016; Zhang et al. 2020).

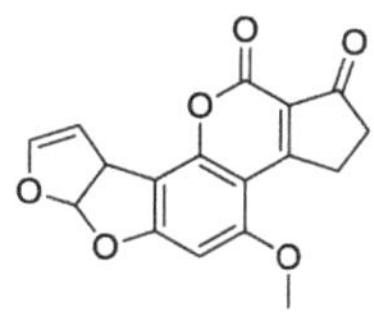

Aflatoxin

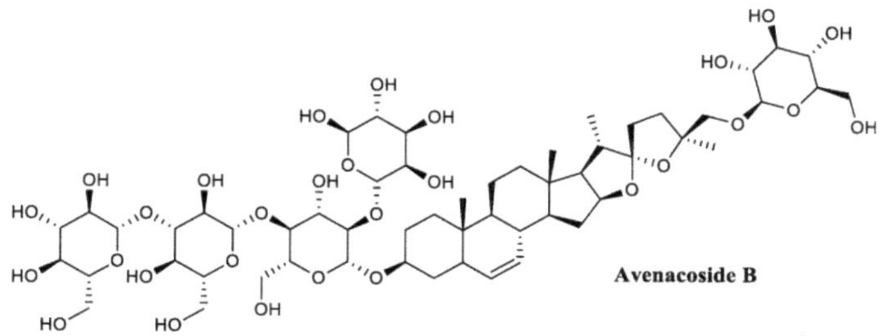

Avenacoside B

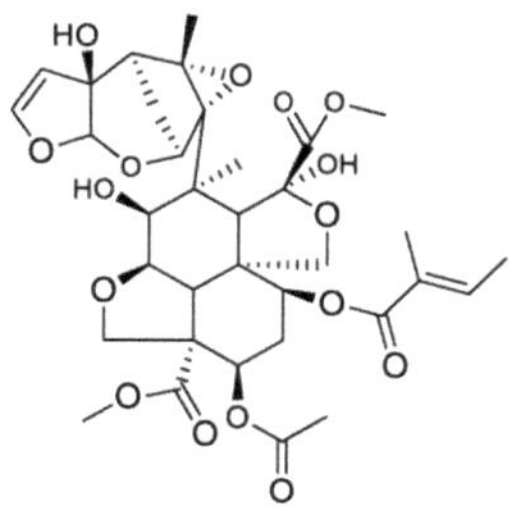

Azadirachtin

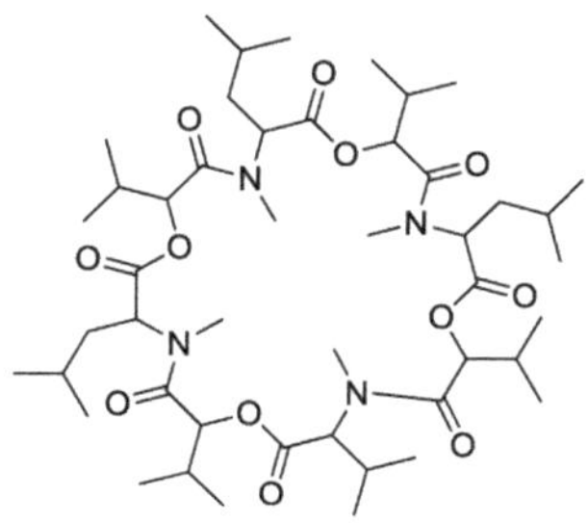

Bassianolide

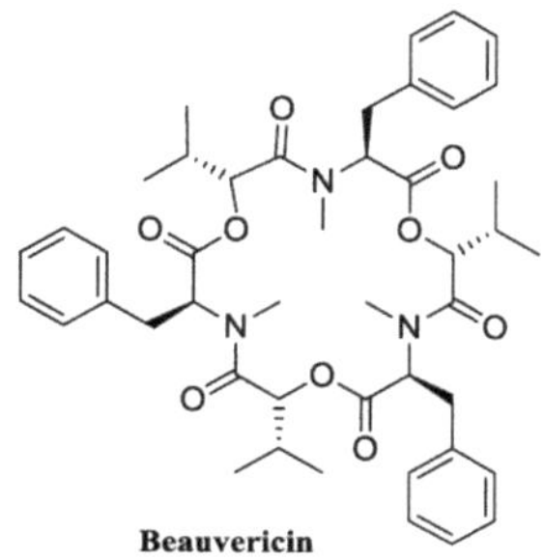

Beauvericin

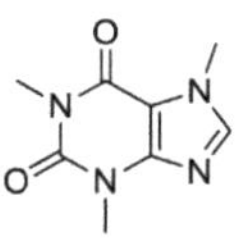

Caffeine

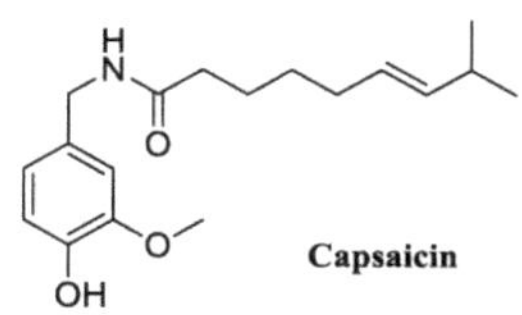

Capsaicin

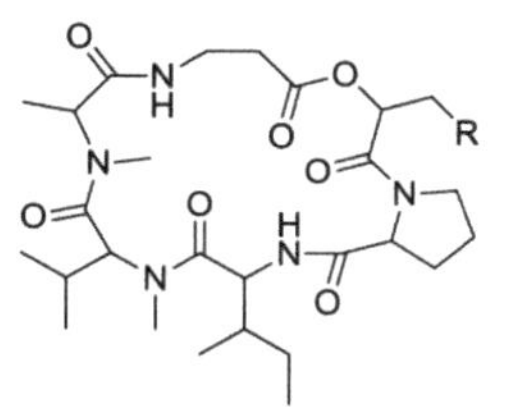

Destruxins:

A: R= HC=CH

B: R= HC-$(CH_3)2$

E: R= 

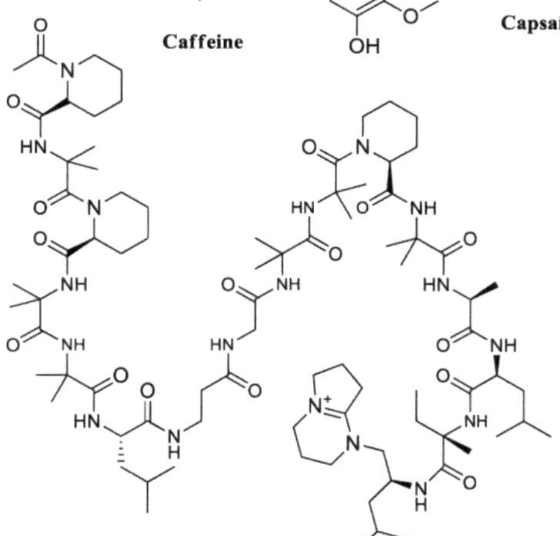

Efrapeptin F

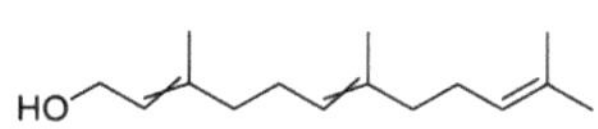

Farnesol

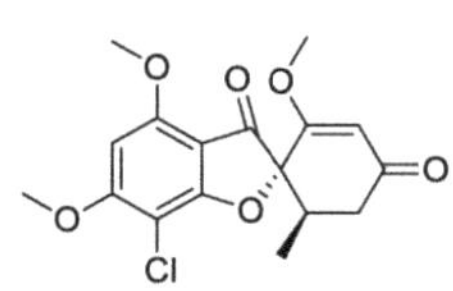

Griseofulvin

Fig. 6.5 Structures of some bioactive compounds produced by entomopathogenic fungi

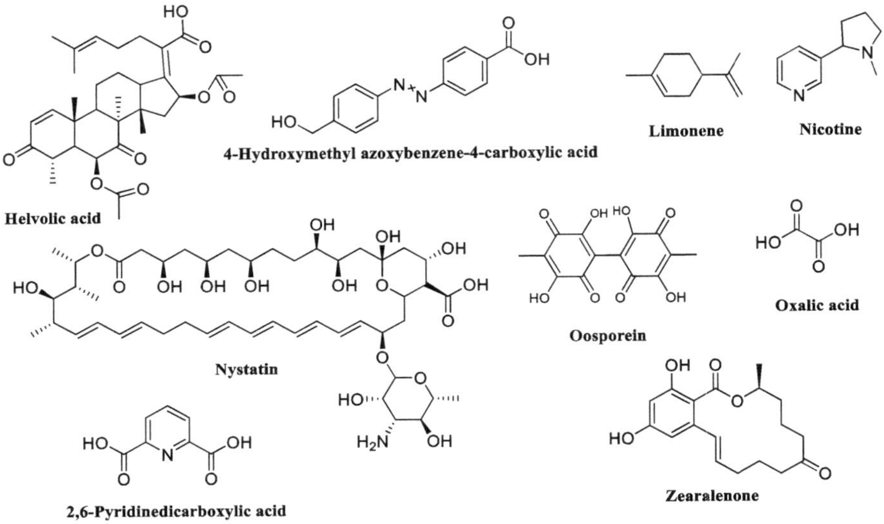

Fig. 6.5 (continued)

The mechanisms through which these bioactive compounds operate in pest control are essential for showcasing their potential as biocontrol agents. Many bioactive compounds exhibit harmful effects on pests and produce antibiotics that fight against bacterial and fungal infections (Patocka 2016; San Juan-Maldonado et al. 2024). In addition, bioactive compounds can modulate insect behavior, immune responses, and critical biological functions. A thorough examination of their roles allows researchers to enhance the efficacy of EPF in pest management, contributing to more resilient agricultural systems that reduce reliance on chemical pesticides and promote ecological balance. A summary of key bioactive compounds produced by EPF and their diverse chemical structures is presented in Table 6.5 and Fig. 6.5. Many EPF also produce antibiotics that protect themselves and the plants they infect from diseases (Isaka et al. 2005; San Juan-Maldonado et al. 2024).

Bioactive compounds from EPF work through different mechanisms that affect insect physiology (Vega et al. 2009; Zhang et al. 2020). These bioactive compounds (Fig. 6.5) can harm pests by causing significant physiological changes and can interact with plants to enhance their defensive abilities. They have been recognized in various studies as critical for the effectiveness of EPF against pests and pathogens, as described below. The ongoing research into these compounds and their interactions with insects and plants remains essential for developing innovative and sustainable pest management strategies that utilize the natural capabilities of EPF.

Aflatoxin (Chemical Class: Polyketide-Derived Mycotoxin) Produced by several EPF species, aflatoxin is known for its insecticidal properties and other bioactivities, which help improve pest management and plant resilience (Zhang et al. 2020). It is a secondary metabolite with a difuranocoumarin backbone, classified within the polyketide group. Its cytotoxic and genotoxic effects are well established, although no IC_{50} value has been reported for its insecticidal activity (Zhang et al. 2020). In toxicological assessments, particularly in food safety, aflatoxin B_1 is considered one of the most potent naturally occurring carcinogens (Abrehame et al. 2023). According to the Codex Alimentarius Commission, the maximum tolerated levels of aflatoxin B_1 in food are typically set between 1 and 20 μg/kg (0.001–0.02 ppm), depending on the commodity and national regulations (FAO 2024).

Avenacoside B (Chemical Class: Steroidal Saponin) A type of glycolipid (saponin) produced by different EPF, avenacoside B acts as a surface-active agent, improving the adhesion and distribution of fungal spores (Morcia et al. 2022). Structurally, it is classified as a bisdesmosidic saponin, meaning it contains two sugar chains attached to the aglycone nuatigenin, one at the C-3 position and another at C-26, making it part of the steroidal glycoside family (Ahmad and Basha 2006).

Azadirachtin (Chemical Class: Tetranortriterpenoid Limonoid) Present in various EPF, this compound is recognized for its insecticidal properties against different pests and serves as a plant growth regulator (Hernández et al. 2012; Otieno et al. 2016). Chemically, azadirachtin belongs to the limonoid class of tetranortriterpenoids, derived from neem (*Azadirachta indica*) and biosynthesized by some EPF species. A study demonstrated its fumigant toxicity against the lesser grain beetle, *R. dominica*, with an IC_{50} of approximately 60 μL/L air, indicating strong insecticidal potency in stored grain environments (Tine et al. 2018).

Bassianolide (Chemical Class: Cyclooligodepsipeptide) Extracted from *B. bassiana*, this cyclooctadepsipeptide is known for its insecticidal activity. It interacts with ion channels at neuromuscular junctions, leading to paralysis and increased mortality in insect hosts (Patocka 2016; Zhang et al. 2020). Structurally, it belongs to the class of cyclooligodepsipeptides, specifically a cyclooctadepsipeptide, composed of alternating amino and hydroxy acid residues forming an eight-membered macrocyclic ring.

Beauvericin (Chemical Class: Cyclic Hexadepsipeptide) This cyclic depsipeptide is produced by *B. bassiana* and *Paecilomyces* spp. It functions as an ionophore, disrupting ion gradients across membranes and inducing apoptosis in insect and mammalian cells (Wang and Xu 2012; Gibson et al. 2014; Patocka 2016; Pedrini 2018; Weng et al. 2019; Zhang et al. 2020; Hong et al. 2024). It is characterized by alternating ester and amide bonds in a macrocyclic structure. It has an IC_{50} of 1.9 μM for antiviral activity against HIV-1 integrase and approximately 10 μM for cytotoxic activity against CHO-K1 cells (Wang and Xu 2012).

Caffeine (Chemical Class: Purine Alkaloid) Found in various EPF, caffeine shows neurotoxic effects that can alter insect feeding behaviors, serving as a natural deterrent. Chemically, caffeine is a methylxanthine, a subclass of purine alkaloids derived from xanthine (Wink 2018; Zhang et al. 2020).

Capsaicin (Chemical Class: Alkaloid) Present in various EPF species, capsaicin alters the feeding behavior of herbivorous insects, acting as a feeding deterrent. It belongs to the vanilloid class of alkaloids, characterized by a vanillyl group linked to a long hydrophobic tail (Patocka 2016; Raghunandan et al. 2024).

Destruxins (Chemical Class: Cyclic Hexadepsipeptides) This group of cyclic depsipeptides derives from *M. anisopliae*. Destruxins A, B, and E inhibit vital cellular processes such as DNA, RNA, and protein synthesis in insect cells. This activity leads to increased mortality among target insects (Gibson et al. 2014; Rodríguez-Romero et al. 2023; San Juan-Maldonado et al. 2024). They are macrocyclic compounds with alternating amino and hydroxy acid residues. Destruxin A is quantified at 6.3 μg/larva in *T. molitor* and *G. mellonella*, indicating potent bioactivity (Fredenhagen et al. 2006).

Efrapeptin F (Chemical Class: Linear Peptidic Antibiotic) Antibiotic peptides produced by various EPF species possess insecticidal properties and inhibit harmful microorganisms. Their dual function is important in managing pests and diseases (Zhang et al. 2020; Rodríguez-Romero et al. 2023). It contains rare amino acids like pipecolic acid and 1-amino-isobutyric acid (Fredenhagen et al. 2006).

Farnesol (Chemical Class: Acyclic Sesquiterpenoid Alcohol) Produced by various EPF species, farnesol possesses antimicrobial properties that aid biological control by inhibiting certain plant pathogens (Patocka 2016; Raghunandan et al. 2024). It is an acyclic isoprenoid alcohol with three isoprene units. This compound shows an MIC of 0.29–75.0 μM against *Cryptococcus neoformans* and *Cryptococcus gattii* (Pammi et al. 2011).

Griseofulvin (Chemical Class: Benzofuran Derivative) This antifungal metabolite, released by multiple EPF species, halts the growth of pathogenic fungi, offering protection against root diseases in plants (Gibson and Krasnoff 1999; Isaka et al. 2005). It features a fused benzene and furan ring system with chlorine and methoxy substitutions (Aplin et al. 1964).

Helvolic Acid (Chemical Class: Fusidane-Type Triterpenoid Antibiotic) EPF synthesize helvolic acid, a fusidane-type triterpenoid with a steroid-like backbone of hydroxyl and carboxyl groups. Helvolic acid enhances fungal efficacy against pests and pathogens (Zhang et al. 2020; Hong et al. 2024). This compound exhibits an MIC of 8.0 μg/mL against *H. pylori* and 15–30 μg/mL against *S. aureus* and *C. albicans* (Li et al. 2005).

4-Hydroxymethylazoxybenzoic Acid (Chemical Class: Azoxybenzoic Acid Derivative) Obtained from *E. virulenta*, this metabolite exhibits insecticidal effects, targeting specific pest populations and contributing to their mortality (Patocka 2016; Raghunandan et al. 2024). This rare compound features an azoxy functional group (–N=N(O)–) attached to a hydroxymethyl-substituted benzoic acid core. It has been reported to exhibit insecticidal activity, particularly against adult blowflies (*Calliphora erythrocephala*) through intra-haemacellular injection, where it accounted for most of the extractable insecticidal effect (Claydon 1978). Despite its potency, the precise mechanism of action remains poorly characterized and warrants further investigation.

Limonene (Chemical Class: Monocyclic Monoterpene) Derived from various EPF, limonene displays antimicrobial properties and promotes plant growth, playing a significant role in pest management strategies (Isaka et al. 2005; San Juan-Maldonado et al. 2024). It plays a significant role in integrated pest management strategies by inhibiting microbial growth and enhancing systemic resistance in plants. Structurally, limonene is a monocyclic monoterpene with two enantiomers, commonly found in citrus oils (Duetz et al. 2003).

Nicotine (Chemical Class: Pyridine Alkaloid) Produced by several EPF species, and it exhibits neurotoxic properties that influence insect behavior and physiology (Wink 2018; Zhang et al. 2020; Farhan et al. 2024). It acts as a potent neurotoxin by targeting insect nicotinic acetylcholine receptors, thereby disrupting neural transmission. This mechanism underlies its function as a natural insecticide. Nicotine has also been investigated for its role in pest deterrence and behavioral modification (Ujváry 1999).

Nystatin (Chemical Class: Polyene Macrolide) Another antifungal compound from various EPF species, nystatin inhibits the growth of fungi, protecting against infections in plants (Patocka 2016; Raghunandan et al. 2024). It provides protective effects against fungal infections in plants (Boros-Majewska et al. 2014). Nystatin shows an MIC_{50} of 0.03–0.06 μg/mL and an MIC_{90} of 0.06–0.23 μg/mL against *C. albicans* (Aljaffary et al. 2023).

Oosporein (Chemical Class: Dibenzoquinone) This dibenzoquinone from *B. bassiana* exhibits insecticidal, antiviral, and antibacterial properties. Oosporein forms in insect cadavers and provides antagonistic effects against pathogens, optimizing nutrient uptake for the fungus (Patocka 2016; Pedrini 2018; Zhang et al. 2020; Hong et al. 2024; San Juan-Maldonado et al. 2024). It accumulates in insect cadavers and contributes to fungal virulence by suppressing host immunity and deterring microbial competitors (García-Estrada et al. 2016).

Oxalic Acid (Chemical Class: Dicarboxylic Acid) Found in *Beauveria* spp. and *M. anisopliae*, oxalic acid compromises the integrity of the insect cuticle, promoting dehydration and weakening host defenses (Weng et al. 2019; Zhang et al. 2020; Rodríguez-Romero et al. 2023; San Juan-Maldonado et al. 2024). Oxalic acid disrupts insect cuticle integrity, promoting dehydration and weakening host defenses. This enhances fungal penetration and virulence (Khachatourians and Qazi 2008).

2,6-Pyridinedicarboxylic Acid (Chemical Class: Pyridine Derivative) Isolated from *Isaria* spp. and *M. anisopliae*, this compound shows insecticidal properties, especially against larvae of significant agricultural pests. It can also enhance plant defenses, making it crucial for sustainable pest management (Patocka 2016; Rodríguez-Romero et al. 2023; Raghunandan et al. 2024; San Juan-Maldonado et al. 2024).

Zearalenone (Chemical Class: Resorcylic Acid Lactone) This bioactive compound is synthesized by multiple EPF. It exhibits hormonal activity that affects both plant and insect physiology, particularly concerning reproduction and overall health (Isaka et al. 2005; San Juan-Maldonado et al. 2024). Its role as a mycoestrogen contributes to the regulation of reproductive functions and developmental pathways in a variety of biological systems (Biesaga-Kos'Cielniak and Filek 2009).

3.1 Mechanisms of Action of Bioactive Compounds

EPF serve as an effective and eco-friendly choice for addressing the issues caused by pesticide resistance. The effectiveness and environment-friendly nature of EPF make them a viable alternative to traditional chemical methods in agriculture. The success of EPF as biological control agents relies on the complex mechanisms of action employed by their diverse bioactive compounds (Pedrini 2018; Quesada-Moraga et al. 2023).

3.1.1 Cell Membrane Disruption

The low-molecular-weight bioactive compounds, such as destruxins (Table 6.5 and Fig. 6.5) and cytochalasins (Fig. 6.6), can disrupt the integrity of insect cell membranes (Ortiz-Urquiza and Keyhani 2013; Ma et al. 2024). This disruption changes the permeability of membranes, leading to osmotic imbalances. As a result, cells experience leakage and dysfunction, ultimately causing cell death. By

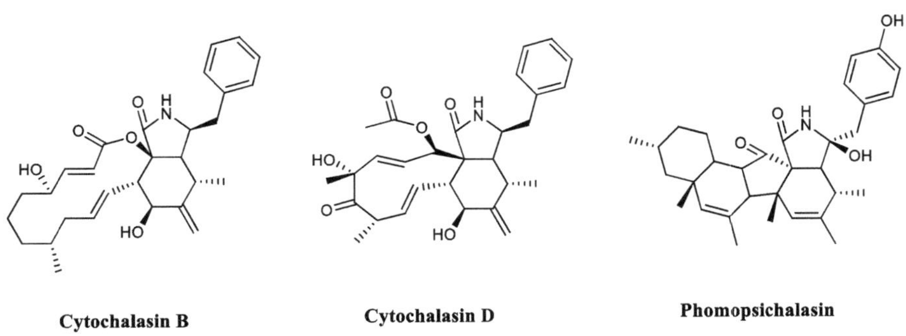

Fig. 6.6 Structure of cytochalasin B and cytochalasin D, and phomopsichalasin

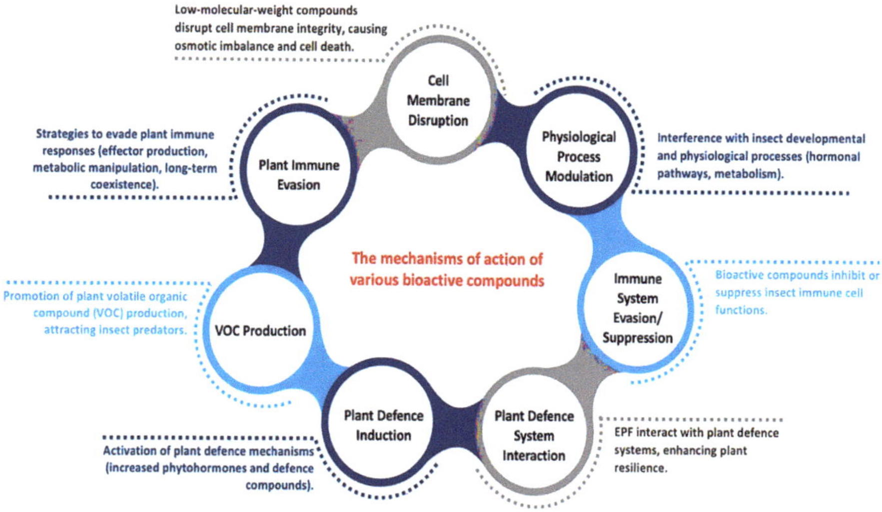

Fig. 6.7 The mechanisms of action of various bioactive compounds

breaching these protective barriers, EPF succeed in establishing infections and promoting successful colonization within hosts (Davari et al. 2018; Mascarin et al. 2024).

Cytochalasins are well known for their interaction with actin filaments, leading to disruptions in the cytoskeleton and influencing various cellular functions such as motility and cell division (Boguś et al. 2021; Lambert et al. 2023). Notably, cytochalasin B and cytochalasin D (Fig. 6.6), sourced from *Helminthosporium dematioideum* and *M. anisopliae*, respectively, have gained attention in cell biology not only for their efficacy in inhibiting actin polymerization but also for their insecticidal properties (Boguś et al. 2021; Lambert et al. 2023). The mechanisms of action of various bioactive compounds are presented in Fig. 6.7.

3.1.2 Modulation of Physiological Processes

Bioactive compounds from EPF interfere with critical developmental and physiological processes within insect hosts. The targeting of hormonal pathways that are essential for growth and reproduction adversely affects the life cycles of pests. This disruption leads to reduced pest populations and improved crop protection (Qu and Wang 2018; Mascarin et al. 2024). The interruption of hormone production and metabolic pathways weakens insects, making them more vulnerable to infections (Davari et al. 2018).

3.1.3 Immune System Evasion and Suppression

EPF employ strategies to avoid detection and attack from the immune systems of insect hosts. Certain secondary bioactive compounds inhibit or suppress the functions of immune cells, neutralizing the defense mechanisms of the host

(Qu and Wang 2018; Mascarin et al. 2024). The understanding of these immune evasion strategies could lead to more effective biocontrol methods that exploit weaknesses within insect immune systems (Davari et al. 2018).

3.2 Interaction with Plant Defense Systems

EPF not only target insect pests but also significantly interact with plant defense mechanisms. This interaction enhances plant resilience and overall health (Ahsan et al. 2024; Aravinthraju et al. 2024). The various bioactive compounds produced by these fungi contribute to their effectiveness against pests and provide beneficial effects on plants (Ahsan et al. 2024; Farhan et al. 2024).

3.2.1 Plant Defense Induction

EPF activate intrinsic defense mechanisms in colonized plants. This activation increases the production of phytohormones and defense-related compounds (Ahsan et al. 2024; Aravinthraju et al. 2024). This biochemical response enhances the capacity of plants to withstand biotic stresses, improving their resistance to pests (Iida et al. 2023; Farhan et al. 2024).

3.2.2 Volatile Organic Compound Production

EPF promote the production of volatile organic compounds (VOCs) in plants, which act as signaling molecules. These VOCs attract natural predators of herbivorous insects, offering indirect protection to the colonized plant and neighboring plants (Crespo et al. 2008; Hung et al. 2015; Khoja et al. 2021).

3.2.3 Immune System Evasion

To establish successful infections, EPF must evade complex plant immune responses (Hernández-Chávez et al. 2017). They achieve this through several strategies as described below.

Effector Production EPF produce effector proteins that can alter the immune responses of host plants, allowing fungi to bypass plant defenses (Hernández-Chávez et al. 2017).

Metabolic Manipulation EPF can adjust metabolic pathways in host plants, influencing the balance of phytohormones. This change creates conditions that favor fungal growth while suppressing plant defenses (Hernández-Chávez et al. 2017).

Long-Term Coexistence EPF can form lasting associations with host plants, which helps them evade defenses and manipulate physiological responses. These relationships provide consistent nutrient sources for fungi and assist plants in adapting to different stresses (Hernández-Chávez et al. 2017).

The relationship between EPF and plants highlights their role as effective biological control agents and beneficial organisms (Ahsan et al. 2024; Aravinthraju et al. 2024). The use of EPF in pest management strategies presents a significant opportunity to decrease reliance on chemical pesticides and adopt more environmentally friendly methods. The various actions of EPF improve pest control efforts while supporting ecosystem health. The increasing demand for sustainable agricultural practices emphasizes the importance of recognizing the role of EPF in promoting biodiversity and maintaining ecological balance (Mantzoukas et al. 2022; Aravinthraju et al. 2024; Shahbaz et al. 2024).

3.2.4 Implications for Crop Protection

Bioactive compounds produced by EPF enhance their efficacy against insect pests and contribute to overall plant health (Mantzoukas et al. 2022; Shahbaz et al. 2024). These features position EPF as essential components in IPM strategies, emphasizing sustainable agricultural methods. The implementation of EPF in pest management systems can improve crop health by stimulating systemic resistance against various pathogens and pests (Mantzoukas et al. 2022; Shahbaz et al. 2024). The ability to boost plant immune responses has significant implications for agriculture, with potential applications in pest control and disease management.

The ongoing exploration of bioactive compounds may lead to the discovery of new compounds that strengthen the roles of EPF in agriculture and ecology (Mantzoukas et al. 2022; Shahbaz et al. 2024). The knowledge of the ecological functions and impacts of EPF and their bioactive compounds is critical for developing effective and environmentally friendly pest management strategies that sustainably support agricultural productivity. Research indicates that EPF can significantly reduce crop losses caused by insect pests, thereby enhancing agricultural sustainability (Mantzoukas et al. 2022; Shahbaz et al. 2024). The use of EPF in crop protection not only targets pests but also promotes plant growth and resilience, making them a valuable tool in modern agriculture.

4 Case Studies Against Pest Species

EPF have demonstrated its effectiveness as a biocontrol agent across various agricultural contexts. Real-world studies illustrate the efficacy of EPF against specific pest species, underscoring their potential as integral components of IPM strategies (Abdel-Raheem and Yassin 2019; Francis et al. 2022). Bioactive compounds produced by fungi such as *B. bassiana* and *M. anisopliae* are crucial for enhancing biological control effectiveness (Rath 2010; Hassan et al. 2024).

Bioactive compounds produced by EPF not only disrupt pest populations but also promote sustainable agricultural practices (Sharma and Sharma 2021; Sharma et al. 2023). Their interaction with plant systems enhances overall health and resilience in the face of biotic stresses, emphasizing their dual role in pest management. Notable case studies of EPF against target pest species are presented in Fig. 6.8. These studies demonstrate that EPF provide substantial benefits beyond direct pest control. They

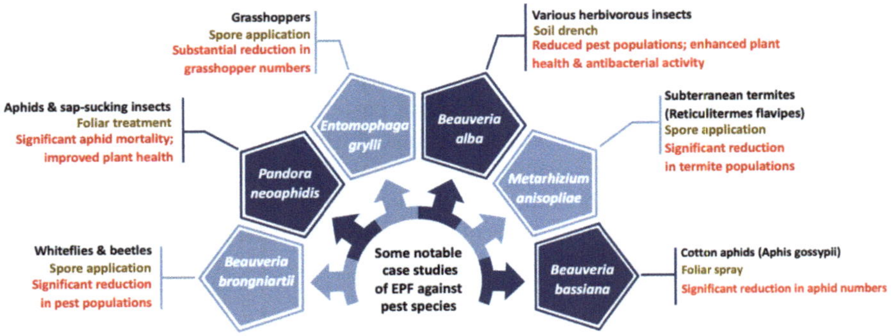

Fig. 6.8 Some notable case studies of entomopathogenic fungi against pest species

enhance overall plant health and resilience by inducing systemic resistance and improving nutrient uptake (Aksoy et al. 2023; Ma et al. 2024). Research indicates that EPF effectively manages insect populations while also benefiting soil health and plant vitality. This dual purpose establishes them as valuable tools in sustainable agriculture.

4.1 *Beauveria bassiana* Against Cotton Aphids

One study demonstrated that *B. bassiana* effectively reduced populations of cotton aphids (*Aphis gossypii*). The fungal spores were applied as a foliar spray, resulting in a significant decrease in aphid numbers within a short time. The conidia attached to the aphids and penetrated their cuticles, causing a systemic infection that led to their death (Abdel-Raheem and Yassin 2019; Francis et al. 2022). This research emphasizes the potential of *B. bassiana* as an essential component of IPM strategies aimed at controlling aphid populations while reducing reliance on synthetic pesticides (Abdel-Raheem and Yassin 2019; Francis et al. 2022). This case study exemplifies how biocontrol with EPF can significantly impact agricultural pest management.

4.2 *Metarhizium anisopliae* Against Subterranean Termites

Research on subterranean termites (*Reticulitermes flavipes*) establishes *M. anisopliae* as an effective biocontrol agent. The field trials confirmed significant reductions in termite populations after applying the fungal spores (Rath 2010; Hassan et al. 2024). This effectiveness results from the enzymatic activity of the EPF, which allows for cuticle penetration and disrupts essential metabolic functions in termites (Rath 2010; Hassan et al. 2024). The capability of *M. anisopliae* to protect wooden structures and crops from termite damage illustrates its potential as a sustainable pest management strategy.

4.3 *Beauveria alba* Against Various Herbivorous Insects

Numerous studies have evaluated the effectiveness of *B. alba* against various herbivorous insects. When applied as a soil drench, this fungus exhibited strong antibacterial activity against *B. subtilis* (Sharma and Sharma 2021; Sharma et al. 2023). The resilience of plants against pest pressures was enhanced by these bioactive compounds produced by the fungus. Research indicates that the bioactive compounds of *B. alba* improve plant health and assist in managing herbivorous populations (Sharma and Sharma 2021; Sharma et al. 2023). This suggests that the application of *B. alba* can provide various benefits and promote plant vigor while directly addressing pest threats (Sharma and Sharma 2021; Sharma et al. 2023).

4.4 *Entomophaga grylli* Against Grasshoppers

A detailed study on *E. grylli* demonstrated its effectiveness in managing grasshopper populations. The application of fungal spores resulted in substantial reductions in grasshopper numbers, showcasing its potential for maintaining ecological balance within agricultural ecosystems (Goettel et al. 1995). This case study highlights that EPF can provide an environmentally friendly solution for controlling grasshopper infestations while supporting overall biodiversity.

4.5 *Pandora neoaphidis* Against Aphids

The efficacy of *P. neoaphidis* was evaluated for controlling aphids and other sap-sucking insects. The foliar treatments induced significant mortality in aphid populations while simultaneously promoting plant health. The dual function of EPF emphasizes its effectiveness in managing pest populations while simultaneously improving the resilience and overall well-being of host plants (Baverstock et al. 2005). These findings suggest that incorporating EPF, such as *P. neoaphidis*, into pest management programs can enhance plant vitality and lead to better overall results.

4.6 *Beauveria brongniartii* Against Insect Pests

Recent studies have shown the effectiveness of *B. brongniartii* as a biocontrol agent for managing insect pests in agriculture. The field trials indicate that applying *B. brongniartii* spores significantly reduces pest populations, including whiteflies and beetles. The fungal conidia attach to the exoskeletons of these pests, penetrate their cuticles, and proliferate in their haemolymph, causing systemic infections that lead to pest mortality (Zimmermann 2007; Sinha et al. 2016).

5 Antimicrobial Activity and Mechanisms

EPF are recognized for their potent antimicrobial activities. These fungi produce a wide variety of antimicrobial compounds that inhibit the growth of pathogenic bacteria and fungi, offering protection not only to themselves but also to their plant hosts. This remarkable ability contributes to the overall health and resilience of host plants, reinforcing the role of EPF as effective biocontrol agents in sustainable agriculture (Sharma et al. 2023; Ma et al. 2024). The multifaceted capabilities of EPF enable significant enhancements in agricultural productivity while promoting environmental sustainability.

EPF produce a diverse array of bioactive compounds, each playing unique roles in managing pests and promoting plant growth (Aksoy et al. 2023; Sharma et al. 2023). This variety highlights the intricate relationships between these fungi, their insect hosts, and the surrounding ecosystems (Aksoy et al. 2023; Sharma et al. 2023). A comprehensive understanding of the types and functions of bioactive compounds is essential for effectively implementing IPM strategies in sustainable agriculture (Sharma et al. 2023; Ma et al. 2024). Moreover, studies have shown that bioactive compounds produced by fungi are vital for their survival and effectiveness as biological control agents (Sharma and Sharma 2021; Aksoy et al. 2023).

Bioactive compounds produced by EPF, shown in Fig. 6.5, provide protective effects and facilitate fungal infection in hosts. In addition to their role in pest control, EPF are significant for colonizing and infecting their host organisms (Sharma et al. 2023; Ma et al. 2024). Their diverse bioactive compounds also help improve plant health and resilience against biological and environmental stresses (Sharma and Sharma 2021; Aksoy et al. 2023). The ongoing research into these diverse bioactive compounds will deepen the understanding of their ecological functions, leading to more effective and environmentally friendly pest management strategies (Sharma et al. 2023; Ma et al. 2024).

5.1 Antimicrobial Activity

The antimicrobial properties of EPF are significant in their roles as biological control agents. These properties demonstrate the ability of EPF to combat pathogens that pose threats to plant health (Shin et al. 2016; Nakashima et al. 2024). Helvolic acid inhibits *H. pylori* (MIC = 8.0 μg/mL), *S. aureus* (MIC = 15–30 μg/mL), and *C. albicans* (MIC = 15–30 μg/mL) (Li et al. 2005).

Beauvericin disrupts ion balance in *C. albicans* and *Aspergillus* spp., with no MIC values reported (Wang and Xu 2012; Tong et al. 2016; Mantzoukas et al. 2022; Verma et al. 2024). Destruxins inhibit protein synthesis in *E. coli* and *S. aureus*, with no MIC values documented (Deka et al. 2021; Mantzoukas et al. 2022). Farnesol shows an MIC of 0.29–75.0 μM against *C. neoformans* and *C. gattii* (Pammi et al. 2011). Nystatin exhibits an MIC_{50} of 0.03–0.06 μg/mL and an MIC_{90} of 0.06–0.23 μg/mL against *C. albicans* (Aljaffary et al. 2023). Antibiotics target *B. subtilis*, while antifungal metabolites inhibit *Fusarium* spp. and *Rhizoctonia* spp. (Mousa and Raizada 2013; Shin et al. 2016; Deka et al. 2021; Mantzoukas et al. 2022; Nakashima et al. 2024). Nonribosomal peptides target *C. albicans*, and pestalotiopins, from *Pestalotiopsis* spp., inhibit *Botrytis cinerea* (Mishra et al. 2012; Mousa and Raizada 2013; Shin et al. 2016; Nakashima et al. 2024).

The diverse antimicrobial activities exhibited by EPF include the production of various antibiotics and antifungal metabolites, as detailed in Table 6.6 and Fig. 6.9. Bioactive compounds serve as crucial mechanisms for protecting the plants they inhabit. The versatility of these antimicrobial bioactive compounds allows EPF to effectively suppress bacterial and fungal pathogens, which are common threats in agricultural settings. In addition to their direct antimicrobial effects, these compounds support plant health by reducing disease incidence and minimizing the need for chemical fungicides (Deka et al. 2021; Mantzoukas et al. 2022). The ability of EPF to produce a range of bioactive compounds that offer protective functions demonstrates their potential in IPM strategies, providing a pathway toward reducing reliance on synthetic chemicals while fostering sustainable agricultural practices (Mantzoukas et al. 2022; Verma et al. 2024).

5.2 Mechanisms of Antimicrobial Action

The mechanisms by which EPF exert their antimicrobial action comprise various strategies specifically aimed at targeting pathogens. A thorough understanding of these mechanisms is essential for optimizing the use of EPF in agricultural pest management and enhancing plant health (Qin et al. 2023; Ma et al. 2024). The detailed mechanisms of antimicrobial action are depicted in Fig. 6.9. Antimicrobial compounds produced by EPF can significantly interfere with the key cellular processes in pathogens that are essential for survival and proliferation (García-Espinoza et al. 2023; Verma et al. 2024). Such interference includes inhibiting cell wall synthesis and compromising membrane integrity. The disruption of cell wall construction is vital for preventing pathogen colonization and infection.

5.3 Implications for Plant Health and Disease Resistance

The bioactivities of EPF promote overall plant health while enhancing resistance to diseases. The production of various bioactive compounds enables EPF to suppress the growth of harmful pathogens, thereby improving the resilience of host plants (Ahsan et al. 2024; Aravinthraju et al. 2024).

5.3.1 Disease Suppression

EPF play a crucial role in disease suppression by producing antimicrobial compounds that inhibit the growth of bacteria, fungi, and other pathogens that may invade plant tissues. Such protective actions are vital for reducing disease incidence, ensuring that plants remain productive under pest and pathogen pressures (Mantzoukas et al. 2022; Iida et al. 2023). The presence of EPF in the rhizosphere

Table 6.6 Antimicrobial activities of various entomopathogenic fungi

Antimicrobial compound	Fungal source	Target pathogen	Function	Reference
Antibiotics	Various EPF	Bacterial pathogens (*B. subtilis*)	Target bacterial pathogens	Mousa and Raizada (2013), Shin et al. (2016), Nakashima et al. (2024)
Antifungal metabolites	Various EPF	Fungal pathogens (*Fusarium*, *Rhizoctonia*)	Inhibit fungal pathogen growth	Mousa and Raizada (2013), Shin et al. (2016), Gouda et al. (2020), Deka et al. (2021), Mantzoukas et al. (2022)
Beauvericin	*B. bassiana*	*C. albicans*, and *Aspergillus* spp.	Disrupt the ion balance in pathogens	Tong et al. (2016), Mantzoukas et al. (2022), Verma et al. (2024)
Destruxins	*M. anisopliae*	*E. coli*, and *S. aureus*	Inhibit protein synthesis in various pathogens	Deka et al. (2021), Mantzoukas et al. (2022)
Helvolic acid	Various EPF	*H. pylori*, *S. aureus*, and *C. albicans*	Antimicrobial activity (MIC = 8.0 μg/mL against *H. pylori*, 15–30 μg/mL against *S. aureus*, and *C. albicans*)	Li et al. (2005)
Farnesol	Various EPF	*C. neoformans* and *C. gattii*	Antimicrobial activity (MIC = 0.29–75.0 μM)	Pammi et al. (2011)
Nonribosomal peptides	Various EPF	*C. albicans*	Insecticidal and antimicrobial properties	Mousa and Raizada (2013), Shin et al. (2016), Nakashima et al. (2024)
Nystatin	Various EPF	*C. albicans*	Antimicrobial activity (MIC_{50} = 0.03–0.06 μg/mL; MIC_{90} = 0.06–0.23 μg/mL)	Aljaffary et al. (2023)
Pestalotiopins	*Pestalotiopsis* spp.	*B. cinerea*	Exhibit antifungal properties	Mishra et al. (2012)

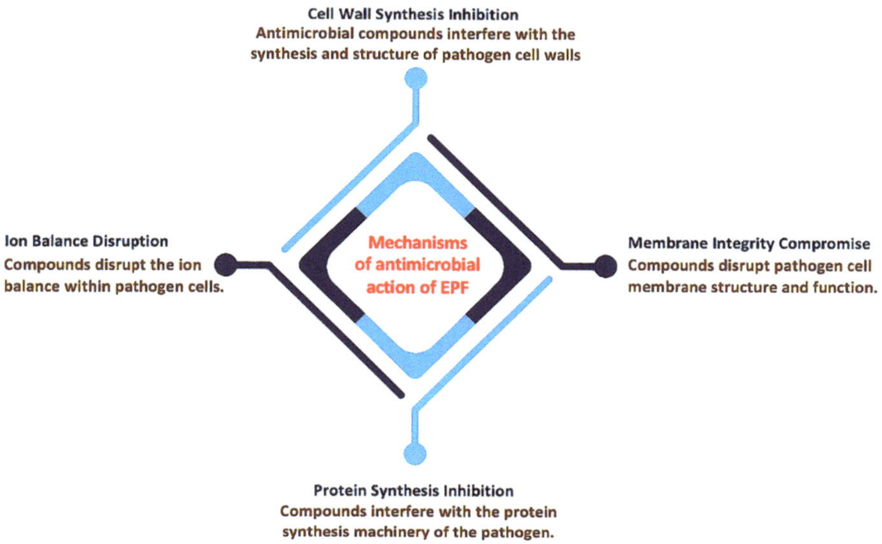

Fig. 6.9 Mechanisms of antimicrobial action of entomopathogenic fungi

contributes to a healthy root environment, essential for plant vitality and resilience toward stress. This aspect becomes especially critical in organic farming systems, where synthetic fungicides are absent, necessitating reliance on natural defenses against diseases.

5.3.2 Enhanced Resistance Mechanisms

EPF enhance the intrinsic defense mechanisms of their host plants. Interactions with these fungi can trigger systemic acquired resistance (SAR) and induce systemic resistance (ISR), empowering plants to respond more effectively to subsequent biotic stressors, such as insect herbivores and pathogens (Mantzoukas et al. 2022; Aravinthraju et al. 2024). An increased production of phytohormones, including salicylic acid and jasmonic acid (Fig. 6.10), mediates these defense responses. By increasing the resistance mechanisms, plants are better able to manage stresses more effectively, which reduces reliance on external inputs and contributes to sustainable agricultural production practices (Mantzoukas et al. 2022; Aravinthraju et al. 2024). The enhancement of stress tolerance can also result in better growth outcomes during adverse environmental conditions, thus increasing agricultural productivity.

5.3.3 Synergistic Effects with Other Beneficial Microorganisms

EPF can interact positively with other microorganisms in the soil, such as beneficial bacteria and mycorrhizal fungi. When EPF colonize plant roots, they enhance the establishment and activity of these beneficial microbes. This interaction improves nutrient acquisition, water retention, and overall plant health (Qayyum et al. 2021; Verma et al. 2024). The combination of EPF and beneficial microorganisms fosters a

Fig. 6.10 Structures of salicylic acid and jasmonic acid

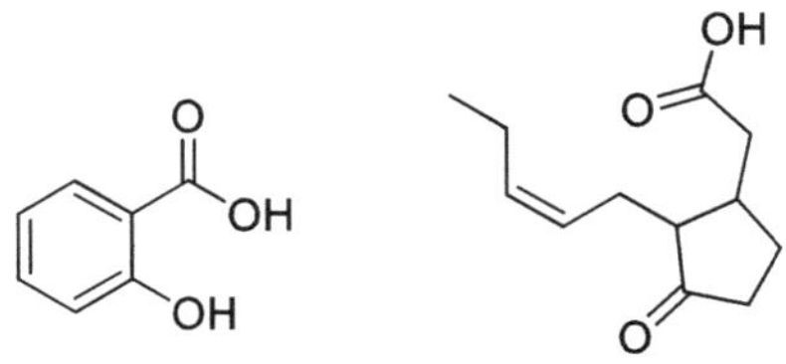

Salicylic acid **Jasmonic acid**

robust ecosystem in the rhizosphere, further promoting plant resilience against pathogens and pests (Salazar-Mendoza et al. 2025).

5.3.4 Contribution to Soil Health

The utilization of EPF positively influences the overall health of the soil. By enhancing microbial diversity within the soil, EPF support a balanced ecosystem that fosters the growth and development of plants. Their ability to degrade organic matter enhances soil fertility, enriching the nutrient profile available for plant uptake (Murindangabo et al. 2024; Vivekanandhan et al. 2024). The bioactivities of EPF not only aid in the control of pest populations but also play a significant role in bolstering the health of plants and improving their resistance to diseases. By utilizing these beneficial interactions, agricultural producers can establish more resilient cropping systems that support sustainable practices.

5.4 Growth-Promoting Effects

EPF contribute actively to plant growth and resilience beyond its effectiveness in managing pests and pathogens. These fungi produce phytohormones and growth-enhancing compounds that significantly influence plant development, including auxins, cytokinins, and gibberellins (Sharma et al. 2023; Aravinthraju et al. 2024). EPF are diverse, easily localized, and have long-lasting effects on insect pests. When inhabiting plants, these fungi alter secondary metabolites, volatile organic compounds, and microbiomes, enhancing plant resistance to pests and diseases and sometimes improving growth (Sharma et al. 2023; Aravinthraju et al. 2024).

5.4.1 Phytohormones and Growth-Enhancing Compounds

EPF possess the ability to produce phytohormones such as auxins, cytokinins, and gibberellins (Fig. 6.11) and other growth-promoting substances that significantly impact plant development. The influence of these phytohormones and growth-enhancing compounds plays a crucial role in agricultural productivity, particularly in stress conditions (Chanclud and Morel 2016; Gonzalez-Guzman et al. 2021).

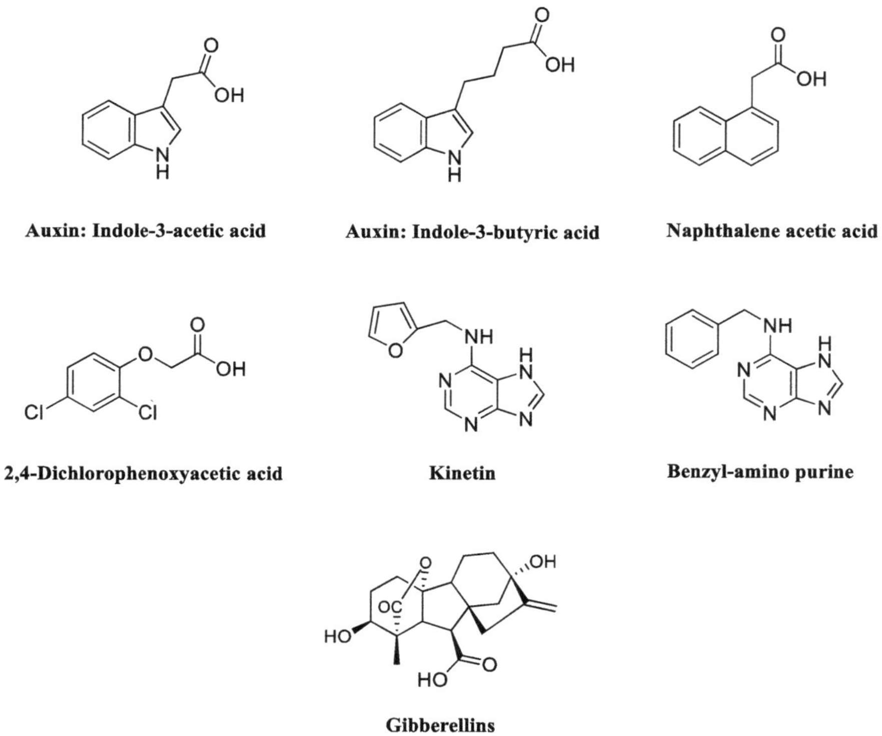

Fig. 6.11 Structures of indole-3-acetic acid, indole butyric acid, naphthalene acetic acid, 2,4-dichlorophenoxyacetic acid, kinetin, benzyl-aminopurine, and gibberellins

Auxins These hormones are essential for root development. Auxins encourage robust root systems that enhance nutrient and water uptake (Gonzalez-Guzman et al. 2021). They promote cell elongation, particularly in shoots, and induce apical dominance. These phenomena occur when the main stem grows taller and faster than side branches. Natural auxins (Fig. 6.11) include indole-3-acetic acid (IAA) and indole butyric acid (IBA), and synthetic auxins (Fig. 6.6), such as naphthalene acetic acid (NAA) and 2,4-dichlorophenoxyacetic acid (2,4-DPA), are widely used in agricultural practices. These compounds are particularly valuable for promoting root development in cuttings and plant tissue culture (Chanclud and Morel 2016; Gonzalez-Guzman et al. 2021). They are widely used to ascertain the minimum inhibitory concentration (MIC) values for these compounds to ensure accuracy and reliability.

Cytokinins These hormones stimulate cell division and differentiation. They delay leaf and flower aging while promoting shoot growth. The most frequently used cytokinins (Fig. 6.11) in plant cell cultures include kinetin and benzylaminopurine (BAP). These hormones are critical for supporting shoot proliferation and improving plant health and longevity (Chanclud and Morel 2016; Gonzalez-Guzman et al. 2021). Recent studies demonstrate that cytokinins may enhance the effectiveness of EPF in promoting plant resilience under pest pressure (Chanclud and Morel 2016; Gonzalez-Guzman et al. 2021).

Gibberellins These hormones stimulate stem elongation and seed germination. They assist in breaking seed dormancy and can increase the size of fruits and flowers. The application of gibberellins contributes to facilitating seed germination and delaying senescence in leaves and fruit. Gibberellins (Fig. 6.11) also support plant resilience against various stresses, thereby contributing to sustainable agricultural practices (Ahsan et al. 2024; Aravinthraju et al. 2024).

The complex relationship between EPF and plant growth is demonstrated by the role of these phytohormones. An understanding of the mechanisms through which EPF produce and influence these compounds provides potential pathways for enhancing crop health and resilience (Ahsan et al. 2024; Aravinthraju et al. 2024). This knowledge strengthens efforts in integrated pest management and reveals the diverse advantages of incorporating EPF in sustainable agriculture.

5.4.2 Benefits in Stress Conditions

Entomopathogenic fungi demonstrate capabilities in enhancing plant resilience under unfavorable conditions, such as drought and salinity stresses. These fungi improve root growth, which allows plants better access to water during drought periods (Samal et al. 2023; Ahsan et al. 2024). In saline environments, EPF assist in improving nutrient uptake and maintaining ion balance, which mitigates the adverse effects of salinity on plant growth.

The ability of EPF to provide antimicrobial protection while promoting plant growth creates a synergistic relationship that boosts plant health, resilience, and agricultural productivity. The bioactivities of EPF contribute not only to pest control but also to fostering sustainable agricultural practices. Agricultural producers can achieve improved crop health, reduced chemical inputs, and enhanced ecosystem stability by taking advantage of these multifaceted benefits (Samal et al. 2023; Ahsan et al. 2024).

6 Agricultural Applications and Biopesticide Development

EPF have gained considerable attention as a promising alternative to chemical pesticides due to their eco-friendly nature and multiple benefits. Their application in agriculture encompasses various strategies aimed at enhancing pest control, improving plant health, and contributing to sustainable farming practices (Irsad et al. 2023; Sharma et al. 2023). The key application methods in agriculture are presented in Table 6.7. The integration of EPF into agricultural systems promotes ecological stability and agricultural productivity.

Table 6.7 Key application methods in agriculture

Application method	Description	Benefit	Reference
Seed treatments	Applying EPF to seeds before planting	Boosts germination and protects young plants	Irsad et al. (2023), Sharma et al. (2023)
Soil amendments	Mixing EPF into the soil	Improves roots and soil health	Sharma et al. (2023), Vivekanandhan et al. (2024)
Foliar sprays	Spraying EPF on plant leaves	Controls pests and supports plant growth	Irsad et al. (2023), Sharma et al. (2023)

The integration of EPF into agricultural practices presents a novel approach to pest management. This approach effectively combines biological control with enhanced plant health. The capability of EPF to naturally combat insect pests while promoting growth and resilience in crops aligns with the principles of sustainable agriculture (Irsad et al. 2023; Sharma et al. 2023). This dual functionality makes EPF a valuable ally in the quest for eco-friendly farming solutions.

6.1 Application Methods in Agriculture

Several application methods maximize the benefits of EPF in agricultural settings. These methods target pest populations directly while supporting the overall vigor of crops, thus leading to healthier and more productive agricultural systems (Irsad et al. 2023; Sharma et al. 2023). Each approach demonstrates the versatility of EPF and their potential to contribute significantly to modern pest management strategies.

6.1.1 Seed Treatments

Seed coating with EPF before planting provides clear benefits by improving germination rates. This process stimulates early plant growth and provides initial protection against pests and diseases. *B. bassiana* has demonstrated considerable effectiveness in seed treatments, offering systemic protection once the plant emerges from the soil (Aravinthraju et al. 2024; Verma et al. 2024). This proactive approach supports early plant establishment and mitigates pest pressures during critical growth stages. Healthier and better crop development results from this method. In addition, seed treatments with EPF can enhance seedling vigor while promoting robust root systems, which increase the resilience of plants to subsequent environmental challenges. By providing a natural defense mechanism at the earliest stages of plant growth, this method decreases reliance on synthetic pesticides. The effective seed treatment strategies also lead to cost savings for farmers and contribute to overall agricultural sustainability.

6.1.2 Soil Amendments

The incorporation of EPF into soil amendments can significantly improve soil health. This method enhances nutrient cycling and promotes the establishment of beneficial microbial communities within the soil ecosystem. Research indicates that applications of EPF to the soil can facilitate root development and improve nutrient acquisition, particularly in nutrient-deficient or stressed environments (Sharma et al. 2023; Vivekanandhan et al. 2024). The improvements in the physical and biological properties of the soil allow for better water retention and drainage, ensuring that plants can access nutrients more effectively. The soil amendments stimulate beneficial interactions between EPF and native soil microorganisms, promoting a balanced agroecosystem. The use of EPF as soil amendments enriches the soil environment and supports integrated pest management by fostering healthy plant growth.

6.1.3 Foliar Sprays

The application of EPF as foliar sprays demonstrates strong insecticidal effects against various pests. Spraying crops with fungal spores allows for effective infection and population control of pests. Studies indicate that foliar sprays of *M. anisopliae* significantly reduce infestations of insect pests while enhancing overall crop health (Irsad et al. 2023; Sharma et al. 2023). This method is particularly efficient for managing foliar-feeding insects, such as aphids and caterpillars, which can cause substantial damage to crops. The use of EPF via foliar sprays offers a targeted approach to pest management that minimizes harm to beneficial insects and the surrounding environment (Irsad et al. 2023; Sharma et al. 2023). The quick action of these fungal spores provides immediate relief from pest pressures, which supports farmers in maintaining crop quality and yield during critical growing periods.

6.2 Biopesticide Development

The integration of EPF into commercial biopesticide products serves as a focal point for enhancing pest management strategies in sustainable agriculture (Bamisile et al. 2021a; Verma et al. 2024). The development of these biopesticides involves several key components, as highlighted in Fig. 6.12.

6.2.1 Advances in Formulation Technology

Recent advancements in the formulation of biopesticides have significantly improved the stability and efficacy of EPF-based products. Techniques such as microencapsulation and the use of inert carrier materials protect fungal spores from environmental degradation while maximizing delivery to target sites (Iida et al. 2023; Verma et al. 2024). Microencapsulation shields EPF from adverse conditions, including UV radiation, temperature extremes, and fluctuations in humidity (Iida et al. 2023; Verma et al. 2024). Such advancements enhance the viability of EPF during storage and application, leading to more effective pest management solutions.

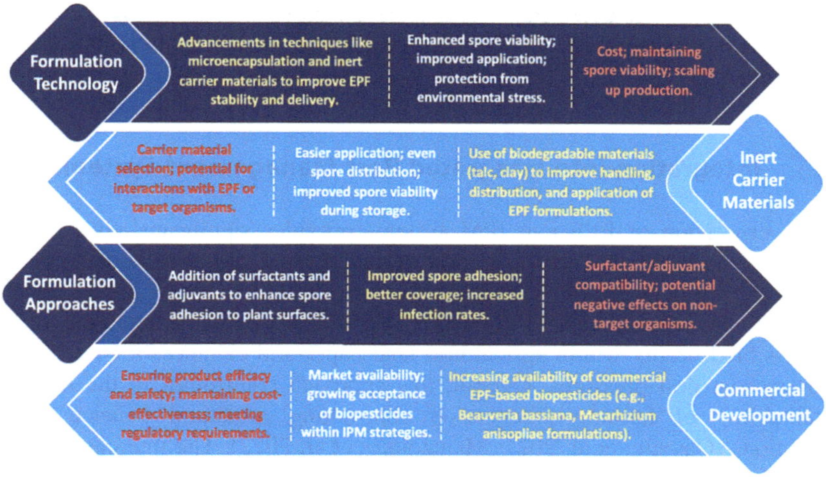

Fig. 6.12 Key aspects of entomopathogenic fungi biopesticide development

6.2.2 Inert Carrier Materials

Biodegradable materials, such as talc and clay, frequently enhance the handling, distribution, and application of EPF-based formulations (Bamisile et al. 2021a). These inert carriers provide a conducive environment for the fungi during storage while maintaining their viability. They also assist in evenly dispersing fungal spores over the intended application area, thereby increasing the chances of successful colonization and effective infection of target pests (Bamisile et al. 2021b).

6.2.3 Innovation in Formulation Approaches

Innovations in formulation approaches include incorporating surfactants and adjuvants to enhance the adhesion of fungal spores to plant surfaces. Additives that improve the spreading and wetting properties of the application solution allow for uniform coverage of target plants, thus increasing the likelihood of effective pest management (Sabbahi et al. 2022; Parveen 2023). This consideration ensures greater adherence of fungal spores to plant surfaces, enhancing their ability to infect and control pest populations.

6.2.4 Commercial Development of Biopesticides

The commercial development of biopesticides based on EPF is progressing, with several products already available on the market. Notable examples include formulations based on *B. bassiana* and *M. anisopliae*, which are recognized for their broad-spectrum insecticidal properties (Qayyum et al. 2021; Vivekanandhan et al. 2024). These biopesticides are gaining traction in various agricultural contexts, reflecting a growing appreciation for biological control agents within IPM strategies (Qayyum et al. 2021; Vivekanandhan et al. 2024). The ongoing research efforts focus on optimizing the effectiveness and safety of these products through extensive field trials that assess their efficacy against target pest populations while evaluating their safety for non-target organisms, including beneficial insects and soil microorganisms.

6.3 Integration into Integrated Pest Management Systems

The incorporation of EPF into IPM systems provides significant advantages for sustainable agriculture. These fungi act as natural antagonists to pest populations while reducing dependency on chemical pesticides, thus fostering a balanced ecosystem (Qayyum et al. 2021; Mantzoukas et al. 2022). The integration of EPF supports biodiversity within agricultural landscapes and enhances the resilience of crops against pest pressures. Overall, this approach contributes to the long-term sustainability of farming practices by promoting a natural balance in pest populations (Qayyum et al. 2021; Mantzoukas et al. 2022). The integration of EPF into IPM is summarized in Fig. 6.13.

6.3.1 Contributing to Biodiversity

The implementation of biocontrol agents like EPF supports biodiversity within agroecosystems. These fungi contribute to the sustenance of beneficial organisms, such as pollinators and natural pest predators, which play crucial roles in maintaining ecological balance and promoting crop health (Meyling and Eilenberg 2007; Mantzoukas et al. 2022). The presence of EPF enhances ecosystem resilience by

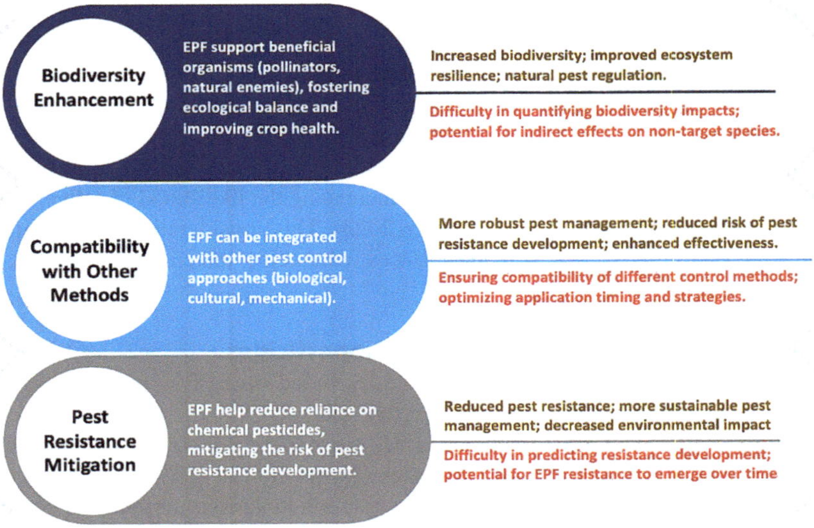

Fig. 6.13 The integration of entomopathogenic fungi into integrated pest management

creating habitats that support diverse species, which are essential for sustaining natural pest regulation and improving overall agricultural productivity (Meyling and Eilenberg 2007; Mantzoukas et al. 2022).

6.3.2 Compatibility with Other Control Methods

The compatibility of EPF with other pest control methods, such as chemical insecticides and biological agents, enhances the overall effectiveness of IPM strategies. Studies have shown that EPF can be combined with reduced doses of insecticides to achieve effective pest control while minimizing the negative impacts on non-target organisms (Sharma et al. 2023; Zhou et al. 2024). This integrated approach allows for the use of multiple control methods in a complementary manner, reducing the reliance on chemical pesticides and promoting sustainable pest management (Sharma et al. 2023; Zhou et al. 2024).

6.3.3 Resilience to Pest Resistance

The repeated application of chemical pesticides can lead to the development of pest resistance over time. The inclusion of EPF as part of IPM strategies presents an alternative approach to pest control that reduces selection pressures contributing to resistance (Qayyum et al. 2021; Zhou et al. 2024). The application of EPF maintains the long-term effectiveness of pest management programs while promoting ecologically sound solutions.

7 Research Advances with Bioactive Compounds

Recent research has significantly expanded the potential applications of EPF in agricultural systems. Novel insights into the interactions between EPF, plants, and soil microorganisms have revealed their capacity to influence plant physiology, nutrient dynamics, and stress responses. These findings underscore the role of EPF as a key component in sustainable pest management strategies (Samal et al. 2023; Mascarin et al. 2024). Advances in the identification and application of EPF-derived bioactive compounds have opened new avenues for biopesticide development. These compounds exhibit diverse biological activities, including insecticidal, antimicrobial, and plant growth-promoting effects. Table 6.8 summarizes recent approaches and innovations in EPF-based biopesticide development.

7.1 Plant-Microbe Interactions

Research has identified how EPF interact with plants, influencing their growth and responses to stress. Studies indicate that EPF enhance nutrient uptake and bioavailability through complex signaling pathways (Rasool et al. 2021; Ahsan et al. 2024). These pathways facilitate beneficial interactions within the rhizosphere, where fungi engage with rhizobacteria and other beneficial soil microbes, creating a synergistic environment that supports plant health. In addition, these interactions improve plant

Table 6.8 Approaches for EPF biopesticide development

Approach	Description	Advantage	Challenge	Reference
Formulation studies	Improving EPF product delivery, stability, and shelf life	Better product performance and longer shelf life	High cost and maintaining spore viability	Bamisile et al. (2021a), Sharma et al. (2023), Vivekanandhan et al. (2024)
Field trials	Testing EPF performance in real farming conditions	Real-world data on safety and effectiveness	Time and resource-intensive	Bamisile et al. (2021a), Mantzoukas et al. (2022), Qin et al. (2023), Sharma et al. (2023)
Genomic research	Identifying genes for virulence and compound production	Selection of strong strains for pest control	Complex analysis and data interpretation	Zhao et al. (2016), Harith-Fadzilah et al. (2021), Altimira et al. (2022), Mantzoukas et al. (2022)
Metabolomic analysis	Finding key compounds for pest and disease control	Enhanced production of useful compounds	Advanced tools are needed for compound identification	Mantzoukas et al. (2022), Mascarin et al. (2024)
Synthetic biology	Modifying EPF traits using genetic tools	Tailored strains for specific needs	Ethical and regulatory concerns	Mantzoukas et al. (2022), Mascarin et al. (2024)

resilience to biotic and abiotic stresses while promoting growth by enhancing nutrient absorption and utilization. Alterations in phytohormone levels, typically induced by EPF, result in increased root biomass and overall plant vigor, enabling plants to adapt more effectively to their environment. By promoting these beneficial relationships, EPF contribute to a more sustainable agricultural ecosystem by enhancing crop productivity and reducing dependency on chemical fertilizers. This multifaceted role emphasizes the potential of EPF not only as biocontrol agents but also as vital contributors to nutrient cycling and soil health (Rasool et al. 2021; Ahsan et al. 2024).

7.2 Pathogenic Mechanisms

A thorough understanding of the genetic and biochemical pathways involved in EPF infection processes provides critical insights into their mechanisms of action. This knowledge enhances their efficacy as biological control agents. Current studies focus on virulence factors, enzyme production, and host recognition mechanisms, aiming to optimize EPF for pest control while ensuring safety for non-target organisms (Sharma and Sharma 2021; Qin et al. 2023). Genomic research continues to identify novel genes associated with virulence traits, supporting the development of enhanced fungal strains specifically designed for pest management (Sharma and Sharma 2021; Qin et al. 2023).

7.3 Potential for Biopesticide Development

The growing demand for sustainable agricultural solutions has driven interest in developing effective biopesticides from EPF (Bamisile et al. 2021b; Emaru et al. 2024). As agricultural practices evolve in response to environmental concerns and food security needs, EPF-based biopesticides offer a viable alternative to synthetic pesticides. These biological control agents can reduce chemical inputs, promote biodiversity, and enhance soil health. The versatility and effectiveness of EPF in controlling a wide range of pests make them attractive candidates for integration into modern pest management strategies. With advancements in biotechnology and agricultural sciences, researchers are exploring various approaches to improve the efficacy, safety, and longevity of EPF formulations, as outlined in Table 6.8. These efforts highlight the role of EPF in achieving sustainable agricultural systems.

7.3.1 Formulation Studies

Further research is needed to enhance the effectiveness of EPF-based biopesticides through advanced formulation technologies. These formulations aim to improve product delivery, stability, and shelf life. One promising approach involves micro-encapsulation, which protects fungal spores from environmental stresses and enhances their viability (Bamisile et al. 2021a; Vivekanandhan et al. 2024).

7.3.2 Field Trials

Large-scale field trials are essential for evaluating the effectiveness and safety of specific EPF strains under real-world conditions. These trials provide valuable data on optimal environmental conditions, dosage, and application frequency (Bamisile et al. 2021a; Qin et al. 2023). The understanding of the interactions between EPF, beneficial insects, soil health, and plant biodiversity is crucial to ensure that biopesticide applications do not disrupt key ecological components (Mantzoukas et al. 2022).

7.4 Integration of Biotechnology and Molecular Methods

The integration of advanced biotechnological and molecular techniques has transformed the study and application of EPF. These innovations enable researchers to manipulate and enhance the bioactive properties of EPF, improving their effectiveness in agricultural systems (Harith-Fadzilah et al. 2021; Altimira et al. 2022).

7.4.1 Genomic Research

Genomic sequencing of EPF provides a powerful platform for identifying novel genes associated with virulence and the production of bioactive compounds. This genetic information allows for the selection and development of fungal strains with enhanced pest control capabilities (Zhao et al. 2016; Harith-Fadzilah et al. 2021). Furthermore, genetic mapping reveals traits related to host specificity and environmental adaptability, which are essential for designing targeted solutions for specific crops and pest species (Altimira et al. 2022; Mantzoukas et al. 2022).

7.4.2 Metabolomic Analysis

Metabolomic studies play a critical role in identifying and characterizing the bioactive compounds produced by EPF. These compounds are responsible for a range of biological activities, including insecticidal and antimicrobial effects. By understanding the metabolic profiles of EPF, researchers can enhance the production of desirable compounds, thereby improving pest control efficacy and promoting plant health (Mantzoukas et al. 2022; Mascarin et al. 2024). This research direction is vital for optimizing the therapeutic potential of EPF, as the synergistic effects of multiple compounds can significantly boost their overall effectiveness.

7.4.3 Synthetic Biology

Synthetic biology offers innovative tools for engineering EPF with enhanced traits, such as increased resistance to environmental stressors and improved production of beneficial metabolites. Techniques like CRISPR-Cas9 enable precise genome editing, allowing for the targeted enhancement of virulence factors and other desirable characteristics (Mantzoukas et al. 2022; Mascarin et al. 2024). These advancements support the development of next-generation biocontrol agents tailored for specific agricultural needs.

The continued exploration of EPF through biotechnological and molecular approaches presents significant opportunities for sustainable pest management. By deepening the understanding of their ecological roles, infection mechanisms, and bioactive compounds, researchers can develop more effective biopesticides and integrated pest management strategies that align with the goals of sustainable agriculture.

8 Discussion

EPF represent a promising strategy for sustainable agriculture, offering biological control methods that align with ecological goals. As viable alternatives to chemical pesticides, EPF enhance pest control while minimizing environmental impact (Bamisile et al. 2021a; Ahsan et al. 2024). Their diverse mechanisms of action establish their effectiveness against a wide range of insect pests. Moreover, their role in promoting plant health and resilience contributes to a holistic approach to agricultural sustainability (Gutiérrez Cárdenas et al. 2024).

The ability of EPF to produce a variety of bioactive compounds is central to their biocontrol efficacy. These compounds disrupt essential physiological functions in insect hosts, leading to mortality. For instance, destruxins and beauvericin (Table 6.5 and Fig. 6.5) interfere with neural signaling and metabolic processes, ultimately proving fatal to insect targets (Wang and Xu 2012; Mantzoukas et al. 2022). Beyond pest control, these compounds also enhance nutrient uptake and protect plants from pathogens, reinforcing the role of EPF in integrated pest and disease management (Patocka 2016; Deka et al. 2021).

Recent studies highlight the benefits of integrating EPF into agricultural practices. Seed coating with EPF has been shown to improve germination rates and establish early-stage protection against pests. Field trials confirm the effectiveness of this approach and its potential to support sustainable farming systems (Rasool et al. 2021; Mantzoukas et al. 2023). Additionally, soil inoculation with EPF enriches the microbiome, enhances nutrient cycling, and fosters beneficial microbial interactions in the rhizosphere, contributing to robust plant growth (Liu et al. 2022; Ahsan et al. 2024).

Despite these advantages, several challenges hinder the widespread adoption of EPF. Environmental variability, such as fluctuations in temperature and humidity, can affect fungal viability and efficacy (Tomar and Yadav 2023). The complex interactions between EPF and their insect hosts require further investigation to optimize their use. Moreover, the stability of formulated products remains a concern, as fungal spores may lose viability under suboptimal storage or application conditions. Research emphasizes the importance of formulation stability for successful field deployment (Mantzoukas et al. 2022; Rodríguez-Romero et al. 2023).

Advancements in biotechnology and molecular biology offer promising solutions to these challenges. Genomic studies have elucidated the genetic basis of virulence traits, enabling the development of enhanced fungal strains tailored for specific pest targets (Zhao et al. 2016). Innovations in formulation technologies, such as microencapsulation and the use of inert carriers, can improve the stability and delivery of EPF products (Paschapur et al. 2021; Liu et al. 2022). Advanced ecological research will further inform integrated pest management strategies that fully utilize the benefits of EPF (Chaurasiya et al. 2021; Gonzalez-Guzman et al. 2021).

The broad spectrum of bioactive compounds produced by EPF, including antibacterial and antifungal agents, underscores their versatility in promoting agricultural sustainability (Rodríguez-Romero et al. 2023). These compounds not only target pests but also strengthen plant defenses against pathogens. Notable examples, such as destruxins, beauvericin, oxalic acid, and antimicrobial peptides, play critical roles in enhancing plant resilience to biotic stressors (Gibson et al. 2014; Weng et al. 2019; Farhan et al. 2024). This dual functionality positions EPF as both pest control agents and key contributors to plant health.

The integration of EPF into agricultural systems reflects a broader shift toward sustainable practices that prioritize ecological balance and reduce reliance on synthetic inputs. Their multifaceted benefits, including enhanced crop resilience, biodiversity promotion, and support for global food security, highlight their value in modern agriculture (Bamisile et al. 2021a; Liu et al. 2022). Future research and collaboration among scientists, industry stakeholders, and policymakers will be essential to fully realize the potential of EPF. Through the strategic use of their bioactive compounds, EPF can support integrated pest management. These approaches improve plant health, reduce pest pressure, and strengthen ecosystem resilience.

9 Conclusion

This chapter has outlined the significance of EPF and their bioactive compounds in sustainable agriculture. EPF offer a reliable alternative to synthetic pesticides. These fungi provide biological control of insect pests and contribute to plant health. The effectiveness of EPF depends on its mechanisms of action. Bioactive compounds such as destruxins and beauvericin disrupt essential functions in pests. These compounds also strengthen plant defenses and improve resilience to environmental stress. The integration of EPF into farming systems reflects a shift toward ecological balance and reduced chemical input. The benefits of EPF include improved crop resilience, increased biodiversity, and support for food security. The development of stable formulations and efficient delivery systems remains a priority for practical application. Advances in formulation and delivery methods are required to ensure consistent results under field conditions. Research and collaboration among scientists, industry, and policymakers will support further progress. The use of EPF-derived compounds can support integrated pest management. These strategies reduce pest pressure, improve plant health, and promote sustainable agricultural outcomes.

Acknowledgments Lutfun Nahar gratefully acknowledges the support from the European Regional Development Fund (Project ENOCH #CZ.02.1.01/0.0/0.0/16_019/0000868), the Czech Science Foundation (Project #23-05474S), and the Chinese Academy of Sciences (PIFI Project #2025PVA0074). We thank Seyedeh Mahsa Hosseini Nozari for her assistance with all graphical work.

References

Abdel-Raheem MA, Yassin SA (2019) Pathogenicity of entomopathogenic fungi against cotton aphid, *Aphis gossypii* under laboratory conditions. Acad J Entomol 12(2):44–48. https://idosi.org/aje/12(2)19/3.pdf

Abrehame S, Manoj VR, Hailu M, Chen YY, Lin YC et al (2023) Aflatoxins: source, detection, clinical features and prevention. Processes 11(1):204. https://doi.org/10.3390/pr11010204

Afifah L, Saputro NW (2020) Growth and viability of entomopathogenic fungus *Beauveria bassiana* (Balsamo) Vuillemin in different alternative media. IOP Conf Ser Earth Environ Sci 468(1):012037. https://doi.org/10.1088/1755-1315/468/1/012037

Ahmad VU, Basha A (eds) (2006) Spectroscopic data of steroid glycosides: stigmastanes, furostanes, spirtostanes: volume 2. Springer US, Boston. https://doi.org/10.1007/978-0-387-39572-2_352

Ahsan SM, Injamum-Ul-Hoque M, Das AK, Rahman MM, Mollah MMI et al (2024) Plant–entomopathogenic fungi interaction: recent progress and future prospects on endophytism-mediated growth promotion and biocontrol. Plants 13(10):1420. https://doi.org/10.3390/plants13101420

Aksoy A, Dmour SM, Qaralleh H, Alwaneen WS, Farsi RM et al (2023) An in vitro antimicrobial activity of *Beauveria bassiana* secondary metabolites. Pak J Agris Sci 60(3):331–336. https://doi.org/10.21162/PAKJAS/23.36

Aljaffary M, Jang H, Alomeir N, Zeng Y, Alkhars N et al (2023) Effects of nystatin oral rinse on oral *Candida* species and *Streptococcus mutans* among healthy adults. Clin Oral Investig 27(7):3557–3568. https://doi.org/10.1007/s00784-023-04969-5

Altimira F, Arias-Aravena M, Jian L, Real N, Correa P et al (2022) Genomic and experimental analysis of the insecticidal factors secreted by the entomopathogenic fungus *Beauveria pseudobassiana* RGM 2184. J Fungi 8(3):253. https://doi.org/10.3390/jof8030253

Aplin RT, Billek G, Grove JF, Halsall TG, Keller-Schierlein W et al (1964) Griseofulvin and some analogues. In: Fortschritte der Chemie Organischer Naturstoffe/Progress in the chemistry of organic natural products/Progrès dans la Chimie des substances Organiques Naturelles, vol 22. Springer, Vienna, pp 203–264. https://doi.org/10.1007/978-3-7091-7145-5_4

Aravinthraju K, Shanthi M, Murugan M, Srinivasan R, Maxwell LA et al (2024) Endophytic entomopathogenic fungi: their role in enhancing plant resistance, managing insect pests, and synergy with management routines. J Fungi 10(12):865. https://doi.org/10.3390/jof10120865

Bamisile BS, Akutse KS, Siddiqui JA, Xu Y (2021a) Model application of entomopathogenic fungi as alternatives to chemical pesticides: prospects, challenges, and insights for next-generation sustainable agriculture. Front Plant Sci 12:741804. https://doi.org/10.3389/fpls.2021.741804

Bamisile BS, Siddiqui JA, Akutse KS, Ramos Aguila LC, Xu Y (2021b) General limitations to endophytic entomopathogenic fungi use as plant growth promoters, pests and pathogens biocontrol agents. Plants. 10(10):2119. https://doi.org/10.3390/plants10102119

Baverstock J, Elliot SL, Alderson PG, Pell JK (2005) Response of the entomopathogenic fungus *Pandora neoaphidis* to aphid-induced plant volatiles. J Invertebr Pathol 89(2):157–164. https://doi.org/10.1016/j.jip.2005.05.006

Biesaga-Kos'cielniak J, Filek M (2009) Occurrence and physiology of zearalenone as a new plant hormone. In: Lichtfouse E (ed) Sociology, organic farming, climate change and soil science. Sustainable agriculture reviews, vol 3. Springer, Dordrecht, pp 419–435. https://doi.org/10.1007/978-90-481-3333-8_15

Boguś MI, Wrońska AK, Kaczmarek A, Boguś-Sobocińska M (2021) In vitro screening of 65 mycotoxins for insecticidal potential. PLoS One 16(3):e0248772. https://doi.org/10.1371/journal.pone.0248772

Boros-Majewska J, Salewska N, Borowski E, Milewski S, Malic S et al (2014) Novel Nystatin A 1 derivatives exhibiting low host cell toxicity and antifungal activity in an in vitro model of oral candidosis. Med Microbiol Immunol 203:341–355. https://doi.org/10.1007/s00430-014-0343-4

Butt TM, Coates CJ, Dubovskiy IM, Ratcliffe NA (2016) Entomopathogenic fungi: new insights into host–pathogen interactions. Adv Genet 94:307–364. https://doi.org/10.1016/bs.adgen.2016.01.006

Chanclud E, Morel JB (2016) Plant hormones: a fungal point of view. Mol Plant Pathol 17(8): 1289–1297. https://doi.org/10.1111/mpp.12393

Charnley AK, St Leger RJ (1991) The role of cuticle-degrading enzymes in fungal pathogenesis in insects. In: Cole GT, Hoch HC (eds) The fungal spore and disease initiation in plants and animals. Springer US, Boston, pp 267–286. https://doi.org/10.1007/978-1-4899-2635-7_12

Chaurasiya DK, Singh S, Kumar M, Singh AP, Sahni S (2021) Bio-control of insect pests through entomopathogenic fungi: a novel approach. In: Singh HK (ed) Current research and innovations in plant pathology, vol 27, pp 77–99. https://doi.org/10.22271/ed.book.1163

Chen W, Xie W, Cai W, Thaochan N, Hu Q (2021) Entomopathogenic fungi biodiversity in the soil of three provinces located in Southwest China and first approach to evaluate their biocontrol potential. J Fungi 7(11):984. https://doi.org/10.3390/jof7110984

Claydon N (1978) Insecticidal secondary metabolites from entomogenous fungi: *Entomophthora virulenta*. J Invertebr Pathol 32(3):319–324. https://doi.org/10.1016/0022-2011(78)90195-7

Crespo R, Pedrini N, Juárez MP, Dal Bello GM (2008) Volatile organic compounds released by the entomopathogenic fungus *Beauveria bassiana*. Microbiol Res 163(2):148–151. https://doi.org/10.1016/j.micres.2006.03.013

Davari A, Skinner M, Parker BL (2018) Cell electrofusion to improve efficacy and thermotolerance of the entomopathogenic fungus, *Beauveria bassiana*. J Appl Microbiol 125(5):1482–1493. https://doi.org/10.1111/jam.14031

Deka B, Baruah C, Babu A (2021) Entomopathogenic microorganisms: their role in insect pest management. Egypt J Biol Pest Control 31:1–8. https://doi.org/10.1186/s41938-021-00466-7

Dillon R, Charnley A (1990) Initiation of germination in conidia of the entomopathogenic fungus, *Metarhizium anisopliae*. Mycol Res 94(3):299–304. https://doi.org/10.1016/S0953-7562(09)80353-5

Duetz WA, Bouwmeester H, Van Beilen JB, Witholt B (2003) Biotransformation of limonene by bacteria, fungi, yeasts, and plants. Appl Microbiol Biotechnol 61:269–277. https://doi.org/10.1007/s00253-003-1221-y

Emaru A, Nyaanga JG, Mwanarusi S (2024) Use of Entomopathogenic fungi as biopesticides to manage insect pest: a review. Am J Appl Sci 21:1–14. https://doi.org/10.3844/ajassp.2024.1.14

FAO W (2024) Codex Alimentarius Commission. General standard for contaminants and toxins in food and feed (CXS 193-1995, Rev. 2024). Food and Agriculture Organization of the United Nations. Retrieved from https://www.fao.org/fao-who-codexalimentarius/thematic-areas/contaminants/en/

Farhan M, Pan J, Hussain H, Zhao J, Yang H et al (2024) Aphid-resistant plant secondary metabolites: types, insecticidal mechanisms, and prospects for utilization. Plants. 13(16):2332. https://doi.org/10.3390/plants13162332

Francis F, Fingu-Mabola JC, Ben Fekih I (2022) Direct and endophytic effects of fungal entomopathogens for sustainable aphid control: a review. Agriculture (Basel) 12(12):2081. https://doi.org/10.3390/agriculture12122081

Fredenhagen A, Molleyres LP, Böhlendorf B, Laue G (2006) Structure determination of neoefrapeptins A to N: peptides with insecticidal activity produced by the fungus *Geotrichum candidum*. J Antibiot 59(5):267–280. https://doi.org/10.1038/ja.2006.38

García-Espinoza F, García M, Quesada-Moraga E, Yousef-Yousef M (2023) Entomopathogenic fungus-related priming defense mechanisms in cucurbits impact *Spodoptera littoralis* (Boisduval) fitness. Appl Environ Microbiol 89(8):e00940-23. https://doi.org/10.1128/aem.00940-23

García-Estrada C, Cat E, Santamarta I (2016) *Beauveria bassiana* as biocontrol agent: formulation and commercialization for pest management. In: Singh H, Sarma B, Keswani C (eds) Agriculturally important microorganisms: commercialization and regulatory requirements in Asia, pp 81–96. https://doi.org/10.1007/978-981-10-2576-1_5

Gibson DM, Krasnoff SB (1999) Exploring the potential of biologically active compounds from plants and fungi. In: Cutler HG, Cutler SJ (eds) Biologically active natural products: agrochemicals. Routledge, pp 231–242

Gibson DM, Donzelli BG, Krasnoff SB, Keyhani NO (2014) Discovering the secondary metabolite potential encoded within entomopathogenic fungi. Nat Prod Rep 31(10):1287–1305. https://doi.org/10.1039/c4np00054d

Gielen R, Ude K, Kaasik A, Põldmaa K, Teder T et al (2024) Entomopathogenic fungi as mortality agents in insect populations: a review. Ecol Evol 14(12):e70666. https://doi.org/10.1002/ece3.70666

Goettel MS, Johnson DL, Inglis GD (1995) The role of fungi in the control of grasshoppers. Can J Bot 73(S1):71–75. https://doi.org/10.1139/b95-227

Gonzalez-Guzman A, Raya-Diaz S, Sacristán D, Yousef M, Sánchez-Rodríguez AR et al (2021) Effects of entomopathogenic fungi on durum wheat nutrition and growth in the field. Eur J Agron 128:126282. https://doi.org/10.1016/j.eja.2021.126282

Gouda SA, Hassanein NM, Rizk SA, Foad PF, Rabeea OE (2020) Evaluation of a potential biocontrol rhizosphere *Fusarium moniliforme* against *Curvularia spicifera* the causal agent of rice leaf spot and utility DFT study to evaluate one constituent in GC/MS. Curr Sci Int 9(1):57–66. https://doi.org/10.36632/csi/2020.9.1.7

Gutiérrez Cárdenas OG, Cortez Madrigal H, Garzón A (2024) Use and impact of endophytic entomopathogenic fungi: their potential in the context of agricultural sustainability. Biotecnia 26:e1986. https://doi.org/10.18633/biotecnia.v26.1986

Han P, Gong Q, Fan J, Abbas M, Chen D et al (2022) Destruxin a inhibits scavenger receptor B mediated melanization in *Aphis citricola*. Pest Manag Sci 78(5):1915–1924. https://doi.org/10.1002/ps.6809

Harith-Fadzilah N, Abd Ghani I, Hassan M (2021) Omics-based approach in characterising mechanisms of entomopathogenic fungi pathogenicity: a case example of *Beauveria bassiana*. J King Saud Univ Sci 33(2):101332. https://doi.org/10.1016/j.jksus.2020.101332

Hassan A, Li Z, Zhou X, Mo J, Huang Q (2024) Termite management by entomopathogenic fungi: recent advances and future prospects. Curr Res Biotechnol 7:100183. https://doi.org/10.1016/j.crbiot.2024.100183

Hernández MM, Martínez-Villar E, Peace C, Pérez-Moreno I, Marco V (2012) Compatibility of the entomopathogenic fungus *Beauveria bassiana* with flufenoxuron and azadirachtin against *Tetranychus urticae*. Exp Appl Acarol 58:395–405. https://doi.org/10.1007/s10493-012-9594-1

Hernández-Chávez MJ, Pérez-García LA, Niño-Vega GA, Mora-Montes HM (2017) Fungal strategies to evade the host immune recognition. J Fungi 3(4):51. https://doi.org/10.3390/jof3040051

Hong S, Shang J, Sun Y, Tang G, Wang C (2024) Fungal infection of insects: molecular insights and prospects. Trends Microbiol 32(3):302–316. https://doi.org/10.1016/j.tim.2023.09.005

Hung R, Lee S, Bennett JW (2015) Fungal volatile organic compounds and their role in ecosystems. Appl Microbiol Biotechnol 99:3395–3405. https://doi.org/10.1007/s00253-015-6494-4

Iida Y, Higashi Y, Nishi O, Kouda M, Maeda K et al (2023) Entomopathogenic fungus *Beauveria bassiana*–based bioinsecticide suppresses severity of powdery mildews of vegetables by inducing the plant defense responses. Front Plant Sci 14:1211825. https://doi.org/10.3389/fpls.2023.1211825

Irsad SM, Haq E, Mohamed A, Rizvi PQ et al (2023) Entomopathogen-based biopesticides: insights into unraveling their potential in insect pest management. Front Microbiol 14:1208237. https://doi.org/10.3389/fmicb.2023.1208237

Isaka M, Kittakoop P, Kirtikara K, Hywel-Jones NL, Thebtaranonth Y (2005) Bioactive substances from insect pathogenic fungi. Acc Chem Res 38(10):813–823. https://doi.org/10.1021/ar040247r

Islam W, Adnan M, Shabbir A, Naveed H, Abubakar YS et al (2021) Insect-fungal-interactions: a detailed review on entomopathogenic fungi pathogenicity to combat insect pests. Microb Pathog 159:105122. https://doi.org/10.1016/j.micpath.2021.105122

Khachatourians GG, Qazi SS (2008) Entomopathogenic fungi: biochemistry and molecular biology. In: Brakhage AA, Zipfel PF (eds) Human and animal relationships. The Mycota, vol 6. Springer, Berlin, pp 33–61. https://doi.org/10.1007/978-3-540-79307-6_3

Khoja S, Eltayef KM, Baxter I, Myrta A, Bull JC et al (2021) Volatiles of the entomopathogenic fungus, *Metarhizium brunneum*, attract and kill plant parasitic nematodes. Biol Control 152:104472. https://doi.org/10.1016/j.biocontrol.2020.104472

Kryukov VY, Glupov VV (2023) Special issue on "Entomopathogenic fungi: ecology, evolution, adaptation": an editorial. Microorganisms 11(6):1494. https://doi.org/10.3390/microorganisms11061494

Kurniawan I, Afandhi A, Tarno H (2022) Colony growth, sporulation, and viability of entomopathogenic fungus *Beauveria bassiana* (Balsamo) Vuillemin on various agricultural waste as growing media. Res J Life Sci 9(2):77–90. https://doi.org/10.21776/ub.rjls.2022.009.02.4

Lambert C, Schmidt K, Karger M, Stadler M, Stradal TE et al (2023) Cytochalasans and their impact on actin filament remodeling. Biomolecules 13(8):1247. https://doi.org/10.3390/biom13081247

Li Y, Song Y, Liu J, Ma Y, Tan R (2005) Anti-*Helicobacter pylori* substances from endophytic fungal cultures. World J Microbiol Biotechnol 21:553–558. https://doi.org/10.1007/s11274-004-3273-2

Liu Y, Yang Y, Wang B (2022) Entomopathogenic fungi *Beauveria bassiana* and *Metarhizium anisopliae* play roles of maize (*Zea mays*) growth promoter. Sci Rep 12(1):15706. https://doi.org/10.1038/s41598-022-19899-7

Lu HL, Leger RS (2016) Insect immunity to entomopathogenic fungi. Adv Genet 94:251–285. https://doi.org/10.1016/bs.adgen.2015.11.002

Ma M, Luo J, Li C, Eleftherianos I, Zhang W et al (2024) A life-and-death struggle: interaction of insects with entomopathogenic fungi across various infection stages. Front Immunol 14: 1329843. https://doi.org/10.3389/fimmu.2023.1329843

Mannino MC, Davyt-Colo B, Pedrini N (2021) Toxic secondary metabolites and virulence factors expression by entomopathogenic fungi during insect infection and potential impact as a tool for pest management. In: Khan MA, Ahmad W (eds) Microbes for sustainable insect pest management. Hydrolytic enzyme & secondary metabolite–volume 2, pp 121–134. https://doi.org/10.1007/978-3-030-67231-7_6

Mantzoukas S, Kitsiou F, Natsiopoulos D, Eliopoulos PA (2022) Entomopathogenic fungi: interactions and applications. Encyclopedia 2(2):646–656. https://doi.org/10.3390/encyclopedia2020044

Mantzoukas S, Papantzikos V, Katsogiannou S, Papanikou A, Koukidis C et al (2023) Biostimulant and bioinsecticidal effect of coating cotton seeds with endophytic *Beauveria bassiana* in semi-field conditions. Microorganisms 11(8):2050. https://doi.org/10.3390/microorganisms11082050

Mascarin GM, Shrestha S, De Carvalho Barros Cortes MV, Ramirez JL, Dunlap CA et al (2024) CRISPR-Cas9-mediated enhancement of *Beauveria bassiana* virulence with overproduction of oosporein. Fungal Biol Biotechnol 11(1):21. https://doi.org/10.1186/s40694-024-00190-5

Meyling NV, Eilenberg J (2007) Ecology of the entomopathogenic fungi *Beauveria bassiana* and *Metarhizium anisopliae* in temperate agroecosystems: potential for conservation biological control. Biol Control 43(2):145–155. https://doi.org/10.1016/j.biocontrol.2007.07.007

Mishra A, Gond SK, Kumar A, Sharma VK, Verma SK et al (2012) Sourcing the fungal endophytes: a beneficial transaction of biodiversity, bioactive natural products, plant protection and nanotechnology. In: Satyanarayana T, Johri B (eds) Microorganisms in sustainable agriculture and biotechnology. Springer, Dordrecht, pp 581–612. https://doi.org/10.1007/978-94-007-2214-9_26

Mondal S, Baksi S, Koris A, Vatai G (2016) Journey of enzymes in entomopathogenic fungi. Pac Sci Rev A Nat Sci Eng 18(2):85–99. https://doi.org/10.1016/j.psra.2016.10.001

Morcia C, Piazza I, Ghizzoni R, Delbono S, Felici B et al (2022) In search of antifungals from the plant world: the potential of saponins and *Brassica* species against *Verticillium dahliae* Kleb. Horticulturae 8(8):729. https://doi.org/10.3390/horticulturae8080729

Mousa WK, Raizada MN (2013) The diversity of anti-microbial secondary metabolites produced by fungal endophytes: an interdisciplinary perspective. Front Microbiol 4:65. https://doi.org/10.3389/fmicb.2013.00065

Murindangabo YT, Kopecký M, Perná K, Konvalina P, Bohatá A et al (2024) Relevance of entomopathogenic fungi in soil–plant systems. Plant Soil 495(1):287–310. https://doi.org/10.1007/s11104-023-06325-8

Nakashima M, Mitaka Y, Inagaki T, Matsuura K (2024) An antifungal compound secreted by termite workers, phenylacetic acid, inhibits the growth of both termite egg-mimicking fungus and entomopathogenic fungi. Insect Soc 71(2):221–232. https://doi.org/10.1007/s00040-024-00966-3

Ortiz-Urquiza A, Keyhani NO (2013) Action on the surface: entomopathogenic fungi versus the insect cuticle. Insects 4(3):357–374. https://doi.org/10.3390/insects4030357

Otieno J, Pallmann P, Poehling HM (2016) The combined effect of soil-applied azadirachtin with entomopathogens for integrated management of western flower thrips. J Appl Entomol 140(3): 174–186. https://doi.org/10.1111/jen.12242

Pammi M, Liang R, Hicks JM, Barrish J, Versalovic J (2011) Farnesol decreases biofilms of *Staphylococcus epidermidis* and exhibits synergy with nafcillin and vancomycin. Pediatr Res 70(6):578–583. https://doi.org/10.1203/PDR.0b013e318232a984

Parveen SS (2023) Entomopathogenic fungi as a sustainable biopesticide: current status and future prospects. J Entomol Res 47(4):666–671. https://doi.org/10.5958/0974-4576.2023.00122.6

Paschapur A, Subbanna AR, Singh AK, Jeevan B, Stanley J et al (2021) Unraveling the importance of metabolites from entomopathogenic fungi in insect pest management. In: Khan MA, Ahmad W (eds) Microbes for sustainable insect pest management. Hydrolytic enzyme & secondary metabolite–volume 2, pp 89–120. https://doi.org/10.1007/978-3-030-67231-7_5

Patocka J (2016) Bioactive metabolites of entomopathogenic fungi *Beauveria bassiana*. Mil Med Sci Lett 85(2):80. https://doi.org/10.31482/mmsl.2016.015

Pedrini N (2018) Molecular interactions between entomopathogenic fungi (Hypocreales) and their insect host: perspectives from stressful cuticle and hemolymph battlefields and the potential of dual RNA sequencing for future studies. Fungal Biol 122(6):538–545. https://doi.org/10.1016/j.funbio.2017.10.003

Qayyum MA, Saeed S, Wakil W, Nawaz A, Iqbal N et al (2021) Diversity and correlation of entomopathogenic and associated fungi with soil factors. J King Saud Univ Sci 33(6):101520. https://doi.org/10.1016/j.jksus.2021.101520

Qin Y, Liu X, Peng G, Xia Y, Cao Y (2023) Recent advancements in pathogenic mechanisms, applications and strategies for entomopathogenic fungi in mosquito biocontrol. J Fungi 9(7): 746. https://doi.org/10.3390/jof9070746

Qu S, Wang S (2018) Interaction of entomopathogenic fungi with the host immune system. Dev Comp Immunol 83:96–103. https://doi.org/10.1016/j.dci.2018.01.010

Quesada-Moraga E, Garrido-Jurado I, González-Mas N, Yousef-Yousef M (2023) Ecosystem services of entomopathogenic ascomycetes. J Invertebr Pathol 201:108015. https://doi.org/10.1016/j.jip.2023.108015

Raghunandan B, Dave A, Baria PR, Manjari (2024) Applications of bioactive compounds from fungal entomopathogens. In: Deshmukh SK, Sridhar KR (eds) Entomopathogenic fungi: prospects and challenges. Springer, pp 453–478. https://doi.org/10.1007/978-981-97-5991-0_18

Rasool S, Vidkjær NH, Hooshmand K, Jensen B, Fomsgaard IS et al (2021) Seed inoculations with entomopathogenic fungi affect aphid populations coinciding with modulation of plant secondary metabolite profiles across plant families. New Phytol 229(3):1715–1727. https://doi.org/10.1111/nph.16979

Rath AC (2010) The use of entomopathogenic fungi for control of termites. Biocontrol Sci Tech 10(5):563–581. https://doi.org/10.1080/09583150075001637O

Rodríguez-Romero H, Rodríguez-Peláez L, Reyes-Castro A, Tlakaele Notario-Rendón O, González-Peréz M et al (2023) Perspective chapter: secondary metabolites of entomopathogens as biotechnological tools for the biological control of agricultural insect pests. In: Insecticides—advances in insect control and sustainable pest management. IntechOpen. https://doi.org/10.5772/intechopen.111465

Sabbahi R, Hock V, Azzaoui K, Saoiabi S, Hammouti B (2022) A global perspective of entomopathogens as microbial biocontrol agents of insect pests. J Agr Food Res 10:100376. https://doi.org/10.1016/j.jafr.2022.100376

Salazar-Mendoza P, Magalhães DM, Pec M, Azevedo KE, Delalibera JI et al (2025) Endophytic entomopathogenic fungus, individually and in combination with rhizobacteria, enhances resistance in wild and cultivated tomatoes to *Tuta absoluta*. J Pest Sci 98:731–745. https://doi.org/10.1007/s10340-024-01854-2

Samal I, Bhoi TK, Majhi PK, Murmu S, Pradhan AK et al (2023) Combatting insects mediated biotic stress through plant associated endophytic entomopathogenic fungi in horticultural crops. Front Plant Sci 13:1098673. https://doi.org/10.3389/fpls.2022.1098673

San Juan-Maldonado N, Rodrí S, Angel-Cuapio A, Vá N, Barranco-Florido JE (2024) Secondary metabolites of entomopathogenic fungi, biological alternative for the control of agricultural pests and disease: present and perspectives. Adv Enzyme Res 12(1):1–15. https://doi.org/10.4236/aer.2024.121001

Shahbaz M, Palaniveloo K, Tan YS, Palasuberniam P, Ilyas N et al (2024) Entomopathogenic fungi in crops protection with an emphasis on bioactive metabolites and biological activities. World J Microbiol Biotechnol 40(7):217. https://doi.org/10.1007/s11274-024-04022-x

Sharma R, Sharma P (2021) Fungal entomopathogens: a systematic review. Egypt J Biol Pest Control 31:1–13. https://doi.org/10.1186/s41938-021-00404-7

Sharma A, Sharma S, Yadav PK (2023) Entomopathogenic fungi and their relevance in sustainable agriculture: a review. Cogent Food Agric 9(1):2180857. https://doi.org/10.1080/23311932.2023.2180857

Shin TY, Bae SM, Woo SD (2016) Screening and characterization of antimicrobial substances originated from entomopathogenic fungi. J Asia Pac Entomol 19(4):1053–1059. https://doi.org/10.1016/j.aspen.2016.09.008

Sinha KK, Choudhary AK, Kumari P (2016) Entomopathogenic fungi. In: Omkar (ed) Ecofriendly pest management for food security. Elsevier, pp 475–505. https://doi.org/10.1016/B978-0-12-803265-7.00015-4

Sun R, Hong B, Reichelt M, Luck K, Mai DT et al (2023) Metabolism of plant-derived toxins from its insect host increases the success of the entomopathogenic fungus *Beauveria bassiana*. ISME J 17(10):1693–1704. https://doi.org/10.1038/s41396-023-01480-3

Tine S, Halaimia A, Chechoui J, Tine-Djebbar F (2018) Fumigant toxicity and repellent effect of azadirachtin against the lesser grain beetle, *Rhyzopertha dominica* (F.) (Col.: Bostrichidae). In: Kallel A, Ksibi M, Ben Dhia H, Khélifi N (eds) Recent advances in environmental science from the Euro-Mediterranean and surrounding regions. EMCEI 2017. Advances in science, technology & innovation. Springer, Cham. https://doi.org/10.1007/978-3-319-70548-4_125

Tomar P, Yadav AN (2023) Entomopathogenic microorganisms as biocontrol agents for agricultural sustainability. Biocatal Agric Biotechnol 62:103431. https://doi.org/10.1016/j.bcab.2024.103431

Tong Y, Liu M, Zhang Y, Liu X, Huang R et al (2016) Beauvericin counteracted multi-drug resistant *Candida albicans* by blocking ABC transporters. Synth Syst Biotechnol 1(3):158–168. https://doi.org/10.1016/j.synbio.2016.10.001

Ujváry I (1999) Nicotine and other insecticidal alkaloids. In: Yamamoto I, Casida JE (eds) Nicotinoid insecticides and the nicotinic acetylcholine receptor. Springer, Tokyo, pp 29–69. https://doi.org/10.1007/978-4-431-67933-2_2

Valero-Jiménez CA, Van Kan JA, Koenraadt CJ, Zwaan BJ, Schoustra SE (2017) Experimental evolution to increase the efficacy of the entomopathogenic fungus *Beauveria bassiana* against malaria mosquitoes: effects on mycelial growth and virulence. Evol Appl 10(5):433–443. https://doi.org/10.1111/eva.12451

Vega FE, Goettel MS, Blackwell M, Chandler D, Jackson MA et al (2009) Fungal entomopathogens: new insights on their ecology. Fungal Ecol 2(4):149–159. https://doi.org/10.1016/j.funeco.2009.05.001

Verma ML, Kumar A, Chintagunta AD, Samudrala PJK, Bardin M et al (2024) Microbial production of biopesticides for sustainable agriculture. Sustainability 16(17):7496. https://doi.org/10.3390/su16177496

Vivekanandhan P, Swathy K, Sarayut P, Patcharin K (2024) Biology, classification, and entomopathogen-based management and their mode of action on *Tuta absoluta* (Meyrick) in Asia. Front Microbiol 15:1429690. https://doi.org/10.3389/fmicb.2024.1429690

Wang Q, Xu L (2012) Beauvericin, a bioactive compound produced by fungi: a short review. Molecules 17(3):2367–2377. https://doi.org/10.3390/molecules17032367

Weng Q, Zhang X, Chen W, Hu Q (2019) Secondary metabolites and the risks of *Isaria fumosorosea* and *Isaria farinosa*. Molecules 24(4):664. https://doi.org/10.3390/molecules24040664

Wink M (2018) Plant secondary metabolites modulate insect behavior-steps toward addiction? Front Physiol 9:364. https://doi.org/10.3389/fphys.2018.00364

Zhang L, Fasoyin OE, Molnár I, Xu Y (2020) Secondary metabolites from hypocrealean entomopathogenic fungi: novel bioactive compounds. Nat Prod Rep 37(9):1181–1206. https://doi.org/10.1039/c9np00065h

Zhao H, Lovett B, Fang W (2016) Genetically engineering entomopathogenic fungi. Adv Genet 94:137–163. https://doi.org/10.1016/bs.adgen.2015.11.001

Zhou W, Arcot Y, Medina RF, Bernal J, Cisneros-Zevallos L et al (2024) Integrated pest management: an update on the sustainability approach to crop protection. ACS Omega 9(40):41130–41147. https://doi.org/10.1021/acsomega.4c06628

Zimmermann G (2007) Review on safety of the entomopathogenic fungi *Beauveria bassiana* and Beauveria brongniartii. Biocontrol Sci Tech 17(6):553–596. https://doi.org/10.1080/09583150701309006

Production and Application of Enzymes from Entomopathogenic Fungi in Biological Control

7

Amanda do Carmo Alves and Filippe Elias de Freitas Soares

Abstract

Biological pest control has gained increasing attention as a sustainable alternative to the excessive use of chemical pesticides, which pose risks such as insect resistance, environmental contamination, and threats to human health. Entomopathogenic fungi, particularly from the genera *Beauveria*, *Metarhizium*, and *Ophiocordyceps*, represent promising biocontrol agents due to their specificity in infecting and eliminating target pests, minimizing effects on non-target organisms. A key factor in their pathogenicity is the secretion of hydrolytic enzymes, including chitinases, proteases, and lipases, which degrade vital host structures such as the insect cuticle, enabling successful infection and colonization. This chapter reviews the production, optimization, and application of these fungal enzymes in the development of biopesticides, covering molecular mechanisms and advanced production and purification methods, as well as innovative formulation strategies such as encapsulation and enzyme cocktails. Challenges related to culture medium optimization, enzyme stability, and industrial scalability are also discussed. Emphasizing the potential of fungal enzymes as eco-friendly biocatalysts, this work advocates the integration of scientific advances, cutting-edge technologies, and sustainable agricultural practices to promote enzymatic biopesticides as a vital component of integrated pest management strategies. Furthermore, continued development in this field can contribute to reducing agriculture's environmental impact and fostering more resilient and efficient production systems.

A. do Carmo Alves · F. E. de Freitas Soares (✉)
Department of Chemistry, Universidade Federal de Lavras, Minas Gerais, Brazil
e-mail: filippe.soares@ufla.br

S. K. Deshmukh et al. (eds.), *Frontiers in Entomopathogenic Fungi*,
https://doi.org/10.1007/978-981-95-6529-0_7

Keywords

Chitinase · Biocontrol · Biopesticides · Protease

1 Introduction

Given the growing concern about environmental impacts and the search for sustainable agricultural systems, the development of green technologies has become the focus of intense scientific research. In Brazil, the growth of agricultural areas and the intensification of cultivation have led to an increase in pests, as well as the indiscriminate use of chemical pesticides. This scenario has worsened over the years, causing problems such as the development of resistance in insects, environmental contamination, and risks to human health (Begg et al. 2017; Siqueira et al. 2021).

Biological control has emerged as an effective form of management that employs the use of living organisms to regulate the incidence of pests, diseases, and weeds. Different categories exist, each offering unique strategies for sustainable pest management. Classical biological control is a strategy in which parasitoids, predators, or pathogens are used to control a population that is out of balance. These natural enemies are introduced to establish themselves permanently in the area (Heimpel and Cock 2018; Van Lenteren et al. 2018). A good example is the use of the wasp *Cotesia flavipes* to control the sugarcane borer (*Diatraea saccharalis*) (Mahmoud et al. 2011). Augmentative biological control consists of the large-scale release of natural enemies to control a population, aiming to achieve a balance between the target population and the ecosystem. However, this type of control acts temporarily, requiring release at various times of the year according to the needs of the area (Collier and Van Steenwyk 2004). For example, the release of *Trichogramma* spp. to control lepidoptera in corn crops has proven to be effective when integrated into integrated pest management (IPM) (Parra 2014).

Conservative biological control, on the other hand, is a sustainable management strategy that seeks to preserve, increase, and optimize the performance of natural enemies that are already present in a given area through adjustments to the environment and agricultural practices (Gurr et al. 2017). Unlike classical biological control, which involves the introduction of exotic agents or augmentative control based on large-scale releases, this approach assumes that agricultural ecosystems already harbor a diversity of beneficial organisms that, when managed correctly, can provide effective pest control (Begg et al. 2017). Studies show that the adoption of flower strips in crops increases the abundance of ladybugs (*Coccinellidae*) and syrphids, reducing aphid infestations (Tschumi et al. 2016; Hatt et al. 2018; Tougeron et al. 2023).

Biological control strategies are based on the action of biological control agents (BCAs). Among these agents, notable examples include parasitoids, such as wasps from the Braconidae family; predators, such as lady beetles (Coccinellidae) and predatory mites (Phytoseiidae); and entomopathogenic microorganisms, including bacteria (*Bacillus thuringiensis*), viruses (baculoviruses), and fungi (*Beauveria* and *Metarhizium*). Among these groups, entomopathogenic fungi have stood out due to their effectiveness in controlling various agricultural pests, acting as natural pathogens of insects and mites. Furthermore, these fungi possess the ability to infect and eliminate insect pests in a specific manner, reducing adverse effects on non-target organisms (Vega et al. 2009; Ortiz-Urquiza et al. 2015).

Entomopathogenic fungi serve as the foundation for developing mycopesticides, which are fungi-based microbial pesticides used to control various agricultural pests. In the current Brazilian agricultural landscape, 330 microbial insecticides are marketed, 68% of which are derived from entomopathogenic fungi, highlighting the growth of biological solutions (Fig. 7.1) (Brazil, Ministry of Agriculture, Livestock and Supply 2025).

However, the sustainable expansion of the mycopesticide market in Brazil depends critically on continuous improvement in product quality. Low-efficacy mycopesticides, in addition to compromising results in the field, can pose risks to human and environmental health. To ensure safety and efficiency, it is essential to adopt strict quality control standards, from the selection of fungal isolates to formulation, packaging, and storage. Despite the numerous advantages of mycopesticides, the emergence of resistance in insect pest populations remains a growing challenge, similar to what is observed with conventional agrochemicals. This resistance may occur through mechanisms such as modifications to the insect cuticle, reduced fungal recognition, or the activation of more efficient immune responses. According to Qi et al. (2023), insects have evolved multiple defense strategies against entomopathogenic fungi, including physical barriers, innate immune responses, and adaptive behavioral changes. In this context, the search for technological innovations has guided research toward enzymes produced by these fungi, such as chitinases and proteases, which are capable of degrading key insect structures. These enzymes represent the next frontier in the development of high-quality biopesticides, which is the central theme of this chapter.

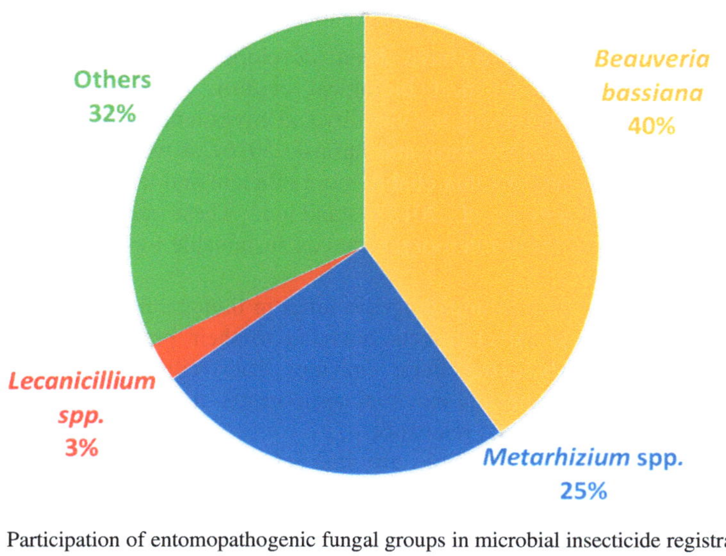

Fig. 7.1 Participation of entomopathogenic fungal groups in microbial insecticide registrations in the AGROFIT system

2 Mechanisms of Action and Enzymes

More than 1000 species of filamentous fungi and yeasts establish a variety of ecological interactions, including symbiosis, commensalism, and parasitism with diverse host organisms. According to Elsherbiny et al. (2023), many vascular plants, such as *Zea mays* (maize), *Triticum aestivum* (wheat), and *Glycine max* (soybean), form mutualistic associations with mycorrhizal and endophytic fungi that enhance nutrient acquisition and stress tolerance. Additionally, Elkhateeb et al. (2021) describe intricate symbiotic relationships between insects and fungi, including leaf-cutter ants (*Atta* spp.) cultivating *Leucoagaricus* fungi, ambrosia beetles (*Xyleborus* spp.) harboring ambrosial fungi, and entomopathogenic fungi such as *Metarhizium* and *Beauveria* species that parasitize insect pests.

They belong mostly to *Hypocreales* (Ascomycota) and *Entomophthorales* (Entomophthoromycota). The most studied and widely used genera in various commercial formulations are *Beauveria*, *Metarhizium*, *Ophiocordyceps*, and *Lecanicillium* (de Faria and Wraight 2007; Sung et al. 2007). Some species formerly classified under the genus *Cordyceps* have been reclassified, with many entomopathogenic fungi now placed in the genus *Ophiocordyceps* (Sung et al. 2007). *Beauveria bassiana* is very effective against stink bugs such as *Euschistus heros* (Sosa-Gómez et al. 2020) and also against whiteflies (*Bemisia tabaci*) (de Faria and Wraight 2007). *Metarhizium anisopliae* is used in the management of beetles such as *Anthonomus grandis* (cotton boll weevil) (Nussenbaum and Lecuona 2012) and is also employed against disease vectors such as *Aedes aegypti* (Scholte et al. 2004). *Lecanicillium fumosoroseum* (formerly *Isaria fumosorosea*) has been widely marketed for the control of psyllids (*Diaphorina citri*) (Zare and Gams 2001), and *Metarhizium rileyi* (formerly *Nomuraea rileyi*) has proven effective against caterpillars such as *Spodoptera frugiperda* (Sosa-Gómez et al. 2020). In addition, *Lecanicillium lecanii* is effective against aphids and mealybugs (Vega et al. 2009).

Technological innovations have been developed to enhance the use of entomopathogenic fungi in pest biological control. Among them, improved formulations aimed at increasing the viability and environmental stability of conidia under adverse conditions (Mascarin and Jaronski 2016), as well as the development of controlled-release systems that enable more efficient and targeted application of biological agents (Lacey et al. 2015), stand out. These advances significantly strengthen the role of these microorganisms as sustainable tools in integrated pest management.

The effectiveness of entomopathogenic fungi in biological control is not solely due to their ability to infect insects but also to their production of bioactive compounds, particularly extracellular enzymes that facilitate host invasion and degradation. These enzymes are essential components of the infection process and also hold significant industrial relevance.

Enzymes are a group of biomolecules that act as biological catalysts in various metabolic processes. According to Orlandelli et al. (2012), enzymes exhibit both intracellular and extracellular activity, performing vital functions in biological systems. Their ability to accelerate chemical reactions without being consumed in the process makes them valuable for industrial applications (Nelson and Cox 2014). Fungi stand out as efficient producers of extracellular enzymes, a characteristic that facilitates their collection and use on a large scale (Dey et al. 2022). There is a wide range of enzymes produced by fungi, including amylases, proteases, lipases, cellulases, and chitinases. *Paecilomyces* and *Beauveria* are particularly efficient in the production of hydrolytic enzymes such as keratinases and chitinases (Fernandes 2012; Alves et al. 2020). Brazil, due to its rich biodiversity and abundance of organic matter, offers ideal conditions for the production of these enzymes from low-cost substrates. The mechanism of enzymatic catalysis consists of accelerating chemical reactions by reducing the activation energy, that is, the amount of energy required for the reaction to occur. This process is possible thanks to the presence of specific active sites in the enzyme structure, where the substrate binds and the reaction is facilitated (Nelson and Cox 2014). Enzymatic activity is highly sensitive to environmental conditions, so the optimal temperature and pH ranges for the use of each enzyme must always be considered. This specificity makes strict control of processing conditions very important when using enzymes in industrial applications.

3 Infection Process and the Role of Enzymes

Infection by entomopathogenic fungi is a sequential process that begins with the adhesion of conidia to the insect cuticle, mediated by hydrophobic interactions and adhesion proteins. Environmental factors such as humidity and temperature influence both adhesion and spore germination (Holder and Keyhani 2005). The cuticle, the insect's main physical barrier, consists of an outer lipid layer (epicuticle) and an inner matrix of chitin and proteins (procuticle) (Andersen 2010).

To penetrate this barrier, fungi use a combination of mechanical pressure, exerted through appressoria, and the secretion of extracellular enzymes. The degradation process occurs in stages: first, lipases break down the hydrophobic waxy layer (Pedrini et al. 2013); then, proteases digest the protein matrix, while chitinases hydrolyze the chitin polymers (St Leger et al. 1996). This synergistic action weakens the cuticle, allowing the fungal penetration and releasing of essential nutrients for its establishment (Wang and St Leger 2007). The efficiency of this process varies among species. *Beauveria bassiana* and *Metarhizium* spp. directly influence their virulence. After penetration, the fungus invades the hemocoel, where it multiplies and produces toxins (such as destruxins) to suppress the host's immunity (Ortiz-Urquiza et al. 2015). The insect dies due to septicemia and the action of toxic metabolites (Fig. 7.2) (Vega et al. 2009). Finally, the fungus emerges from the cadaver to produce new conidia, restarting the infection cycle, as these conidia spread with the wind (Lacey et al. 2015). Given the central importance of enzymes in this process, the following topics will explore the role of chitinases, proteases, and lipases in the success of infection.

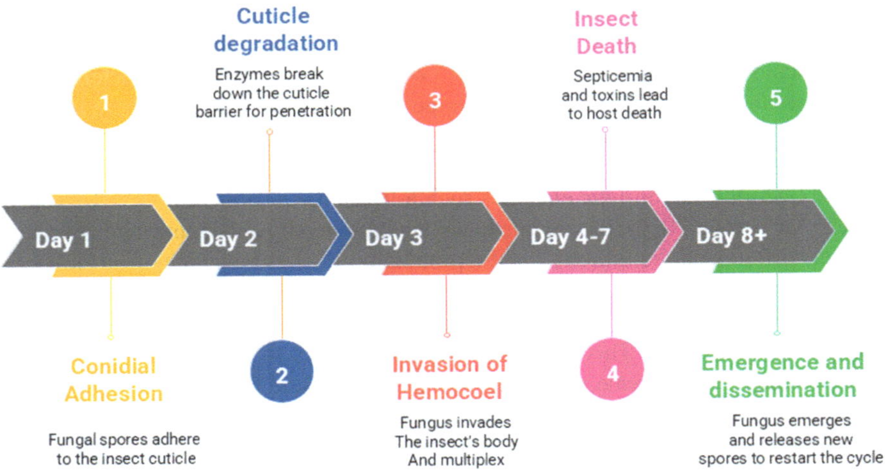

Fig. 7.2 Summary of days involved in the infection process of entomopathogenic fungi in insects

3.1 Chitinases

Chitinases (EC 3.2.1.14) are enzymes essential for the pathogenicity of entomopathogenic fungi, responsible for degrading chitin, a structural polysaccharide in the cuticle of insects. Chitinases are classified into families GH18 and GH19 (Glycosyl Hydrolases) based on their molecular structure. In most entomopathogenic fungi, GH18 predominates. This family is subdivided into

- Endochitinases (EC 3.2.1.14): They act on β-1,4 bonds within the chitin chain.
- Exochitinases (EC 3.2.1.52): Release monomers (N-acetylglucosamine) from the ends of the chain.
- Chitobiases (EC 3.2.1.29): Catalysis of chitin oligomer degradation into smaller units.

The genes that encode chitinases in *Metarhizium* spp. are *Mechit1* and *Mechit2*. The main function of chitinases expressed by *Mechit1* during the infection process is related to penetration of the insect cuticle, while *Mechit2* acts in the degradation of chitin into soluble oligomers, facilitating nutrient absorption. In addition, mutations that can occur in these genes can lead to a 60–70% reduction in the virulence of the fungus (St Leger et al. 2011). These genes are very important during the process, but under normal conditions they are suppressed.

For *B. bassiana*, the genes that encode chitinases are *Bbchit1* and *Bbchit2*, which belong to the GH18 family. *Bbchit1* plays an important role in cuticle penetration and evasion of the insect's immune system. Δ*Bbchit1* mutants have been shown to have 50% reduced virulence against *S. frugiperda* (Jin et al. 2011). *Bbchit2* enzyme exhibits hydrolytic activity on complex chitinous substrates, including protein-associated chitin, such as those present in the cuticle of the whitefly (*Bemisia tabaci*), indicating its participation in the virulence of the fungus.

3.2 Proteases

Proteases (EC 3.4) are a group of enzymes that have been extensively studied due to their biotechnological applications in the pharmaceutical and food industries (Choi et al. 2015; Souza et al. 2015). They are grouped into five classes according to the active amino acid residue at the catalytic site.

Serine proteases (EC 3.4.21) are enzymes that have a highly reactive serine residue at their catalytic site, which is essential for cleaving peptide bonds in proteins. Examples of these enzymes include trypsin, chymotrypsin, elastase, and subtilisin (Han et al. 2017). Among them, subtilisins are widely studied because they are produced by entomopathogenic fungi (Muszewska et al. 2017). Research conducted by Bagga et al. (2004) demonstrated the existence of more than 15 copies of genes that express secreted serine proteases of the proteinase K subfamily (subtilisin, S8A) in *M. anisopliae.*

Another class of proteases is metalloproteases (EC 3.4.24), which are enzymes that depend on metal ions (Zn^{2+}, Mn^{2+}) for catalytic activity. Fungi such as *Aspergillus* and *Beauveria* secrete metalloproteases to degrade the proteins of the host's extracellular matrix (Vertyporokh and Wojda 2017). Cysteine proteases (EC 3.4.22), on the other hand, have a cysteine residue at the active site. These proteases are widely found in various organisms, including pathogens such as *Candida albicans*, where they play an important role in the invasion of human tissues through the degradation of extracellular matrix components (Naglik et al. 2004; Moyes et al. 2015).

Aspartic proteases (EC 3.4.23) use aspartic acid residues in their active site to catalyze the hydrolysis of peptide bonds, acting optimally at acidic pH. These enzymes are produced by various filamentous fungi, especially species of the genus *Aspergillus*, such as *A. niger*, *A. flavus*, *A. oryzae*, *A. fumigatus*, and *A. terreus*. Among them, *A. niger* is widely used in the industrial production of aspartic proteases (Wei et al. 2023). On the other hand, threonine proteases (EC 3.4.25) have a threonine residue at the active site and are essential components of the proteasome, a protein complex responsible for the degradation of ubiquitinated proteins in eukaryotes (Seemüller et al. 1995). In organisms such as *Saccharomyces cerevisiae* and *A. nidulans*, proteasomes play a crucial role in protein regulation, participating in processes such as cell cycle control and stress response (Shishodia and Shankar 2020).

In this scenario, proteases expressed by *Metarhizium* spp. and *B. bassiana* are of great importance, since they act in the degradation of the insect cuticle, allowing penetration and colonization of the host. *B. bassiana* expresses a variety of serine proteases and metalloproteases. PR1-like subtilisins act by degrading the proteins of the insect cuticle protein layer, facilitating fungal penetration (Ortiz-Urquiza et al. 2015). Their expression is strongly regulated by environmental factors such as pH and nutrient availability. In addition, these enzymes act in later stages of infection, aiding in the spread of the fungus in the host's hemocoel.

Metarhizium spp. secrete several proteases that act by degrading cuticle components and suppressing the host's immunity. Among the main ones are alkaline proteases Pr1 (serine proteases), which are effective in cleaving cuticle proteins such as resilin and sclerotin and are considered markers of virulence (Wang and St Leger 2007). The cysteine proteases Mcp1 complement the action of Pr1 by degrading specific structural components of the cuticle (Screen et al. 1997). Zinc-dependent metalloproteases (Mpr1/Mpr2), on the other hand, inhibit melanization and the host's immune response by degrading antimicrobial peptides (Thompson 2022). The expression of these enzymes is strongly regulated during the infectious process, reflecting metabolic adaptations to the different barriers encountered in the host (Screen and St Leger 2000).

3.3 Lipases

Lipases (triacylglycerol hydrolases, EC 3.1.1.3) are enzymes that catalyze the hydrolysis of triacylglycerols into fatty acids and glycerol, in addition to participating in esterification, transesterification, and interesterification reactions in non-aqueous media (Jaeger and Eggert 2002). These enzymes are widely distributed in nature and are found in animals, plants, and microorganisms, including entomopathogenic fungi such as *B. bassiana* and *M. anisopliae* (Pedrini et al. 2013).

Lipases are enzymes widely distributed in nature, especially in fungi and bacteria, and are classified into five main families based on their structural and functional properties (Arpigny and Jaeger 1999). Family I includes fungal lipases, such as those from *Rhizopus* and *Aspergillus*. Family II comprises bacterial lipases, with representatives from the genera *Pseudomonas* and *Burkholderia*. Family III consists of lipases found in Gram-positive bacteria, exhibiting distinct catalytic characteristics. Family IV includes lipases widely used in biotechnology, such as the lipase from *Candida antarctica*, which holds great industrial importance. Finally, Family V encompasses bacterial lipases and esterases with specific catalytic mechanisms (Sarmah et al. 2018; Arpigny and Jaeger 1999). Detailed knowledge of these families aids in selecting the appropriate enzyme for various industrial applications, including biodiesel synthesis, detergent production, and food processing.

From a biological point of view, microbial lipases, especially those from fungi, are valued for their versatility and stability under industrial conditions (Hasan et al. 2006). In the case of *B. bassiana*, genes such as *Bblip1* and *Bblip2* encode lipases and play an important role in the fungal virulence. Deletion of the *Bblip2* gene has been shown to significantly reduce pathogenicity, and its expression is induced by lipid sources and alkaline pH (Jin et al. 2011). *B. bassiana* genome includes about 20 to 30 genes annotated as lipases, many of which are expressed during infection (Xiao et al. 2012). In *M. anisopliae*, lipases also play a central role in insect infection, especially during cuticle penetration. *Malip1* gene encodes an extracellular lipase whose expression is intensified in the presence of *Manduca sexta* cuticle. Its deletion resulted in reduced virulence and lower penetration efficiency (Wang and St Leger 2007). Another gene, *Malip2*, is regulated during the early stages of infection and is associated with the degradation of insect epicuticle lipids (Wang and St Leger 2005).

4 Enzyme Production

Efficient production of enzymes by entomopathogenic fungi requires the integration of multiple factors, from the selection of specific substrates to the strict control of cultivation conditions (Mishra et al. 2020). Successful production also depends on the formulation of culture media, which should be similar to the natural components found in target hosts, such as chitin, cuticular proteins, and lipids.

Physicochemical parameters such as pH, temperature, aeration, and agitation must also be optimized for each fungal species, considering their metabolic and physiological needs. Standardization of these protocols not only ensures the reproducibility of results but also lays the foundation for a successful transition to industrial scales. This standardization is a challenge for biopesticide formulations, where cost-effectiveness and final product stability are key determinants (Wang and St Leger 2006). The standardization of these processes can also be advantageous, since when well developed, it can use the integration of low-cost strategies, such as the use of agro-industrial waste and systematic evaluation of stability and enzymatic activity under different conditions, to reduce costs.

The choice of carbon source is of paramount importance for the optimization of culture media. Although the use of simple sugars such as glucose and sucrose may promote rapid growth, it can repress the synthesis of hydrolytic enzymes due to carbon catabolite repression (CCR) (Jin et al. 2011). This phenomenon is well described in several microorganisms, including filamentous fungi such as *Aspergillus oryzae*, where the presence of glucose represses the production of amylolytic enzymes (Tanaka and Gomi 2021).

To mitigate the effects of CCR, strategies such as using reduced glucose concentrations (below 2%) or replacing it with polysaccharides after the logarithmic growth phase can be effective. The substrate must also be chosen with the utmost caution; colloidal chitin, for example, could accentuate the induction of chitinases (Ren et al. 2022). A study by Baratto et al. (2006) showed that *Beauveria bassiana* produces up to three times more chitinase when grown in a medium containing 1% chitin, highlighting the inductive effect of chitin on enzyme production. The use of rice bran or insect waste can stimulate the expression of proteases in entomopathogenic fungi. Screen et al. (2001) showed that *Metarhizium anisopliae* produces the serine-protease Pr1, important for virulence, when grown in medium supplemented with beetle cuticle. Pr1 is initially produced as a precursor and then matured. It is structurally similar to other serine proteases and has proved to be highly effective in degrading the cuticle and its structural proteins, outperforming proteinase K by a significant margin.

Nitrogen sources are also important factors, and the use of organic nitrogen, such as peptones and casein, is essential for protease production (Geisseler and Horwath 2008). Inorganic nitrogen (ammonium sulfate) has also been used in *Metarhizium* strains for low-cost enzyme production (Li et al. 2021). In addition, it has been proven that supplementation with minerals such as Ca^{2+} and Mg^{2+} ions increases the thermal stability of proteases, and Zn^{2+} modulates the activity of metalloproteases in *M. anisopliae* (Qazi and Khachatourians 2007).

5 Critical Environmental Factors

Enzyme production is also influenced by critical environmental factors, which regulate both microbial growth and the expression of key enzymes. For most strains, temperatures between 25 and 28 °C must be rigorously controlled, as variations above 30 °C can inhibit sporulation and reduce chitinase production by up to 50%, as observed in *B. bassiana* (Fernandes et al. 2015).

The medium's pH significantly influences the expression of cuticle-degrading enzymes in *Metarhizium anisopliae*. Studies show that acidic conditions around pH 5 favor chitinase production (optimal at ≈ 5), while alkaline conditions near pH 8 enhance Pr1 and other serine-protease expression. Other enzymes such as metalloproteases and aminopeptidases peak at neutral to slightly alkaline pH (6–8) (St Leger et al. 1996). In addition, the presence of substrates like colloidal chitin or yeast extract strongly induces chitinase production in *M. anisopliae* cultures (Dhar and Kaur 2011). Aeration is also an important variable. Ideal values are around 1–2 vvm (air volume/medium volume/min) in bioreactors, and agitation at 150–200 rpm is essential to prevent pellet formation, which occurs due to hyphal agglomerates. These agglomerates limit nutrient and oxygen transfer, negatively impacting lipase and protease production. Fine-tuning these parameters, combined with real-time monitoring, can increase enzyme productivity up to threefold, as demonstrated in *B. bassiana* cultures with automated pH and temperature control (Baratto et al. 2006).

5.1 Optimization Strategies

In the context of culture media optimization, the most used strategies are two-stage cultivation (Scheidt et al. 2009), statistical experimental design (Ladeira et al. 2010), and the use of alternative substrates (Paiva-Guimarães et al. 2019). These techniques have shown promise in overcoming challenges such as scalability and costs. The integration of these techniques can significantly increase enzyme productivity while reducing waste and production costs. During the first phase of two-stage cultivation (24–48 h), a glucose-rich medium (20 g/L) promotes exponential biomass growth, providing a rapid energy source and favoring cell multiplication. In the second phase, replacement with specific inducers, such as chitin (1%) or insect residues, stimulates the production of key enzymes like chitinases, proteases, and lipases, which are essential for degrading the cuticle and other host components (Lopes et al. 2023). This strategy allows for the optimization of both fungal biomass and enzyme production, increasing enzyme efficiency and yield.

Another strategy is systematic optimization via response surface methodology (RSM), which has revolutionized the standardization of enzymatic processes. RSM-based strategies have been successfully applied to optimize variables such as moisture content and yeast extract concentration, resulting in significant yields. In the research by Dhar and Kaur (2011), *B. bassiana* cultivation optimized using RSM achieved a maximum production of 28.8×10^9 conidia/g with mixed substrates. The study by Yasmeen et al. (2013) demonstrated that optimizing the production of ligninolytic enzyme from *Schizophyllum commune* through RSM allowed for the identification of ideal solid-state fermentation (SSF) conditions, resulting in high enzyme activity.

The growing demand for sustainable processes has driven research into alternative substrates. According to Doroški et al. (2022), the use of food waste as a substrate for *Pleurotus ostreatus* cultivation contributed to sustainable waste management and the production of high-quality food. The study by Sala et al. (2021) shows that agro-industrial residues such as rice husks, apple pomace, potato peels, and brewer's spent grain can be used as effective substrates for producing conidia of fungi with biopesticidal potential, such as *Beauveria* and *Trichoderma harzianum*, reaching concentrations exceeding 10^9 conidia g^{-1} of dry matter.

Although cultivation optimization is fundamental for maximizing enzyme production, the effectiveness of these biomolecules equally depends on robust extraction and purification methods. The proper selection of techniques to preserve enzyme activity, ensure adequate purity, and enable industrial applications is a challenge. In this context, approaches such as salt precipitation, affinity chromatography, and aqueous two-phase systems have stood out in the recovery of fungal enzymes, which will be detailed in the following sections.

5.2 Extraction and Purification of Enzymes

Obtaining fungal enzymes with preserved activity and high purity is essential for biotechnological applications, requiring efficient extraction and purification strategies (Singh et al. 2022; Hamrouni et al. 2025). Conventional methods (salt precipitation, ultrafiltration) and innovative approaches (ionic liquid-assisted extraction, aqueous two-phase systems) have been explored to maximize yield, preserve enzymatic stability, and reduce operational costs (Nadar et al. 2018). The choice of enzyme extraction method critically depends on the enzyme localization (intra- or extracellular) and scalability feasibility, factors that demand cost-benefit analysis in prior studies (Juturu and Wu 2014). We present a detailed analysis of the most effective techniques for purifying fungal enzymes, highlighting methods with industrial scaling potential, advantages, and key technological challenges to overcome in the following sections.

5.2.1 Centrifugation

In this method, after fungal growth in liquid medium, cells are separated using a centrifuge. The sample is subjected to high centrifugal forces (typically between 3000 and 10,000 × g) at low temperatures, causing cells and solid particles in the culture medium to sediment at the tube bottom, forming a pellet. The supernatant contains enzymes and metabolites produced by the fungi, which can then be collected for subsequent steps. This supernatant-derived sample is called the cell-free crude extract.

This method is rapid, preserves enzymatic activity by avoiding thermal or chemical stress, and can be scaled for industrial processes, though it requires robust equipment and incurs significant costs for large volumes. For intracellular enzymes, centrifugation is often combined with prior cell lysis steps to separate cellular debris from the crude enzymatic extract.

5.2.2 Tangential Filtration

Tangential filtration (cross-flow filtration) is an efficient strategy for recovering fungal enzymes. Unlike conventional filtration, where flow occurs perpendicular to the filter, in tangential filtration the culture medium is circulated parallel to a semipermeable membrane (typically with 0.1–0.45 μm pores). This allows enzymes and small molecules to pass through (permeate), while cells and larger particles are retained (retentate) (Orr et al. 2012).

This method is highly scalable, avoids rapid membrane clogging (fouling), and can be coupled with continuous fermentation systems, making it ideal for industrial processes. Furthermore, tangential flow filtration preserves enzyme activity as it does not involve harsh chemical agents or high temperatures and is often combined with subsequent ultrafiltration steps for the concentration and purification of enzymes of interest (Cheison et al. 2006). A process was patented by Andersson et al. (2022) describing the separation of enzymes produced by *Trichoderma reesei* using a combination of a filter press and tangential flow filtration with ceramic membranes. This method resulted in high yields, with enzyme concentrations reaching up to 200 g/L after ultrafiltration. Process efficiency was attributed to the ceramic membranes' ability to retain fungal particles while allowing the desired enzymes to pass through.

5.3 Purification Techniques

After extraction, purification aims to isolate specific enzymes with minimal activity loss for industrial, pharmaceutical, and environmental applications. Different techniques are employed according to the enzymes' physicochemical properties, such as molecular size, surface charge, and substrate affinity. Classical methods, such as salt precipitation (ammonium sulfate), ultrafiltration, and chromatography, are widely used on a laboratory and industrial scale, while modern approaches, such as separation by functional membranes and hybrid systems, have gained prominence for their efficiency and sustainability.

5.3.1 Ammonium Sulfate Precipitation

Ammonium sulfate precipitation is a classical and efficient method for enzyme recovery. This process is based on the "salting-out" principle, where high salt concentrations (generally between 40 and 80% saturation) reduce enzyme solubility, leading to their selective precipitation (Wingfield 2017). This technique is particularly advantageous due to its simplicity, low cost, and ability to preserve enzyme activity, making it ideal for initial purification steps in industrial processes. For example, Illuri et al. (2021) applied ammonium sulfate precipitation followed by dialysis to partially purify ligninolytic enzymes from basidiomycete fungi, achieving significant increases in enzyme purity and specific activity. The precipitation technique is particularly advantageous due to its simplicity, low cost, and ability to preserve enzyme activity, making it ideal for initial purification steps in industrial processes.

Despite its wide applicability, the method presents challenges. The residual presence of $(NH_4)_2SO_4$ can interfere with analyses, requiring additional steps such as dialysis or ultrafiltration, as shown in previous studies (Janson 2011; Bedade and Pawar 2023). Other studies also highlight that combining it with techniques like ultrafiltration or ion-exchange chromatography can significantly improve the final purity of enzyme preparations (Ahmed et al. 2024; Bekavac et al. 2024). Furthermore, rigorous control of pH, generally between 7 and 8, and temperature, around 4 ° C, is very important to maintain enzyme stability throughout the process.

5.3.2 Ultrafiltration

Ultrafiltration (UF) has stood out as an efficient technique for enzyme purification. It uses membranes with specific pores to retain proteins while allowing smaller components, such as salts and low molecular weight residues, to pass through. Besides its effectiveness in separating and concentrating enzymes, ultrafiltration is considered a sustainable technology, as it avoids the use of organic solvents, consumes less energy compared to other separation techniques, and allows for the recovery and reuse of water and valuable process components, significantly contributing to the reduction of waste and environmental impacts.

In UF, different types of membranes are used to meet the specific needs of each application. Polymeric membranes, such as those made of polysulfone (PES), polyacrylonitrile, regenerated cellulose, or cellulose acetate, are the most common due to their high rigidity, thermal resistance, and mechanical stability (Balakrishnan and Agarwal 1996; Yang and Tong 1997; Saxena et al. 2009). These membranes are widely used in protein separation and particle removal in biotechnology and the food industry. In addition to these, inorganic membranes, such as those made of nanoporous carbon, offer greater chemical and thermal resistance, enabling their use in more aggressive conditions (Shah et al. 2007). There are also ceramic membranes, which stand out for their robust mechanical support and stability under high temperatures and with strong chemical solutions, making them suitable for processes requiring rigorous cleaning and corrosion resistance (Ali et al. 2018). The work of Mondal et al. (2019) highlights that the enzyme purification process of *B. bassiana* can be efficiently performed using ceramic membranes in two sequential steps: ultrafiltration and nanofiltration. This approach not only improves enzyme yield and purity but also enables resource reuse during the process.

This diversity of materials provides specific options according to requirements for resistance, durability, and chemical compatibility, expanding the application possibilities of UF in different areas. Nevertheless, the combination of ultrafiltration with the absence of toxic solvents positions this technique as a key technology for green biotechnology, as highlighted in the review by Thompson et al. (2023) on sustainable enzymatic processes. Thus, ultrafiltration is established as an effective and environmentally responsible strategy for enzyme purification, offering high efficiency in macromolecule separation and promoting more sustainable practices in the biotechnology industry.

5.3.3 Chromatography

Chromatography is one of the most powerful and versatile techniques for the purification of fungal enzymes, allowing a high degree of purity and recovery of enzymatic activity. It is based on physicochemical principles such as ionic interactions, molecular affinity, hydrophobicity, and size. With the advancement of specific resins and automated systems, chromatography has become indispensable in both research and industrial enzyme production, offering scalable and reproducible solutions for biotechnological applications. The following sections will detail the main chromatographic techniques used in the purification of fungal enzymes, highlighting their operating principles, specific applications, and comparative advantages.

Ion-Exchange Chromatography

Ion-exchange chromatography (IEX) stands out as one of the most versatile and efficient techniques for purifying fungal enzymes, being particularly valuable in the biopesticide production process. The principle of the method is based on the electrostatic attraction between the charged functional groups of the enzymes and the chromatographic column matrix, which can be functionalized with anionic (e.g., DEAE-Sepharose) or cationic (e.g., CM-Sepharose) groups (Linskens et al. 1963; Acikara 2013). This selectivity allows not only for the effective purification of enzymes such as chitinases, glucanases, and proteases but also for the concentration of diluted samples, making the process especially advantageous for processing large volumes of fermentation broth.

Despite its advantages, IEX presents challenges that require careful optimization to ensure maximum process efficiency. The technique's sensitivity to pH and ionic strength variations requires the use of specific buffers, such as Tris-HCl or phosphate, and controlled salt gradients (NaCl or KCl) for enzyme elution. Furthermore, the presence of ionic contaminants in the crude extract can interfere with the binding of enzymes to the resin, reducing purification capacity. To overcome these limitations, modern resins have been developed, such as those combining ion exchange with hydrophobic interactions, like Capto MMX (Yang et al. 2023). These advancements have been particularly useful in purifying enzymes with similar properties, such as laccases and peroxidases, which often coexist in fungal extracts and require highly resolute separation methods.

The future of IEX in the purification of fungal enzymes for biopesticides looks promising, with research focused on developing more selective resins and continuous flow purification strategies. The integration of IEX with other chromatographic techniques, such as size-exclusion chromatography (SEC) or affinity chromatography, has shown exceptional results in obtaining ultra-pure enzymes (Chakraborty et al. 2022). Moreover, combining IEX with advanced analytical techniques, like mass spectrometry, allows for detailed characterization of purified enzymes, ensuring the quality and consistency of the final product (Murisier et al. 2021). These advances not only optimize the purification process but also reduce costs and increase the commercial viability of enzymatic biopesticides, reinforcing the central role of IEX in sustainable biotechnology.

Size-Exclusion Chromatography

Size-exclusion chromatography (SEC) is a fundamental technique for the final purification of enzymes, based on the principle of separation by differences in molecular weight (Porath and Flodin 1959). In this method, enzymes are fractionated as they pass through a column containing porous microspheres of an inert polymer (e.g., Sephadex or Sephacryl). Larger molecules, such as protein aggregates or high molecular weight contaminants, are excluded from the pores and elute first, while enzymes of interest, with a size compatible with the matrix pores, have their path retarded, resulting in efficient separation.

An example of this technique's use was described by Fang et al. (2005), who demonstrated its effectiveness in purifying chitinases from *B. bassiana*. In this study, Sephadex G-100 resin allowed for the efficient separation of the chitinase *Bbchit1*, with an estimated molecular weight of 33 kDa. This approach demonstrated the effectiveness of size exclusion in obtaining an enzyme with a high degree of purity for biochemical characterization and subsequent functional application.

Despite its advantages, SEC has limitations in large-scale applications. The low yield and high cost of traditional resins have led to the development of new matrices, such as Superdex resins, which combine higher loading capacity with better resolution (Hellberg et al. 1996; Herkersdorf et al. 2021). These advancements have significantly expanded the applications of SEC in the enzyme industry, particularly for the production of high-value-added biocatalysts.

Affinity Chromatography

This is a highly specific method that uses immobilized ligands. The purification of fungal enzymes using a His-tag is based on the construction of a recombinant expression system for the enzyme of interest, which is fused to a sequence of 6–10 histidines. After genetic transformation of the fungus, the tagged enzyme is expressed and secreted into the culture medium or extracted by cell lysis.

For the purification of the enzyme of interest, selective capture is performed by immobilized metal affinity chromatography (IMAC). When the crude extract is applied to a column containing resin with metal ions (such as nickel or cobalt), the His-tag specifically binds to these ions through coordination with the imidazole groups of the histidines (Salem et al. 2024). Contaminating proteins, with no affinity for the resin, are removed by washing, while the target enzyme remains adsorbed. Finally, the purified protein is eluted with a solution containing imidazole (which competes for binding with the metal ions) or by pH alteration, resulting in a highly purified enzyme preparation (>95% purity) in a single step (Freydank et al. 2008; Salem et al. 2024). This approach is particularly advantageous for genetically modified fungi, as it overcomes challenges such as the presence of cell wall polysaccharides that interfere with conventional purification methods.

Hydrophobic Interaction Chromatography

Hydrophobic Interaction Chromatography (HIC) is a technique based on the interaction between hydrophobic regions on the protein surface and alkyl or aryl groups immobilized on the column matrix (Hofstee 1973). The structural complexity and functional diversity of fungal lipases, such as those from *Rhizopus oryzae*, often require refined techniques to separate and stabilize isoenzymes with subtle differences. Ramachandran et al. (2015) demonstrated that immobilizing *R. oryzae* lipase on hierarchical mesoporous silica supports not only enhanced its catalytic efficiency in the transesterification of rice bran oil but also revealed the enzyme's isoform-specific behavior, indicating that support interactions can differentially stabilize and expose distinct enzymatic conformations. The great advantage of HIC lies in its ability to operate under native conditions, preserving enzyme structure and activity, as observed for proteases from *Aspergillus* spp. (Kumar et al. 2021).

The application of HIC in industrial processes has advanced significantly with the development of new resins and optimized protocols. According to Fischer et al. (1995), combining HIC with pretreatment by ammonium sulfate precipitation increases the yield in the purification of enzymes produced by *Thermomyces lanuginosus* by 22.5 times. These data highlight the potential of the combined strategy for obtaining thermotolerant microbial enzymes with high purification efficiency.

Hydrophilic Interaction Chromatography

Hydrophilic interaction chromatography (HILIC) is a liquid separation technique that can be used for the purification of fungal enzymes with polar characteristics, such as chitinases, proteases, and glucanases. In this approach, a polar stationary phase is used along with a mobile phase predominantly composed of acetonitrile, with a small proportion of water or aqueous buffer (Alpert 1990). The enzymes present in the fungal extract interact with the stationary phase through hydrophilic forces, being gradually eluted as the proportion of water in the mobile phase increases. This technique is especially advantageous for separating proteins soluble in organic solvents and with high surface polarity, in addition to being compatible with subsequent analyses by mass spectrometry. Due to its selectivity and efficiency, HILIC represents a valuable alternative to traditional purification techniques, allowing the recovery of active enzymes with high purity. Although HIC and HILIC are opposing techniques, they share complementary principles in the purification of fungal enzymes. This duality is particularly useful, as enzymes often present both hydrophobic domains (e.g., active sites of lipases) and hydrophilic residues (e.g., glycosylation in glycosidases).

In summary, the methods for extracting and purifying fungal enzymes play a fundamental role in obtaining efficient, high-quality biocatalysts. The appropriate choice of techniques, such as salt precipitation, chromatography, and ultrafiltration is essential to ensure enzyme recovery and purity while maintaining their biological activity. With these well-established processes, it is possible to advance to the formulation stage of enzymatic biopesticides, where the preservation of activity and enzyme stability becomes a critical factor for the success of practical application.

6 Techniques for Preserving Enzyme Activity

Preserving enzyme activity is a central challenge in the development of effective biopesticides, as enzymes are biomolecules sensitive to environmental factors such as temperature, humidity, UV radiation, and oxidation. To overcome these limitations, various techniques have been employed, each with specific protection mechanisms and ideal application scenarios.

6.1 Lyophilization

Lyophilization (or freeze-drying) has established itself as the preferred method for preserving enzymes in biopesticide formulations, with its effectiveness proven by several scientific studies. The process consists of three main stages: freezing the enzymatic material (−40 °C to −80 °C), sublimation of water (primary phase) carried out under vacuum (0.1–0.01 mBar), and removal of residual water (secondary phase), resulting in a dry and stable powder. The process efficiently preserves the three-dimensional conformation of fungal enzymes such as chitinases and β-1,3-glucanases (Zhang et al. 2023).

Studies show that the addition of cryoprotectants during lyophilization can significantly improve enzyme stability. Lyophilization has also been effective in preserving fungal cultures. Several studies support the use of lyophilization as a reliable method for preserving spores of filamentous fungi. Ibatsam et al. (2012) demonstrated that *Penicillium* strains maintained full viability and morphological characteristics for up to 3 months after lyophilization, even in the absence of cryoprotectants. The authors confirmed that colony morphology and growth rates remained consistent throughout the storage period. Similarly, Rodriguez et al. (2019) observed that strains from the *Nigri* section of *Aspergillus* preserved their biotechnological potential and showed only minimal phenotypic changes after lyophilization and storage under accelerated conditions. These findings reinforce the suitability of lyophilization for the long-term preservation of sporulating fungal cultures. Although the initial investment in equipment is considerable, lyophilization represents a strategic and effective step for the conservation of fungal enzymes, ensuring the maintenance of biological activity, stability, and viability of preparations for later use.

6.2 Encapsulation

Encapsulation of fungal enzymes has proven to be a promising strategy to overcome stability and efficacy challenges in biopesticide formulations. It consists of enveloping biologically active molecules (such as enzymes, cells, or bioactive compounds) in protective materials, forming microscopic structures. This process creates a physical barrier that protects enzymes against degrading environmental factors, such as high temperatures, UV radiation, pH variations, and oxidation.

Encapsulation protects enzymes using three key mechanisms: (1) a physical barrier using polymeric matrices (alginate, chitosan) that isolates them from the external environment; (2) controlled release triggered by stimuli such as pH or pathogen enzymes; and (3) synergy with adjuvants (trehalose, nanoparticles) that increase stability and efficiency. For encapsulation, different techniques can be employed. Among polymer-based techniques, ionic gelation forms gels upon contact with ions, creating microspheres that protect enzymes. It is ideal for heat-sensitive bioactive compounds, allowing controlled release through matrix degradation. It is a technique that offers advantages such as simplicity, low cost, and biocompatibility, making it ideal for heat-sensitive systems. Encapsulation in liposomes and micelles are techniques that improve the stability, solubility, and targeted delivery of active ingredients, potentiating action against agricultural pests. Liposomes are spherical vesicles made of phospholipids that encapsulate enzymes or microorganisms in their aqueous core, offering protection against degradation and controlled release. Micelles are nanometric structures formed by amphiphilic molecules like PLGA, ideal for carrying hydrophobic compounds such as essential oils in biopesticide formulations (Lagreca et al. 2020). In addition to natural polymers, advanced systems such as liposomes and polymeric micelles have been explored to improve the delivery and stability of fungal enzymes. These systems also allow for the synergistic combination of enzymes with other antifungal agents, such as fungal secondary metabolites, potentiating their action against plant pathogens.

Among drying-based encapsulation techniques, spray-drying can be successfully employed in biopesticide formulation. In this process, a solution containing enzymes and protective polymers such as maltodextrin, alginate, or chitosan is atomized into fine droplets and rapidly dried with hot air (Grewal 2002). This technique offers high productivity and industrial scalability, allowing simultaneous encapsulation with adjuvants (e.g., silica nanoparticles for UV protection). Despite the use of hot air, typically between 150 °C and 220 °C at the inlet, the instantaneous evaporation of the droplets minimizes enzyme exposure to degradative temperatures.

A variation is spray-freeze-drying, where atomized droplets are instantly frozen in liquid nitrogen (−196 °C) and then lyophilized. It can be used for preserving thermosensitive enzymes. Finally, spray-chilling involves dispersing enzymes in a lipid matrix, such as waxes or vegetable fats, and atomizing them into a cold chamber, where they solidify. It is ideal for heat-sensitive compounds and has a temperature-stimulated release system (melts in the agricultural environment), releasing the active ingredients only when exposed to ambient temperatures (Tao et al. 2023). These variations allow encapsulation to be adapted to the specific thermodynamic characteristics of each bioactive compound, expanding the spectrum of applications in biological control. Recent studies demonstrate that encapsulation systems in polymeric matrices, such as calcium alginate and chitosan, can protect enzymes like chitinases and glucanases against degradation by environmental factors, such as UV radiation and pH variations (Zhang et al. 2023). These systems not only preserve enzyme activity but also allow for controlled release into the environment, increasing the biopesticide's persistence and efficacy.

Despite advances, challenges such as production cost and scalability still limit the large-scale application of these technologies. Recent works indicate that optimizing encapsulation methods, such as the electrospray technique, can reduce costs and improve particle uniformity (Felizatti et al. 2021). Furthermore, enzyme engineering for greater intrinsic stability, combined with encapsulation, emerges as an innovative approach. These advancements highlight the potential of encapsulation to revolutionize the development of sustainable and efficient enzymatic biopesticides.

7 Development of Formulations

Formulations of microbiological agents, such as entomopathogenic fungi and bacteria, represent technological systems to enable the application of these organisms in agricultural management. According to de Faria and Wraight (2007), these formulations are developed to protect microorganisms against adverse environmental factors, such as UV radiation and desiccation, in addition to facilitating their application in the field. The main formulations used can be classified as wettable powder (WP), suspension concentrate (SC), oil dispersion (OD), oil-in-water emulsion (EW), water-dispersible granules (WG), and more recently, microencapsulated (CS), based on their physicochemical characteristics, raw materials used, and application method. Recent studies highlight that the selection of the appropriate formulation is critical for maintaining cell viability, optimizing dispersion in the target environment, and ensuring the product's biological efficacy (Keswani et al. 2016).

According to data from Agrofit (Ministry of Agriculture, Livestock, and Food Supply 2025), 330 products classified as microbiological insecticides were registered in Brazil by 2025. Among them, 55 are based on *Bacillus* spp., 133 on *B. bassiana*, 9 on *Isaria* spp., and 83 on *Metarhizium* spp.. Although the products registered under the name *Isaria* spp. retain this designation, the species they contain are currently classified as *Lecanicillium* spp., reflecting updated taxonomic revisions (Zare and Gams 2001). Formulations in which these products are marketed include dry powder, wettable powder, granules, emulsifiable oil, suspension concentrate, and other forms adapted to ensure the viability and efficacy of biological agents in the field (Fig. 7.3). These formulations are developed to preserve the integrity of microorganisms and facilitate their application, contributing to the sustainable control of agricultural pests with less environmental impact. In the following topics, each type of formulation will be detailed to understand its applications, advantages, and limitations.

7.1 Suspension Concentrate

Suspension concentrate (SC) is one of the most widely used formulations in biological and chemical control due to its stability and ease of application. It is composed of solid particles of the active ingredient, such as fungal spores or bacterial crystals, suspended in a liquid medium, which is usually water with

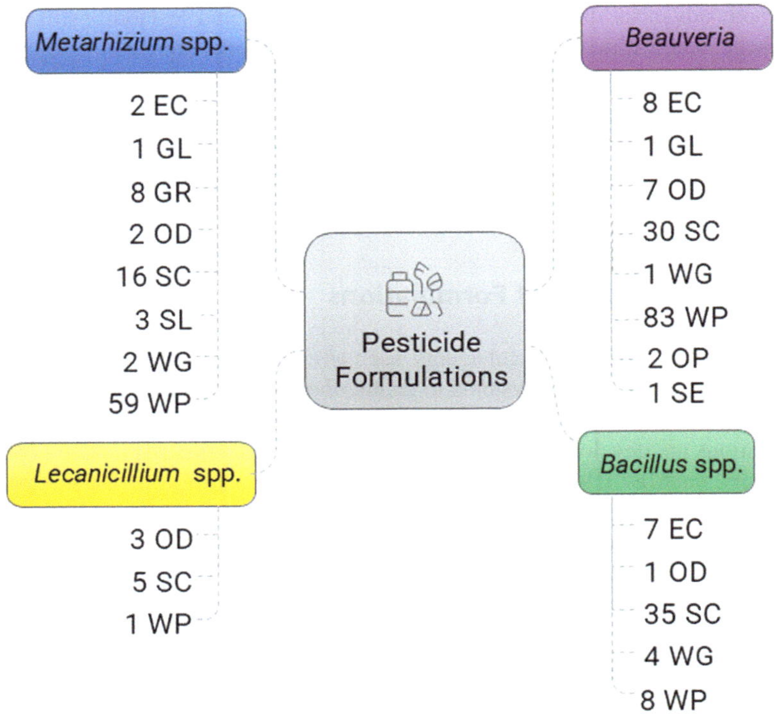

Fig. 7.3 Quantity of biological products marketed in different formulations, according to the number of registered products per type. Includes: emulsifiable concentrate (EC), emulsifiable gel (GL), oil dispersion (OD), powder in oil (PO), suspension concentrate (SC), suspo-emulsion (SE), water-dispersible granule (WG), wettable powder (WP), soluble concentrate (SL), and granule (GR). (Source: Agrofit, 2025s)

adjuvants (de Faria and Wraight 2007). Unlike wettable powders (WP), SCs do not settle quickly, reducing the risk of clogging spray nozzles. Furthermore, their composition allows for the incorporation of UV protectants and foliar adhesives, increasing the bioagent's persistence in the field (Hernandez-Tenorio et al. 2022). The SC formulations are compatible with different application methods, including conventional spraying and ultra-low volume (ULV) systems, which are common in drones or portable equipment (García-Estrada et al. 2016). One of the main advantages is lower abrasiveness; in frequent applications, dry powders can corrode metal components of agricultural equipment, while SCs prolong their lifespan (Tikar and Prakash 2017).

A common problem in SCs is sedimentation during storage, where solid particles settle at the bottom of the container, requiring agitation before use. To prevent this, thickeners, which are anti-settling agents like xanthan gum or modified cellulose, are added to keep the suspension homogeneous. In practice, SCs of *B. bassiana* and *M. anisopliae* have demonstrated superior efficacy in humid climates, where adhesion to leaves is critical (García-Estrada et al. 2016).

7.2 Wettable Powder

The wettable powder (WP) formulation is a dry formulation composed of fine particles of the active ingredient (such as spores or fungal cells), inert carriers (e.g., clay, silica), and adjuvants (wetting agents, dispersants, and anti-settling agents) (Sivakumar et al. 2020). This formulation requires dilution in water for application, forming a stable suspension that can be sprayed onto plants or soil. The WP formulations are widely used in the biological control of pests, especially for entomopathogenic fungi like *B. bassiana* and *M. anisopliae* (Ali 2016). However, there are limiting factors such as dispersibility, which is the ability of particles to distribute uniformly in water, and foliar adhesion, which consists of the product's fixation to the plant surface, that are critical challenges for the efficacy of wettable powders (WP) (Brar et al. 2006; Wraight et al. 2016). In practice, poorly dispersed particles can clog spray nozzles and require constant agitation. Furthermore, the product can run off the leaf with rain or irrigation; thus, repeated applications are necessary, increasing costs for the farmer.

To overcome these challenges, it is necessary to improve particle dispersibility and adhesion by incorporating advanced polymers or surfactants. The use of non-ionic surfactants like hydrophobic silica can reduce particle agglomeration, ensuring homogeneity in the spray mixture, and the use of PVA (polyvinyl alcohol) and surfactant–humectant adjuvants can increase foliar retention and improve formulation efficacy (Köcher and Kocur 1993; Polli et al. 2022). Additionally, WP formulations have been adapted for ultra-low volume (ULV) application systems and precision agriculture, aiming to reduce waste and increase biological efficacy (FAO 2022; Zhang et al. 2023). Despite competition from more modern formulations, WP remains relevant due to its cost-effectiveness and compatibility with traditional agricultural infrastructures (de Faria and Wraight 2007; Tikar and Prakash 2017).

7.3 Oil Dispersion

Oil dispersion (OD) stands out as one of the most effective formulations for microbiological agents, especially under adverse environmental conditions. In this formulation, solid particles of the active ingredient (such as fungal spores or bacterial cells) are uniformly dispersed in an oily base, which can be composed of vegetable oils like soybean or cottonseed or mineral oils (de Faria and Wraight 2007; Bejarano and Puopolo 2020).

The main advantage of OD lies in its excellent protection against environmental factors, such as UV radiation and desiccation, due to the physical barrier formed by the oil around the particles. Furthermore, the formulation's viscosity promotes greater adhesion to foliar surfaces, reducing losses from runoff and increasing the product's persistence on the plant (Keswani et al. 2016). This characteristic is particularly valuable in perennial crops or in regions with frequent rainfall, where product retention is a constant challenge. The compatibility of the OD formulation with ultra-low volume (ULV) application systems makes it ideal for use in drones and precision sprayers, optimizing coverage and reducing the volume of spray mixture needed (Sun et al. 2024).

Comparative studies have shown that conidia not formulated in mineral oil, when exposed to 45 °C for 4 hours, exhibited a significant reduction in germination rate and appressoria formation on the cuticle of *Rhipicephalus sanguineus*, in contrast to conidia protected by the oily formulation (Ortíz-Catón et al. 2013). Furthermore, the addition of oils to fungal formulations significantly improves the adhesion of conidia to the body surface of arthropods. This phenomenon occurs because oils impart lipophilic properties to the formulation, facilitating the attachment and penetration of conidia through the characteristic hydrophobic cuticle of arthropods (Prior et al. 1988; Bateman et al. 1993). Recent research focuses on developing more accessible vegetable oils and incorporating smart adjuvants to further improve formulation stability and dispersion, consolidating OD as one of the most promising technologies for the future of biopesticides (FAO 2022).

7.4 Oil-in-Water Emulsion

The oil-in-water emulsion (EW) consists of a stable dispersion of oil droplets containing the active ingredient in an aqueous phase, stabilized by surfactants and emulsifiers. This technology is especially advantageous for biological products, such as enzymes or microorganisms, as it combines the protection offered by oil against desiccation and UV radiation with the ease of application in aqueous systems. The study conducted by Rondelli et al. (2011) demonstrated that aqueous emulsions (EWs) based on vegetable oils, such as castor oil, improve foliar adhesion and the persistence of biopesticides, such as those formulated with *B. bassiana*. However, challenges such as phase separation during storage require the use of thickening agents like xanthan gum and high-efficiency emulsifiers (Tan and McClements 2021).

The EW formulations have also been adopted for the delivery of hydrolytic enzymes like lipases and chitinases. In the work of Liu et al. (2023), a lipase from *Candida* sp. was encapsulated using water-in-oil Pickering emulsions and showed greater stability and activity compared to traditional methods. Innovations in nanostructured EWs, where oil droplets are reduced to nanometric sizes, have increased stability and foliar penetration, reducing the risk of phytotoxicity (Jamir et al. 2024). These advancements position EWs as a promising alternative to traditional formulations (like SCs), especially in organic farming where synthetic solvents are avoided (FAO 2022).

7.5 Water-Dispersible Granules

Water-dispersible granules (WG) formulations are widely used in biopesticides due to their efficiency, safety, and ease of application. These granules are designed to disperse rapidly in water, forming a homogeneous suspension, which facilitates spraying in crops. Unlike liquid formulations, WGs exhibit greater stability, reducing problems such as sedimentation and degradation of active biological ingredients, as in bacteria, fungi, or plant extracts (Butu et al. 2022). Furthermore, this formulation minimizes exposure to organic solvents, making it safer for the applicator and the environment (Knowles 2008).

One of the main advantages of WGs is the absence of dust during handling, reducing the risks of inhalation and contamination. Additionally, this formulation allows for the combination of multiple biological agents and adjuvants, such as surfactants and UV protectants, which increase product efficacy. The study conducted by Alves et al. (2002) demonstrated that WG formulations of *M. anisopliae* maintain high conidial viability even after long storage periods, due to the protection provided by inert components. WG technology is also adaptable to different production scales, facilitating its adoption in both family farming and large plantations (IBMA 2021).

Despite the advantages, the development of WGs for biopesticides faces challenges, such as selecting carriers and dispersants compatible with living microorganisms. Residual moisture and heat during granulation can affect the viability of biological agents, requiring optimized processes, such as wet or dry granulation. Recent research explores the use of biopolymers like alginate as sustainable alternatives to improve adhesion and controlled release. With advances in nanotechnology and encapsulation, WG formulations tend to become even more efficient, driving sustainable agriculture (OECD 2022).

7.6 Microencapsulation

The microencapsulated formulation (capsule suspension, CS) consists of microscopic particles that envelop active ingredients, such as enzymes, in a polymeric or lipid matrix, forming a stable suspension in water of 1–100 μm (Chaturvedi and Sharma 2024; Hazra and Purkait 2019). This technology was developed to overcome common challenges in biopesticides, such as rapid degradation by environmental factors (UV light, temperature, and humidity) and uncontrolled release into the environment (Gupta et al. 2022). By encapsulating enzymes like chitinases, proteases, or glucanases, it is possible to protect them until they reach the target, ensuring greater efficiency and durability in the field. Furthermore, CS allows the combination of different active agents in a single system, such as enzymes with essential oils or pheromones, broadening their spectrum of action. The release of the active ingredient in CS formulations can be controlled by different mechanisms, such as capsule degradation by humidity, pH, or specific enzymes. This targeted release reduces product waste and minimizes effects on non-target organisms, such as pollinators and natural enemies of pests. Another advantage is greater stability during storage, as the capsule structure protects enzymes against denaturation, prolonging their shelf life compared to conventional liquid formulations. The CS formulations have been applied in agricultural pest control, such as in the use of microcapsules containing enzymes like chitinase and protease, which degrade the insect exoskeleton or fungal cell wall, increasing the efficacy of biological control (Cortés et al. 2017; Mejías and Roy 2019). However, advances in biodegradable materials, such as chitosan and PLGA (poly(lactic-co-glycolic acid)), can improve the sustainability of these formulations (Campos et al. 2015; Lagreca et al. 2020).

In addition to microformulations, nanoformulations represent an innovative frontier in biopesticide development, offering unique advantages such as greater delivery efficiency and molecular-scale interaction with biological targets. These formulations are characterized by colloidal systems in the 1–100 nm range that exhibit distinct thermodynamic and kinetic properties. The high surface area/volume ratio enhances the solubilization of hydrophobic compounds, while the quantum size effect in metallic nanoparticles like Ag and SiO_2 amplifies their biocidal activity.

Despite challenges, such as the high cost of production and the need for specific regulation for microencapsulation of biopesticides (Chaturvedi and Sharma 2024), micro- and nanoformulations represent an innovative solution for Integrated Pest Management, combining high efficacy with reduced ecological impacts (Knowles 2008; Kumar et al. 2021).

8 Challenges and Advancements in Second-Generation Biopesticides

Despite the continuous and promising growth of the biopesticide market in Brazil, several obstacles still limit its effective consolidation and expansion. One of the main challenges lies in the competitiveness of these biological products against conventional chemical insecticides, especially regarding pest control efficacy. In this context, strengthening research and developing more effective, economically viable, and adaptable formulations for different cultivation conditions becomes essential (Copping and Menn 2000; Glare et al. 2012).

Given these limitations, the so-called second generation of biopesticides, composed of natural metabolites produced by microorganisms or their synthetic analogs, gains prominence. These compounds exhibit direct bioactive action on biological targets, offering promising alternatives to the use of living organisms as control agents (Marrone 2019). Within this category, extracellular enzymes, such as chitinases, proteases, and lipases, stand out, often applied in combined formulations like enzyme cocktails to potentiate insecticidal or antifungal action through synergism (Lorito et al. 2010; St Leger et al. 1996; Alves et al. 2020).

One of the main advantages of enzyme cocktails is the possibility of individual component optimization, allowing fine adjustments in formulations according to the type of pest or specific environmental conditions. Furthermore, these enzymes exhibit greater stability during storage and application, which can be an important differential compared to the use of living microorganisms (Gullino et al. 2020; Mascarin and Jaronski 2016). As an example, the study by Alves et al. (2020) describes the development of a process to obtain an enzyme cocktail with insecticidal activity from *B. bassiana*. The enzymes were produced by solid-state fermentation (SSF) and concentrated by ultrafiltration, resulting in a rich extract with proven efficacy against insect pests. These results reinforce the potential of enzymatic formulations as sustainable and efficient alternatives to chemical insecticides.

Despite advances, industrial-scale production still faces technical and economic hurdles, such as defining accessible fermentation substrates and compatibility with other agricultural inputs. Moreover, regulatory and commercial barriers persist; in many countries, including Brazil, there are still no specific regulations for enzymatic biopesticides, which hinders the registration, approval, and commercialization of these products. Farmers' resistance to adopting little-known technologies and the perception of inferior efficacy are also relevant challenges.

Various strategies are being explored to overcome these limitations. Enzymes can be incorporated into traditional fungicides and insecticides as synergistic additives, contributing to increasing their efficacy and reducing the necessary doses, thus minimizing environmental impacts. Another innovative approach involves the use of functionalized microbial biofilms, which allow for the controlled and prolonged release of enzymes into the soil, extending their persistence and reducing the frequency of reapplications. These innovations open new perspectives for the use of enzymes in integrated pest management, contributing to more sustainable, safe, and efficient agriculture.

9 Conclusion

Use of extracellular enzymes produced by entomopathogenic fungi represents an innovative and promising strategy in the biological control of agricultural pests. Throughout this chapter, it has been discussed in detail how these enzymes, especially chitinases and proteases, can act synergistically in degrading the insect cuticle, promoting an effective and selective insecticidal effect. For these enzymes to be viable in practical applications, the standardization and optimization of production processes are essential, from the selection of the microorganism and culture medium to the fermentation, extraction, and concentration stages of bioactive compounds. The standardization of these processes is fundamental to ensure reproducibility, stability, and scalability, aspects indispensable for the future commercialization of biopesticides based on enzymatic extracts.

Furthermore, the importance of understanding and appropriately selecting the types of formulation to be employed was discussed, as this directly influences the efficacy, stability, and applicability of enzymatic products in the field. The correct formulation can increase enzyme stability, facilitate application, and promote controlled release, which are decisive factors for product performance. It is concluded, therefore, that the application of enzymes as biocontrol agents is a promising alternative but demands a rigorous technical approach in standardization, production, and formulation, coupled with regulatory advancements and awareness actions in the field. By overcoming these challenges, enzymatic biopesticides have the potential to play an important role in more sustainable, safe, and technologically advanced agriculture.

References

Acikara ÖB (2013) Ion-exchange chromatography and its applications. Column Chromatogra 10: 55744. https://doi.org/10.5772/55744

Ahmed T, Juhász A, Bose U, Terefe NS, Colgrave ML (2024) Research trends in production, separation, and identification of bioactive peptides from fungi–a critical review. J Funct Foods 119:10634. https://doi.org/10.1016/j.jff.2024.106343

Ali SS (2016) Preparation and evaluation of three laboratory formulations of the entomopathogenic fungus *Beauveria bassiana*. Egypt J Biol Pest Control 26(1):107–110

Ali MB, Hamdi N, Rodriguez MA, Mahmoudi K, Srasra E (2018) Preparation and characterization of new ceramic membranes for ultrafiltration. Ceram Int 44(2):2328–2335. https://doi.org/10.1016/j.ceramint.2017.10.199

Alpert AJ (1990) Hydrophilic-interaction chromatography for separation of peptides, nucleic acids, and other polar compounds. J Chromatogr 499:177–196. https://doi.org/10.1016/S0021-9673(00)96972-3

Alves RT, Bateman RP, Gunn J, Prior C, Leather SR (2002) Effects of different formulations on viability and medium term storage of *Metarhizium anisopliae* conidia. Neotrop Entomol 31:91–99. https://doi.org/10.1590/S1519-566X2002000100013

Alves EA, Schmaltz S, Tres MV, Zabot GL, Kuhn RC et al (2020) Process development to obtain a cocktail containing cell wall degrading enzymes with insecticidal activity from *Beauveria bassiana*. Biochem Eng J 156:107484. https://doi.org/10.1016/j.bej.2019.107484

Andersen SO (2010) Insect cuticular sclerotization: a review. Insect Biochem Mol Biol 40(3): 166–178. https://doi.org/10.1016/j.ibmb.2009.10.007

Andersson C, Tjerneld F, Hedenström M (2022) Method for separating enzymes from microorganisms (U.S. Patent No. 11,274,122 B2). US Patent and Trademark Office

Arpigny JL, Jaeger KE (1999) Bacterial lipolytic enzymes: classification and properties. Biochem J 343(1):177–183. https://doi.org/10.1042/bj3430177

Bagga S, Hu G, Screen SE, Leger RJS (2004) Reconstructing the diversification of subtilisins in the pathogenic fungus *Metarhizium anisopliae*. Gene 324:159–169. https://doi.org/10.1016/j.gene.2003.09.031

Balakrishnan M, Agarwal GP (1996) Protein fractionation in a vortex flow filter. II: separation of simulated mixtures. J Membr Sci 112(1):75–84

Baratto CM, Dutra V, Boldo JT, Leiria LB, Vainstein MH et al (2006) Isolation, characterization, and transcriptional analysis of the chitinase chi2 gene (DQ011663) from the biocontrol fungus *Metarhizium anisopliae* var. anisopliae. Curr Microbiol 53:217–221. https://doi.org/10.1007/s00284-006-0078-6

Bateman RP, Carey M, Moore DE, Prior C (1993) The enhanced infectivity of *Metarhizium flavoviride* in oil formulations to desert locusts at low humidities. Ann Appl Biol 122(1): 145–152

Bedade DM, Pawar SS (2023) Advances in enzyme purification techniques: challenges and solutions. J Biotechnol Bioprocess Eng 28(2):145–158

Begg GS, Cook SM, Dye R, Ferrante M, Franck P et al (2017) A functional overview of conservation biological control. Crop Prote 97:145–158. https://doi.org/10.1016/j.cropro.2016.11.008

Bejarano A, Puopolo G (2020) Bioformulation of microbial biocontrol agents for a sustainable agriculture. In: De Cal A, Melgarejo P, Magan N (eds) How research can stimulate the development of commercial biological control against plant diseases. Springer, pp 275–293. https://doi.org/10.1007/978-3-030-53238-3_16

Bekavac N, Benković M, Jurina T, Valinger D, Gajdoš Kljusurić J et al (2024) Advancements in aqueous two-phase systems for enzyme extraction, purification, and biotransformation. Molecules 29(16):3776. https://doi.org/10.3390/molecules29163776

Brar SK, Verma M, Tyagi RD, Valéro JR (2006) Recent advances in downstream processing and formulations of *Bacillus thuringiensis* based biopesticides. Process Biochem 41(2):323–342. https://doi.org/10.1016/j.procbio.2005.07.015

Brazil, Ministry of Agriculture, Livestock and Supply (2025) AGROFIT: Phytosanitary Pesticides System. Available at: http://agrofit.agricultura.gov.br/agrofit_cons/principal_agrofit_cons. Accessed on: 21 May 2025.

Butu M, Rodino S, Butu A (2022) Biopesticide formulations-current challenges and future perspectives. In: Rakshit A, Meena VS, Abhilash PC, Sarma BK, Singh HB, Fraceto L, Parihar M, Singh AK (eds) *Bi*opesticides volume 2: advances in bio-inoculants. Woodhead Publishing, pp 19–29. https://doi.org/10.1016/B978-0-12-823355-9.00010-9

Campos EVR, de Oliveira JL, Fraceto LF, Singh B (2015) Polysaccharides as safer release systems for agrochemicals. Agron Sustain Dev 35:47–66. https://doi.org/10.1007/s13593-014-0263-0

Chakraborty A, Puja R, Bose K (2022) Protein purification by ion exchange chromatography. In: Bose K (ed) Textbook on cloning, expression and purification of recombinant proteins, Springer Nature Singapore, pp 173–198. https://doi.org/10.1007/978-981-16-4987-5_7

Chaturvedi P, Sharma P (2024) A review on microencapsulation as method of drug delivery. BIO Web of Conf 86:01033. https://doi.org/10.1051/bioconf/20248601033

Cheison SC, Wang Z, Xu SY (2006) Hydrolysis of whey protein isolate in a tangential flow filter membrane reactor: II. Characterisation for the fate of the enzyme by multivariate data analysis. J Membr Sci 286(1–2):322–332. https://doi.org/10.1016/j.memsci.2006.10.019

Choi JM, Han SS, Kim HS (2015) Industrial applications of enzyme biocatalysis: current status and future aspects. Biotechnol Adv 33(7):1443–1454. https://doi.org/10.1016/j.biotechadv.2015.02.014

Collier T, Van Steenwyk R (2004) A critical evaluation of augmentative biological control. Biol Control 31(2):245–256. https://doi.org/10.1016/j.biocontrol.2004.05.001

Copping LG, Menn JJ (2000) Biopesticides: a review of their action, applications and efficacy. Pest Manag Sci 56(8):651–676. https://doi.org/10.1002/1526-4998(200008)56:8%3C651::AID-PS201%3E3.0.CO;2-U

Cortés P, Cano-Sarabia M, Colom J, Otero J, Maspoch D et al (2017) Nano/Micro formulations for bacteriophage delivery. In: Azeredo M, Sillankorva S (eds) Bacteriophage therapy: from lab to clinical practice. Springer, pp 271–283. https://doi.org/10.1007/978-1-4939-7395-8_20

de Faria MR, Wraight SP (2007) Mycoinsecticides and mycoacaricides: a comprehensive list with worldwide coverage and international classification of formulation types. Biol Control 43(3): 237–256. https://doi.org/10.1016/j.biocontrol.2007.08.001

Dey P, Rangarajan V, Singh J, Nayak J, Dilip KJ (2022) Current perspective on improved fermentative production and purification of fungal cellulases for successful biorefinery applications: a brief review. Biomass Convers Biorefin 1–29. https://doi.org/10.1007/s13399-020-01227-6

Dhar P, Kaur G (2011) Response surface methodology for optimizing process parameters for the mass production of *Beauveria bassiana* conidiospores. Afr J Microbiol Res 5(17):2399–2406. https://doi.org/10.5897/AJMR10.088

Doroški A, Klaus A, Režek Jambrak A, Djekic I (2022) Food waste originated material as an alternative substrate used for the cultivation of oyster mushroom (*Pleurotus ostreatus*): a review. Sustainability 14(19):12509. https://doi.org/10.3390/su141912509

Elkhateeb W, Karunarathna SC, Daba GM (2021) Fungi and insects as models of extraordinary symbiosis. Stud Fungi 6(1):469–479. https://doi.org/10.5943/sif/6/1/36

Elsherbiny EA, Selim MAE, Elattaapy AM (2023) Plant growth-promoting fungi for growth improvement and resistance induction. In: Rashad YM, Baka ZAM, Moussa TAA (eds) Plant mycobiome. Springer, Cham, pp 267–288. https://doi.org/10.1007/978-3-031-28307-9_11

Fang W, Leng B, Xiao Y, Jin K, Ma J et al (2005) Cloning of *Beauveria bassiana* chitinase gene Bbchit1 and its application to improve fungal strain virulence. Appl Environ Microbiol. 71(1): 363–370. https://doi.org/10.1128/AEM.71.1.363-370.2005

FAO (Food and Agriculture Organization) (2022) Guidelines for registration of enzymatic biopesticides. FAO

Felizatti AP, Manzano RM, Rodrigues IMW, da Silva MFDGF, Fernandes JB et al (2021) Encapsulation of *B. bassiana* in biopolymers: improving microbiology of insect pest control. Front Microbiol 12:704812

Fernandes EG, Valério HM, Feltrin T, Sand STVD (2012) Variability in the production of extracellular enzymes by entomopathogenic fungi grown on different substrates. Braz J Microbiol 43:827–833. https://doi.org/10.1590/S1517-83822012000200049

Fernandes ÉK, Rangel DE, Braga GU, Roberts DW (2015) Tolerance of entomopathogenic fungi to ultraviolet radiation: a review on screening of strains and their formulation. Curr Genet 61:427–440. https://doi.org/10.1007/s00294-015-0492-z

Fischer L, Scheckermann C, Wagner F (1995) Purification and characterization of a thermotolerant beta-galactosidase from *Thermomyces lanuginosus*. Appl Environ Microbiol 61(4):1497–1501. https://doi.org/10.1128/aem.61.4.1497-1501.1995

Freydank AC, Brandt W, Dräger B (2008) Protein structure modeling indicates hexa histidine-tag interference with enzyme activity. Proteins Struct Funct Bioinform 72(1):173–183

García-Estrada C, Cat E, Santamarta I (2016) *Beauveria bassiana* as biocontrol agent: Formulation and commercialization for pest management. In: Singh HB, Sharma B, Keswani C (eds) Agriculturally important microorganisms. Springer, Singapore, pp 81–96. https://doi.org/10.1007/978-981-10-2576-1_5

Geisseler D, Horwath WR (2008) Regulation of extracellular protease activity in soil in response to different sources and concentrations of nitrogen and carbon. Soil Biol Biochem 40(12): 3040–3048

Glare TR, Caradus JR, Gelernter WD, Jackson TA, Keyhani NO et al (2012) Have biopesticides come of age? Trends Biotechnol 30(5):250–258. https://doi.org/10.1016/j.tibtech.2012.01.003

Grewal PS (2002) Formulation and application technology. In: Gaugler R (ed) Entomopathogenic nematology. CABI Publishing, pp 265–287

Gullino ML, Albajes R, Nicot PC (eds) (2020) Integrated pest and disease management in greenhouse crops (2nd ed., Plant Pathology in the 21st Century, Vol. 9). Springer Nature. https://doi.org/10.1007/978-3-030-22304-5

Gupta A, Tripathy DB, Kumar G, Agarwal P, Ghosal A (2022) Nanopesticides, nanoherbicides, and nanofertilizers. CRC Press, p 178

Gurr GM, Wratten SD, Snyder WE (2017) Biodiversity and insect pests: key issues for sustainable management. Wiley-Blackwell, p 360

Hamrouni R, Regus F, Farnet Da Silva AM, Orsiere T, Boudenne JL et al (2025) Current status and future trends of microbial and nematode-based biopesticides for biocontrol of crop pathogens. Crit Rev Biotechnol 45(2):333–352. https://doi.org/10.1080/07388551.2024.2370370

Han Z, Kautto L, Nevalainen H (2017) Secretion of proteases by an opportunistic fungal pathogen *Scedosporium aurantiacum*. PLoS One 12(1):e0169403. https://doi.org/10.1371/journal.pone.0169403

Hasan F, Shah AA & Hameed A (2006) Influence of culture conditions on lipase production by Bacillus sp.FH5. Ann. Microbiol. 56, 247–252. https://doi.org/10.1007/BF03175013

Hatt S, Boeraeve F, Francis F (2018) Spatial diversification of agroecosystems to enhance biological control and other regulating services. Agron Sustain Dev 38(3):30. https://doi.org/10.1016/j.scitotenv.2017.11.296

Hazra DK, Purkait A (2019) Role of pesticide formulations for sustainable crop protection and environment management: a review. J Pharmacogn Phytochem 8(2):686–693

Heimpel GE, Cock MJW (2018) Shifting paradigms in the history of classical biological control. Biol Control 63(1):27–37. https://doi.org/10.1007/s10526-017-9841-9

Hellberg U, Ivarsson JP, Johansson BL (1996) Characteristics of Superdex prep grade media for gel filtration chromatography of proteins and peptides. Process Biochem 31(2):163–178

Herkersdorf S, Krüger T, Wein P, Löffler S, Fontaine T et al (2021) Bacterial cell wall degrading enzymes induce basidiomycete natural product biosynthesis. Environ Microbiol 23(8): 4360–4371. https://doi.org/10.1111/1462-2920.15621

Hernandez-Tenorio F, Miranda AM, Rodríguez CA, Giraldo-Estrada C, Sáez AA (2022) Potential strategies in the biopesticide formulations: a bibliometric analysis. Agronomy 12(11):2665. https://doi.org/10.3390/agronomy12112665

Hofstee BHJ (1973) Hydrophobic affinity chromatography of proteins. Anal Biochem 52(2): 430–448

Holder DJ, Keyhani NO (2005) Adhesion of the entomopathogenic fungus *Beauveria bassiana* to substrata. Appl Environ Microbiol 71(9):5260–5266. https://doi.org/10.1128/AEM.71.9.5260-5266.2005

Ibatsam K, Rukhsana B, Nasim G (2012) Preservation of *Penicillium* species by lyophilization. Afr J Food Agric Nutr Dev 12(3):6055–6064

IBMA (International Biocontrol Manufacturers Association) (2021) Guidelines for WG formulations of biocontrol agents, 2nd edn. IBMA

Illuri R, Kumar M, Eyini M, Veeramanikandan V, Almaary KS et al (2021) Production, partial purification and characterization of ligninolytic enzymes from selected basidiomycetes mushroom fungi. Saudi J Biol Sci 28(12):7207–7218. https://doi.org/10.1016/j.sjbs.2021.08.026

Jaeger KE, Eggert T (2002) Lipases for biotechnology. Curr Opin Biotechnol 13(4):390–397. https://doi.org/10.1016/s0958-1669(02)00341-5

Jamir Y, Bhushan M, Sanjukta R, Robindro Singh L (2024) Plant-based essential oil encapsulated in nanoemulsions and their enhanced therapeutic applications: an overview. Biotechnol Bioeng 121(2):415–433. https://doi.org/10.1002/bit.28590

Janson JC (2011) Protein purification: principles, high resolution methods, and applications, 3rd edn. Wiley

Jin K, Luo Z, Jiang X, Zhang Y, Zhou Y et al (2011) Carbon catabolite repressor gene *BbCre1* influences carbon source uptake but does not have a big impact on virulence in *Beauveria bassiana*. J Invertebr Pathol 106(3):400–406. https://doi.org/10.1016/j.jip.2010.11.008

Juturu V, Wu JC (2014) Microbial cellulases: engineering, production and applications. Renew Sustain Energy Rev 33:188–203. https://doi.org/10.1016/j.rser.2014.01.077

Keswani C, Bisen K, Singh V, Sarma BK, Singh HB (2016) Formulation technology of biocontrol agents: present status and future prospects. In: Arora N, Mehnaz S, Balestrini R (eds) Bioformulations: for sustainable agriculture. Springer, pp 35–52

Knowles A (2008) Recent developments of safer formulations of agrochemicals. Environmentalist 28:35–44. https://doi.org/10.1007/s10669-007-9045-4

Köcher H, Kocur J (1993) Influence of wetting agents on the foliar retention, uptake and herbicidal activity of glufosinate on barley and barnyard. Pestic Sci 37(1):45–53. https://doi.org/10.1002/ps.2780370107

Kumar P, Pandhi S, Mahato DK, Kamle M, Mishra A (2021) Bacillus-based nano-bioformulations for phytopathogens and insect-pest management. Egypt J Biol Pest Control 31(1):1–11

Lacey LA, Grzywacz D, Shapiro-Ilan DI (2015) Insect pathogens as biological control agents: back to the future. J Invertebr Pathol 132:1–41

Ladeira SA, Andrade MVV, Delatorre AB, Perez VH, Martins MLL (2010) Use of agro-industrial residues for protease production by thermophilic *Bacillus* sp. in submerged fermentation: medium optimization using experimental design technique. Quim Nova 33:324–328. https://doi.org/10.1590/S0100-40422010000200018

Lagreca E, Onesto V, Di Natale C (2020) Recent advances in the formulation of PLGA microparticles for controlled drug delivery. Prog Biomater 9:153–174. https://doi.org/10.1007/s40204-020-00139-y

Li J, Xie J, Zeng D, Liang L, Liu W et al (2021) Effective control of *Frankliniella occidentalis* by *Metarhizium anisopliae* CQMa421 under field conditions. J Pest Sci 94(1):111–117. https://doi.org/10.1007/s10340-020-01223-9

Linskens HF, Tracey MV, Beiss U, Bendall F, Björk W et al (1963) Purification of enzymes by ion exchange chromatography. In: Linskens HF (ed) Modern methods of plant analysis / moderne methoden der Pflanzenanalyse. Springer, pp 393–394. https://doi.org/10.1007/978-3-642-94878-7_23

Liu L, Wei J, Ho KM, Chiu KY, Ngai T (2023) Capsules templated from water-in-oil Pickering emulsions for enzyme encapsulation. J Colloid Interface Sci 629:559–568

Lopes IR, Canellas ALB, de Oliveira BFR, Laport MS (2023) Marine microorganisms: a reservoir of biotechnologically interesting hydrolases. Rev Biol 22(1):32–46. https://doi.org/10.11606/issn.1984-5154.v22p32-46

Lorito M, Woo SL, Harman GE, Monte E (2010) Translational research on Trichoderma: from 'omics to the field. Annu Rev Phytopathol 48:395–417

Mahmoud AMA, De Luna-Santillana EJ, Rodríguez-Pérez MA, Guo X (2011) Parasitism by the endoparasitoid *Cotesia flavipes* induces cellular immunosuppression and enhances susceptibility of the sugarcane borer (*Diatraea saccharalis*) to *Bacillus thuringiensis*. J Insect Sci 11(119): 1–15. https://doi.org/10.1673/031.011.11901

Marrone PG (2019) Pesticidal natural products—status and future potential. Pest Manag Sci 75(9): 2325–2340. https://doi.org/10.1002/ps.5433

Mascarin GM, Jaronski ST (2016) The production and uses of *Beauveria bassiana* as a microbial insecticide. World J Microbiol Biotechnol 32(11):177

Mejías JC, Roy K (2019) In-vitro and in-vivo characterization of a multi-stage enzyme-responsive nanoparticle-in-microgel pulmonary drug delivery system. J Control Release 316:393–403. https://doi.org/10.1016/j.jconrel.2019.09.012

Mishra P, Mishra J, Dwivedi SK, Arora NK (2020) Microbial enzymes in biocontrol of phytopathogens. In: Arora N, Mishra J, Mishra V (eds) Microbial enzymes: roles and applications in industries. Springer, pp 259–285. https://doi.org/10.1007/978-981-15-1710-5_10

Mondal S, Berihun M, Kovacs M, Vatai G, Koris A (2019) Partial refinement of fungal chitinase (*Beauveria bassiana*) with multistage membrane filtration. Acta Aliment 48(3):277–287. https://doi.org/10.1556/066.2019.48.3.1

Moyes DL, Richardson JP, Naglik JR (2015) *Candida albicans*-epithelial interactions and pathogenicity mechanisms: scratching the surface. Virulence 6(4):338–346. https://doi.org/10.1080/21505594.2015.1012981

Murisier A, Duivelshof BL, Fekete S, Bourquin J, Schmudlach A et al (2021) Towards a simple on-line coupling of ion exchange chromatography and native mass spectrometry for the detailed characterization of monoclonal antibodies. J Chromatogr A 1655:462499. https://doi.org/10.1016/j.chroma.2021.462499

Muszewska A, Stepniewska-Dziubinska MM, Steczkiewicz K, Pawlowska J, Dziedzic A et al (2017) Fungal lifestyle reflected in serine protease repertoire. Sci Rep 7(1):9147. https://doi.org/10.1038/s41598-017-09644-w

Nadar SS, Rao P, Rathod VK (2018) Enzyme assisted extraction of biomolecules as an approach to novel extraction techniques: a review. Food Res Int 108:309–330. https://doi.org/10.1016/j.foodres.2018.03.006

Naglik J, Albrecht A, Bader O, Hube B (2004) *Candida albicans* proteinases and host/pathogen interactions. Cell Microbiol 6(10):915–926

Nelson DL, Cox MM (2014) Lehninger principles of biochemistry, 6th edn. WH Freeman

Nussenbaum AL, Lecuona RE (2012) Selection of *Beauveria bassiana* sensu lato and *Metarhizium anisopliae* sensu lato isolates as microbial control agents against the boll weevil (*Anthonomus grandis*) in Argentina. J Invertebr Pathol 110(1):1–7. https://doi.org/10.1016/j.jip.2012.01.010

OECD (Organisation for Economic Co-operation and Development) (2022) Guidance document on WG formulations. OECD Publishing

Orlandelli RC, Specian V, Pamphile JA (2012) Fungal enzymes: production and biotechnological applications. Semin: Ciênc Agrár 33(1):123–136

Orr V, Scharer J, Moo-Young M, Honeyman CH, Fenner D et al (2012) Simultaneous clarification of *Escherichia coli* culture and purification of extracellularly produced penicillin G acylase using tangential flow filtration and anion-exchange membrane chromatography (TFF-AEMC). J Chromatogr B 900:71–78. https://doi.org/10.1016/j.jchromb.2012.05.039

Ortíz-Catón M, Alatorre-Rosas R, Valdivia-Bernal R, Ortíz-Catón A, Medina-Torres R et al (2013) Efecto de la temperatura y humedad relativa sobre el desarrollo de los hongos entomopatógenos [Effect of temperature and relative humidity on the development of entomopathogenic fungi]. Rev Bio Ciênc 1(2):1–12. https://doi.org/10.15741/revbio.01.02.05

Ortiz-Urquiza A, Luo Z, Keyhani NO (2015) Improving mycoinsecticides for insect biological control. Appl Microbiol Biotechnol 99(3):1057–1068. https://doi.org/10.1007/s00253-014-6270-x

Paiva-Guimarães AGL, Freire KRL, Santos SFM, Almeida AF, Sousa ACB (2019) Alternative substrates for conidiogenesis of the entomopathogenic fungus *Beauveria bassiana* (Bals) Vuillemin (Deuteromycotina: Hyphomycetes). Braz J Biol 80:133–141. https://doi.org/10.1590/1519-6984.195711

Parra JRP (2014) Biological control in Brazil: an overview. Scientia Agricola 71:420–429. https://doi.org/10.1590/0103-9016-2014-0167

Pedrini N, Ortiz-Urquiza A, Huarte-Bonnet C, Zhang S, Keyhani NO (2013) Targeting of insect epicuticular lipids by the entomopathogenic fungus *Beauveria bassiana*: Hydrocarbon oxidation within the context of a host-pathogen interaction. Front Microbiol 4:24

Polli EG, Alves GS, de Moraes JG, Kruger GR (2022) Influence of surfactant–humectant adjuvants on physical properties, droplet size, and efficacy of glufosinate formulations. Agrosyst Geosci Environ 5(1):e20230. https://doi.org/10.1002/agg2.20230

Porath J, Flodin P (1959) Gel filtration: a method for desalting and group separation. Nature 183(4676):1657–1659. https://doi.org/10.1038/1831657a0

Prior C, Jollands P, Le Patourel G (1988) Infectivity of oil and water formulations of Beauveria bassiana (Deuteromycotina: Hyphomycetes) to the cocoa weevil pest Pantorhytes plutus (Coleoptera: Curculionidae). J Invertebr Pathol 52(1):66–72. https://doi.org/10.1016/0022-2011(88)90103-6

Qazi SS, Khachatourians GG (2007) Hydrated conidia of *Metarhizium anisopliae* release a family of metalloproteases. J Invertebr Pathol 95(1):48–59. https://doi.org/10.1016/j.jip.2006.12.002

Qi H, Zhang D, Shan L, Chen G, Zhang B (2023) Advances in the mechanisms of entomopathogenic fungi infecting insect hosts and the defense strategies of insects. Biodivers Sci 31(11):Article 23273. https://doi.org/10.17520/biods.2023273

Ramachandran P, Narayanan GK, Gandhi S, Basak A (2015) *Rhizopus oryzae* lipase immobilized on hierarchical mesoporous silica supports for transesterification of rice bran oil. Appl Biochem Biotechnol 175(5):2332–2346. https://doi.org/10.1007/s12010-014-1432-y

Ren XB, Dang YR, Liu SS, Huang KX, Qin QL et al (2022) Identification and characterization of three chitinases with potential in direct conversion of crystalline chitin into N, N′--diacetylchitobiose. Mar Drugs 20(3):165. https://doi.org/10.3390/md20030165

Rodriguez R, Santos C, Simões MF, Soares C, Santos C et al (2019) Polyphasic, including MALDI-TOF MS, evaluation of freeze-drying long-term preservation on *Aspergillus* (section *Nigri*) strains. Microorganisms 7(9):291

Rondelli VM, Pratissoli D, Polanczyk RA, Marques EJ, Sturm GM et al (2011) Association of castor bean oil with *Beauveria bassiana* for diamondback moth control. Pesqui Agropecu Bras 46:212–214. https://doi.org/10.1590/S0100-204X2011000200014

Sala A, Vittone S, Barrena R, Sánchez A, Artola A (2021) Scanning agro-industrial wastes as substrates for fungal biopesticide production: Use of *Beauveria bassiana* and *Trichoderma harzianum* in solid-state fermentation. J Environ Manage 295:113113

Salem K, Elgharbi F, Ben Hlima H, Alghamdi OA, Perduca M, Sayari A et al (2024) His-tag effect on biochemical properties of *Bacillus subtilis* US572 α-amylase produced in *Escherichia coli*: application of the recombinant enzyme in breadmaking. Food Biotechnol 38(2):134–158. https://doi.org/10.1080/08905436.2024.2345768

Sarmah N, Borah S, Bora T (2018) Microbial lipases: sources, production and applications. Microb Pathog 117:149–162

Saxena A, Tripathi BP, Kumar M, Shahi VK (2009) Membrane-based techniques for the separation and purification of proteins: an overview. Adv Colloid Interface Sci 145(1–2):1–22

Scheidt GN, da Silva ALL, Dronk AG, Biasi LA, Arakaki AH et al (2009) Multiplicação in vitro de *Oncidium leucochilum* (Orchidaceae) em diferentes sistemas de cultivo [In vitro multiplication of *Oncidium leucochilum* (Orchidaceae) in different cultivation systems]. Biociências 17(1): 82–85

Scholte EJ, Knols BGJ, Takken W (2004) Infection of *Aedes aegypti* with entomopathogenic fungi. J Invertebr Pathol 86(3):151–154

Screen SE, St Leger RJ (2000) Cloning, expression, and substrate specificity of a fungal chymotrypsin: evidence for lateral gene transfer from an actinomycete bacterium. J Biol Chem 275(9): 6689–6694. https://doi.org/10.1074/jbc.275.9.6689

Screen SE, St Leger RJ, Cooper RM (1997) The role of *Metarhizium anisopliae* protease in insect pathogenesis. J Invertebr Pathol 70(2):85–91

Screen SE, Hu G, St Leger RJ (2001) Regulation of subtilisin-like protease production in *Metarhizium anisopliae*. Mycol 105(6):705–712

Seemüller E, Lupas A, Stock D, Löwe J, Huber R et al (1995) Proteasome from *Thermoplasma acidophilum*: a threonine protease. Science 268(5210):579–582. https://doi.org/10.1126/science.7725107

Shah TN, Foley HC, Zydney AL (2007) Development and characterization of nanoporous carbon membranes for protein ultrafiltration. J Membr Sci 295(1–2):40–49

Shishodia SK, Shankar J (2020) Proteomic analysis revealed ROS-mediated growth inhibition of *Aspergillus terreus* by shikonin. J Proteomics 224:103849. https://doi.org/10.1016/j.jprot.2020.103849

Singh G, Kapoor T, Kaur H (2022) Current status and future perspective on enzyme involving in biocontrol of plant pathogen. J Res Appl Sci Biotechnol 1(1):12–17. https://doi.org/10.55544/jrasb.1.1.3

Siqueira HAA, Martins AJ, Valadares-Inglis MC (2021) Resistência a inseticidas e desafios para o manejo de pragas no Brasil [Insecticide resistance and challenges for pest management in Brazil]. Pesqui Agropec Bras 56:e02144

Sivakumar T, Renganathan PBP, Sanjeevkumar K (2020) Bio efficacy of bio-nematon (Paecilomyces lilacinus 1.15% wp) against root-knot nematode (*Meloidogyne incognita*) in cucumber crop. Plant Arch 20(2):3805–3810

Sosa-Gómez DR, Del Pino M, Jelkmann W (2020) Efficacy of *Beauveria bassiana* against *Euschistus heros* (Hemiptera: Pentatomidae). Pest Manag Sci 76(2):675–682

Souza PM, Almeida AF, Bittencourt ML (2015) Proteases from fungi: current applications and future prospects. Food Technol Biotechnol 53(1):1–14

St Leger RJ, Joshi L, Roberts DW (1996) Ambient pH is a major determinant in the expression of cuticle-degrading enzymes and hydrophobin by the entomopathogenic fungus *Metarhizium anisopliae*. Appl Environ Microbiol 62(2):716–722. https://doi.org/10.1128/AEM.64.2.709-713.1998

St Leger RJ, Wang C, Fang W (2011) New insights into the mechanisms of fungal pathogenesis in insects. Trends Microbiol 19(6):256–262

Sun Z, Zhao R, Yu M, Liu Y, Ma Y et al (2024) Enhanced dosage delivery of pesticide under unmanned aerial vehicle condition for peanut plant protection: tank-mix adjuvants and formulation improvement. Pest Manag Sci 80(3):1632–1644. https://doi.org/10.1002/ps.7895

Sung GH, Hywel-Jones NL, Sung JM, Luangsa-ard JJ, Shrestha B et al (2007) Phylogenetic classification of *Cordyceps* and the clavicipitaceous fungi. Stud Mycol 57:5–59. https://doi.org/10.3114/sim.2007.57.01

Tan C, McClements DJ (2021) Application of advanced emulsion technology in the food industry: a review and critical evaluation. Foods 10(4):812

Tanaka M, Gomi K (2021) Induction and repression of hydrolase genes in *Aspergillus oryzae*. Front Microbiol 12:677603. https://doi.org/10.3389/fmicb.2021.677603

Tao R, You C, Qu Q, Zhang X, Deng Y et al (2023) Recent advances in the design of controlled-and sustained-release micro/nanocarriers of pesticide. Environ Sci Nano 10(2):351–371. https://doi.org/10.1039/D2EN00446A

Thompson MW (2022) Regulation of zinc-dependent enzymes by metal carrier proteins. Biometals 35(2):187–213. https://doi.org/10.1007/s10534-022-00373-w

Thompson R, Wilson E, Brown A (2023) Green processes for enzyme production and purification. Trends Biotechnol 41(2):156–170

Tikar S, Prakash S (2017) Fly ash-based *Bacillus thuringiensis israelensis* formulation: an ecofriendly approach. Indian J Med Res 146(6):680–682. https://doi.org/10.4103/ijmr.IJMR_1679_16

Tougeron K, Ferrais L, Gardin P, Lateur M, Hance T (2023) Flower strips increase the control of rosy apple aphids after parasitoid releases in an apple orchard. Ann Appl Biol 182(2):245–256. https://doi.org/10.1111/aab.12816

Tschumi M, Albrecht M, Entling MH, Jacot K (2016) High effectiveness of tailored flower strips in reducing pests and crop plant damage. Proc R Soc B-Biol Sci 283(1823):20152290. https://doi.org/10.1098/rspb.2015.1369

Van Lenteren JC, Bolckmans K, Köhl J (2018) Biological control using invertebrates and microorganisms: plenty of new opportunities. BioControl 63(1):39–59. https://doi.org/10.1007/s10526-017-9801-4

Vega FE, Goettel MS, Blackwell M (2009) Fungal entomopathogens: new insights on their ecology. Fungal Ecol 2(4):149–159

Vertyporokh L, Wojda I (2017) Expression of the insect metalloproteinase inhibitor IMPI in the fat body of *Galleria mellonella* exposed to infection with *Beauveria bassiana*. Acta Biochim Pol 64(2):273–278. https://doi.org/10.18388/abp.2016_1376

Wang C, St Leger RJ (2005) Developmental and transcriptional responses to host and nonhost cuticles by the specific locust pathogen *Metarhizium anisopliae* var. *acridum*. Eukaryot Cell 4(5):937–947

Wang CS, St Leger RJ (2006) A collagenous protective coat enables Metarhizium anisopliae to evade insect immune responses. Proc Natl Acad Sci USA 103:6647–6652. https://doi.org/10.1073/pnas.0601951103

Wang C, St Leger RJ (2007) The *Metarhizium anisopliae* perilipin homolog MPL1 regulates lipid metabolism, appressorial turgor pressure, and virulence. J Biol Chem 282(29):21110–21115. https://doi.org/10.1074/jbc.M609592200

Wei M, Chen P, Zheng P, Tao X, Yu X et al (2023) Purification and characterization of aspartic protease from *Aspergillus niger* and its efficient hydrolysis applications in soy protein degradation. Microb Cell Fact 22(1):42. https://doi.org/10.1186/s12934-023-02047-9

Wingfield PT (2017) Protein precipitation using ammonium sulfate. Curr Protoc Protein Sci 88(1): 4.1.1–4.1.9. https://doi.org/10.1002/0471140864.psa03fs13

Wraight SP, Filotas MJ, Sanderson JP (2016) Comparative efficacy of emulsifiable-oil, wettable-powder, and unformulated-powder preparations of *Beauveria bassiana* against the melon aphid *Aphis gossypii*. Biocontrol Sci Technol 26(7):894–914. https://doi.org/10.1080/09583157.2016.1157851

Xiao G, Ying SH, Zheng P, Wang ZL, Zhang S et al (2012) Genomic perspectives on the evolution of fungal entomopathogenicity in *Beauveria bassiana*. Sci Rep 2(1):483. https://doi.org/10.1038/srep00483

Yang MC, Tong JH (1997) Loose ultrafiltration of proteins using hydrolyzed polyacrylonitrile hollow fiber. J Membr Sci 132(1):63–71. https://doi.org/10.1016/S0376-7388(97)00038-0

Yang L, Li H, Zhang T (2023) Agro-industrial residues as substrates for enzyme production. Waste Biomass Valori 14(5):1567–1580

Yasmeen Q, Asgher M, Sheikh MA, Nawaz H (2013) Optimization of ligninolytic enzymes production through response surface methodology. Bioresour 8(1):944–956. https://doi.org/10.15376/biores.8.1.944-968

Zare R, Gams W (2001) A revision of *Verticillium* sect. *Prostrata* and *Paecilomyces* sect. *Isarioidea* to several anamorphic genera, including *Lecanicillium*. Mycol Res 105(12): 1419–1429. https://doi.org/10.1127/nova.hedwigia/72/2001/329

Zhang Z, Zhao Z, Huang K, Liang Z (2023) Acid-resistant enzymes: the acquisition strategies and applications. Appl Microbiol Biotechnol 107(20):6163–6178. https://doi.org/10.1007/s00253-023-12702-1

Mycotoxins in Entomopathogenic Fungi 8

Duong Minh Lam, Le Thi Tuoi, and Ha Huy Cong

Abstract

The biosphere hosts a vast diversity of organisms, including more than 155,000 described fungal species, of which thousands are recognized as entomopathogenic fungi. These fungi possess unique mechanisms that allow them to overcome host immune defenses, often through the production of specialized compounds known as mycotoxins. Mycotoxins have promising applications in biological control as well as in medicine and pharmacies. However, effective utilization requires comprehensive knowledge of their properties. This chapter reviews several notable mycotoxins, such as beauvericin, bassiatin, beauverolides, brevianamide, cyclosporin, myriocin, destruxins, efrapeptins, enniatin, fumonisins, fusarin, iso-isariin, lateritin, leucinostatins, metarhizin A, oosporein, pentostatin, rhizoferrin, swainsonine, and trichothecenes that are derived from commonly occurring entomopathogenic fungi. We highlight recent advances in understanding their diversity, bioactivity, biosynthetic pathways, mechanisms of action, detection techniques, detoxification strategies, and potential applications in agriculture and medicine. The findings presented aim to offer a comprehensive overview of the mycotoxins produced by entomopathogenic fungi and their diverse significance.

D. M. Lam (✉) · Le Thi Tuoi
Faculty of Biology, Hanoi National University of Education, Hanoi, Vietnam

Institute of Natural Sciences, Hanoi National University of Education, Hanoi, Vietnam
e-mail: lamdm@hnue.edu.vn

Ha Huy Cong
Faculty of Biology, Hanoi National University of Education, Hanoi, Vietnam

S. K. Deshmukh et al. (eds.), *Frontiers in Entomopathogenic Fungi*,
https://doi.org/10.1007/978-981-95-6529-0_8

Keywords

Biocontrol · Mycotoxin monitoring · Structure and biosynthesis · Toxicity · Toxic mechanisms

1 Introduction

Fungi have drawn much attention from mycologists over the last three decades. The number of described species has rapidly increased from 80,000 in 2000 to over 155,000 at present (Catalogue of Life). However, the estimated number of fungal species ranges from 2.2 to 3.8 million (Hyde et al. 2020), highlighting the vast unexplored diversity that remains to be studied. Despite their ecological and economic importance, fungi also pose significant risks due to their toxicity to humans, animals, and ecosystems. Many well-known examples involve cereals such as peanuts, corn, wheat, and rice being colonized or contaminated by fungi. Both general and entomopathogenic fungi (EPF) can infect crops and agricultural products, leading to reduced value, spoilage, and toxicity when consumed by humans or animals (Mutalov et al. 2025).

Mycotoxins are toxic secondary metabolites produced by fungi during fungal growth and development. These compounds exhibit harmful effects on animals and humans, with thousands of distinct mycotoxins typically produced by specific fungal species. A prominent example is aflatoxin, produced by *Aspergillus flavus*, which is associated with carcinogenic effects in humans (Jallow et al. 2021).

Entomopathogenic fungi are those capable of infecting, colonizing, and killing insect hosts. They employ diverse invasion strategies, including the secretion of hydrolytic enzymes and the production of toxic secondary metabolites. This chapter explores key aspects of mycotoxins produced by EPF, including their taxonomy, chemical diversity, biosynthetic pathways, modes of action, detoxification strategies, detection techniques, and agricultural and medicinal applications.

2 Mycotoxin-Producing Entomopathogenic Fungi

The number of identified mycotoxins has increased, with thousands of toxic metabolites now known to be produced by fungi. Entomopathogenic fungi, a specialized group capable of killing their insect hosts, often generate a wide array of bioactive metabolites with toxic effects. De Wint et al. (2024) introduced EntomoFun 1.0, an entomopathogenic fungi and insect host global database that includes 1791 records of nearly 400 species and their insect hosts, including geographic distribution data. Members of the order Hypocreales are commonly recognized as entomopathogenic, but other fungal taxa have also recently been identified as insect pathogens and evaluated for potential biocontrol use. For instance, *Fusarium verticillioides* has shown pathogenicity against the grasshopper *Tropidacris collaris* (Pelizza et al. 2011); *Trichoderma longibrachiatum* has been reported to infect new hosts *Aphis craccivora* (Ibrahim et al. 2011) and *Bemisia tabaci* (Anwar et al. 2016); *T. hamatum* infects *Spodoptera littoralis* (Lana et al. 2023); *Alternaria murispora* parasitizes the ladybird *Diaspis echinocacti* (El Aalaoui et al. 2025); and *Aspergillus nidulans* and *A. fumigatus* isolated from *Sitophilus oryzae* showed toxicity against the common stored-cereal product beetles *Rhyzopertha dominica*, *Tribolium castaneum,* and *Trogoderma granarium* (Khan et al. 2024). The recently isolated strain *Fusarium oxysporum* ZZ-L1 is a potential agent for biological control of the red palm weevil *Rhynchophorus ferrugineus* (Pu et al. 2023).

Despite promising applications, some of these newly categorized EPF raise concerns due to their known production of toxic compounds. For example, *F. verticillioides* is notorious for producing fumonisin B1, a mycotoxin commonly detected in maize (Blacutt et al. 2018). More than 30 species of *Fusarium* have been reported infecting insects and considered entomopathogenic species (da Silva Santos et al. 2020).

Duong and Van (2025) reviewed over 250 secondary metabolites from 56 EPF species and highlighted their diverse biological activities. Several are mycotoxins, including beauvericin, cyclosporine, destruxin, isaridin, isariotins, oosporein, and reseotoxin. Beauvericin (BEA) and enniatins (ENNs) are structurally related cyclic hexadepsipeptides, characterized by alternating N-methyl L-amino acid and D-2-hydroxyisovaleric acid residues. These are frequently found in *B. bassiana* and various EPF (Liuzzi et al. 2017; Van et al. 2021; Huang et al. 2024). Cyclosporin A and myriocin are well-known bioactive compounds with pharmaceutical applications (Patocka et al. 2021; Pradier et al. 2019). Leucinostatin A, a lipopeptide antibiotic, was first isolated from *Penicillium lilacinum* (present name *Purpureocillium lilacinum*) (Arai et al. 1973) and inhibits human prostate cancer growth (Kawada et al. 2010). It serves as a model compound for the development of the synthetic drug lefleuganan (Rimle et al. 2025). Table 8.1 lists some mycotoxins, fungal species, and toxicities.

3 Mycotoxins from Entomopathogenic Fungi: Structure and Biosynthesis

Mycotoxins produced by entomopathogenic fungi are remarkably diverse in structure and biosynthesized through a range of distinct biosynthetic pathways (Duong and Van 2025). The primary biosynthetic routes include polyketide synthases (PKS), nonribosomal peptide synthetases (NRPS), terpene synthesis, and several hybrid mechanisms such as PKS-NRPS systems. Additionally, siderophore formation pathways contribute to the formation of iron chelators such as rhizoferrin. Table 8.2 presents several biosynthetic pathways and corresponding mycotoxins commonly associated with entomopathogenic fungi.

Table 8.1 Mycotoxins from entomopathogenic fungi, their toxicities and applications

Mycotoxin	Entomopathogenic fungal species	Toxicity/effect/application	Reference
Bassiatin 7-hydroxybassiatin	*Beauveria bassiana* K-717 and *Cordyceps cicadae* JXCH-1	Platelet aggregation inhibitor	Kagamizono et al. (1995); Fan et al. (2023)
Beauvericin	*Beauveria bassiana*, *Cordyceps fumosorosea*, *C. cateniannulata* CPA14V and *Fusarium oxysporum*	Nematicidal activities against *Bursaphelenchus xylophilus;* BEA reduced blood cell density in *Bombyx mori*, causing blood cells cluttering, increasing melanization of the monocytic lymphoma cells U-937 and the promyelocytic leukemia cells HL-60. Inhibited progesterol, estradiol, CYP11A1, and CYP19A1 gene expression therefore indirectly affect reproduction of cattle that consume contaminated feed; impairs development, fertility, and life span in the nematode *Caenorhabditis elegans,* accompanied by increased germ cell apoptosis and lipofuscin accumulation	Albonico et al. (2017); Büchter et al. (2020); Wu et al. (2021); Van et al. (2021); Sánchez-Gómez et al. (2023); Calò et al. (2004)
Beauverolides C, F, I, ja, L, M, and N	*Cordyceps fumosorosea* and *Isaria fumosorosea*	Cholesterol acyltransferase inhibitors; Strong lethal activity against *Diaphorina citri* nymphs and adults, with 70, 80, and 100% mortality after 2 and 3 days of application, respectively	Qasim et al. (2020); Šimčíková et al. (2020)
Brevianamide F	*Cordyceps fumosorosea*	Antioxidant activity, mycotoxin	Wu et al. (2021)
Cyclosporine A	*Tolypocladium inflatum*	Immunosuppression, nephrotoxicity, hepatotoxicity, neurotoxicity, myocardial toxicity	Patocka et al. (2021)
Deoxynivalenol (DON)	*Fusarium pernambucanum* and *F. caatingaense*	Hepatotoxicity, nephrotoxicity, immunotoxicity, gastrointestinal toxicity, neurotoxicity, genetic toxicity and carcinogenicity; male infertility	Maciel et al. (2021); Dai et al. (2025)
Destruxin A, B, D and E	*Metarhizium robertsii* and *Metarhizium* spp.	Inhibition of the hemocytin-mediated hemolymph immunity of host insects Control of adults of *Anastrepha obliqua* Macquart (Diptera: Tephritidae), main pest of mangos	Lozano-Tovar et al. (2023); Wang et al. (2024)

(continued)

Table 8.1 (continued)

Mycotoxin	Entomopathogenic fungal species	Toxicity/effect/application	Reference
Efrapeptins	*Tolypocladium* spp.	Complex V inhibitor in the mitochondrial electron transport chain	Masuma et al. (2009); Wallace and Starkov (2000)
Enniatin 29 derives (ENNs)	*Fusarium* spp. and *Verticillium* sp.	Antibacterial, antihelmintic, antifungal, herbicidal, and insecticidal, potent mammalian cytotoxic activity in several mammalian cell lines	Prosperini et al. (2017)
Fumonisins	*Fusarium solani, F. verticillioides* and *Tolypocladium cylindrosporum*	Insecticidal activities against grasshoppers/sugar beet root maggot pupae	Majumdar et al. (2008); Pelizza et al. (2011); Zabalgogeazcoa et al. (2018)
Fusarin C	*Fusarium moniliforme* (present name: *F. fujikuroi)* and *Metarhizium anisopliae*	Inhibited the activation of macrophages by macrophage activation factor and the cytotoxic activity of activated macrophages; highly mutagenic; potentially carcinogenic (induces esophageal and forestomach carcinomas in DBA mice and Wistar rats)	Li et al. (1990); Krasnoff et al. (2006); Niehaus et al. (2014)
Fusarin-like NG391	*Metarhizium robertsii*	Impairs nucleic acid formation in K-562 leukemia cells	Donzelli et al. (2010); Bohnert et al. (2013)
5-methoxymethyl-1H- pyrrole-2-carboxaldehyde	*Purpureocillium lavendulum*	Nematicidal activity on *Meloidogyne incognita*	Bao et al. (2022)
Myriocin	*Isaria sinclarii*	Immunosuppression of mouse lymphocytes	Fujita et al. (1994)
Iso-isariin B, D and E	*B. felina*	Insecticidal against *Sitophilus* spp. with an LD50 value of 10 μg/mL	Langenfeld et al. (2011)
Lateritin	*Gibberella lateritium* and *Isaria cicadae*	Inhibitor of acyl-CoA: cholesterol acyltransferase	Hasumi et al. (1993); He et al. (2019)
Leucinostatins	*Paecilomyces lilacinus* (present name: *Purpureocillium lilacinum*)	Complex V inhibitor in the mitochondrial electron transport chain; Nematicides	Park et al. (2004); Girardi et al. (2022)
Metarhizin A	*Metarhizium flavoviride*	Complex IV inhibitor in the mitochondrial electron transport chain; anti-proliferative activity on insect and mammalian cells	Katou et al. (2014)

(continued)

Table 8.1 (continued)

Mycotoxin	Entomopathogenic fungal species	Toxicity/effect/application	Reference
Oosporein	*Beauveria bassiana* and *B. caledonica*	Insecticidal, antibiotic, antiviral, and antifungal; mycotoxin causes gout in chickens	Fan et al. (2017); Pegram and Wyatt (1981); McNamara et al. (2019)
Pentostatin	*Cordyceps militaris*	Maternal toxicity in rats and rabbits	Dostal et al. (1991)
Rhizoferrin	*Trichoderma hamatum*	Insecticide against *Spodoptera littoralis*	Lana et al. (2023)
Swainsonine	*Metarhizium anisopliae* and *Metarhizium* spp.	Golgi α mannosidase II and lysosomal α-mannosidase inhibitions and locoism	Broquist (1985); Singh and Kaur (2014)
Trichothecenes	*Beauveria bassiana; Cordyceps confragosa* and *Isaria tenuipes*	Inhibit eukaryotic protein synthesis; induce oxidative stress, cell cycle arrest, and apoptosis	Proctor et al. (2018)
T-2 and HT-2	*Conidiobolus coronatus*	Toxic to *Galleria mellonella* larvae and Sf9 cells	Paszkiewicz et al. (2016)
Zearalenone (ZEA)	*Fusarium* spp.	Hepatic injury, DNA damage, immunosuppression and intestinal inflammation	Chen et al. (2023)
Combination of secondary metabolites	*Cladosporium cladosporioides*	Insecticidal; activate AST. ALT, ACP ALP, inhibit AChE and GST of *Spodoptera frugiperda*	Abdullah et al. (2024)
Combination of secondary metabolites	*Verticillium lecanii*	Insecticidal; activate AST. ALT, ACP ALP, inhibit AChE and GST of *Spodoptera frugiperda*	Abdullah et al. (2024)

3.1 Polyketides

Polyketides are a diverse class of natural products assembled via successive Claisen condensation reactions involving small carboxylic acid monomers. This process is catalyzed by polyketide synthases (PKSs), which are categorized into three structural classes. The majority of fungal PKSs belong to type I, which are characterized as large, multifunctional enzymes comprising multiple catalytic domains within a single polypeptide chain (Molnar et al. 2010).

Type I fungal PKSs function iteratively, using the same enzymatic domains for chain elongation. The three core domains essential for this process include acyltransferase (AT), acyl carrier protein (ACP), and ketosynthase (KS) (Zhang et al. 2020). The AT domain is responsible for loading malonyl-derived extender units onto the PKS assembly line, the ACP domain serves as a tether for

Table 8.2 Mycotoxin biosynthesis pathways

Biosynthetic pathway	Class of natural compound	Compound	Reference
NRPS	Cyclic nonribosomal (depsi)peptides	Beauvericin	Huang et al. (2024)
		Enniatins	Liuzzi et al. (2017)
		Destruxins	Wang et al. (2012a)
PKS	Dibenzoquinone	Oosporein	Feng et al. (2015)
	Linear polyketide backbone	Fumonisins	Zhu et al. (2006)
	Resorcylic acid lactone	Zearalenone	Gaffoor and Trail (2006)
PKS and NRPS	Linear nonribosomal peptides	Leucinostatins	Wang et al. (2016)
		Beauverolides	Yin et al. (2020)
		Efrapeptins	Niu et al. (2020)
	Cyclic nonribosomal (depsi)peptides	Cyclosporins	Yang et al. (2018)
PKS-NRPS hybrid	Linear polyketide backbone	Fusarins	Huang et al. (2024)
	Indolizidine alkaloid	Swainsonine	Cook et al. (2017); Luo et al. (2020)
	Other	Tenuazonic acid	Yun et al. (2015)
Siderophore	Carboxylate-type siderophore	Rhizoferrin	Carroll et al. (2017)
Terpene	Diterpene	Metarhizins A	Kikuchi et al. (2009)
	Sesquiterpenoid	Trichothecene	Gao et al. (2025b)

intermediates and building blocks, and the KS domain catalyzes the key condensation reaction between the growing polyketide chain and incoming malonyl units.

In addition to these core domains, fungal type I PKSs may include auxiliary catalytic domains, such as methyltransferases (MeT), which introduce methyl groups during chain assembly, and a thioesterase (TE) domain responsible for releasing the completed polyketide—either in linear or cyclic form. Structural diversity among polyketides arises from variations in type, order, and stereochemistry of monomer units; differing β-keto group reduction; and the cyclization strategies employed during product release. Furthermore, accessory enzymes that modify the polyketide backbone post-synthesis contribute significantly to the rich chemical complexity of these fungal metabolites (Molnar et al. 2010) (Fig. 8.1).

Oosporein, a symmetrical 1,4-bibenzoquinone derivative, was first isolated from the insect-pathogenic fungus *Beauveria bassiana* (el-Basyouni and Vining 1966) and later identified in *B. brongniartii* (Feng et al. 2015). Its biosynthesis is initiated by the polyketide synthase OpS1, which catalyzes the production of orsellinic acid from acetyl-CoA and malonyl-CoA. This intermediate is then modified by enzymes encoded by a dedicated gene cluster comprising seven genes (*opS1-opS7*). Within this cluster, the hydroxylase OpS4 converts orsellinic acid into benzenetriol, which is further oxidized by the putative dioxygenase OpS7 to form benzenetetrol. Finally, the catalase OpS5 catalyzes the dimerization of benzenetetrol, resulting in the formation of the final oosporein molecule (Fig. 8.2) (Feng et al. 2015).

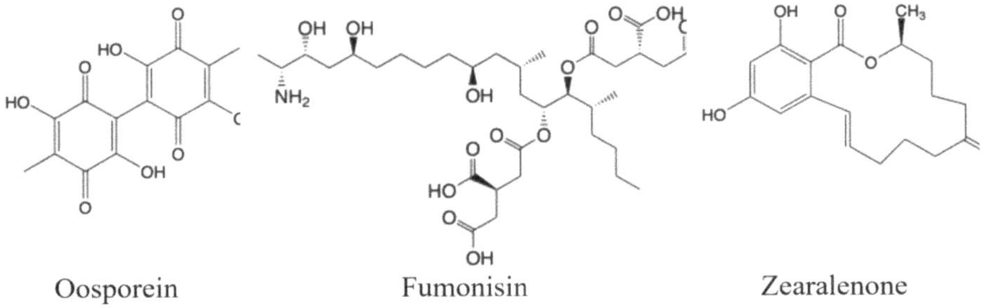

Fig. 8.1 Polyketide mycotoxins

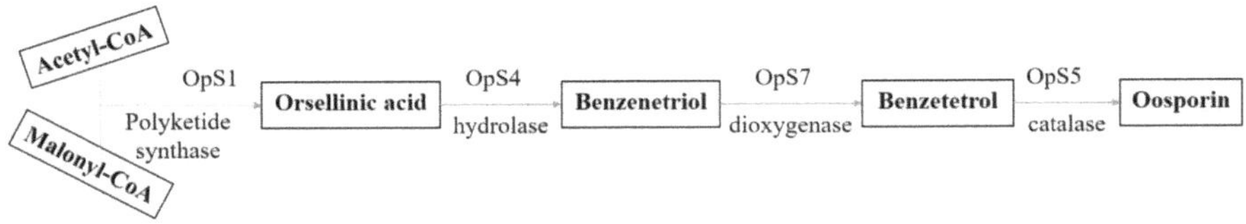

Fig. 8.2 Oosporein biosynthetic pathway in *Beauveria bassiana*

Fumonisins produced by *Fusarium verticillioides* feature a linear polyketide-derived carbon backbone (Fig. 8.1). Their biosynthesis is governed by a 15-gene cluster known as the *fum* cluster, which includes a single polyketide synthase gene, *fum1* (Proctor et al. 2003). This gene is predicted to encode a multi-domain PKS possessing seven functional domains: acyl carrier protein (ACP), acyltransferase (AT), β-ketoacyl reductase (KR), β-ketoacyl synthase (KS), dehydratase (DH), enoylreductase (ER), and methyltransferase (MT) (Fig. 8.3) (Proctor et al. 1999; Zhu et al. 2006).

The biosynthetic process begins with the action of PKS FUM1, which catalyzes the formation of a nonaketide intermediate incorporating two methyl groups (Fig. 8.3). This intermediate is then condensed with a CoA derivative, followed by the addition of L-alanine—a reaction mediated by the aminotransferase FUM8 to form an amine-containing structure. This intermediate is subsequently oxidized by FUM6 and reduced by the ketoreductase FUM13. Two successive oxidation reactions, both catalyzed by FUM2, generate a triol structure. Following oxidation, the triol undergoes esterification at two hydroxyl groups using the CoA derivatives of a tricarboxylic acid, which is believed to originate from citrate reduction. A final oxidative modification completes the pathway, yielding fumonisin B1 as the end product (Alexander et al. 2009).

Zearalenone (ZEN) is another notable polyketide mycotoxin synthesized by various *Fusarium* species and is commonly detected in wheat and other cereal grains (Ingram et al. 2025). Its biosynthesis requires two polyketide synthase genes, *zea1* and *zea2*. Biochemical evidence indicates that the ZEN scaffold is assembled from one acetyl-CoA starter unit and eight malonyl-CoA extender units. The structural arrangement and domain functions of the ZEN polyketide synthases mirror other

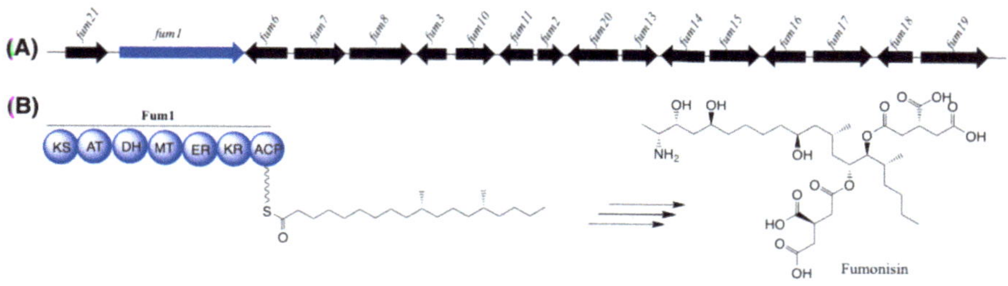

Fig. 8.3 (**a**) *Fusarium verticillioides fum* gene cluster responsible for fumonisin biosynthesis (Gil-Serna et al. 2020). (**b**) Schematic representation of the PKS FUM1 domain architecture involved in fumonisin production (Zhu et al. 2006)

iterative fungal type I PKSs, contributing to the production of estrogenic mycotoxins with significant health implications (Gaffoor and Trail 2006).

3.2 Nonribosomal Peptides

Fungal nonribosomal peptides represent a diverse group of secondary metabolites synthesized by nonribosomal peptide synthetases (NRPSs). These large, modular enzymes are essential for producing structurally complex and functionally versatile peptides. Each NRPS module typically comprises three core domains: (1) Adenylation (A) domain—selects and activates a specific monomer, commonly an amino acid, hydroxy acid, or keto acid. (2) Thiolation (T) domain, also known as the peptidyl carrier protein (PCP)—covalently tethers the activated monomer or peptide intermediate via a thioester bond. (3) Condensation (C) domain—catalyzes peptide bond formation between the growing peptide chain and the newly introduced monomer.

Beyond these core domains, many NRPS modules incorporate auxiliary domains such as N-methyltransferase, epimerase, and oxidase, which contribute to structural diversity by modifying the incorporated residues. Additionally, some NRPSs contain specialized cyclization domains, which facilitate the formation of heterocyclic structures through intramolecular reactions—further enriching the chemical complexity of nonribosomal peptides (Zhang et al. 2020) (Fig. 8.4).

Beauvericin is a cyclic depsipeptide synthesized via a nonribosomal, thiol-templated mechanism. Its biosynthesis is catalyzed by BEA synthase, an NRPS. A deeper genetic understanding has been achieved in *F. proliferatum* LF061, where *Agrobacterium* AGL-1-mediated targeted gene knockouts were used to elucidate the structure and function of the *bea* gene cluster (Huang et al. 2024).

In this strain, BEA biosynthesis is governed by a two-gene cluster comprising an NRPS gene (*bea1*) and a gene encoding ketoisovalerate reductase (Zobel et al. 2016). The *bea1* gene encodes a multifunctional NRPS (BEA1), with domain architecture essential for substrate recognition and modification (Fig. 8.5). Within module 1, the adenylation domain (A1) specifically recognizes D-hydroxyisovaleric

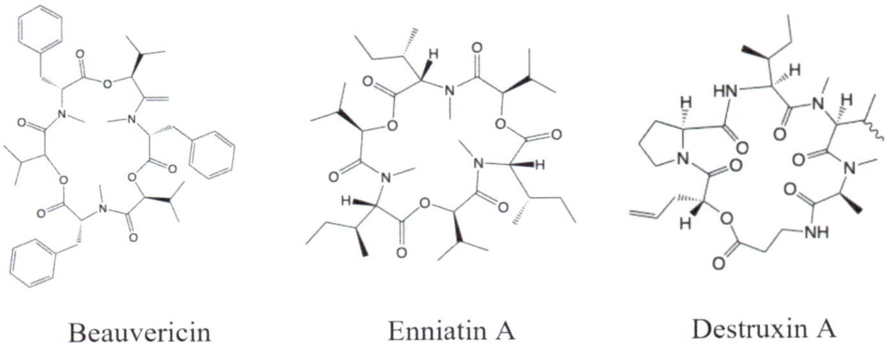

Fig. 8.4 Cyclic nonribosomal (depsi)peptide mycotoxins

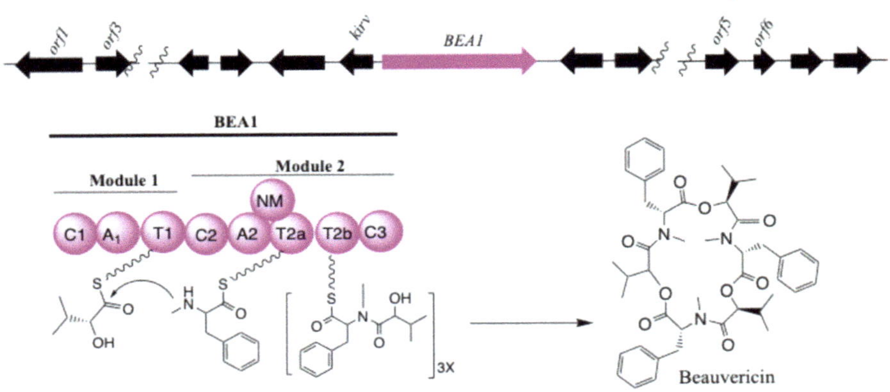

Fig. 8.5 The *bea* gene cluster and BEA biosynthesis in *F. proliferatum* LF061 (Huang et al. 2024)

acid (D-Hiv) and activates it for transfer to the thiolation domain (T1) via a thioester bond. In module 2, the adenylation domain (A2) activates L-phenylalanine (L-Phe), which is then sequentially transferred to the twin thiolation domains T2a and T2b. The NRPS also contains an integrated N-methyltransferase domain, which methylates the L-Phe residue during elongation (Niehaus et al. 2016). This biosynthetic route exemplifies the modular flexibility of NRPS and serves as a model system for studying the assembly of cyclic depsipeptides (Huang et al. 2024).

Destruxins (DTXs) are cyclohexadepsipeptides synthesized via a nonribosomal mechanism, primarily catalyzed by the multifunctional synthetase DtxS1. This enzyme comprises six adenylation domains, two of which exhibit substrate flexibility, enabling the incorporation of different amino acids required for the formation of DTX-B and its structural analogs (Fig. 8.6) (Wang et al. 2012a). Following initial assembly, DTX-B undergoes structural diversification through the action of the cytochrome P450 enzyme DtxS2, which mediates a series of oxidation reactions, each yielding a distinct destruxin derivative. Additional enzymes involved in the

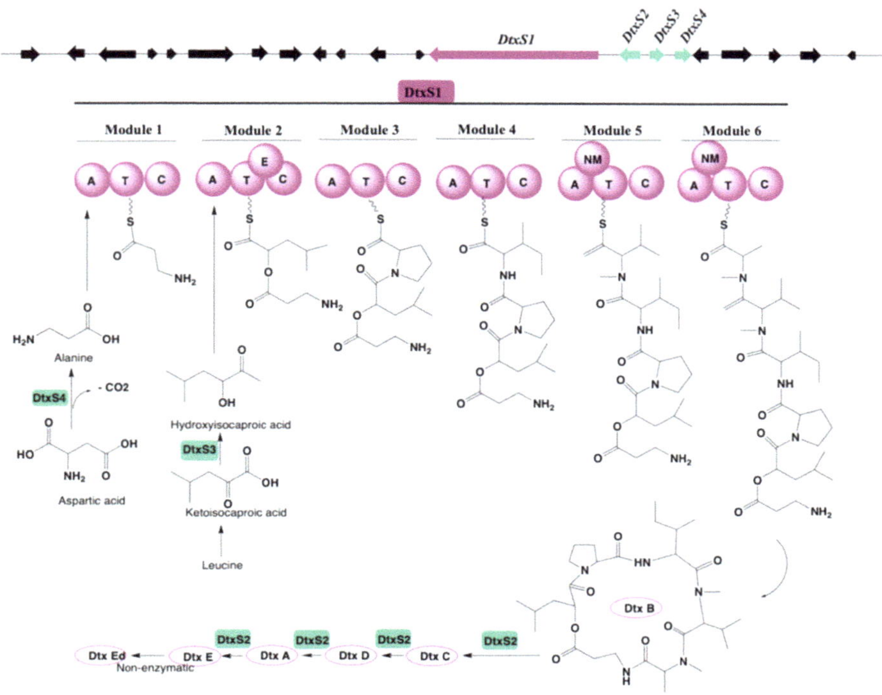

Fig. 8.6 Destruxin biosynthetic gene cluster and DtxS1 and DtxS2 domains involved in destruxin synthesis (Wang et al. 2012a)

destruxin biosynthetic pathway include DtxS3, an aldo-keto reductase, contributing to the modification of biosynthetic intermediates, and DtxS4, an aspartic acid decarboxylase, providing the terminal precursor required for initiating the DTX biosynthetic assembly line (Wang et al. 2012a).

3.3 Polyketide-Nonribosomal Peptide Hybrids

Several mycotoxins produced by EPF are biosynthesized through hybrid polyketide synthase–nonribosomal peptide synthetase (PKS-NRPS) pathways. These include both cyclic and linear structures, depending on the domain organization and enzymatic processes involved. Notable cyclic hybrids include beauverolides and cyclosporine, while leucinostatin and efrapeptins represent linear forms.

Cyclosporine A is a cyclic undecapeptide assembled by a large, modular NRPS SimA. SimA comprises 11 semi-independent modules, each containing homologous domains responsible for the sequential activation, modification, and condensation of 11 amino acid residues (Fig. 8.7) (Weber and Leitner 1994; Yang et al. 2018). Modules 2, 3, 4, 5, 7, 8, 10, and 11 contain integrated N-methyltransferase domains

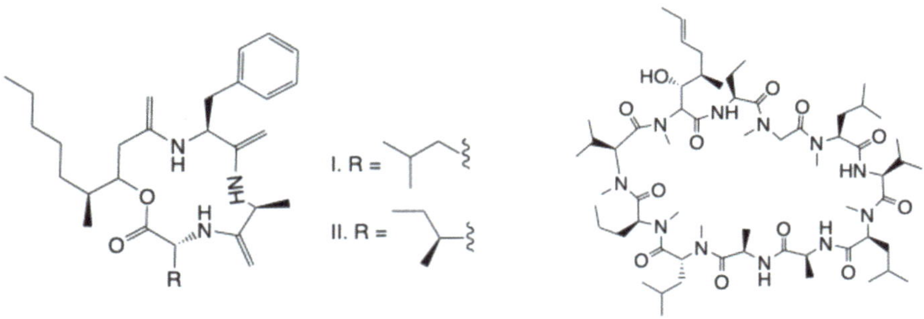

Beauverolides   Cyclosporin A

Fig. 8.7 Cyclic polyketide-nonribosomal (depsi)peptide mycotoxins

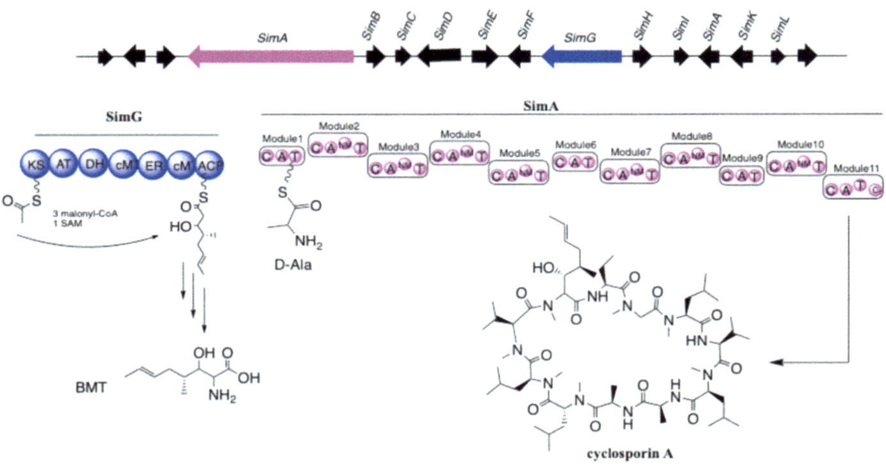

Fig. 8.8 Biosynthetic gene cluster of cyclosporin A in *Tolypocladium inflatum* (Sang et al. 2024)

in addition to the canonical condensation, adenylation, and thiolation domains. This modular arrangement reflects the structure of cyclosporine A, which contains eight N-methylated amino acid residues (Fig. 8.8). The presence of eight N-methylation domains underscores the essential role of N-methylation in supporting the conformation rigidity, bioactivity, and successful cyclization of the undecapeptide structure. A terminal condensation domain, located at the C-terminus of the final module, is thought to catalyze the macrocyclization step that yields the final cyclic structure of cyclosporine A and its derivatives.

The *sim* biosynthetic gene cluster includes 11 additional open reading frames (ORFs) implicated in the complete biosynthesis of cyclosporine A (Yang et al. 2018). Among them, SimG, a type I iterative PKS, plays a key role in producing the non-proteinogenic building block (4R)-4-[(E)-2-butenyl]-4-methyl-L-threonine

(BMT). Targeted disruption or deletion of *simG* abolishes cyclosporin A production in *Tolypocladium inflatum,* emphasizing its indispensable role in the biosynthetic process (Sang et al. 2024).

Yin et al. (2020) reported the production of beauveriolides (BVDs) in three species of entomopathogenic fungi. Among these, *Beauveria bassiana* and *B. brongniartii* were found to synthesize distinct analogs of BVDs, reflecting species-specific biosynthetic adaptations. Genomic analysis revealed a conserved four-gene biosynthetic cluster responsible for BVD production across all three fungi. This gene cluster comprises a polyketide synthase (PKS), which synthesizes 3-hydroxy fatty acids (FAs) of varying chain lengths, and an NRPS, which assembles the peptide moiety. The coordinated activity of the enzymes leads to the formation of the BVDs (Fig. 8.9) (Yin et al. 2020).

Leucinostatin A is a linear peptide composed of nine amino acid residues and an unusual amino acid at the C-terminal end, all linked via peptide bonds (Isogai et al. 1992; Li et al. 2024b). Its structure incorporates polyketide and peptide moieties, suggesting a biosynthetic origin that involves either a PKS, an NRPS, or a hybrid PKS-NRPS system. It is hypothesized that a reducing PKS produces the 4-methylhex-2-enoic acid component, while an NRPS assembles the peptide portion (Wang et al. 2016). Biosynthetic studies indicate that the PKS-derived polyketide intermediate is transferred to the NRPS LcsA through the coordinated actions of two accessory enzymes: LcsD, an acyl-CoA ligase that converts the intermediate into a CoA thioester, and LcsE, a thioesterase that hydrolyzes the thiol bond, enabling the transfer of the polyketide chain to the NRPS system. The central NRPS enzyme LcsA is composed of ten modules, each comprising condensation (C), adenylation (A), and peptidyl carrier protein (PCP) domains. This enzyme is specifically involved in leucinostatin biosynthesis in both *Purpureocillium lilacinum* and *Tolypocladium ophioglossoides* (Wang et al. 2016).

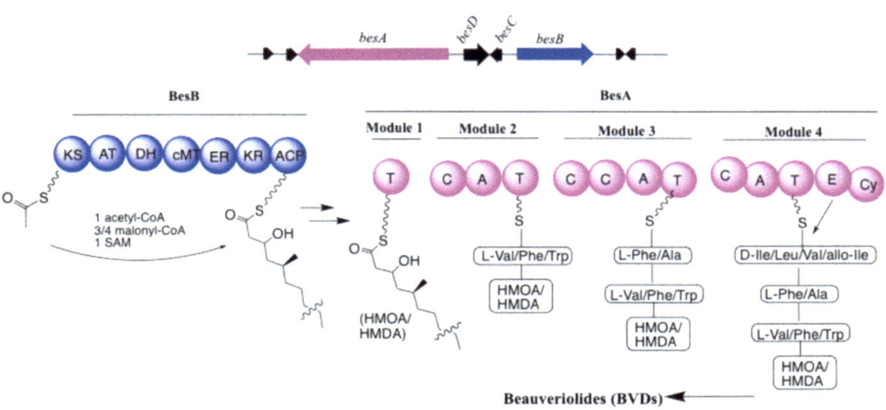

Fig. 8.9 Biosynthesis gene cluster of BDVs in *B. bassiana* and the domains of PKS (BesB) and modules of NRPS (BesA) involved in BDVs synthesis (Yin et al. 2020)

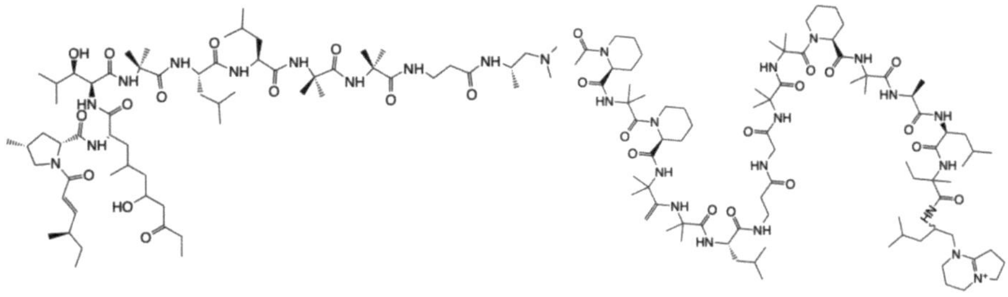

Fig. 8.10 Linear polyketide-nonribosomal peptide mycotoxins

Efrapeptins (Fig. 8.10) are a class of linear peptides produced by the entomopathogenic fungus *Tolypocladium niveum*. These compounds are characterized by the distinctive structure of a bicyclic amine moiety at the C-terminal end. Efrapeptins exhibit potent insecticidal properties, primarily through the inhibition of mitochondrial ATPase activity, which disrupts ATP production and energy metabolism in target organisms (Sakuda and Kimura 2010).

Fusarin C belongs to the broader acyl-tetramic acid family of secondary metabolites, which are found across a range of organisms, including fungi, myxomycetes, and marine sponges. As a representative member of the substituted 2-pyrrolidinone metabolite group, fusarin C was first identified in *Fusarium moniliforme* and is now known to be produced by several *Fusarium* species, including *F. graminearum*, *F. oxysporum*, *F. verticillioides*, and *F. fujikuroi* (Song et al. 2004; Thrane et al. 2004; Diaz-Sanchez et al. 2012; Kleigrewe et al. 2012; Huang et al. 2024). The biosynthesis of fusarin C is mediated by a hybrid PKS-NRPS complex enzyme encoded by the *fus1* gene. This atypical iterative type I PKS, fused with NRPS modules, initiates the biosynthetic pathway by condensing one acetyl-CoA starter with six malonyl-CoA extender units to form a heptaketide intermediate (Fig. 8.11). The intermediate is subsequently coupled to homoserine via an amide bond and released from the enzyme complex as an alcohol intermediate (Yang et al. 2021). Subsequent tailoring steps involve a series of gene-encoded enzymes. The *fus8* gene encodes a cytochrome P450 monooxygenase that catalyzes hydroxylation at the C-20 position, producing 20-hydroxy-fusarin. The *fus2* encodes a putative α/β hydrolase, facilitating the formation of the 2-pyrrolidone ring. This ring is further modified via epoxidation. The *fus8* may also mediate an additional oxidation at C-20 post-cyclization, generating carboxy-fusarin C. Finally, the *fus9* encodes a carboxyl methyltransferase that converts carboxy-fusarin C to fusarin C through a terminal methylation reaction. In contrast, the biosynthetic pathway responsible for producing fusarin A, a structurally distinct member of the fusarin family featuring a cyclized tetrahydrofuran-conjugated pyrrolidone ring, remains unresolved (Yang et al. 2021).

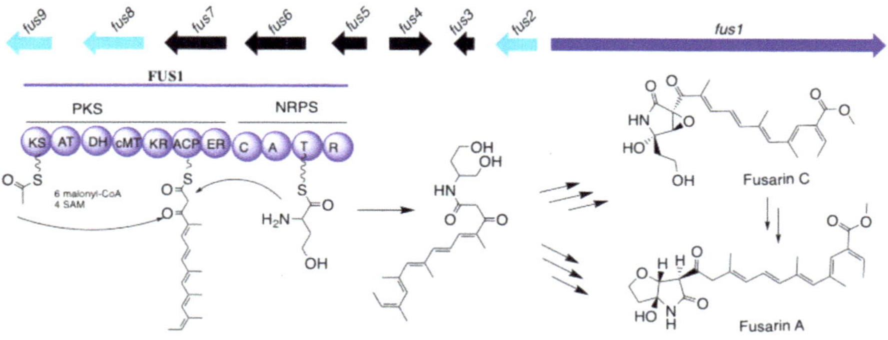

Fig. 8.11 Biosynthetic gene cluster of fusarins from *F. graminearum* and the function of hybrid PKS-NRPS enzyme FUS1 (Huang et al. 2024; Yang et al. 2021)

Swainsonine (SW) is an indolizidine produced by entomopathogenic fungi such as *Metarhizium* species. In *Metarhizium robertsii*, its biosynthesis is governed by a multibranched PKS-NRPS pathway. Central to this process is the *swnK* gene, which encodes a hybrid NRPS–PKS enzyme responsible for multiple biosynthetic intermediates through both ketoreductase domain functionality and the domain-skipping mechanism. The biosynthetic route initiates with the conversion of lysine into pipecolic acid, a key precursor in SW formation. This transformation is mediated by a combination of enzymes encoded within the SW biosynthetic gene cluster, along with additional contribution of non-clustered enzymes such as lysine cyclodeaminase. The hybrid NRPS–PKS enzyme then assembles at least three distinct intermediates as part of the SW pathway (Luo et al. 2020).

3.4 Terpene

Trichothecenes represent a large and structurally related class of sesquiterpenoid mycotoxins predominantly produced by *Fusarium* species, especially *F. graminearum*, and are frequently detected in cereal grains. More recently, they have also been identified in several entomopathogenic fungi, including *Akanthomyces lecanii, Beauveria bassiana, Cordyceps cicadae*, and *Isaria tenuipes* (Proctor et al. 2018). Despite their structural variability, all trichothecenes share a defining cyclic sesquiterpene core. The central feature of these molecules is the 12,13-epoxytrichothec-9-ene (EPT) scaffold (Fig. 8.12), a tricyclic core that serves as the biosynthetic backbone. Functional diversity among trichothecenes arises from the specific substitution patterns attached to this core.

Although the formation of the EPT scaffold was once thought to occur spontaneously, a recent study by Gao et al. (2025b) has revealed an enzymatically regulated mechanism. The findings demonstrated that effective biosynthesis of the EPF core requires the acetylation of the C11 hydroxyl group on trichothecene precursors (e.g., isotrichodiol and isotrichotriol) by the O-acetyltransferase Tri3. This acetylation

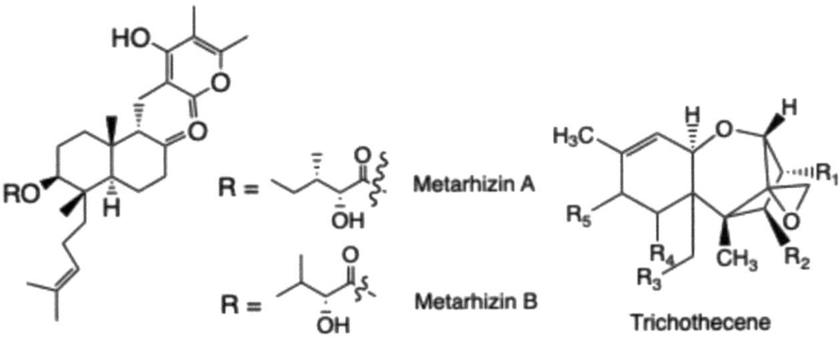

Fig. 8.12 Terpene-derived mycotoxins

enhances the leaving group potential, facilitating subsequent allylic rearrangements in aqueous environments. In the absence of further enzymatic control, these intermediates preferentially convert into shunt products trichodiol and trichotriol. The enzyme Tri14, a dedicated cyclase, catalyzes the crucial oxide bridge formation that finalizes the EPT ring system. This energetically demanding step is key to maintaining the structural integrity and biosynthetic fidelity of trichothecene production (Gao et al. 2025b).

3.5 Siderophore

Siderophores are small, high-affinity iron-chelating molecules produced by microorganisms to sequester iron from their environment. Among them, rhizoferrin is a representative example, synthesized via a nonribosomal peptide synthetase-independent siderophore (NIS) pathway. In this pathway, enzymes catalyze the condensation of citric acid with either a polyamine or an amino alcohol to form the siderophore (Fig. 8.13). *Trichoderma hamatum* has been reported to produce a substantial amount of rhizoferrin upon contact with *Spodoptera littoralis,* suggesting that this siderophore may serve as a virulence factor contributing to the fungus's pathogenicity against the leafworm (Lana et al. 2023).

In *Rhizopus delemar*, an opportunistic human pathogen, the biosynthesis of rhizoferrin has been investigated using comparative genomics. The genome of *R. delemar* harbors homologous sequences to the bacterial NIS protein SfnaD, which is essential for the biosynthesis of staphyloferrin A in *Staphylococcus aureus.* This protein features an N-terminal IucA/IucC domain, characteristic of siderophore biosynthesis enzymes, and a C-terminal FhuF-like domain, which is conserved among ferric iron reductase transporters (Carroll et al. 2017).

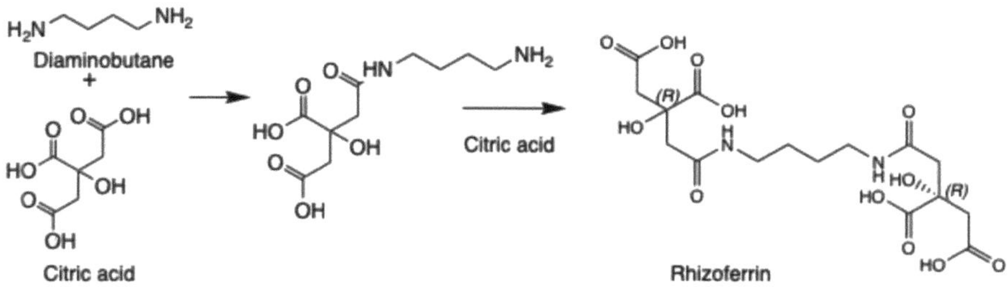

Fig. 8.13 Proposed biosynthetic pathway for fungal rhizoferrin (Carroll et al. 2017)

4 Toxic Mechanisms of Entomopathogenic Toxins

4.1 Mechanisms of Virulence Against Insects

Fungi, like all living organisms, have evolved specialized strategies for defense and survival in challenging environments. Entomopathogenic fungi have developed a diverse arsenal of virulence mechanisms that enable them to overcome insect immune defenses and establish successful infections. Central to their virulence are fungal secondary metabolites—including alkaloids, cyclodepsipeptides, peptide antibiotics, and polyketides—that specifically target host physiological and immunological systems. These compounds not only facilitate fungal penetration and colonization but also suppress host immune responses, disrupt cellular integrity, and alter insect behavior or physiology. By interfering with key processes such as hemocyte aggregation, calcium signaling, mitochondrial respiration, and membrane stability, EPF's toxins synergistically weaken the host and promote fungal proliferation. This section explores the molecular strategies by which EPF deploy these bioactive metabolites, with an emphasis on their roles in pathogenesis and their potential in biocontrol applications.

4.1.1 Depsipeptides

Among destruxin homologs, destruxin A (DTX-A) and destruxin B (DTX-B) are notable for their biological activities. Despite differing structurally only in vinyl and ethyl groups in DXT-A and DXT-B, respectively, DTX-A exhibits significantly higher insecticidal potency, while DTX-B demonstrates more pronounced antitumor activity. These differences are attributed to their differential binding affinities to insect hemocyte ATPase, where DTX-B shows a 100-fold higher affinity compared to DTX-A (Chen et al. 2014). This contrast underscores how subtle structural variations in depsipeptides can dramatically alter their targets and modes of action.

In *Metarhizium* spp., DTX-A suppresses silkworm immune response by targeting multiple proteins in the host hemolymph. Proteomic analysis identified 48 DTX-A-binding proteins, among which hemocytin exhibits the strongest binding affinity. By interacting with multiple sites on hemocytin, DTX-A induces conformational changes that block hemocyte aggregation. Furthermore, it disrupts hemocytin's interaction with actin A3 and gelsolin—key components of the vesicle transport machinery—thereby inhibiting the formation and release of hemocytin-storing vesicles. This impairment significantly reduces hemocytin secretion and hemocyte aggregation (Wang et al. 2024).

The DTX-A also alters cytoskeletal architecture, impairing phagocytosis and encapsulation, which are essential defense mechanisms in insects (Vey et al. 2002; Fan et al. 2013). Wang et al. (2023) demonstrated that DTX-A disrupts Bm12 cell morphology by decreasing attachment capacity, reducing cellular motility, and diminishing microvilli density. Mechanistically, DTX-A binds to several cytoskeletal proteins, including actin, α-actinin, and α-tubulin, which are integral to microfilament and microtubule structure. Additionally, DTX-A interacts with the myosin heavy chain, dynein heavy chain, and microtubule-actin cross-linking factor 1, all of which are essential for intracellular transport and cytoskeletal organization. Through inhibition of these components, DTX-A destabilizes cytoskeletal homeostasis and impairs structural integrity (Wang et al. 2023).

Beyond structural interference, DTX-A also affects protein synthesis. Wang et al. (2020) reported that DTX-A binds to seven aminoacyl-tRNA synthetases, enzymes responsible for catalyzing amino acid attachment to tRNA molecules during translation. These interactions hamper the elongation phase of protein synthesis. Further studies revealed that DTX-A has an affinity for components of the translational machinery, including elongation factor 1 delta (EF-1δ), elongation factor 1-alpha (EF-1α), translation elongation factor 2 (TEF-2), and eukaryotic translation initiation factor 3 (eIF3) (Wang et al. 2023). These findings suggest that DTX-A plays a regulatory role in host protein biosynthesis, ultimately contributing to immune suppression and pathogenesis.

Destruxin A has been shown to interfere with protein degradation by targeting the ubiquitin-proteasome pathway, thereby impairing the removal of misfolded or damaged proteins and disrupting cellular homeostasis. In addition, DTX-A acts on endoplasmic reticulum (ER) components such as ATP synthase, calnexin, and the COP complex, all of which are integral to vesicular transport from the ER to the Golgi apparatus and to proper protein folding (Molinari et al. 2004; Yin et al. 2021). DTX-A also interacts with peptidases, including prolyl oligopeptidase and protein disulfide isomerase precursor, enzymes that may contribute to detoxifying DTX-A through hydrolysis or nutrient conversion. Notably, the high affinity of prolyl oligopeptidase for DTX-A is likely due to the presence of a proline residue in the toxin's structure. Beyond protein-level interactions, DTX-A may influence nuclear-cytoplasmic transport and affect transcriptional regulation, leading cumulatively to alterations in cell morphology, organelle dysfunction, and ultimately, cell death (Wang et al. 2023).

Studies in *Bombyx mori* showed that DTX-A alters hemocyte morphology, inducing changes including cell aggregation, membrane shrinkage, and the formation of surface hollows at a low concentration of 10 μL of 12.5 μg/mL. The effect intensifies with prolonged exposure (1–24 hours) and increasing concentrations (25–200 μg mL-1), results in failed phagocytosis, a key host defense response

(Fan et al. 2013). Destruxin A also exhibits strong binding affinity for transmembrane protein 214 and transport proteins in *B. mori,* contributing to increased larval mortality upon treatment (Yin et al. 2021). Furthermore, DTX-A interacts with three modification-related proteins: calreticulin (BmCRT), dipeptidyl peptidase III (BmDPP3), and protein disulfide isomerase A5 (BmPDIA5). Among these, DTX-A inhibits BmDPP3 activity but does not significantly affect BmPDIA5. Notably, DTX-A upregulates BmDPP3 while downregulating BmCRT. Both BmCRT and BmPDIA5 function as molecular ER chaperones. BmCRT assists protein repair and folding, whereas BmPDIA5 facilitates disulfide bond rearrangement (Hashimoto et al. 2008; Yin et al. 2022). BmDPP3, a zinc-dependent metalloprotease, is involved in protein metabolism (Podversnik et al. 2022). Destruxin A was found to interact with 22 hemolymph proteins in *B. mori*, with juvenile hormone binding protein (JHBP) identified as the most sensitive target. JHBPs play an important role in hormone signal transmission related to insect development. When bound by destruxin A, its signal transmission function is changed, leading to disrupted metamorphosis and inhibited growth (Hu et al. 2023a).

Destruxins modulate insect immunity through multiple mechanisms. DTXs have been shown to suppress prophenoloxidase (PPO) activity and downregulate antimicrobial peptide (AMP) gene expression, primarily through interactions with immunophilins (Wang et al. 2012b; Han et al. 2022). In *Drosophila*, DTX-A may target the nervous system, although its toxicity can be mitigated by a cut peptide derived from Baramicin A, a protective immune protein (Huang et al. 2023). Physiologically, DTX-A disrupts ion regulation by inhibiting V-type H^+ ATPases in excretory organs without altering intracellular ATP production. This suggests that DTX targets ion transport systems (Ruiz-Sanchez et al. 2010). Additionally, DTXs impair cellular immune responses by inhibiting nodule formation (Peng et al. 2005) and disrupting phagocytosis (Pal et al. 2007).

DTXs also modulate calcium signaling by activating insect muscle calcium channels, influencing contractility (Samuels et al. 1988; Dumas et al. 1996). A laser scanning confocal microscopy (LSCM) study showed that DTX-A and DTX-B induced rapid Ca^{2+} influx in *Exolontha serrulata*, *B. mori*, and *Spodoptera litura* hemocytes. This influx was not blocked by known calcium channel inhibitors such as 2-APB and U73122. Both DTX-A (690 μmol/L) and DTX-B (20 μmol/L) significantly reduced intracellular H^+ levels, indicating disruption of proton gradients crucial for immune signaling and cellular homeostasis (Chen et al. 2014).

Other cyclopeptides produced by EPF also contribute to virulence, such as beauvericin (BEA), beauveriolides (BVDs), and cyclosporin A (CsA). BEA and BVDs, both synthesized by *Beauveria bassiana* (Xu et al. 2008; Yin et al. 2020), exhibit broad-spectrum insecticidal activity. BEA, in particular, is also produced by *Cordyceps fumosorosea* (*Isaria fumosorosea*), *C. cateannulata*, *Fusarium semitectum*, *F. moniliforme*, and various Basidiomycota plant pathogens (Wang and Xu 2012; Qasim et al. 2018; Van et al. 2021) and acts as an ionophore that forms metal ion complexes (Nkyaagye et al. 2024). Its toxic mechanisms include reactive oxygen species (ROS) generation, mitochondrial dysfunction, disruption of hormonal signaling, and elevation of intracellular Ca^{2+}, contributing to host cytotoxicity (Ferrer et al. 2009; Ortiz-Urquiza and Keyhani 2016). CsA, produced by *Tolypocladium inflatum* (Yang et al. 2018), exerts immunosuppressive effects by binding to peptidyl-prolyl cis-trans isomerase (cyclophilins), which are conserved across organisms (Fischer et al. 1989). In *T. inflatum*, a luster-encoded cyclophilin, SimC, enhances CsA's antifungal properties and may directly suppress host immune responses during infection (Yang et al. 2018).

Another potent depsipeptide, bassianolide, is produced by *Akanthomyces lecanii* (formerly *Lecanicillium lecanii*) (Ravindran et al. 2018) and strains of *B. bassiana* isolated from *B. mori* pupae cadavers (Xu et al. 2009). Like other cyclopeptides, bassianolide acts as an ionophore with strong insecticidal activity (Kanaoka et al. 1978). Recent studies showed that *Lecanicillium muscarium*-derived bassianolide can disrupt acetylcholine receptors in insect muscle tissues, leading to reduced acetylcholine-esterase activity in whiteflies and contributing to its toxicity (Ali et al. 2017; Villegas-Mendoza et al. 2019).

4.1.2 Oosporein and Zearalenone

In addition to cyclopeptides, *Beauveria bassiana* produces benzoquinone-derived secondary metabolites such as oosporein (OSN) and 2-pyridone tenellin (TEN), both of which play notable roles in enhancing fungal virulence (Feng et al. 2015; Chen et al. 2021). Oosporein exhibits antibacterial activity by suppressing the growth of gut bacteria in the insect host, facilitating fungal colonization within the insect hemocoel, and improving systemic infection efficiency (Fan et al. 2017; Wei et al. 2017). Meanwhile, tenellin exhibits iron-chelating properties, enabling *B. bassiana* to outcompete co-infecting microorganisms by depriving them of this essential nutrient (Jirakkakul et al. 2015).

The alkaloid swainsonine (SW), a known inhibitor of Golgi α-mannosidase II, accumulates in *Metarhizium* spp. conidia and may serve an ecological function by deterring predators such as birds or scavengers from consuming infected insects (Luo et al. 2020). Compounds with diverse bioactivities, including insecticidal and immunosuppressive molecules, have also been isolated from *Isaria* (the anamorphic name of *Cordyceps*) species, further underscoring the sophisticated chemical arsenal that EPF employ to manipulate and invade insect immune responses (Zhang et al. 2019).

Beyond their direct insecticidal roles, EPF-produced siderophores also enhance microbial competitiveness by sequestering iron and other essential micronutrients, thereby suppressing competing microbes and reshaping insect defensive ectomicrobiotas (Krasnoff et al. 2020; Chen et al. 2021). These findings illustrate the ecological complexity of fungus–insect interactions and the position of EPF as key players in multipartite networks involving not only the host insect but also its microbiome and surrounding microbial communities (Hong et al. 2023).

4.1.3 Leucinostatins and Efrapeptins

Leucinostatins are a family of peptide-based natural products with broad-spectrum biological activities. Their primary action involves targeting cellular membranes. Early studies demonstrated that leucinostatin A functions as an ionophore, facilitating cation transport across mitochondrial membranes (Ishiguro and Arai 1976; Csermely et al. 1994; Fresta et al. 2000). In model systems such as liposomes, leucinostatin A associates within the lipid bilayers in a concentration-dependent manner, altering membrane fluidity and forming pores that disrupt membrane integrity (Fresta et al. 2000). More recent studies have expanded the mechanistic understanding of leucinostatins. Brand et al. (2021) found that leucinostatin A induces transbilayer scrambling of phospholipids, increasing external exposure of normal cytoplasmic lipids like phosphatidylinositol. This disruption of phospholipid asymmetry affects membrane structure and enhances substrate accessibility to external enzymes, highlighting membrane scrambling as a key element of its biological activity (Brand et al. 2021).

In addition to membrane disruption, leucinostatins are potent inhibitors of mitochondrial oxidative phosphorylation. Shima et al. (1990) showed that leucinostatins A and B bind to the F_0 subunit of H^+-ATPase, blocking proton translocation essential for ATP synthesis. At lower concentrations ($\leq$300 nM), they act as phosphoryl transfer inhibitors, while at higher levels, they function as uncouplers through a protonophoric mechanism, leading to rapid ATP depletion and metabolic collapse. These effects are enhanced by co-treatment with mitochondrial inhibitors like venturicidin, DCCD, and oligomycin, reinforcing their potential as energy-targeting bioagents (Shima et al. 1990). The combination of ionophoric, uncoupling, and membrane-scrambling properties of leucinostatins contributes to their cytotoxicity and makes them promising candidates for bioinsecticides and antiparasitic agents. Alongside other EPF-produced metabolites such as destruxins, beauvericins, tenellin, and bassianolide, leucinostatins represent a critical tool in the fungal biochemical reservoir for disrupting host physiology and outcompeting microbial rivals in dynamic ecological niches.

Among the diverse peptide-based compounds produced by EPF, efrapeptins stand out as another potent class of toxins. Primarily synthesized by soil-dwelling hyphomycetes such as *Beauveria nivea* and *Tolypocladium niveum* (syn. *Tolypocladium inflatum*), efrapeptins exist as complex mixtures of peptide antibiotics (Niu et al. 2020). They demonstrate strong insecticidal activity against several pests such as *Leptinotarsa decemlineata* (Bandani and Butt 1999), suggesting a significant role in EPF virulence. Functionally, efrapeptins are known for their strong ATPase inhibitory activity, specifically targeting mitochondrial oxidative phosphorylation. Mechanistic studies showed that efrapeptins bind to the F1 catalytic subunit of mitochondrial ATP synthase, disrupting ATP production and leading to bioenergetic failure and cell death (Goettel et al. 2005; Junge and Nelson 2015).

4.1.4 Zearalenone

Zearalenone (ZEN) is a nonsteroidal mycotoxin that structurally mimics the estrogenic 17-β-estradiol. Due to its conformational similarity, ZEN can bind to estrogen receptors, thereby eliciting endocrine-disrupting effects. Exposure to ZEN and its derivatives has been linked to a range of hormonal disturbances, including precocious puberty, estrogenization, and reproductive disorders in both humans and animals. Although ZEN is not acutely lethal, its persistent effects on endocrine regulation raise considerable concerns regarding its toxicological significance and long-term health impacts (Kharayat and Singh 2018).

In summary, fungal secondary metabolites such as beauvericins, beauverolides, destruxins, efrapeptins, leucinostatins, and zearalenone reflect the sophisticated biochemical strategies developed by entomopathogenic fungi to invade insect hosts, modulate immunity, and secure ecological dominance. These compounds exhibit a diverse spectrum of mechanisms, ranging from immunity suppression to energy-metabolism disruption to hormonal mimicry and microbiome modulation, underscoring their evolutionary role in pathogenicity. Continued research into their molecular targets, biosynthetic pathways, and ecological functions holds great promise for developing novel biopesticides and advancing sustainable pest management strategies.

4.2 In Vitro Cytotoxic Mechanisms

4.2.1 Beauvericins and Enniatins

A growing body of research has investigated the cytotoxicity of mycotoxins using *in vitro* models to elucidate their molecular targets and mechanisms of cellular disruption. Among these, beauvericin and enniatins cyclic depsipeptides produced by *Beauveria bassiana* and other fungi have emerged as potent disruptors of lysosomal stability and mitochondrial function. Enniatin B1 (ENN B1), commonly found in cereal-based products, has been shown to compromise lysosomal membrane integrity in mouse embryonic fibroblasts (MEFs). Wild-type MEFs demonstrated greater sensitivity to ENN B1 than cells lacking LAMP-1 and LAMP-2, two lysosome-associated membrane proteins. This suggests that ENN B1 preferentially targets LAMP-2, leading to lysosomal alkalinization, membrane permeabilization, and disruption of chaperone-mediated autophagy (CMA), and releasing proteins such as Hsp70 and Hsp90 into the cytosol (Oliveira et al. 2020).

Both ENN B and BEA have been reported to elevate lysosomal pH and stimulate the release of cathepsin B, a lysosomal protease known to trigger apoptosis. Cytosolic translocation of cathepsin B activates caspase cascades, which initiate programmed cell death. Selective inhibition of cathepsin B significantly reduced BEA and ENN B cytotoxicity, underscoring its key role in the apoptotic process (Aufy et al. 2023). These findings highlight lysosomal damage as a shared mechanism among ionophoric mycotoxins.

Beauvericin (BEA) also demonstrated broad anticancer potential, with cytotoxic effects observed in diverse human cancer cell lines, including C6 glioma, MDA-MB-231, HeLa, HCT-15, LoVo, and U251. In C6 glioma cells, BEA induces dose-dependent apoptosis via activation of caspase-3 and caspase-9, while suppressing Bcl-2, an anti-apoptotic gene. Additionally, BEA disrupts actin polymerization, alters cytoskeletal integrity, and modulates phosphorylation of signaling proteins such as Src kinase and the transcription factor STAT3. It directly binds to the C-terminal region, thereby inhibiting Src kinase activity and attenuating the Src/STAT3 signaling axis, a pathway implicated in cancer progression (Kim et al. 2023).

In human SH-SY5Y neuroblastoma cells, exposure to mycotoxins such as deoxynivalenol, fumonisin B1, enniatin A, and beauvericin (BEA) caused mitochondrial membrane dysfunction and significantly reduced cell viability, consistent with activation of the intrinsic apoptosis pathway. Notably, binary mixtures of ENN and BEA often showed antagonistic interactions, whereas naturally contaminated matrices (e.g., animal feed or milk) exhibited enhanced toxicity, likely due to synergistic effects among multiple compounds and matrix factors (Pérez-Fuentes et al. 2021). The findings emphasize the need to evaluate mycotoxin toxicity under real-world exposure conditions, where additive or unpredictable interactions may significantly elevate health risks.

4.2.2 Metarhizin and Fumonisins

Metarhizin A, a secondary metabolite isolated from *Metarhizium flavoviride*, exhibits strong cytotoxic activity against insect and mammalian cell lines. It demonstrates potent cytotoxic effects in human breast cancer MCF-7 cells and shows similar efficacy across other tumor cell types. A notable cellular response to metarhizin A exposure is the selective upregulation of p21, a cyclin-dependent kinase inhibitor involved in cell cycle arrest. Additionally, treatment with metarhizin A significantly increases intracellular reactive oxygen species (ROS) levels and activates the ERK1/2 signaling pathway, indicating engagement of the mitogen-activated protein kinase cascade.

At the mitochondrial level, Metarhizin A acts as a powerful inhibitor of cytochrome c oxidase (complex IV), completely suppressing oxygen consumption and impairing mitochondrial respiration. These effects are particularly lethal under nutrient-limited or metabolically stressed conditions, where the compound induces cell death by simultaneously impairing ATP production and promoting oxidative damage. Cytochrome c oxidase has been identified as a key molecular target, supported by evidence from biochemical assays using rat liver mitochondrial extracts (Katou et al. 2014).

The fumonisins, particularly fumonisin B1 (FB1), structurally resemble sphingolipids and are believed to target sphingolipid biosynthesis. FB1 inhibits ceramide synthase, leading to the accumulation of sphinganine and depletion of complex sphingolipid. These disturbances compromise membrane integrity, impair cell signaling, and contribute to hepatic and neuronal toxicity. Experimental studies have linked FB1 exposure to developmental defects, such as neural tube malformations, emphasizing its similar effects in humans (Kharayat and Singh 2018).

These findings demonstrate that fungal toxins exert cytotoxic effects through multiple intracellular targets, including lysosomes, mitochondria, signaling proteins, and nucleic acid pathways. Their diverse mechanisms of action underscore the importance of further mechanistic studies to better assess their health risks and to explore their potential as novel therapeutics.

4.3 Monitoring and Detecting Mycotoxins

Detection of mycotoxins relies on diverse approaches that target either the toxins themselves or their biosynthetic markers. Molecular, biochemical, and analytical techniques are employed depending on the objectives, whether to identify the producing organism, the biosynthetic gene cluster of enzymes, or to quantify the toxins in biological matrices.

4.4 PCR-Based Methods

The advent of the polymerase chain reaction (PCR) has revolutionized molecular biology, enabling rapid amplification of specific DNA sequences. In the context of mycotoxin detection, PCR is widely applied to identify genes within biosynthetic gene clusters, particularly those encoding nonribosomal peptide synthetases and polyketide synthases. For example, Zhang et al. (2013) designed primer pairs EA1F/EA1R and EA2F/EA2R targeting adenylation domains in NRPSs involved in beauvericin synthesis. Amplification of a 750 bp fragment using these primers indicated the presence of NRPS genes homologous to enniatin synthetase of *Fusarium sambucinum* (former name *Gibberella pulicaris)* (Zhang et al. 2013). More primer pairs and PCR techniques have been developed and successfully applied in mycotoxin detection (Chehri and Godini 2017; Natarajan et al. 2025).

4.5 Lateral Flow Assays

Lateral flow assays (LFAs) are rapid, user-friendly immunochromatographic tests capable of qualitative or semi-quantitative detection in a single-strip test within 60 minutes. These assays are increasingly used in on-site mycotoxin screening due to their portability and simplicity (Pedreira et al. 2025). Deoxynivalenol (DON) was successfully detected using a gold-based LFA at a concentration of 250–500 μg/kg and 1000–2000 μg/kg (Kolosova et al. 2008).

Song et al. (2014) introduced a new lateral immunoassay for the detection of ZEA and DON in cereal samples based on their specific antibody class. The visual detection limit was 1.6 μg/kg for ZEN and 10 μg/kg for DON, with 81% and 100% agreement with LC/MS/MS results, respectively. Chen et al. (2016) developed a multiplex LFA for ZEN detection in corn, rice, and peanuts, achieving a visual detection limit of 50 μg/kg and a quantitative detection limit as low as 0.42–0.46 μg/kg through an optimized antibody-gold nanoparticle system (Chen et al. 2016). Zhu et al. (2021) applied up-converting phosphor technology-based lateral flow (UPT-LF) for detecting aflatoxin B_1, ZEN, and DON in feed samples. Although UPT-LF was less sensitive than HPLC-MS/MC or ELISA, it enabled reliable on-site detection with limits of 3 μg/kg (aflatoxin B_1), 50 μg/kg (ZEA), and 200 μg/kg (DON). Zhang et al. (2025) enhanced LFA performance by coupling it with immunomagnetic separation. This approach enabled sensitive detection of ZEN and DON at 2.77 and 20.99 μg/kg in maize and 1.92 and 19.44 μg/kg in wheat, respectively.

4.6 Liquid Chromatography

High-performance liquid chromatography (HPLC) is one of the most widely used methods for analyzing the chemical composition of complex samples. To accurately identify specific mycotoxins, it is essential to reduce the background noise caused by co-eluting compounds. Sample preparation plays a critical role in achieving sensitive and specific detection. The choice of extraction and cleanup method depends on the physicochemical properties of the mycotoxins and the matrix being analyzed. Common techniques include solid-phase extraction (SPE), solid-liquid extraction, ultrasonic-assisted extraction (UAE), dispersive liquid-liquid microextraction (DLLME), and the use of immunoaffinity columns (IACs). Commonly used extraction solvents are methanol, acetonitrile, H_2O, acetic acid, and formic acid, alone or in combination. For analytical separation, different HPLC and Ultra HPLC (UHPLC) systems are employed, often coupled with fluorescent detectors, mass spectrometry (MS), and triple quadrupole MS (LC-MS/MS) systems (Huang et al. 2014; De Ruyck et al. 2020; Delgado-Povedano et al. 2025).

The aflatoxin M1, ochratoxin A, zearalenone, and α-zearalenol in milk were extracted using UAE with acetonitrile and SPE via Oasis HLB cartridges. The treated samples were then applied for UHPLC coupled with electrospray ionization triple quadrupole tandem MS, achieving quantification limits of 0.003–0.015 μg/L, with a recovery rate exceeding 87.0% (Huang et al. 2014).

Delgado-Povedano et al. (2025) developed a screening method for 144 mycotoxins and metabolites, building on the targeted method described by De Ruyck et al. (2020). The new method combined UHPLC-MS/MS data (experimental results and predicted retention times) with UNIFI software containing curated MS/MS spectra. When applied to human serum tainted with 41 known mycotoxins, including BEA, ENNs, fumonisins, trichothecenes, and ZEN, the screening method successfully detected fumonisins and trichothecene A, while the targeted method detected fumonisins, trichothecene A and B, and ZEN (Delgado-Povedano et al. 2025). These results demonstrate the growing utility of high-throughput LC-MS/MS workflows in mycotoxin biomonitoring.

5 Mycotoxin Removal by Microorganisms

While physical and chemical strategies such as thermal treatment, ozonation, ultrasonification, and magnetic nanostructures have been explored for mycotoxin removal, microbial detoxification presents a promising biological solution, especially for entomopathogenic mycotoxins.

Although ubiquitous and harmful to insect hosts, EPF and their mycotoxins are controlled by other biological agents, primarily bacteria and yeasts. These microbes use enzymatic or metabolic pathways to neutralize or transform toxins into less harmful derivatives.

Numerous bacterial strains, including *Nocardia glubera, Rhodococcus fascians,* and *R. erythropolis*, have been patented for degrading BEA from *Fusarium graminearum*-infected wheat kernels (Duvick and Rood 2000). *Lactobacillus* (*L. animalis, L. casei, L. casei-casei, L.* plantarum, *L. rhamnosus,* and *L. ruminus*), *Bifidobacterium* (*B. adolescents, B. bifidum, B. breve*, and *B. longum*), *Salmonella* (*S. fecalis* and *S. termofilus*), and *Eubacterium crispatus* have also shown the ability to degrade BEA by 66.5–83.1% of the 5 mg/L initial concentration after 48 hours (Meca et al. 2012). For enniatins (ENN A, A1, B, and B1), degradation efficiency varied widely from 5.2 ± 2.5% by *Bifidobacterium breve* CECT4839 to 99.5 ± 2.0% by *Streptococcus thermophilus* CECT986. *Saccharomyces cerevisiae* (CECT27, CECT28, CECT31, CECT33, and CECT74) also exhibited 92% degradation of the tested ENNs (Roig et al. 2013).

De-epoxification of trichothecenes, particularly deoxynivalenol (DON), has been observed in microbial communities from chicken and fish intestines (Young et al. 2007; Guan et al. 2009), as well as soil bacteria, including *Serratia*, *Clostridium*, *Citrobacter*, *Enterococcus*, *Stenotrophomonas,* and *Streptomyces* (Islam et al. 2012). The combination of the 6 bacterial taxa (mainly *Serratia*) completely transformed 50 μg/ml of DON after 60, 72, 72, and 96 hours in aerobic, aerobic-anaerobic, anaerobic-aerobic, and anaerobic growth conditions, respectively (Islam et al. 2012). *Sphingomonas* S3–4 totally degraded 100 μg/mL of DON to 3-oxo-DON and 3-epi-DON within 72 hours, with the activated enzyme identified as AKR18A1. The responsible enzyme of the *Sphingomonas* S3–4 strain was determined to be AKR18A1, an aldo/keto reductase. The enzyme remains active across pH 7–11 and temperatures of 10–50 °C, optimal at pH 10.6 and 45 °C (He et al. 2017). *Enterococcus faecalis* YC12-C10 and *Morganella morganii* YC12-C3 from crucian fish intestines degraded DON into a new, less toxic compound, 3-deoxy-6-demethanol-DON. The two strains were used to degrade 150 μg kg − 1 DON supplemented in *Coix lacryma-jobi* seed, achieving degradation efficacy of 68.89% and 59.94%, respectively (An et al. 2025).

There have been varieties of reported microbial enzymes that target zearalenone (ZEN). *Neurospora crassa* encodes ZENC, a lactonase. The recombinant ZENC in *Pichia pastoris* exhibited high activity on ZEN (530.4 U/mg). This protein has potential application in degrading ZEN from agricultural products (Bi et al. 2018). *Exophiala spinifera* produces a new ZEN-degrading enzyme (ZHD_LD) that has 60.15% homology with ZHD101 and is most active at 50 °C, pH 7.0–10.0 (Gao et al. 2022). Hu et al. (2023b) found *Aeromicrobium* expresses ZenH, a new ZEN-degrading enzyme (ZenH), which hydrolyzes ZEN to a new and less toxic compound (*S*,*E*)-4-hydroxy-2-(10-hydroxy-6-oxoundec-1-en-1-yl)-7-oxabicyclo [4.2.0]octa-1,3,5-trien-8-one under optimal conditions of 55 °C, pH 7.0 (Hu et al. 2023a). *Gordonia hydrophobica* HAU421, a soil spiramycin-resistant bacterium, produces lactonohydrolase GhZH, capable of hydrolyzing ZEN in beer and corn steep liquor by 53.2% and 74.0%, respectively (Chen et al. 2025).

Various *Lactobacillus* species (*L. rhamnosus, L. plantarum, L. sakei KTU05–6, L. ruminis, L. casei,* and *L. animalis*) can degrade multiple mycotoxins, including BEA, patulin, trichothecene (DON, T-2), and ZENs (El-Nezami et al. 2002; Hatab et al. 2012; Meca et al. 2012; Hawar et al. 2013; Juodeikiene et al. 2018). *Lactobacillus brevis* utilizes a combination of enzymes, including aryl-alcohol dehydrogenase, carboxylesterase, and cell surface components such as lipoteichoic acid, to bind or convert DON and ZEN (Ghamsari et al. 2025).

Yeasts such as *Trichosporon mycotoxinovorans* (Molnar et al. 2004), *Pichia ohmeri* (Coelho et al. 2008), *Rhodosporidium kratochvilovae* (Castoria et al. 2011), *Saccharomyces cerevisiae* (Meca et al. 2013), *Kodameae ohmeri* (Dong et al. 2015), *Pichia caribbica* (Zheng et al. 2018), and *Rhodotorula mucilaginosa* (Li et al. 2019) exhibit growth inhibitions of EPF and degradation of various mycotoxins. Additionally, EPF themselves can participate in detoxification. *Cordyceps fumosorosea*, *C. farinosa*, and *C. tenuipes* (Nowak et al. 2023), and *Metarhizium anisopliae* (Nowak et al. 2024) effectively transformed ZEN, suggesting a dual role as both producers and mitigators of toxic compounds.

6 Entomopathogenic Mycotoxin Applications

Understanding the biosynthetic and ecological aspects of EPF enables the strategic application of their bioactive compounds in agriculture and medicine. Historically used as biocontrol agents and remedies, EPF-derived toxins are gaining attention as environmentally friendly tools for pest management and beyond.

6.1 Applications of Entomopathogenic Mycotoxins in Agriculture

Leucinostatin, produced by *Paecilomyces lilacinus,* shows strong nematicidal activity. For instance, leucinostatins cause 100% mortality in nematode juveniles at 100 ppm within 12 hours and 74% mortality at 10 ppm after 24 hours, although egg viability remains unaffected (Park et al. 2004).

A newly isolated compound from *Purpureocillium lavendulum* YMF1.683, 5-(methoxymethyl)-1H-pyrrole-2-carboxaldehyde, showed attraction to *Meloidogyne incognita* at 5 ppm, a strong negative chemotropism effect at 40 ppm, and >98% mortality at 400 ppm after 96 hours (Bao et al. 2022). Notably, the strain was effective even in a $CdCl_2$-contaminated soil (100 mg/kg Cd^{2+}). The fungus resisted high concentrations of Cd^{2+} by increasing secretion of Cd^{2+}, upregulating autophagy, enhancing antioxidant defenses, and reducing the Cd^{2+} uptake in tomato roots and fruit (Li et al. 2024a). *Purpureocilium lilacinum* SR38, adapted to various living conditions in soil and endophytic in tomato, produces leucinostatins. The strain is significantly toxic to *Nacobbus aberrans,* offering potential for plant growth promotion and nematode control in biostimulant formulation (Girardi et al. 2022).

The host specificity of *Metarhizium anisopliae* was investigated, revealing different host species. The results showed that *M. anisopliae* exhibited differential virulence among insect orders: *Apis mellifera* (Hymenoptera: Apidae), *Gryllus assimilis* (Orthoptera: Gryllidae), *Spodoptera frugiperda* (Lepidoptera: Noctuidae), and *Tenebrio molitor* (Coleoptera: Tenebrionidae). The fungus was most lethal to *T. molitor,* showing 100% mortality. Transcriptomic analysis revealed 10,103 expressed genes during early infection stages. On wings of *T. molitor*, blastospores produced appressoria, which transformed orsellinic acid to a toxic, aromatic compound. Genes encoding destruxin synthase and MAD1 adhesin were highly upregulated, suggesting targeted virulence strategies and potential use of destruxins and MAD1 adhesin for controlling *T. molitor* (Iwanicki et al. 2025).

Deoxynivalenol (DON) is toxic to humans and animals. *Tenebrio molitor* survival was not affected by DON at the concentrations of 4.9, 8, 16, and 25 mg/g wheat bran, but larval body weight was reduced at the highest concentration. DON exposure in *T. molitor* larvae significantly upregulated superoxide dismutase (SOD) activity, indicating a stress response, though catalase activity remained unchanged. DON significantly reduced larval movement distance and time, though not always dose-dependent, and only the highest concentration slowed the average (Janković-Tomanić et al. 2019).

6.2 Potential Application of Entomopathogenic Mycotoxins in Medicine

Natural products are diverse and precious sources for drug discovery, including entomopathogenic mycotoxins. Many of these secondary metabolites exhibit potent antitumor, anticancer, antibacterial, antiviral, and immunosuppressive activities, making them promising leads in pharmaceutical research.

Anticancer potential has been demonstrated in compounds such as BEA (Kim et al. 2023; Juan-García et al. 2024; Patra et al. 2024), metarhizin A (Katou et al. 2014), furacin C (Li et al. 1990), ENNs (Prosperini et al. 2017), and swainsonine (Ren et al. 2017). Immunosuppressive activities were reported for cyclosporin (Fischer et al. 1989; Yang et al. 2018; Patocka et al. 2021), myriocin (Fujita et al. 1994), DON (Maciel et al. 2021; Dai et al. 2025), swainsonine (Zhang et al. 2019), and leucinostatin A (Csermely et al. 1994). BEA and ENNs possessed antibacterial activities (Prosperini et al. 2017; Fan et al. 2017), especially against multidrug-resistant bacteria (Bonilla et al. 2024). Antifungal activities were shown in ENNs (Prosperini et al. 2017), oosporein (Pegram and Wyatt 1981), and cyclosporin (Yang et al. 2018). Oosporein and BEA exhibited antiviral activities (Pegram and Wyatt 1981; Al Khoury et al. 2022). Recently, leucinostatin A derivatives have shown antiprotozoal activity (Rimle et al. 2025). The research findings bring potential applications of these mycotoxins in drug development.

Several studies have evaluated the toxicological profile and therapeutic potential of *B. bassiana* and its metabolites in murine models. Intramuscular injection of spores (2×10^5 and 2×10^8 spores/ml) resulted in transient lesions that healed rapidly within 12–96 hours depending on dose (Semalulu et al. 1992). Westwood et al. isolated four genes encoding proteins Bb-Eno1, Bb-f2, Bb-Ald, and Bb-Hex from *B. bassiana*, which cause allergies in humans. The recombinant proteins Bb-Ald and Bb-Hex bind to IgE derived from fungal allergenic patients (Westwood et al. 2006). In contrast, BEA demonstrated therapeutic promise by alleviating asthma symptoms in mice. It reduced inflammatory substances and stimulated the apoptosis of lymphocytic cells. The density of neutrophils, eosinophils, lymphocytes, and macrophages decreased in alveolar and bronchial fluid. In addition, BEA exhibits a dose-dependent stimulating effect (>1 μM), upregulating proteins involved in apoptosis, such as Bax and caspase-3, and downregulating Bcl-2, which plays a role in preventing cell apoptosis (Zhang et al. 2016). Acute toxicity of *B. bassiana* extract and spore preparations in mice was $LD_{50} > 5000$ mg/kg and $>1.1 \times 10^{10}$ conidia/kg body weight, respectively. White blood cell counts increased overall, but neutrophils, eosinophils, monocytes, and lymphocytes decreased. Despite these changes, no lethal effect was observed, and all treatment groups showed significantly increased body weight and decreased male average brain somatic index (Farag et al. 2022). According to the European Food Safety Authority, there has not yet been any convincing evidence that BEA induces chromosomal disruption (EFSA 2024). Nevertheless, intake limits for BEA and ENNs have not been established due to significant gaps in toxicokinetic and long-term safety data (Behr et al. 2025).

Swainsonine has been studied for over 40 years (Hino et al. 1985). Clinical testing reported that swainsonine reduced tumors and improved symptoms such as smaller tumors, less coughing, and easier breathing (Goss et al. 1994). Additionally, swainsonine was found to enhance the efficacy of 5-fluorouracil in colorectal cancer cells (Hamaguchi et al. 2007), potentiate cisplatin activity in Ehrlich ascites carcinoma cells, diminishing ascites volumes and mortality (Santos et al. 2011), and synergize with paclitaxel against hepatocellular carcinoma (You et al. 2012).

Mechanistically, swainsonine inhibits human promyelocytic cell proliferation by interfering with the S-phase of the cell cycle (Singh and Kaur 2014). Swainsonine and its derivatives, such as 8-epi-L-SWN and C-6 fluorinated D-swainsonine analogs, inhibited α-mannosidase, α-L-rhamnosidase, and antileukaemic activities (Gao et al. 2025a; Shimadate et al. 2024; Singh and Kaur 2014). Swainsonine increases intestinal permeability and intestinal microbiota's LPS gene expression (Lei et al. 2024). On the other side, swainsonine toxicity was reduced in the presence of trehalose in renal cells (Zhang et al. 2024). Transcriptomic analysis revealed that swainsonine mediated the gene expression of coding and non-coding RNAs in rat renal tubular epithelial cells, offering insight into potential detoxification mechanisms (Liu et al. 2025). Tang et al. (2025) clarified the mechanisms of swainsonine in renal tubular epithelial cells. It activates endoplasmic reticulum stress, unfolded protein response and autophagy by upregulating the expression of GRP78, XBP1s, LC3-II/I, and ATG5 (Tang et al. 2025). These findings underscore the pharmaceutical versatility of EPF-derived mycotoxins. However, further research is essential to refine the dose-response relationship, elucidate the mechanisms of action, and ensure safety and efficacy before clinical application.

7 Conclusion

Entomopathogenic fungi represent a distinct group of fungi due to their ability to grow and develop on insect hosts. To invade and colonize insect bodies, these fungi possess specialized traits, including hydrolytic enzymes and bioactive secondary metabolites, that enable them to circumvent the insect immune system. Insects, the most diverse group of organisms on Earth, are matched by the host specificity of EPF. As a result, EPF are remarkably diverse and rich in both species richness and the range of metabolites they produce. The unique biological traits of certain EPF species have been exploited in biological control strategies. Several genera, such as *Metarhizium* and *Beauveria,* are widely used in commercially available biocontrol products. With increasing taxonomic exploration, the number of identified EPF species continues to grow. Recently, the focus has expanded from fungal species to their mycotoxins due to their promising applications in agriculture and medicine and advances in sequencing technology, genome analysis, proteomics, and transcriptomics.

Despite the relatively small number of EPF-mycotoxins characterized to date, these compounds exhibit a broad spectrum of biological activities, including insecticidal, nematicidal, anti-protozoal, antibacterial, antifungal, antiviral, antioxidant, anticancer, immunosuppressive, pro-apoptotic, and plant growth-promoting effects, demonstrating their vast potential for practical application.

Although recent studies have provided new insights into EPF and their toxins, current knowledge remains limited when compared to the full extent of their diversity, complexity, and potential utility. Thus, further research demands interdisciplinary collaboration and should focus on (1) discovering novel EPF species in unexplored ecosystems and monitoring known and emerging EPF species to understand ecological shifts. (2) Elucidating the genome and biosynthetic pathways responsible for secondary metabolites, particularly mycotoxins. (3) Investigating gene regulatory networks under varying environmental conditions to predict and direct the biosynthesis of desired compounds. (4) Characterizing EPF ecological interactions and evaluating their practical applications. (5) Examining mycotoxin mechanisms of action and effective dosage across various cell types to inform the development of agricultural and medical products. (6) Assessing the metabolic transformation and biodegradability of mycotoxins to support environmentally sustainable detoxification strategies. In conclusion, EPF and their mycotoxins represent a rich and promising frontier for scientific research with broad implications for biotechnology, agriculture, and medicine.

References

Abdullah RR, Abd El-Wahab AH, Abd El-Salam SA (2024) Insecticidal activity and possible modes of action of secondary metabolites of some fungal strains and wild plants as natural pesticides against *Spodoptera frugiperda*. Beni-Suef Univ J Basic Appl Sci 13(1):9. https://doi.org/10.1186/s43088-024-00467-z

Al Khoury C, Bashir Z, Tokajian S, Nemer N, Merhi G et al (2022) In silico evidence of beauvericin antiviral activity against SARS-CoV-2. Comput Biol Med 141:105171. https://doi.org/10.1016/j.compbiomed.2021.105171

Albonico M, Schutz LF, Caloni F, Cortinovis C, Spicer LJ (2017) In vitro effects of the *Fusarium* mycotoxins fumonisin B1 and beauvericin on bovine granulosa cell proliferation and steroid production. Toxicon 128:38–45. https://doi.org/10.1016/j.toxicon.2017.01.019

Alexander NJ, Proctor RH, McCormick SP (2009) Genes, gene clusters, and biosynthesis of trichothecenes and fumonisins in *Fusarium*. Toxin Rev 28(2–3):198–215. https://doi.org/10.1080/15569540903092142

Ali S, Zhang C, Wang Z, Wang XM, Wu JH et al (2017) Toxicological and biochemical basis of synergism between the entomopathogenic fungus *Lecanicillium muscarium* and the insecticide matrine against *Bemisia tabaci* (Gennadius). Sci Rep 7(1):46558. https://doi.org/10.1038/srep46558

An J, Chen Y, Zhou S, Gao Y, Yang C et al (2025) Identification and characterization of *Morganella morganii* strain YC12-C3 and *Enterococcus faecalis* strain YC12-C10 and elucidation of its deoxynivalenol-degrading potential. Mycotoxin Res 41(1):113–126. https://doi.org/10.1007/s12550-024-00568-9

Anwar W, Subhani MN, Haider MS, Shahid AA, Mushatq H et al (2016) First record of *Trichoderma longibrachiatum* as entomopathogenic fungi against *Bemisia tabaci* in Pakistan. Pak J Phytopathol 28(2):287–294. https://doi.org/10.5555/20173096518

Arai T, Mikami Y, Fukushima K, Utsumi T, Yazawa K (1973) A new antibiotic, leucinostatin, derived from *Penicillium lilacinum*. J Antibiot 26(3):157–161. https://doi.org/10.7164/antibiotics.26.157

Aufy M, Abdelaziz RF, Hussein AM, Topcagic N, Shamroukh H et al (2023) Impact of enniatin B and beauvericin on lysosomal cathepsin B secretion and apoptosis induction. Int J Mol Sci 24(3):Article 3. https://doi.org/10.3390/ijms24032030

Bandani AR, Butt TM (1999) Insecticidal, Antifeedant and growth inhibitory activities of efrapeptins, metabolites of the mungus *Tolypocladium*. Biocontrol Sci Tech 9(4):499–506. https://doi.org/10.1080/09583159929460

Bao ZX, Liu R, Li CQ, Pan XR, Zhao PJ (2022) Pathogenicity and metabolites of *Purpureocillium lavendulum* YMF1. 00683 against *Meloidogyne incognita*. Pathogens 11(7):795. https://doi.org/10.3390/pathogens11070795

Behr AC, Fæste CK, Azqueta A, Tavares AM, Spyropoulou A et al (2025) Hazard characterization of the mycotoxins enniatins and beauvericin to identify data gaps and improve risk assessment for human health. Arch Toxicol 99:1791–1841. https://doi.org/10.1007/s00204-025-03988-3

Bi K, Zhang W, Xiao Z, Zhang D (2018) Characterization, expression and application of a zearalenone degrading enzyme from *Neurospora crassa*. AMB Express 8:1–10. https://doi.org/10.1186/s13568-018-0723-z

Blacutt AA, Gold SE, Voss KA, Gao M, Glenn AE (2018) *Fusarium verticillioides*: advancements in understanding the toxicity, virulence, and niche adaptations of a model mycotoxigenic pathogen of maize. Phytopathology® 108(3):312–326. https://doi.org/10.1094/PHYTO-06-17-0203-RVW

Bohnert M, Dahse HM, Gibson DM, Krasnoff SB, Hoffmeister D (2013) The fusarin analog NG-391 impairs nucleic acid formation in K-562 leukemia cells. Phytochem Lett 6(2):189–192. https://doi.org/10.1016/j.phytol.2013.01.001

Bonilla JNV, Florido EB, Partida AH, Alquicira EP, Loera O (2024) Interaction of beauvericin in combination with antibiotics against methicillin-resistant *Staphylococcus aureus* and *Salmonella typhimurium*. Toxicon 243:107713. https://doi.org/10.1016/j.toxicon.2024.107713

Brand M, Wang L, Agnello S, Gazzola S, Gall FM et al (2021) Antiprotozoal structure–activity relationships of synthetic leucinostatin derivatives and elucidation of their mode of action. Angew Chem Int Ed 60(28):15613–15621. https://doi.org/10.1002/anie.202102153

Broquist HP (1985) The indolizidine alkaloids, slaframine and swainsonine: contaminants in animal forages. Annu Rev Nutr 5(1):391–409. https://doi.org/10.1146/annurev.nu.05.070185.002135

Büchter C, Koch K, Freyer M, Baier S, Saier C et al (2020) The mycotoxin beauvericin impairs development, fertility and life span in the nematode *Caenorhabditis elegans* accompanied by increased germ cell apoptosis and lipofuscin accumulation. Toxicol Lett 334:102–109. https://doi.org/10.1016/j.toxlet.2020.09.016

Calò L, Fornelli F, Ramires R, Nenna S, Tursi A et al (2004) Cytotoxic effects of the mycotoxin beauvericin to human cell lines of myeloid origin. Pharmacol Res 49(1):73–77. https://doi.org/10.1016/j.phrs.2003.07.002

Carroll CS, Grieve CL, Murugathasan I, Bennet AJ, Czekster CM et al (2017) The rhizoferrin biosynthetic gene in the fungal pathogen *Rhizopus delemar* is a novel member of the NIS gene family. Int J Biochem Cell Biol 89:136–146. https://doi.org/10.1016/j.biocel.2017.06.005

Castoria R, Mannina L, Durán-Patrón R, Maffei F, Sobolev AP et al (2011) Conversion of the mycotoxin patulin to the less toxic desoxypatulinic acid by the biocontrol yeast *Rhodosporidium kratochvilovae* strain LS11. J Agric Food Chem 59(21):11571–11578. https://doi.org/10.1021/jf203098v

Chehri K, Godini R (2017) Detection of nivalenol and deoxynivalenol chemotypes produced by *Fusarium graminearum* species complex isolated from barley in Iran using specific PCR assays. J Pl Prot Res 57(3):237–242. https://doi.org/10.1515/jppr-2017-0033

Chen XR, Hu QB, Yu XQ, Ren SX (2014) Effects of destruxins on free calcium and hydrogen ions in insect hemocytes. Insect Sci 21(1):31–38. https://doi.org/10.1111/1744-7917.12028

Chen Y, Chen Q, Han M, Zhou J, Gong L et al (2016) Development and optimization of a multiplex lateral flow immunoassay for the simultaneous determination of three mycotoxins in corn, rice and peanut. Food Chem 213:478–484. https://doi.org/10.1016/j.foodchem.2016.06.116

Chen B, Sun Y, Li S, Yin Y, Wang C (2021) Inductive production of the iron-chelating 2-Pyridones benefits the producing fungus to compete for diverse niches. MBio 12(6):e0327921. https://doi.org/10.1128/mbio.03279-21

Chen Z, Wang F, Zhang W, Zhou S, Wen D et al (2023) Chronic exposure to zearalenone induces intestinal inflammation and oxidative injury in adult *Drosophila melanogaster* midgut. Ecotoxicol Environ Saf 251:114555. https://doi.org/10.1016/j.ecoenv.2023.114555

Chen X, Li Z, Zhang X, Zheng H, Lv H et al (2025) Insights into the zearalenone degradation performance and pathway by *Gordonia hydrophobica* HAU421 and characterization of a novel lactonohydrolase involved. Int J Biol Macromol 296:139631. https://doi.org/10.1016/j.ijbiomac.2025.139631

Coelho A, Celli M, Ono ES, Hoffmann F, Pagnocca F et al (2008) Patulin biodegradation using *Pichia ohmeri* and *Saccharomyces cerevisiae*. World Mycotoxin J 1(3):325–331. https://doi.org/10.3920/WMJ2008.1040

Cook D, Donzelli BGG, Creamer R, Baucom DL, Gardner DR et al (2017) Swainsonine biosynthesis genes in diverse symbiotic and pathogenic fungi. G3 Genes|Genomes|Genetics 7(6):1791–1797. https://doi.org/10.1534/g3.117.041384

Csermely P, Radics L, Rossi C, Szamel M, Ricci M et al (1994) The nonapeptide leucinostatin A acts as a weak ionophore and as an immunosuppressant on T lymphocytes. Biochim Biophys Acta 1221(2):125–132. https://doi.org/10.1016/0167-4889(94)90004-3

da Silva Santos AC, Diniz AG, Tiago PV, de Oliveira NT (2020) Entomopathogenic *Fusarium* species: a review of their potential for the biological control of insects, implications and prospects. Fungal Biol Rev 34(1):41–57. https://doi.org/10.1016/j.fbr.2019.12.002

Dai C, Hao Z, Liu D, Wang Z, Conti GO et al (2025) Deoxynivalenol exposure-related male reproductive toxicity in mammals: molecular mechanisms, detoxification and future directions. Environ Int:109478. https://doi.org/10.1016/j.envint.2025.109478

De Ruyck K, Huybrechts I, Yang S, Arcella D, Claeys L et al (2020) Mycotoxin exposure assessments in a multi-center European validation study by 24-hour dietary recall and biological fluid sampling. Environ Int 137:105539. https://doi.org/10.1016/j.envint.2020.105539

De Wint FC, Nicholson S, Koid QQ, Zahra S, Chestney-Claassen G et al (2024) EntomoFun 1.0: a global database of entomopathogenic fungi and associations with their hosts [dataset]. Dryad. https://doi.org/10.5061/dryad.1zcrjdg17

Delgado-Povedano M, Maris E, Kellner N, Mulisa G, Gámiz-Gracia L et al (2025) Liquid chromatography-tandem mass spectrometry for the determination of multiple mycotoxins in serum through suspect screening and targeted approaches: advancing human mycotoxin biomonitoring. Microchem J 208:112562. https://doi.org/10.1016/j.microc.2024.112562

Diaz-Sanchez V, Avalos J, Limon MC (2012) Identification and regulation of fusA, the polyketide synthase gene responsible for fusarin production in *Fusarium fujikuroi*. Appl Environ Microbiol 78(20):7258–7266. https://doi.org/10.1128/AEM.01552-12

Dong X, Jiang W, Li C, Ma N, Xu Y et al (2015) Patulin biodegradation by marine yeast *Kodameae ohmeri*. Food Addit Contam Part A Chem Anal Control Expo Risk Assess 32(3):352–360. https://doi.org/10.1080/19440049.2015.1007090

Donzelli BGG, Krasnoff SB, Churchill ACL, Vandenberg JD, Gibson DM (2010) Identification of a hybrid PKS–NRPS required for the biosynthesis of NG-391 in *Metarhizium robertsii*. Curr Genet 56:151–162. https://doi.org/10.1007/s00294-010-0288-0

Dostal LA, Brown S, Bleck J, Anderson JA (1991) Developmental toxicity of pentostatin (2′--deoxycoformycin) in rats and rabbits. Teratology 44(3):325–334. https://doi.org/10.1002/tera.1420440309

Dumas C, Matha V, Quiot JM, Vey A (1996) Effects of destruxins, cyclic depsipeptide mycotoxins, on calcium balance and phosphorylation of intracellular proteins in lepidopteran cell lines. Comp Biochem Physiol C: Pharmacol Toxicol Endocrinol 114(3):213–219. https://doi.org/10.1016/0742-8413(96)00041-2

Duong ML, Van NTT (2025) Secondary metabolites and potential applications of *Cordyceps* and allies. In: Deshmukh SK, Sridhar KR (eds) *Cordyceps* and allied species. Springer Nature Singapore, Singapore, pp 119–163. https://doi.org/10.1007/978-981-97-6345-0_6

Duvick J, Rood TA (2000) U.S. Patent No. 6,117,668. Washington, DC: U.S. Patent and Trademark Office. https://patents.google.com/patent/US6117668A/en

EFSA Panel on Contaminants in the Food Chain (CONTAM), Knutsen HK, Åkesson A, Bampidis V, Bodin L et al (2024) Genotoxicity of beauvericin. EFSA J 22(10):e9031. https://doi.org/10.2903/j.efsa.2024.9031

El Aalaoui M, Rammali S, Bencharki B, Sbaghi M (2025) Integrated biological control of *Diaspis echinocacti* (Bouché) on *Opuntia ficus-indica* (L.) mill (Cactaceae) using predatory ladybirds and fungal pathogens. Crop Prot 187:106950. https://doi.org/10.1016/j.cropro.2024.106950

el-Basyouni SH, Vining LC (1966) Biosynthesis of oosporein in *Beauveria bassiana* (Bals.) Vuill. Can J Biochem 44(5):557–565. https://doi.org/10.1139/o66-067

El-Nezami H, Polychronaki N, Salminen S, Mykkänen H (2002) Binding rather than metabolism may explain the interaction of two food-grade *Lactobacillus* strains with zearalenone and its derivative ά-zearalenol. Appl Environ Microbiol 68(7):3545–3549. https://doi.org/10.1128/AEM.68.7.3545-3549.2002

Fan JQ, Chen XR, Hu QB (2013) Effects of destruxin A on hemocytes morphology of *Bombyx mori*. J Integr Agric 12(6):1042–1048. https://doi.org/10.1016/S2095-3119(13)60482-7

Fan Y, Liu X, Keyhani NO, Tang G, Pei Y et al (2017) Regulatory cascade and biological activity of *Beauveria bassiana* oosporein that limits bacterial growth after host death. Proc Natl Acad Sci USA 114(9):E1578–E1586. https://doi.org/10.1073/pnas.1616543114

Fan J, Liu P, Zhao K, Chen HP (2023) Three previously undescribed metabolites from *Cordyceps cicadae* JXCH-1, an entomopathogenic fungus. Nat Prod Bioprospect 13(1):46. https://doi.org/10.1007/s13659-023-00410-2

Farag AAG, Ali SS, Ibrahim AA, El-Saadany HM (2022) Acute oral toxicity of two products from a microbial pest control agent (*Beauveria bassiana*) on physiological status aspects of male albino rats. Egypt J Hosp Med 86(1):943–951. https://www.ajol.info/index.php/ejhm/article/view/223971

Feng P, Shang Y, Cen K, Wang C (2015) Fungal biosynthesis of the bibenzoquinone oosporein to evade insect immunity. Proc Natl Acad Sci USA 112(36):11365–11370. https://doi.org/10.1073/pnas.1503200112

Ferrer E, Juan-García A, Font G, Ruiz MJ (2009) Reactive oxygen species induced by beauvericin, patulin and zearalenone in CHO-K1 cells. Toxicol In Vitro 23(8):1504–1509. https://doi.org/10.1016/j.tiv.2009.07.009

Fischer G, Wittmann-Liebold B, Lang K, Kiefhaber T, Schmid FX (1989) Cyclophilin and peptidyl-prolyl cis-trans isomerase are probably identical proteins. Nature 337(6206): 476–478. https://doi.org/10.1038/337476a0

Fresta M, Ricci M, Rossi C, Furneri PM, Puglisi G (2000) Antimicrobial nonapeptide leucinostatin A-dependent effects on the physical properties of phospholipid model membranes. J Colloid Interface Sci 226(2):222–230. https://doi.org/10.1006/jcis.2000.6816

Fujita T, Inoue K, Yamamoto S, Ikumoto T, Sasaki S et al (1994) Fungal metabolites. Part 11. A potent immunosuppressive activity found in *Isaria sinclairii* metabolite. J Antibiot 47(2): 208–215. https://doi.org/10.7164/antibiotics.47.208

Gaffoor I, Trail F (2006) Characterization of two polyketide synthase genes involved in zearalenone biosynthesis in *Gibberella zeae*. Appl Environ Microbiol 72(3):1793–1799. https://doi.org/10.1128/AEM.72.3.1793-1799.2006

Gao H, Lu D, Xing M, Xu Q, Xue F (2022) Excavation, expression, and functional analysis of a novel zearalenone-degrading enzyme. Folia Microbiol 67(4):633–640. https://doi.org/10.1007/s12223-022-00967-4

Gao FT, Zhang M, Shimadate Y, Kato A, Li YX et al (2025a) Enantiomeric C-6 fluorinated swainsonine derivatives as highly selective and potent inhibitors of α-mannosidase and α-l-rhamnosidase: design, synthesis and structure-activity relationship study. Eur J Med Chem 282: 117031. https://doi.org/10.1016/j.ejmech.2024.117031

Gao J, Liu D, Nguyen C, McCormick SP, Proctor RH et al (2025b) Biosynthesis of the central tricyclic skeleton of trichothecene mycotoxins. J Am Chem Soc 147(12):10331–10338. https://doi.org/10.1021/jacs.4c16973

Ghamsari FA, Ghadaksaz A, Ebrahimi MT, Varzaneh MB, Iranbakhsh A et al (2025) An integrated study on the interactions of *Lactobacillus brevis* components with deoxynivalenol and zearalenone. Process Biochem 154:12–21. https://doi.org/10.1016/j.procbio.2025.04.005

Gil-Serna J, Vazquez C, Patino B (2020) Genetic regulation of aflatoxin, ochratoxin A, trichothecene, and fumonisin biosynthesis: a review. Int Microbiol 23(1):89–96. https://doi.org/10.1007/s10123-019-00084-2

Girardi NS, Sosa AL, Etcheverry MG, Passone MA (2022) In vitro characterization bioassays of the nematophagous fungus *Purpureocillium lilacinum*: evaluation on growth, extracellular enzymes, mycotoxins and survival in the surrounding agroecosystem of tomato. Fungal Biol 126(4):300–307. https://doi.org/10.1016/j.funbio.2022.02.001

Goettel MS, Eilenberg J, Glare T (2005) Entomopathogenic fungi and their role in regulation of insect populations. In: Gilbert LI (ed) Comprehensive molecular insect science. Elsevier, pp 361–405. https://doi.org/10.1016/B0-44-451924-6/00088-0

Goss PE, Baptiste J, Fernandes B, Baker M, Dennis JW (1994) A phase I study of swainsonine in patients with advanced malignancies. Cancer Res 54(6):1450–1457

Guan S, He J, Young JC, Zhu H, Li XZ et al (2009) Transformation of trichothecene mycotoxins by microorganisms from fish digesta. Aquaculture 290(3–4):290–295. https://doi.org/10.1016/j.aquaculture.2009.02.037

Hamaguchi J, Nakagawa H, Takahashi M, Kudo T, Kamiyama N et al (2007) Swainsonine reduces 5-fluorouracil tolerance in the multistage resistance of colorectal cancer cell lines. Mol Cancer 6: 1–9. https://doi.org/10.1186/1476-4598-6-58

Han P, Gong Q, Fan J, Abbas M, Chen D et al (2022) Destruxin A inhibits scavenger receptor B mediated melanization in *Aphis citricola*. Pest Manag Sci 78(5):1915–1924. https://doi.org/10.1002/ps.6809

Hashimoto S, Okada K, Imaoka S (2008) Interaction between bisphenol derivatives and protein disulphide isomerase (PDI) and inhibition of PDI functions: requirement of chemical structure for binding to PDI. J Biochem 144(3):335–342. https://doi.org/10.1093/jb/mvn075

Hasumi K, Shinohara C, Iwanaga T, Endo A (1993) Lateritin, a new inhibitor of acyl-CoA: cholesterol acyltransferase produced by *Gibberella lateritium* IFO 7188. J Antibiot 46(12): 1782–1787. https://doi.org/10.7164/antibiotics.46.1782

Hatab S, Yue T, Mohamad O (2012) Removal of patulin from apple juice using inactivated lactic acid bacteria. J Appl Microbiol 112(5):892–899. https://doi.org/10.1111/j.1365-2672.2012.05279.x

Hawar S, Vevers W, Karieb S, Ali BK, Billington R et al (2013) Biotransformation of patulin to hydroascladiol by *Lactobacillus plantarum*. Food Control 34(2):502–508. https://doi.org/10.1016/j.foodcont.2013.05.023

He WJ, Zhang L, Yi SY, Tang XL, Yuan QS et al (2017) An aldo-keto reductase is responsible for *Fusarium* toxin-degrading activity in a soil *Sphingomonas* strain. Sci Rep 7(1):9549. https://doi.org/10.1038/s41598-017-08799-w

He Y, Zhang W, Peng F, Lu R, Zhou H et al (2019) Metabolomic variation in wild and cultured *Cordyceps* and mycelia of *Isaria cicadae*. Biomed Chromatogr 33(4):e4478. https://doi.org/10.1002/bmc.4478

Hino M, Nakayama O, Tsurumi Y, Adachi K, Shibata O et al (1985) Studies of an immunomodulator, swainsonine I. Enhancement of immune response by swainsonine in vitro. J Antibiot 38(7): 926–935. https://doi.org/10.7164/antibiotics.38.926

Hong S, Sun Y, Chen H, Zhao P, Wang C (2023) Fungus–insect interactions beyond bilateral regimes: the importance and strategy to outcompete host ectomicrobiomes by fungal parasites. Curr Opin Microbiol 74:102336. https://doi.org/10.1016/j.mib.2023.102336

Hu H, Yin X, Pang S, Jiang Y, Weng Q et al (2023a) Mechanism of destruxin a inhibits juvenile hormone binding protein transporting juvenile hormone to affect insect growth. Pestic Biochem Physiol 197:105654. https://doi.org/10.1016/j.pestbp.2023.105654

Hu J, Wang G, Hou M, Du S, Han J et al (2023b) New hydrolase from *Aeromicrobium* sp. HA for the biodegradation of zearalenone: identification, mechanism, and application. J Agric Food Chem 71(5):2411–2420. https://doi.org/10.1021/acs.jafc.2c06410

Huang LC, Zheng N, Zheng BQ, Wen F, Cheng JB et al (2014) Simultaneous determination of aflatoxin M1, ochratoxin A, zearalenone and α-zearalenol in milk by UHPLC–MS/MS. Food Chem 146:242–249. https://doi.org/10.1016/j.foodchem.2013.09.047

Huang J, Lou Y, Liu J, Bulet P, Cai C et al (2023) A Toll pathway effector protects *Drosophila* specifically from distinct toxins secreted by a fungus or a bacterium. Proc Natl Acad Sci U S A 120(12):e2205140120. https://doi.org/10.1073/pnas.2205140120

Huang Z, Zhu W, Bai Y, Bai X, Zhang H (2024) Non-ribosomal peptide synthetase (NRPS)-encoding products and their biosynthetic logics in *Fusarium*. Microb Cell Fact 23(1):93. https://doi.org/10.1186/s12934-024-02378-1

Hyde KD, Jeewon R, Chen YJ, Bhunjun CS, Calabon MS et al (2020) The numbers of fungi: is the descriptive curve flattening? Fungal Divers 103:219–271. https://doi.org/10.1007/s13225-020-00458-2

Ibrahim HYE, Salam AM, Abdel-Mogib M, El-nagar ME, Salem HA et al (2011) Survey of entomopathogenic fungi naturally infecting cowpea aphid, *Aphis craccivora*. KOCH J Plant Prot Pathol 2(12):1063–1070. https://doi.org/10.21608/jppp.2011.86637

Ingram B, Marin S, Kiaitsi E, Magan N, Verheecke-Vaessen C et al (2025) *Fusarium graminearum* and zearalenone in wheat: a water activity–temperature model. Fungal Biol 129(4):101572. https://doi.org/10.1016/j.funbio.2025.101572

Ishiguro K, Arai T (1976) Action of the peptide antibiotic leucinostatin. Antimicrob Agents Chemother 9(6):893–898. https://doi.org/10.1128/aac.9.6.893

Islam R, Zhou T, Christopher Young J, Goodwin PH, Peter Pauls K (2012) Aerobic and anaerobic de-epoxydation of mycotoxin deoxynivalenol by bacteria originating from agricultural soil. World J Microbiol Biotechnol 28:7–13. https://doi.org/10.1007/s11274-011-0785-4

Isogai A, Nakayama J, Takayama S, Kusai A, Suzuki A (1992) Structural elucidation of minor components of peptidyl antibiotic P168s (leucinostatins) by tandem mass spectrometry. Biosci Biotechnol Biochem 56(7):1079–1085. https://doi.org/10.1271/bbb.56.1079

Iwanicki NSA, Gotti IA, Delalibera JI, Licht HHDF (2025) Host-specific patterns of virulence and gene expression profiles of the broad-host-range entomopathogenic fungus *Metarhizium anisopliae*. J Invertebr Pathol 209:108242. https://doi.org/10.1016/j.jip.2024.108242

Jallow A, Xie H, Tang X, Qi Z, Li P (2021) Worldwide aflatoxin contamination of agricultural products and foods: from occurrence to control. Compr Rev Food Sci Food Saf 20(3): 2332–2381. https://doi.org/10.1111/1541-4337.12734

Janković-Tomanić M, Petković B, Todorović D, Vranković J, Perić-Mataruga V (2019) Physiological and behavioral effects of the mycotoxin deoxynivalenol in *Tenebrio molitor* larvae. J Stored Prod Res 83:236–242. https://doi.org/10.1016/j.jspr.2019.07.008

Jirakkakul J, Cheevadhanarak S, Punya J, Chutrakul C, Senachak J et al (2015) Tenellin acts as an iron chelator to prevent iron-generated reactive oxygen species toxicity in the entomopathogenic fungus *Beauveria bassiana*. FEMS Microbiol Lett 362(2):1–8. https://doi.org/10.1093/femsle/fnu032

Juan-García A, Ilie AM, Juan C, Martínez L (2024) Evaluating the combined and individual cytotoxic effect of beauvericin, enniatin B and ochratoxin a on breast cancer cells, leukemia cells, and fresh peripheral blood mononuclear cells. Toxicol In Vitro 99:105890. https://doi.org/10.1016/j.tiv.2024.105890

Junge W, Nelson N (2015) ATP Synthase. Annu Rev Biochem 84:631–657. https://doi.org/10.1146/annurev-biochem-060614-034124

Juodeikiene G, Bartkiene E, Cernauskas D, Cizeikiene D, Zadeike D et al (2018) Antifungal activity of lactic acid bacteria and their application for *Fusarium* mycotoxin reduction in malting wheat grains. LWT 89:307–314. https://doi.org/10.1016/j.lwt.2017.10.061

Kagamizono T, Nishino E, Matsumoto K, Kawashima A, Kishimoto M et al (1995) Bassiatin, a new platelet aggregation inhibitor produced by *Beauveria bassiana* K-717. J Antibiot 48(12): 1407–1413. https://doi.org/10.7164/antibiotics.48.1407

Kanaoka M, Isogai A, Murakoshi S, Ichinoe M, Suzuki A et al (1978) Bassianolide, a new insecticidal cyclodepsipeptide from *Beauveria bassiana* and *Verticillium lecanii*. Agric Biol Chem 42(3):629–635. https://doi.org/10.1080/00021369.1978.10863029

Katou Y, Endob N, Suzuki T, Yua J, Kikuchi H et al (2014) Metarhizin A suppresses cell proliferation by inhibiting cytochrome c oxidase activity. Life Sci 103:1–7. https://doi.org/10.1016/j.lfs.2014.03.023

Kawada M, Inoue H, Ohba S, Masuda T, Momose I et al (2010) Leucinostatin A inhibits prostate cancer growth through reduction of insulin-like growth factor-I expression in prostate stromal cells. Int J Cancer 126(4):810–818. https://doi.org/10.1002/ijc.24915

Khan HAA, Khan T, Iqbal N, Anwar W, Ali M (2024) Control efficacy of isolates of *Aspergillus nidulans* and *Aspergillus fumigatus* against three destructive insect pests of stored grain. J Stored Prod Res 108:102357. https://doi.org/10.1016/j.jspr.2024.102357

Kharayat BS, Singh Y (2018) Mycotoxins in foods: mycotoxicoses, detection, and management. In: Holban AM, Grumezescu AM (eds) Microbial contamination and food degradation. Academic Press, pp 395–421. https://doi.org/10.1016/B978-0-12-811515-2.00013-5

Kikuchi H, Hoshi T, Kitayama M, Sekiya M, Katou Y et al (2009) New diterpene pyrone-type compounds, metarhizins A and B, isolated from entomopathogenic fungus, *Metarhizium flavoviride* and their inhibitory effects on cellular proliferation. Tetrahedron 65(2):469–477. https://doi.org/10.1016/j.tet.2008.11.014

Kim HG, Sung NY, Kim JH, Cho JY (2023) *In vitro* anti-cancer effects of beauvericin through inhibition of actin polymerization and Src phosphorylation. Phytomedicine 109:154573. https://doi.org/10.1016/j.phymed.2022.154573

Kleigrewe K, Aydin F, Hogrefe K, Piecuch P, Bergander K et al (2012) Structure elucidation of new fusarins revealing insights in the rearrangement mechanisms of the *Fusarium* mycotoxin fusarin C. J Agric Food Chem 60(21):5497–5505. https://doi.org/10.1021/jf3009469

Kolosova AY, Sibanda L, Dumoulin F, Lewis J, Duveiller E et al (2008) Lateral-flow colloidal gold-based immunoassay for the rapid detection of deoxynivalenol with two indicator ranges. Anal Chim Acta 616(2):235–244. https://doi.org/10.1016/j.aca.2008.04.029

Krasnoff SB, Sommers CH, Moon YS, Donzelli BG, Vandenberg JD et al (2006) Production of mutagenic metabolites by *Metarhizium anisopliae*. J Agric Food Chem 54(19):7083–7088. https://doi.org/10.1021/jf061405r

Krasnoff SB, Howe KJ, Heck ML, Donzelli BGG (2020) Siderophores from the Entomopathogenic Fungus *Beauveria bassiana*. J Nat Prod 83(2):296–304. https://doi.org/10.1021/acs.jnatprod.9b00698

Lana M, Simón O, Velasco P, Rodríguez VM, Caballero P et al (2023) First study on the root endophytic fungus *Trichoderma hamatum* as an entomopathogen: development of a fungal bioinsecticide against cotton leafworm (*Spodoptera littoralis*). Microbiol Res 270:127334. https://doi.org/10.1016/j.micres.2023.127334

Langenfeld A, Blond A, Gueye S, Herson P, Nay B et al (2011) Insecticidal cyclodepsipeptides from *Beauveria felina*. J Nat Prod 74(4):825–830. https://doi.org/10.1021/np100890n

Lei L, Deng D, Xu W, Yue M, Wu D et al (2024) Increased intestinal permeability and lipopolysaccharide contribute to swainsonine-induced systemic inflammation. Ecotoxicol Environ Saf 284:116912. https://doi.org/10.1016/j.ecoenv.2024.116912

Li M, Jiang Y, Bjeldanes LF (1990) Carcinogenicity of fusarin C isolated from *Fusarium moniliforme*. Chin J Cancer Res 2:1–5. https://doi.org/10.1007/BF02997222

Li X, Tang H, Yang C, Meng X, Liu B (2019) Detoxification of mycotoxin patulin by the yeast *Rhodotorula mucilaginosa*. Food Control 96:47–52. https://doi.org/10.1016/j.foodcont.2018.08.029

Li X, Liang LM, Hua ZB, Zhou XK, Huang Y et al (2024a) Eco-friendly management of *Meloidogyne incognita* in cadmium-contaminated soil by using nematophagous fungus *Purpureocillium lavendulum* YMF1. 683: efficacy and mechanism. Environ Res 244:117930. https://doi.org/10.1016/j.envres.2023.117930

Li Z, Jiao Y, Ling J, Zhao J, Yang Y et al (2024b) Characterization of a methyltransferase for iterative N-methylation at the leucinostatin termini in *Purpureocillium lilacinum*. Commun Biol 7(1):757. https://doi.org/10.1038/s42003-024-06467-0

Liu Y, Zhang S, Zhai Y, Zhang Q, Wu K et al (2025) Transcriptomic profiling provide insights into swainsonine-induced toxic responses and activation of the aryl hydrocarbon receptor pathway. Food Chem Toxicol 202:115556. https://doi.org/10.1016/j.fct.2025.115556

Liuzzi VC, Mirabelli V, Cimmarusti MT, Haidukowski M, Leslie JF et al (2017) Enniatin and beauvericin biosynthesis in *Fusarium* species: production profiles and structural determinant prediction. Toxins 9(2). https://doi.org/10.3390/toxins9020045

Lozano-Tovar MD, Ballestas Álvarez KL, Sandoval-Lozano LA, Palma Mendez GM, Barrera-Cubillos GP (2023) Study on the insecticidal activity of entomopathogenic fungi for the control of the fruit fly (*Anastrepha obliqua*), the main pest in mango crop in Colombia. Arch Microbiol 205(3):83. https://doi.org/10.1007/s00203-023-03405-2

Luo F, Hong S, Chen B, Yin Y, Tang G et al (2020) Unveiling of swainsonine biosynthesis via a multibranched pathway in fungi. ACS Chem Biol 15(9):2476–2484. https://doi.org/10.1021/acschembio.0c00466

Maciel MD, do Amaral AC, da Silva TD, Bezerra JD, de Souza-Motta CM et al (2021) Evaluation of mycotoxin production and phytopathogenicity of the entomopathogenic fungi *Fusarium caatingaense* and *F. pernambucanum* from Brazil. Curr Microbiol 78:1218–1226. https://doi.org/10.1007/s00284-021-02387-y

Majumdar A, Boetel MA, Jaronski ST (2008) Discovery of *Fusarium solani* as a naturally occurring pathogen of sugar beet root maggot (Diptera: Ulidiidae) pupae: prevalence and baseline susceptibility. J Invertebr Pathol 97:1–8. https://doi.org/10.1016/j.jip.2007.05.003

Masuma R, Shiomi K, Omura S (2009) Helminth electron transport inhibitors produced by fungi. In: Anke T, Weber D (eds) The Mycota XV—physiology and genetics. Springer, Berlin, Heidelberg, pp 247–271. https://doi.org/10.1007/978-3-642-00286-1_12

Mc Namara L, Dolan SK, Walsh JM, Stephens JC, Glare TR et al (2019) Oosporein, an abundant metabolite in *Beauveria caledonica*, with a feedback induction mechanism and a role in insect virulence. Fungal Biol 123(8):601–610. https://doi.org/10.1016/j.funbio.2019.01.004

Meca G, Ritieni A, Mañes J (2012) Reduction in vitro of the minor *Fusarium mycotoxin* beauvericin employing different strains of probiotic bacteria. Food Control 28(2):435–440. https://doi.org/10.1016/j.foodcont.2012.04.002

Meca G, Zhou T, Li XZ, Ritieni A, Mañes J (2013) Ciclohexadespipeptide beauvericin degradation by different strains of *Saccharomyces cerevisiae*. Food Chem Toxicol 59:334–338. https://doi.org/10.1016/j.fct.2013.06.010

Molinari M, Eriksson KK, Calanca V, Galli C, Cresswell P et al (2004) Contrasting functions of calreticulin and calnexin in glycoprotein folding and ER quality control. Mol Cell 13(1):125–135. https://doi.org/10.1016/s1097-2765(03)00494-5

Molnar O, Schatzmayr G, Fuchs E, Prillinger H (2004) *Trichosporon mycotoxinivorans* sp. nov., a new yeast species useful in biological detoxification of various mycotoxins. Syst Appl Microbiol 27(6):661–671. https://doi.org/10.1078/0723202042369947

Molnar I, Gibson DM, Krasnoff SB (2010) Secondary metabolites from entomopathogenic hypocrealean fungi. Nat Prod Rep 27(9):1241–1275. https://doi.org/10.1039/c001459c

Mutalov A, Kalandarov P, Kannazarova Z, Gazieva R, Shavazov K (2025) Comprehensive strategies for managing stored grains: tackling insect infestation and quality loss: a review. J Stored Prod Res 112:102624. https://doi.org/10.1016/j.jspr.2025.102624

Natarajan S, Bucur D, Kildea S, Doohan F (2025) Digital PCR assays for quantifying trichothecene-producing *Fusarium* species, including *Fusarium langsethiae*, *F. poae*, and *F. sporotrichioides*, in oats. Anal Bioanal Chem 417(13):2957–2969. https://doi.org/10.1007/s00216-025-05840-0

Niehaus EM, Diaz-Sanchez V, von Bargen KW, Kleigrewe K, Humpf HU et al (2014) Fusarins and fusaric acid in fusaria. In: Biosynthesis and molecular genetics of fungal secondary metabolites. Springer New York, New York, pp 239–262. https://doi.org/10.1007/978-1-4939-1191-2_11

Niehaus EM, Studt L, von Bargen KW, Kummer W, Humpf HU et al (2016) Sound of silence: the beauvericin cluster in *Fusarium fujikuroi* is controlled by cluster-specific and global regulators mediated by H3K27 modification. Environ Microbiol 18(11):4282–4302. https://doi.org/10.1111/1462-2920.13576

Niu X, Thaochan N, Hu Q (2020) Diversity of linear non-ribosomal peptide in biocontrol fungi. J Fungi 6(2):61. https://doi.org/10.3390/jof6020061

Nkyaagye E, Limbach MN, Do TD (2024) Molecular selectivity in the binding of alkali metals, alkaline earth metals, first-row transition metals, and lanthanides with cyclic depsipeptides. J Phys Chem B 128(5):1209–1219. https://doi.org/10.1021/acs.jpcb.3c08385

Nowak M, Soboń A, Bernat P, Różalska S (2023) Entomopathogenic fungi of the genus *Cordyceps* biotransform zearalenone- metabolomic and proteomic backgrounds. Int Biodeterior Biodegrad 179:105572. https://doi.org/10.1016/j.ibiod.2023.105572

Nowak M, Bernat P, Różalska S (2024) Mutual interaction of the entomopathogenic and endophytic fungus *Metarhizium anisopliae* with zearalenone as a native component of crude *Fusarium* extract. Sci Rep 14(1):22493. https://doi.org/10.1038/s41598-024-73022-6

Oliveira CAF, Ivanova L, Solhaug A, Fæste CK (2020) Enniatin B1-induced lysosomal membrane permeabilization in mouse embryonic fibroblasts. Mycotoxin Res 36(1):23–30. https://doi.org/10.1007/s12550-019-00366-8

Ortiz-Urquiza A, Keyhani NO (2016) Molecular genetics of *Beauveria bassiana* infection of insects. Adv Genet 94:165–249. https://doi.org/10.1016/bs.adgen.2015.11.003

Pal S, St Leger RJ, Wu LP (2007) Fungal peptide destruxin A plays a specific role in suppressing the innate immune response in *Drosophila melanogaster*. J Biol Chem 282(12):8969–8977. https://doi.org/10.1074/jbc.M605927200

Park JO, Hargreaves JR, McConville EJ, Stirling GR, Ghisalberti EL et al (2004) Production of leucinostatins and nematicidal activity of Australian isolates of *Paecilomyces lilacinus* (Thom) Samson. Lett Appl Microbiol 38(4):271–276. https://doi.org/10.1111/j.1472-765X.2004.01488.x

Paszkiewicz M, Tyma M, Ligęza-Żuber M, Włóka E, Boguś MI et al (2016) Trichothecenes production by entomopathogenic fungus *Conidiobolus coronatus*. Adv Toxicol Toxic Effects 1(1):007–014. https://www.medsciencegroup.us/articles/ATTE-1-102.pdf

Patocka J, Nepovimova E, Kuca K, Wu W (2021) Cyclosporine A: chemistry and toxicity–a review. Curr Med Chem 28(20):3925–3934. https://doi.org/10.2174/0929867327666201006153202

Patra A, Ghosh SS, Saini GK (2024) Exploring potential molecular targets and therapeutic efficacy of beauvericin in triple-negative breast cancer cells. Comput Biol Chem 112:108154. https://doi.org/10.1016/j.compbiolchem.2024.108154

Pedreira J, Rivas L, Comenge J, Skouridou V, Camprubí-Ferrer D et al (2025) A comprehensive review of competitive lateral flow assays over the past decade. Lab Chip 25:2578–2608. https://doi.org/10.1371/journal.pone.0250250

Pegram RA, Wyatt RD (1981) Avian gout caused by oosporein, a mycotoxin produced by *Chaetomium trilaterale*. Poult Sci 60(11):2429–2440. https://doi.org/10.3382/ps.0602429

Pelizza SA, Stenglein SA, Cabello MN, Dinolfo MI, Lange CE (2011) First record of *Fusarium verticillioides* as an entomopathogenic fungus of grasshoppers. J Insect Sci 11(1):70. https://doi.org/10.1673/031.011.7001

Peng KC, Huang HS, Tzeng YM, Liu SY (2005) Circular dichroism analysis of destruxins from *Metarhizium anisopliae*. J Biochem Biophys Methods 62(1):41–50. https://doi.org/10.1016/j.jbbm.2004.08.001

Pérez-Fuentes N, Alvariño R, Alfonso A, González-Jartín J, Gegunde S et al (2021) Single and combined effects of regulated and emerging mycotoxins on viability and mitochondrial function of SH-SY5Y cells. Food Chem Toxicol 154:112308. https://doi.org/10.1016/j.fct.2021.112308

Podversnik H, Jha S, Macheroux P, Breinbauer R (2022) Design and synthesis of efficient fluororethylene-peptidomimetic inhibitors of dipeptidyl peptidase III (DPP3). Bioorg Med Chem 67:116831. https://doi.org/10.1016/j.bmc.2022.116831

Pradier A, Papaserafeim M, Li N, Rietveld A, Kaestel C et al (2019) Small-molecule immunosuppressive drugs and therapeutic immunoglobulins differentially inhibit NK cell effector functions in vitro. Front Immunol 10:556. https://doi.org/10.3389/fimmu.2019.00556

Proctor RH, Desjardins AE, Plattner RD, Hohn TM (1999) A polyketide synthase gene required for biosynthesis of fumonisin mycotoxins in *Gibberella fujikuroi* mating population A. Fungal Genet Biol 27(1):100–112. https://doi.org/10.1006/fgbi.1999.1141

Proctor RH, Brown DW, Plattner RD, Desjardins AE (2003) Co-expression of 15 contiguous genes delineates a fumonisin biosynthetic gene cluster in *Gibberella moniliformis*. Fungal Genet Biol 38(2):237–249. https://doi.org/10.1016/s1087-1845(02)00525-x

Proctor RH, McCormick SP, Kim HS, Cardoza RE, Stanley AM et al (2018) Evolution of structural diversity of trichothecenes, a family of toxins produced by plant pathogenic and entomopathogenic fungi. PLoS Pathog 14(4):e1006946. https://doi.org/10.1371/journal.ppat.1006946

Prosperini A, Berrada H, Ruiz MJ, Caloni F, Coccini T et al (2017) A review of the mycotoxin enniatin B. Front Public Health 5:304. https://doi.org/10.3389/fpubh.2017.00304

Pu YC, Zheng ZW, Ding CH, Chen XD (2023) Development of potential microbial agents with two new entomopathogenic fungal strains to control the red palm weevil *Rhynchophorus ferrugineus* (Olivier) (Coleoptera: Curculionidae). Egypt J Biol Pest Control 33(1):107. https://doi.org/10.1186/s41938-023-00754-4

Qasim M, Lin Y, Dash CK, Bamisile BS, Ravindran K et al (2018) Temperature-dependent development of Asian citrus psyllid on various hosts, and mortality by two strains of *Isaria*. Microb Pathog 119:109–118. https://doi.org/10.1016/j.micpath.2018.04.019

Qasim M, Islam SU, Islam W, Noman A, Khan KA et al (2020) Characterization of mycotoxins from entomopathogenic fungi (*Cordyceps fumosorosea*) and their toxic effects to the development of Asian citrus psyl lid reared on healthy and diseased citrus plants. Toxicon 188:39–47. https://doi.org/10.1016/j.toxicon.2020.10.012

Ravindran K, Sivaramakrishnan S, Hussain M, Dash CK, Bamisile BS et al (2018) Investigation and molecular docking studies of bassianolide from *Lecanicillium lecanii* against *Plutella xylostella* (Lepidoptera: Plutellidae). Comp Biochem Physiol C Toxicol Pharmacol 206:65–72. https://doi.org/10.1016/j.cbpc.2018.03.004

Ren Z, Song R, Wang S, Quan H, Yang L et al (2017) The biosynthesis pathway of swainsonine, a new anticancer drug from three endophytic fungi. J Microbiol Biotechnol 27(11):1897–1906. https://doi.org/10.4014/jmb.1709.09003

Rimle L, Pliatsika D, Arnold N, Kurth S, Kaiser M et al (2025) Dissecting structural requirements of leucinostatin A derivatives for antiprotozoal activity and mammalian toxicity. J Med Chem 68(4):4237–4258. https://doi.org/10.1021/acs.jmedchem.4c01989

Roig M, Meca G, Ferrer E, Mañes J (2013) Reduction of the enniatins A, A1, B, B1 by an in vitro degradation employing different strains of probiotic bacteria: identification of degradation products by LC–MS–LIT. Toxicon 70:44–53. https://doi.org/10.1016/j.toxicon.2013.04.001

Ruiz-Sanchez E, Orchard I, Lange AB (2010) Effects of the cyclopeptide mycotoxin destruxin A on the malpighian tubules of *Rhodnius prolixus* (Stål). Toxicon 55(6):1162–1170. https://doi.org/10.1016/j.toxicon.2010.01.006

Sakuda S, Kimura M (2010) Toxins of microorganisms. In: Mander L, Liu HW (eds) Comprehensive natural products II, Chemistry and biology, vol 4, pp 411–455. https://doi.org/10.1016/B978-008045382-8.00102-7

Samuels RI, Reynolds SE, Charnley AK (1988) Calcium channel activation of insect muscle by destruxins, insecticidal compounds produced by the entomopathogenic fungus *Metarhizium anisopliae*. Comp Biochem Physiol C: Comp Pharmacol 90(2):403–412. https://doi.org/10.1016/0742-8413(88)90018-7

Sánchez-Gómez T, Harte SJ, Zamora P, Bareyre M, Díez JJ et al (2023) Nematicidal effect of *Beauveria* species and the mycotoxin beauvericin against pinewood nematode *Bursaphelenchus xylophilus*. Front For Glob Change 6:1229456. https://doi.org/10.3389/ffgc.2023.1229456

Sang M, Feng P, Chi LP, Zhang W (2024) The biosynthetic logic and enzymatic machinery of approved fungi-derived pharmaceuticals and agricultural biopesticides. Nat Prod Rep 41(4):565–603. https://doi.org/10.1039/d3np00040k

Santos FM, Latorre AO, Hueza IM, Sanches DS, Lippi LL et al (2011) Increased antitumor efficacy by the combined administration of swainsonine and cisplatin in vivo. Phytomedicine 18(12):1096–1101. https://doi.org/10.1016/j.phymed.2011.06.005

Semalulu SS, MacPherson JM, Schiefer HB, Khachatourians GG (1992) Pathogenicity of *Beauveria bassiana* in mice. J Veterin Med Ser B 39(1–10):81–90. https://doi.org/10.1111/j.1439-0450.1992.tb01141.x

Shima A, Fukushima K, Arai T, Terada H (1990) Dual inhibitory effects of the peptide antibiotics leucinostatins on oxidative phosphorylation in mitochondria. Cell Struct Funct 15(1):53–58. https://doi.org/10.1247/csf.15.53

Shimadate Y, Oshima S, Kasamatsu N, Yamamoto S, Taguchi A et al (2024) Synthesis of 8-epi-l-swainsonine, related C6 alkylated derivatives and their α-l-rhamnosidase inhibition. Tetrahedron Lett 139:155008. https://doi.org/10.1016/j.tetlet.2024.155008

Šimčíková D, Tůma P, Jegorov A Jr, Šimek P, Heneberg P (2020) Rapid methods for the separation of natural mixtures of beauverolides, cholesterol acyltransferase inhibitors, isolated from the fungus *Isaria fumosorosea*. J Sep Sci 43(5):962–969. https://doi.org/10.1002/jssc.201901084

Singh D, Kaur G (2014) The antileukaemic cell cycle regulatory activities of swainsonine purified from *Metarhizium anisopliae* fermentation broth. Nat Prod Res 28(22):2044–2047. https://doi.org/10.1080/14786419.2014.919287

Song Z, Cox RJ, Lazarus CM, Simpson TT (2004) Fusarin C biosynthesis in *Fusarium moniliforme* and *Fusarium venenatum*. Chembiochem 5(9):1196–1203. https://doi.org/10.1002/cbic.200400138

Song S, Liu N, Zhao Z, Njumbe Ediage E, Wu S et al (2014) Multiplex lateral flow immunoassay for mycotoxin determination. Anal Chem 86(10):4995–5001. https://pubs.acs.org/doi/abs/10.1021/ac500540z

Tang L, Zhang S, Zhai Y, Wu K, Wang P et al (2025) Endoplasmic reticulum stress regulates swainsonine-induced the autophagy in renal tubular epithelial cells through UPR signaling pathway. Sci Total Environ 981:179616. https://doi.org/10.1016/j.scitotenv.2025.179616

Thrane U, Adler A, Clasen PE, Galvano F, Langseth W et al (2004) Diversity in metabolite production by *Fusarium langsethiae*, *Fusarium poae*, and *Fusarium sporotrichioides*. Int J Food Microbiol 95(3):257–266. https://doi.org/10.1016/j.ijfoodmicro.2003.12.005

Van NTT, Lam DT, Minh PTH, Le VTT, Hoan BV et al (2021) Biological activities of extracts and beauvericin from *Cordyceps cateniannulata* CPA14V. Int J Agric Technol 17(6):2449–2460

Vey A, Matha V, Dumas C (2002) Effects of the peptide mycotoxin destruxin E on insect haemocytes and on dynamics and efficiency of the multicellular immune reaction. J Invertebr Pathol 80(3):177–187. https://doi.org/10.1016/s0022-2011(02)00104-0

Villegas-Mendoza JM, Mireles-Martínez M, Rodríguez Castillejos G, Santiago-Adame R, Rosas-García NM (2019) Evaluación de la toxicidad de un extracto metanólico de *Beauveria bassiana* en larvas de *Spodoptera exigua* y *Spodoptera frugiperda*. Southwest Entomol 44(3):715–722. https://doi.org/10.3958/059.044.0316

Wallace KB, Starkov AA (2000) Mitochondrial targets of drug toxicity. Annu Rev Pharmacol Toxicol 40:353–388. https://doi.org/10.1146/annurev.pharmtox.40.1.353

Wang Q, Xu L (2012) Beauvericin, a bioactive compound produced by fungi: a short review. Molecules 17(3):2367–2377. https://doi.org/10.3390/molecules17032367

Wang B, Kang Q, Lu Y, Bai L, Wang C (2012a) Unveiling the biosynthetic puzzle of destruxins in *Metarhizium* species. Proc Natl Acad Sci USA 109(4):1287–1292. https://doi.org/10.1073/pnas.1115983109

Wang L, Fang Q, Zhu J, Wang F, Rean Akhtar Z et al (2012b) Molecular cloning and functional study of calreticulin from a lepidopteran pest, *Pieris rapae*. Dev Comp Immunol 38(1):55–65. https://doi.org/10.1016/j.dci.2012.03.019

Wang G, Liu Z, Lin R, Li E, Mao Z et al (2016) Biosynthesis of antibiotic leucinostatins in bio-control fungus *Purpureocillium lilacinum* and their inhibition on *Phytophthora* revealed by genome mining. PLoS Pathog 12(7):e1005685. https://doi.org/10.1371/journal.ppat.1005685

Wang J, Weng Q, Yin F, Hu Q (2020) Interactions of destruxin A with silkworms' arginine tRNA synthetase and Lamin-C proteins. Toxins 12(2):137. https://doi.org/10.3390/toxins12020137

Wang J, Weng Q, Zhang K, Hu Q (2023) Binding proteins of destruxin A from *Metarhizium* against insect cell. BMC Microbiol 23:96. https://doi.org/10.1186/s12866-023-02843-8

Wang J, Hu H, Pang S, Yin X, Cao B et al (2024) Destruxin A inhibits the hemocytin-mediated hemolymph immunity of host insects to facilitate *Metarhizium* infection. Cell Rep 43(2):113686. https://doi.org/10.1016/j.celrep.2024.113686

Weber G, Leitner E (1994) Disruption of the cyclosporin synthetase gene of *Tolypocladium niveum*. Curr Genet 26(5–6):461–467. https://doi.org/10.1007/BF00309935

Wei G, Lai Y, Wang G, Chen H, Li F et al (2017) Insect pathogenic fungus interacts with the gut microbiota to accelerate mosquito mortality. Proc Natl Acad Sci USA 114(23):5994–5999. https://doi.org/10.1073/pnas.1703546114

Westwood GS, Huang SW, Keyhani NO (2006) Molecular and immunological characterization of allergens from the entomopathogenic fungus *Beauveria bassiana*. Clin Mol Allergy 4:1–11. https://doi.org/10.1186/1476-7961-4-12

Wu J, Yang B, Xu J, Cuthbertson AG, Ali S (2021) Characterization and toxicity of crude toxins produced by *Cordyceps fumosorosea* against *Bemisia tabaci* (Gennadius) and *Aphis craccivora* (Koch). Toxins 13(3):220. https://doi.org/10.3390/toxins13030220

Xu Y, Orozco R, Wijeratne EMK, Gunatilaka AAL, Stock SP et al (2008) Biosynthesis of the cyclooligomer depsipeptide beauvericin, a virulence factor of the entomopathogenic fungus *Beauveria bassiana*. Chem Biol 15(9):898–907. https://doi.org/10.1016/j.chembiol.2008.07.011

Xu Y, Orozco R, Kithsiri Wijeratne EM, Espinosa-Artiles P, Leslie Gunatilaka AAL et al (2009) Biosynthesis of the cyclooligomer depsipeptide bassianolide, an insecticidal virulence factor of *Beauveria bassiana*. Fungal Genet Biol 46(5):353–364. https://doi.org/10.1016/j.fgb.2009.03.001

Yang X, Feng P, Yin Y, Bushley K, Spatafora JW et al. (2018) Cyclosporine biosynthesis in *Tolypocladium inflatum* benefits fungal adaptation to the environment. MBio 9(5):10.1128/mbio.01211–18. 10.1128/mbio.01211-18

Yang Q, Wang W, Lin Y, Lin Y, Tang Z et al (2021) Characterization of a carboxyl methyltransferase in *Fusarium graminearum* provides insights into the biosynthesis of fusarin A. Org Biomol Chem 19(30):6638–6643. https://doi.org/10.1039/d1ob01010g
Yin Y, Chen B, Song S, Li B, Yang X et al (2020) Production of diverse beauveriolide analogs in closely related fungi: a rare case of fungal chemodiversity. mSphere 5(5):e00667. https://doi.org/10.1128/msphere.00667-20
Yin F, Xiao M, Berestetskiy A, Hu Q (2021) The *Metarhizium anisopliae* toxin, destruxin A, interacts with the SEC23A and TEME214 proteins of *Bombyx mori*. J Fungi 7(6):460. https://doi.org/10.3390/jof7060460
Yin X, Peng H, Weng Q, Hu Q, Wang J (2022) Interaction of destruxin A with three silkworm proteins: BmCRT, BmDPP3, and BmPDIA5. Molecules 27(22):7713. https://doi.org/10.3390/molecules27227713
You N, Liu W, Wang T, Ji R, Wang X et al (2012) Swainsonine inhibits growth and potentiates the cytotoxic effect of paclitaxel in hepatocellular carcinoma in vitro and in vivo. Oncol Rep 28(6): 2091–2100. https://doi.org/10.3892/or.2012.2035
Young JC, Zhou T, Yu H, Zhu H, Gong J (2007) Degradation of trichothecene mycotoxins by chicken intestinal microbes. Food Chem Toxicol 45(1):136–143. https://doi.org/10.1016/j.fct.2006.07.028
Yun CS, Motoyama T, Osada H (2015) Biosynthesis of the mycotoxin tenuazonic acid by a fungal NRPS–PKS hybrid enzyme. Nat Commun 6(1):8758. https://doi.org/10.1038/ncomms9758
Zabalgogeazcoa I, Alvarez A, Herrero N, de Aldana BRV (2018) Production of fumonisins by endophytic strains of *Tolypocladium cylindrosporum* and its relation to fungal virus infection. Mycotoxin Res 34:49–57. https://doi.org/10.1007/s12550-017-0298-6
Zhang T, Zhuo Y, Jia X, Liu J, Gao H et al (2013) Cloning and characterization of the gene cluster required for beauvericin biosynthesis in *Fusarium proliferatum*. Sci China Life Sci 56(7): 628–637. https://doi.org/10.1007/s11427-013-4505-1
Zhang J, Zhou X, Zhu J (2016) *Beauveria* attenuates asthma by inhibiting inflammatory response and inducing lymphocytic cell apoptosis. Oncotarget 7(46):74557. https://doi.org/10.18632/oncotarget.12958
Zhang X, Hu Q, Weng Q (2019) Secondary metabolites (SMs) of *Isaria cicadae* and *Isaria tenuipes*. RSC Adv 9(1):172–184. https://doi.org/10.1039/C8RA09039D
Zhang L, Fasoyin OE, Molnar I, Xu Y (2020) Secondary metabolites from hypocrealean entomopathogenic fungi: novel bioactive compounds. Nat Prod Rep 37(9):1181–1206. https://doi.org/10.1039/c9np00065h
Zhang S, Yin H, Zhang Y, Zhu Y, Zhu X et al (2024) Autophagic-lysosomal damage induced by swainsonine is protected by trehalose through activation of TFEB-regulated pathway in renal tubular epithelial cells. Chem Biol Interact 394:110990. https://doi.org/10.1016/j.cbi.2024.110990
Zhang L, Zhang G, Chen J, Luo K, Zhai W et al (2025) Sensitive lateral flow immunoassay based on integrated green pretreatment for detection of three mycotoxins in maize and wheat. Food Biosci 64:105908. https://doi.org/10.1016/j.fbio.2025.105908
Zheng X, Li Y, Zhang H, Apaliya MT, Zhang X et al (2018) Identification and toxicological analysis of products of patulin degradation by *Pichia caribbica*. Biol Control 123:127–136. https://doi.org/10.1016/j.biocontrol.2018.0.4.019
Zhu X, Yu F, Bojja RS, Zaleta-Rivera K, Du L (2006) Functional replacement of the ketosynthase domain of FUM1 for the biosynthesis of fumonisins, a group of fungal reduced polyketides. J Ind Microbiol Biotechnol 33(10):859–868. https://doi.org/10.1007/s10295-006-0137-9
Zhu F, Zhang B, Zhu L (2021) An up-converting phosphor technology-based lateral flow assay for rapid detection of major mycotoxins in feed: comparison with enzyme-linked immunosorbent assay and high-performance liquid chromatography-tandem mass spectrometry. PLoS One 16(4):e0250250. https://doi.org/10.1371/journal.pone.0250250
Zobel S, Boecker S, Kulke D, Heimbach D, Meyer V et al (2016) Reprogramming the biosynthesis of cyclodepsipeptide synthetases to obtain new enniatins and beauvericins. Chembiochem 17(4):283–287. https://doi.org/10.1002/cbic.201500649

Pharmaceutical Applications of Metabolites of Entomopathogenic Fungi

9

Abdelsamed I. Elshamy, Ahmed Sabt, Rania F. Ahmed, Rehab F. Taher, Mai M. Elghonemy, Asmaa S. Abd Elkarim, Fathalla Ayoob, Yusuke Kasai, and Hiroshi Imagawa

Abstract

Entomopathogenic fungi (EPF), which were once mainly recognized as insect killers in the natural environment, are increasingly gaining attention as potent friends in biotechnology, medicine, and sustainable agriculture. The amazing path of these fungal creatures is examined in this chapter, from their historical use in pest management to their recent discovery as abundant producers of pharmacologically powerful chemicals. We study the synthesis of a wide range of beneficial secondary metabolites by species like *Beauveria*, *Cordyceps*, *Isaria* (Family: Cordycipitaceae), and *Metarhizium* (Family: Clavicipitaceae), from the perspectives of both contemporary molecular research and ancient ethnomedicine. These include compounds with specific antibacterial, antifungal, antimalarial, and enzyme-inhibitory properties, as well as those with notable anticancer action, such as beauvericin, cycloaspeptides, and trichocaranes. EPF are a nexus where human creativity and nature's inventiveness meet, since they are both

A. I. Elshamy (✉)
Chemistry of Natural Compounds Department, National Research Centre, Dokki, Giza, Egypt

Faculty of Pharmaceutical Sciences, Tokushima Bunri University, Yamashiro-cho, Tokushima, Japan
e-mail: ai.el-shamy@nrc.sci.eg

A. Sabt · R. F. Ahmed · R. F. Taher · M. M. Elghonemy
Chemistry of Natural Compounds Department, National Research Centre, Dokki, Giza, Egypt

A. S. Abd Elkarim · F. Ayoob
Chemistry of Tanning Materials and Leather Technology Department, National Research Centre, Dokki, Giza, Egypt

Y. Kasai · H. Imagawa
Faculty of Pharmaceutical Sciences, Tokushima Bunri University, Yamashiro-cho, Tokushima, Japan

S. K. Deshmukh et al. (eds.), *Frontiers in Entomopathogenic Fungi*,
https://doi.org/10.1007/978-981-95-6529-0_9

environmentally benign biocontrol agents and a source of therapeutic potential. This chapter emphasizes their increasing importance in tackling urgent issues in agriculture and global health and imagines a time when fungal metabolites will drive interdisciplinary discoveries that will bring ecosystems, industry, and medicine together in previously unheard-of ways.

Keywords

EPF · Ethnomedicine · Anticancer · EFP-metabolites · Cycloaspeptides

1 Introduction

Entomopathogenic fungi (EPF) are a group of fungi that parasitize and infect insects, acting as natural biological control agents in ecosystems. These fungi, which belong to several taxonomic families, including Cordycipitaceae and Clavicipitaceae, are widely distributed in nature and play significant roles in regulating insect populations (Zhang et al. 2020; Elshamy et al. 2024; Rajput et al. 2024). The parasitism of insects by these fungi is facilitated through a combination of virulence factors, such as the production of enzymes that degrade the insect's cuticle, as well as the release of toxic secondary metabolites (Zhang et al. 2020; Rajput et al. 2024).

The potential applications of EPF as biocontrol agents are vast, particularly in agricultural settings. They offer an environmentally friendly alternative to chemical pesticides, which can have detrimental effects on non-target organisms, human health, and the environment. EPF have been successfully used in the control of pest insects such as locusts, termites, and aphids and are now integrated into integrated pest management (IPM) programs (Rajput et al. 2024). Furthermore, the study of EPF has provided significant insights into fungal biology, host-pathogen interactions, and evolutionary processes. Understanding the molecular mechanisms of pathogenesis, including the identification of virulence factors, host immune responses, and environmental factors influencing fungal growth, is critical for enhancing the effectiveness of these fungi as biocontrol agents (Bihal et al. 2023).

The pharmaceutical potential of secondary metabolites derived from EPF is an area of growing interest (Zhang et al. 2020). These compounds exhibit a variety of bioactivities, including antimicrobial, anticancer, anti-inflammatory, and immunomodulatory effects. The complexity and diversity of these metabolites, which include alkaloids, polyketides, and peptides, suggest that they could serve as valuable sources of novel drugs for the treatment of a variety of human diseases. Furthermore, the growing concern over antibiotic resistance and the need for new pharmacological agents has underscored the importance of exploring natural product sources like EPF for new and effective treatments (Hama et al. 2019; Zhang et al. 2020; Elshamy et al. 2024). Research into the isolation, characterization, and mechanism of action of these secondary metabolites is crucial for harnessing their full therapeutic potential. As the pharmaceutical industry faces increasing challenges in drug discovery, EPF represent an untapped reservoir of bioactive compounds that

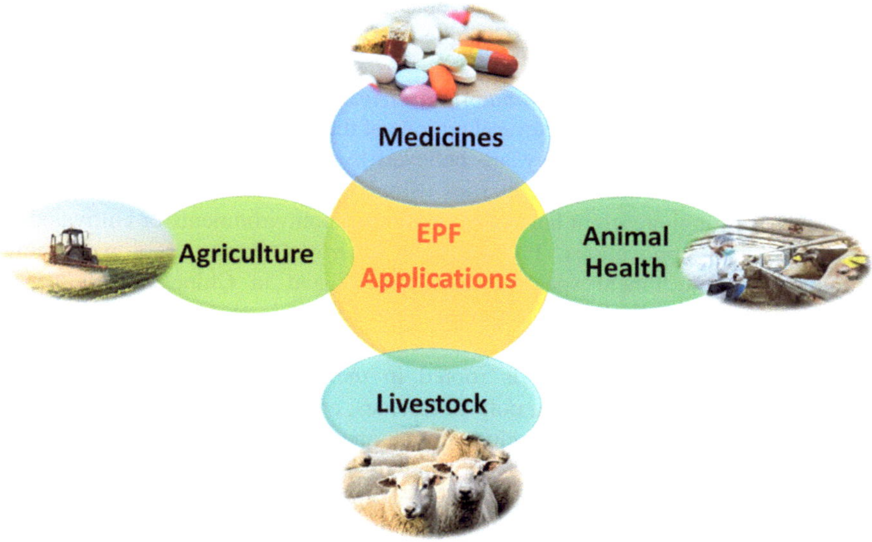

Fig. 9.1 A systematic diagram of the different applications of EPF

could lead to the development of innovative therapeutic agents (Singh and Kumar 2023). This exploration of secondary metabolites from EPF not only enhances our understanding of fungal biology but also paves the way for the future application of these compounds in pharmaceutical development, providing new solutions for combating diseases and enhancing human health (Paschapur et al. 2021).

EPF and their metabolites hold substantial promise across multiple fields, from sustainable pest management in agriculture to enhancing animal health and providing new avenues in medicine (Fig. 9.1). Their versatility, ecological safety, and efficacy position them as valuable tools in the development of alternative, eco-friendly solutions to current challenges in pest control and disease management.

2 Traditional Medicinal Uses

EPF are a diverse group of microorganisms that primarily act as pathogens to insects and other arthropods. These fungi have garnered significant interest due to their potential therapeutic properties, especially in traditional medicine. Historically, various cultures have recognized the medicinal value of these fungi, which are often employed in the treatment of a wide range of ailments, including infections, inflammation, and general health maintenance (Chandler 2017; Elshamy et al. 2024).

Some EPF have demonstrated potential as sources of bioactive compounds with therapeutic effects. Fungal metabolites exhibit a variety of pharmacological activities, including antibacterial, antifungal, and anticancer properties. Specific EPF-like *Beauveria* and *Metarhizium* species have been investigated for their potential in drug discovery, offering novel approaches to treating infections, inflammatory diseases, and even cancer (Chandler 2017; Singh and Kumar 2023; Elshamy et al. 2024).

EPF, such as *Beauveria bassiana*, *Metarhizium anisopliae*, and *Cordyceps* spp., have been traditionally used in several regions of the world, particularly in East Asia and parts of Africa, where they are believed to possess immune-boosting, anti-inflammatory, and anticancer properties (Singh and Kumar 2023; Elshamy et al. 2024). These fungi are often harvested from natural environments or cultivated for their medicinal applications. For example, *Cordyceps sinensis*, a well-known entomopathogenic fungal species, has been utilized in Chinese medicine for centuries to enhance vitality, improve lung function, and promote longevity (Paterson 2008; Chandler 2017).

The medicinal use of EPF is rooted in their ability to produce bioactive compounds, including alkaloids, polysaccharides, and peptides, that exhibit various pharmacological effects. These include immunomodulatory, antioxidant, antimicrobial, and anticancer activities. While modern scientific research is increasingly validating these traditional uses, the full scope of their therapeutic potential remains a subject of ongoing investigation (Wadhwa et al. 2024). As interest in alternative and complementary therapies grows, the exploration of EPF for medicinal purposes continues to expand, offering promising avenues for novel drug discovery and natural health solutions (Paterson 2008; Singh and Kumar 2023; Elshamy et al. 2024).

3 Pharmaceutical Applications

EPF produce a wide range of bioactive metabolites that have garnered significant interest due to their potential applications in agriculture, medicine, and biotechnology. The bioactive compounds isolated from EPF encompass various classes of natural products, including polyketides, alkaloids, terpenoids, and peptides. These metabolites often exhibit antimicrobial, anticancer, antiviral, and immunosuppressive activities, making them valuable candidates for the development of novel therapeutic agents and pest control strategies (Molnár et al. 2010; Paschapur et al. 2021).

The complexity and diversity of these metabolites arise from the intricate biosynthetic pathways employed by the fungi in response to environmental stimuli, such as insect host interaction, nutrient availability, and stress conditions. Research into the molecular mechanisms of these bioactive metabolites has led to the discovery of novel compounds with potential commercial value (Pedrini 2022; Hyde et al. 2024).

The therapeutic promise of EPF-derived metabolites is expanding, with studies showing their efficacy in combating human pathogens and modulating immune responses. Moreover, their low environmental impact and specificity to target organisms highlight the growing importance of these metabolites in sustainable agricultural practices. However, further research is required to fully understand the mechanisms of action, safety profiles, and potential side effects of these bioactive compounds before they can be widely implemented in commercial applications (Molnár et al. 2010; Wadhwa et al. 2024). The bioactive metabolites derived from EPF represent a promising source of novel compounds with diverse biological activities. Continued exploration of these fungi and their secondary metabolites will likely yield new insights into their ecological roles and open new avenues for their application in various fields, from pest control to pharmaceutical development (Alves et al. 2025).

Metabolites isolated from EPF exhibit a range of bioactivities that make them valuable in biotechnology and pest control (Paschapur et al. 2021). These metabolites include secondary metabolites such as peptides, polyketides, and alkaloids, which often display antimicrobial, insecticidal, and antifungal properties. Some metabolites, like destruxins produced by *Metarhizium* spp., have potent insecticidal effects, targeting specific insect pests by disrupting their cellular processes (Chandler 2017). Other metabolites may possess antitumor, antiviral, or antioxidant activities, highlighting their potential for use in medicine and agriculture, as summarized in Figs. 9.2, 9.3, 9.4, 9.5, and 9.6.

3.1 Some Cytotoxic Active Metabolites

Cycloaspeptides F (**1**) and G (**2**) (Fig. 9.2), two cyclic pentapeptides isolated from *Isaria farinosa* XJC04-CT-303, have stepped into the spotlight for their anticancer potential. Against MCF7 breast cancer cells, these compounds posted GI_{50} values of 18.7 μM and 15.2 μM, putting them nearly on par with the established chemotherapeutic agent 5-fluorouracil (5-Fu), which clocked in at 15.0 μM. Their activity extended to HeLa cervical cancer cells as well, though with milder potency, GI_{50} values of 49.8 μM and 30.4 μM, in contrast to 5-Fu's sharper 10.0 μM effect. Still, these results mark cycloaspeptides F (**1**) and G (**2**) as promising leads in the search for novel anticancer agents (Zhang et al. 2009).

The 8-hydroxyherbarin (**3**) (Fig. 9.2) that was isolated from *Cordyceps bifusispora* demonstrated substantial cytotoxicity toward all cell lines, including CCRF-CEM, CEM-ADR5000, MDA-MB-231-pcDNA3, HCT116 (p53+/+), and HEK293 cells, with the respective IC_{50} values of 1.35 μM, 3.01 μM, 6.05 μM, 13.68 μM (HCT116), and 14.52 μM (HEK293) (Elshamy et al. 2023). Cardinalisamides B (4) and C (5) (Fig. 9.2), two intriguing compounds derived from the fungus *Blackwellomyces roseostromatus* BCC56290, have shown remarkable pancytotoxic effects across a broad panel of human cancer cell lines, including lung (A549), breast (MCF-7), ovarian (SKOV-3), prostate (PC-3), epidermoid (A431), and endocervical adenocarcinoma (KB3.1), as well as mouse fibroblasts (L929). Impressively, these compounds exhibited IC_{50} values ranging from just 2.2 to 13.9 μM (Phutthacharoen et al. 2024). Ascherxanthone A (**6**), a symmetrical tetrahydroxanthone dimer, was isolated from *Aschersonia* sp. BCC 8401, an entomopathogenic fungus. This compound exhibited cytotoxic properties against *Plasmodium falciparum* K1 with an IC_{50} value of 0.20 μg/mL (Isaka et al. 2005).

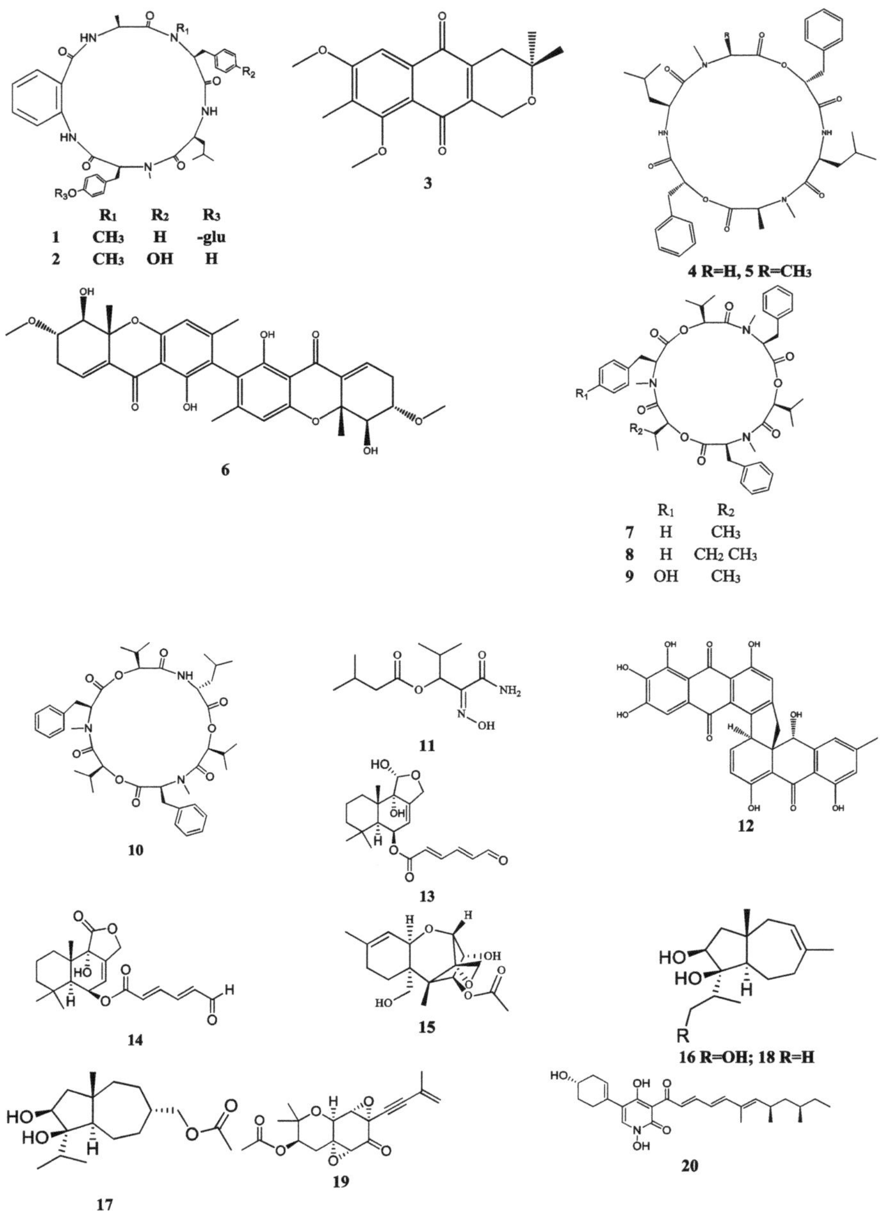

Fig. 9.2 Some examples of the isolated cytotoxic active metabolites

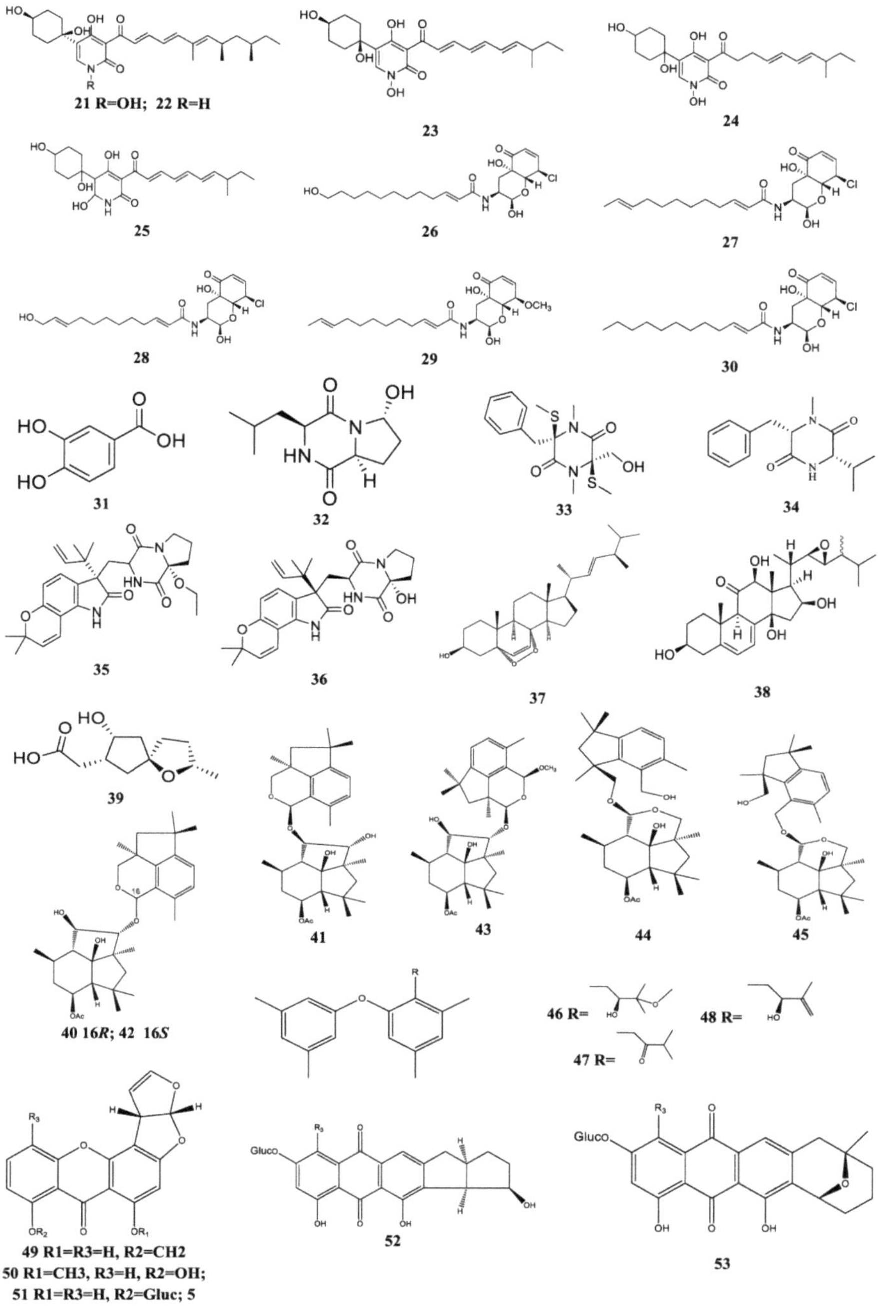

Fig. 9.2 (continued)

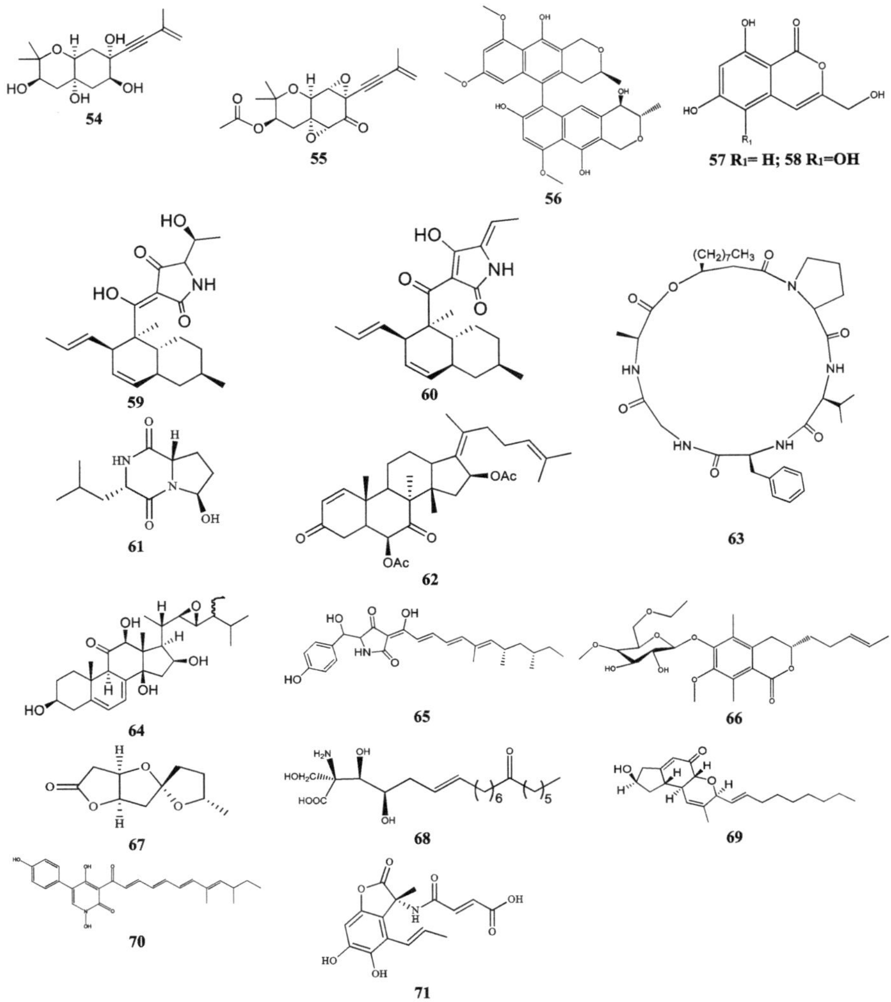

Fig. 9.3 Some examples of the isolated antimicrobial active metabolites

Beauvericin (**7**) (Fig. 9.2), a cyclopeptide derived from various EPF, has been shown to exhibit potent inhibitory effects on the proliferation of PANC-1 cancer cells, with an IC50 value of 4.8 μM. Furthermore, this compound has been reported to suppress PANC-1 cell migration by upregulating E-cadherin gene expression while downregulating N-cadherin and Snail gene expression in a dose-dependent manner, within the concentration range of 0.1–1 μM (Tao et al. 2015; Kuca et al. 2018; Yahagi et al. 2020). This compound demonstrated cytotoxic effects by inhibiting the growth of both normal and cancerous HepG2 cell lines, with an IC_{50} of 2.40 μM for the HepG2 cells and a significantly higher IC50 for multi-drug-

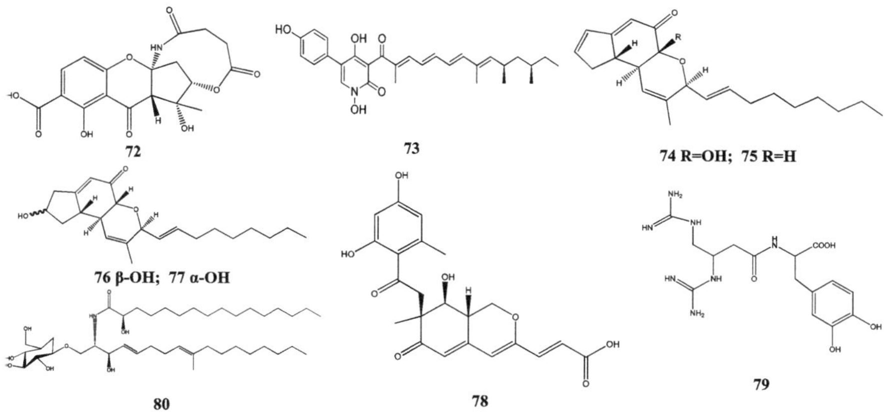

Fig. 9.4 Some examples of the isolated PTP1B-inhibitory active metabolites, 79 and 80, with different bioactivities

resistant (MDR)-HepG2 cells, which was 25 times greater than the IC_{50} observed with doxorubicin, the reference drug (Wang and Xu 2012). Among all compounds isolated from *Paecilomyces cicadae*, this was the only one to show substantial growth inhibition in human lung (A549) and human leukemia (HL-60) cell lines, with IC_{50} values of 5.995 μM and 5.800 μM, respectively (Wang et al. 2014). Furthermore, this compound was reported to effectively suppress the proliferation of A549 non-small-cell lung cancer cells through the activation of the mitogen-activated protein kinase (MAPK) signaling pathway (Lu et al. 2016).

Beauvericin A (**8**), J (**9**), and E (**10**) (Fig. 9.2), cyclopeptides derived from the wild co-culture of *Paecilomyces cicadae* fruit bodies and insect specimens (ID: 2012081827), have been shown to possess significant anticancer properties. These three compounds demonstrated strong inhibitory effects against HepG2 and HepG2/ADM cell lines, with IC_{50} values ranging from 2.40 to 14.48 μM, and were found to be 25 times more sensitive than the reference drug, doxorubicin (Wang et al. 2014). Additionally, aspergilliamide (**11**) (Fig. 9.2), identified in the primordia of *Paecilomyces cicadae*, was tested for brine shrimp toxicity, revealing a lethal concentration (LC_{50}) of 71.09 nM (Zhang et al. 2018). Torrubiellin B (**12**) (Fig. 9.2), a compound extracted from *Torrubiella* sp. BCC 28517, demonstrated potent cytotoxic activity against cancer cell lines KB, NCI-H187, and MCF-7, boasting IC_{50} values of 0.48, 0.20, and 3.20 mg/mL^{-1}, respectively, outperforming even doxorubicin hydrochloride in certain assays (IC_{50}: 0.30, 0.045, and 6.47 mg/mL^{-1}) (Isaka et al. 2012). Asperflavinoid C (**13**) and ustusolate E (**14**) (Fig. 9.2), extracted from the co-culture of *Beauveria felina* KMM 4639 and *Aspergillus carneus* KMM 4638, demonstrated strong inhibition of human breast cancer MCF-7 cell viability with an IC_{50} of 10 μM. They induced caspase-dependent apoptosis and caused cell cycle arrest in the G2/M phase of MCF-7 cells (Zhuravleva et al. 2022). Additionally, 4-β-acetoxyscirpendiol (**15**) (Fig. 9.2), a trichothecene compound sourced from the fruit body of *Paecilomyces tenuipes* and *Isaria*

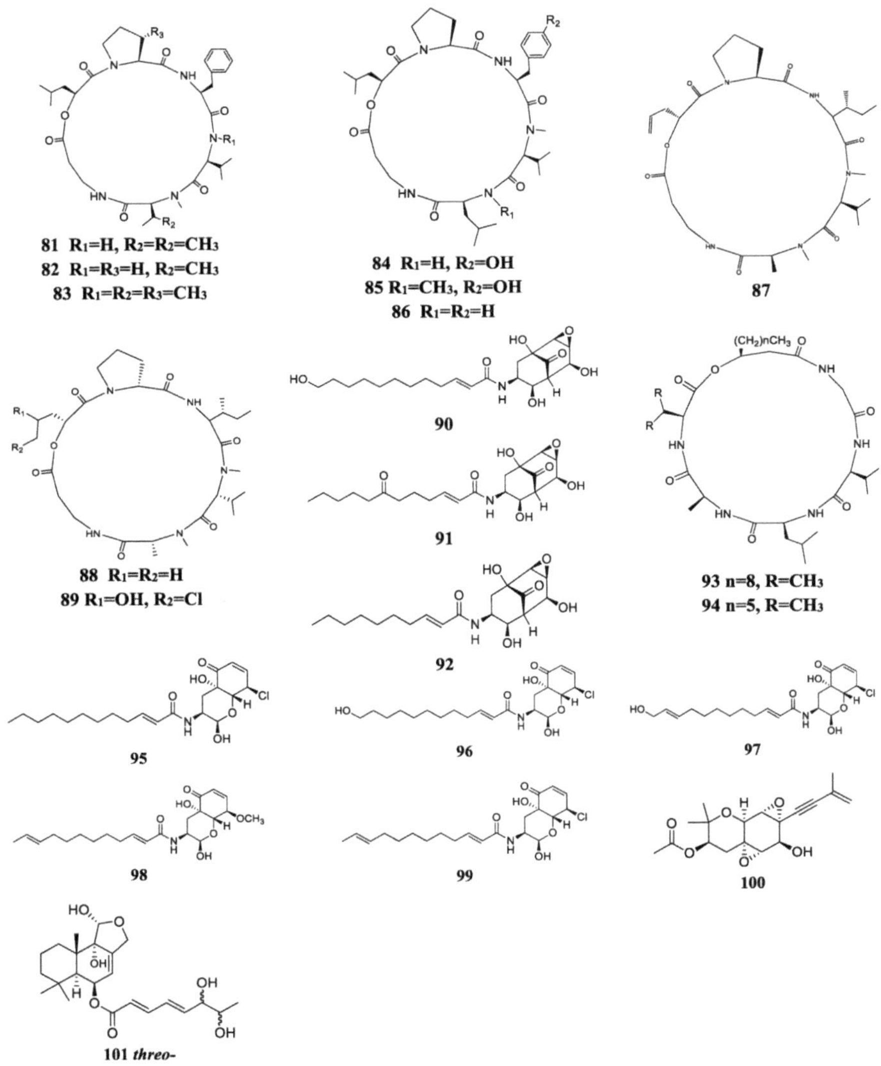

Fig. 9.5 Some examples of the isolated metabolites with different bioactivities

japonica, showed notable apoptosis-inducing effects across various human cancer cell lines, including leukemia (HL-60) cells (Nam et al. 2001; Oh et al. 2001).

The three compounds—trichocaranes E (**16**), F (**17**), and CAF-603 (**18**) (Fig. 9.2), isolated from *Isaria fumosorose* ACCC 37775—have demonstrated strong inhibitory effects on a range of cancer cell lines, including MDA, MCF-7, SKOV-3, HeLa, A549, and HepG2. Their IC_{50} values span from 0.1 to 6.0 μg/ml. Specifically, these trichocaranes exhibited notable cytotoxicity with IC50 values against these cell lines of (0.1, 2.5, 1.0, 2.3, 1.4, and 1.9 μg/mL), (0.89, 4.38, 1.46,

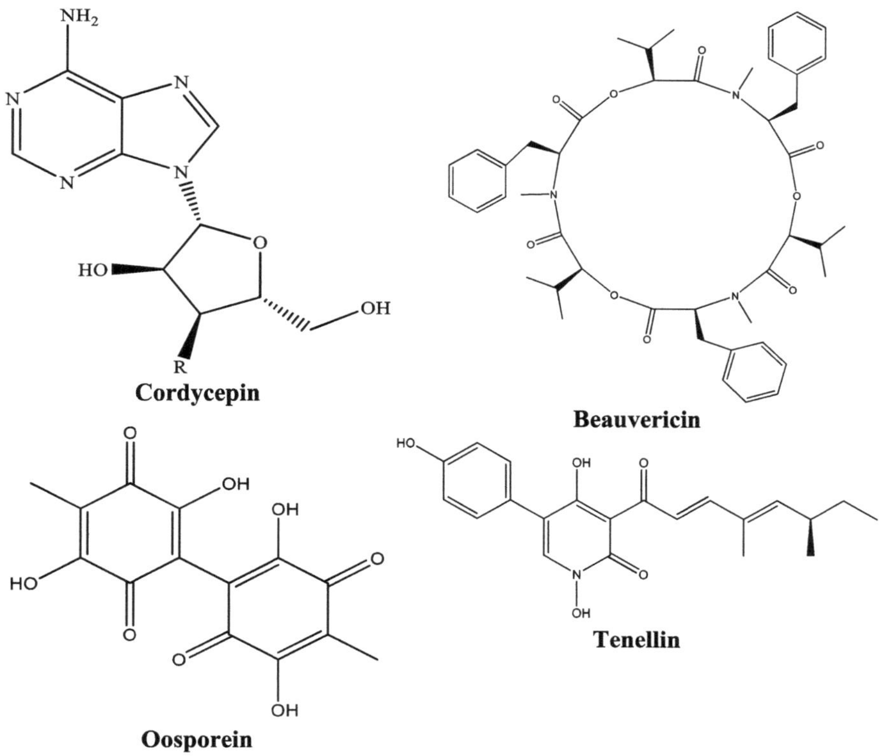

Fig. 9.6 Some EPF-metabolites with recent preclinical studies

4.57, 1.66, and 3.66 μg/mL), and (1.12, 6.00, 1.57, 3.35, 1.32, and 3.44 μg/mL), respectively (Zhang et al. 2018). Oxirapentyn A (**19**) (Fig. 9.2), extracted from a marine sediment-associated *Isaria feline* (DC.) Fr, displayed moderate inhibitory effects against the T-47D, SK-Mel-5, and SK-Mel-28 cancer cell lines, with IC_{50} values of 25, 19, and 17 μM, respectively. At a concentration of 400 μM, Oxirapentyn A (**19**) exhibited cytotoxicity of 34%, 16%, and 35% against T-47D, SK-Mel-5, and SK-Mel-28 cells, respectively, when compared to cisplatin (positive control), which had IC_{50} values of 360, 252, and 140 μM for these same cells (Smetanina et al. 2012). The two compounds, militarinone E (**20**) and militarinone A (**21**) (Fig. 9.2), extracted from cultures of *Cordyceps*-infecting fungus *Isaria farinose*, demonstrated significant cytotoxic activity against A549 cell lines, with IC_{50} values of 1.59 and 1.75 μM, respectively. These values were notably lower than that of the reference drug cisplatin, which had an IC_{50} of 4.17 μM (Ma et al. 2011).

The compound (+)-*N*-deoxymilitarinone A (**22**) (Fig. 9.2), derived from *Isaria farinosa* RCEF 0097, has been shown to stimulate neurite outgrowth in PC12 cells at concentrations of 33 and 100 μM, while exhibiting cytotoxic effects on human neuroblastoma IMR-32 cells at 100 μM (Cheng et al. 2004). Additionally, the two polysaccharides isolated from *Isaria farinosa* B05, WIPS1, and WEPS1, with molecular weights of 42 kDa and 208 kDa, respectively, demonstrated notable anticancer and antioxidant activities in the in vivo model using mice (Jiang et al. 2008). The isolated pyridone alkaloids JBIR-130 (**23**), JBIR-131 (**24**), and JBIR-132 (**25**) (Fig. 9.2), from *Isaria* sp. (NBRC 104353), have been demonstrated to be effective inhibitors of the proliferation of human lung cancer cells (A549) with corresponding IC_{50} values of 87, 53, and 4125 mM following a 48-hour incubation period (Hosoya et al. 2013). It has been reported that isariotins G–J (**26–29**) (Fig. 9.2), which were isolated from *Isaria tenuipes* BCC 21283 and 231,112, are cytotoxic to KB cells, with IC_{50} values ranging from 2.98 to 8.26 μg/ml. Nevertheless, the above compounds suppressed the proliferation of MCF-7, NCI-H187, and Vero cell lines with varying IC50 values of 4.37–18.56 μg/ml, 0.64–44.86 μg/ml, and 1.27–26.97 μg/ml, independently (Bunyapaiboonsri et al. 2011). Isariotin F (**30**) (Fig. 9.2), isolated from *Isaria tenuipes* BCC 12625, has demonstrated cytotoxic activity against several human cancer cell lines. These include African green monkey kidney fibroblast (Vero), breast cancer (BC), small-cell lung cancer (NCI-H187), and oral epidermoid carcinoma (KB), with IC_{50} values of 15.8, 2.4, 1.6, and 2.9 μM, respectively (Bunyapaiboonsri et al. 2009).

Researchers found that protocatechuic acid (**31**) (Fig. 9.2), isolated from *Paecilomyces cicadae* strain F0004, effectively suppressed the growth of MCF-7 and Jurkat cancer cells, with IC_{50} values of 5.97 mM and 3.15 mM, respectively (Dung et al. 2021). Meanwhile, cordysinin A (**32**) (Fig. 9.2), a diketopiperazine compound also derived from *Paecilomyces cicadae,* demonstrated notable antiproliferative effects on human glioma cell lines U87-MG and U25 (Lu et al. 2017; Ye et al. 2017). The two macrolides, (4*R*,10*R*)-4-hydroxy-10-methyloxecane-2,8-dione (**33**) and (3*S*)-6-Benzyl-3-isopropyl-1-methylpiperazine-2,5-dione (**34**) (Fig. 9.2), both derived from *Isaria tenuipes*, demonstrated moderate anticancer activity in MTT assays. They inhibited the growth of prostate cancer cell lines 22RV1 and DU-145, with respective inhibition rates of 37.8%, 38.6%, 32.5%, and 40.6% (Yang et al. 2019). The compound, 17-O-ethylnotoamide M (**35**) (Fig. 9.2), obtained from the co-culture of *Aspergillus sulphureus* and *Isaria feline*, has been shown to inhibit the growth of human prostate cancer cells (22Rv1). At a concentration of 10 μM, it reduced colony formation by 25%. Additionally, notoamide M (**36**) (Fig. 9.2) demonstrated a more significant effect, decreasing colony formation by 55% (Afiyatullov et al. 2018). Ergosterol peroxide (**37**) (Fig. 9.2), isolated from *Paecilomyces cicadae*, demonstrated notable cytotoxicity against several cancer cell lines—P-388, KB, A549, and HT-29—with ED_{50} values of 0.4, 2.1, 2.7, and 1.4 μg/mL, respectively (He et al. 2018). A highly oxygen-rich sterol known as JBIR-14 (**38**) (Fig. 9.2), derived from *Isaria* sp. NBRC 104353, has been found to counteract growth suppression caused by dynAP in Mad2-deficient mutant yeast, showing activity at a 25 μM concentration (Ueda et al. 2010). Meanwhile, researchers isolated a compound called (2*S*,5*S*,7*S*)-3a-hydroxyl-exogonic acid (**39**) (Fig. 9.2) from *Isaria cateniannulata*, which demonstrated mild cytotoxicity against HeLa cells, with an IC50 value of 80.5 μg/mL (Yang et al. 2012).

In a fascinating breakthrough, six unearthed botryane meroterpenoid ethers—dubbed hypocriols A-F (**40–45**) (Fig. 9.2)—emerged from the culture extract of *Trichoderma* sp. EC1–35 (formerly *Hypocrea*), revealing potent cytotoxic effects against a lineup of formidable human cancer cell lines: HeLa, HCT116 (colorectal carcinoma), A549, and MCF-7 (Ren et al., 2016). The reported compounds produced from the co-cultivation of *Aspergillus sulphureus* and *Isaria feline*, diorcinol J (**46**), B (**47**), and D (**48**) (Fig. 9.2), were reported to have cytotoxic activity against Ehrlich carcinoma cells with ED_{50} values of 37.6, 95.1, and 98.1 μM (Zhuravleva et al. 2016). In a study of the scale insect fungi *Aschersonia coffeae* BCC 2871273 and *Aschersonia marginata* BCC 2872176, five purified compounds—xanthones (**49–51**) and anthraquinones **52–53** (Fig. 9.2)—stood out for their cytotoxic punch against the NCI-H187 cancer cell line, with IC_{50} values of 1.17, 12.93, 2.86, 8.06, and 5.12 mg/mL, respectively (for comparison, ellipticine clocked in at 1.31 mg/mL). However, xanthones (**49–51**) showed a double edge, also exhibiting toxicity toward Vero cells (African green monkey kidney fibroblasts), with IC_{50} values of 0.34, 1.40, and 0.33 mg/mL—slightly more potent than ellipticine's 1.29 mg/mL (Kornsakulkarn et al. 2012).

3.2 Some Antimicrobial Active Metabolites

In the study of Smetanina et al. (2012), oxirapentyn D (**54**) and oxirapentyn A (**55**) (Fig. 9.3), isolated from *Isaria felina* (DC.) Fr., were reported to exhibit weak inhibitory effects against *Staphylococcus aureus* and *Bacillus subtilis* at minimum inhibitory concentrations (MICs) of 140 and 150 μM, respectively, compared with nitrofungin as a reference antimicrobial drug with MICs for the two strains at 116 μM. Torrubiellin B (**12**) (Fig. 9.2), isolated from *Torrubiella* sp. BCC 28517, demonstrated remarkable bioactivity, exhibiting potent antifungal effects against *Candida albicans* with an IC_{50} of 1.66 mg/mL, in comparison to the reference drug amphotericin B (IC_{50} = 0.072 mg/mL). Additionally, it showed strong antibacterial activity against *Bacillus cereus*, with an IC_{50} of 6.25 mg/mL, relative to vancomycin hydrochloride (IC_{50} = 4.0 mg/mL) (Isaka et al. 2012).

Beauvericin (**7**) (Fig. 9.2), which was isolated from *Cordyceps tenuipes* NBRC 111630, was shown to have antibacterial activity against *Staphylococcus aureus* and *Bacillus subtilis*, with MICs of 6.25 and 12.5 μM, respectively. Furthermore, the documented docking results showed that beauvericin (**7**) had strong binding affinities against the HK and MurE proteins, with $\Delta G = -8.585$ kcal/mol and -8.021 kcal/mol, respectively (Yoneyama et al. 2022). In a recent study by Phutthacharoen et al. (2024), the compounds ES-242-9 (**56**) (Fig. 9.3) and cardinalisamide C (**5**) (Fig. 9.2), extracted from *Blackwellomyces roseostromatus* BCC56290, showcased noteworthy nematocidal activity, causing corrected mortality rates of 51.1% and 53.6%, respectively, against *Caenorhabditis elegans* at a concentration of 100 μg/mL.

Among the compounds isolated from *Conoideocrella tenuis*, 6,8-dihydroxy-3-hydroxymethylisocoumarin (**57**) (Fig. 9.3) emerged as a moderate inhibitor against both *Herpes simplex* virus-1 and *Mycobacterium tuberculosis* H37Ra. Meanwhile, crude extracts from strain BCC 44534 showed notable biological activity—demonstrating antimalarial effects against *Plasmodium falciparum* K1 (IC_{50} = 9.53 mg/mL) and cytotoxicity toward the NCI-H187 human small-cell lung cancer line (IC_{50} = 4.60 mg/mL). Interestingly, these bioactivities did not persist in the purified compounds, with one exception: isocoumarin (**58**) (Fig. 9.3), which displayed weak cytotoxicity against both NCI-H187 (IC_{50} = 27.7 mg/mL) and Vero cells (IC_{50} = 9.6 mg/mL), when compared to the positive control ellipticine (IC_{50} = 4.0 and 1.9 mg/mL, respectively) (Saepua et al. 2018). The two compounds isolated from *Isaria farinose*, paecilosetins A (**59**) and C (**60**) (Fig. 9.3), were reported to show bacterial growth inhibitory actions against the methicillin-sensitive and methicillin-resistant strains of *S. aureus* (MSSA and MRSA) at MIC 1–2 μg/mL (Brel et al. 2020). The isolated peptide, isarfelin A (**61**), from *Isaria feline*, showed inhibition against *Rhizoctonia solani* at an IC_{50} of 3.1 $\mu g/mL^{-1}$, without any inhibitory actions toward other bacterial strains, including *B. subtilis*, *E. coli,* and *S. aureus* (Guo et al. 2005).

The compound that emerged as a promising antimicrobial agent, torrubiellin B (**12**) (Fig. 9.2), was isolated from the cultures of *Torrubiella* sp. BCC 28517. Notably, it demonstrated antimalarial activity against *Plasmodium falciparum*, with an IC_{50} value of 0.33 mg/mL^{-1}—while not as potent as the reference drug mefloquine (IC_{50} = 0.013 mg/mL^{-1}), it still showed significant bioactivity worthy of further exploration (Isaka et al. 2012). Helvolic acid (**62**) (Fig. 9.3), isolated from the cultures of *Ophiocordyceps coccidiicola* NBRC 100683, was documented to show significant antibacterial effects, broadening its therapeutic promise against pathogens like those responsible for human African trypanosomiasis (Ganaha et al. 2016). In a similar vein, another compound, 5α,8α-epidioxy-(22*E*,24*R*)-ergosta-6,22-dien-3β-ol (**63**) (Fig. 9.3), isolated from *Aschersonia luteola* BCC 31749, demonstrated low cytotoxicity: IC_{50} values of 5.6 mM for KB cells and 15 mM for MCF-7 cells, once again paling in comparison to doxorubicin IC_{50} of 4.9 mM (Jinming et al. 2001).

Among the intriguing metabolites uncovered from fungi, JBIR-14 (**64**) (Fig. 9.3), sourced from *Isaria* sp. NBRC 104353, stood out—albeit modestly—for its biological activity. It showed only weak cytotoxicity against KB cells, with an IC_{50} of 10 mM (for comparison, doxorubicin hydrochloride, used as the positive control, exhibited a much stronger IC_{50} of 0.27 mM) (Isaka et al. 2011; Isaka et al. 2012; Ueda et al. 2010). Militarinone B (**65**) (Fig. 9.3), isolated from the cultures of *Cordyceps*-colonizing fungus *I. farinose*, showed inhibitory actions against *S. aureus* (CGMCC 1.2465) and *Streptococcus pneumoniae* (CGMCC 1.1692), at IC_{50} of 2.11 and 2.60 μM, respectively, compared to the ampicillin reference drug with IC_{50} values of 0.06 and 1.08 μM, respectively. Moreover, it was found to have weak inhibitory actions against *C. albicans* CGMCC 2.2086 (IC_{50}: 43.5) (Ma et al. 2011). From the solid cultures of a distinct *Metarhizium robertsii* strain (No. DTH12–10), researchers isolated acyl isocoumarin glycosides (**66**) (Fig. 9.3), which demonstrated potent antibacterial action against *Pseudomonas aeruginosa*—not by brute force, but by cunningly disrupting its quorum sensing machinery, thereby curbing biofilm development and secretory activity (Tian et al. 2016).

Cephalosporolide F (**67**) (Fig. 9.3), isolated from *Isaria tenuipes*, was reported to show moderate inhibition against the nematode *Panagrellus redivivus* at a concentration of 2.5 mg mL^{-1} with a mortality ratio of 79.0% (Yang et al. 2019). The amino acid myriocin (**68**) (Fig. 9.3), isolated from *Paecilomyces cicadae* (Zhang et al. 2018) and *Isaria sinclairii* ATCC24400 (Fujita et al. 1994), was reported to have significant antifungal activity against *C. albicans*. However, some reports have shown that this compound showed high toxicity with LD_{50} of 5–10 and 2–5 mg kg^{-1} in acute toxicity studies in mice and rats, respectively (Kluepfel et al. 1972). Du et al. (2014) found that desmethylisaridin C1 (**69**) (Fig. 9.3) derived from *Isaria felina* EN-135 exerted strong antibacterial activity against *E. coli* with an MIC value of 8 μg/mL. Farinosone B (**70**) (Fig. 9.3) from the culture of *Paecilomyces farinose* RCEF 0101 was reported to have the ability to inhibit *B. subtilis* and *S. aureus* and moderate toxicity against *Artemia salina* brine shrimp larvae (Cheng et al. 2004; Liu et al. 2015). Fumimycin (**71**) (Fig. 9.3), a product of the *Paecilomyces cicadae*, was reported to have an antibacterial effect against the growth of *S. aureus* at an IC_{50} of 4.1 μM via inhibition of peptide deformylase (Kwon et al. 2007; Lu et al. 2017).

3.3 Some Protein Phosphatase 1B Inhibition (PTP1B) Inhibitory Active Metabolites

Unearthed from *Cordyceps gracilioides*, Cryptosporioptide A (**72**) (Fig. 9.4), a uniquely ring-contracted xanthone featuring an unusual *N*-succinyl aminal bridge, revealed its biochemical prowess by inhibiting the protein tyrosine phosphatases PTP1B, SHP2, CDC25B, and SHP1 in vitro, with activity measured in the low millimolar range (5–8 mM) (Wei et al. 2015). The reported data showed that fumosorinone (**73**) (Fig. 9.4), isolated from *Isaria fumosorosea,* inhibited protein tyrosine phosphatase 1B (PTP1B) at an IC_{50} value of 14.04 μM (Chen et al. 2018; Liu et al. 2015). Beauvericin (**7**) (Fig. 9.2), isolated from several *Isaria* species, was found to have inhibitory action against PTP1B at an IC_{50} of 0.59 μM (Zhang et al. 2018). Metabolites of *Isaria tenuipes* RCEF 37776 were chemically characterized by Chen et al. to afford the isolation of penostatin A (**74**), B (**75**), C (**76**), and J (**77**) (Fig. 9.4) (Chen et al. 2014). The protein phosphatase 1B inhibitory actions of these four penostatins were evaluated, and the results demonstrated that they have strong PTP1B inhibitory effects at IC_{50} values ranging from 0.37 to 43.6 μM. Penostatin C (**76**) and penostatin J (**77**) (Fig. 9.4), were found to have significant actions via selectivity between PTP1B and the protein tyrosine phosphatases (LAR) with IC_{50} values of 1.0 and 1.6 μM, respectively (Chen et al. 2014). Pinophilin C (**78**) (Fig. 9.4), an intriguing azaphilone dihydroxymethylbenzoate ester obtained from the rare fungus *Cordyceps gracilioides*, has been shown to suppress the activity of several key protein tyrosine phosphatases, including PTP1B, SHP2, CDC25B, LAR, and SHP1, with inhibition observed in the 3–8 μM concentration range (Wei et al. 2015).

3.4 Some Metabolites with Other Different Biological Activities

Sakakura et al. (2009) explored the antioxidant prowess of hanasanagin (**79**) (Fig. 9.4), a pseudo-di-peptide isolated from *Isaria japonica*, employing the DPPH assay to assess its free radical quenching abilities. Their findings revealed a notable EC_{50} of 8.1 μM—making it 2.6 times more effective than ascorbic acid. Intriguingly, hanasanagin (**79**) did not stop there: it also demonstrated superoxide dismutase (SOD)-mimicking behavior, clocking in at 80 SOD units per micromole, nearly doubling the SOD-like activity of ascorbic acid (1.9-fold higher), marking it as a remarkably potent antioxidant agent. Phiocordylongiiside A (**80**) (Fig. 9.5), a cerebroside extracted from the n-butanol-soluble fraction of *Ophiocordyceps longissima* NBRC 106965, demonstrated potent antioxidant activity, exhibiting a strong DPPH-free radical scavenging effect with an IC_{50} value of 55.72 mg/mL (Elshamy et al. 2020). Cardinalisamides B (**4**) and C (**5**) (Fig. 9.2), natural products sourced from *Blackwellomyces roseostromatus* BCC56290, were first spotlighted in the literature by Phutthacharoen et al. (2024) for their nearly identical potency against *Trypanosoma brucei*, showing in vitro IC_{50} values of 12.8 and 12.5 μM, respectively.

Chung and colleagues spotlighted the anti-inflammatory potential of cyclodepsipeptides derived from *Beauveria felina*, revealing their suppressive impact on immune cell activity. Specifically, desmethylisaridin E (**81**) (Fig. 9.5) curbed superoxide anion production with an IC_{50} of 10.00 μM, while desmethylisaridin C2 (**82**) (Fig. 9.5) effectively dampened elastase release, achieving inhibition at an IC_{50} of 10.01 μM (Chung et al. 2013). In a study by Du et al. (2014), four compounds—isaridin E (**83**), isaridin G (**84**), desmethylisaridin G (**85**), and desmethylisaridin C1 (**86**) (Fig. 9.5), isolated from *Beauveria felina* EN-135—were put to the test for their antibacterial prowess. Each showed varying degrees of effectiveness against *E. coli*, with minimum inhibitory concentrations (MICs) of 16, 64, 64, and 8 μg/mL, respectively. For comparison, the standard antibiotic chloramphenicol demonstrated superior potency with an MIC of just 4 μg/mL. From a pigment-deficient mutant of *Ophiocordyceps coccidiicola* NBRC 100683 emerged a quartet of potent natural products—destruxin A (**87**), destruxin B (**88**), destruxin E chlorohydrin (**89**) (Fig. 9.5), and helvolic acid (**62**) (Fig. 9.3)—each demonstrating formidable in vitro antitrypanosomal activity against *Trypanosoma brucei brucei* GUTat3.1. Their IC_{50} values, at 0.33, 0.16, 0.061, and 5.08 mg/mL, respectively, rivaled and even surpassed that of the reference drug suramin (IC_{50} = 1.58 mg/mL) (Ganaha et al. 2016).

Furthermore, compounds xanthone **49** and anthraquinones **52** and **53** (Fig. 9.2), isolated from *Aschersonia coffeae* BCC 2871273 and *A. marginata* BCC 2872176, were also identified as possessing mild antimalarial properties, with IC_{50} values of 8.54, 3.88, and 1.60 mg/mL, respectively. These activities, while modest, stand in sharp contrast to the potent benchmark set by dihydroartemisinin, the positive control, which exhibits an IC_{50} of just 0.0005 mg/mL (Kornsakulkarn et al. 2012). In the documented studies, 5α,8α-epidioxy-(22*E*,24*R*)-ergosta-6,22-dien-3β-ol (**63**) (Fig. 9.3)—sourced from *Aschersonia luteola* BCC 31749—was found to possess only modest antimalarial activity against the *Plasmodium falciparum* K1, with an IC_{50} of 9.8 mM. This potency pales in comparison to that of dihydroartemisinin, the positive control, which demonstrated an IC_{50} of 0.0044 mM (Jinming et al. 2001; Isaka et al. 2011, 2012).

Isariotins A (**90**), C (**91**), and D (**92**) (Fig. 9.5), metabolites derived from *Isaria tenuipes* BCC 7831, have demonstrated notable activity against *Mycobacterium tuberculosis* H37Ra, with minimum inhibitory concentrations (MICs) of 486, 488, and 544 μM, respectively (Haritakun et al. 2007). Meanwhile, from *Isaria felina* EN-135, isaridin A (**93**) (Fig. 9.5) exhibited antispasmodic effects against *Plasmodium falciparum* 3D7 strain, achieving an IC_{50} of 230 μmol/L, whereas its counterpart isaridin B (**94**) (Fig. 9.5) proved even more potent, with an IC_{50} of 118 μmol/L (Sabareesh et al. 2007). From *Isaria tenuipes* BCC 12625, the compound isariotin F (**95**) (Fig. 9.5) emerged as a notable antimalarial agent, displaying inhibitory activity against *Plasmodium falciparum* K1 with an IC_{50} of 5.1 μM (Bunyapaiboonsri et al. 2009). In an extended study, Bunyapaiboonsri et al. (2011) uncovered additional isariotins, namely G (**96**), F (**95**), I (**97**), and J (**98**) (Fig. 9.5), isolated from *Isaria tenuipes* strains BCC 21283 and 23,112, which exhibited promising antimalarial efficacy, with IC_{50} values spanning 2.10 to 5.51 mg/ml. In contrast, isariotin H (**99**) (Fig. 9.5) demonstrated significantly weaker activity, achieving only 50% inhibition at a higher concentration of 10 mg/ml.

Anisimov and his team successfully extracted oxirapentyn A (**19**) (Fig. 9.2), oxirapentyn B (**100**), and oxirapentyn E (**101**) (Fig. 9.5) from *Isaria felina* KMM 4639. These compounds were then put to the test for their ability to boost root growth in the seedlings of several key crops—including barley (*Hordeum vulgare* L.), buckwheat (*Fagopyrum esculentum* Moench), corn (*Zea mays* L.), soy (*Glycine max* (L.) Merr.), and wheat (*Triticum aestivum* L.). Notably, oxirapentyns A (**19**) (Fig. 9.2) and B (**100**) (Fig. 9.5) demonstrated significant root-promoting activity in barley and wheat across various concentrations. Meanwhile, oxirapentyn E (**101**) (Fig. 9.5) showed marked stimulatory effects on the root development of maize, soy, and wheat, suggesting a broad-spectrum potential for agricultural enhancement (Anisimov et al. 2014). Yurchenko et al. (2013) uncovered a striking phytoregulatory effect from oxirapentyn E (**101**), a compound derived from *Isaria feline*, demonstrating that even at vanishingly small doses—ranging from 10^{-5} to an astonishing 10^{-17} M—the compound most potently stimulated rootlet growth in *Zea mays* (corn) and *Hordeum vulgare* (barley), with peak bioactivity occurring at these ultralow concentrations.

Xu et al. (2018) uncovered two rare and structurally intriguing polysaccharides from *Paecilomyces cicadae* Miquel, boasting hefty molecular weights of 30.9 and 555.3 kDa and rich in sulfates and uronic acids. These compounds stood out for their robust anti-inflammatory prowess, curbing the proliferation of RAW264.7 cells without a trace of cytotoxicity within a defined concentration window. Even more compelling, they markedly boosted macrophage viability while ramping up the production of nitric oxide (NO), interleukin-6 (IL-6), and tumor necrosis factor-alpha (TNF-α), spotlighting their potential as immune-enhancing agents.

3.5 Some Recent Medicinal Applications of Metabolites

3.5.1 Medicinal and Pharmaceutical Preclinical Applications

Cordycepin (Fig. 9.6), a 3′-deoxyadenosine analog, was initially isolated from the sporoderm-broken spores of *Paecilomyces cicadae*, a species known for its extensive ethnopharmacological use in East Asian traditional medicine. This nucleoside derivative has garnered significant scientific interest due to its multifaceted bioactivities, including pronounced antioxidant, antineoplastic (antitumor), immunomodulatory, antidepressant, and antimetastatic properties (Tuli et al. 2013; Olatunji et al. 2016). Mechanistically, cordycepin has been shown to enhance male reproductive function by upregulating steroidogenesis in Leydig cells, both in vitro and in vivo, via activation of the cAMP-dependent protein kinase A (PKA) signaling cascade, thereby increasing testosterone biosynthesis (Chen et al. 2017). In a landmark regulatory milestone, cordycepin was designated in 2007 by the U.S. Food and Drug Administration (FDA) as an investigational therapeutic agent for terminal deoxynucleotidyl transferase (TdT)-positive acute lymphoblastic leukemia (ALL). Subsequently, from 2008 to 2010, a combination therapy involving cordycepin and pentostatin underwent phase I/II clinical evaluation for efficacy and safety in patients with TdT-positive leukemia. Beyond hematologic malignancies, cordycepin has also entered clinical use in China, where it is prescribed for managing non-small-cell lung carcinoma (NSCLC) and diabetic nephropathy associated with type 2 diabetes mellitus, reflecting its expanding therapeutic potential across oncological and metabolic disease spectra (Lu et al. 2016; Wang et al. 2018).

The cyclic hexadepsipeptide beauvericin (Fig. 9.6), which originates from *Beauveria bassiana*, might seem weak, but it is a powerful ionophore that targets cancer cells with lethal accuracy. Through both intrinsic and extrinsic apoptotic mechanisms, it triggers cell death, oxidative stress, and ion homeostasis disruption. Strong cytotoxic efficacy against a variety of malignancies, including colon, leukemia, and glioblastoma, is demonstrated by preclinical research. Its potential to overcome medication resistance is particularly encouraging, despite the fact that it is still in the lab stage. Its toxicity to healthy cells is a worry, though, and delivery and dosage methods will be major obstacles in the future (Heilos et al. 2017; Patra et al. 2024).

Primary red pigment, oosporein of *Beauveria caledonica* (Fig. 9.6), has a sharp kick and a remarkable range of antibacterial action, especially against Gram-positive bacteria (Mc Namara et al. 2019). Its capacity to damage bacterial membranes and prevent the formation of biofilms—two processes essential to antibiotic resistance—was demonstrated in preclinical research. It has a promising multitarget profile and is also being investigated for anti-inflammatory and anticancer effects (Mc Namara et al. 2019; Jen et al. 2024). Tenellin (Fig. 9.6) is a yellow polyketide that was extracted from *Beauveria neobassiana* and has a milder effect. Although its antimicrobial potency is not as strong by itself, it is becoming more popular when used in combination therapy since it can strengthen the effects of current antibiotics and even show anti-quorum sensing qualities, which may disarm bacteria without killing them and thereby lower resistance (Toshe et al. 2024).

The antiviral field has recently focused on destruxins, a class of cyclodepsipeptides made by several EPF. In early research, these fungal metabolites, which have historically been investigated for their insecticidal and anticancer effects, have demonstrated unexpected efficacy against a variety of viruses (Pedras et al. 2002; Wu et al. 2013). According to preclinical research, destruxins may disturb the machinery of the host cell or specifically target viral proteins to interrupt viral reproduction cycles. For example, via altering host immunological signaling and gene expression, destruxin B has been shown to limit hepatitis B virus (HBV) multiplication in vitro. Additionally, there is evidence of potential protection against HIV and dengue virus, although the data is still preliminary (Yeh et al. 2012; Dornetshuber-Fleiss et al. 2013). Destruxins are unique possibilities in the hunt for broad-spectrum antiviral medicines because of their dual bioactivity, which combines immunomodulatory potential with antiviral activities. However, before moving further with extensive clinical studies, their cytotoxicity and pharmacokinetics require improvement (Hyde et al. 2019).

3.5.2 Agricultural Applications

In agricultural settings, EPF are increasingly used as biocontrol agents to manage insect pests, reducing the reliance on chemical pesticides. EPF such as *Beauveria bassiana*, *Metarhizium anisopliae*, and *Isaria fumosorosea* have proven effective against a wide range of insect pests, including aphids, whiteflies, and soil-dwelling larvae. Their use in integrated pest management (IPM) strategies enhances crop yield while minimizing environmental damage and the development of pesticide resistance (Vivekanandhan et al. 2024a, b).

4 Natural Pest Control in Agriculture

4.1 Traditional Agricultural Practices

EPF have long been utilized by farmers as a natural pest management tool in different regions of the globe. Fungi are frequently found in soil or in gardens and agricultural fields organically. Farmers recognized these fungi for their potential to manage insect pests, even though they were not usually identified or investigated properly (Vivekanandhan et al. 2024a, b).

4.2 Insect Control

EPF, particularly *Beauveria bassiana*, *Metarhizium anisopliae*, and *Isaria fumosorosea*, have been utilized for fighting widespread crop pests such as locusts, aphids, and beetles in places where access to synthetic pesticides is restricted. The fungi infiltrate the insects, resulting in illness and death without the need for harmful substances (Kisaakye et al. 2021; Irsad et al. 2023).

4.3 Soil Inoculation

Some ancient methods involved inoculating soils with fungal spores in order to create a biological barrier that effectively prevented pests. Unaware of it, farmers may have depended on natural EPF populations to defend crops from a variety of pest dangers (Tuininga et al. 2014; Chandler 2017).

5 Use in Livestock and Animal Health

EPF show promise in veterinary applications, particularly in the treatment of ecto-parasitic infestations such as mites, lice, and ticks, which affect livestock and companion animals. Fungal biocontrol agents have been explored to manage these external parasites, offering a more sustainable and less toxic alternative to chemical treatments. Moreover, EPF could help mitigate the spread of zoonotic diseases by targeting insect vectors that transmit pathogens to animals and humans (Ebani and Mancianti 2021).

5.1 Animal Protection

EPF have historically been used in some areas to manage livestock-harming pests like flies, lice, and ticks. These fungi can help lessen the load of ectoparasites by infecting the insects that parasitize farm animals. In many cultures, this use improved the health of animals, despite the fact that it may have been informal and unfounded in scientific understanding (Bihal et al. 2023; Rajput et al. 2024).

6 Examples of Common EPF in Cultural Traditional Use

6.1 *Beauveria bassiana*

Beauveria bassiana is an entomopathogenic fungus belonging to Clavicipitaceae within *Hypocreales*. This fungus is widely distributed in nature, occurring in diverse habitats, including soils, decaying organic matter, and as an endophyte in plants. *B. bassiana* is known for its pathogenicity toward a wide range of insect pests, making it a valuable biocontrol agent in IPM strategies. The fungus infects its hosts through the attachment of conidia (asexual spores) to the insect cuticle, followed by germination and the penetration of the host's exoskeleton. Once inside, *B. bassiana* grows throughout the insect's body, leading to tissue damage, systemic infection, and eventual host mortality (Jalinas et al. 2022; Sui et al. 2023). The application of *B. bassiana* as a biopesticide is particularly appealing due to its host specificity, low environmental toxicity, and ability to reduce chemical pesticide use. Moreover, *B. bassiana* exhibits potential in controlling a broad spectrum of pests, including those that affect crops, forests, and stored products. Research into its mechanisms of

Fig. 9.7 *Beauveria bassiana* targets a broad spectrum of pest control

pathogenesis, host range, and ecological impact continues to advance, highlighting the fungus as a promising agent in sustainable pest management approaches (Meyling and Eilenberg 2007; Wang et al. 2021). This fungus targets a wide variety of insect pests, including aphids, beetles, and flies (Fig. 9.7). It has a history of use in many cultures, particularly in China and parts of Africa, as a natural way to protect crops from insect damage (Li et al. 2024).

6.2 *Metarhizium anisopliae*

Metarhizium anisopliae is an entomopathogenic fungus that belongs to Clavicipitaceae. It is widely recognized for its ability to infect and kill a broad spectrum of insect pests, making it an important biocontrol agent in IPM systems. The fungus primarily targets insects by adhering to their exoskeleton, germinating, and infiltrating the host's body. The fungus eventually causes death through a combination of mechanical damage and the release of toxic metabolites. *M. anisopliae* has shown efficacy against numerous agricultural pests, including weevils, locusts, and beetles, and has been investigated for use in controlling vector-borne diseases, such as malaria and dengue, by targeting disease-carrying mosquitoes (Sharma et al. 2023; Vivekanandhan et al. 2024a, b). Beyond its role in pest control, *M. anisopliae* is a subject of significant interest due to its genetic diversity, potential for bioinsecticide development, and environmental

Fig. 9.8 *Metarhizium anisopliae* targets a soil-dwelling pest

sustainability. Studies have focused on understanding its pathogenicity, ecology, and molecular mechanisms, including the synthesis of various secondary metabolites and enzymes. Furthermore, the fungus's ability to adapt to different environmental conditions, along with its relatively low risk to non-target organisms, positions *M. anisopliae* as a promising candidate for sustainable agriculture and pest management solutions (Paschapur et al. 2021; de Miranda et al. 2024). Another widely used entomopathogenic fungus, *M. anisopliae*, has been recognized for its ability to infect a broad range of insects, especially soil-dwelling pests like termites and weevils (Fig. 9.8). In some African and Southeast Asian farming communities, this fungus was applied to control pests in agriculture, particularly where other pest management methods were unavailable (Zimmermann 2007; Deka et al. 2021).

6.3 *Isaria fumosorosea*

Isaria fumosorosea is a filamentous fungal species within Cordycipitaceae, known for its entomopathogenic properties. It is a widely studied microorganism with significant relevance in biological pest control, particularly for the management of insect pests in agriculture and forestry. Originally isolated from various insect hosts, this fungus has been shown to infect and kill a range of arthropods, including

Fig. 9.9 *Isaria fumosorosea* targets whiteflies, aphids, and thrips

important pest species like whiteflies, aphids, and thrips (Fig. 9.9) (Lopes et al. 2017; Xu et al. 2017). *I. fumosorosea* produces asexual conidia (spores), which are the primary infectious agents. These spores attach to the exoskeleton of host insects, germinate, and invade the host's body, eventually leading to death through a combination of tissue invasion and the release of toxins. The fungus is also known for its ability to adapt to various environmental conditions, making it a promising candidate for biocontrol in diverse agro-ecosystems (Mora et al. 2017; Nermuť et al. 2020). Research into the molecular mechanisms behind its pathogenicity and host specificity has provided valuable insights into its potential as a natural pesticide. Furthermore, *I. fumosorosea* has been explored for its role in IPM, reducing the reliance on chemical pesticides and promoting sustainable agricultural practices. Its efficacy, safety profile, and environmental compatibility continue to make it an important subject of study in the field of microbial biocontrol (Lei et al. 2021; Irsad et al. 2023).

7 Conclusion

EPF are at the intersection of biomedical and ecological innovation, providing a wealth of biologically active compounds that go far beyond their conventional use in pest management. The various metabolic characteristics of EPF, including

Beauveria, Metarhizium, Cordyceps, and *Isaria*, provide an intriguing synergy between medicine and nature, as this chapter has looked at. These fungi produce metabolites with substantial medicinal potential, such as immunological modulators, enzyme inhibitors, anticancer and antibacterial agents, in addition to offering environmentally friendly substitutes for conventional pesticides.

Future study on EPF will likely involve more thorough genomic and metabolomic analyses, which should reveal new molecules with previously unheard-of bioactivities. Their pharmacological significance can be further enhanced by sophisticated synthetic biology and fermentation technologies, which allow for the scalable synthesis of uncommon metabolites. Furthermore, finding therapeutically viable compounds could be accelerated by combining high-throughput screening and developing drug platforms alongside conventional ethnomedical insights.

EPF are positioned to play a revolutionary role as the need for innovative therapies and environmentally sustainable biotechnologies increases globally. To fully realize the promise of these adaptable fungi, interdisciplinary cooperation among mycologists, pharmacologists, agronomists, and indigenous knowledge holders will be crucial. EPF are ultimately a prime example of how traditional biological practices may stimulate innovative approaches to environmental sustainability, agriculture, and healthcare.

References

Afiyatullov SS, Zhuravleva OI, Antonov AS, Berdyshev DV, Pivkin MV et al (2018) Prenylated indole alkaloids from co-culture of marine-derived fungi *Aspergillus sulphureus* and *Isaria felina*. J Antibiot 71:846–853. https://doi.org/10.1038/s41429-018-0072-9

Alves V, Zamith-Miranda D, Frases S, Nosanchuk JD (2025) Fungal metabolomics: a comprehensive approach to understanding pathogenesis in humans and identifying potential therapeutics. J Fungi 11:93. https://doi.org/10.3390/jof11020093

Anisimov MM, Chaikina EL, Smetanina OF, Afiyatullov SS (2014) Oxirapentyns A, B and E from the marine-derived strain of *Isaria felina* KMM 4639 as stimulators of initial stages of development of agricultural plants. Nat Prod Commun 9:1934578X1400900628. https://doi.org/10.1177/1934578X1400900628

Bihal R, Al-Khayri JM, Banu AN, Kudesia N, Ahmed FK et al (2023) Entomopathogenic fungi: an eco-friendly synthesis of sustainable nanoparticles and their nanopesticide properties. Microorganisms 11:1617. https://doi.org/10.3390/microorganisms11061617

Brel O, Touré S, Levasseur M, Lechat C, Pellissier L et al (2020) Paecilosetin derivatives as potent antimicrobial agents from *Isaria farinosa*. J Nat Prod 83:2915–2922. https://doi.org/10.1021/acs.jnatprod.0c00444

Bunyapaiboonsri T, Yoiprommarat S, Intereya K, Rachtawee P, Hywel-Jones NL et al (2009) Isariotins E and F, spirocyclic and bicyclic hemiacetals from the entomopathogenic fungus *Isaria tenuipes* BCC 12625. J Nat Prod 72:756–759. https://doi.org/10.1021/np800702c

Bunyapaiboonsri T, Yoiprommarat S, Srisanoh U, Choowong W, Tasanathai K et al (2011) Isariotins G–J from cultures of the Lepidoptera pathogenic fungus *Isaria tenuipes*. Phytochem Lett 4:283–286. https://doi.org/10.1016/j.phytol.2011.04.018

Chandler D (2017) Basic and applied research on entomopathogenic fungi. In: Lacey LA (ed) Microbial control of insect and mite pests. From theory to practice. Elsevier, pp 69–89

Chen YP, Yang CG, Wei PY, Li L, Luo DQ et al (2014) Penostatin derivatives, a novel kind of protein phosphatase 1B inhibitors isolated from solid cultures of the entomogenous fungus *Isaria tenuipes*. Molecules 19:1663–1671. https://doi.org/10.3390/molecules19021663

Chen YC, Chen YH, Pan BS, Chang MM, Huang BM (2017) Functional study of *Cordyceps sinensis* and cordycepin in male reproduction: a review. J Food Drug Anal 25:197–205. https://doi.org/10.1016/j.jfda.2016.10.020

Chen C, Xue T, Fan P, Meng L, Wei J et al (2018) Cytotoxic activity of Shp2 inhibitor fumosorinone in human cancer cells. Oncol Lett 15:10055–10062. https://doi.org/10.3892/ol.2018.8593

Cheng Y, Schneider B, Riese U, Schubert B, Li Z et al (2004) Farinosones A– C, neurotrophic alkaloidal metabolites from the entomogenous deuteromycete *Paecilomyces f arinosus*. J Nat Prod 67:1854–1858. https://doi.org/10.1021/np049761w

Chung YM, El-Shazly M, Chuang DW, Hwang TL, Asai T et al (2013) Suberoylanilide hydroxamic acid, a histone deacetylase inhibitor, induces the production of anti-inflammatory cyclodepsipeptides from *Beauveria felina*. J Nat Prod 76:1260–1266. https://doi.org/10.1021/np400143j

de Miranda RPR, Soares TKA, Castro DP, Genta FA (2024) General aspects, host interaction, and application of *Metarhizium* sp. in arthropod pest and vector control. Front Fungal Biol 5:1456964. https://doi.org/10.3389/ffunb.2024.1456964

Deka B, Babu A, Peter AJ, Kumhar KC, Sarkar S et al (2021) Potential of the entomopathogenic fungus, *Metarhizium anisopliae* sl in controlling live-wood eating termite, Microtermes obesi (Holmgren)(Blattodea: Termitidae) infesting tea crop. Egypt J Biol Pest Control 31:1–8. https://doi.org/10.1186/s41938-021-00477-4

Dornetshuber-Fleiss R, Heffeter P, Mohr T, Hazemi P, Kryeziu K et al (2013) Destruxins: fungal-derived cyclohexadepsipeptides with multifaceted anticancer and antiangiogenic activities. Biochem Pharmacol 86:361–377. https://doi.org/10.1016/j.bcp.2013.05.022

Du FY, Zhang P, Li XM, Li CS, Cui CM et al (2014) Cyclohexadepsipeptides of the isaridin class from the marine-derived fungus *Beauveria felina* EN-135. J Nat Prod 77:1164–1169. https://doi.org/10.1021/np4011037

Dung NC, Ngoc HT, Ninh PTM, Phu DH, Hiep DM, Suong NK et al (2021) Studying on cytotoxic activity of ethyl acetate extracts and isolated substances from cultured Isaria cicadae F0004 in Vietnam against the MCF-7 cell lines and Jurkat cell lines. IOP Conference Series: Earth and Environmental Science: IOP Publishing.

Ebani VV, Mancianti F (2021) Entomopathogenic fungi and bacteria in a veterinary perspective. Biology 10:479. https://doi.org/10.3390/biology10060479

Elshamy AI, Yoneyama T, Van Trang N, Son NT, Okamoto Y et al (2020) A new cerebroside from the entomopathogenic fungus *Ophiocordyceps longiissima*: structural-electronic and antioxidant relations. Experimental and DFT calculated studies. J Mol Struct 1200:127061. https://doi.org/10.1016/j.molstruc.2019.127061

Elshamy AI, Mohamed TA, Yoneyama T, Noji M, Ban S et al (2023) Bifusicoumarins A-D: cytotoxic 3*S*-dihydroisocoumarins from the entomopathogenic fungus *Cordyceps bifusispora* (NBRC 108997). Phytochemistry 212:113743. https://doi.org/10.1016/j.phytochem.2023.113743

Elshamy AI, Essa AF, Sabt A, Abd El-Razek MH, Gaara AH et al (2024) *Isaria* entomopathogenic fungi: chemistry and biological potency. Stud Nat Prod Chem 81:415–465. https://doi.org/10.1016/B978-0-443-15628-1.00006-4

Fujita T, Inoue K, Yamamoto S, Ikumoto T, Sasaki S et al (1994) Fungal metabolites. Part 11. A potent immunosuppressive activity found in *Isaria sinclairii* metabolite. J Antibiot 47:208–215. https://doi.org/10.7164/antibiotics.47.208

Ganaha M, Yoshii K, Ōtsuki Y, Iguchi M, Okamoto Y et al (2016) In vitro antitrypanosomal activity of the secondary metabolites from the mutant strain IU-3 of the insect pathogenic fungus *Ophiocordyceps coccidiicola* NBRC 100683. Chem Pharm Bull 64:988–990. https://doi.org/10.1248/cpb.c16-00220

Guo Y, Liu Q, Ng T, Wang H (2005) Isarfelin, a peptide with antifungal and insecticidal activities from *Isaria felina*. Peptides 26:2384–2391. https://doi.org/10.1016/j.peptides.2005.05.020

Hama M, Elshamy AI, Yoneyama T, Kasai Y, Yamamoto H et al (2019) New alkaloidal metabolites from cultures of entomopathogenic fungus *Cordyceps takaomontana* NBRC 101754. Fitoterapia 139:104364. https://doi.org/10.1016/j.fitote.2019.104364

Haritakun R, Srikitikulchai P, Khoyaiklang P, Isaka M (2007) Isariotins A–D, alkaloids from the insect pathogenic fungus *Isaria tenuipes* BCC 7831. J Nat Prod 70:1478–1480. https://doi.org/10.1021/np070291q

He L, Shi W, Liu X, Zhao X, Zhang Z (2018) Anticancer action and mechanism of ergosterol peroxide from *Paecilomyces cicadae* fermentation broth. Int J Mol Sci 19:3935. https://doi.org/10.3390/ijms19123935

Heilos D, Rodríguez-Carrasco Y, Englinger B, Timelthaler G, Van Schoonhoven S et al (2017) The natural fungal metabolite beauvericin exerts anticancer activity *in vivo*: a pre-clinical pilot study. Toxins 9:258. https://doi.org/10.3390/toxins9090258

Hosoya T, Takagi M, Shin-ya K (2013) New pyridone alkaloids JBIR-130, JBIR-131 and JBIR-132 from *Isaria* sp. NBRC 104353. J Antibiot 66:235–238. https://doi.org/10.1038/ja.2012.106

Hyde KD, Xu J, Rapior S, Jeewon R, Lumyong S et al (2019) The amazing potential of fungi: 50 ways we can exploit fungi industrially. Fungal Divers 97:1–136. https://doi.org/10.1007/s13225-019-00430-9

Hyde KD, Baldrian P, Chen Y, Thilini Chethana K, De Hoog S et al (2024) Current trends, limitations, and future research in the fungi? Fungal Divers 125:1–71. https://doi.org/10.1007/s13225-023-00532-5

Irsad, Shahid M, Haq E, Mohamed A, Rizvi PQ et al (2023) Entomopathogen-based biopesticides: insights into unraveling their potential in insect pest management. Front Microbiol 14:1208237. https://doi.org/10.3389/fmicb.2023.1208237

Isaka M, Palasarn S, Kocharin K, Saenboonrueng J (2005) A cytotoxic xanthone dimer from the entomopathogenic fungus *Aschersonia* sp. BCC 8401. J Nat Prod 68:945–946

Isaka M, Palasarn S, Supothina S, Komwijit S, Luangsa-ard JJ (2011) Bioactive compounds from the scale insect pathogenic fungus *Conoideocrella tenuis* BCC 18627. J Nat Prod 74:782–789. https://doi.org/10.1021/np058028h

Isaka M, Palasarn S, Tobwor P, Boonruangprapa T, Tasanathai K (2012) Bioactive anthraquinone dimers from the leafhopper pathogenic fungus *Torrubiella* sp. BCC 28517. J Antibiot 65:571–574. https://doi.org/10.1038/ja.2012.76

Jalinas J, Lopez-Moya F, Marhuenda-Egea FC, Lopez-Llorca LV (2022) *Beauveria bassiana* (Hypocreales: Clavicipitaceae) volatile organic compounds (VOCs) repel *Rhynchophorus ferrugineus* (Coleoptera: Dryophthoridae). J Fungi 8:843. https://doi.org/10.3390/jof8080843

Jen KG, Yi EYW, Yi SKX, Quan TY (2024) Bioactive compounds of *Cordyceps* and their potential pharmaceutical avenues. In: Advances in *Cordyceps* research. CRC Press, pp 192–229

Jiang Y-H, Jiang X-l, Wang P, Mou HJ, Hu XK et al (2008) The antitumor and antioxidative activities of polysaccharides isolated from *Isaria farinosa* B05. Microbiol Res 163:424–430. https://doi.org/10.1016/j.micres.2006.07.002

Jinming G, Lin H, Jikai L (2001) A novel sterol from Chinese truffles *Tuber indicum*. Steroids 66: 771–775. https://doi.org/10.1016/S0039-128X(01)00105-2

Kisaakye J, Fourie H, Coyne D, Cortada L, Masinde S et al (2021) Evaluation of the entomopathogenic potential of *Beauveria bassiana*, *Metarhizium anisopliae* and *Isaria fumosorosea* for management of cosmopolites sordidus Germar (Coleoptera: Curculionidae). Agriculture (Basel) 11:1290. https://doi.org/10.3390/agriculture11121290

Kluepfel D, Bagli J, Baker H, Charest MP, Kudelski A et al (1972) Myriocin, a new antifungal antibiotic from *Myriococcum albomyces*. J Antibiot 25:109–115. https://doi.org/10.7164/antibiotics.25.109

Kornsakulkarn J, Saepua S, Srichomthong K, Supothina S, Thongpanchang C (2012) New mycotoxins from the scale insect fungus *Aschersonia coffeae* Henn. BCC 28712. Tetrahedron 68:8480–8486. https://doi.org/10.1016/j.tet.2012.07.059

Kuca K, Patocka J, Wu Q (2018) Biological activities of beauvericin, a mini review. Mini Rev Med Chem 19(3):206–214. https://doi.org/10.2174/1389557518666180928161808

Kwon YJ, Sohn MJ, Zheng CJ, Kim WG (2007) Fumimycin: a peptide deformylase inhibitor with an unusual skeleton produced by *Aspergillus fumisynnematus*. Org Lett 9:2449–2451. https://doi.org/10.1021/ol0703231

Lei Y, Hussain A, Guan Z, Wang D, Jaleel W et al (2021) Unraveling the mode of action of *Cordyceps fumosorosea*: potential biocontrol agent against *Plutella xylostella* (Lepidoptera: Plutellidae). Insects 12:179. https://doi.org/10.3390/insects12020179

Li XL, Zhang JJ, Li DD, Cai XY, Qi YX et al (2024) Toxicity of *Beauveria bassiana* to *Bactrocera dorsalis* and effects on its natural predators. Front Microbiol 15:1362089. https://doi.org/10.3389/fmicb.2024.1362089

Liu L, Zhang J, Chen C, Teng J, Wang C et al (2015) Structure and biosynthesis of fumosorinone, a new protein tyrosine phosphatase 1B inhibitor firstly isolated from the entomogenous fungus *Isaria fumosorosea*. Fungal Genet Biol 81:191–200. https://doi.org/10.1016/j.fgb.2015.03.009

Lopes RS, Lima G, Correia MTS, da Costa AF, Lima EÁLA et al (2017) The potential of Isaria spp. as a bioinsecticide for the biological control of *Nasutitermes corniger*. Biocontrol Sci Tech 27: 1038–1048. https://doi.org/10.1080/09583157.2017.1380163

Lu CL, Lin HI, Chen BF, Jow GM (2016) Beauvericin-induced cell apoptosis through the mitogen-activated protein kinase pathway in human nonsmall cell lung cancer A549 cells. J Toxicol Sci 41:429–437. https://doi.org/10.2131/jts.41.429

Lu Y, Luo F, Cen K, Xiao G, Yin Y et al (2017) Omics data reveal the unusual asexual-fruiting nature and secondary metabolic potentials of the medicinal fungus *Cordyceps cicadae*. BMC Genomics 18:1–15. https://doi.org/10.1186/s12864-017-4060-4

Ma C, Li Y, Niu S, Zhang H, Liu X et al (2011) *N*-hydroxypyridones, phenylhydrazones, and a quinazolinone from *Isaria farinosa*. J Nat Prod 74:32–37. https://doi.org/10.1021/np100568w

Mc Namara L, Dolan SK, Walsh JM, Stephens JC, Glare TR et al (2019) Oosporein, an abundant metabolite in Beauveria caledonica, with a feedback induction mechanism and a role in insect virulence. Fungal Biol 123:601–610. https://doi.org/10.1016/j.funbio.2019.01.004

Meyling NV, Eilenberg J (2007) Ecology of the entomopathogenic fungi *Beauveria bassiana* and *Metarhizium anisopliae* in temperate agroecosystems: potential for conservation biological control. Biol Control 43:145–155. https://doi.org/10.1016/j.biocontrol.2007.07.007

Molnár I, Gibson DM, Krasnoff SB (2010) Secondary metabolites from entomopathogenic Hypocrealean fungi. Nat Prod Rep 27:1241–1275. https://doi.org/10.1039/C9NP00065H

Mora MAE, Castilho AMC, Fraga ME (2017) Classification and infection mechanism of EPF. Arq Inst Biol 84:e0552015. https://doi.org/10.1590/1808-1657000552015

Nam KS, Jo YS, Kim YH, Hyun JW, Kim HW (2001) Cytotoxic activities of acetoxyscirpenediol and ergosterol peroxide from *Paecilomyces tenuipes*. Life Sci 69:229–237. https://doi.org/10.1016/S0024-3205(01)01125-0

Nermuť J, Konopická J, Zemek R, Kopačka M, Bohatá A et al (2020) Dissemination of *Isaria fumosorosea* Spores by *Steinernema feltiae* and *Heterorhabditis bacteriophora*. J Fungi 6:359. https://doi.org/10.3390/jof6040359

Oh GS, Hong KH, Oh H, Pae HO, Kim IK et al (2001) 4-Acetyl-12, 13-epoxyl-9-trichothecene-3, 15-diol isolated from the fruiting bodies of *Isaria japonica* Yasuda induces apoptosis of human leukemia cells (HL-60). Biol Pharm Bull 24:785–789. https://doi.org/10.1248/bpb.24.785

Olatunji OJ, Feng Y, Olatunji OO, Tang J, Ouyang Z et al (2016) Cordycepin protects PC12 cells against 6-hydroxydopamine induced neurotoxicity via its antioxidant properties. Biomed Pharmacother 81:7–14. https://doi.org/10.1016/j.biopha.2016.03.009

Paschapur A, Subbanna A, Singh AK, Jeevan B, Stanley J et al (2021) Unraveling the importance of metabolites from entomopathogenic fungi in insect pest management. In: Khan MA, Ahmad W (eds) Microbes for sustainable Insect pest management: hydrolytic enzyme & secondary metabolite, vol 2, pp 89–120

Paterson RRM (2008) *Cordyceps*–a traditional Chinese medicine and another fungal therapeutic biofactory? Phytochemistry 69:1469–1495. https://doi.org/10.1016/j.phytochem.2008.01.027

Patra A, Arora A, Ghosh SS, Kaur Saini G (2024) Beauvericin reverses epithelial-to-mesenchymal transition in triple-negative breast cancer cells through regulation of notch signaling and autophagy. ACS Pharmacol Transl Sci 7:2878–2893. https://doi.org/10.1021/acsptsci.4c00370

Pedras MSC, Zaharia LI, Ward DE (2002) The destruxins: synthesis, biosynthesis, biotransformation, and biological activity. Phytochemistry 59:579–596. https://doi.org/10.1016/S0031-9422(02)00016-X

Pedrini N (2022) The entomopathogenic fungus *Beauveria bassiana* shows its toxic side within insects: expression of genes encoding secondary metabolites during pathogenesis. J Fungi 8: 488. https://doi.org/10.3390/jof8050488

Phutthacharoen K, Llanos-López NA, Toshe R, Noisripoom W, Khonsanit A et al (2024) Bioactive bioxanthracene and cyclodepsipeptides from the entomopathogenic fungus *Blackwellomyces roseostromatus* BCC56290. Antibiotics 13:585. https://doi.org/10.3390/antibiotics13070585

Rajput M, Sajid MS, Rajput NA, George DR, Usman M et al (2024) EPF as alternatives to chemical acaricides: challenges, opportunities and prospects for sustainable tick control. Insects 15:1017. https://doi.org/10.3390/insects15121017

Ren F., Zhu S., Bo Wang, Li L., Liu X., Su R., Yongsheng Che Y. Hypocriols A–F et al (2016) Heterodimeric Botryane Ethers from Hypocrea sp., an Insect-Associated Fungus. Journal of Natural Products 79(7), 1848–1856. https://doi.org/10.1021/acs.jnatprod.6b00394

Sabareesh V, Ranganayaki R, Raghothama S, Bopanna M, Balaram H et al (2007) Identification and characterization of a library of microheterogeneous cyclohexadepsipeptides from the fungus *Isaria*. J Nat Prod 70:715–729. https://doi.org/10.1021/np060532e

Saepua S, Kornsakulkarn J, Somyong W, Laksanacharoen P, Isaka M et al (2018) Bioactive compounds from the scale insect fungus *Conoideocrella tenuis* BCC 44534. Tetrahedron 74: 859–866. https://doi.org/10.1016/j.tet.2018.01.004

Sakakura A, Shioya K, Katsuzaki H, Komiya T, Imamura T et al (2009) Isolation, structural elucidation and synthesis of a novel antioxidative pseudo-di-peptide, Hanasanagin, and its biogenetic precursor from the *Isaria japonica* mushroom. Tetrahedron 65:6822–6827. https://doi.org/10.1016/j.tet.2009.06.069

Sharma A, Sharma S, Yadav PK (2023) Entomopathogenic fungi and their relevance in sustainable agriculture: a review. Cogent Food Agric 9:2180857. https://doi.org/10.1080/23311932.2023.2180857

Singh VK, Kumar A (2023) Secondary metabolites from endophytic fungi: production, methods of analysis, and diverse pharmaceutical potential. Symbiosis 90:111–125. https://doi.org/10.1007/s13199-023-00925-9

Smetanina OF, Yurchenko AN, Afiyatullov SS, Kalinovsky AI, Pushilin MA et al (2012) Oxirapentyns B–D produced by a marine sediment-derived fungus *Isaria felina* (DC.) Fr. Phytochem Lett 5:165–169. https://doi.org/10.1016/j.phytol.2011.12.002

Sui L, Lu Y, Zhou L, Li N, Li Q et al (2023) Endophytic *Beauveria bassiana* promotes plant biomass growth and suppresses pathogen damage by directional recruitment. Front Microbiol 14:1227269. https://doi.org/10.3389/fmicb.2023.1227269

Tao YW, Lin YC, She ZG, Lin MT, Chen PX et al (2015) Anticancer activity and mechanism investigation of beauvericin isolated from secondary metabolites of the mangrove endophytic fungi. Anti Cancer Agents Med Chem 15:258–266. https://doi.org/10.2174/1871520614666140825112255

Tian JF, Li PJ, Li XX, Sun PH, Gao H et al (2016) New antibacterial isocoumarin glycosides from a wetland soil derived fungal strain *Metarhizium anisopliae*. Bioorg Med Chem Lett 26:1391–1396. https://doi.org/10.1016/j.bmcl.2016.01.074

Toshe R, Charria-Girón E, Khonsanit A, Luangsa-Ard JJ, Khalid SJ et al (2024) Bioprospection of tenellins produced by the entomopathogenic fungus *Beauveria neobassiana*. J Fungi 10:69. https://doi.org/10.3390/jof10010069

Tuininga AR, Miller JL, Morath SU, Daniels TJ, Falco RC et al (2014) Isolation of entomopathogenic fungi from soils and *Ixodes scapularis* (Acari: Ixodidae) ticks: prevalence and methods. J Med Entomol 46:557–565. https://doi.org/10.1603/033.046.0321

Tuli HS, Sharma AK, Sandhu SS, Kashyap D (2013) Cordycepin: a bioactive metabolite with therapeutic potential. Life Sci 93:863–869. https://doi.org/10.1016/j.lfs.2013.09.030

Ueda JY, Kunoh T, Sekigawa M, Wada SI, Mukai Y, Ohta S, Sasaki R, Mizukami T, Takagi M, Shin-Ya K et al (2010) JBIR-14, a highly oxygenated ergostane, from *Isaria* sp. NBRC 104353. J Antibiot 63:139–141. https://doi.org/10.1038/ja.2010.3

Vivekanandhan P, Alford L, Krutmuang P (2024a) Role of entomopathogenic fungi in sustainable agriculture. Frontiers Media SA, p. 1504175. https://doi.org/10.3389/fmicb.2024.1504175

Vivekanandhan P, Swathy K, Sarayut P, Patcharin K (2024b) Classification, biology and entomopathogenic fungi-based management and their mode of action against *Drosophila* species (Diptera: Drosophilidae): a review. Front Microbiol 15:1443651. https://doi.org/10.3389/fmicb.2024.1443651

Wadhwa K, Kapoor N, Kaur H, Abu-Seer EA, Tariq M et al (2024) A comprehensive review of the diversity of fungal secondary metabolites and their emerging applications in healthcare and environment. Mycobiology:1–53. https://doi.org/10.1080/12298093.2024.2416736

Wang Q, Xu L (2012) Beauvericin, a bioactive compound produced by fungi: a short review. Molecules 17:2367–2377. https://doi.org/10.3390/molecules17032367

Wang J, Zhang DM, Jia JF, Peng QL, Tian HY et al (2014) Cyclodepsipeptides from the ascocarps and insect-body portions of fungus *Cordyceps cicadae*. Fitoterapia 97:23–27. https://doi.org/10.1016/j.fitote.2014.05.010

Wang X, Gong X, Li P, Lai D, Zhou L (2018) Structural diversity and biological activities of cyclic depsipeptides from fungi. Molecules 23:169. https://doi.org/10.3390/molecules23010169

Wang H, Peng H, Li W, Cheng P, Gong M (2021) The toxins of *Beauveria bassiana* and the strategies to improve their virulence to insects. Front Microbiol 12:705343. https://doi.org/10.3389/fmicb.2021.705343

Wei PY, Liu LX, Liu T, Chen C, Luo DQ et al (2015) Three new pigment protein tyrosine phosphatases inhibitors from the insect parasite fungus *Cordyceps gracilioides*: terreusinone A, pinophilin C and cryptosporioptide A. Molecules 20:5825–5834. https://doi.org/10.3390/molecules20045825

Wu CC, Chen TH, Liu BL, Wu LC, Chen YC et al (2013) Destruxin B isolated from entomopathogenic fungus *Metarhizium anisopliae* induces apoptosis via a Bcl-2 family-dependent mitochondrial pathway in human nonsmall cell lung cancer cells. Evid Based Complement Alternat Med 2013:548929. https://doi.org/10.1155/2013/548929

Xu J, Xu X, Shakeel M, Li S, Wang S et al (2017) The EPF *Isaria fumosorosea* plays a vital role in suppressing the immune system of *Plutella xylostella*: RNA-Seq and DGE analysis of immunity-related genes. Front Microbiol 8:1421. https://doi.org/10.3389/fmicb.2017.01421

Xu Z, Yan X, Song Z, Li W, Zhao W et al (2018) Two heteropolysaccharides from *Isaria cicadae* Miquel differ in composition and potentially immunomodulatory activity. Int J Biol Macromol 117:610–616. https://doi.org/10.1016/j.ijbiomac.2018.05.164

Yahagi H, Yahagi T, Furukawa M, Matsuzaki K (2020) Antiproliferative and antimigration activities of beauvericin isolated from *Isaria* sp. on pancreatic cancer cells. Molecules 25:4586. https://doi.org/10.3390/molecules25194586

Yang XL, Li ZZ, Zhang S, Yu H, Zhu HJ et al (2012) A new spirocyclic compound from the liquid culture of entomogenous fungus Isaria cateniannulata. J Asian Nat Prod Res 14:1093–1096. https://doi.org/10.1080/10286020.2012.707650

Yang NN, Jiang N, Ma QY, Kong FD, Xie QY et al (2019) Chemical study of the strain *Cordyceps* spp. from cell fusion between *Cordyceps militaris* and *Cordyceps cicadae*. J Asian Nat Prod Res 21:449–455. https://doi.org/10.1080/10286020.2018.1451518

Ye X, Chai W, Lian XY, Zhang Z (2017) Novel propanamide analogue and antiproliferative diketopiperazines from mangrove *Streptomyces* sp. Q24. Nat Prod Res 31:1390–1396. https://doi.org/10.1080/14786419.2016.1253079

Yeh CT, Rao YK, Ye M, Wu WS, Chang TC et al (2012) Preclinical evaluation of destruxin B as a novel Wnt signaling target suppressing proliferation and metastasis of colorectal cancer using non-invasive bioluminescence imaging. Toxicol Appl Pharmacol 261:31–41. https://doi.org/10.1016/j.taap.2012.03.007

Yoneyama T, Elshamy AI, Yamada J, El-Kashak WA, Kasai Y et al (2022) Antimicrobial metabolite of *Cordyceps tenuipes* targeting MurE ligase and histidine kinase via in silico study. Appl Microbiol Biotechnol 106:6483–6491. https://doi.org/10.1007/s00253-022-12176-7

Yurchenko A, Smetanina O, Khudyakova YV, Kirichuk N, Chaikina E et al (2013) New oxirapentyn E from marine isolate of the fungus *Isaria felina*. Chem Nat Compd 49:857–860. https://doi.org/10.1007/s10600-013-0764-0

Zhang Y, Liu S, Liu H, Liu X, Che Y (2009) Cycloaspeptides F and G, cyclic pentapeptides from a cordyceps-colonizing isolate of *Isaria farinosa*. J Nat Prod 72:1364–1367. https://doi.org/10.1021/np900205m

Zhang X, Hu Q, Weng Q (2018) Secondary metabolites (SMs) of *Isaria cicadae* and *Isaria tenuipes*. RSC Adv 9:172–184. https://doi.org/10.1039/C8RA09039D

Zhang L, Fasoyin OE, Molnár I, Xu Y (2020) Secondary metabolites from hypocrealean EPF: novel bioactive compounds. Nat Prod Rep 37:1181–1206. https://doi.org/10.1039/C9NP00065H

Zhuravleva O, Kirichuk N, Denisenko V, Dmitrenok P, Yurchenko E et al (2016) New diorcinol J produced by co-cultivation of marine fungi *Aspergillus sulphureus* and *Isaria felina*. Chem Nat Compd 52:227–230. https://doi.org/10.1007/s10600-016-1601-z

Zhuravleva OI, Belousova EB, Oleinikova GK, Antonov AS, Khudyakova YV et al (2022) Cytotoxic drimane-type sesquiterpenes from co-culture of the marine-derived fungi *Aspergillus carneus* KMM 4638 and *Beauveria felina* (= *Isaria felina*) KMM 4639. Mar Drugs 20:584. https://doi.org/10.3390/md20090584

Zimmermann G (2007) Review on safety of the entomopathogenic fungus *Metarhizium anisopliae*. Biocontrol Sci Tech 17:879–920. https://doi.org/10.1080/09583150701593963

Part III

Agricultural Applications

Recent Advances in Entomopathogenic Fungi as Biocontrol Agents in the Management of Insect Pests in Agriculture

10

Pranaba Nanda Bhattacharyya, Bharat Chandra Nath, Parismita Sarma, Jafrin Farha Hussain, Bhaskar Sarma, Dhruba Jyoti Borgohain, Anurag Kashyap, Dusmanta Kumar Sut, and Liza Handique Bhattacharyya

Abstract

Agriculture faces significant environmental and health risks due to rapid and unlawful use of toxic chemical pesticides, which harm native flora and fauna, degrade soil quality, and interfere with essential biological activities, leading to fertility loss, pesticide resurgence, and global warming. Climate-smart insect pest management in agriculture, utilizing the entomopathogenic fungi has gained significant prominence as an effective non-chemical biological approach, owing

P. N. Bhattacharyya (✉)
Department of Botany, Nanda Nath Saikia College, Jorhat, Assam, India

Institutional Biotech Hub (IBT-HUB), Nanda Nath Saikia College, Jorhat, Assam, India

B. C. Nath · A. Kashyap
Department of Plant Pathology, Assam Agricultural University, Jorhat, Assam, India

P. Sarma
Department of Botany, Rangia College, Rangia, Assam, India

J. F. Hussain
Department of Zoology, Nanda Nath Saikia College, Jorhat, Assam, India

B. Sarma
Department of Botany, Dhemaji College, Dhemaji, Assam, India

D. J. Borgohain
C. S. Central Library, Nanda Nath Saikia College, Jorhat, Assam, India

D. K. Sut
Department of Mathematics, Nanda Nath Saikia College, Jorhat, Assam, India

L. H. Bhattacharyya
Department of Botany, Jagannath Barooah University, Jorhat, Assam, India

S. K. Deshmukh et al. (eds.), *Frontiers in Entomopathogenic Fungi*,
https://doi.org/10.1007/978-981-95-6529-0_10

to its minimal or nil toxicity issues, enhanced bioefficacy, extended shelf life, and less risk of environmental contamination. Multidimensional pest immobilization activities mediated by entomopathogenic fungi are accomplished by diverse action mechanisms of several mycotoxins and enzymes produced by this biocontrol agents, resulting in significant mycopathogenesis. In contrast to synthetic pesticides, entomopathogenic fungi do not compromise with essential soil nutrients like carbon and nitrogen, rendering them more environmentally acceptable and sustainable. In addition to its potential as a microbial biocontrol agent against major agricultural insect pests, this category of pathogenic fungus may persist for extended periods of time in adverse topography, climate, and soil. The integration of advanced technologies, including the genetic engineering, myconanoformulations, artificial intelligence, and the Internet of Things, alongside real-time data analysis and fuzzy logic, has recently revolutionized entomopathogenic research in agriculture, leading to sustainable crop protection. Additionally, bibliometrics has been an excellent tool in examining the trends, patterns, research gaps, and progress in entomopathogen-mediated plant protection in agriculture.

Keywords

Bibliometrics · Climate-smart agriculture · Insect pest management · Genetic engineering · Internet of things · Myconanoformulations

1 Introduction

Agriculture, the practice of cultivating natural resources to continue human life, civilization, and sustained economic growth, has been the foundation for over 10,000 years (Bhattacharyya et al. 2025). It encompasses the cultivation of different economically important crops (EICs), livestock farming, and deals to address prominent sustainable practices to ensure sufficient food security, economic stability, and environmental health. However, there are reports of considerable damages in the farming sector, mainly due to various abiotic and biotic interferences (Teshome et al. 2020), including anthropogenic factors, thereby creating significant risks in quality food production, public health, safety, and security. Primary factors contributing to losses in agriculture include major devastations caused by the interference of pests, pathogens, nematodes, climate crises, salinity, drought, loss of arable land and land degradation, and frequent issues of weed proliferations within the cultivated areas (Mahanta et al. 2023; Shahbazi et al. 2025). A report by Dar et al. (2021) and Sarma et al. (2023a) predicts that global warming and climate variability will increase the frequency, intensity, and duration of abiotic stressors in agriculture, leading to impact the global agricultural productivity, quality, and food security. The rising human population necessitates more food and nutritional demands, eventually generating an upgrade in agricultural production and quality worldwide. As per Seijger et al. (2025), insufficient resource management results in the degradation and depletion of arable land, while climate variability intensifies major pest populations (Subedi et al. 2023), potentially predicting to diminish agricultural productivity by 25% over the next 50 years (Seijger et al. 2025).

Besides, a significant alteration in atmospheric composition has also been observed, characterized by an increase in atmospheric carbondioxide (CO_2) levels (Friedlingstein et al. 2022). Elevated CO_2 concentration (ECC) results in reduced water loss via transpiration, enhanced stomatal openings, and elevated photosynthetic rates (Sarma et al. 2023a). ECC may modify plant nutrient concentrations, especially the proteins, hence influencing the plant's defense systems against devastating insect pests. ECC diminishes the buildup of defense-related hormone jasmonic acid (JA) while increasing the synthesis of salicylic acid (SA) in plants (Li and Ahammed 2023). ECC may influence the rate of food intake, growth, fertility, and population density of major insect pests in agriculture. For example, ECC enhanced the eating behavior and reproduction abilities of the Japanese beetle, *Popillia japonica* Newman (Scarabaeidae: Coleoptera) and the western corn rootworm, *Diabrotica virgifera* (Coleoptera: Chrysomelidae) (Subedi et al. 2023). Extended drought followed by intense precipitaion may increase the populations of Oriental armyworm, *Mythimna separata* (Walker) (Lepidoptera: Noctuidae), since drought adversely affects the activity and abundance of this pest's natural predators (Subedi et al. 2023). Variations in precipitation levels considerably influence the insect pest biology and dispersal mechanism, hence affecting worldwide agricultural productivity.

Due to global warming, insects are anticipated to migrate towards the poles and higher altitudes, hence increasing the risks associated with agriculture, such as those caused by migrating locusts. Nonetheless, the effects of climate change on insect pests are significantly context-dependent and exhibit variations with changes in topography, thus, influencing a rational change, while others observe a decline in generational frequency. *Icerya aegyptiaca* (Douglas) (Hemiptera: Coccoidea: Monophlebidae) is anticipated to occupy a larger habitable range in Africa, South America, and Asia, while *I. purchasi* Mask (Hemiptera: Coccoidea: Monophlebidae) is forecasted to prevail in South America, Asia, and Europe (Liu and Shi 2020). Mukhopadhyay and Mondal (2017) have examined the potential impact of climate-intensive variables on insect infestation in tea. It has been acknowledged that a rain-fed agricultural system, marked by alternating wet and dry periods, can significantly facilitate major pest infestations (Bhattacharyya et al. 2024a), thereby posing a considerable risk to achieving adequate food production and quality maintenance, globally (Kakoti et al. 2023). Deka et al. (2021a) have identified the prevailing microclimate and food supplies as significant variables, leading to possible outbreaks of insect pests in tea, including the infestations of nematodes and mites. *Biston suppressaria* (tea looper), *Ectropis obliqua*, *Empoasca onukii* (tea green leafhopper), *Entypotrachelus meyeri* (tea weevil), *Euwallacea fornicatus* (tea shot hole borer), *Fiorinia theae*, *Gyropsylla spegazziniana* (Paraguay tea ampul), *Helopeltis theivora* (tea mosquito bug), *Hyposidra talaca* (black inchworm), *Microtermes obesi* (wheat termite), and *Oligonychus coffea* (tea red spider mite) have been recognized as significant tea pests (Huang et al. 2008; Roy et al. 2015; Formentini et al. 2016; Baruah and Deka 2017; Bhattacharyya et al. 2023).

Insect pest occurrences in agriculture are associated with significant damages, leading to substantial losses in crop production, quality deterioration, and overall economic crisis (Bhattacharyya et al. 2024b). Worldwide food production must rise up to 70% by 2050 to satisfy the increasing population demands, with pests contributing significant losses, responsible for 20%–40% of the worldwide food supply (Savary et al. 2019). Lahlali et al. (2024) assert a significant threat to vital plant components such as leaves, stems, and underground parts from destructive insect pests and pathogens, aggravated by favorable climate and soil. Soil-borne agricultural pests could become challenging issues if extreme weather and temperature situations prevail, causing significant losses to microbial antagonists present in soil. Addressing the sustainability in soil fertility and protecting agriculture from the attack of devastating plant pests and pathogens, thus, requires substantial technological and climate-smart perspectives. Furthermore, invasive alien species (IAS) pose serious risks to global biodiversity conservation, since their presence typically results in huge economic losses to natural ecosystems such as agriculture, forestry, and aquatic life. Ectothermic species, such as insect pests, can undergo temperature adaptation by altering behavioral features that control their energy metabolism.

Post-harvest crop losses are other challenges of significant concern, leading to global food waste and threatening food security (Debebe 2022; Saikia et al. 2025b). These losses refer to the reduction in the quantity or quality of crops between the time they are harvested and consumed. Causes can range from improper storage, transportation issues, pest infestations, mechanical damage, or even poor handling practices. It is noteworthy to mention that underdeveloped countries face severe food security challenges due to uncontrolled measures and lack of public awareness in pest management, requiring more projected interest to ensure sustainable food security (Junaid and Gokce 2024). To meet the growing demand for food security and safety against the insect pest devastations in agriculture, applications of chemical pesticides have become more popular and formidable options amongst the farmers. Keeping the trust in agrochemical pest management, Rahaman et al. (2018) have indicated the potential of chemical pest management as a primary means to protect agriculture against major insect pest devastations. Between 1870 and 1945, farmers began to use inorganic-based substances as the primary means of pest management in agriculture. Sheail (1991) stated that copper- and sulfur-containing compounds were frequently employed to eradicate infectious pests, including the microbiological pests in agriculture. The need for applications of different synthetic pesticides increased only after the end of World War II. As of 2019, around 4.19 million metric tons of insecticides were used globally. In terms of pesticide consumption, China is reported as in the top with 1.76 million metric tons, followed by US with 408,000 tons, Brazil with 377,000 tons, and Argentina with 204,000 tons.

Twenty percent of emerging nations, including Vietnam, Cambodia, and Laos, are classified as major users of pesticides, according to the World Health Organization's forecast for the increase in pesticide consumption in Southeast Asia each year (Pathak et al. 2022). India is recognized as a prominent manufacturer of pesticides in the Asian subcontinent, producing over 90,000 tons annually, which include organochlorine pesticides such as dichlorodiphenyltrichloroethane (DDT) and benzene hexachloride (BHC) (Pathak et al. 2022). Zitko (2003) recognized DDT, β-hexachlorocyclohexane (β-HCH), dieldrin, endrin, aldrin, chlordane, captan, and parathion as the primary pesticides utilized in agriculture. Agricultural experts define pesticides as chemical formulations used to eliminate or control pests, primarily utilized in crop protection. Mada et al. (2014) emphasized the significance of agrochemicals in agriculture, since they protect the crops from detrimental pests. Cooper and Dobson (2007) reported that insecticides contribute to reducing agricultural losses through diminishing the probability of severe damage from various insect pests and pathogens. Damalas and Koutroubas (2016) suggests that agrochemicals have dual characteristics in agriculture, namely inducing toxicity in plants and adversely affecting pest population dynamics.

Although the use of pesticides is advocated for increased crop protection, the rapid and unlawful exploitation of chemical substances has generated numerous adverse effects on non-target organisms, including the humans and animals, additionally creating an enormous threat to soils, plants, and ecosystems (Bhattacharyya et al. 2024c). Pesticide exposure to humans may occur by ingestion, inhalation, or dermal contact. Nicolopoulou-Stamati et al. (2016) assert that pesticides may be metabolized, excreted, stored, or bioaccumulated in the bodies of humans and other animals. Pesticides, according to researchers, have the potential to interfere with and even impede a number of essential biological processes in living things. These include protein synthesis, cell membrane formation, and the manufacturing of vital cellular organelles. Pesticide exposure is associated to several human ailments, including the causes of allergies, asthma, cancer, hormonal disturbances, and hypersensitivity reactions. Agrochemicals are known to degrade the soil quality (Bhattacharyya et al. 2024c), disrupt beneficial microbial population dynamics (Bhattacharyya et al. 2024d), and severely affect the functions of essential soil enzymes (Bhattacharyya et al. 2023, 2024a; Borah et al. 2024). Mahanta et al. (2023) have demonstrated a link between the adverse impacts of pesticide usage and rapid decline in soil fertility, resulting in heightened toxicity. Bhattacharyya et al. (2024e) linked the toxic effects of chemical substances to a major decline in microbial population numbers and their functional attributes. The persistent presence of chemical toxicity in soil presents significant risks, including those related to maximum residue limits (MRLs) in agriculture (Bhattacharyya and Sarmah 2018).

The widespread use of chemical pesticides in the tea ecosystem, as noted by Deka et al. (2021a), causes soil and groundwater contamination, depletes beneficial microbiomes, and diminishes natural predators, while also contributing to the resurgence and emergence of pest resistance and residue accumulation in the finished product. The residual activity and persistence of pesticides in soil are influenced by various essential elements, including the chemical composition and active ingredients of the pesticides, as well as soil properties and the prevailing temperature and topography. Moreover, there is a lack of documentation on the cataloguing of approved molecules and agrochemicals that delineate bioefficacy standards as confirmed by authorized regulatory bodies, especially in relation to managing harmful insect pest populations in agriculture. Consequently, while new molecules undergo expedited testing to achieve requisite certification and field evaluation approval from various governmental and non-governmental organizations (NGOs), recently articulated climate-smart approaches in agriculture emphasize the promotion of biosafety protocols for enduring sustainability in agroecosystems.

Reducing the use of harmful agrochemicals and transitioning towards the validated and sanctioned use of alternative, non-chemical, and biological methods is crucial for ensuring safe, sustainable, and environmentally sound agriculture (Bhattacharyya and Jha 2012; Roy et al. 2020; Mitra et al. 2021; Herrmann et al. 2025). By reducing emissions of greenhouse gases (GHGs), increasing agricultural output, and adjusting to and fortifying agricultural food security systems, climate-smart agriculture (CSA) seeks to promote sustainable agricultural practices. Recently, CSA aims to promote more sustainable development and resistance against climate crises while simultaneously maintaining sufficient food safety security, and plant protection for enhanced sustainable agriculture (Lipper and Building 2018). One important component of CSA is climate-smart pest management (CSPM), which advocates for minimizing the use of toxic agrochemicals in field and promotes for an environmentally sustainable pest control strategy. Reducing the chemical intervention in agriculture and promoting a more sustainable pest management strategy are, thus, considered crucial components of the CSPM program. Application of microbial consortium-mediated bioformulations has tremendous promise in advancing the rhizosphere through enhanced nutrient mining and key enzyme activities (Saikia et al. 2025b).

CSPM is seen as a contemporary method that may facilitate the transformation and reinvention of sustainable agriculture systems to attain global food security and safety. CSPM strategies seek to anticipate changes in pest dynamics induced by climate change using rigorous pest surveillance and pest-infestation forecasting systems for prompt diagnosis and adoption of effective risk management protocols. Effective interventions in CSPM primarily rely on ecologically sustainable pest management practices to enhance agricultural productivity. This shift includes the use of transgenic plants for pest resistance, although the use efficiency of genetically modified crops (GMC) is not found to be suitable in all instances, due to the constant evolution of insects (Wheelis 2002) and other ethical issues. Microbial biopesticides, such as endophytic entomopathogenic fungi (EPF), represent an alternative, non-chemical, eco-friendly solution against the toxic pesticides with a growing market of approximately US$ 7.7 billion, expanding at a rate of 14.1% annually (Sharma et al. 2020). According to Mamun et al. (2014) and Bhattacharyya et al. (2024e), the potential of endophytic EPF against insect pests in agriculture, including in tea, has been widened owing to their less or no toxicity issues, improved bioefficacy, shelf life, and reduced chances of environmental contamination, ensuring their success as one of the sustainable components of integrated pest management (IPM). According to most of the workers (Kumhar et al. 2020; Deka et al. 2021b), EPF represents one of the most preferable and effective pest management tactics in plant protection under CSA, as they generate an environmentally friendly farming system without causing any detrimental effect on non-target organisms, including the human beings.

Among the recorded 100 genera and 1000 species, the genus *Metarhizium* serves as one of the most pertinent microbial biocontrol agents owing to its cosmopolitan nature and ability to infect a wide range of hosts with diverse modes of action and pathogenicity (Bhattacharyya et al. 2024b, e). According to entomologists, EPF offer several advantages over conventional chemical pesticides that are basically toxic in nature. These microorganisms are highly effective against primary insect pests and are increasingly being used in organic farming throughout the world (Bhattacharyya et al. 2024a, e). Mongkolsamrit et al. (2020) have examined the advantages of utilizing *M. anisopliae* over conventional insecticides in the North-eastern region, as it results in little negative consequences for non-target organisms, including the silkworm industry. Unlike synthetic pesticides, EPFs never disrupt the key nutrient contents in soil, such as carbon and nitrogen, making them environmentally more suitable and sustainable (Ownley et al. 2008). Additionally, EPF have their abilities to withstand adverse topography, climate, and soil for more than a year; thus, they are considered preferable non-chemical and sustainable alternatives against major pest infestations under CSA. García-Estrada et al. (2016) assert that the use of *B. bassiana* can successfully reduce insect pests and mitigate plant damage while additionally encouraging long-term sustainability in farming. As EPFs are considered safe for humans and non-target organisms and pose minimal environmental risks, they have become an effective component of IPM. EPF are host-specific, effective in small quantities, and even degrade quickly, reducing the need for frequent field reapplications (Sandhu et al. 2012). EPFs are also less prone to developing insect resistance, further enhancing their sustainability in CSPM (Islam et al. 2021). The aim of the current review is to emphasize the potential of EPF as a prominent biocontrol agent for managing devastating insect pests in agriculture.

Entomopathogen-mediated pest immobilization may be accomplished by the modes of action of several mycotoxins produced by EPF (Bhattacharyya et al. 2024e). The EPF favors conditions with elevated moisture and humidity levels. Consequently, these fungi thrive in damp and humid situations, particularly in the tropical and subtropical regions of the world. The current article highlights different modes of action in EPF-mediated pest management and additionally discusses diverse tactics to improve the bioefficacy and virulence of the EPF and further strengthen the concepts of CSA using efficacious microbial strains under diverse topography and multilocational field situations. Besides, scopes for using genetic engineering in improving the effectiveness of EPF have been addressed to strengthen the concept of CSA.

According to molecular biologists, genetic engineering can assist in producing efficient EPF strains that would be more resilient to extreme environmental situations, allowing them to be used in a wider range of agricultural settings under diverse soil, climate, and topography. Myconanoformulation using pertinent EPF to promote the use efficiency of this microbial biocontrol agent consequently enhanced the scope of EPF-mediated pest management (Vega 2018). Nanoformulations provide numerous benefits, such as elevated surface area, adjustable characteristics, and the capacity to produce reactive oxygen species (ROS) upon extreme light exposure conditions, thereby playing a key role in remediating metal-polluted soil and water bodies (Dullah et al. 2025). Bibliometrics serves as a powerful tool for analyzing the trends, patterns, and research developments (Bhattacharyya et al. 2024d) in plant protection, including in addressing the recent trends and patterns in EPF-mediated pest management. By quantifying recent and up-to-date publications, citations, and the evolution of research topics or interests over time, bibliometrics provide suitable insights into key research areas, influential authors, and emerging technologies in EPF-mediated pest management, such as the use efficiency of genetic engineering, transgenics, and EPF-based nanoformulation. The data eventually assist to shape the future directions in CSA for improving pest control and crop protection strategies, ensuring a more sustainable approach to agricultural practices.

Incorporating advanced technologies like artificial intelligence (AI) and the internet of things (IoT) has recently revolutionized insect-pest management and crop protection strategies to another level. IoT-enabled devices and sensors can monitor essential environmental factors, including the temperature, humidity, and soil moisture, as well as strategic effects of biotic interferences that influence the rapid pest development and succession trends. Even combined with AI, Scientometrics and IoT can process vast datasets to predict pest outbreaks, identify invasive species, and eventually assist in pest control effectiveness in real-time challenges. AI-based models, particularly those using machine learning (ML) and deep learning (DL) technologies, can optimize pest management by adjusting treatment schedules and dosages according to real-time data analysis. Additionally, fuzzy logic, when adopted in pest management, can handle uncertainties in pest behavior and prevailing environmental conditions, thereby enlightening the researchers while improving the decision-making targets within the IPM.

2 Insect Pest Challenges in Agriculture

Agriculture has been identified as one of the significant sectors enriching the integrity of a country's economy (Tiwari 2024; Saikia et al. 2025a). It was around 12,000 years ago when humanity shifted from roaming hunter-gatherer societies to the process of agricultural establishments, thus facilitating the Neolithic Revolution. Since the inception of agriculture, pests and their interactions with crops have consistently posed significant threats to farmers (Mukhtar et al. 2023). Pests are insects in general but can also be referred to as any form of living organism that may cause significant harm to humans or human concerns. Diverse phytopathogens, weeds, viruses, fungi, nematodes, and protozoa are more commonly included under agricultural pests. According to Ballal et al. (2022) and Bhattacharyya et al. (2024a, e), insect pests are considered as major biotic challenges in agriculture, resulting in considerable crop losses, impaired crop quality, and consequently elevated production costs.

Table 10.1 Emerging insect pests in EICs

Emerging insect pests in agriculture	Common name	Target host
Bactrocera spp.	Fruit fly	Diverse fruits and vegetables
Bemisia tabaci Gennadius	Whitefly	*Gossypium* spp. and *Nicotiana tabacum* L.
Casmara patrona (Lepidoptera: Oecophoridae)	Large stem borer	*C. sinensis* (L.) Kuntza
Ceratovacuna lanigera Zehntner	Sugarcane woolly aphid	*Saccharum officinarum* L.
Helicoverpa armigera Hübner	American bollworm	*Cajanus cajan* (L.) Huth
Maconellicoccus hirsutus (Pseudococcidae: Hemiptera)	Mealy bug	*C. cajan* (L.) Huth
Macrosiphum miscanthi (*Sitobion miscanthi*) Takahashi	Indian grain aphid	*Triticum aestivum* L., *Hordeum vulgare* L., *Avena sativa* L.
Maruca vitrata Fabricius	Spotted pod borer	*C. cajan* (L.) Huth and *Vigna unguiculata* (L.) Walp.
Nilaparvata lugens Stal	Brown plant hopper	*Oryzae sativa* L.
Plutella xylostella L.	Diamondback moth	*Brassica oleracea* L.
Sesamia inferens Walker	Pink stem borer	*Sorghum bicolor* (L.) Moench
Spodoptera exigua Hübner	Beet army worm	*Cicer arietinum* L.
Zeuzera coffeae Nietner	Common red borer	*Camellia sinensis* (L.) Kuntze

Source: Ghosh and Samanta (2003), Sharma (2016), and Mukhtar et al. (2023)

According to an estimate, insect pests cause diverse crop losses ranging from 18 to 20% in Indian agriculture, where pests like mealybug (Family: Pseudococcidae Heymons), beet armyworm (*Spodoptera exigua*, Hübner), and pink stem borer (*Sesamia inferens*, Walker) pose serious challenges by threatening the food security and economic stability of human beings (Mukhtar et al. 2023; Tiwari 2024). In recent years, pest species like American bollworm (*Helicoverpa armigera*), brown plant hopper (*Nilaparvata lugens*), diamondback moth (*Plutella xylostella*), pink stem borer (*Sesamia inferens*), whitefly (*Bemisia tabaci*), and wheat aphid (*Macrosiphum miscanthi*) have emerged as key threats to economically important crops (EICs) like wheat, barley, oats, sorghum, pigeon pea, and rice (Mukhtar et al. 2023). Emerging insect pests in agriculture, along with the infested crop species, have been represented in Table 10.1.

It is noteworthy to mention that the untimely and unwanted emergence of insect pests in agriculture has eventually led to considerable losses, and the farmers face tremendous difficulties in maintaining the quality of crops and yield economics. A sudden influx of insect pests adversely impacts the agricultural economics, crop productivity, and general biodiversity (Tiwari 2024). Household animal treasures, including the livestock and cattle feed, are subject to regular disturbances due to damage of grasslands infested by pest populations. Adverse impacts have also been observed due to severe housefly infestations in cattle that increase bacterial contamination in milk (Gioia et al. 2022). Mukhtar et al. (2023) have addressed the challenges incurred due to anthropogenic activities, such as excessive and illegal use of agrochemicals, irrational farming practices, hybrid variety introduction, and so on, that eventually contribute to the advancement of major insect pest infestations in agriculture. These advancements, in turn, have a negative consequence on pest resistance, secondary outbreaks, and pest resurgences.

Thus, considering the devastation of insect pests in agriculture, discussing its suitable and timely management options becomes a significant challenge under CSA perspectives. Bhattacharyya et al. (2024b) and Bhattacharyya et al. (2024e) have addressed for the implementation of healthy CSPM approaches in agriculture, like the recommended use of plant protection formulations (PPFs), effective biological control agents, including the potential use of EPF, and adoption of good agricultural practices (GAP). Additionally, Tiwari (2024) has focused on suitable farmer-friendly IPM strategies and other mechanical controls, including the adoption of pheromones and light traps for systematic and orderly management of devastating pest populations and their outbreaks in CSA. The following sections aim to address the risks and challenges associated with chemical-based pest management, besides, highlighting diverse eco-friendly, non-chemical options of insect pest management, especially focusing on the improved use efficiency of EPF-mediated insect pest management to promote biodiversity and long-term agricultural sustainability.

3 Agrochemical-Based Pest Management: Risks and Challenges

Management of agricultural pests using diverse agrochemicals is broadly defined as agrochemical-based pest management (He et al. 2019; Borah et al. 2024). Pest management in agriculture involving the applications of diverse agrochemicals began in the early twentieth century, though its origins can be traced even further back. According to agricultural fraternities, agrochemicals can reduce agro-pest infestations, including the damages caused by major insect-pests, phytopathogen damage, growth of unwanted weeds, and nematodes (Al-Ani et al. 2025). According to Singh et al. (2024), chemical pesticides provide promising results in reducing pest populations, thereby assisting in preventing crop losses, improving yields, and controlling the plant damage caused by other invasive species.

The modern era of pesticide commenced after the discovery of high-efficiency pesticides such as dichlorodiphenyltrichloroethylene (DDT) and organophosphates. According to research conducted by Manyilizu (2019), global use of pesticides had increased manyfold by the second half of the nineteenth century. Synthetic insecticides, including the organochlorides such as DDT and benzene hexachloride (BHC), organophosphates (parathion, malathion, and azinophos methyl), phenoxyacetic acids (2,4-dichlorophenoxyacetic acid, 2-methyl-4-chlorophenoxyacetic acid, and 2,4,5-trichlorophenoxyacetic acid), captan, carbamates (aldicarb, carbofuran, oxamyl, and methomyl), neonicotinoids (imidacloprid, acetamiprid, clothianidin, and nitenpyram), and glyphosate came into force for their applications after World War II (Khan and Ahmad 2019). Prior to which the chemicals used to protect the plants from insect pests were identified as lead (Pb), horticultural oils, lime, sulfur, nicotine, soaps (Karlik 2003), and certain other inorganic elements such as copper (Cu) and sulfur (S) that were earlier used against significant fungal pathogens (Sheail 1991). At the onset of the twentieth century, the utilization of synthetic chemicals, including the copper salts, nicotine extracts, and arsenic compounds, gained significant prominence in various agroclimatic regions. Current report indicates the existence of thousands of pesticides designed for plant protection and agricultural enhancement. As mentioned above, globally around 4.19 million tons of pesticides are used, with China being the largest consumer at 1.76 million tons, followed bt US, Brazil and Argentina. India is also a significant player in pesticide production and consumption, as agrochemicals are increasingly utilized to improve crop yield (Aktar et al. 2009). There are more than 900 registered pesticides, and approximately two-thirds of them are employed in agricultural practices worldwide. Thus, agrochemicals are rapidly being used in farming systems to increase crop yield and enhanced plant protection.

Although chemical pesticides are reportedly being utilized in agriculture to protect the plants from diverse pest infestations, they are also known to alter the physicochemical properties of soil. Changes have been observed in natural soil color that has frequently changed to a brown-white color with large-scale and rapid applications of agrochemicals. Conversion of ferric ion to ferrous ion is an important detrimental chemical reaction to justify the quality of healthy soil. Salts of ferrous ions are more soluble, and hence, their uptake becomes easier for the plants and, therefore, plays an important role in maintaining the soil fertility. Overuse of pesticides in fields can dramatically reduce the conversion of ferric ions to ferrous ions that have toxic effects in soil, resulting in deterioration of soil health and porosity. Pesticides also alter the acidity of soil and, subsequently, result in severe degradation of the cultivated land. Ammonification of arginine is greatly reduced to about 70%–90% upon rapid application of pesticides consecutively for three years.

Besides, the ligating sites of pesticides form complexes with metal ions in agricultural lands. Thiophanate methyl can interact quickly with Cu and nickel (Ni) metal ions, whereas carbofuran forms complexes with iron (Fe) and manganese (Mn) ions (Kaur et al. 2017). The soil further loses its essential nutrients and generates undesirable porosity, leading to extensive soil degradation and erosion activities. Apart from alteration of physical properties of soil, chemical pesticides also negatively affect the native soil microbiomes and organic matter content (OMC) in soil. Formation of complexes with metal ions usually restricts their availability in soil and to the microorganisms inhabiting the soil, thus damaging the natural integrity of healthy soil (Rajak et al. 2023). Functional attributes of beneficial soil microorganisms respond differently to chemical pesticides, causing unfavorable changes, leading to a decline in the populations of beneficial soil microbiota. It has been documented that pesticides may disrupt key biochemical reactions and protein profiling properties of agriculturally important bacteria (AIB). Shahid and Khan (2022) have addressed the decline of cellular integrity of native microbiota due to chemical toxicity, thereby affecting microbial population size, cell permeability, shelf life, and viability.

Additionally, pesticides can enter into the aquatic ecosystem through agricultural runoffs, leaching activities, rapid application in home gardens, and industrial runoffs. It is noteworthy to mention that chemical pesticides, particularly the insecticides, are highly toxic to aquatic organisms. They end up as sediments in aquatic bodies and bioaccumulate in aquatic organisms, which later appears as a phenomenon of biomagnification at different trophic levels. Soil, plants, and ecosystems suffer tremendous challenges from the inappropriate and illegal application of toxic pesticides, despite the fact that these chemicals, too, play essential roles in farming, ensuring crop protection and yield enhancement. The development of insecticide resistance in the target insect species can be caused by the excessive use of pesticides. This, in turn, requires the use of larger concentrations of insecticides, leading to higher production costs and disrupting the natural insecticidal characteristics. Toxicity generation in soil, plant, and environment due to rapid and undesirable use of chemical molecules is represented in Fig. 10.1.

Exposure to insecticides, herbicides, and fungicides at adequate quantities adversely impacts the non-target organisms, including the human beings, affecting their survival, behavior, fecundity rates, and various other biological and metabolic processes. While pointing out the toxicity of chemical pesticides, the United Nations Human Rights Council (UNHRC) has advocated that about two hundred thousand people die from exposure to pesticides, most of them being from the underdeveloped or lower-income countries (Nag and Gite 2020), thus indicating the justified use of chemicals in agriculture. Many pesticide users are, however, unaware of the toxicity induced due to chemical exposure as well as high doses and frequent usages of

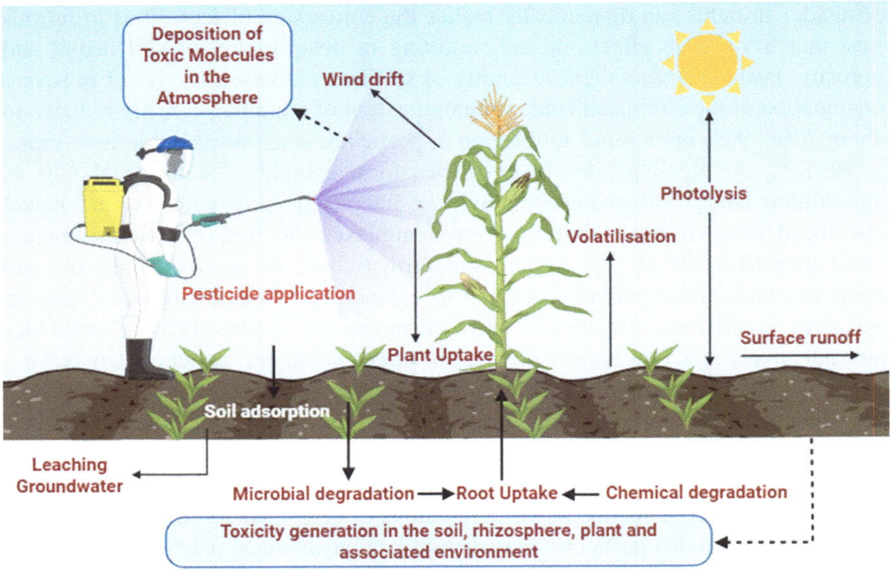

Fig. 10.1 Risks associated to agrochemical toxicity in soil, rhizosphere, plant, and environment

chemicals. Farmers who are well aware of the risks encounter difficulty in translating them into effective agro-practices (Damalas et al. 2006), resulting in the hasty, irrational, uncontrolled, and unscientific use of pesticides, posing substantial threats to soil, plant, and ecosystem. Besides human beings, several other organisms are also adversely affected due to pesticide exposure and related toxicity, including the fruit flies, earthworms, amphibians, and birds, resulting in their morbidity and mortality (Mandi et al. 2020). The negative impacts of pesticides in non-target organisms include oxidative stress, organ damage, reproductive failure, and neuronal degeneration (Rajak et al. 2018). Negative consequences exerted by pesticides affecting different organ systems and animal physiology have been depicted in Table 10.2.

Given that the potential risks associated with pesticide exposure exceed their advantages due to rapid and unjustified application, it is imperative to establish stringent chemical toxicity monitoring systems. These systems suggest incorporating advanced techniques such as HPLC (high-performance liquid chromatography) and GC-MS (gas chromatography-mass spectrometry) to accurately quantify pesticide levels in soil, water, and agricultural products, thereby regulating their overuse and related toxicity. IPM has emerged as an environmentally friendly, sustainable alternative for pest control in agricultural fields, incorporating environmentally safer alternatives, such as biopesticides, pheromones, and biological control agents, thereby diminishing the reliance on toxic chemical pesticides (Kashyap et al. 2024) in agriculture.

4 Integration of Entomopathogenic Fungi as a Key IPM Component

Although a farmer is required to rely on conventional pest management approaches to manage pest infestations under specific circumstances, there also appear to rely on other non-chemical and ecologically sustainable alternatives (Sarmah et al. 2020). Adhikari (2022) asserts that effective and enduring pest management solutions in agroecosystems rely fundamentally on a thorough knowledge and understanding of the appropriate strategy, pest behavior, biology, and ecological integration. Thus, there appear various approaches to manage insect pests in agriculture, such as cultural, mechanical or physical, biological, chemical and genetic control approaches, and ultimately an integrated approach, more commonly known as IPM, that involves an integration of cultural, biological, mechanical, and judicial use of plant protection formulations (PPFs) (Desneux et al. 2010; Sarmah et al. 2020). Physical or mechanical control strategies employed against the insect pests result in insect elimination, disrupt their typical behavior, or modify the environment to deter the pest activity (Adhikari 2022).

This type of methodologies possesses negligible or minimal adverse environmental implications and generates nil residue on agricultural products. An efficient insect pest control necessitates an integration of conventional and cutting-edge techniques, enabling our marginal and small-scale farmers (SSFs) to tackle diverse pest challenges while fostering biodiversity, environmental sustainability, and enduring

Table 10.2 Diverse pesticides and toxicity concerns to different target hosts and nature of damage

Chemical pesticides	Target organisms (hosts)	Nature of damage on non-target organ system
Aldrin/Dieldrin	Insect pests of corn and cotton	Affects organ development, causes significant damages to gastrointestinal, hepatic, neurological, and reproductive system
Atrazine	Weeds	Affects body weight, causes disruption in cardiovascular, hepatic, renal, and reproductive system
Chlordane	Household insect pests as well as insect pests in gardens and agricultural fields	Affects different developmental stages of hematological, neurological, and hepatic system
Chlordecone	Insect pests of tobacco, ornamental shrubs, bananas, and citrus trees	Affects liver, kidneys, thyroid, and reproductive health
2,4-Dichlorophenoxyacetic acid (2,4-D)	Weeds (land/water conditions)	Loss in body weight, occurrence of cardiovascular diseases, imbalance in hormone production, hematological disorders, hepatic diseases, disruption of neurological system, renal damage, and respiratory disorders
DDT	Used to reduce the insect pest populations in agriculture and mosquitoes, causing malaria	Causes cancer, affects hepatic, neurological and reproductive systems. Also affects different developmental stages related to organ development
Endrin	Insects, rodents and birds	Affects the hepatic and neurological system
Endosulfan	Insects of food and non-food crops	Affects developmental stages of organs, disrupts immunology, and neurology
Glyphosate	Weeds in farms, home gardens, and lawns	Causes loss in body weight, damages to gastrointestinal, hepatic, renal, and respiratory system
Heptachlor/heptachlor epoxide	House-hold insects and food crops (specially corn)	Affects the development of vital organs, disrupts hepatic, immunological, and reproductive system
Hexachlorobenzene	Fungus of seeds of onion, sorghum, wheat, and other grains	Causes cancer, affects developing organs and damages to hepatic, neurological, and reproductive system
Methoxychlor	Flies, mosquitoes, chiggers, cockroaches, and other insects	Affects development of neurological and reproductive systems

(continued)

Table 10.2 (continued)

Chemical pesticides	Target organisms (hosts)	Nature of damage on non-target organ system
Methylene chloride	No specific target, important ingredient in different pesticide products	Causes different forms of cancer, and causes imbalances in neurological system
Pentachlorophenol	Fungus, insects, weeds	Responsible in causing cancer, affects in vital organ development, and damages to hepatic system
Pyrethrins and pyrethroids	Insects of pets and livestocks	Causes developmental toxicity to vital organs, gastrointestinal damage. Creates toxicity to neurological, and reproductive system in children
Toxaphene	Insect pests of cotton and other crops	Disrupts immunological and neurological system

agricultural viability (Tiwari 2024). IPM encompasses an array of sustainable solutions aimed at controlling or declining insect pest populations in agriculture which would otherwise compromise crop yield and the quality of finished products or food grains, thereby affecting human health, safety, and security. IPM in agriculture provides a sustainable strategy for the control of insect pests that frequently impose damage on agriculture (Roy et al. 2020). Methods involving crop rotation, intercropping, and the selection of pest-resistant varieties are among the good cultural practices (GCP) usually adopted in IPM that eventually assist in the disruption of pest life cycles, pressure minimization, and enhancement in crop resilience (Islam et al. 2016). According to the workers, adoption of GCP in IPM plays a significant role in modifying the crop environment to make it less conducive to diverse pest infestations. While lowering the collateral damage linked to broad-spectrum pesticide applications, mechanical control methods, such as the use of light traps, barriers, and physical removal techniques, can be regarded as sustainable, non-chemical alternatives for managing insect pest populations in IPM under precision agriculture (Tiwari 2024).

The significance of IPM lies in its capacity to tackle the intricate challenges associated with insect pest infestations while preserving environmental integrity and fostering long-term agricultural sustainability concurrently. Desneux et al. (2010) highlighted the impending necessity of IPM in contemporary agriculture, considering the negative consequences associated with exposure of chemical pesticides on crops. IPM also assists in the reduction of selective pressures on pest populations and mitigates the risk of pesticide resistance (Tiwari 2024). Harnessing ecological interactions that regulate pest populations in natural ecosystems also brings significance in highlighting IPM-based pest management in agriculture (Ogori et al. 2022). Tiwari (2024) has asserted the integration of multiple pest management strategies in IPM to achieve effective and enduring pest control. The incorporation of beneficial insects such as ladybugs and parasitic wasps promotes an ever-changing equilibrium between the pests and their natural predators, thus diminishing overreliance on toxic chemicals and pesticides.

EPF are a broad range of natural biocontrol agents, contributing significantly to the management of insect pest populations in agriculture (Islam et al. 2021; Bhattacharyya et al. 2024a, b, e). This group of biological control agents does this through infecting or inhibiting the growth, reproduction, and proliferation of devastating insect pests. Due to their safe, easy-to-use, non-chemical, and sustainable pest management tactics, this microbial tool is becoming more popular as a broad-spectrum microbial biopesticide against toxic chemicals in IPM. Researchers have established the use efficiency of EPF as pertinent microbial biocontrol agents (Bhattacharyya et al. 2023), establishing beneficial microbial population dynamics in soil and plant endosphere through reducing the activity and populations of soil-borne pests and pathogens, thereby assisting in the building of beneficial crop microbiomes. For instance, in the tea ecosystem, the efficient utilization of EPF as microbial biopesticides has improved plant defense mechanisms (Bhattacharyya et al. 2024c), potentially resulting in the creation of novel tea germplasms capable to withstand against various biotic and abiotic stressors (Zhang et al. 2019).

As there appear toxicity challenges related to the administration of *Bacillus thuringiensis* and nuclear polyhedrosis virus (NPV) biopesticides (Fernandez-Chapa et al. 2019), these biopesticides are currently being restricted for field experimentations in tea plantations in Northeast India (Bhattacharyya et al. 2023). In this perspective, EPF shows possibilities as a non-chemical, more ecologically friendly solution against toxic chemicals when it comes to controlling insect pest populations in agriculture (Bhattacharyya et al. 2024e). Bhattacharyya et al. (2023) have emphasized the significance of *Beauveria bassiana*, an entomopathogenic fungus, in effective management of major insect pest infestations in tea, therefore, elucidating recent advancements in EPF-mediated insect pest management. EPFs, such as *B. bassiana* and *Metarhizium anisopliae*, are effective against the ancient looper pest in tea (*Biston suppressaria*), the emerging looper (*Hyposidra talaca*), the tea mosquito bug (TMB) (*Helopeltis theivora*), and the red spider mite (*Oligonychus cofeae*) respectively (Bhattacharyya et al. 2023, 2024b). Researchers have evaluated that EPFs can effectively combat against a wide variety of arthropod pests. This makes them a potentially useful tool for developing environmentally friendly agricultural systems, instead of exploiting toxic chemicals (Bhattacharyya et al. 2024e). *B. bassiana* is renowned for its abilities to release a wide spectrum of mycotoxins and enzymes, which may be capable of impeding the key metabolic processes of destructive pest populations (Bhattacharyya et al. 2023). They are reported to release efficient enzymes like lipases, proteases, and phospholipases, which are believed to have tremendous potential in causing pest mortality (Moharram et al. 2021).

Likewise, EPFs are also effective against a wide range of insect pests, including the aphids, whiteflies, caterpillars, and termites. A classic example can be attributed to the application of *B. bassiana* that has successfully reduced aphid and whitefly infestations in vegetable crops. Groden (1999) has reported the use potential of *B. bassiana* in the control of soil-borne insects like the May beetle in Europe and the

Argentine stem borer in New Zealand and stated that this fungus is effective in targeting foliage-feeding pests. The efficacy of *B. bassiana* spray applications against coffee berry borer pests has also been reported (Wraight et al. 2021), and they were able to infect more than 700 insect species (Mantzoukas et al. 2022). The use of *M. anisopliae* in the control of termite populations in sugarcane cultivation has led to the contribution of approximately a 40% reduction in the use of chemical pesticides (Mesquita et al. 2023; Rani et al. 2024).

EPFs have several benefits, including their cost-effectiveness, high efficiency, safety for beneficial organisms (Bhattacharyya et al. 2024a), lesser residual effect in the environment (Bhattacharyya et al. 2023), and enhanced biodiversity in ecosystems due to rapid applications (Islam et al. 2021; Bhattacharyya et al. 2024b, e), which render them advantageous over chemical pesticides (Deka et al. 2021b). Wang et al. (2002) assert that *B. bassiana* is a superior biological control agent to traditional broad-spectrum chemical insecticides, owing to its significant reproductive potential and virulence, targeted mechanism of action, relatively short generation time, ability to diminish chemical residues in soil and the environment, and prolonged shelf life, even in the absence of a specific host. EPFs are non-toxic and, therefore, never generate any harmful residues in crops, soil, and water (Groden 1999). They are considered safe for pollinators like bees and butterflies and other beneficial insects because certain beneficial insects have exerted a natural tolerance to such fungus. The benefits of EPF in IPM are reported to have significant contributions to adopt long-term pest control modulations in several agricultural systems (Islam et al. 2021; Moharram et al. 2021). Recent advancements in targeted genome editing methodologies, such as clustered regularly interspaced short palindromic repeats (CRISPR), have enabled the scientists to develop genetically modified plant germplasms with enhanced adaptability to climate resilience and pathogen surveillance (Zaidi et al. 2020).

The current review sought to highlight the potential of EPF as significant microbial biocontrol agents for the management of catastrophic insect pests in agriculture. Kumhar et al. (2020) assert that pest immobilization can be accomplished through the mechanisms of various mycotoxins released by EPF. According to Sharma et al. (2023), the EPF prefers to enjoy high wetness and humidity conditions thus thrives in damp, humid environments, such as tropical and subtropical climates. In the light of above, the current section has been focused on highlighting the EPF as an integral component of IPM to enhance sustainability in CSA. Recent advancements in entomopathogenesis and behavioral defenses, alongside investigations into the molecular, genetic, and physiological mechanisms underlying entomopathogen-mediated insect pest control in agriculture, ought to be understood while promoting microbe-based technology under CSA. This evaluation encompasses application requirements, dose, viable spore load, and concentrations (cfu/mL), and compatibility issues pertinent to the effective utilization of EPF as a significant microbial biocontrol agent within IPM. The study examines the adaptation and long-term prospects of EPF concerning climate crises and substantial topographical variations.

5 EPF-Mediated Enhancement in Plant Resilience

EPFs are microbial biocontrol agents that exert direct or indirect influences on insect pests (Hu and Leger 2002; Yousef et al. 2017, 2018; Garrido-Jurado et al. 2020; Bhattacharyya et al. 2023, 2024a, e). EPF shows their uniqueness in several ways, such as acquiring the nutrients from soil and making them available for plant utilization. Due to incorporation, endophytic microbes are known to play a significant role in inducing systemic and acquired resistance in plants (Bhattacharyya and Jha 2012). Endophytic EPFs are considered as fascinating microorganisms that can establish symbiotic relationships with the host plants and have been reported to improve ecological adaptability. They can do it through boosting the plants against adverse drought, salinity, and extreme temperature situations, thereby assisting to improve nutrient allocations and modulating physiological responses. It may be mentioned that most of the EPFs are basically soil dwellers (Mantzoukas and Eliopoulos 2020) that form an important component of beneficial soil microbiome. Experts have suggested the potential of EPF as endophytes that play a vital role in mitigating the biotic and abiotic stresses in plants under climate change scenarios. Combatting the insect-mediated biotic stress through plant-associated endophytic EPF in horticultural crops has been investigated by Samal et al. (2023). EPFs are known to change the plant's physiochemical properties, like the volatile emission profile (VEP), and thereby assist in eliciting the production of secondary metabolites to regulate insect pest defense in agriculture. The workers have identified several enzymes, including the chitinases, laccases, amylases, and cellulases, produced by EPF that confer crucial plant resilience characteristics. Fungal endophytes have demonstrated the abilities to act as substantial biocontrol agents through infecting sucking pest complexes, lepidopteran larvae, and other cosmopolitan insects, resulting in major mycopathogenesis and contributing to effective plant protection mechanism. Through effective colonization of plant tissues and rapid insect infection mechanisms, endophytic EPFs establish a tripartite relationship wherein the endophytic fungi, the host, and the insect exchange their nutrients (Bhattacharyya et al. 2024a, e). Various adhesion molecules, including MAD1 in *Metarhizium robertsii* and hydrophobins in *Beauveria* spp., have been documented to facilitate effective spore attachment to the insect cuticle, leading to rapid pest infection and mortality. In order to ensure proper food allocation, EPFs are known to generate significant antibacterial substances in the dead cadaver, thereby decreasing microbial competition (Fan et al. 2017). *M. robertsii* rhizosphere proliferation and root competency are facilitated by the existence of raffinose and sucrose in root exudates. Additional aspects have been added to the EPF-mediated plant defense system through rendering sucrose and raffinose available for EPF growth and colonization. EPF may offer comprehensive protection to plants against phytophagous insects, either directly via the synthesis of toxic chemical compounds or indirectly through the release of volatile organic compounds (VOCs) to attract natural enemies. Many EPFs, such as *Metarhizium* sp. Sorokin (Hypocreales: Clavicipitaceae), *Verticillium lecanii*, *Paecilomyces* spp., and *B. bassiana*, can inhabit the plant rhizosphere or establish endophytic relationships with the plants, thereby eliciting significant plant resilience against biotic and abiotic stress-related factors. *B. bassiana*, an EPF, can limit the damage caused by devastating insect pests in agriculture, including the lepidopteran cob- and stemborers, *Ostrinia nubilalis,* and *Sesamia calamistis.* Hampson (Lepidoptera: Noctuidae) in maize (Bing and Lewis 1991; Cherry et al. 2004); the South American tomato pinworm, *Tuta absoluta* (Lepidoptera: Gelechiidae) (Powell et al. 2009), the tomato fruitworm *Helicoverpa zea* Boddie (Lepidoptera: Noctuidae) in tomato (Toffa et al. 2021); the tomato fruit worm, *Helicoverpa armigera* Hübner (Lepidoptera: Noctuidae) (Mantzoukas et al. 2022); the banana weevil, Cosmopolites sordidus Germar (Coleoptera: Curculionidae) in banana (Akello et al. 2008); the poppy stem gall wasp, *Iraella luteipes* Thompson (Hymenoptera: Cynipidae) in opium poppy (Quesada-Moraga et al. 2009); and the stem weevil *Apion corchori* Marshall (Coleoptera: Curculionidae) in white jute (Biswas et al. 2013), respectively. Sphaeropsidin A (SphA) is a pimarane diterpene recognized for its larvicidal and phagodeterrent properties against lepidopteran insects, including the *Spodoptera littoralis* (Andolfi et al. 2014). SphA was identified to exhibit contact and oral toxicity towards the chewing lepidopteran, *S. littoralis*, contingent upon its capacity for in vivo SphA synthesis (Di Lelio et al. 2022). An in vitro investigation demonstrates the sublethal impacts of partially purified protein extracted from *Lecanicillium lecanii*, Zimmerman, linked to solanaceous plants like tomato, on the green peach aphid, *Myzus persicae* (Sulzer). Figure 10.2 represents EPF-mediated enhancement in plant resilience under CSA, leading to prominent myco-biocontrol activities and pest mortality, thereby playing an immense role in plant protection.

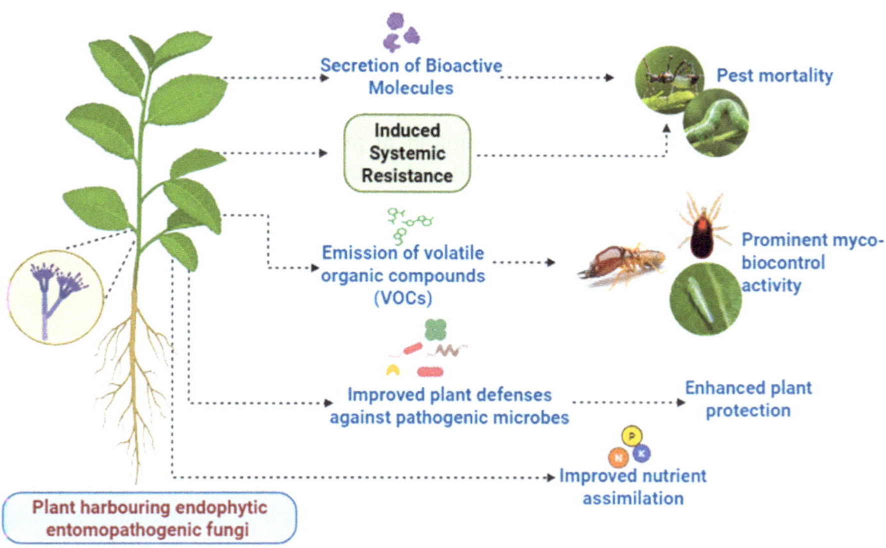

Fig. 10.2 Endophytic EPF-mediated enhancement in plant nutrient assimilation and crop protection

Furthermore, RT-qPCR expression studies of pivotal genes linked to the SA and JA pathways demonstrated upregulation, indicating substantial adverse effects on *M. persicae* survival and fertility patterns. The protein derived from *L. lecanii* significantly upregulated the SA-associated genes including the PR1, BGL2, AOS, PAL, LOX, and AOC, suggesting an augmentation of induced systemic resistance (ISR) in plants. This indicates the necessity for its purification and characterization as a novel bioactive molecule against the aphids and other phloem-feeding insect pests (Hanan et al. 2020). *Trichoderma harzianum* Tr6 is also considered a potent inducer of ISR that stimulates the immune system in plants and adversely affects the host preference and egg production abilities of *Trialeurodes vaporariorum* (Aldaghi et al. 2021). In response to threats like insect pests, plants trigger a cascade of responses that prime their defense-related (DR-genes) genes and increase their resistance to subsequent attacks. Inoculation of tomato plants by endophytic EPF, *B. bassiana* strain GHA, and *M. anisoplae* F01 affected the development of leaf miner larva and caused lower oviposition. Ghaffari et al. (2025) have indicated the involvement of plant immune response genes, leading to the development of significant plant resilience.

Endophyte treatment was documented to diminish the populations of *Aphis fabae* and *Aphis pisum* nymphs by 1.6–14.6 and 3.7–11.0, respectively, whereas endophyte seed treatment enhanced seedling survival by 20–100% relative to the control intervention (Akello and Sikora 2012). The individual or combined inoculation of *Rhizophagus intraradices*, an arbuscular mycorrhizal fungus (AMF), and *B. bassiana*, an entomopathogenic fungus, led to elevated concentrations of monoterpenes and sesquiterpenes in tomato plants infected by the beet armyworm (*Spodoptera exigua* Hübner), signifying an enhanced terpenoid-mediated defense response in the host plants (Shrivastava et al. 2015). Mycorrhizal symbiosis influences reactive oxygen metabolism and membrane lipid peroxidation pathways, therefore promoting the capacity for osmotic adjustment and antioxidant synthesis in plant roots and foliage (Bhattacharyya and Jha 2015).

In addition to influencing several biological processes, climate change has the potential to influence both the top-down and bottom-up aspects of agroecosystems through interactions among tritrophs (Bhattacharyya et al. 2016). The diversity, abundance, and function of microbes associated with plants can be impacted by changes in community structure and interspecies connections brought about by climate change (Boukal et al. 2019). In order to increase their resistance to diverse biotic and abiotic stressors, including growth augmentation, plants preferentially engage with their microbiome (Rodriguez and Durán 2020). Sui et al. (2020) have investigated the relationships between temperature-mediated EPF interactions, as well as temperature-induced physiological alterations in *Zea mays*. There appear to be significant differences in growth behavior, spore production abilities, and conidia survival rates among the potent EPF strains at different abiotic stress-related factors, thereby signifying the influences of abiotic stressors such as temperature, salinity, drought, and nutritional stresses in EPF-mediated plant resilience. EPF, such as *M. anisoplae* and *M. brunneum*, have been identified for their significance in protecting the plants against salt and enhancing soil fertility (Khan et al. 2012;

Sanchez-Rodriguez et al. 2015). Treatment with *B. bassiana* is known to protect *Zea mays* and *Quercus* from adverse drought (Kuzhuppillymyal-Prabhakarankutty et al. 2020). At higher air temperatures, maize positively influenced the conidia yield, germination rate, and pathogenicity of *B. bassiana*, whereas the growth and biological properties of *B. bassiana* remained unaffected. This implies that higher air temperatures may cause the plants' and EPFs' interactions to change from mutualism to commensalism. Additionally, EPFs' capacity to rapidly adjust and "transfer" resistance to their hosts may encourage plants to adapt to climate change. This could be interpreted as a positive feedback loop and better utilized to breed plants with a higher colonization potential of EPFs in order to withstand, adapt to, and mitigate climate change. EPF may hasten climate change reactions in agricultural and wild plant communities, necessitating measures to enhance EPF-facilitated plant growth and resistance. EPF's potential to improve plant resilience against abiotic stresses like heat, salt, and drought adds a new dimension to their relationship with host plants. This possibility deserves further exploration in the agricultural sector, where it might serve as a replacement for EPF dissemination in flood-prone applications and reduce pest or disease pressures in the context of climate change scenarios.

It is noteworthy to mention that either as endophytes or as rhizospheric fungi, EPFs provide protective effects against diverse phytopathogens, including the bacteria, fungi, viruses, and parasitic nematodes. EPF has demonstrated a protective effect against agricultural pests when administered via stem injection, seed dressing, and root or soil drenching methods (Quesada-Moraga et al. 2023). *B. bassiana* has been demonstrated to augment gene expression associated with plant defense, stimulate the synthesis of proteins pertinent to phytopathogenesis, and activate systemic immunity, thereby providing adequate crop resistance through the synthesis of anti-insect metabolites such as peramine, loline, and isoflavonoids, as well as the activation of the JA cascade (Tomilova et al. 2021; Batool et al. 2022; Gupta et al. 2022). Endophytic EPFs have also been found to enhance plants' photosynthetic efficiency and increase gene expression related to iron (Raya-Díaz et al. 2017; Gupta et al. 2022; Moustaka et al. 2022; Zitlalpopoca-Hernandez et al. 2022). This, eventually, activates both enzymatic and non-enzymatic defense reactions and induces the expression of genes that are induced by auxin (Ahmad et al. 2020; Tomilova et al. 2021; Zitlalpopoca-Hernandez et al. 2022). Combating the crops against significant pests through induction of plant growth-promoting (PGP) abilities in crops, this group of fungi plays a vital role in the maintenance of ecological integrity in agriculture. Due to their application as inundative sprays for short-term pest management, EPFs are typically vulnerable to several environmental variables. Upon effective establishment as endophytes in plants, EPF can offer enduring and effective protective strategies against destructive pests and pathogens. Integrating the omics approach, including detailed studies on proteomics, genomics, transcriptomics, and metabolomics, would be helpful to understand the EPF-mediated host plant-insect interaction leading to enhanced plant defense.

6 EPF: Biology and Mechanism of Action

EPFs are fascinating organisms that can act as natural pathogens to diverse insects, causing mortality. This group of natural bioagents plays a crucial role in mitigating insect populations under CSPM. EPF may be heterotrophic, eukaryotic microorganisms that are unicellular or multicellular (filamentous) and possess the capabilities to reproduce sexually, asexually, or both, consequently producing various infective propagules (Bahadur 2018). Hypocreales, Onygenales (the species Ascosphaera), Entomophthorales, and Neozygitales are recognized as the prominent fungal orders with the highest population density of EPF (Sung et al. 2008). EPFs play multifaceted roles as endophytic biocontrol agents, causing them to antagonize the plant pathogens, thereby boosting sustainability in plants through an array of mechanisms like production of PGP substances, facilitating rhizosphere microbial status and vigorous colonization (Mantzoukas et al. 2022). According to Bruck (2009), in addition to their function as pest regulators through the induction of epizootics, EPFs exhibit multitrophic level functions, including the colonization of internal plant tissues, rhizosphere colonization, and enhancement of plant fitness through promoting PGP traits. They are hostile to crucial phytopathogens because they can take up space and nutrients and cause the target plants to develop systemic resistance (ISR) (Rodriguez et al. 2009).

Research conducted by Ownley et al. (2008) indicates that treating the plant seedlings with *B. bassiana* leads to endophytic proliferation in their seedlings, potentially enhancing their protection against the phytopathogenic fungi, *Rhizoctonia solani* and *Pythium myriotylum*, respectively. The workers further reported that the applications of plant seedlings to fungal endophytes are known to exhibit mycoparasitism and induce effects like antagonistic races for nutrition and antibiosis against major pest infestations (Mantzoukas et al. 2022). Microbial infection by EPF causes substantial damages to insect pests, affecting their overall physiology, biochemistry and behavior, including the feeding deterrence, pest inertia, and alterations in oviposition, which are likely consequences of the secondary metabolites produced by the antagonistic fungi (Shahid et al. 2012). Some common EPF strains along with their target insect pests are tabulated in Table 10.3.

EPFs have their abilities to control a wide range of insect pest stages, including the larva to adult. It is reported that out of a total of 31 orders of devastating insect pests, 20 orders could be significantly infected by EPF (Islam et al. 2021). EPF primarily belongs to the phylum Ascomycota and are usually found in soil and plant surfaces. Certain key ecological aspects of this biocontrol agent are listed in Table 10.4.

According to Klingen et al. (2002), the degree of fungal adhesion and ability to penetrate the insect exoskeleton could be used to determine the level of antagonism of EPF infection. Microbial biocontrol agents, including the EPF, have the abilities to produce a wide range of extracellular enzymes like lipase, phospholipase, protease, chitinase, and esterase, leading to complete degradation of the insect integuments (Jaronski and Mascarin 2017). The hydrolysis of ester bonds of lipoproteins, lipids, and waxes present in the interior regions of the insect integument

Table 10.3 Common EPFs along with their target pests

EPF genera	Key target pest
Beauveria bassiana	White flies, aphids, caterpillars, beetles
Isaria fumosorosea	Whiteflies, thrips, mealy bugs
Lecanicillium lecanii	Whiteflies, aphids, scale insects
Metarhizium anisopliae	Termites, locusts, mosquitoes, soil pests
Nomuraea rileyi	Armyworms, boll worms

Table 10.4 Key ecological aspects of entomopathogenic fungi

Key ecological aspects	Detailed information
Kingdom	Fungi
Phylum	Ascomycota
Habitat	Soil, plant, and animal surfaces
Mode of action	Primarily parasitic in nature
Enzymes in action during mycoinfection	Mainly chitinase, may also include protease and lipase
Reproduction	Conidia or spores
Target species	Aphids, whiteflies, caterpillars, termites, mosquitoes, and beetles
Mode of application	Spray formulation, seed treatment, and soil application

is carried out by lipases in EPF, and strains like as *M. anisopliae* and *B. bassiana* have been demonstrated to be among the finest producers of this enzyme (Golzan et al. 2023). Phospholipases in EPF are a group of enzymes that degrade the phospholipids of an insect pest's cuticle (Moharram et al. 2021). Proteases are produced in great quantities by EPF. This enzyme normally gets activated after the exoskeleton of the insect pest is broken down by the enzymatic action of other molecules (Moharram et al. 2021). Insect proteases break down the epicuticle's proteinaceous components, and additional proteolytic degradation of soluble proteins releases the amino acids that sustain EPF (Wang et al. 2002). Fungal chitinases act synergistically with the protease enzyme systems and assists in degradation of insect cuticle. Moharram et al. (2021) have indicated the prominence of these groups of enzymes as pertinent virulence factors of EPF. Table 10.5 represents some of the common enzymes associated with explaining the EPF-mediated insect pest infection leading to pest mortality.

EPF usually follows a parasitic life cycle, where they invade or kill the insect hosts before producing new spores. EPF could be well differentiated from other microbial biocontrol agents and recommended organic tactics by their mode of action, which is typically a combination of mechanical penetration, nutrient depletion, and toxin production. EPF usually breaks into the insect's exoskeleton, followed by the consumption of the insect's internal resources, thereby releasing secondary metabolites that eventually weaken the insects, leading to their death.

Table 10.5 Common enzymes associated with EPF-mediated pest management

Enzymes	Source of origin and secretion	Effects on target organ	Reference
Chitinases (endo and exo)	*Metarhizium anisopliae*	Complete degradation of the insect chitin	Duo-Chuan (2006)
Lipases	*M. brunneum*	Degrade the insect pest's cuticle/epicuticle	Santi et al. (2010)
Phospholipases	*Beauerisa bassiana* and *M. anisopliae*	Assists in damaging the lipid profile of an insect pest's cuticle	Moharram et al. (2021)
Proteases	*B. bassiana, M. anisopliae,* and *Ophiocordyceps. Sinensis*	Hydrolysis of cuticular proteins and modifies the cuticle surface	Shah et al. (2005)

Different modes of EPF-mediated insect pest management can be explained as follows: a. attachment to the fungal spores (here, at the beginning, conidia, or fungal spores get attached to the insect's exoskeleton; b. spore germination (conidia absorbs moisture and starts to germinate); c. exoskeleton (cuticle) penetration (this process initiates as the infecting fungi releases chitinase enzyme, thereby assists in degradation of insect cuticle following penetration within the target host); d. invasion and growth of the fungus (the fungus grows and multiplies inside the target hosts, causing disruption of its conventional metabolism; e. insect pest mortality (the insect dies within the initial 3–7 days of microbial infection due to release of fungal toxins and nutrient depletion); f. toxicity release and further production of fungal spores (the dead insect releases new fungal spores that spreads to other nearby pests). Different stages of mycoinfection in insect pests resulting in mycopathogenesis have been highlighted in Fig. 10.3. Viable spore load (approx. 1×10^8 colony-forming units/mL), optimum virulence, and preferred shelf life are considered as limiting factors contributing to enhanced mycoinfection in insect pests in agriculture.

7 Recent Advances in EPF-Mediated Insect Pest Management in Agriculture

7.1 Genetic Engineering and Biotechnological Intervention

Entomopathogens play crucial roles as prominent biological control agents against significant pest infestations in agriculture. However, the applications of EPF in pest management have received certain limitations due to their environmental instability, high cost for large-scale applications, virulence and prominent viability issues during field applications and storage (Bhattacharyya et al. 2024b). Thus, recent advancements in biotechnology tools and techniques are seen to be essential to increase their efficiency in plant protection against devastating insect pests. Since the onset of the twenty-first century, scientific investigations has primarily been focused

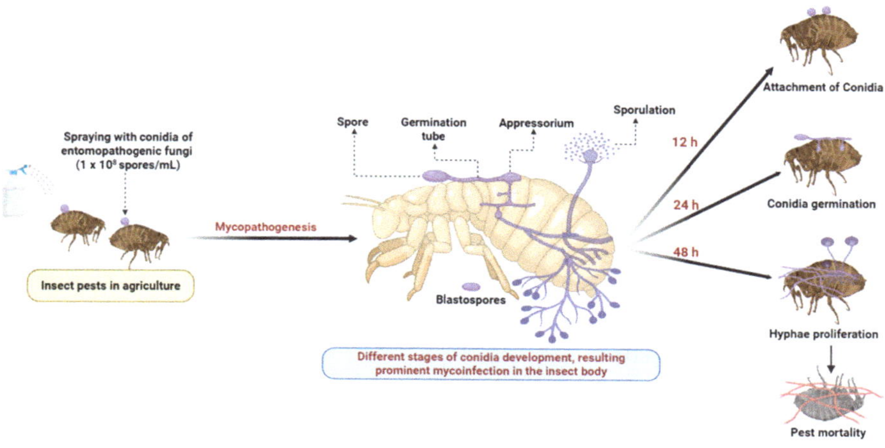

Fig. 10.3 Different stages of mycopathogenesis in EPF-mediated insect pest management

on the genetic engineering of fungal endophytes to augment crop yields and bolster plant resilience against devastating insect pests. Genetic engineering has provided multidimensional approaches to overcome the shortcomings associated with EPF-mediated plant protection in agriculture, where gene manipulation-related approaches have gained prominent significance to enhance the stress tolerance in crops against pest infestations (Fang and St Leger 2012).

Researchers have proposed breakthroughs in gene modification that could significantly enhance fungal virulence against various stresses. Du et al. (2022) have indicated the role of gene manipulation in increasing the viabilities and shelf life of microbial conidia during pest infection. As conidia are major infectious units of EPF, therefore, conidial production is considered an important factor contributing to the long-term efficacy of entomopathogens (Holder et al. 2007). Workers have indicated the significance of core transcription factors (TFs) as potential candidates for genetic improvement of EPF. These TFs are believed to be vital parts of key regulatory pathways during microbe-mediated pest management. Overexpression of the *MaSom1* gene in the cAMP/PKA pathway of *M. acridum* led to the increase in conidiation, stress tolerance, and augmented virulence in the pathogen (Du et al. 2022). Recent research has focused on developing recombinant EPF strains with improved viability and virulence, thanks to the development of RNA interference (RNAi) technology and the introduction of gene modification as a strategy to EPF (Wang and St Leger 2007). The RNAi technology has shown promising role in the field of insect pest control using EPF. A common fungal entomopathogen, *Isaria fumosorosea*, pathogenic to whitefly, has been genetically engineered to express a specific *TLR7* gene from dsRNA of whitefly.

The procedure silenced the TLR gene and increased the virulence of *I. fumosorosea* against whitefly. Nair et al. (2025) have discussed the significance of CRISPR-Cas9 technology in increasing the virulence and spore life (colony-forming unit/mL) of mycopathogens. The work assisted in eliminating the *Bbsmr1* gene, which downregulates oosporin production, resulting in enhanced virulence of the entomopathogen. These aspects of genetic engineering will lead to the introduction of novel, efficient strains in the field of biocontrol. The introduction of recombinant endophytic genes as significant biocontrol agents has gained prominence in advancing the sustainability in plant protection. Recombinant endophytic microorganisms release anti-pest proteins for the control of significant insect pest populations and can efficiently interact with and colonize the host plants (Fadiji and Babalola 2020).

7.2 Nanoformulations in Improving Limitations of EPF in Pest Management

Nanotechnology is an advancement in sustainable agriculture (Thakur and Yadav 2025) that assists in developing environmentally stable, cost-effective and resilient crop yield, besides protection efficiencies. It offers an excellent method for synthesizing appropriate nanoparticles (1–100 nm in size) (Dullah et al. 2025) that can be utilized as nanopesticides to eliminate or hinder undesirable pest infestations in agriculture. Researchers have recently gained popularity in using microorganisms as nanoformulations that could substantially assist in promoting a more cost-effective, easy, and reliable tool in CSA. For instance, myconanofungicides have been introduced in plant protection against fungal pathogenesis. EPF is increasingly regarded as a desirable option for nanoparticle synthesis due to its high yields, capacity for substantial protein production, simplicity of manipulation, and minimal environmental risks (Gong et al. 2023). A variety of organisms, both targeted and non-targeted, have been shown to be susceptible to the synthesis of copper nanoparticles (CuNPS) using an aqueous extract of *M. robertsii*, including *Aedes aegypti*, *Anopheles stephensi*, *Artemia nauplii*, and *A. salina*, *Culex quinquefasciatus*, *Eudrilus eugeniae*, *E. andrei*, *Tenebrio monitor*, respectively.

Nowadays, entomopathogen-based silver nanoparticles (AgNPs) have been practiced as potent biopesticides to kill or deter the insect pests. AgNPs are commonly utilized in agroforestry and medicine owing to their multifaceted properties. AgNPs produced by mycogenic processes are employed in controlling pest populations due to their antifeedant, larvicidal properties, and environmental compatibility. Treatment with AgNPs has decelerated insect growth and extended the larval duration by up to 4 days relative to larvae nourished with a fungal cell-free extract and untreated castor leaves. A recent investigation by Kamil et al. (2017) has evaluated the impact of AgNPs generated by insect pathogenic fungus, *B. bassiana,* on mustard aphids. The experiment involved bioefficacy testing of 25 *B. bassiana* isolates against mustard aphids. Based on the mortality rate of the mustard aphid, the EPF isolates were divided into two groups: Group I (0–50%; 20 isolates) and Group II (>50%; 20 isolates) (51–100%; 5 isolates) (Kamil et al. 2017).

According to the workers, there appear potent opportunities in using AgNPs in agricultural pest control. This study reports one of the pioneering works on synthesizing nanosilver formulations utilizing the EPF. Nanoparticles mediated by

Trichoderma viride were shown to exhibit pupicidal activity and up to 100% antifeedant action against first, second, and third instar larvae of *Helicoverpa armigera* at a concentration of 100 ppm. *H. armigera* third-instar larvae exposed to nanoparticles showed an increase in glutathione-S-transferase activity and a decrease in glucosidase enzyme level (Chinnaperumal et al. 2018). AgNP-loaded fungal metabolite nanoconjugates exhibit high mortality against *Spodoptera litura* (Namasivayam and Bharani 2021). The nanoconjugate showed up to 100%, 100%, 91.2%, 84.3%, 78.4%, and 71.3% mortality against 1st, 2nd, 3rd, 4th, 5th, and 6th instar larvae of *S. litura*, respectively (Namasivayam and Bharani 2021). It is additionally reported that nanoparticles made by *M. anisopliae* have significant insecticidal properties. Yosri et al. (2018) demonstrated that fungal nanoparticles had a 22% larvicidal activity against *Galleria mellonella*, but when combined with gamma-irradiated *Mycoplasma anisopliae*, which had higher activities for amylase, nitrate reductase, protease, and lipase, the synergistic interaction led to an 82% larval mortality rate.

Biopolymer-based formulations are emerging as innovative approaches for the smart delivery of EPF as bioinsecticides, presenting a viable strategy in IPM. Recent advancements include the use efficiency of various biopolymers such as cellulose derivatives, chitosan, alginate, and starch for creating bioformulations of EPF (Batista et al. 2017). Biopolymers can enhance the insect-infecting capabilities of EPF through prolonging the shelf life and conidia survival during field applications. They also improve the interaction between EPF and target insects, leading to better stability, adherence, and persistence of fungal spores on the insect cuticle, thereby increasing EPF's efficacy as bioinsecticides.

Microbial encapsulation, further, provides major landmarks in improving the efficiency of EPF in pest infection, leading to insect mortality. Microencapsulation and nanoencapsulation entail the integration of fungi into diminutive polymeric spheres or capsules that serve as protective barriers around the EPF, thereby safeguarding them from environmental stressors such as ultraviolet radiation and desiccation. Fungal conidia stability and field survival prospects in multilocational field trials are both improved by encapsulation. Concerns regarding fungal adherence and sporulation on the target insect's cuticle, bioproduct shelf life, and preservation of EPF's global efficacy are additional aspects to consider when using microbial bioinsecticides, such as EPF, for agricultural insect pest management (Saberi Riseh et al. 2022).

Chitosan-based bioformulations are a promising approach for developing bioinsecticides, particularly for EPF. Chitosan, a biopolymer obtained from chitin found in crustaceans and insects (Kou et al. 2021), enhances the efficacy, persistence, and adherence of these formulations to insects. Eco-friendly, biodegradable, and non-toxic, chitosan formulations like pellets, alginate beads, and hydrogels have demonstrated notable effectiveness against target insects, positioning them as viable alternatives to chemical pesticides.

Research conducted by Wu et al. (2022) demonstrated that bioformulations containing chitosan were effective in controlling the western flower thrips, *Plutella xylostella* (diamondback moth), by delivering *M. anisopliae* spores. Chitosan nanoparticles have been developed for the delivery of *Trichoderma asperellum* spores and have demonstrated prominent efficacy in controlling three significant soil-borne fungal pathogenesis caused by *Fusarium oxysporum*, *Sclerotium rolfsii*, and *Rhizoctonia solani* (Boruah and Dutta 2021), respectively.

Additionally, alginate-based bioformulations have garnered significant attention in CSPM for enhancing the efficacy of EPF as sustainable bioinsecticides. Alginate, a biopolymer from brown seaweed, consists of linear copolymers of β-D-mannuronic acid and α-L-guluronic acid (Wang et al. 2019). Formulations containing alginate-EPF are considered as safe for humans and the environment because they are non-toxic to non-target organisms and never leave behind any toxic residues (Gheorghita Puscaselu et al. 2020). The targeted delivery of EPF against numerous invasive agricultural insect pests has just been rendered possible with the development of several bioformulations based on alginate, including alginate gels and beads. By shielding the fungal spores from environmental stresses, alginate-based bioformulations increase the bioefficacy of EPF through improving the spores' persistence on target insects and influencing their shelf life and viability. Sarma et al. (2023b) demonstrated that alginate gel bioformulations containing *M. anisopliae* spores significantly enhanced the fungal spores' persistence and dynamism. For the purpose of maintaining *Trichoderma asperellum* submerged conidia and microsclerotia (MS), an additional formulation based on alginate had been developed. Alginate beads containing spores were part of the formulation; they were kept at 8, 25, and 35 °C for up to 120 days. For freeze-dried beads kept at 8 °C, the formulation successfully protected the conidia and MS concentration (Lopes et al. 2020). The sugarcane borer, *Diatraea saccharalis*, can be controlled by delivering spores of *B. bassiana* and *M. anisopliae* through an alginate bead bioformulations. In a study conducted by Wenzel et al. (2016), it was discovered that alginate beads could shield the spores from harmful elements including ultraviolet light and drying out, hence enhancing their bioinsecticide activity and prolonging their shelf life. Since the fungus were still alive after exposure, the alginate bioformulation shielded the conidia from the radiation for up to 48 hours. Commercially accessible alginate-based formulations for targeted administration of EPF include the product Algibio, which contains alginate beads infused with *B. bassiana* spores. Algibio has been utilized for the management of many insect pests, including the citrus blackfly (*Aleurocanthus woglumi*) and the coffee berry borer (*Hypothenemus hampei*). According to Friuli et al. (2023), Algibio has demonstrated significant efficacy and an extended impact on the target insects.

Starch, a biopolymer from various plant sources like corn, potato, and rice, is increasingly being used in developing microbial bioformulations to enhance the effectiveness of EPF (Bertoft 2017). Specifically, starch pellets and microspheres encapsulating *B. bassiana* spores (Friuli et al. 2023) have been developed to improve spore delivery, persistence, and adherence on target hosts (insects), resulting in higher mortality rates and efficacy. These formulations also protect the fungal spores from adverse environmental stressors, such as UV radiation and desiccation, thus extending their shelf life and overall performance.

According to Singh et al. (2020), the microspheres serve as a physical barrier, protecting the fungal spores and maintaining their shelf life and viability. To augment the protection and efficacy of starch-based formulations, supplementary coatings have been invented for the pellets or microspheres. These coatings may comprise biodegradable polymers, surfactants, or bioadhesive elements that promote adherence to target insects, thereby facilitating controlled release qualities, or improve the compatibility of microbial bioformulations against particular crops. Advanced encapsulation technologies, including spray drying, fluidized bed coating, and electrostatic encapsulation, have been investigated to enhance the formulation of starch-based bioformulations.

These techniques enable precise control over particle size, morphology, and encapsulation efficiency, enhancing mycoformulations' uniformity and performance. Further, the addition of surfactants, stabilizers, synergistic agents, and other bioactive compounds improves the efficacy and biocontrol properties of fungal spores in starch-based formulations (Vassilev et al. 2020).

7.3 Role of IoT, AI, and Image Capturing Technology in Improving the EPF-Mediated Pest Management in Agriculture

Recently, with the rapid development of information and communication technology (ICT), the smart plant protection (SPP) strategy is globally becoming a novel interest for agricultural fraternities (Huang and Shu 2021; Nayagam et al. 2023). The approach has the advantages of being sustainable, intelligent, and unmanned. The effective application of EPF-based pest control solutions necessitates precise monitoring and timely interventions. To improve the effectiveness and efficiency of EPF-mediated insect-pest management, cutting-edge technologies including the IoT, AI, and image-capturing technology are essential. The concept of IoT refers to the connection of a set of objects, devices, humans, animals, or machines in which all elements can communicate and integrate with each other (Wang et al. 2023). In fact, IoT is a connected network consisting of several entities to integrate the physical and cyber worlds with the addition of data processing capabilities and AI. The emergence of IoT and digital transformation has enabled the intelligent regulation of the physicochemical properties of agroecosystems, such as soil moisture and temperature, alongside the monitoring of bioproduct growth post-field applications. Furthermore, it allows special focus on preventive measures to identify devastating insect pests in agriculture (Berahmand et al. 2023). These real-time datasets serve as critical inputs for optimizing the application potential of alternative non-chemical and biological pest control strategies.

For instance, the application efficacy of EPF is highly contingent upon specific environmental factors such as humidity and temperature, which influence the germination of fungal conidia, spore concentration, and virulence. IoT sensors can rapidly monitor these parameters and relay data to centralized platforms, thereby facilitating the precise timing and localized application of EPF. Besides, pest activity can be tracked using the IoT sensors, designed to detect the frequency of pest performances through various techniques, such as motion detection or the use of pheromone traps. By enabling on-demand data transmission, IoT systems can enhance decision-making processes, ensuring the judgements of the best working conditions of microbial biocontrol agents, including the EPFs, if they are applied only when the pest populations reach critical thresholds, thus minimizing unnecessary interventions and optimizing the process of application standards of microbial biological control agents.

AI, through ML and DL algorithms, is a transformative tool for pest management, enabling effective detection, classification, and quantification of herbivorous pests using data inputs like images or videos. Research by Li et al. (2023) and Yuan et al. (2023) highlights AI's role in enhancing decision-making processes dependent on big data analysis, allowing machines to derive insights from extensive datasets and apply them to real-world situations (Jarrahi 2018). This capability enables AI to navigate complex environments better than traditional decision-making models. AI techniques can be leveraged to analyze large-scale data generated by IoT sensors and image-capturing technologies, enabling the automated identification of pest species, prediction of pest outbreaks, and optimization of EPF application schedules for repeated use in agriculture. By processing historical data on pest behavior, weather patterns, and environmental conditions, AI models can predict the likelihood of pest infestations and suggest preventive measures based on statistical patterns and probabilistic reasoning activities.

Furthermore, AI-based classification algorithms can enhance pest identification through analysis of visual data captured by image recognition systems. ML, a category of AI, trains algorithms to make predictions based on established data. In contrast, DL, a subcategory of ML, uses suitable networks with multiple layers to learn and detect features and patterns (Dolatabadian et al. 2025). DL algorithms, specifically convolutional neural networks (CNNs), can be trained on annotated datasets to detect and classify the pest species and behavioral changes associated with them with high accuracy, even under varying environmental conditions, as designed in multilocational field experimentations. This predictive capacity allows the farmers to anticipate pest outbreaks and proactively deploy the EPF-based interventions, thereby reducing crop damage and minimizing reliance on agrochemicals. The "DeepAgro" platform by AgShift employs DL algorithms to analyze crop images for identifying pest infestations and pathogen outbreaks (Khan and Ullah 2022). Recently, the AI-driven YOLOv3 algorithm has been adopted for pest categorization, while CNN techniques automatically count insect numbers on target plants. Agroforestry pest outbreaks can be foreseen through the use of computer vision, ML, and DL methods by analyzing a variety of meteorological records and data. In contemporary farming methods, ML tools such as Decision Tree (DT), Support Vector Machine (SVM), Naïve Bayes (NB), and Random Forest (RF), and DL methods such as CNN, Deep Convolutional Neural Networks (DCNN), Long Short-Term Memory (LSTM), and Deep Belief Network (DBN) are frequently employed for pest prediction and related crop protection purposes (Chithambarathanu and Jeyakumar 2023). As part of IPM strategies, a new Android app can detect patterns of fall armyworm infestation in maize. The application software achieves a validation accuracy of up to 93.47% and a training accuracy of 98.47% through the use of Deep Convolutional Neural Networks (DCNNs) and transfer learning, respectively, to detect pest-infested areas with high accuracy (Kariyanna and Sowjanya 2024). The integration of ICT, including the application of drones, satellites, and ground-based cameras, has eventually revolutionized the pest-monitoring practices in precision agriculture. These systems are capable of capturing high-resolution images that provide detailed insights into crop health, pest activity, and overall field performances. Image-based pest detection coupled with AI-based image recognition algorithms could facilitate rapid identification of pest infestations and quantification of pest densities at various spatial scales and diversified infestation patterns.

Drones, equipped with multispectral or hyperspectral imaging sensors, offer the advantage of rapidly surveying large areas and capturing real-time images of crops. These images can then be processed using AI algorithms to detect the pest-induced damage or the presence of specific pest species, even at early developmental stages. Image analysis can also be useful to assess the health of the plants and to detect stress symptoms caused by pest damage or pathogen infestations. By integrating image-based pest detection with IoT systems, farmers gain access to a comprehensive monitoring framework that substantially enhances the pest management decisions and thereby could improve the application timing and dose specificity of EPFs. Despite the promising potential of these technological advancements, several challenges still remain in IoT while designing the proper module of EPF-mediated insect pest management in agriculture. One of the primary barriers while adopting such IT systems is the high initial cost of deploying IoT sensors, AI systems, and ICT, particularly in resource-constrained farming regions and for the small and marginal farmers. Furthermore, the variability in the effectiveness of EPF across different environmental conditions like soil, climate, and topography requires advanced scientific research to optimize their application strategies, taking into account local ecological factors, integrity, and pest dynamics. Additionally, the integration of diverse technological components into a unified system requires the development of standardized protocols, communication frameworks, and data management platforms. Data privacy and security concerns, particularly with IoT devices, also necessitate the implementation of robust cybersecurity measures to safeguard sensitive and timely information for better adoption of CSA perspectives. Future research should prioritize facilitating the development of better, more economical, and more scalable technologies suitable for widespread adoption by small and marginal farmers.

7.4 Integration of Fuzzy Logic in EPF-Mediated Pest Management in Agriculture

Fuzzy logic offers a reliable approach for addressing uncertainty and imprecision in agricultural decision-making processes. Unlike traditional binary logic systems that operate on rigid, deterministic rules, fuzzy logic systems model uncertainty and offer more flexible and approximate reasoning. In the context of pest management, fuzzy logic can synthesize multiple inputs, including the environmental conditions, pest population dynamics, and effectiveness of biocontrol agents, thereby generating actionable recommendations based on a range of possibilities rather than a single deterministic outcome. For example, fuzzy logic can be used to determine the optimal quantity and timing of EPF applications by processing continuous data streams from IoT sensors and image recognition technology. It can consider a variety of variables, such as pest density, weather conditions, and fungal virulence, providing a more nuanced decision-making process that accounts for the complexity and variability under field situations.

The implementation of a Fuzzy inference system allows for the dynamic adjustment of pest control tactics in response to varying environmental conditions, hence improving the adaptability and accuracy of EPF-based interventions. Efforts must focus on improving the accuracy and specificity of EPF applications, further integrating diverse data streams, and redefining fuzzy logic models to enhance decision-making in precision agriculture. Jiang et al. (2022) have suggested the integration of fuzzy logic in ecological pest forecasting in agriculture. According to subject experts, fuzzy logic can precisely forecast the population dynamics of agricultural pests through handling the inherent uncertainties, vagueness, and complexity of ecological systems. In contrast to conventional boolean logic, which depends on binary true/false values, fuzzy logic accommodates partial truths, depicted by membership functions with values ranging from 0 to 1. This makes it well suited for modeling imprecise and dynamic agricultural data, such as weather conditions, pest populations, and crop health, to support climate-resilient IPM strategies.

Accompanied by tabulated values and explanations for membership functions, fuzzy logic pretends to explain ecological pest forecasting, even in a more defined way; thus, scientists can anticipate pest activity and recommend timely interventions. Several steps need to be taken care of while defining the role of fuzzy logic in CSA, such as:

(a) *Handling Uncertainty and Vagueness*: Agricultural systems are influenced by several biotic and abiotic variables like temperature, humidity, and pest activity, which are often imprecise or subject to variation. Fuzzy logic models these variables using linguistic terms (e.g., "high," "moderate," "low") and assigns degrees of membership, enabling robust decision-making despite incomplete data and sources (Zadeh 1965).
(b) *Weather-Based Diagnosis and Pest Forecasting*: Ecological pest forecasting relies on meteorological data, including the temperature, humidity, and leaf wetness duration, to predict the pest or pathogen outbreaks. Fuzzy logic analyzes these variables to predict risks, thereby aiding farmers in optimizing pesticide application (Siraj and Arbaiy 2006) and adopting non-chemical, sustainable pest management strategies, including the use of pheromone traps, light traps, ML models, and evaluating the bioefficacy of microbial biocontrol agents such as EPF.

(c) *Decision Support Systems*: Fuzzy logic integrates ecological expertise and site-specific data into rule-based systems (e.g., "IF temperature is high AND humidity is high, THEN pest risk is high"). These systems provide actionable insights for IPM, balancing pest management (Adriaenssens et al. 2004) through adoption of sustainable expertise including the use efficiency of EPF.

(d) *Optimizing IPM Strategies*: By integrating complex relationships between the pests, crops, and environmental factors, fuzzy logic supports precise interventions, such as determining the timing and exact pesticide applications or introducing the biological control agents, including the EPF, for enhanced sustainability, thereby assisting in minimizing the chemical hazards in agriculture (Mikail et al. 2024).

(e) *Adaptability to Non-Stationary Data*: Agricultural systems are considered as more dynamic, with non-stationary patterns. For instance, the prevalence of shifting climate conditions influencing the productivity and plant protection in agriculture. Fuzzy logic, especially non-stationary fuzzy time series, adapts to these changes, improving forecast accuracy over time (Silva et al. 2020).

7.4.1 Fuzzy Logic Realities in CSA

For instance, we may consider a fuzzy Logic system that takes humidity and temperature into account to predict whether wheat will be infested with aphids. In order to influence IPM choices, the system generates the level of pest crisis. The standard fuzzy logic rule-based for aphid risk forecasting has been represented in Table 10.6.

Table 10.6 Aphid risk forecasting based on standard Fuzzy logic rules[a]

Fuzzy logic rule	Temperature	Humidity	Pest risk (output)
1	Low	Low	Low
2	Low	Moderate	Low
3	Low	High	Moderate
4	Moderate	Low	Low
5	Moderate	Moderate	Moderate
6	Moderate	High	High
7	High	Low	Moderate
8	High	Moderate	High
9	High	High	High

[a] The table defines nine rules combining the temperature and humidity levels to predict pest risks. For example, Rule 6 states, "IF temperature is moderate AND humidity is high, THEN pest risk is high," reflecting conditions favorable for aphid proliferation. These rules are derived from ecological knowledge and field observations, processed by fuzzy logic to handle partial memberships. In contrast, the degree to which an input is a part of a fuzzy set is defined by a membership function (MF), which can take values between zero (no membership) and one (full membership). In the aphid risk example, MFs are defined for temperature, humidity, and pest risk

7.4.2 Mathematical Steps in Fuzzy Logic Process

(a) *Fuzzification*: It converts crisp inputs (e.g., temperature = 25 °C, humidity = 70%) into Fuzzy sets using membership functions (e.g., "low," "moderate," "high").
(b) *Rule Evaluation*: It is based on applying predefined IF-THEN rules based on expert knowledge (e.g., "IF temperature is high AND humidity is high, THEN pest risk is high").
(c) *Inference*: This process combines rules to compute the fuzzy output (pest risk level).
(d) *Defuzzification*: It converts the Fuzzy output into a crisp value (e.g., pest risk = 0.75) to inform decisions, such as applying a biological control agent.

7.4.3 Examples of Membership Functions

Temperature (°C):

Low: 10–18 °C.
Moderate: 16–24 °C.
High: 22–30 °C.

Humidity (%):

Low: 30–50%.
Moderate: 45–65%.
High: 60–80%.

Pest Risk:

Low: 0–0.3.
Moderate: 0.25–0.6.
High: 0.551.0 MFs are often triangular or trapezoidal for simplicity.

For temperature:

Low: $\mu(x) = 1$ at 10 °C, decreases linearly to 0 at 18 °C.
Moderate: $\mu(x) = 0$ at 16 °C, peaks at 1 at 20 °C, decreases to 0 at 24 °C.
High: $\mu(x) = 0$ at 22 °C, increases to 1 at 26 °C, remains 1 beyond 30 °C.

8 Research Data Integration Using Scientometrics

Research publications within the period 1984 to 2025 on the selected topics like agriculture, human civilization, biotic stress in plants, abiotic stress in plants, agrochemical toxicity, IPM, CSA, EPF, biocontrol, PGP, AI in crop protection, IoT in pest management, plant resilience, EPF and pest management and myconanoformulations, microbial virulence, shelf life, viability of entomopathogens, *Beauveria bassiana*, *Metarhizium anisopliae*, bioefficacy, microbial inoculum, biocontrol agents, fuzzy logics, indexed in the premium Scopus database are plotted in the 2D graph with computer software for data analysis, MS-Excel, and Biblioshiny R, with a linear and exponential trendline depicted with the value of r^2. There are a total of 1190 research papers published in the global perspective. To determine if the growth pattern is exponential, the Price Law (Price 1963) is used, as demonstrated by Okoroiwu et al. (2020). The linear trendline is expressed by the equation $y = 2.7606x—5505.2$, while the exponential trendline is represented by the equation $y = 0.1529x^2–610.05x + 608,663$ (Fig. 10.4). This is accomplished by computing the correlation coefficient between the exponential and linear curves. A growth pattern is considered exponential if the correlation coefficient of the exponential curve (r_1) is larger than that of the linear curve (r_2), indicating that the price law is satisfied. The first exponential coefficient (r_1) is 0.8632, while the second (r_2) is 0.6348. Since $r_1 > r_2$, the Price law is fulfilled and the growth rate is recorded as exponential. Figure 10.5 depicts the publication count for the period 1984 to 2025.

8.1 Analysis of Prominent Themes and Sub-Themes

The consideration of keywords for current data analysis is made through using the following three key aspects: i) identification of keywords with the highest frequency of occurrence ii) year-wise growth of keywords, and iii) data visualization of the trending topics. All of the data analyses were performed using VOSviewer and Biblioshiny-R tools. A developed network of keyword co-occurrence provides a link

Fig. 10.4 Chronological growth of research publications during the period 1984–2025

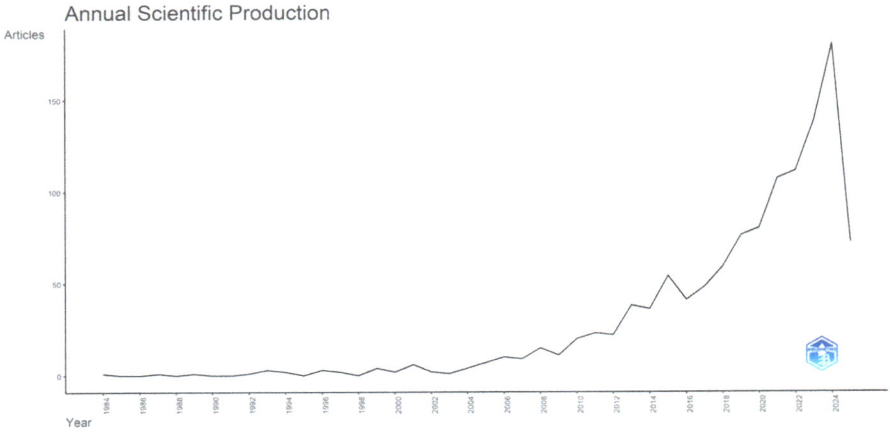

Fig. 10.5 Year-wise growth of research publications

between the two keywords. It may be stated that the first aspect of this data analysis will reveal the areas where maximum research is done and the areas where significant contribution is still required. Thus, this analysis will be useful in determining the areas of predominance in the subject. Fig. 10.6 illustrates the word cloud of significant terms that frequently occur in the research publications within this domain. The font size of the words are closely correlated with their frequency of occurrence. Keywords such as oxidative stress, toxicity, metabolism, biotic stress, bioaccumulation, ecotoxicology, physiological stress, gene expression, and enzyme activity appear to be significant.

Figure 10.7 was generated using Biblioshiny R, employing designated graphical procedures, including the "keyword plus" field and frequency of occurrences (FoO) per year, without a confidence interval, highlighting the top 10 prominent keywords based on their maximum frequency of occurrences. Oxidative stress (FoO = 372), metabolism (273), cadmium (218), enzyme activity (171), biotic stress (159), and toxicity (474) are identified as certain prominent keywords.

Figure 10.8 is generated using VOSviewer by employing the rainbow color method. The font size of the term and the intensity of the color are proportional to the FoO of the keyword. The FoO value is elevated if the hue of the node denoting a keyword is more intense. The deepest color in the center signifies the toxicity of keywords such as oxidative stress, abiotic stress, biotic stress, silicon, transcriptomics, antioxidants, and heavy metals, which are located at the core. Moving away from the core, the FoO decreases. Keywords in the green region just outside the core are microplastics, boron toxicity, metal toxicity, nanotechnology, genomics, multiple stressors, toxic effects, and crop protection. In the periphery, the blue-colored region has multiple keywords like combined toxicity, inflammation, biotic and abiotic stress, crop protection, genome editing, bioavailability, nanobiotechnology, lipid peroxidation, toxic effects, detoxification, free radicals,

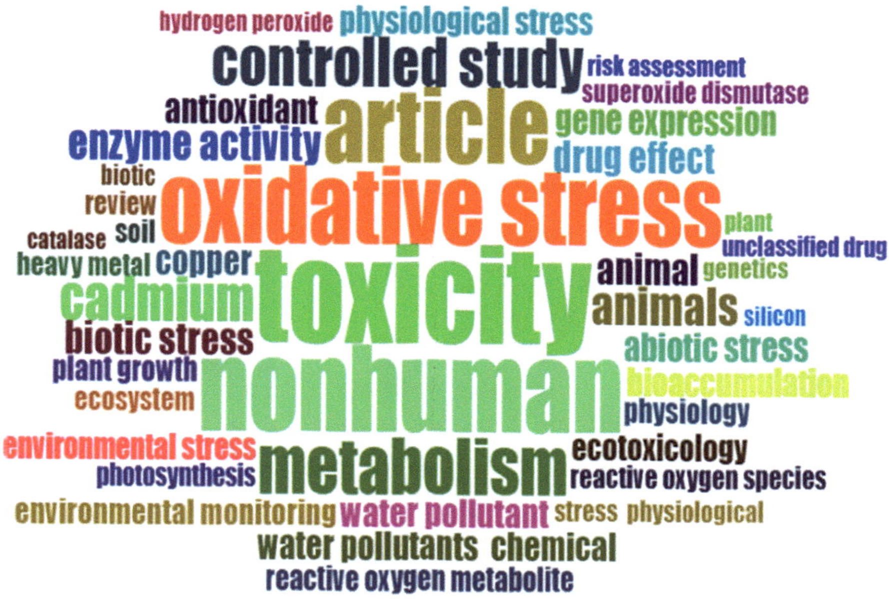

Fig. 10.6 Word cloud of prominent keywords

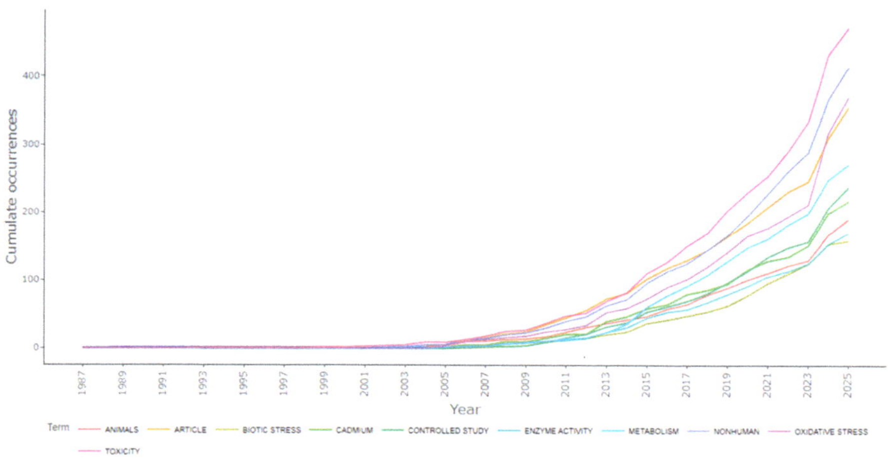

Fig. 10.7 Year-wise FoO of prominent keywords

boron toxicity, and pesticides. It is noteworthy to indicate that the mentioned areas are directly or indirectly related to the core areas. Another depiction by Fig. 10.9 displays the prominent keywords with FoO and their percentage.

Figure 10.10 is developed with the Biblioshiny R that represents the trending areas of research in this field. Keywords like bivalve, anthropogenic, Mytilus, drug

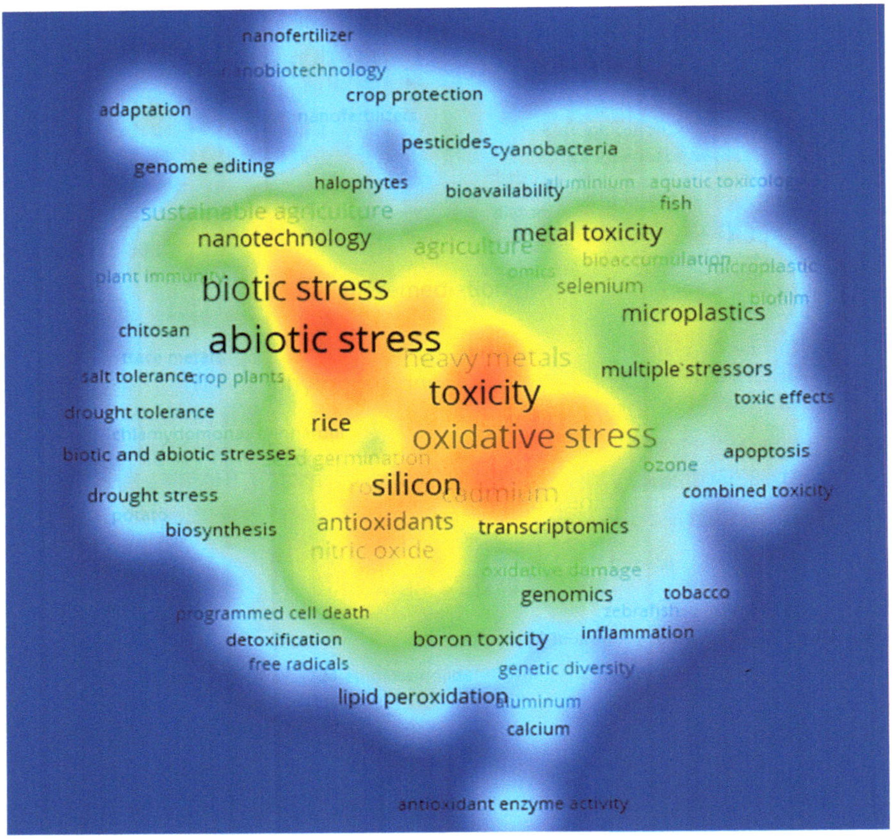

Fig. 10.8 Construction of keyword density map. The deeper the color, the more the value of FoO

effect, microplastic, superoxide dismutase, nanoparticle, oxidative stress, and environmental stress are recognized as the most trending areas during the period 2019 to 2024. However, the emergence of keywords like microplastics, bivalve, and nanoparticle in 2023–24 implicates future perspectives of research in these vital areas of science for enhancing sustainability in plant protection.

8.2 Bradford's Law of Scattering to Identify the Core Journals

This law is fundamental in bibliometrics for identifying core journals in a subject area. It states that when scientific journals are arranged by article productivity, they can be divided into a nucleus of core periodicals and several zones containing an equal number of articles, following a ratio of 1: n: n^2 (Bradford 1934; Garfield 1980). Bradford's distribution law indicates there should be 184 articles in each zone of the 552 journals analyzed (Table 10.7). However, the distribution is not uniform, with a

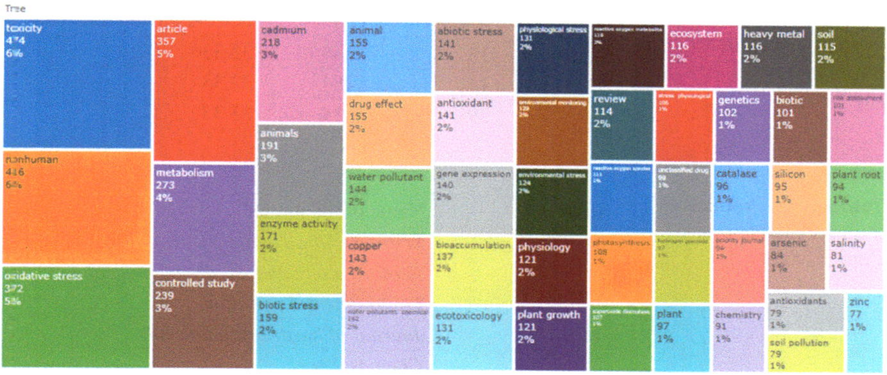

Fig. 10.9 The tree maps of prominent keywords. It displays the highly occurring keywords with FoO and the percentage of occurrence

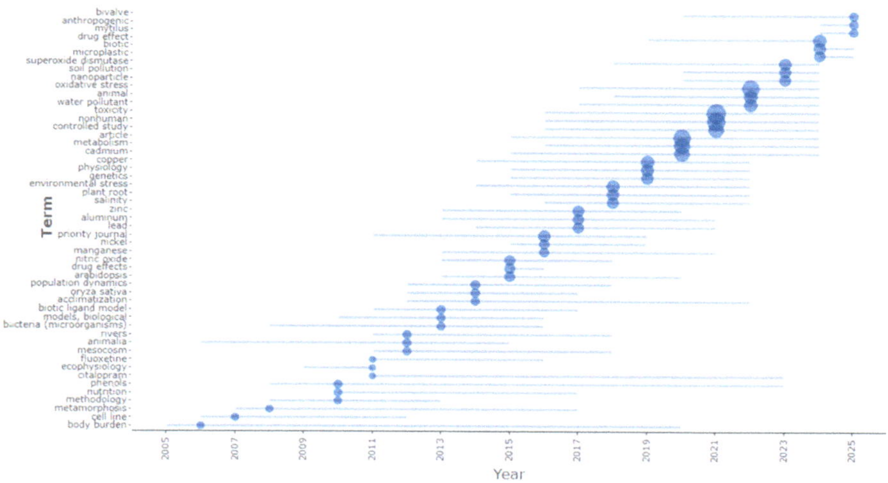

Fig. 10.10 Trending topics in research on crop protection. The consideration is made by developing a minimum word frequency of 5 and a number of words per year of 5

calculated multiplier of approximately 8 for the first two zones. Zone 3 should have 1128 journals, but only 391 are observed, indicating a mismatch. Despite this, the law successfully pinpointed 18 core journals that focus on the main areas. But interestingly, moving from Zone 1 to the peripheral Zones, i.e., 2 and 3, the number of journals increases manifold, but the number of articles decreases. These journals do not produce articles directly related to the subject or publish articles that are of low standards (Borgohain et al. 2021). The journals in Zone 1 (core) are listed in Table 10.8, and the journals in the core are depicted in Fig. 10.11.

Table 10.7 Distribution of papers and number of journals as per Bradford's law

Zone	Journal	NP	Percent among total papers
1	18	184	33.33
2	141	184	33.33
3	391	184	33.33

Table 10.8 Prominent journals and their ranks in Zone 1 (core)

Ranks	Name of the journal
1	*Science of the Total environment*
2	*Ecotoxicology and environmental safety*
3	*Environmental pollution*
4	*Journal of hazardous materials*
5	*Chemosphere*
6	*Frontiers in plant science*
7	*Aquatic toxicology*
8	*Environmental science and pollution research*
9	*Plants*
10	*International journal of molecular sciences*
11	*Plant physiology and biochemistry*
12	*Environmental science and technology*
13	*Environmental toxicology and chemistry*
14	*Marine pollution bulletin*
15	*Journal of environmental management*
16	*PLOS ONE*
17	*Acta Physiologiae Plantarum*
18	*Journal of experimental botany*

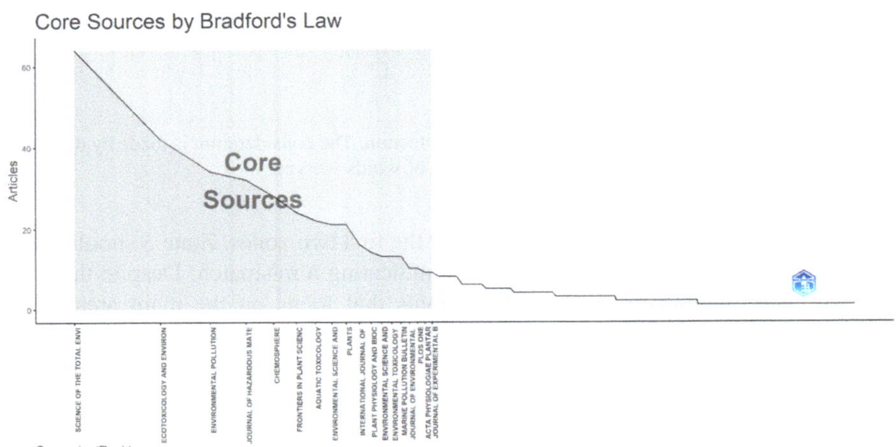

Fig. 10.11 Core journals in this area of research with Bradfrod's law

9 Conclusions

Multidimensional pest infestations in agriculture ultimately result from favorable microclimate and topography, as well as the rapid and illegal applications of toxic chemicals and pesticides. This eventually leads to substantial damages in plant health followed by losses in food production and reflects increased toxicity in the soil, water, and ecosystem. It has been shown that abrupt use of high-dosage chemical pesticides cause significant damage in plants through oxidative damage (e.g., production of reactive oxygen species), reduce beneficial crop microbiome populations, and interfere with key soil enzyme activities. This leads to a decline in the plant's energy-generating systems and disrupts key hormonal balance, which weakens the plant's natural defenses over time. The continual persistence of chemical toxicity in soil poses considerable hazards in soil, generating maximum residue limits in agriculture. Numerous human health issues, including the immune system disorders, neurological damage, hormone disturbance, hypersensitivity, cancer, and so on, have been linked to pesticide exposure to chemical toxicity.

Reducing the environmental impact of toxic agrochemicals and transitioning towards the alternative, non-chemical, and sustainable biological technologies are essential for ensuring the agro-ecosystem as safe, sustainable, and environmentally sound. An expanding market of recently introduced and improved biological products, which can be used either alone, in rotation, or in conjunction with chemical pesticides. The development of microbial biopesticides that combat a range of significant insect pests is essential for managing agricultural ecosystems, promoting sustainable food production, and protecting the environment and public health. Entomopathogenic fungi is a safe, affordable, superior, non-chemical, and excellent sustainable pest management tool extensively employed in agriculture. While penetrating through the natural apertures, pest ingestions, or sometimes their entry through the pest integuments, these virulent pathogens kill or deter target insects. Different kinds of toxin complexes are released by entomopathogenic fungi that exhibit prominent insecticidal, antibacterial, and antimycotic properties, thereby assisting the farmers to overcome multidimensional insect pest problems in agriculturally important crops. Despite the promising potential of this microbial biocontrol agent in insect pest management, several challenges still remain with this category of microbial biocontrol agent during field applications. These may be listed as unpredictable effectiveness, time-intensive factors, adverse effects on non-target organisms, survival challenges and limited efficiencies under changing climate and topography, restricted range in pest management (unique mode of action), limited virulence, and challenges in maintaining the shelf life and viability that have posed significant limitations in addressing the prominence of EPF in insect pest management in agriculture.

It may be stated that by introducing or modifying specific genes, researchers may be able to enhance the shelf life, virulence, adaptability, and environmental resistance of entomopathogenic fungi, thereby making this microbial pest management tool more effective as well as attractive in climate smart agriculture. Genetic modifications can increase the production of effective enzymes or toxins that specifically target the insect pests, thereby assisting in enhancing the microbial mode of action and efficiency. Alongside genetic engineering, the use of nano-entomopathogen formulation has shown tremendous promise in improving the stability, shelf life, and virulence. By incorporating entomopathogen into nanoparticles, researchers may now create appropriate myconanoformulations that shield fungi from environmental threats such as ultraviolet radiation, desiccation, or dehydration, thus prolonging their shelf life, viability, and virulence. Microbial biocontrol agents are typically more costly to produce than synthetic insecticides. We need a more deeper understanding of the bio-functionality of entomopathogenic fungi in various agroecosystem services, as well as the effects of climate change and prevailing agronomic practices. Additionally, the integration of internet of things, artificial intelligence, and fuzzy logics into a cohesive pest management system offers an unprecedented opportunity for optimizing the entomopathogen-mediated pest control in agriculture. By combining an internet of things-enabled real-time environmental monitoring system with artificial intelligence-driven pest detection and prediction models, farmers are empowered to make informed, data-driven decisions. Image recognition technology, machine learning and deep learning methodologies further enhances pest detection accuracy, while fuzzy logic allows for developing flexible decision-making models based on the dynamic interplay of environmental and pest-related factors.

This holistic approach ensures that EPFs are applied precisely when and where they are most effective, thereby minimizing unnecessary pesticide use and optimizing resource allocation. Besides, the integration of these modern technologies enables the automation of pest management processes, reducing labor costs, and enhancing the scalability of entomopathogen-based interventions, further strengthening the concept of adoption of climate-resilient pest management strategies in agriculture. Future research ought to concentrate on learning more about how pests and entomopathogens interact in plants and how to make microbial biocontrol treatments more effective and easier to use under field conditions. In order to achieve desired success, governments, and non-government organizations along with proper discussions with agricultural stakeholders, must establish and execute new biopesticide policies that would promote the production, registration, and utilization of environmentally friendly pesticides in agriculture. Contemporary regulatory frameworks in industrialized nations mandating farmers to adopt integrated pest management standards are likely to boost the microbial biopesticide industry.

Acknowledgments The corresponding author extends sincere gratitude to the principal of NNS College, Titabar, Jorhat, Assam, India, for offering logistical support. Additionally, the corresponding author acknowledges the Department of Biotechnology (DBT), Government of India, for allocating research funds for the Institutional Biotech Hub (BT/NER/143/SP44344/2021) at NNS College, as part of the NER Biotech Hub Programme. We deeply gratitude Zubeen Garg, the Humming King of the World, for demonstrating his dedication through multiple acts of kindness, his advocacy for social justice, and his conviction in the possibility of unification transcending religious and societal divisions.

References

Adhikari U (2022) Insect pest management: mechanical and physical techniques. RFNA 3(1): 48–53. https://doi.org/10.26480/rfna.01.2022.48.53

Adriaenssens V, De Baets B, Goethals PLM, De Pauw N (2004) Fuzzy rule-based models for decision support in ecosystem management. Sci Total Environ 319(1–3):1–12. https://doi.org/10.1016/S0048-9697(03)00433-9

Ahmad I, del Jiménez-Gasco MM, Luthe DS, Shakeel SN et al (2020) Endophytic *Metarhizium robertsii* promotes maize growth, suppresses insect growth, and alters plant defense gene expression. Biol Control 114:104167. https://doi.org/10.1016/j.biocontrol.2019.104167

Akello J, Sikora R (2012) Systemic acropedal influence of endophyte seed treatment on *Acyrthosiphon pisum* and *Aphis fabae* offspring development and reproductive fitness. Biol Control 61(3):215–221. https://doi.org/10.1016/j.biocontrol.2012.02.007

Akello J, Dubois T, Coyne DL, Kyamaniwa S (2008) Endophytic *Beauveria bassiana* in banana (Musa spp.) reduces banana weevil (*Cosmopolites sordidus*) fitness and damage. Crop Prot 27(11):1437–1441. https://doi.org/10.1016/j.cropro.2008.07.003

Aktar MW, Sengupta D, Chowdhury A (2009) Impact of pesticides use in agriculture: their benefits and hazards. Interdiscip Toxicol 2:1–12. https://doi.org/10.2478/v10102-009-0001-7

Al-Ani LKT et al (2025) *Trichoderma* a holistic fungus with biocontrol potential for agricultural diseases. In: Shaaban M (ed) Soils and sustainable agriculture. Frontier studies in soil science. Springer, Cham. https://doi.org/10.1007/978-3-031-91114-9_37

Aldaghi M, Allahyari H, Hosseininaveh V, Behboudi K (2021) Effect of *Trichoderma harzianum* Tr6 in inducing resistance in tomato against *Trialeurodes vaporariorum* (Hem: Aleyodidae). Plant Prot (Scientific J Agriculture) 44(3):107–117. https://doi.org/10.22055/PPR.2021.17128

Andolfi A, Maddau L, Basso S, Linaldeddu BT, Cimmino A et al (2014) Diplopimarane, a 20-nor-ent-pimarane produced by the oak pathogen *Diplodia quercivora*. J Nat Prod 77(11): 2352–2360. https://doi.org/10.1021/np500258r

Bahadur A (2018) Entomopathogens: role of insect pest management in crops. Trends Hortic 1:1–9. https://doi.org/10.24294/th.v1i4.833

Ballal CR, Sreedevi K, Salini S, Gupta A, Amala U et al (2022) Biodiversity of agriculturally important insects: status, issues, and challenges. In: Kaur S, Batish D, Singh H, Kohli R (eds) Biodiversity in India: status, issues and challenges. Springer, Singapore. https://doi.org/10.1007/978-981-16-9777-7_12

Baruah N, Deka AC (2017) Myco-biocontrol of red spider Mite (*Oligonychus coffeae*, Nietner) using *Metarhizium anisopliae*. Int J Res Appl Sci Eng Technol 5:1800–1804. https://doi.org/10.22214/IJRASET.2017.10264

Batista DPC, de Oliveira IN, Ribeiro ARB, Fonseca EJS, Santos-Magalhães NS et al (2017) Encapsulation and release of *Beauveria bassiana* from alginate–bentonite nanocomposite. RSC Adv 7:26468–26477. https://doi.org/10.1039/C7RA02185B

Batool R, Umer MJ, Wang Y, He K, Shabbir MZ et al (2022) Myco-synergism boosts herbivory-induced maize defense by triggering antioxidants and phytohormone signaling. Front Plant Sci 13:790504. https://doi.org/10.3389/fpls.2022.790504

Berahmand K, Mohammadi M, Sheikhpour R, Li Y, Xu Y (2023) WSNMF: weighted symmetric nonnegative matrix factorization for attributed graph clustering. Neurocomputing 566:127041. https://doi.org/10.1016/j.neucom.2023.127041

Bertoft E (2017) Understanding Starch Structure: Recent Progress. Agronomy 7:56. https://doi.org/10.3390/agronomy7030056

Bhattacharyya PN, Jha DK (2012) Plant growth-promoting rhizobacteria (PGPR): emergence in agriculture. World J Microbiol Biotechnol 28:1327–1350. https://doi.org/10.1007/s11274-011-0979-9

Bhattacharyya PN, Jha DK (2015) Mycorrhizal symbiosis in the formation of antioxidant compounds. In: Dubey NK (ed) Plants as a source of natural antioxidants. CAB International, Oxfordshire, pp 252–281. https://doi.org/10.1079/9781780642666.0000

Bhattacharyya PN, Sarmah SR (2018) The role of microbes in tea cultivation. In: Sharma VS, Gunasekare MTK (eds) Global tea science: current status and future needs. Burleigh Dodds Science Publishing, Cambridge. https://doi.org/10.1201/9781351114516

Bhattacharyya PN, Goswami MP, Bhattacharyya LH (2016) Perspective of beneficial microbes in agriculture under changing climatic scenario: a review. J Phytol 8:26–41. https://doi.org/10.19071/jp.2016.v8.3022

Bhattacharyya PN, Sarmah SR, Roy S, Sarma B, Nath BC et al (2023) Perspectives of *Beauveria bassiana*, an entomopathogenic fungus for the control of insect-pests in tea [*Camellia sinensis* (L.) O. Kuntze]: opportunities and challenges. Int J Trop Insect Sci 43:1–19. https://doi.org/10.1007/s42690-022-00932-1

Bhattacharyya PN, Baishya B, Borgohain DJ, Sarma B, Bhattacharyya LH et al (2024a) Unveiling the virulence and stress tolerance of *Beauveria bassiana* against major insect pests in tea with an emphasis on Bibliometrics using data mining tools. In: Deshmukh SK, Sridhar KR (eds) Entomopathogenic fungi. Springer, Singapore, pp 247–298. https://doi.org/10.1007/978-981-97-5991-0_11

Bhattacharyya PN, Sarma B, Sarmah SR, Nath BC, Borchetia S et al (2024b) Entomopathogen-based biological control of looper pests (old looper, *Biston (=Buzura) suppressaria* and emerging looper, *Hyposidra Talaca*): an in vitro sustainable approach for tea pest management. Int J Trop Insect Sci 44:1713–1727. https://doi.org/10.1007/s42690-024-01268-8

Bhattacharyya PN, Islam NF, Sarma B, Nath BC, Al-Ani LK et al (2024c) *Frankia*-actinorhizal symbiosis: a non-chemical biological assemblage for enhanced plant growth, nodulation and reclamation of degraded soils. Symbiosis 92:1–26. https://doi.org/10.1007/s13199-023-00956-2

Bhattacharyya PN, Sandilya SP, Sarma B, Pandey AK, Dutta J et al (2024d) Biochar as soil amendment in climate-smart agriculture: opportunities, future prospects, and challenges. J Soil Sci Plant Nutr 24:135–158. https://doi.org/10.1007/s42729-024-01629-9

Bhattacharyya PN, Nath BC, Sarma B, Al-Ani LK, Borgohain DJ et al (2024e) Interaction of *Metarhizium anisopliae* against emergent insect pest problems in the north-eastern tea industry. In: Deshmukh SK, Sridhar KR (eds) Entomopathogenic fungi. Springer, Singapore, pp 319–357. https://doi.org/10.1007/978-981-97-5991-0_13FAO

Bhattacharyya PN, Sarma B, Bhattacharyya LH (2025) A textbook of biofertilizers. As per FYUGP (NEP) syllabus, 1st edn. Publisher Mahaveer Publications, pp 1–174

Bing LA, Lewis LC (1991) Suppression of *Ostrinia nubilalis* (Hübner) (Lepidoptera: Pyralidae) by Endophytic *Beauveria bassiana* (Balsamo) Vuillemin. Environ Entomol 20:1207–1211. https://doi.org/10.1093/ee/20.4.1207

Biswas C, Dey P, Satpathy S, Satya P, Mahapatra BS (2013) Endophytic colonization of white jute (*Corchorus capsularis*) plants by different *Beauveria bassiana* strains for managing stem weevil (*Apion corchori*). Phytoparasitica 41:17–21. https://doi.org/10.1007/s12600-012-0257-x

Borah D, Bhattacharyya PN, Nath BC, Chetia R, Islam NF et al (2024) Synthesis and antagonistic evaluation of fluorinated imidazolium salt [1-(2,6-diisopropylphenyl)-3-(2-fluoro- benzyl)-1H-imidazol-3-ium bromide] against significant phytopathogens in agriculture. J Fluor Chem 274: 110259. https://doi.org/10.1016/j.jfluchem.2024.110259

Borgohain DJ, Verma MK, Nazim M, Sarkar M (2021) Application of Bradford's law of Scattering and Leimkuhler model to information science literature. Collnet J Scientomet 15(1):197–212. https://doi.org/10.1080/09737766.2021.1943041

Boruah S, Dutta P (2021) Fungus mediated biogenic synthesis and characterization of chitosan nanoparticles and its combine effect with *Trichoderma asperellum* against *Fusarium oxysporum*, *Sclerotium rolfsii* and *Rhizoctonia solani*. Indian Phytopathol 74:81–93. https://doi.org/10.1007/s42360-020-00289-w

Boukal DS, Bideault A, Carreira BM, Sentis A (2019) Species interactions under climate change: connecting kinetic effects of temperature on individuals to community dynamics. Curr Opin Insect Sci 35:88–95. https://doi.org/10.1016/j.cois.2019.06.014

Bradford S (1934) Sources of information on specific subjects. Engineering 37(3550):85–86. https://doi.org/10.1177/016555158501000406

Bruck DJ (2009) Fungal entomopathogens in the rhizosphere. In: Roy HE, Vega FE, Chandler D, Goettel MS, Pell J, Wajnberg E (eds) The ecology of fungal Entomopathogens. Springer, Dordrecht, pp 103–112. https://doi.org/10.1007/978-90-481-3966-8_8

Cherry AJ, Banito A, Djegui D, Lomer C (2004) Suppression of the stem-borer *Sesamia calamistis* (Lepidoptera; Noctuidae) in maize following seed dressing, topical application and stem injection with African isolates of Beauveria bassiana. Int J Pest Manag 50:67–73. https://doi.org/10.1080/09670870310001637426

Chinnaperumal K, Govindasamy B, Paramasivam D, Dilipkumar A, Dhayalan A et al (2018) Bio-pesticidal effects of *Trichoderma viride* formulated titanium dioxide nanoparticle and their physiological and biochemical changes on *Helicoverpa armigera* (Hub.). Pestic Biochem Physiol 149:26–36. https://doi.org/10.1016/j.pestbp.2018.05.005

Chithambarathanu M, Jeyakumar MK (2023) Survey on crop pest detection using deep learning and machine learning approaches. Multimed Tools Appl 11:1–34. https://doi.org/10.1007/s11042-023-15221-3

Cooper J, Dobson H (2007) The benefits of pesticides to mankind and the environment. Crop Prot 26(9):1337–1348. https://doi.org/10.1016/j.cropro.2007.03.022

Damalas CA, Koutroubas SD (2016) Farmers' exposure to pesticides: toxicity types and ways of prevention. Toxics 4(1):1. https://doi.org/10.3390/toxics4010001

Damalas CA, Georgiou EB, Theodorou MG (2006) Pesticide use and safety practices among Greek tobacco farmers: a survey. Int J Environ Health Res 16:339–348. https://doi.org/10.1080/09603120600869190

Dar MH, Bano DA, Waza SA, Zaidi NW, Majid A et al (2021) Abiotic stress tolerance-progress and pathways of sustainable rice production. Sustainability 13(4):2078. https://doi.org/10.3390/su13042078

Debebe S (2022) Post-harvest losses of crops and its determinants in Ethiopia: tobit model analysis. Agric Food Secur 11:13. https://doi.org/10.1186/s40066-022-00357-6

Deka B, Babu A, Peter AJ, Pandey AK, Kumhar KC et al (2021a) *Beauveria bassiana*: as a potential microbial biocontrol agent for tea mosquito bug, *Helopeltis theivora* Waterhouse (Hemiptera: Miridae) in Dooars and Darjeeling, India. Egypt J Biol Pest Control 31:1–9. https://doi.org/10.1186/s41938-021-00478-3

Deka B, Baruah C, Babu A (2021b) Entomopathogenic microorganisms: their role in insect pest management. Egypt J Biol Pest Control 31:121. https://doi.org/10.1186/s41938-021-00466-7

Desneux N, Wajnberg E, Wyckhuys KAG (eds) (2010) Biological control: perspective. Springer

Di Lelio L, Salvatore MM, Della Greca M, Mahamedi AE, Alves A et al (2022) Defensive mutualism of endophytic fungi: effects of sphaeropsidin an against a model lepidopteran pest. Chem Proc 10(1):42. https://doi.org/10.3390/IOCAG2022-12216

Dolatabadian A, Neik TX, Danilevicz MF, Upadhyaya SR, Batley J et al (2025) Image-based crop disease detection using machine learning. Plant Pathol 74:18–38. https://doi.org/10.1111/ppa.14006

Du Y, Xia Y, Jin K (2022) Enhancing the biocontrol potential of the entomopathogenic fungus in multiple respects via the overexpression of a transcription factor gene MaSom1. J Fungi 8(105): 1–16. https://doi.org/10.3390/jof8020105

Dullah S, Gogoi R, Bhattacharyya PN, Borah D, Kumari R et al (2025) Current perspectives of nanomaterials in wastewater treatment for sustainable environment. In: Agarwal N, Shah MP, Solanki VS, Singh N (eds) Nanomaterials in wastewater research. Adv Wastewater Res. Springer, Singapore. https://doi.org/10.1007/978-981-96-3908-3_17

Duo-Chuan L (2006) Review of fungal chitinases. Mycopathologia 161:345–360. https://doi.org/10.1007/s11046-006-0024-y

Fadiji AE, Babalola OO (2020) Exploring the potentialities of beneficial endophytes for improved plant growth. Saudi J Biol Sci 27:3622–3633. https://doi.org/10.1016/j.sjbs.2020.08.002

Fan Y, Liu X, Keyhani NO, Tang G, Pei Y et al (2017) Regulatory cascade and biological activity of *Beauveria bassiana* oosporein that limits bacterial growth after host death. Proc Natl Acad Sci U S A 114(9):1578–1586. https://doi.org/10.1073/pnas.1616543114

Fang W, St Leger RJ (2012) Enhanced UV resistance and improved killing of malaria mosquitoes by photolyase transgenic Entomopathogenic fungi. PLoS One 7:43069. https://doi.org/10.1371/journal.pone.0043069

Fernandez-Chapa D, Ramírez-Villalobos J, Galán-Wong L (2019) Toxic potential of *Bacillus thuringiensis*: an overview. In: Protecting Rice Grains in the Post-Genomic Era. IntechOpen. https://doi.org/10.5772/intechopen.85756

Formentini MA, Alves LF, Schapovaloff ME (2016) Insecticidal activity of neem oil against *Gyropsylla spegazziniana* (Hemiptera: Psyllidae) nymphs on Paraguay tea seedlings. Braz J Biol 76:951–954. https://doi.org/10.1590/1519-6984.04915

Friedlingstein P, O'Sullivan M, Jones MW, Andrew RM, Gregor L et al (2022) Global carbon budget. Earth Syst Sci Data 14(11):4811–4900. https://doi.org/10.5194/essd-14-4811-2022

Friuli M, Pellegrino R, Lamanna L, Nitti P, Madaghiele M et al (2023) Materials engineering to help Pest control: a narrative overview of biopolymer-based Entomopathogenic fungi formulations. J Fungi 9:918. https://doi.org/10.3390/jof9090918

García-Estrada C, Cat E, Santamarta I (2016) *Beauveria bassiana* as biocontrol agent: formulation and commercialization for pest management. In: Singh H, Sarma B, Keswani C (eds) Agriculturally important microorganisms. Springer, Singapore. https://doi.org/10.1007/978-981-10-2576-1_5

Garfield E (1980) Bradford's law and related statistical literature. Current Contents 19:40–45

Garrido-Jurado I, Montes-Moreno D, Sanz-Barrionuevo P, Quesada-Moraga E (2020) Delving into the causes and effects of entomopathogenic endophytic *Metarhizium brunneum* foliar application-related mortality in *Spodoptera littoralis* larvae. Insects 11:429. https://doi.org/10.3390/insects11070429

Ghaffari S, Karimi J, Cheniany M, Seifi A, Loverodge J et al (2025) Endophytic entomopathogenic fungi enhance plant immune responses against tomato leaf miner. J Invertebr Pathol 209:108270. https://doi.org/10.1016/j.jip.2025.108270

Gheorghita Puscaselu R, Lobiuc A, Dimian M, Covasa M (2020) Alginate: from food industry to biomedical applications and management of metabolic disorders. Polymers 12:2417. https://doi.org/10.3390/polym12102417

Ghosh I, Samanta R (2003) TEAPEST: an expert system for insect pest management in tea. Appl Eng Agric 19(5):619–625. https://doi.org/10.13031/2013.15309

Gioia G, Freeman J, Sipka A, Santisteban C, Wieland M et al (2022) Pathogens associated with houseflies from different areas within a New York State dairy. JDS Commun 3(4):285–290. https://doi.org/10.3168/jdsc.2021-0200

Golzan SR, Talaei-Hassanloui R, Homayoonzadeh M, Safavi SA (2023) Role of cuticle-degrading enzymes of *Beauveria bassiana* and *Metarhizium anisopliae* in virulence on *Plodia interpunctella* (Lepidoptera, Pyralidae) larvae. J Asia Pac Entomol 26(2):102038. https://doi.org/10.1016/j.aspen.2023.102038

Gong D, Sun L, Li X, Zhang W, Zhang D et al (2023) Micro/nanofabrication, assembly, and actuation based on microorganisms: recent advances and perspectives. Small Structures 2023:2200356. https://doi.org/10.1002/sstr.202200356

Groden E (1999) Proceedings of the New England vegetable and berry growers conference and trade show, Sturbridge, pp 313–315

Gupta R, Keppanan R, Leibman-Markus M, Rav-David D, Elad Y et al (2022) The Entomopathogenic fungi *Metarhizium brunneum* and *Beauveria bassiana* promote systemic immunity and confer resistance to a broad range of pests and pathogens in tomato. Phytopathology® 112(4):784–793. https://doi.org/10.1094/PHYTO-08-21-0343-R

Hanan A, Basit A, Nazir T, Majeed MZ, Qiu D (2020) Anti-insect activity of a partially purified protein derived from the entomopathogenic fungus *Lecanicillium lecanii* (Zimmermann) and its putative role in a tomato defense mechanism against green peach aphid. J Invertebr Pathol 170: 107282. https://doi.org/10.1016/j.jip.2019.107282

He H, Liu L, Munir S, Bashir NH, Wang Y et al (2019) Crop diversity and pest management in sustainable agriculture. J Integr Agric 18(9):1945–1952. https://doi.org/10.1016/S2095-3119(19)62689-4

Herrmann L, Bhattacharyya PN, Lesueur D (2025) Soil health must be included in one health. In: Roger F, Olive M-M, Peyre M, Pfeiffer D, Zinsstag J (eds) One health atlas. Quæ, CABI, pp 74–75. https://doi.org/10.35690/978-2-7592-4027-2

Holder DJ, Kirkland BH, Lewis M, Keyhani NO (2007) Surface characteristics of the entomopathogenic fungus *Beauveria (Cordyceps) bassiana*. Microbiology 153:3448–3457. https://doi.org/10.1099/mic.0.2007/008524-0

Hu G, Leger RJS (2002) Field studies using a recombinant mycoinsecticide (*Metarhizium anisopliae*) reveal that it is rhizosphere competent. Appl Environ Microbiol 68:6383–6387. https://doi.org/10.1128/AEM.68.12.6383-6387.2002

Huang K, Shu L (2021) Grand challenges in sustainable and intelligent phytoprotection. Front Plant Sci 12. https://doi.org/10.3389/fpls.2021.755510

Huang ZH, Feng MG, Chen XX, Liu SS (2008) Pathogenic fungi and parasitoids of aphids present in air captures of migratory alates in the low-latitude plateau of Yunnan, China. Environ Entomol 37:1264–1271. https://doi.org/10.1603/0046-225x

Islam MS, Rahman MM, Paul NK (2016) Arsenic-induced morphological variations and the role of phosphorus in alleviating arsenic toxicity in rice (*Oryza sativa* L.). Plant Sci Arch 1(1):1–10. https://doi.org/10.1016/j.jia.2024.07.022

Islam W, Adnan M, Shabbir A, Naveed H, Abubakar YS et al (2021) Insect-fungal-interactions: a detailed review on entomopathogenic fungi pathogenicity to combat insect pests. Microb Pathog 159:105122. https://doi.org/10.1016/j.micpath.2021.105122

Jaronski ST, Mascarin GM (2017) Mass production of fungal entomopathogens. In: Lacey LA (ed) A. Academic Press, Microbial control of insect and mite pests, pp 141–155. https://doi.org/10.1016/B978-0-12-803527-6.00009-3

Jarrahi MH (2018) Artificial intelligence and the future of work: human-AI symbiosis in organizational decision making. Bus Horiz 61:577–586. https://doi.org/10.1016/j.bushor.2018.03.007

Jiang JA, Syue CH, Wang CH, Liao MS, Shieh JS et al (2022) Precisely forecasting population dynamics of agricultural pests based on an interval type-2 fuzzy logic system: case study for oriental fruit flies and the tobacco cutworms. Precis Agric 23:1302–1332. https://doi.org/10.1007/s11119-022-09886-3

Junaid M, Gokce A (2024) Global agricultural losses and their causes. Bull Biol Allied Sci Res 1: 66. https://doi.org/10.54112/bbasr.v2024i1.66

Kakoti B, Deka B, Roy S, Babu A (2023) The scale insects: its status, biology, ecology and management in tea plantations. Front Insect Sci 2:1048299. https://doi.org/10.3389/finsc.2022.1048299

Kamil D, Prameeladevi T, Ganesh S, Prabhakaran N, Nareshkumar R et al (2017) Green synthesis of silver nanoparticles by entomopathogenic fungus *Beauveria bassiana* and their bioefficacy against mustard aphid (*Lipaphis erysimi* Kalt.). Indian J Exp Biol 5:555–561

Kariyanna B, Sowjanya M (2024) Unravelling the use of artificial intelligence in management of insect pests. Smart Agric Technol 8:100517. https://doi.org/10.1016/j.atech.2024.100517

Karlik JF (2003) Insects and other animals. Insecticides. In: Roberts AV (ed) Encyclopedia of rose science. Elsevier, pp 460–466. https://doi.org/10.1016/B0-12-227620-5/001579

Kashyap U, Garg S, Arora P (2024) Pesticide pollution in India: environmental and health risks, and policy challenges. Toxicol Rep 9(13):101801. https://doi.org/10.1016/j.toxrep.2024.101801

Kaur S, Kumar V, Chawla M, Cavallo L, Poater A et al (2017) Pesticides curbing soil fertility: effect of complexation of free metal ions. Front Chem 5:43. https://doi.org/10.3389/fchem.2017.00043

Khan MA, Ahmad W (2019) Synthetic chemical insecticides: environmental and agro contaminants. In: Khan M, Ahmad W (eds) Microbes for sustainable insect pest management.

Sustainability in plant and crop protection. Springer, Cham. https://doi.org/10.1007/978-3-030-23045-6_1

Khan MK, Ullah MO (2022) Deep transfer learning inspired automatic insect pest recognition. Proceedings of the 3rd international conference on computational sciences and technologies, Mehran University of Engineering and Technology, Jamshoro, Pakistan, pp. 17–19

Khan AL, Hamayun M, Khan SA, Kang SM, Shinwari ZK et al (2012) Pure culture of *Metarhizium anisopliae* LHL07 reprograms soybean to higher growth and mitigates salt stress. World J Microbiol Biotechnol 28(4):1483–1494. https://doi.org/10.1007/s11274-011-0950-9

Klingen I, Eilenberg J, Meadow R (2002) Effects of farming system, field margins and bait insect on the occurrence of insect pathogenic fungi in soils. Agric Ecosyst Environ 91(1–3):191–198. https://doi.org/10.1016/S0167-8809(01)00227-4

Kou S, Peters LM, Mucalo MR (2021) Chitosan: a review of sources and preparation methods. Int J Biol Macromol 169:85–94. https://doi.org/10.1016/j.ijbiomac.2020.12.005

Kumhar KC, Babu A, Arulmarianathan JP, Deka B, Bordoloi M et al (2020) Role of beneficial fungi in managing diseases and insect pests of tea plantation. Egypt J Biol Pest Control 30(1): 1–9. https://doi.org/10.1186/s41938-020-00270-9

Kuzhuppillymyal-Prabhakarankutty L, Tamez-Guerra P, Gomez-Flores R, Rodriguez-Padilla MC, Ek-Ramos MJ (2020) Endophytic *Beauveria bassiana* promotes drought tolerance and early flowering in corn. World J Microbiol Biotechnol 36:47. https://doi.org/10.1007/s11274-020-02823-4

Lahlali R, Taoussi M, Laasli SE, Gachara G, Ezzouggari R et al (2024) Effects of climate change on plant pathogens and host-pathogen interactions. Crop and Environment 3(3):159–170. https://doi.org/10.1016/j.crope.2024.05.003

Li Z, Ahammed GJ (2023) Salicylic acid and jasmonic acid in elevated CO_2-induced plant defense response to pathogens. J Plant Physiol 286:154019. https://doi.org/10.1016/j.jplph.2023.154019

Li S, Yuan Z, Peng R, Leybourne D, Xue Q et al (2023) An effective farmer-centred mobile intelligence solution using lightweight deep learning for integrated wheat Pest management. J Ind Inf Integr 42:100705. https://doi.org/10.1016/j.jii.2024.100705

Lipper L, Building DZ (2018) A short history of the evolution of the climate smart agriculture approach and its links to climate change and sustainable agriculture debates. In: Climate smart agriculture: building resilience to climate change. Springer, Cham

Liu Y, Shi J (2020) Predicting the potential global geographical distribution of two *Icerya* species under climate change. Forests 11(6):684. https://doi.org/10.3390/f11060684

Lopes ARO, Locatelli GO, Barbosa RM, Lobo Junior M, Moura Mascarin G et al (2020) Preparation, characterisation and cell viability of encapsulated *Trichoderma asperellum* in alginate beads. J Microencapsul 37:270–282. https://doi.org/10.1080/02652048.2020.1729884

Mada DA, Hussaini ID, Adamu IG (2014) Study on impact of annual post-harvest losses of grain and post-harvest technology in Ganye Southern Adamawa State-Nigeria. IOSRN J Eng 4(5): 15–20

Mahanta K, Bhattacharyya PN, Sharma AK, Rajkhowa D, Lesueur D et al (2023) Residue and soil dissipation kinetics of chloroacetanilide herbicides on rice (*Oryzae sativa* L.) and assessing the impact on soil microbial parameters and enzyme activity. Environ Monit Assess 195:910. https://doi.org/10.1007/s10661-023-11513-1

Mamun MSA, Ahmed M, Paul SK (2014) Integrated approaches in tea pest management for sustainable tea production. In: Proceedings of the workshop on tea production technology updated, vol 24, pp 18–32. Dhaka: 24 December 2014, Organized by Bangladesh Tea Research Institute, Srimangal, Moulvibazar and Krishi Gobeshona Foundation, BARC Campus

Mandi M, Khatun S, Rajak P, Mazumdar A, Roy S (2020) Potential risk of organophosphate exposure in male reproductive system of a non-target in- sect model Drosophila melanogaster. Environ Toxicol Pharmacol 74:103308. https://doi.org/10.1016/j.etap.2019.103308

Mantzoukas S, Eliopoulos PA (2020) Endophytic Entomopathogenic fungi: a valuable biological control tool against plant pests. Appl Sci 10:360. https://doi.org/10.3390/app10010360

Mantzoukas S, Kitsiou F, Natsiopoulos D, Eliopoulos PA (2022) Entomopathogenic fungi: interactions and application. Encyclopaedia 2:646–656. https://doi.org/10.3390/encyclopedia2020044

Manyilizu WB (2019) Pesticides, anthropogenic activities, history and the health of our environment: lessons from Africa. In: Larramendy M, Soloneski S (eds) Pesticides – use and misuse and their impact in the environment. IntechOpen. https://doi.org/10.5772/in-techopen.82600

Mesquita E, Hu S, Lima TB, Golo PS, Bidochka MJ (2023) Utilization of *Metarhizium* as an insect biocontrol agent and a plant bioinoculant with special reference to Brazil. Front Fungal Biol 4: 1276287. https://doi.org/10.3389/ffunb.2023.1276287

Mikail N, Çığ A, Gılıcova T, Yagubova Y (2024) Fuzzy expert system applications in plant cultivation. In: Mammadov FS, Aliev RA, Kacprzyk J, Pedrycz W (eds) International conference on smart environment and green technologies—ICSEGT2024. ICSEGT. Lecture notes in networks and systems, vol 1251. Springer, Cham. https://doi.org/10.1007/978-3-031-81567-6_42

Mitra B, Chowdhury AR, Dey P, Hazra KAK, Hossain A et al (2021) Use of agrochemicals in agriculture: alarming issues and solutions. In: Bhatt R, Meena RS, Hossain A (eds) Input use efficiency for food and environmental security. Springer, Singapore. https://doi.org/10.1007/978-981-16-5199-1_4

Moharram AM, Abdel-Galil FA, Hafez WMM (2021) On the enzymes' actions of entomopathogenic fungi against certain indigenous and invasive insect pests. Egypt J Biol Pest Control 31. https://doi.org/10.1186/s41938-021-00397-3

Mongkolsamrit S, Khonsanit A, Thanakitpipattana D, Tasanathai K, Noisripoom W et al (2020) Revisiting Metarhizium and the description of new species from Thailand. Stud Mycol 95:171–251. https://doi.org/10.1016/j.simyco.2020.04.001

Moustaka J, Meyling NV, Hauser TP (2022) Root-associated entomopathogenic fungi modulate their host plant's photosystem II photochemistry and response to herbivorous insects. Molecules 27(1):207. https://doi.org/10.3390/molecules27010207

Mukhopadhyay M, Mondal TK (2017) Cultivation, improvement, and environmental impacts of tea. In: Oxford University Press, Oxford, UK. https://doi.org/10.1093/acrefore/9780199389414.013.373

Mukhtar Y, Shankar U, Wazir ZA (2023) Prospects and challenges of emerging insect Pest problems vis-à-vis climate change. Indian J Ecol 50(4):1096–1103. https://doi.org/10.55362/IJE/2023/4019

Nag PK, Gite LP (2020) Pesticides and chemical toxicity—challenges in farming. In: Human-centered agriculture. Design science and innovation. Springer, Singapore. https://doi.org/10.1007/978-981-15-7269-2_10

Nair A, Rao AS, Surabhi MA, Gnanika M, More SS (2025) Unravelling fungal pathogenesis: advances in CRISPR-Cas9 for understanding virulence and adaptation. Fungal Genet Biol 179: 104006. https://doi.org/10.1016/j.fgb.2025.104006

Namasivayam SKR, Bharani RSA (2021) Biocompatible silver nanoparticles-loaded fungal metabolites nanoconjugate (agnp–fm) preparation for the noteworthy pesticidal activity. Natl Acad Sci Lett 44:511–517. https://doi.org/10.1007/s40009-021-01044-z

Nayagam MG, Vijayalakshmi B, Somasundaram K, Mukunthan MA, Yogaraja CA et al (2023) Control of pests and diseases in plants using IOT technology. Meas Sensors 26:100713. https://doi.org/10.1016/j.measen.2023.100713

Nicolopoulou-Stamati P, Maipas S, Kotampasi C, Stamatis P, Hens L (2016) Chemical pesticides and human health: the urgent need for a new concept in agriculture. Front Public Health 18(4): 148. https://doi.org/10.3389/fpubh.2016.00148

Ogori AF, Eke MO, Girgih TA, Abu JO (2022) Influence of Aduwa (*Balanites aegyptiaca. del*) meal protein enrichment on the proximate, phytochemical, functional and sensory properties of Ogi. Acta Bot Plantae 3(3):22–35. https://doi.org/10.5281/zenodo.7458578

Okoroiwu HU, Munoz FL, Montero FP (2020) Bibliometric analysis of global sickle cell disease research from 1997 to 2017. Hemato, Transfus Cell Ther 44(2):1–12. https://doi.org/10.1016/j.htct.2020.09.156

Ownley BH, Griffin MR, Klingeman WE, Gwinn KD, Moulton JK et al (2008) *Beauveria bassiana*: Endophytic colonization and plant disease control. J Invertebr Pathol 98(3):267–270. https://doi.org/10.1016/J.JIP.2008.01.010

Pathak VM, Verma VK, Rawat BS, Kaur B, Babu N et al (2022) Current status of pesticide effects on environment, human health and it's eco-friendly management as bioremediation: a comprehensive review. Front Microbiol 13:962619. https://doi.org/10.3389/fmicb.2022.962619

Powell WA, Klingeman WE, Ownley BH, Gwinn KD (2009) Evidence of endophytic *Beauveria bassiana* in seed-treated tomato plants acting as a systemic entomopathogen to larval *Helicoverpa zea* (Lepidoptera: Noctuidae). J Entomol Sci 44(4):391–396. https://doi.org/10.18474/0749-8004-44.4.39

Price D (1963) Little science, big science. Columbia University Press, New York

Quesada-Moraga E, Munoz-Ledesma FJ, Santiago-Alvarez C (2009) Systemic protection of *Papaver somniferum* L. against *Iraella luteipes* (Hymenoptera: Cynipidae) by an endophytic strain of *Beauveria bassiana* (Ascomycota: Hypocreales). Environ Entomol 38(3):723–730. https://doi.org/10.1603/022.038.0324

Quesada-Moraga E, Garrido-Jurado I, González-Mas N, Yousef-Yousef M (2023) Ecosystem services of entomopathogenic ascomycetes. J Invertebr Pathol 201:108015. https://doi.org/10.1016/j.jip.2023.108015

Rahaman MM, Islam KS, Jahan M (2018) Rice farmers' knowledge of the risks of pesticide use in Bangladesh. J Health Pollut 8(20):181203. https://doi.org/10.5696/2156-9614-8.20.181203

Rajak P, Khatun S, Dutta M, Mandi M, Roy S (2018) Chronic exposure to acephate triggers ROS-mediated injuries at organismal and sub-organismal levels of *Drosophila melanogaster*. Toxicol Res 7(5):874–887. https://doi.org/10.1039/c8tx00052b

Rajak P, Roy S, Ganguly A, Mandi M, Dutta A et al (2023) Agricultural pesticides—friends or foes to biosphere? Journal of Hazardous Materials Advances 10:100264. https://doi.org/10.1016/j.hazadv.2023.100264

Rani D, Krishnamma SK, Rani JS (2024) Entomopathogenic fungi, *Metarhizium anisopliaeas* a chemical substitute for termite pest management in sugarcane. J Environ Biol 45:235–242. https://doi.org/10.1002/ps.2780370410

Raya-Díaz S, Sanchez-Rodrıguez AR, Segura-Fernandez JM, del Campillo MC, Quesada-Moraga E (2017) Entomopathogenic fungi-based mechanisms for improved Fe nutrition in sorghum plants grown on calcareous substrates. PLoS One 12(10):e0185903. https://doi.org/10.1371/journal.pone.0185903

Rodriguez R, Durán P (2020) Natural holobiome engineering by using native extreme microbiome to counteract the climate change effects. Front Bioeng Biotechnol 8. https://doi.org/10.3389/fbioe.2020.00568

Rodriguez R, White J Jr, Arnold A, Redman R (2009) Fungal endophytes: diversity and functional roles. New Phytol 182:314–330. https://doi.org/10.1111/j.1469-8137.2009.02773.x

Roy S, Muraleedharan N, Mukhapadhyay A, Handique G (2015) The tea mosquito bug, *Helopeltis theivora* Waterhouse (Heteroptera: Miridae): its status, biology, ecology and management in tea plantations. Int J Pest Manag 61(3):179–197. https://doi.org/10.1080/09670874.2015.1030002

Roy S, Prasad AK, Neave S, Bhattacharyya PN, Nagpal A et al (2020) Nonchemical based integrated management package for live-wood eating termites in tea plantations of North-East India. Int J Trop Insect Sci 40:435–440. https://doi.org/10.1007/s42690-019-00095-6

Saberi Riseh R, Hassanisaadi M, Vatankhah M, Soroush F, Varma RS (2022) Nano/microencapsulation of plant biocontrol agents by chitosan, alginate, and other important biopolymers as a novel strategy for alleviating plant biotic stresses. Int J Biol Macromol 222:1589–1604. https://doi.org/10.1016/j.ijbiomac.2022.09.278

Saikia R, Bortamuly M, Konwar NJ, Bhattacharyya PN (2025a) Exploring the creative nexus (Vol I) Ecological integrity and environmental sustainability of North-East India: a part of Indo-

Burma mega biodiverse hotspot region. Published by NNS College Teachers' Unit, Titabar, Purbayon Publisher, pp 1–368

Saikia S, Nath BC, Bhattacharyya PN, Gogoi B (2025b) Beneficial effect of microbial bioformulations on nutrient availability, microbial biomass carbon and key enzymatic activities in citrus orchard soil. Indian J Hortic 82(2):165–170. https://doi.org/10.58993/ijh/2025.82.2.6

Samal I, Bhoi TK, Majhi PK, Murmu S, Pradhan AK et al (2023) Combatting insects mediated biotic stress through plant associated endophytic entomopathogenic fungi in horticultural crops. Front Plant Sci 13:1098673. https://doi.org/10.3389/fpls.2022.1098673

Sanchez-Rodriguez AR, Campillo MCD, Quesada Moraga E (2015) *Beauveria bassiana*: an entomopathogenic fungus alleviates Fe chlorosis symptoms in plants grown on calcareous substrates. Sci Hortic 197:193–202. https://doi.org/10.1016/j.scienta.2015.09.029

Sandhu SS, Sharma AK, Beniwal V, Goel G, Batra P et al (2012) Myco-biocontrol of insect pests: factors involved, mechanism, and regulation. J Pathog 2012:126819. https://doi.org/10.1155/2012/12681

Santi L, Beys da Silva WO, Berger M, Guimarães JA, Schrank A et al (2010) Conidial surface proteins of *Metarhizium anisopliae*: source of activities related with toxic effects, host penetration and pathogenesis. Toxicon 55:874–880. https://doi.org/10.1016/j.toxicon.2009.12.012

Sarma B, Kashtoh H, Lama Tamang T, Bhattacharyya PN, Mohanta YK et al (2023a) Abiotic stress in rice: visiting the physiological response and its tolerance mechanisms. Plants 12(23):3948. https://doi.org/10.3390/plants12233948

Sarma BD, Puzari KC, Dutta P, Pandey AK (2023b) An alginate-based encapsulation enhances shelf life and bioactivity of the entomopathogenic fungus, *Metarhizium anisopliae*. Egypt J Biol Pest Control 33:69. https://doi.org/10.1186/s41938-023-00714-y

Sarmah SR, Bhattacharyya PN, Barooah K (2020) Microbial biocides – prominent alternatives of chemicals in tea disease management. J Biol Control 34(2):144–152. https://doi.org/10.18311/jbc/2020/22689

Savary S, Willocquet L, Pethybridge SJ, Esker P, McRoberts N et al (2019) The global burden of pathogens and pests on major food crops. Nat Ecol Evol 3:430–439. https://doi.org/10.1038/s41559-018-0793-y

Seijger C, Urfels A, Christoforidou M, Hellegers P, Borghuis G et al (2025) More food, but less land and water for nature: why agricultural productivity gains did not materialize. Agric Water Manage 307:109229. https://doi.org/10.1016/j.agwat.2024.109229

Shah FA, Wang CS, Butt TM (2005) Nutrition influences growth and virulence of the insect-pathogenic fungus *Metarhizium anisopliae*. FEMS Microbiol Lett 251:259–266. https://doi.org/10.1016/j.femsle.2005.08.010

Shahbazi F, Shahbazi S, Nadimi M, Paliwal J (2025) Losses in agricultural produce: a review of causes and solutions, with a specific focus on grain crops. J Stored Prod Res 111:102547. https://doi.org/10.1016/j.jspr.2025.102547

Shahid M, Khan MS (2022) Ecotoxicological implications of residual pesticides to beneficial soil bacteria: a review, pesticide biochemistry and physiology. Pestic Biochem Physiol 188:Article ID 105272. https://doi.org/10.1016/j.pestbp.2022.105272

Shahid A, Rao Q, Bakhsh A, Husnain T (2012) Entomopathogenic fungi as biological controllers: new insights into their virulence and pathogenicity. Arch Biol Sci 64:21–42. https://doi.org/10.2298/ABS1201021S

Sharma HC (2016) Climate Change vis-à-vis Pest management. In: Conference on National Priorities in plant health management (organised by plant protection Association of India, NBPGR, Hyderabad), Tirupati, India

Sharma L, Bohra N, Rajput VD, Quiroz-Figueroa FR, Singh RK et al (2020) Advances in Entomopathogen isolation: a case of bacteria and fungi. Microorganisms 9(1):16. https://doi.org/10.3390/microorganisms9010016

Sharma A, Sharma S, Yadav PK (2023) Entomopathogenic fungi and their relevance in sustainable agriculture: a review. Cogent Food Agric 9(1):2180857. https://doi.org/10.1080/23311932.2023.2180857

Sheail J (1991) The regulation of pesticides use: an historical perspective. In: Innovation and environmental risks. Belhaven Press, London, UK, pp 38–46

Shrivastava G, Ownley BH, Augé RM, Toler H, Dee M et al (2015) Colonization by arbuscular mycorrhizal and endophytic fungi enhanced terpene production in tomato plants and their defense against a herbivorous insect. Symbiosis 65(2):65–74. https://doi.org/10.1007/s13199-015-0319-1

Silva PCL, Sadaei HJ, Ballini R, Guimarães FG (2020) Probabilistic forecasting with fuzzy time series. IEEE Trans Fuzzy Syst 28(8):1771–1784

Singh A, Dhiman N, Kar AK, Singh D, Purohit MP et al (2020) Advances in controlled release pesticide formulations prospects to safer integrated pest management and sustainable agriculture. J Hazard Mater 385:121525. https://doi.org/10.1016/j.jhazmat.2019.121525

Singh A, Shraogi N, Verma R, Saji J, Kar AK et al (2024) Challenges in current pest management practices: navigating problems and a way forward by integrating controlled release system approach. Chem Eng J 498:154989. https://doi.org/10.1016/j.cej.2024.154989

Siraj F, Arbaiy N (2006) Integrated pest management system using fuzzy expert system. Semantic Scholar

Subedi B, Poudel A, Aryal S (2023) The impact of climate change on insect pest biology and ecology: implications for pest management strategies, crop production, and food security. J Agric Food Res 14:100733. https://doi.org/10.1016/j.jafr.2023.100733

Sui L, Zhu H, Xu W, Guo Q, Wang L et al (2020) Elevated air temperature shifts the interactions between plants and endophytic fungal entomopathogens in an agroecosystem. Fungal Ecol 47: 100940. https://doi.org/10.1016/j.funeco.2020.100940

Sung GH, Poinar GO, Spatafora JW (2008) The oldest fossil evidence of animal parasitism by fungi supports a cretaceous diversification of fungal-arthropod symbioses. Mol Phylogenet Evol 49: 495–502. https://doi.org/10.1016/j.ympev.2008.08.028

Teshome DT, Zharare GE, Naidoo S (2020) The threat of the combined effect of biotic and abiotic stress factors in forestry under a changing climate. Front Plant Sci 11:601009. https://doi.org/10.3389/fpls.2020.601009

Thakur N, Yadav AN (2025) Nanotechnology in agriculture: a review on precision farming and sustainable crop production. BioNanoSci 15:243. https://doi.org/10.1007/s12668-025-01861-2

Tiwari AK (2024) Insect pests in agriculture identifying and overcoming challenges through IPM. Arch Curr Res Int 24(3):124–130. https://doi.org/10.9734/acri/2024/v24i3651

Toffa J, Loko YLE, Kpindou OKD, Zanzana K, Adikpeto J et al (2021) Endophytic colonization of tomato plants by *Beauveria bassiana* Vuillemin (Ascomycota: Hypocreales) and leaf damage in *Helicoverpa armigera* (Hübner) (Lepidoptera: Noctuidae) larvae. Egypt J Biol Pest Control 31(82):1–9. https://doi.org/10.1186/s41938-021-00431-4

Tomilova OG, Kryukova NA, Efimova MV, Kovtun IS, Kolomeichuk LV et al (2021) Early physiological response of potato plants to Entomopathogenic fungi under hydroponic conditions. Horticulturae 7:217. https://doi.org/10.3390/horticulturae7080217

Vassilev N, Vassileva M, Martos V, Garcia del Moral LF, Kowalska J et al (2020) Formulation of microbial inoculants by encapsulation in natural polysaccharides: focus on beneficial properties of carrier additives and derivatives. Front Plant Sci 11:270. https://doi.org/10.3389/fpls.2020.00270

Vega FE (2018) The use of fungal entomopathogens as endophytes in biological control: a review. Mycologia 110(1):4–30. https://doi.org/10.1080/00275514.2017.1418578

Wang C, St Leger RJ (2007) A scorpion neurotoxin increases the potency of a fungal insecticide. Nat Biotechnol 25:1455–1456. https://doi.org/10.1038/nbt1357

Wang C, Typas MA, Butt TM (2002) Detection and characterization of Pr1 virulent gene deficiencies in the insect pathogenic fungus *Metarhizium anisopliae*. FEMS Microbiol Lett 213(2):251–255. https://doi.org/10.1111/j.1574-6968.2002.tb11314.x

Wang B, Wan Y, Zheng Y, Lee X, Liu T et al (2019) Alginate-based composites for environmental applications: a critical review. Crit Rev Environ Sci Technol 49:318–356. https://doi.org/10.1080/10643389.2018.1547621

Wang T, Zhang L, Xu N, Alharbi KH (2023) Adaptive critic learning for approximate optimal event-triggered tracking control of nonlinear systems with prescribed performances. Int J Control. https://doi.org/10.1080/00207179.2023.2250880

Wenzel I, Forim M, Silva M, Fernandes J, Filho A (2016) Effect of ultraviolet radiation on fungi *Beauveria bassiana* and *Metarhizium anisopliae*, pure and encapsulated, and bio-insecticide action on *Diatraea saccharalis*. Adv Entomol 04:151–162. https://doi.org/10.4236/ae.2016.43016

Wheelis M (2002) Biological warfare at the 1346 siege of Caffa. Emerg Infect Dis 8(9):971. https://doi.org/10.3201/eid0809.010536

Wraight SP, Galaini-Wraight S, Howes RL, Castrillo LA, Griggs MH et al (2021) Efficacy of *Beauveria bassiana* strain GHA spray applications against coffee berry borer *Hypothenemus hampei* on Hawai'i Island. Biol Control 161:104587. https://doi.org/10.1016/j.biocontrol.2021.104587

Wu J, Du C, Zhang J, Yang B, Cuthbertson AGS et al (2022) Synthesis of *Metarhizium anisopliae*–chitosan nanoparticles and their pathogenicity against *Plutella xylostella* (Linnaeus). Microorganisms 10(1):1. https://doi.org/10.3390/microorganisms10010001

Yosri M, Abdel-Aziz MM, Sayed RM (2018) Larvicidal potential of irradiated myco-insecticide from *Metarhizium anisopliae* and larvicidal synergistic effect with its mycosynthesized titanium nanoparticles (TiNPs). J Radiat Res Appl Sci 11:328–334. https://doi.org/10.1016/j.jrras.2018.06.001

Yousef M, Garrido-Jurado I, Ruíz-Torres M, Quesada-Moraga E (2017) Reduction of adult olive fruit fly populations by targeting preimaginals in the soil with the entomopathogenic fungus Metarhizium brunneum. J Pest Sci 90:345–354. https://doi.org/10.1007/s10340-016-0779-y

Yousef M, Aranda-Valera E, Quesada-Moraga E (2018) Lure-and-infect and lure-and kill devices based on *Metarhizium brunneum* for spotted wing drosophila control. J Pest Sci 91:227–235. https://doi.org/10.1007/s10340-017-0874-8

Yuan Z, Li S, Peng R, Leybourne D, Yang P et al (2023) Pest DSS: an integrated decision support system for sustainable Pest management in agriculture. In: 32nd international symposium on industrial electronics. Helsinki, Finland, pp 1–6

Zadeh LA (1965) Fuzzy sets. Inf Control 8(3):338–353. https://doi.org/10.1016/S0019-9958(65)90241-X

Zaidi SSA, Mahas A, Vanderschuren H, Mahfouz MM (2020) Engineering crops of the future: CRISPR approaches to develop climate-resilient and disease-resistant plants. Genome Biol 21:289. https://doi.org/10.1186/s13059-020-02204-y

Zhang Z, Feng X, Wang Y, Xu W, Huang K et al (2019) Advances in research on functional genes of tea plant. Gene 711:143940. https://doi.org/10.1016/j.gene.2019.143940

Zitko V (2003) Chlorinated pesticides: Aldrin, DDT, Endrin, Dieldrin, Mirex. In: Fiedler, H. (eds) persistent organic pollutants. The handbook of environmental chemistry, 30. Springer, Berlin, Heidelberg. https://doi.org/10.1007/10751132_4

Zitlalpopoca-Hernandez G, Pozo MJ, Hauser TP, Meyling NV (2022) Combined effects of root-associated entomopathogenic and mycorrhizal fungi on the foliar pathogen *Botrytis cinerea* in tomato. Biol Control 175:105034. https://doi.org/10.1016/j.biocontrol.2022

Native Entomopathogenic Fungi from Mexico: Biological Protection of *Agave* spp. Against *Scyphophorus acupunctatus*

11

Rosa Elvira López-López, Alfonso Vásquez-López, Carlos I. Cortes Martínez, Esperanza Loera-Alvarado, Eduardo Alfonso Treviño-López, and Carlos Granados-Echegoyen

Abstract

The agave weevil (*Scyphophorus acupunctatus*, Coleoptera: Dryophthoridae) is the primary pest of agave crops in Mexico, significantly affecting their productivity and commercial value. The extensive use of chemical insecticides can cause issues such as pest resistance and adverse environmental effects, highlighting the need for sustainable and effective management alternatives. This study aimed to identify entomopathogenic fungi native to the state of Oaxaca that are pathogenic to *S. acupunctatus* adults collected from *Agave potatorum*. Using the insect-trap method with *Tenebrio molitor* larvae, we collected soil samples from six municipalities and isolated strains of *Beauveria* and *Metarhizium*. We evaluated these strains in a laboratory setting using conidial suspensions of 1×10^6 and 1×10^8 conidia/mL, applying both immersion and spray techniques to adult *S. acupunctatus* specimens. We obtained 22 fungal isolates; seven caused $\geq 80\%$ mortality 5 days after immersion. Using the spray technique, six isolates produced $>60\%$ mortality 11 days post-inoculation. Morphological and molecular techniques identified *Beauveria bassiana* and *Metarhizium brunneum* as the most effective fungi.

R. E. López-López · A. Vásquez-López · C. I. Cortes Martínez
Instituto Politécnico Nacional, CIIDIR Oaxaca, Santa Cruz Xoxocotlán, Oaxaca, Mexico

E. Loera-Alvarado
SECIHTI-Universidad Autónoma Chapingo (CRUCO), Lomas del Valle, Morelia, Michoacán, Mexico

E. A. Treviño-López
Centro de Investigación en Química Aplicada, San José de los Cerritos, Saltillo, Coahuila, Mexico

C. Granados-Echegoyen (✉)
SECIHTI-Instituto Politécnico Nacional, CIIDIR Oaxaca, Santa Cruz Xoxocotlán, Oaxaca, Mexico

S. K. Deshmukh et al. (eds.), *Frontiers in Entomopathogenic Fungi*,
https://doi.org/10.1007/978-981-95-6529-0_11

Keywords

Agave · *Beauveria* · Entomopathogenic · Metarhizium

1 Introduction

In Mexico, agaves have played a vital role both economically and culturally for many indigenous communities. For centuries, they have served as sources of food, alcoholic beverages, medicine, and various other purposes (García-Mendoza 2007). With 33 species, 13 of which are unique to Oaxaca, the country boasts the largest collection of agave species. Mezcal production uses many of these species as raw materials (García-Mendoza and Franco-Martínez 2018). The so-called Mezcal Region encompasses the municipalities of Villa Sola de Vega, Miahuatlán, Yautepec, Santiago Matatlán, Tlacolula, Ocotlán, Ejutla, and Zimatlán de Álvarez, all of which are renowned for their high concentration of maguey plants (Illsey et al. 2009). Mezcal production in Oaxaca typically involves cultivated varieties, such as espadín (*Agave angustifolia*), sierra negra (*A. americana var. oaxacensis*), and tobasiche (*A. karwinskii*), along with wild species like mexicano (*A. rhodacantha*), jabalí (*A. convallis*), tepeztate (*A. marmorata*), and the highly prized tóbala (*A. potatorum*) (García-Mendoza 2012).

The Asparagaceae family comprises nine genera native to the Americas: *Agave*, *Beschorneria*, *Furcraea*, *Hesperaloe*, *Hesperoyucca*, *Manfreda*, *Polianthes*, *Prochnyanthes*, and *Yucca*. This family includes approximately 330 species that range from the northern edges of Canada to Bolivia and Paraguay (García-Mendoza et al. 2017). *Agave* ranges from the southern United States through Colombia and Venezuela, also inhabiting Curaçao, Trinidad and Tobago, various Caribbean islands, the Bahamas, and Aruba. Mexico, Cuba, Guatemala, and the United States exhibit the highest diversity, with around 200 recognized species of *Agave* (García-Mendoza and Franco-Martínez 2018). Mexico has 159 of these species, 119 of which are endemic to the country (García-Mendoza and Franco-Martínez 2018). These species are primarily in the states of Jalisco, Guanajuato, Nayarit, Michoacán, Tamaulipas, Oaxaca, Guerrero, Durango, San Luis Potosí, Zacatecas, and Puebla (SAGARPA 2017) (Table 11.1).

2 Economic Importance of *Agave* in Mexico

The mezcal and tequila industries generate over 100,000 direct and indirect jobs each year. Between 2012 and 2015, the export value of mezcal increased by 322%, rising from USD 263,768 to USD 19,796,627. Mexico exports this alcoholic beverage to 47 countries, with the United States as the largest buyer, accounting for 61.13% of exports (COMERCAM 2022). The leading states in mezcal production are Durango, Zacatecas, Tamaulipas, San Luis Potosí, Guanajuato, Michoacán, Guerrero, Puebla, and Oaxaca, which together produce 14,165,505 L. Oaxaca dominates this

Table 11.1 Distribution of *Agave* species in Mexico (García-Mendoza and Franco-Martínez 2018)

Taxon name	Distribution[a]
Agave americana	Oaxaca, Guerrero, Jalisco, Morelos, Mexico, Michoacan, Puebla
A. angustifolia	Oaxaca, Guerrero, Jalisco
A. opplanata	Chihuahua, Durango, Hidalgo, Mexico, Oaxaca, Zacatecas, Puebla
A. atrovirens	Oaxaca, Puebla y Veracruz
A. chiapensis	Chiapas, Oaxaca
A. convallis	Oaxaca
A. cupreata	Michoacan, Guerrero, Oaxaca
A. guiengola	Endemic from Oaxaca
A. isthmensis	Chiapas, Oaxaca
A. karwinskii	Oaxaca, Puebla
A. kavandivi	Endemic from Oaxaca
A. kerchovei	Oaxaca, Puebla
A. marmota	Oaxaca, Puebla
Agave nizandensis	Endemic from Oaxaca
A. potatorum	Oaxaca, Puebla
A. rhodacantha	From Sonora to Oaxaca
A. salmiana	Oaxaca, Durango, Coahuila, Chiapas, Puebla, Veracruz
A. macroacantha	Oaxaca, Puebla

[a]These are 13 of the 32 states that make up the United Mexican States

production, contributing 91.3% of the total volume, or 12,890,609.55 L. Mezcal ranks as the third most exported product from Oaxaca, following beer and coffee. The state accounts for 93% of the country's total mezcal output and is renowned for its high quality. Demand for craft spirits has surged by over 300% in the past 5 years. Byproducts from both the distillation process and the agave plant, such as food products, animal feed, pharmaceuticals, insulin, syrups, and substrates, further enhance the industry's value (COMERCAM 2022). However, the industry faces significant challenges from droughts, landslides, and particularly pests, which can reduce yields by as much as 33.9%.

Plant health defines a pest as any animal population that feeds on, lays eggs in, tunnels through, sucks from, or otherwise disrupts plants or their products. This interference can lead to economic losses, diminish commercial quality, or pose a phytosanitary risk. Pests include insects, mites, nematodes, rodents, mollusks, birds, and even wild vertebrates when their populations surpass the economic damage threshold—the point at which the costs of damage exceed the costs of control measures (Paschapur et al. 2025) (Table 11.2).

Similarly, we can define a plant disease as any physiological, biochemical, or structural change caused by a biological agent (like fungi, bacteria, viruses, or phytoplasmas) or an abiotic agent (such as nutritional deficiencies, pollution, or water stress). These changes reduce the plant's vigor, yield, or commercial value. People use the term to refer to infectious diseases; these involve a pathogen that can invade, reproduce, and spread within or between plants (Mukherjee et al. 2025) (Table 11.3).

Table 11.2 Common *Agave* pests (Pérez and Del Real Laborde 2007; Ramos 2013)

Group	Pest
Borers	Agave leaf borer (*Agathymus rethon*) White maguey worm (*Acentrocneme hesperiaris*) Agave weevil (*Scyphophorus acupunctatus*) Agave funeral beetle (Cerambicid, *Acanthoderes funeraria*) Heart borer (unidentified coleopteran) Cutworms (various lepidopteran species)
Foliage	Mealybug (*Pseudococcus agavis*) Armored scale (*Acutaspis agavis*) Agave bug (*Caulatops agavis*) Grasshoppers (various species) Cochineal insect (*Dactylopius* sp.)
Root feeders	Rhinoceros beetle (*Oryctes nasicornis*) White grub (*Anomala* sp., *Cyclocephala* sp., and *Phyllophaga* sp.)

Table 11.3 Main *Agave* diseases and their economic impact

Disease	Causal agent	Economic impact
Agave wilt	*Fusarium oxysporum*	Up to 35% crop loss
Gray leaf spot	*Cercospora agavicola*	Up to 20% crop loss
Piña rot	*Ceratocystis paradoxa*	Causes the death of up to 23% of plants
Soft rot of the agave heart	*Pectobacterium carotovorum* subsp. *carotovorum*	Up to 64.2% crop loss
Agave wilt	*Fusarium verticillioides*	Up to 30% crop loss

3 Biological Pest Control

Biological control (BC) is a sustainable and environmentally friendly strategy for reducing or eliminating pest populations by utilizing their natural enemies instead of synthetic chemicals. This approach depends on organisms such as viruses, nematodes, entomopathogenic fungi, bacteria, protozoa, predatory insects, and parasitoids, which can naturally regulate pest populations and maintain them below economic damage thresholds (Pérez and Del Real Laborde 2007).

A core principle of biological control is that using living organisms leaves no toxic residues in the environment and maintains the agroecosystem's biodiversity, unlike chemical pesticides. In agave cultivation, BC is significant because native natural enemies, such as entomopathogenic fungi, are often already present in the soil and play a crucial role in managing insect pests.

According to Ramos (2013), one of the most critical factors for successful BC is avoiding the excessive use of agrochemicals. These chemicals not only target pests but also harm beneficial species, disrupting the ecological balance. Therefore, integrated pest management strategies should prioritize biological agents, promoting the conservation of beneficial species while minimizing harm to the environment and human health.

Among the most extensively studied and used microorganisms in biological pest control are entomopathogenic fungi (EPF), particularly *Metarhizium anisopliae* and *Beauveria bassiana*, both belonging to the order Hypocreales and the family Clavicipitaceae. Rigorous testing across various cropping systems has shown these fungi to be effective against pests in several insect orders, including Coleoptera, Lepidoptera, Hemiptera, and Thysanoptera (León-Martínez et al. 2019).

Mexico's agricultural sector has seen the development of commercial bioinsecticide formulations based on fungi, including products containing *Verticillium lecanii* and *Paecilomyces fumosoroseus*. These formulations successfully controlled various pests, including *Schistocerca piceifrons*, *Schistocerca pallens*, *Schistocerca nitens*, *Melanoplus*, *Boopedum*, *Mermiria*, *Sphenarium*, *Brachystola*, *Toxoptera citricida*, *Maconellicoccus*, *Hypothenemus hampei* (coffee borer), *Anthonomus grandis* (cotton boll weevil), *Phyllophaga* spp. (white grubs), *Cosmopolites sordidus* (banana weevil), and *Leptinotarsa decemlineata* (Colorado potato beetle).

One of the primary advantages of BC is its specificity. It targets only certain groups of insects, avoiding harm to beneficial species, pollinators, and other non-target organisms. Biological agents do not contaminate soil, water, or harvested products. A single application often suffices, as these organisms establish themselves, reproduce, and persist in the environment, providing long-term protection. Another significant benefit is that pests do not develop genetic resistance, making BC an essential tool for integrated pest management (García-García et al. 2015).

4 Entomopathogenic Fungi (EPF)

EPF are a group of fungi that infect and kill a diverse array of insects and arthropods, making them crucial agents in biological control. Unlike pathogens such as viruses or bacteria, EPF can directly penetrate the host insect's cuticle with no ingestion (Lacey et al. 2001; Schapovaloff et al. 2015).

Their mode of action relies significantly on their capacity to produce a diverse array of extracellular enzymes. Lipases serve as the initial enzymes by breaking down the insect's outer lipid layer, facilitating fungal entry. Once inside, additional enzymes, such as proteases, chitinases, and phospholipases, degrade the insect's structural proteins, chitin, and lipids (Moharram et al. 2021). This combination of enzymes is crucial for the fungus's pathogenicity and its ability to overcome the host defenses.

EPF also deploy additional mechanisms that enhance its effectiveness. These include the production of secondary metabolites with antibiotic or cytotoxic effects, such as antimicrobial peptides or compounds that suppress the insect's immune system. Some EPF can also restrict access to essential micronutrients within the host, stunting its growth and facilitating fungal colonization (Song et al. 2023).

These fungi inhabit both agricultural and natural ecosystems, reducing pest populations without disrupting ecological balance, harming the environment, or endangering human health, thus providing valuable ecosystem services (Meyling and Eilenberg 2007). We can promote sustainable farming systems by combining these practices with other agroecological practices like intercropping, live barriers, and conservation of natural enemies.

4.1 Insects Infected by EPF

People commonly refer to fungal diseases caused by EPF in insects as muscardinosis, a term used to describe infections caused by *B. bassiana*. The mummification of the host insect, which becomes enveloped in a cotton-like mass composed of fungal mycelium and spores, characterizes these infections. The color of this fungal mass varies depending on the species, leading to designations such as "green muscardine" for *M. anisopliae* (Pacheco et al. 2019).

Research has shown that EPF often exhibit high specificity toward particular insect genera or species. For instance, Gkounti et al. (2015) discovered that isolates of *B. bassiana*, *M. anisopliae*, and *Isaria fumosorosea*, obtained directly from insect hosts, demonstrated high virulence against species within the same genus. This finding suggests a co-evolutionary relationship between the pathogen and its host. It is crucial to grasp this relationship to enhance biological control strategies.

EPF have proven effective against the agave weevil (*Scyphophorus acupunctatus*). Field surveys by Velázquez et al. (2006) documented natural infection and mummification of adult *S. acupunctatus* by *B. bassiana* in agave ecosystems, indicating the presence of this biological control agent. Aquino-Bolaños et al. (2023) demonstrated through laboratory trials that both *B. bassiana* and *M. anisopliae* are effective in controlling this pest. These findings support the use of EPF as a complementary tool in integrated management programs for the agave weevil.

4.1.1 *Beauveria bassiana*

Beauveria bassiana is an entomopathogenic fungus that belongs to the family Clavicipitaceae, the order Hypocreales, and the phylum Ascomycota. Its effectiveness as a bioinsecticide in various agricultural systems has been the subject of extensive study. Some of the primary pests it targets include the coffee berry borer (*Hypothenemus hampei*), the diamondback moth (*Plutella xylostella*), the banana weevil (*Cosmopolites sordidus*), and the chili pepper weevil (*Capsicum annuum* L.) (Monzón 2008; Fang-min et al. 2023).

Insects infected by *B. bassiana* develop a distinctive white, cotton-like covering composed of fungal mycelium and conidia that envelops their entire bodies. This fungus naturally occurs in soils across most regions of the world, and researchers have used it to develop commercial formulations. *Cordyceps bassiana* is the name for its sexual form; however, in practical applications and most scientific literature, it is referred to by its anamorphic name (Pavone 2021a).

Beauveria bassiana's mode of action involves both physical and enzymatic penetration of the insect cuticle, as well as the release of toxins that compromise the host's immune system. One of the most significant of these toxins is beauvericin, a cyclopeptide compound that disrupts cellular ion homeostasis and promotes apoptosis, allowing the fungus to establish a systemic infection. Other notable toxins include beauverolides, bassianolides, and isarolides, along with organic acids such as oxalic acid and insecticidal pigments like tenellin and bassianin (Raghunandan et al. 2024). We divide the life cycle of *B. bassiana* into two main stages: (1) Pathogenic Phase: Fungal conidia attach to the surface of the insect, germinate, and produce a germ tube that penetrates the cuticle. Once inside the hemocoel, the fungus multiplies as blastospores and colonizes the internal organs. (2) Saprophytic Phase: After the insect dies, the fungus consumes the remaining tissues and emerges from the cadaver, producing new spores that spread and start new infections (Spychała et al. 2024).

In addition to cuticular penetration, oral infection by *B. bassiana* has also been demonstrated as a primary route of infection. This broadens its applicability and establishes it as a promising candidate for various biological formulations.

4.1.2 *Metarhizium anisopliae*

Metarhizium anisopliae (Clavicipitaceae, Hypocreales, and Ascomycota) is a widely distributed entomopathogenic fungus used in biological control. Studies have shown this fungus effectively controls several major agricultural pests, including the southern green stink bug (*Nezara viridula*), the diamondback moth (*Plutella xylostella*), the fall armyworm (*Spodoptera frugiperda*), the onion thrips (*Thrips tabaci*), *Aeneolamia varia*, *Blattella germanica* (the German cockroach), *Ceratitis capitata* (the Mediterranean fruit fly), *Diatraea saccharalis*, *Galleria mellonella*, *Helicoverpa armigera*, and *Hypothenemus hampei* (Monzón 2008).

M. anisopliae covers infected insects with white mycelium, which then turns green as the fungus sporulates. This characteristic is why it is associated with "green muscardine." This cosmopolitan fungus belongs to Sordariomycetes and reproduces asexually through the formation of conidia.

One notable feature of *M. anisopliae* is its ability to form symbiotic relationships with plants. It can function as an endophyte, colonizing roots and establishing beneficial associations. This interaction promotes plant growth, enhances nutrient uptake, and provides protection against soil pathogens. Pavone (2021b) explains that the entomopathogenic and endophytic roles are not mutually exclusive; rather, they interact and support one another.

The infection process resembles that of *B. bassiana*, beginning with the germination of spores on the insect cuticle, followed by penetration facilitated by physical pressure and enzymatic activity, including proteases, chitinases, and lipases. Once inside the hemocoel, the fungus proliferates and produces toxins that kill the insect. It then transitions to a saprophytic phase, emerging from the cadaver to sporulate and continue its life cycle (Haq et al. 2022).

The symbiotic relationship between plants and fungi involves metabolic exchange. During infection, the fungus releases organic nitrogen, which the host plant can absorb, while the fungus benefits from the carbon compounds produced by the plant through photosynthesis. This plant-fungus symbiosis may represent an evolutionary adaptation that optimizes nutrient exchange and supports ecosystem stability in agricultural systems (Kariman et al. 2018).

5 *Scyphophorus acupunctatus*

The agave weevil, *Scyphophorus acupunctatus* Gyllenhal (Coleoptera: Dryophthoridae), is classified as follows:

Kingdom: Animalia
Phylum: Arthropoda
Class: Insecta
Order: Coleoptera
Family: Dryophthoridae
Genus: *Scyphophorus*
Species: *Scyphophorus acupunctatus*

This pest's significant damage to the plant's central core, or "piña," where its immature stages develop, makes it one of the most destructive threats to agave cultivation systems (Cuervo-Parra et al. 2019).

5.1 Insect Life Cycle

The life cycle of *S. acupunctatus* comprises four stages: egg, larva, pupa, and adult. The cream-white eggs, measuring 1.3–1.7 mm in length and 0.5–0.7 mm in width, are elongated to ovoid. Oviposition occurs at the base of the agave plant, where conditions are more humid and sheltered. Hatching takes place 5 days after oviposition (Estrada et al. 2010).

The legless larvae are robust and cream-colored, measuring 1–3 cm in length and 9–9.5 mm in width. The abdominal segment curves upward, and the dark brown head measures 4.0–4.5 mm in width, exhibiting no external pigmentation. The larvae undergo 11 instars, with their development taking between 50 and 90 days, depending on environmental conditions. Once larval development is complete, pupation begins. The pupa appears pale yellow and gradually darkens as pigmentation develops. This stage lasts between 9 and 10 days, during which the pupae measure 15–19 mm in length (Figueroa-Castro et al. 2013).

Adults are matte black, measuring 15–18 mm in length, with an elongated face and a distinctive curved snout used for boring into plant tissue. Although the adults possess wings, these are nonfunctional, and the insect exhibits limited dispersal through flight. Under laboratory conditions at 27 °C, the average lifespan of adults is 105 days (Cuervo-Parra et al. 2019). After hatching, the larva burrows into the central core of the agave, creating internal galleries that facilitate the entry of phytopathogenic microorganisms and accelerate the decay of plant tissue.

5.2 Hosts for *Scyphophorus acupunctatus*

The agave weevil is a polyphagous pest that exhibits a strong preference for species within *Agave* (family: Asparagaceae) and, to a lesser extent, for species in Cactaceae. Its primary host plants include species of significant economic and cultural value, particularly those used in the production of alcoholic beverages such as tequila and mezcal, as well as for fiber production and ornamental purposes. Reported agave host species of *S. acupunctatus* include:

Agave tequilana F.A.C. Weber, *A. fourcroydes* Lem., *A. sisalana* Perrine, *A. salmiana* Otto ex Salm-Dyck, *A. lechuguilla* Torr., *A. americana* L., *A. angustifolia* Haw., *A. americana* L., *A. shawii* Engelm., *A. cubensis* Jacq., *A. attenuata* Salm-Dyck, *A. pes-velutina* Trel., *A. atrovirens* Karw. ex Salm-Dyck, *A. longisepala* Tod., *A. palmeri* Engelm., and *A. ferdinandi*-regis.

Their chemical characteristics, including soluble sugar and volatile compound levels, and morphological and anatomical traits facilitating oviposition and larval development, may explain *S. acupunctatus*'s affinity for these species (Bolaños et al. 2014; Romo 2015; Cuervo-Parra et al. 2019).

5.3 Damage Caused by *S. acupunctatus*

This pest poses a significant threat to agave crops because of the direct damage inflicted by its larvae, as well as its role in facilitating secondary bacterial and fungal infections, which speed up the degradation of plant tissues. The larvae bore into the core or stem of the plant, creating galleries and disrupting vascular tissues. This results in symptoms such as wilting, necrosis, and structural collapse (Cuervo-Parra et al. 2019). Oviposition occurs near the leaf apex or at the base of the leaves. After hatching, the larvae burrow into the plant and feed. Hardened necrotic tissue often encircles the resulting galleries. Sometimes, the plant exhibits intense red discoloration, which may serve as a chemical defense response or result from opportunistic pathogen infection.

The damage is severe in young plants; however, *S. acupunctatus* prefers agave plants that are older than 4 years, likely because of the higher sugar concentration that attracts adult insects. Plants aged 1–3 years, as well as young shoots, are also susceptible. Reports indicate that in major production areas like Jalisco, *S. acupunctatus* has damaged 80%–92% of *A. tequilana* cores, causing substantial economic losses for the tequila industry (Solís-Aguilar et al. 2001).

5.4 Management of *S. acupunctatus*

Historically, the management of this pest has depended on chemical strategies, particularly the application of synthetic insecticides through foliar spraying, targeted irrigation, or soil drenching. Commonly used products in agricultural fields include malathion (Malathion 1000 CE), fipronil (Regent 200 SC), endosulfan (Thiodan 35 CE), and methomyl (Lannate). According to in Mexico, as well as reports by Solís-Aguilar et al. (2000), methyl azinphos (Gusathion M20) and methyl parathion (Folidol M50) have shown over 95.7% effectiveness in controlling adult weevils.

Table 11.4 Biological control agents used against *Scyphophorus acupunctatus* (Gkounti et al. 2015)

Beetles (Coleoptera)	Wasps (Hymenoptera)	Nematodes	Fungi
Hololepta quadridentata	*Cyclaulacidea* sp.	*Steinernema carpocapsae*	*Beauveria bassiana*
Phileurus valgus	*Ectatomma ruidum*	*Steinernema felitae*	*Metarhizium anisopliae*
	Odontomachus bauri	*Heterorhabditis bacteriophora*	*Isaria fumosorosea*

Isazofos 5% G, Carbofuran 5% G, and Phorate 15% G (Thimet 15G) caused high mortality rates (exceeding 88%) in treated larvae.

However, the ongoing use of chemical products presents significant environmental, health, and ecological risks and contributes to the development of resistance in pest populations. As a result, more sustainable management strategies are being promoted, particularly biological control methods. Specifically, microbial control using entomopathogenic fungi, bacteria, or other antagonistic organisms that can infect and eliminate the agave weevil without leaving toxic residues has gained attention (Table 11.4).

6 Bio-Potential of Native Entomopathogenic Fungi

6.1 Soil Sample Collection

We collected soil samples from six representative municipalities in Oaxaca state between June and December 2022. Mixteca Region: San Miguel el Grande and Tlaxiaco (17°02′45″ N, 97°37′10″ W; 2460 m above sea level). Sierra Sur Region: San Jerónimo Sosola and Nochixtlán (17°23′34.37″ N, 97°02′34.48″ W; 1930 m above sea level). Sierra Sur Region: San Jerónimo Coatlán, Miahuatlán (16°14′00″ N, 96°52′00″ W; 1740 m above sea level) and El Potrero, Villa Sola de Vega (16°20′52″ N, 97°00′46″ W; 1237 m above sea level). Valles Centrales Region: Santa Catarina Minas, Ocotlán (16°46′45″ N, 96°36′56″ W; 1562 m above sea level) and San Gabriel, Etla (17°12′57″ N, 96°46′00″ W; 1780 m above sea level).

In each location, a 1-hectare plot featuring cultivated *Agave* spp. We selected plants on each 1-hectare plot, which contained cultivated *Agave* spp. Soil collection adhered to the “five-spot” sampling pattern (INIFAP 2012). At each sampling point, we excavated a 0.030 m^3 volume (45 × 45 × 15 cm). We then homogenized the soil and took a 1 kg sub-sample from each point. We combined the five subsamples from each plot and homogenized them to create a final 1 kg composite sample.

6.2 Isolation of Entomopathogenic Fungi

We isolated entomopathogenic fungi from soil using the insect baiting technique with healthy *Tenebrio molitor* larvae (Zimmermann 1986). We divided each composite soil sample into four 250 g subsamples and placed them in transparent plastic containers. We added ten *T. molitor* larvae to each container and moistened the soil to field capacity. We then covered the containers with a muslin cloth secured by a rubber band.

We labeled the containers and maintained them at 22 ± 2 °C under natural light for 7 days. Every 3 days, we added 10 mL of sterile distilled water to sustain moisture levels. After the incubation period, we removed any larvae showing signs of mycosis (i.e., visible mycelium) and placed them in moist chambers to encourage fungal growth. We transferred the mycelium from each infected larva to Petri dishes containing potato dextrose agar (PDA) medium. They then incubated the plates at 22 ± 2 °C for 3 days (Zimmermann 1986; Hernández-Velázquez et al. 2011).

6.3 Purification and Preservation of EPF

We purified fungal isolates using the single-spore culture technique (Ayala-Zermeño et al. 2012). We grew cultures on potato dextrose agar (PDA) at 26 °C until abundant sporulation appeared. We prepared agar-water plates, let them solidify, and then inoculated them with 50 μL of a conidial suspension adjusted to 1×10^3 conidia/mL. After 12 h of incubation, we identified germinated conidia under a light microscope using a 40× objective. Subsequently, we excised 2 mm^2 sections and transferred them to fresh PDA plates for further growth and incubation at 26 °C. We preserved EPF on a slanted PDA covered with sterile mineral oil (Humber 1997). We sterilized the oil for 2 h each day over four consecutive days. We filled glass tubes with 15 mL of PDA, let them solidify for 24 h, and then inoculated them with fungal culture disks. We submerged the fungal culture under 1 cm of sterile mineral oil after 8 days of growth. Then, we sealed and wrapped the tubes in two layers of paraffin film to serve as an antimicrobial barrier.

6.4 Conidial Viability

A microculture technique on the water-agar (WA) medium was used to evaluate the percentage of germinated conidia in vitro. We placed WA squares measuring 1.5 × 2.0 cm on glass microscope slides and inoculated them with 20 μL of a conidial suspension containing 1×10^6 conidia/mL. We then incubated the slides in a moist chamber at 22 ± 2 °C. After 12 h, we assessed germination using a Carl Zeiss light microscope at 100× magnification. Three replicates were prepared for each fungal isolate, with 100 conidia counted per replicate. We calculated conidial viability using the following formula.

$$\%V = (V/T) \times 100,$$

where $\%V$ = viability percentage, V = the number of germinated conidia, T = the total number of conidia observed, including both germinated and non-germinated conidia.

6.5 Morphological and Molecular Characterization of Entomopathogenic Fungi

6.5.1 Macroscopic Characteristics and Radial Growth Rate

We cultured EPF on a 2% PDA medium in constant darkness at 22 ± 2 °C for 7 days. We recorded colony color, texture, and radial mycelial growth (cm/day) according to the criteria established by Tegene et al. (2021). We placed mycelium disks (Ø 0.4 cm) in the center of 90 × 15 mm Petri dishes containing PDA. We then incubated the plates in the dark at 22 ± 2 °C for 16 days. We prepared four replicates for each fungal isolate and measured radial growth every 24 h using digital calipers. We calculated the growth rate using the formula provided by Al-Ogaidi and Al-Temeemy (2024):

$$\text{GR} = (\text{Final growth} - \text{initial growth})/\text{incubation time}.$$

6.5.2 Morphometric Characterization

We conducted morphometric analysis using microcultures incubated for 8 days at 22 ± 2 °C in a moist chamber. We measured 100 conidia and 100 conidiophores for each fungal isolate. The recorded parameters included shape, color, length, and width (Bischoff et al. 2009; Rehner et al. 2011). We visualized and measured the structures using an AxioCam camera and Zen Lite 2012 software.

6.5.3 Molecular Characterization

We cultured fungi on PDA for 15 days. We extracted DNA from 0.5 mg of mycelium using the Quick-DNA Kit (Zymo Research). We assessed the quality of the extracted DNA using a NanoDrop spectrophotometer. For *Beauveria* isolates, we amplified the Bloc region using the B22U and B822L primers (Rehner et al. 2006, 2011). For *Metarhizium* isolates, we targeted the EF1-α gene using the EF1T and EF2T primers (Bischoff et al. 2009).

We sent the PCR products to Macrogen Inc. in South Korea for sequencing. We edited the sequences using BioEdit v7.1.9 and aligned them with Clustal W. We constructed a phylogenetic tree based on maximum likelihood analysis using the Tamura-Nei model with 1000 bootstrap replicates using BLAST (Basic Local Alignment Search Tool). We also performed Bayesian inference using MrBayes v3.2 (Ronquist et al. 2012; Cabrera-Mora et al. 2019).

6.6 Verification of Pathogenicity in *Tenebrio molitor* Larvae

Using *T. molitor* larvae, we assessed the pathogenicity of EPF isolates. Each experimental unit comprised a group of ten larvae. We tested 44 treatments, combining fungal isolates and two conidial concentrations (4×10^6 and 6×10^5 conidia/mL), with three replicates per treatment. We included negative controls (larvae sprayed with sterile distilled water) and positive controls (larvae treated with a commercial *B. bassiana* formulation) at the same concentrations.

The larvae were surface-sterilized with a 0.5% sodium hypochlorite solution for 3 min, rinsed with sterile distilled water for 2 min, and dried on sterile paper. We then placed them in Petri dishes and sprayed 2 mL of the corresponding conidial suspension onto them. During the experiment, we incubated the larvae at 22 ± 2 °C under controlled humidity and fed them bread.

We monitored mortality every 48 h over 7 days. We transferred dead larvae to moist chambers maintained at 22 ± 2 °C to facilitate fungal development, confirming the fungus as the cause of death. We adjusted the observed mortality using Abbott's formula (1925) to account for natural mortality in the control group:

$$\text{Corrected mortality } (\%) : [(\text{MT} - \text{Mt})/(100 - \text{Mt})] \times 100.$$

Where: MT = Mortality in the treatment group, Mt = mortality rate in the control group. We selected isolates that caused 80% or greater mortality for a second evaluation phase involving adult *S. acupunctatus*. We applied the isolates at concentrations of 4×10^6 and 6×10^5 conidia/mL, corresponding to the following locations: San Miguel: 1SM, 2SM, 3SM, 4SM, San Jerónimo Coatlán: 1SJ, 2SJ, 3SJ, San Jerónimo Sosola: 1MA, 1N, Etla: 1E, 2E, 3E, 4E, 5E, 6E, 7E, 8E, Santa Catarina, Minas: 1M, 2M, 3M, El Potrero: 1S, 2S. Controls included: Commercial *B. bassiana* (4×10^6 and 6×10^5 conidia/mL) formulation and treatment with sterile distilled water.

6.7 Pathogenicity in Adult *S. acupunctatus*

We evaluated fungal isolates that caused ≥80% mortality in *T. molitor* larvae in *S. acupunctatus* adults. We collected the adults from *A. potatorum* fields where the plants were 4–7 years old and showed visible core rot symptoms. The fields were in the municipalities of El Camarón and Nejapa de Madero (16°33′15″ N, 96°01′36″ W; 1600 m above sea level) and Santa Catarina Minas and Ocotlán (16°46′45″ N, 96°36′56″ W; 1562 m above sea level).

We collected and quarantined 1600 live adult insects in ventilated plastic containers at 22 ± 2 °C. We fed the insects fresh agave pieces every 3 days. We used only the insects that remained healthy for 2 weeks in the experiment. The adults were surface-sterilized with a 0.5% sodium hypochlorite (NaClO) solution for 3 min and then rinsed with sterile distilled water for 2 min.

The experimental design comprised 28 treatments, which were derived from the combination of seven fungal isolates, two conidial concentrations (1×10^6 and 1×10^8 conidia/mL), and two inoculation techniques (spraying and immersion). Each treatment included three replicates, with each experimental unit comprising 15 adults. We also included negative controls (sterile distilled water) and positive controls (commercial formulations of *B. bassiana* and *M. anisopliae*).

We submerged groups of 15 insects in 20 mL of conidial suspension for 1 min for the immersion treatments. For the spraying treatments, we applied 2 mL of conidial suspension to groups of 15 adults using a handheld atomizer. After inoculation, we dried the insects on sterile paper and placed them into 50 mL plastic cups with lids. Each cup contained an agave leaf for the insects to feed on, along with moistened sterile cotton. Then, we incubated the cups at 22 ± 2 °C for 25 days.

Mortality monitoring began on day 10 post-inoculation and continued at 48-h intervals until day 25. We transferred dead insects to moist chambers to confirm fungal sporulation and find out the cause of death. Abbott's formula was used to calculate corrected mortality rates. Isolates that induced ≥80% mortality were classified as highly pathogenic and selected for final morphological and molecular identification. The fungal treatments included 3SM: isolate 3 from San Miguel el Grande; 4SM: isolate 4 from San Miguel el Grande; 2M: isolate 2 from Santa Catarina Minas; 1MA: isolate 1 from San Jerónimo Sosola; 1N: isolate 1 from San Jerónimo Sosola; 8E: isolate 8 from Etla; 2SJ: two isolates from San Jerónimo Coatlán. We used two commercial positive controls: T1+: *B. bassiana* and T2+: *M. anisopliae*. We applied both at concentrations of 1×10^6 and 1×10^8 conidia per milliliter (conidia/mL). The negative control (T−) comprised sterile distilled water.

6.8 Statistical Analysis

We employed a completely randomized design consisting of 40 treatments and three replications. The variables analyzed included the percentage of mortality, time to death, inoculation technique, and conidial concentration. We transformed mortality percentages using the arcsine function. We conducted an analysis of variance (ANOVA) and a Kruskal–Wallis test for mean comparison ($p \leq 0.05$). The statistical analysis was performed using InfoStat/L software.

7 Observations and Discussion

7.1 Isolated EPF

A total of 24 soil samples were collected from six sampling sites. From these samples, 22 fungal isolates were obtained. Table 11.5 shows the locations of the soil samples and the number of fungal isolates from each site.

Previous studies have shown that EPF, such as *M. anisopliae*, can be isolated from soil. For example, Hernández-Domínguez and Guzmán-Franco (2017)

Table 11.5 Number of fungal isolates from soil samples collected in six localities of Oaxaca, Mexico

Locality	Geographic coordinates	Number of fungal isolate
San Gabriel, Etla	17°06′51″ N, 96°46′6.7″ W	8
San Miguel el Grande, Tlaxiaco	17°16′6.5″ N, 97°40′14″ W	4
San Jerónimo Coatlán, Miahuatlán	16°13′49″ N, 96°52′14″ W	3
San Jerónimo Sosola, Nochixtlán	17°21′55″ N, 97°01′56″ W	2
Santa Catarina Minas, Ocotlán	16°46′39″ N, 96°37′0.0″ W	3
El Potrero, Sola de Vega	16°24′23″ N, 97°01′43″ W	2
Total		22

successfully isolated this fungus from soil in a sugarcane field. Similarly, Schapovaloff et al. (2015) reported the isolation of *Purpureocillium lilacinum*, *M. anisopliae*, and *B. bassiana* from the soil. Fernández-Bravo et al. (2021) observed that *Metarhizium* species are prevalent in various types of soil, including both cultivated and uncultivated environments. Their study highlighted *M. brunneum*, *M. robertsii*, and *M. guizhouense* as abundant in agricultural, grassland, and forest soils. Similarly, Cabrera-Mora et al. (2019) identified EPF in diverse soil types, including *B. bassiana*, *M. robertsii*, and *M. brunneum*.

7.2 Pathogenicity Tests Against *Tenebrio molitor* Larvae

Of the 22 fungal strains evaluated, only 20 demonstrated pathogenic activity against *T. molitor* larvae. Table 11.6 presents all 22 fungal isolates, along with the corrected mortality percentages recorded 5 days post-inoculation. We categorized the data in Table 11.6 into four adjusted mortality levels. Level 1: 100%–80%, Level 2: 79.9%–50%, Level 3: 49.9%–20%, Level 4: 19.9%–0%. We applied these levels to both tested conidial concentrations: 4×10^6 and 6×10^5 conidia per milliliter (conidia/mL). At a concentration of 4×10^6 conidia per milliliter (conidia/mL), the distribution was Level 1: 31.81% of isolates, Level 2: 36.36%, Level 3: 27.27%, Level 4: 4.54%. Among the 22 treatments, 21 showed significant differences compared to the controls ($p = 0.05$). The most effective treatments were 3SM, 4SM, 2SJ, 8E, 1M, 1N, and 1MA, all of which resulted in over 90% mortality within 5 days. At a concentration of 6×10^5 conidia/mL, the distribution was Level 1: 31.8%, Level 2: 22.72%, Level 3: 27.72%, Level 4: 27.72%. In this study, 17 treatments showed significant differences compared to the control group ($p = 0.05$). The same seven treatments—3SM, 4SM, 2SJ, 8E, 1M, 1N, and 1MA—again stood out by inducing over 80% mortality within 5 days.

Isolates causing over 80% mortality were selected and evaluated against adult *S. acupunctatus*. These results align with those reported by Ruiz-Vega et al. (2018), who found that *M. anisopliae* (1×10^6 conidia/mL) resulted in 100% mortality in *T. molitor* larvae within 4 days, while *B. bassiana* caused 66.5% mortality after 6 days. Karabörklü et al. (2019) also reported complete mortality in *T. molitor* larvae 168 h after inoculation with *B. bassiana* and *M. anisopliae* (1×10^6 conidia/mL applied via spray).

Table 11.6 Percentage and time of mortality for two conidial doses of EPF tested on *Tenebrio molitor* larvae

Locality	Isolate	Time to death (days)	Corrected mortality (%) at 4×10^6	Corrected mortality (%) at 6×10^5
San Miguel el Grande	1SM	16	10 ab	0 a
	2SM	6	50 bc	30 ab
	3SM	6	100 e	100 f
	4SM	5	100 e	80 de
San Jerónimo Coatlán	1SJ	12	40 bc	10 ab
	2SJ	5	100 e	80 def
	3SJ	6	50 bc	40 bc
San Gabriel Etla	1E	14	20 ab	10 ab
	2E	6	70 bcd	70 bc
	3E	9	70 bcd	70 bc
	4E	10	30 ab	10 ab
	53	10	60 bc	30 ab
	6E	6	40 ab	10 ab
	7E	5	70 bcd	70 cd
	8E	5	100 e	90 ef
Santa Catarina Minas	1M	7	30 ab	20 ab
	2M	7	90 de	100 f
	3M	5	70 bcd	70 bc
San Jerónimo Sosola	1N	7	90 de	80 de
	1MA	7	90 de	80 de
El Potrero, Sola	1S	5	70 bcd	70 bc
	2S	10	20 ab	20 ab
Control +		16	20 ab	20 ab
Control –		0	0 a	0 a

Mean values that do not share a letter are significantly different ($p = 0.05$)

7.3 Pathogenicity and Mortality Against *S. acupunctatus* Adults

7.3.1 Immersion Technique

The fungal isolates evaluated were 8E, 1MA, 2M, 1N, 3SM, 4SM, and 2SJ (Fig. 11.1). Table 11.7 presents the mortality rates of *S. acupunctatus* adults at 11 days post-inoculation (dpi). At a concentration of 1×10^8 conidia/mL, the fungi showed greater efficacy, with a mortality rate of at least 50% recorded at 11 dpi. We observed significant differences ($p = 0.05$) compared to the control group. The most effective isolate was 3SM, which achieved an efficacy of 89%, followed by 2SJ and 8E, both at 82%. Isolates 4SM, 2M, 1N, and 1MA produced mortality rates ranging from 50% to 76%. The commercial strain of *B. bassiana* resulted in a mortality rate of 9%, while *M. anisopliae* caused 4% mortality at 24 days post-inoculation (1×10^8 conidia/mL). We observed no mortality in the sterile water control. At a lower concentration of 1×10^6 conidia/mL, mortality was ≤49% at 11 days post-inoculation (dpi). We incubated dead insects in moist chambers and observed fungal

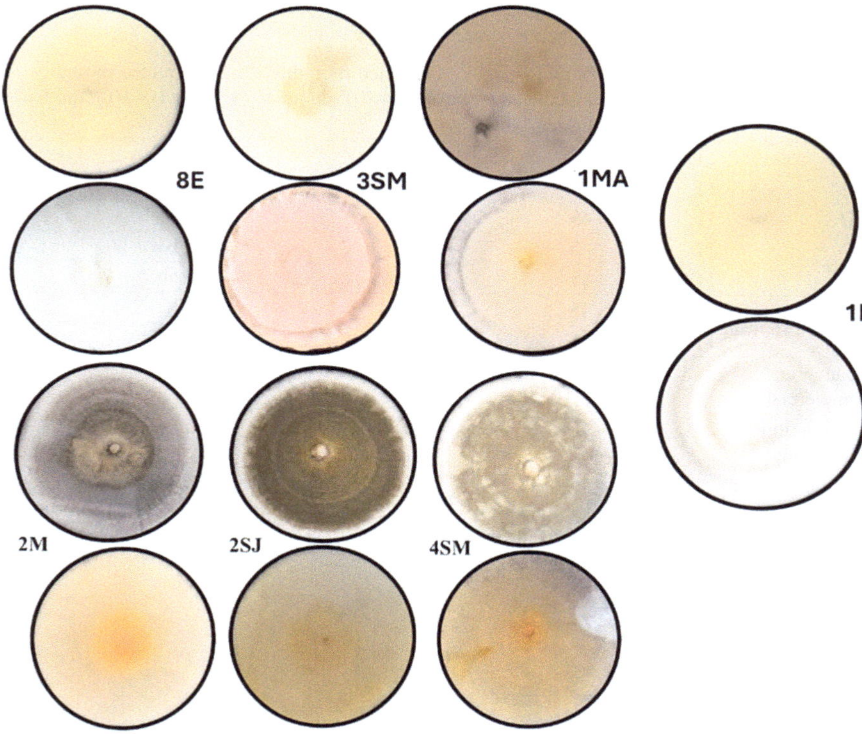

Fig. 11.1 Selected strains used for pathogenicity tests on *S. acupunctatus* adults

sporulation in isolates 3SM, 4SM, 2SJ, 8E, 2M, 1N, and 1MA, but not in the commercial strains (Fig. 11.2).

These results align with those of Asan et al. (2017), who reported that *M. brunneum* induced 86–94% mortality in *Curculio elephas* larvae 12 days post-inoculation. Gkounti et al. (2015) demonstrated that *M. anisopliae*, *B. bassiana*, and *I. fumosorosea* resulted in 100% mortality in larvae and up to 86.67% ± 12% mortality in *S. acupunctatus* adults (1 × 10^8 conidia/mL; immersion). Eckard et al. (2014) observed 83% mortality in *Agriotes obscurus* and *A. lineatus* larvae at 21 days post-inoculation (dpi) using *M. brunneum* (immersion). León-Martínez et al. (2019) noted 78.6% mortality in *Rhynchophorus palmarum* adults treated with *M. anisopliae* (1 × 10^8 conidia/mL). Similarly, Aquino-Bolaños et al. (2023) achieved 100% mortality in *S. acupunctatus* adults within 72 h using oil-based suspensions of *B. bassiana* and *M. anisopliae* (1 × 10^8 conidia/mL).

7.3.2 Spray Application Method

The seven fungi were evaluated (8E, 1MA, 2M, 1N, 3SM, 4SM, and 2SJ), and Table 11.8 presents the results. The EPF were most effective when applied at a concentration of 1 × 10^8 conidia/mL. By day 11 post-inoculation, the adult mortality

Table 11.7 Mortality (%) induced by EPF native to the state of Oaxaca in *S. acupunctatus* adults at 11 days post-inoculation using the immersion technique

Isolate	Time to death (days)	No. of inoculated insects	Corrected mortality (%) 1×10^8 conidia/mL	Corrected mortality (%) 1×10^6 conidia/mL
3SM	11	45	89 d	49 b
4SM	11	45	64 bc	42 b
2SJ	11	45	82 d	47 b
8E	11	45	82 d	47 b
2M	11	45	50 b	29 ab
1N	11	45	76 cd	49 b
1MA	11	45	67 bc	47 b
Beauveria commercial	24	45	9 ab	0a
Metarhizium commercial	24	45	4 ab	0a
Distilled water	–	45	0a	0a

Mean values that do not share a letter are significantly different ($p = 0.05$)

Fig. 11.2 *S. acupunctatus* adults infected by different EPF strains (2M, 1N, 1MA, 2SJ, 4SM, 8E, and 3SM) at 11 dpi using the immersion technique

rate of *S. acupunctatus* had reached at least 50%. All seven isolates exhibited significantly higher mortality rates compared to the positive and negative controls ($p = 0.05$). Isolates 3SM and 1N resulted in the highest mortality rates, at 69% and 67%, respectively. Isolates 4SM, 2M, and 1MA produced mortality rates ranging from 50% to 62%. In contrast, the commercial strains resulted in no mortality (0%).

Table 11.8 Mortality (%) of *S. acupunctatus* adults caused by native entomopathogenic fungi from Oaxaca measured 11 days after spray inoculation

Isolate	Time to death (days)	No. of inoculated insects	Corrected mortality (%) 1×10^8 conidia/mL	Corrected mortality (%) 1×10^6 conidia/mL
3SM	11	45	69 b	49 c
4SM	11	45	60 b	38 b
2SJ	11	45	67 b	49 c
8E	11	45	67 b	44 bc
2M	11	45	50 ab	33 b
1N	10	45	69 b	51 c
1MA	11	45	62 b	49 bc
Beauveria commercial	–	45	0a	0a
Metarhizium commercial	–	45	0a	0a
Distilled water	–	45	0a	0a

Mean values that do not share a letter are significantly different, $p = 0.05$

At a lower concentration of 1×10^6 conidia/mL, the fungi showed reduced effectiveness, with mortality rates not surpassing 49% by day 11. These findings align with those of Gálvez et al. (2017), who reported a mortality rate of 76.66% in *Rhipicephalus (Boophilus) microplus* after 37.3 days using the same conidial concentration of *B. bassiana*. *Metarhizium anisopliae* resulted in a mortality rate of 47% after 10 days at a concentration of 1×10^8 conidia/mL, and 37.75% after 45.5 days at a concentration of 1.3×10^{12} conidia/mL.

Torres de la et al. (2013) reported that spraying *Aeneolamia postica* with soil-derived *M. anisopliae* at a concentration of 1×10^8 conidia/mL resulted in mortality rates ranging from 31.9% to 53.6%. Similarly, Sun et al. (2016) demonstrated that native *M. anisopliae* strains isolated from *R. ferrugineus* induced 100% mortality under laboratory conditions at the same concentration.

7.4 Morphometric Characterization of Entomopathogenic Fungi

The immersion bioassay showed ≥80% mortality; therefore, we selected strains 3SM, 8E, and 2SJ for morphological and molecular characterization.

Strain 3SM After 15 days on PDA, the colony exhibited a cream color with aerial mycelium. The conidia were spherical, globose, and aseptate, forming chains that measured 2 μm × 1 μm (Fig. 11.3). Rehner et al. (2011) classified this strain within *Beauveria*.

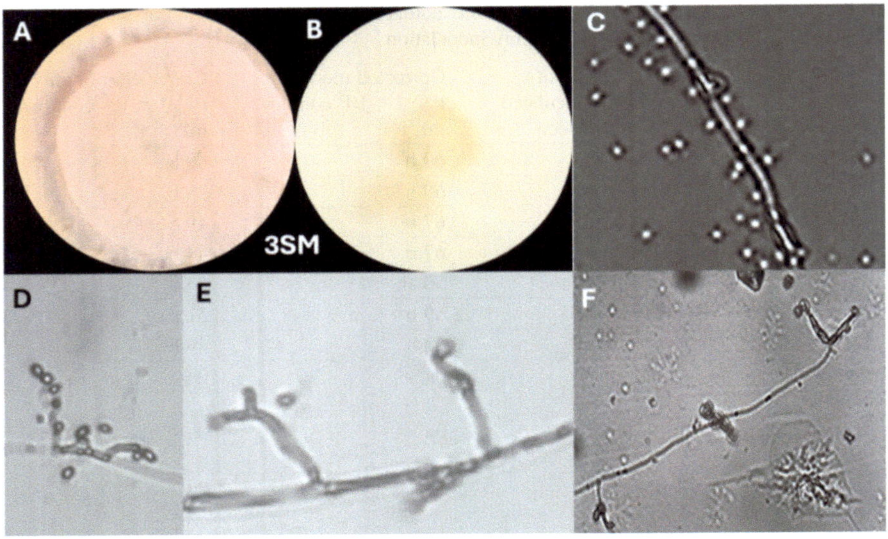

Fig. 11.3 Microscopic characteristics of strain 3SM. (**a**) Top view of the colony on PDA after 15 days of incubation at 22 °C in the dark. (**b**) Reverse view of the colony. (**c**) Conidia. (**d–f**) Conidia and conidiophores

Strain 8E After 15 days on PDA, the colony appeared white with aerial mycelium. The conidia were spherical, globose, and formed in chains measuring 3 μm × 2 μm (Fig. 11.4). We identified it as *Beauveria* sp. (Rehner et al. 2011).

Strain 2SJ On PDA, the colony appeared white after 4 days, transitioning to yellow and then olive green by day 6, with appressed mycelium. The conidia were cylindrical, green, and aseptate, forming chains of 5–7 μm by 2–3 μm. The phialides were green and simple (Fig. 11.5). Based on the descriptions provided by Bischoff et al. (2009), we classify this strain as *Metarhizium* sp.

7.5 Radial Growth Rate

We measured the growth every 24 h over 16 days. We observed significant differences among the strains. During the first 48 h, strain 3SM exhibited a growth rate of 0.21 cm/day, while strains 8E and 2SJ grew at rates of 0.10 cm and 0.18 cm/day. From 96 to 336 h, strain 8E demonstrated the fastest growth. By 384 h, it had completely covered the plate at a remarkable rate of 4.11 cm/day (Table 11.9). These findings align with those reported by Gato Cárdenas et al. (2016), who documented growth rates of 3.10–4.98 mm/day for *M. anisopliae* on PDA over 14 days, and with the results of García-Gutiérrez et al. (2020), who recorded growth rates of 2.62 ± 0.08 cm/day for *M. anisopliae* and 2.76 ± 0.06 cm/day for *B. bassiana*.

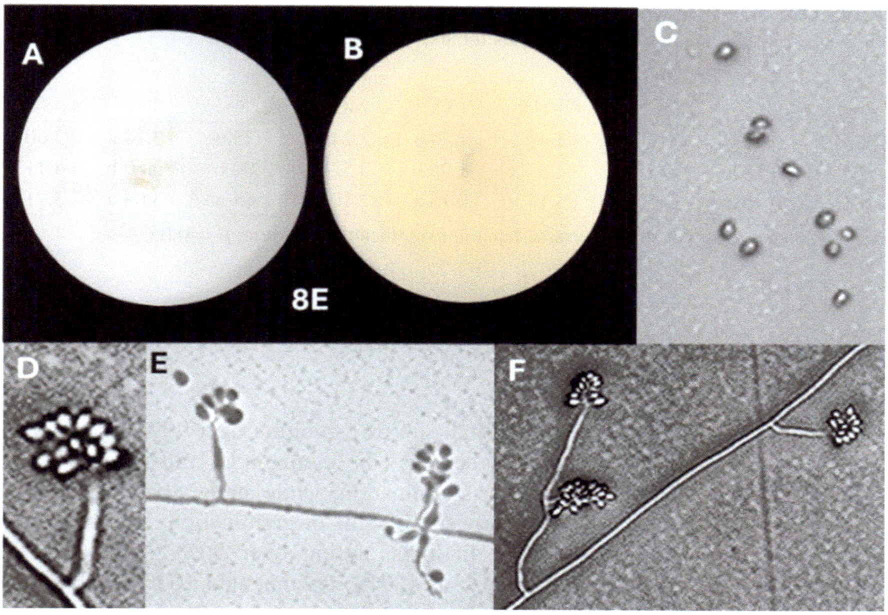

Fig. 11.4 *S. acupunctatus* adults infected with different strains of EPF (2M, 1N, 1MA, 2SJ, 4SM, 8E, and 3SM) after 11 days using the immersion technique. Microscopic characteristics of strain of 8E (**a**) Top view of the colony on PDA after 15 days of incubation at 22 °C in the dark. (**b**) Reverse view of the colony. (**c**) Conidia. (**d–f**) Conidia and conidiophores

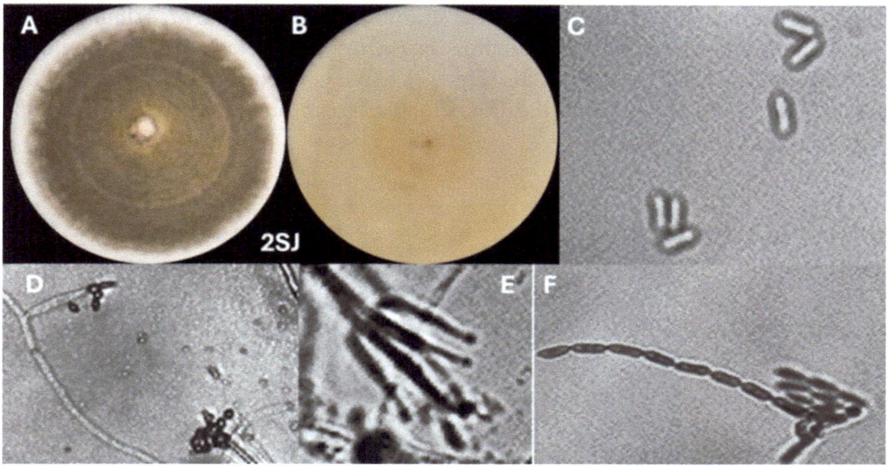

Fig. 11.5 *S. acupunctatus* adults infected with different strains of EPF (2M, 1N, 1MA, 2SJ, 4SM, 8E, and 3SM) 11 days after inoculation using the immersion technique. Microscopic characteristics of strain of 2SJ (**a**) Top view of the colony on PDA after 15 days of incubation at 22 °C in the dark. (**b**) Reverse view of the colony. (**c**) Conidia. (**d–f**) Conidia and conidiophores

Table 11.9 Radial growth rates of three entomopathogenic fungi isolated from *S. acupunctatus*, cultivated on PDA at various time intervals (hours)

	Radial growth (cm/day)							
Strain	48	96	144	192	240	288	336	384
3SM	0.21 b	0.66 ab	1.49 a	1.74 a	2.24 b	2.75 ab	3.14 a	3.60 a
8E	0.18 ab	0.78 b	1.37 a	1.95 b	2.52 c	3.00 b	3.50 b	4.11 b
2SJ	0.10 a	0.57 a	1.14 a	0.15 a	2.10 a	2.64 ab	3.09 a	3.71 a

Mean values that do not share the same letter are significantly different, $p = 0.05$

7.6 Conidial Viability

Germination occurred after 12 h, with all strains exhibiting over 99% germination. We observed significant differences between the commercial and native strains. Strains 3SM and 2SJ achieved 100% germination, while strain 8E reached 99%. The commercial strains, however, showed no signs of germination. Valle-Ramírez et al. (2022) also reported similar findings, noting over 90% germination of *M. anisopliae* after 24 h, and Turcer et al. (2019) documented 100% germination for both *B. bassiana* and *M. anisopliae.*

7.7 Molecular Characterization

We analyzed strains 3SM, 8E, and 2SJ at the molecular level. We found strains 3SM and 8E to be related to *B. bassiana*, while strain 2SJ was related to *M. brunneum*. We constructed a Bayesian phylogenetic tree for *Beauveria* species, incorporating combined ITS, Bloc, B22U, and B822L sequences. We constructed a phylogenetic tree for *Metarhizium* species using the ITS, elongation factor 1-α (EF1-α), EF1T, and EF2T regions (Fig. 11.6).

8 Conclusion

EPF provide a sustainable and promising alternative for the integrated management of economically significant pests, particularly *S. acupunctatus* in Oaxaca, where agave cultivation is essential to the local economy. Isolating, identifying, and evaluating native strains allows for the selection of fungi that exhibit both high virulence and adaptability to regional agro-ecological conditions. This study shows that native entomopathogenic fungi (EPF) strains, particularly those from the genera *Metarhizium* and *Beauveria*, can effectively control adult S. acupunctatus under controlled conditions. Their application reduces reliance on agrochemicals, minimizes environmental impact, and aids in the conservation of beneficial organisms. These findings highlight the significance of integrating EPF into

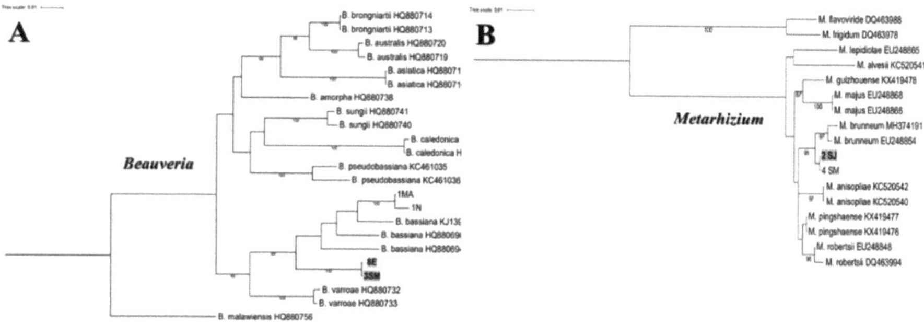

Fig. 11.6 (**a**) Phylogenetic tree of *Beauveria*, (**b**) phylogenetic tree of *Metarhizium*

agroecological biological control programs, establishing a robust foundation for transitioning toward more sustainable, resilient, and environmentally friendly agricultural production systems.

References

Abbott WS (1925) A method of computing the effectiveness of an insecticide. J Econ Entomol 18: 265–267. https://doi.org/10.1093/jee/18.2.265a

Al-Ogaidi M, Al-Temeemy AA (2024) Bio-speckle laser technology in revealing microbial radial growth: a review. Pak J Life Soc Sci 22:1890–1901. https://doi.org/10.57239/PJLSS-2024-22.1.00136

Aquino-Bolaños T, Ortiz-Hernández YD, BautistaCcruz A, Acevedo-Ortíz MA (2023) Viability of entomopathogenic fungi in oil suspensions and their effectiveness against the agave pest Scyphophorus acupunctatus under laboratory conditions. Agronomy 13(6):1468. https://doi.org/10.3390/agronomy13061468

Asan C, Hazir S, Cimen H, Ulug D, Taylor J et al (2017) An innovative strategy for control of the chestnut weevil *Curculio elephas* (Coleoptera: Curculionidae) using *Metarhizium brunneum*. Crop Prot 102:147–153. https://doi.org/10.1016/j.cropro.2017.08.021

Ayala-Zermeño MA, Montesinos-Matias R, Berlanga-Padilla A (2012) Obtención de cultivos monospóricos de hongos entomopatógenos. Centro Nacional de Referencia de Control Biológico, Dirección General de Sanidad Vegetal, SENASICA, Ficha técnica, CB-27

Bischoff JF, Rehner SA, Humber RA (2009) A multilocus phylogeny of the *Metarhizium anisopliae* lineage. Mycologia 100:512–530. https://doi.org/10.3852/07-202

Bolaños TA, Velázquez EP, Hernández UÁ, Gamboa JRD (2014) Host plants of the agave weevil *Scyphophorus acupunctatus* (Gyllenhal) (Coleoptera: Curculionidae) in Oaxaca, Mexico. Southwest Entomol 39:163–169. https://doi.org/10.3958/059.039.0115

Cabrera-Mora JA, Guzmán-Franco AW, Santillán-Galicia MT, Tamayo-Mejía F (2019) Niche separation of species of entomopathogenic fungi within the genera *Metarhizium* and *Beauveria* in different cropping systems in Mexico. Fungal Ecol 39:349–355. https://doi.org/10.1016/j.funeco.2019.02.008

COMERCAM – Consejo Mexicano Regulador de la Calidad del Mezcal AC (2022) Informe estadístico 2023. www.eld-initiative.org

Cuervo-Parra JA, Pérez-España VH, Pérez PAL, Morales-Ovando MA, Arce-Cervantes O et al (2019) *Scyphophorus acupunctatus* (Coleoptera: Dryophthoridae): a weevil threatening the production of agave in Mexico. Fla Entomol 102:1–9. https://doi.org/10.1653/024.102.0101

Eckard S, Ansari MA, Bacher S, Butt TM, Enkerli J et al (2014) Virulence of in vivo and in vitro produced conidia of *Metarhizium brunneum* strains for control of wireworms. J Crop Prot 64: 137–142. https://doi.org/10.1016/j.cropro.2014.06.017

Estrada MEV, Reyes MCH, Ochoa MG, Llanos LA (2010) Determination of the life cycle of *Scyphophorus acupunctatus* (Coleoptera: Curculionidae) under laboratory conditions. Fla Entomol 93:398–402. https://doi.org/10.1653/024.093.0312

Fang-min C, Hsiao-Ling L, Yu-Shin N (2023) Evaluation of potential entomopathogenic fungus *Beauveria bassiana* for controlling the coffee berry borer *Hypothenemus hampei* (Ferrari) (Coleoptera: Curculionidae) in Taiwan. J Asia Pac Entomol 26:102118. https://doi.org/10.1016/j.aspen.2023.102118

Fernández-Bravo M, Gschwend F, Mayerhofer J, Hug A, Widmer F et al (2021) Land-use type drives soil population structures of the entomopathogenic fungal genus *Metarhizium*. Microorganisms 9:1380. https://doi.org/10.3390/microorganisms9071380

Figueroa-Castro P, Solís-Aguilar JF, González-Hernández H, Rubio-Cortés R, Herrera-Navarro EG et al (2013) Population dynamics of *Scyphophorus acupunctatus* (Coleoptera: Curculionidae) on blue agave. Fla Entomol 96:1454–1462. https://doi.org/10.1653/024.096.0425

Gálvez AB, Segura RP, Vázquez AG (2017) Control biológico de *Rhicephalus* (Boophilus) *microplus* con hongos entomopatógenos. CIBA Rev Iberoam Cienc Biol Agropecu 6:33–62. https://doi.org/10.23913/ciba.v6i12.68

García-García MA, Cappello García S, Lesher Gordillo JM, Molina Martínez RF (2015) Hongos entomopatógenos como una alternativa en el control biológico. Kuxulkab 15:21–30. https://doi.org/10.19136/kuxulkab.a15n27.846

García-Gutiérrez C, García-Guajardo MI, Vejar-Cota G, Meza-García L, Chávez-Medina JA (2020) Macromorphology and radial growth of entomopathogenic fungi strains suplement with lepidoteran powder. Rev Colomb Entomol 46:e10164

García-Mendoza AJ (2007) Los agaves de México. Ciencias 87:14–23

García-Mendoza AJ (2012) México país de magueyes. La Jornada Del Campo 53:40–47

García-Mendoza AJ, Franco-Martínez IS (2018) Actualización de la información de las especies y subespecies de magueyes de Oaxaca, con énfasis en las especies mezcaleras. CONABIO, México

García-Mendoza AJ, Sánchez-Teyer LF, Nieto-Sotelo J, Tapia E (2017) Panorama del aprovechamiento de los agaves en México. Red Temática Mexicana Aprovechamiento Integral Sustentable y Biotecnología de los Agaves. Centro de Investigación y Asistencia en Tecnología y Diseño del Estado de Jalisco A.C., Guadalajara, p 302

Gato Cárdenas Y, Márquez Gutiérrez ME, Baró Robaina Y, Porras González Á, Ulloa Martín Y, Quesada Mola Y (2016) Characterization of Cuban isolates of the *Metarhizium anisopliae* species complex with pathogenic activity against Cylas formicarius Fabricius (Coleoptera: Brentidae). Rev Prot Veg 31:50–56

Gkounti VT, Markoyiannaki D, Kontodimas DC (2015) Pathogenicity of indigenous strains of three entomopathogenic fungi to the sisal weevil *Scyphophorus acupunctatus* (Gyllenhal) (Coleoptera: Curculionidae). Hell Plant Prot J 8:46–54. https://doi.org/10.1515/hppj-2015-0007

Haq IU, Khan NA, Sarwar MK (2022) An insight into fungal biology. In: Haq IU, Ijaz S, Khan IA (eds) Phytomycology and molecular biology of plant pathogen interactions. CRC Press, pp 27–50

Hernández-Domínguez C, Guzmán-Franco AW (2017) Species diversity and population dynamics of entomopathogenic fungal species in the genus *Metarhizium*—a spatiotemporal study. Microb Ecol 74:194–206. https://doi.org/10.1007/s00248-017-0942-x

Hernández-Velázquez VM, Cervantes Espíndola Z, Villalobos FJ, García LL, Peña Chora G (2011) Aislamiento de hongos entomopatógenos en suelo y sobre gallinas ciegas (Coleoptera: Melolonthidae) en agroecosistemas de maíz. Acta Zool Mex 27:591–599

Humber RA (1997) Fungi—preservation of cultures. In: Lacey LA (ed) Manual of techniques in insect pathology. Academic Press, pp 269–279. https://doi.org/10.1016/B978-012432555-5/50015-4

Illsey GC, Giovannucci D, Bautista C (2009) La dinámica territorial del estado de Oaxaca: entre la cultura y el comercio. Centro Latinoamericano para el Desarrollo Rural, Grupo de Estudios Ambientales AC, Ciudad de México

INIFAP (2012) Muestreo de suelos y preparación de muestras. www.gobiernofederal.gob.mx, www.sagarpa.gob.mx, www.inifap.gob.mx

Karabörklü S, Altin N, Keskin Y (2019) Native entomopathogenic fungi isolated from Duzce, Turkey and their virulence on the mealworm beetle [*Tenebrio molitor* L. (Coleoptera: Tenebrionidae)]. Philipp Agric Sci 102:82–89. https://www.ukdr.uplb.edu.ph/journal-articles/4395

Kariman K, Barker SJ, Tibbett M (2018) Structural plasticity in root-fungal symbioses: diverse interactions lead to improved plant fitness. PeerJ 6:e6030. https://doi.org/10.7717/peerj.6030

Lacey LA, Frutos R, Kaya HK, Vail P (2001) Insect pathogens as biological control agents: do they have a future? Biol Control 21:230–248. https://doi.org/10.1006/bcon.2001.0938

León-Martínez GA, Campos-Pinzón JC, Arguelles-Cárdenas JH (2019) Pathogenicity and autodissemination of promising entomopathogenic fungi strains on *Rhynchophorus palmarum* L. (Coleoptera: Dryophthoridae). Agron Mesoam 30:631–646. https://doi.org/10.15517/am.v30i3.36184

Meyling NV, Eilenberg J (2007) Ecology of the entomopathogenic fungi *Beauveria bassiana* and *Metarhizium anisopliae* in temperate agroecosystems: potential for conservation biological control. Biol Control 43:145–155. https://doi.org/10.1016/j.biocontrol.2007.07.007

Moharram AM, Abdel-Galil F, Hafez WM (2021) On the enzymes' actions of entomopathogenic fungi against certain indigenous and invasive insect pests. Egypt J Biol Pest Control 31:51. https://doi.org/10.1186/s41938-021-00397-3

Monzón A (2008) Producción y uso de hongos entomopatógenos. Fundación Para el Desarrollo Tecnológico Agropecuario y Forestal de Nicaragua, pp 1–63

Mukherjee R, Ghosh A, Chakraborty C, De JN, Mishra DP (2025) Rice leaf disease identification and classification using machine learning techniques: a comprehensive review. Eng Appl Artif Intell 139:109639. https://doi.org/10.1016/j.engappai.2024.109639

Pacheco HMDL, Reséndiz Martínez JF, Arriola Padilla VJ (2019) Organismos entomopatógenos como control biológico en los sectores agropecuario y forestal de México: una revisión. Rev Mex Cienc For 10(56):4–32. https://doi.org/10.29298/rmcf.v10i56.496

Paschapur AU, Samal I, Ganiger PC, Guru PN, Singh AK (2025) Insect pests of millet crops and their integrated management. In: Joshi DC, Singh AK, Sood S, Prasad M (eds) Minor millets. Springer, Singapore, pp 161–181. https://doi.org/10.1007/978-981-96-4265-6_7

Pavone D (2021a) *Beauveria bassiana*: una alternativa en el manejo integrado de plagas y enfermedades agrícolas. TecnoVita 23:1–5

Pavone D (2021b) *Metarhizium* spp: hongo endófito y patógeno de insectos plaga. TecnoVita 23:1–6

Pérez J, Del Real Laborde JI (2007) Conocimiento y prácticas agronómicas para la producción de Agave Tequilana Weber, en la zona de denominación de origen del tequila. Instituto Nacional de Investigaciones Forestales, Agrícolas y Pecuarias. Centro de Investigación Regional del Pacífico Centro, p 197

Raghunandan BL, Dave A, Baria PR, Manjari (2024) Applications of bioactive compounds from fungal entomopathogens. In: Deshmukh SK, Sridhar KR (eds) Entomopathogenic fungi: prospects and challenges. Springer Nature, Singapore, pp 453–478. https://doi.org/10.1007/978-981-97-5991-0_18

Ramos O (2013) Manual de plagas y enfermedades del agave. Campaña Manejo Fitosanitario de Cultivos Perennes, pp 2–18

Rehner SA, Posada F, Buckley EP, Infante F, Castillo A et al (2006) Phylogenetic origins of African and Neotropical *Beauveria bassiana* s.l. pathogens of the coffee berry borer *Hypothenemus hampei*. J Invertebr Pathol 93:11–21. https://doi.org/10.1016/j.jip.2006.04.005

Rehner SA, Minnis AM, Sung GH, Luangsa-ard JJ, Devotto L et al (2011) Phylogeny and systematics of the anamorphic, entomopathogenic genus *Beauveria*. Mycologia 103:1055–1073. https://doi.org/10.3852/10-302

Romo A (2015) Mexican species of Curculionidae (Insecta: Coleoptera) associated to agaves (Asparagaceae: Agavoideae). Rev Mex Biodiv 83:1025–1035. https://doi.org/10.7550/rmb.30633

Ronquist F, Teslenko M, Van Der Mark P, Ayres DL, Darling A et al (2012) MrBayes 3.2: efficient Bayesian phylogenetic inference and model choice across a large model space. Syst Biol 61:539–542. https://doi.org/10.1093/sysbio/sys029

Ruiz-Vega J, Cortés-Martínez CI, Silva-Rivera ME (2018) Efectos de la Aplicación de Hongos y Nematodos Entomopatógenos sobre la Mortalidad de Larvas de Tenebrio molitor L. (Coleoptera: Tenebrionidae). Entomol Mex 1:156–160

SAGARPA (2017) Agave tequilero y mezcalero. Informe Gobierno de México, p 100

Schapovaloff ME, Alves LF, Urrutia MI, López CC (2015) Ocurrencia natural de hongos entomopatógenos en suelos cultivados con yerba mate (Ilex paraguariensis St. Hil.) en Misiones, Argentina. Rev Argent Microbiol 47:138–142. https://doi.org/10.1016/j.ram.2015.03.005

Solís-Aguilar JF, González-Hernández EH, Valle A, Flores-Mendoza FJ (2000) Control químico del 'picudo del agave' Scyphophorus acupunctatus Gylh. (Coleoptera: Curculionidae) en Jalisco, México. In: Memorias del XXXV Congreso Nacional de Entomología. Soc Mexicana Entomol, Acapulco, pp 679–682

Solís-Aguilar JF, González-Hernández H, Leyva-Vázquez JL, Equihua-Martínez A, Flores-Mendoza FJ, Martínez-Garza Á (2001) Scyphophorus acupunctatus Gyllenhal, plaga del agave tequilero en Jalisco, México. Agrociencia 35:663–670

Song H, Yanlei S, Haimin C, Pengfei Z, Chengshu W (2023) Fungus–insect interactions beyond bilateral regimes: the importance and strategy to outcompete host ectomicrobiomes by fungal parasites. Curr Opin Microbiol 74:102336. https://doi.org/10.1016/j.mib.2023.102336

Spychała K, Piecuch A, Szleszkowski Ł, Kadej M, Ogórek R (2024) Microscopic fungi on the corpse – promising tool requiring further research. Forensic Sci Int 361:112129. https://doi.org/10.1016/j.forsciint.2024.112129

Sun X, Yan W, Qin W, Zhang J, Niu X et al (2016) Screening of tropical isolates of *Metarhizium anisopliae* for virulence to the red palm weevil *Rhynchophorus ferrugineus* Olivier (Coleoptera: Curculionidae). Springer Plus 5:1100. https://doi.org/10.1186/s40064-016-2780-6

Tegene S, Dejene M, Terefe H, Tegegn G, Tena E et al (2021) Evaluation of native *Trichoderma* isolates for the management of sugarcane smut (*Ustilago scitaminea*) in sugar plantations of Ethiopia. Cogent Food Agric 7:1872853. https://doi.org/10.1080/23311932.2021.1872853

Torres de la CM, Cortez MH, Ortiz GF, Cappello GS, de la Cruz PA (2013) Caracterización de aislamientos nativos de *Metarhizium anisopliae* y su patogenicidad hacia Aeneolamia postica, en Tabasco, México. Rev Colomb Entomol 39:40–46

Turcer C, Kushieyev R, Erper I, Ozdemir I, Saruhan I (2019) Efficacy of native isolates of *Metarhizium anisopliae* and *Beauveria bassiana* against the invasive ambrosia beetle *Xylosandrus germanus* (Coleoptera: Curculionidae: Scolytinae). Egypt J Biol Pest Control 29:28. https://doi.org/10.1186/s41938-019-0132-x

Valle-Ramírez SB, Torres-Gutiérrez R, Caicedo-Quinche WO, Abril-Saltos RV, Sucoshañay-Villalba DJ (2022) Aislamiento y caracterización de *Metarhizium* spp. de cultivos de caña de azúcar y su patogenicidad contra *Mahanarva andigena* (Hemiptera: Cercopidae). Cienc Tecnol Agropecu 23:1

Velázquez J, Joly LJ, García JL, Romero Y, González M et al (2006) Enemigos naturales del "picudo del agave" *Scyphophorus acupunctatus* Gyllenhal (Coleoptera: Curculionidae) en el estado Falcón, Venezuela. Entomotropica 21:185–193

Zimmermann G (1986) The "Galleria bait method" for detection of entomopathogenic fungi in soil. J Appl Entomol 102:213–215. https://doi.org/10.1111/j.1439-0418.1986.tb00912.x

Entomopathogenic and Endophytic Fungi: An Alternative for Soil Fertility and Pesticide Degradation in Sustainable Agriculture

12

Hemraj Chhipa and Sunil Kumar Deshmukh

Abstract

Entomopathogenic fungi offer an eco-friendly solution for managing insect pest problems in a wide range of crops. Their insecticidal properties not only help control pests but also contribute essential nutrients to the soil and plants, thereby enhancing soil fertility. This nutrient-recycling ability underscores their potential role in sustainable soil health management. Both entomopathogenic and endophytic fungi produce a variety of enzymes, including cellulase, ligninase, chitinase, pectinase, amylase, and lipase, which facilitate the breakdown of complex organic matter and enhance nutrient availability for other organisms. Additionally, these fungi participate in the mineralization and solubilization of essential macro- and micronutrients in the soil. Numerous studies have reported their capacity to solubilize phosphorus and zinc through the production of organic acids. Moreover, the transfer of nutrients such as nitrogen, carbon, and iron from insect cadavers to plants and soil has also been documented, highlighting their role as nutrient mobilizers. The enrichment of soil with these nutrients supports a more diverse microbial community, enhances soil biodiversity, and improves overall soil physiology and biological activity. The presence of entomopathogenic and endophytic fungi in the soil thus indirectly promotes soil fertility and contributes to better soil health. This chapter discusses the role of these fungi in nutrient mobilization and their enzymatic degradation potential in promoting soil fertility. It also explores the endophytic behavior of certain

H. Chhipa (✉)
College of Agriculture, Kota (Agriculture University, Kota), Kota, Rajasthan, India
e-mail: hemraj@aukota.org

S. K. Deshmukh
Research and Development Division, Greenvention Biotech Pvt. Ltd., Uruli-Kanchan, Maharashtra, India

S. K. Deshmukh et al. (eds.), *Frontiers in Entomopathogenic Fungi*,
https://doi.org/10.1007/978-981-95-6529-0_12

entomopathogenic fungi and their involvement in nutrient translocation between soil and plants.

Keywords

Agro-chemicals · Entomopathogen · Fertility · Nutrient management · Soil health

1 Introduction

Fertile soil forms the foundation for high crop yields and nutrient-rich agricultural produce. Soil serves as a habitat for numerous organisms that play a vital role in maintaining soil fertility and crop health. Microorganisms participate in various soil processes, including nutrient management, pollutant degradation, soil structure management, biogeochemical cycling, and overall soil health improvement (Xu et al. 2024). Soil health is continuously deteriorating due to the unbalanced application of agrochemicals. Excessive use of these agrochemicals not only fails to meet the demands of a growing population but also increases the toxicity in the soil. As a result, soil fertility and the health of the soil are steadily declining in developing countries. The excessive burden of agrochemicals is inhibiting the growth and activity of many beneficial microorganisms that support soil health and high-quality crop production (Bhat et al. 2025). Currently, organic farming and natural farming are being promoted as alternatives to improve soil structure, with most governments encouraging sustainable farming methods. Such farming practices enhance the soil microbial biodiversity and nutrient availability, supporting the survival and activity of beneficial microorganisms. High microbial diversity is an indicator of healthy soil (Pedrinho et al. 2024).

Entomopathogenic fungi are important members of the soil microbial community. Entomopathogenic fungi are a well-known group that helps control insect pests in agricultural systems. The presence of entomopathogenic fungi in soil not only inhibits pest growth but also enhances soil fertility, plant growth, nutrient uptake, and plant immunity (Ciceoi et al. 2021) (Fig. 12.1). More than 700 species of entomopathogenic fungi, belonging to over 100 genera, have been reported across Ascomycota, Basidiomycota, Chytridiomycota, Entomophthoromycota, Microsporidia, and Oomycetes (de Mandal and Passari 2021; Ahsan et al. 2024). Entomopathogenic fungi kill insect pests or other arthropods by attaching to their outer surface as spores, using mucus or glucans present on the host for adhesion. The spores germinate to produce a germ tube and secrete cell wall-degrading enzymes such as proteases, lipases, and chitinases, which facilitate the germ tube's penetration into the host. After entry, entomopathogenic fungi multiply within the host, invading its tissues and proliferating as blastospores and fungal colonies in the hemocoel. The growth of fungal hyphae produces toxins and causes mechanical damage to the host, which together are the primary causes of pest mortality (Sharma et al. 2020; Sisay and Legese 2020). The degradation of insect pests creates opportunities for other microorganisms to colonize the cadavers and release various nutrients into the soil.

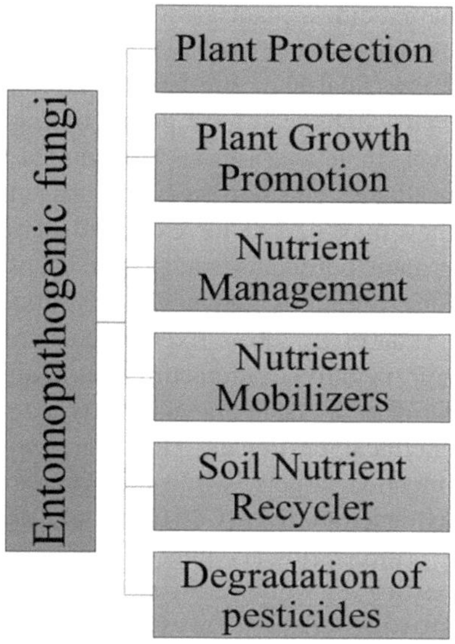

Fig. 12.1 Role of entomopathogenic fungi in soil fertility and agricultural sustainability

Entomopathogenic fungi also provide nutrients to the plant from insects through endophytic interactions. Behie et al. (2017) reported the exchange of insect-derived nitrogen into plants through *Metarhizium robertsii*. *Metarhizium robertsii*, associated with plant roots, supports plant growth by killing soil-borne insects and helps in the translocation of nitrogen to the associated plants of wheat and soybean and carbon exchange with switchgrass. These entomopathogenic fungi also play various roles, including acting as endophytes, antagonists, rhizosphere colonizers, nutrient mobilizers, and plant growth promoters under various ecological conditions (St Leger and Wang 2020). Entomopathogenic fungi are also considered as indicators of soil health due to their roles in plant protection, plant growth promotion, and the maintenance of nutrient uptake (Ciceoi et al. 2021).

However, the roles of entomopathogenic and endophytic fungi in soil nutrient management, organic matter recycling, and soil structure enhancement remain unexplored, and limited research is available (Murindangabo et al. 2024). In this chapter, we reviewed the literature on the roles of entomopathogenic and endophytic fungi in sustainable agriculture. We also discuss the roles of endophytic fungi in enhancing soil fertility and in bioremediation of pollutants.

2 Soil Degradation

Soil is the foundational matrix for food production that sustains all living beings. Rising global food demand is compelling farmers to apply increasing amounts of agrochemicals to boost crop yields. A wide range of agrochemicals—including chemical fertilizers, pesticides, fungicides, herbicides, insecticides, rodenticides, nematicides, plant growth regulators, micronutrient mixtures, and antibiotics—are used to enhance production and protect crops from diseases (Fig. 12.2) (Ahmad et al. 2024; Jalal and Bondarenko 2025). The application of agrochemicals has significantly enhanced food production, particularly during the Green Revolution. However, their overuse has contributed to soil degradation. Excessive use of chemical pesticides and fertilizers can cause these substances to bind to soil particles, where they may persist for extended periods, ranging from months to years. Similarly, organic pollutants, such as persistent organic pollutants (POPs), remain in the soil for long durations and negatively impact soil ecology and the broader environment (Aktar et al. 2009). Excessive use of agrochemicals leads to their persistence in the soil, negatively impacting soil health (Prashar and Shah 2016). It is estimated that farmers in Asian and South African countries apply an average of 6.5–60 kg/ha of chemical pesticides. However, only about 0.1% of these pesticides reach their intended targets, while the remainder persists in the environment as residues for extended periods (Carriger et al. 2006).

Agrochemicals play a key role in controlling pests and pathogens through pesticides, while fertilizers and micronutrient mixtures supply essential nutrients to growing crops. However, excessive use of these chemicals negatively impacts beneficial soil microbes involved in various biogeochemical cycles of minerals and nutrients. Microorganisms secrete a range of enzymes—such as hydrolases, oxidoreductases, and dehydrogenases—that contribute to essential soil enzymatic activities. Overuse of chemical pesticides and fertilizers can inhibit or reduce the

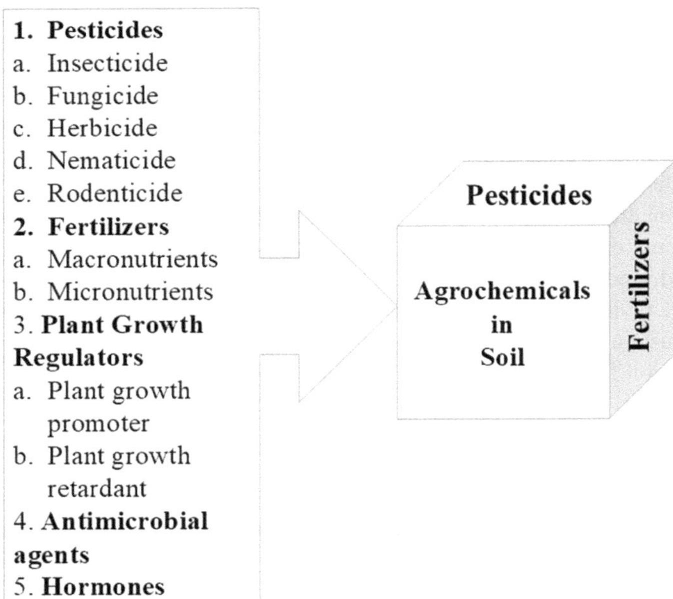

Fig. 12.2 Applications of agrochemicals in agriculture

activity of these enzymes, thereby disrupting soil health and ecosystem functioning (Table 12.1) (Riah et al. 2014). Similarly, prolonged use of copper-based fungicides has been shown to inhibit earthworm populations, which play a vital role in soil structure and organic matter management (Van Zwieten et al. 2004; Eijsackers et al. 2005). Agrochemicals have adverse effects on soil by reducing its water-holding capacity, limiting its ability to absorb rainwater, and increasing surface runoff and erosion, ultimately leading to soil degradation (Gregory et al. 2015). In addition to their environmental impacts, agrochemicals used in crop production can also affect human health and have been linked to various disorders, including cancer, kidney disease, bone deformities, and related health problems (Parks et al. 2016).

Long-term application of agrochemicals also leads to the accumulation of heavy metals and xenobiotic compounds, which can disrupt the food chain and harm the broader environment. Additionally, these substances alter key physicochemical properties of soil, such as pH, moisture content, and organic carbon levels. Prolonged chemical exposure can induce resistance in weeds, pathogens, and pests, indirectly disturbing the ecological balance of soil communities (Lamichhane et al. 2016).

3 Role of Microbes in Restoration of Soil Fertility

The presence of microbes in the soil is a vital indicator of soil health. These microorganisms not only sustain soil fertility but also suppress harmful pathogens and produce bioactive metabolites that enhance plant growth. They play an essential role in decomposing complex organic compounds, thereby supporting nutrient cycling and maintaining soil structure (Ding et al. 2024). However, the excessive use of agrochemicals introduces toxic substances into the soil, which inhibits the growth of beneficial microbial communities and ultimately reduces soil fertility. A decline in microbial populations is closely linked to soil degradation and decreased agricultural productivity.

Microbes are among the most promising alternatives for improving crop productivity and soil fertility. They also contribute significantly to soil structure maintenance. Various microbial groups—such as bacteria, fungi, and algae—enhance soil aggregation by producing extracellular polymeric substances, including mucilage, glycoproteins, and polysaccharides, which help bind soil particles and prevent erosion (Qayyum et al. 2015; Saleem et al. 2019). In addition to their structural functions, soil microbes interact synergistically, mutualistically, or symbiotically with plants and other soil organisms, promoting plant growth and sustaining soil fertility (Mukherjee et al. 2021a, b). These microbes synthesize and release various plant growth-promoting metabolites and hormones into the rhizosphere, which enhance microbial colonization and nutrient mobilization (Mendes et al. 2021). One key hormone is indole acetic acid (IAA), produced via tryptophan-dependent and -independent pathways. IAA enhances microbial colonization and stimulates plant root development (Contreras-Cornejo et al. 2009).

Table 12.1 Enzymatic activities in soil affected by chemical inputs present in soil (Mandal et al. 2020)

Name of agrochemical and pesticide	Affected activity	Reference
Anilofos Benthiocarb Baytan Brominal *Butachlor* Carbendazim *Captain* Carbofuran Cinmethylin Cuprosan 2,4-Dichlorophenoxyacetic acid(2,4-D) Endosulfan Fenvalerate *Funaben* gamma-Hexachlorocyclohexane (γ-HCH) *Imazetapyr* Linuron Methabenzthiazuron Monocrotophos Oxafun Pretilachlor Prometryn Quinalphos Simazine Thiram Terbutryn	The nitrogen fixation activity of microbes decreased and in some cases inhibition of nitrogenase enzyme activity were also observed	Gallori et al. (1991) Omar and Abd-Alla (1992) Chalam et al. (1996, 1997) Di Ciocco (1997) Martınez-Toledo et al. (1998) Hammouda (1999) Singh and Wright (1999) Niewiadomska (2004) Niewiadomska and Klama (2005) Khan et al. (2006) Prasad et al. (2011)
Carbendazim Chlorpyrifos Diazinon 2,4-D Glufosinate ammonium Imidacloprid Mancozeb Mefenoxam Metalaxyl Monocrotophos Nitrapyrin Quinalphos and Lindane	The mineralization of organic phosphorus into inorganic phosphorus by phosphatase enzymes affected	Tu (1981) Ismail et al. (1996) Monkiedje et al. (2002) Madhuri and Rangaswamy (2002) Mayanglambam et al. (2005) Singh and Singh (2005) Srinivasulu et al. (2012)
Benomyl *Captan* Diazinon	Conversion of urea into CO_2 and NH_3 by urease activity was inhibited or increased Decrease nitrate reductase activity	Abdel-Mallek et al. (1994) Chen et al. (2001)

(continued)

Table 12.1 (continued)

Name of agrochemical and pesticide	Affected activity	Reference
Isoproturon Profenofos		Nowak et al. (2004) Ingram et al. (2005)
Azadirachtin Acetamiprid Atrazine Glyphosate Quinalphos	Influence the oxidoreductase enzyme and dehydrogenase activity	Andrea et al. (2000) Mayanglambam et al. (2005) Nweke et al. (2007) Singh and Kumar (2008) Kızılkaya et al. (2012)
Atrazine Carbaryl Paraquat	Hydrolysis of sucrose into fructose and glucose by the invertase enzyme activity affected	Gianfreda et al. (1995) Sannino and Gianfreda (2001)
Metalaxyl Ridomil *Gold Plus*	Affected the β-glucosidase activity (hydrolyzes disaccharides in soil to form β-glucose)	Demanou et al. (2004) Sukul (2006)
Benlate *Captan* Brominal	Hydrolysis of cellulose into D-glucose by Cellulase enzyme	Arinze and Yubedee (2000) Omar and Abdel-Sater (2001)
Cinosulfuron Prosulfuron Thifensulfuron methyl Triasulfuron	Showed stronger detrimental effect on hydrolysis of aryl sulphate by arylsulphatase enzyme	Sofo et al. (2012)

3.1 Role of Microbes in Nutrient Management

Soil microbes also contribute to the solubilization of essential nutrients. For instance, phosphorus—often present in unavailable forms such as rock phosphate and apatite [$Ca_5(PO_4)_3(Cl/F/OH)$], is made bioavailable through microbial secretion of organic acids like citric and oxalic acid (do Nascimento et al. 2021; Vassileva et al. 2022). Similarly, zinc, which exists in forms such as zinc carbonate, zinc sulphate, and zinc oxide, is solubilized through microbial processes like acidification, chelation, and chemical transformation. Organic acids such as gallic and caffeic acid lower the pH, enabling conversion of zinc compounds into plant-absorbable Zn^{2+} ions (Martino et al. 2003).

Iron availability is also enhanced by microbial production of siderophores—specialized compounds that chelate iron from the soil and transport it to plant

Table 12.2 Entomopathogenic and endophytic fungi in the soil

Taxon	Activity	Reference
Aspergillus nomiae	Exhibited pathogenicity against various insect pests	Zhang et al. (2024)
Beauveria bassiana	Remediation of cadmium contamination in soil	Wang et al. (2025)
Beauveria brongniartii	Infect *Spodoptera litura* larva	Vivekanandhan et al. (2025)
Isaria fumosorosea	Biodegradation of p-chloroaniline	Huang et al. (2024)
Lecanicillium lecanii	Antimicrobial activity	Ramteke et al. (2022)
Metarhizium anisopliae	Management of root-knot nematode, *Meloidogyne incognita*	Kassam et al. (2022)
Metarhizium pingshaense	Zinc solubilization and plant growth promotion	Kumar et al. (2024)
Metarhizium robertsii	Bioremediation of mercury	Wu et al. (2022)
Paecilomyces fumosoroseus (presently *Isaria fumosorosea*)	Control agricultural pests and pathogens	Chhipa et al. (2024)
Paecilomyces farinosus (presently *Isaria farinose*)	Biocontrol activity	Moreno-Gavíra et al. (2021)
Purpureocillium sp.	Nematode control and plant growth-promoting fungi	Rigobelo et al. (2024)

roots (Sagar et al. 2021). In addition to nutrient mobilization, soil microbes improve fertility by decomposing organic matter, detoxifying pollutants, and remediating heavy metals through bioremediation processes. These diverse microbial functions are fundamental to sustainable soil health and agricultural productivity (Table 12.2).

4 Involvement of Entomopathogenic and Endophytic Fungi in Soil Fertility

Entomopathogenic and endophytic fungi also contribute to the improvement of soil structure. Entomopathogenic fungi are chemo-organoheterotrophic organisms that rely on organic matter in the soil for their survival. They exhibit various forms of interaction with other microbial communities—ranging from symbiotic and mutualistic to endophytic associations—particularly during the degradation of complex organic matter. In this process, they produce a range of enzymes that break down organic materials and release essential nutrients (Rigobelo 2024). These fungi are known to secrete enzymes such as cellulase, ligase, chitinase, pectinase, amylase, and lipase, which facilitate the decomposition of dead organic matter and contribute to the release of vital nutrients, including phosphorus, carbon, nitrogen, potassium, iron, manganese, copper, and magnesium, into the surrounding soil (Leach et al. 2017). In addition to organic matter decomposition, entomopathogenic fungi also play a role in degrading chemical pesticides and xenobiotic compounds, processes that are crucial for maintaining soil fertility. For instance, species of *Lecanicillium* have demonstrated the ability to degrade the insecticide chlorpyrifos (Swathy et al. 2024). A list of entomopathogenic fungi involved in agrochemical degradation is provided below (Table 12.3).

Table 12.3 Potentials of entomopathogenic and other fungi in agrochemical degradation

Entomopathogenic fungus	Chemical compound and pesticide	Reference
Aspergillus niger	Malathion	Mohapatra et al. (2021)
Aspergillus niger and Trichoderma viride	Chlorpyrifos	Bose et al. (2021)
Bjerkandera sp.	Toxaphene	Prieto et al. (2021)
Calocybe indica	Imidacloprid	Deka et al. (2018)
Cladosporium cladosporioides	Chlorpyrifos	Chen et al. (2012)
Gloeophyllum, Daedalea, and Fomitopsis sp.	DDT	Purnomo et al. (2008)
Metarhizium anisopliae	Chloropyrifos	Abd El-Ghany and Masmali (2016)
Metarhizium anisopliae	Malathion	Abd El-Ghany and Masmali (2016)
Metarhizium brunneum	Nonylphenol	Różalska et al. (2013)
Metarhizium robertsii	Nonylphenol	Nowak et al. (2019)
Metarhizium robertsii	Tributylin	Siewiera et al. (2015)
Metarhizium sp.	Nonylphenol	Różalska et al. (2015)
Penicillium camemberti	Pentachlorophenol (PCP), 2-chlorophenol, and Trichloroacetic acid	Taşeli (2006)
Phanerochaete chrysoporium	Pesticide biodegradation	Kaur et al. (2021)
Phanerochaete chrysosporium and *Phlebia* sp.	Organochlorine heptachlor	Xiao et al. (2011)
Trametes versicolor	1,1,1-Trichloro-2, 2-bis (4-chlorophenyl) ethane (DDT)	Hossain and Anantharaman (2005)
Trichoderma viride	Toxaphene and aldrin	Patil et al. (1972)
Trichoderma sp. *Gc1*	1,1-Dichloro-2,2-bis-(4-chlorophenyl) ethane	Ortega et al. (2011)

4.1 Entomopathogenic and Endophytic Fungi in Nutrient Solubilization and Mineralization

Nutrient mineralization is another key function. According to Murindangabo et al. (2024), entomopathogenic fungi can mechanically and biochemically break down mineral sources such as iron, sulfur, calcium, potassium, and magnesium. These fungi grow on mineral surfaces, solubilize them, and increase their bioavailability. The solubilized nutrients often form complexes with extracellular polymers secreted by fungi and other microbes, leading to biomineral formation (Gadd 2010). Notably, *Metarhizium pingshaense* IISR-EPF-14 has demonstrated high zinc-solubilizing ability and plant growth promotion. It produces a variety of organic acids—such as formic, gluconic, keto-gluconic, malonic, oxalic, tartaric, and succinic acids—that solubilize zinc compounds through acidification and chelation (Mumtaz et al. 2019; Kumar et al. 2024). These acids also aid in phosphorus solubilization, thereby enhancing soil fertility and nutrient availability. Other fungi, such as *Beauveria* spp., have also been shown to solubilize zinc via organic acids like citric, acetic, and oxalic acid. *M. pingshaense* promotes rice and cardamom growth by producing IAA, siderophores, and solubilizing phosphate. Its application increased total chlorophyll content in rice by 48% and in cardamom by 62%, indicating improved iron absorption (Senthil Kumar et al. 2018; Murindangabo et al. 2024).

4.2 Entomopathogenic and Endophytic Fungi in Nutrient Mobilization and Bioremediation

Entomopathogen also involved in nutrient mobilization from insect to plant and soil. Pineda (2014) reported that endophytic fungi such as *B. bassiana* and *M. anisopliae* not only suppress insect pests but also translocate nitrogen from insect cadavers to plants. Similarly, *Trichoderma* spp. enhances plant growth and protects against pests such as thrips (Behie et al. 2012; Muvea et al. 2014). Seed treatment with entomopathogenic fungi like *B. bassiana* and *M. brunneum* has increased wheat grain yield by 63% and 26%, respectively, even in the absence of phosphate and zinc fertilizers (González-Guzmán et al. 2020). These fungi also mobilize nutrients like manganese and iron, thereby activating plant immune responses and improving growth (Raya-Díaz et al. 2017; Sánchez-Rodríguez et al. 2018). Additionally, they degrade pesticide residues through the secretion of enzymes such as esterases, oxidoreductases, and cytochrome P450s, which convert toxic compounds into less harmful forms (Swathy et al. 2024). Specific strains of *Trichoderma* have been shown to degrade toxaphene and aldrin, while *Penicillium camemberti* can degrade PCP and chlorinated compounds. Other effective degraders include *Lecanicillium* sp., *Cladosporium cladosporioides*, *Phanerochaete chrysosporium*, *Phlebia* sp., and *Calocybe indica* (Patil et al. 1972; Hossain and Anantharaman 2005; Purnomo et al. 2008; Xiao et al. 2011; Chen et al. 2012; Abd El-Ghany and Masmali 2016; Deka et al. 2018; Prieto et al. 2021).

Endophytic fungi contribute similarly to soil fertility by producing exopolysaccharides (EPS), exoenzymes, phytohormones, and siderophores, and by solubilizing nutrients like phosphate and fixing nitrogen (Mukherjee et al. 2019, 2020, 2021a, b). Enzymes such as cellulase, pectinase, lipase, urease, and phosphatase help decompose complex organic materials and enhance microbial biomass, nitrogen, and organic carbon levels in the soil (Hou et al. 2020). Moreover, endophytes stimulate soil microbial interactions through the release of chemical signals—proteins, amino acids, sugars, and secondary metabolites—that affect microbial diversity, soil respiration, and plant-microbe communication (Nandy et al. 2020).

Several endophytes, including *Arthrobacter*, *Aspergillus*, *Alternaria*, *Bacillus*, *Pseudomonas*, *Phoma*, and *Microbacterium*, have been reported to tolerate and remediate heavy metals and organic pollutants through immobilization, chelation, and degradation (Li et al. 2012; Sharma et al. 2018; Liu et al. 2021). Chen et al. (2021) observed that leaf endophytes can effectively degrade agrochemicals while supporting microbial biodiversity. Notably, the pesticide N-(3,5-dichlorophenyl) succinimide did not significantly alter microbial diversity in the phyllosphere of *Nicotiana tabacum*. *Metarhizium robertsii* demonstrates dual functionality in its entomopathogenic and endophytic stages. It transfers nitrogen from insect hosts to *Phaseolus vulgaris* while absorbing carbon from the plant, enhancing soil fertility and microbial interactions (Behie et al. 2012, 2017). Similar nutrient exchanges were observed with *B. bassiana*, *M. guizhouense*, *M. brunneum*, and *M. robertsii* in crops such as green bean, soybean, switchgrass, and wheat, resulting in increased productivity (Behie and Bidochka 2014). Entomopathogenic fungi (*B. bassiana*) also enhance soil water-holding capacity and root development, as observed in cotton (Dara and Dara 2019). They influence microbial community dynamics, sometimes suppressing or stimulating populations such as actinomycetes, nitrogen fixers, cellulose degraders, and spore-formers. *B. bassiana* has been reported to increase these populations by 7–30 times compared to untreated soil (Draganova et al. 2008).

Entomopathogenic and endophytic fungi are crucial for maintaining soil fertility, enhancing nutrient availability, supporting plant growth, and promoting microbial biodiversity. Their multifunctional role in nutrient management, biocontrol, and bioremediation makes them key agents for sustainable soil and crop management. However, further research is needed to fully unlock their potential and optimize their application across diverse agroecosystems.

5 Conclusion and Future Prospects

The increasing global population has driven a higher demand for food, leading to intensified farming practices and a continuous rise in agrochemical use. While these practices have boosted crop production, they have also contributed to the steady decline of soil health. In response, there has been growing interest in organic farming and the role of soil microorganisms as sustainable alternatives to maintain and restore soil quality. The use of agrochemicals in agriculture is deteriorating soil health by negatively impacting soil functions, physiology, and beneficial microbial communities, ultimately disrupting the soil environment and ecological balance. Excessive application of agrochemicals introduces toxicity to non-target soil flora and fauna, reducing microbial activity and overall soil fertility.

Entomopathogenic and endophytic fungi are emerging as essential components of sustainable agriculture. These fungi not only degrade chemical pollutants but also contribute to improved soil fertility and provide protection against pests and pathogens. However, only a limited number of entomopathogenic fungi have been studied in this context, and further research is needed to explore their untapped potential.

The diverse climatic conditions across agricultural zones necessitate the identification and adaptation of local fungal strains. Environmental factors significantly influence the survivability and effectiveness of entomopathogenic and endophytic fungi in the soil. Therefore, exploring and utilizing local strains can enhance their performance under region-specific conditions, increasing the success rate of field applications.

Currently, only a few effective fungal strains are commercially available, and their inconsistent performance across different climates has reduced farmers' confidence in microbe-based agricultural inputs. Developing microbial consortia that include local entomopathogenic or endophytic fungi, along with other agriculturally beneficial microorganisms, represents a promising but challenging approach. Such consortia are likely to be more effective than single-strain applications. To promote microbe-based inputs in sustainable farming, future research must focus on understanding microbial interactions within these consortia. A well-formulated product based on entomopathogenic or endophytic fungi would be eco-friendly and highly beneficial for both plant protection and soil fertility restoration.

We can conclude that entomopathogenic and endophytic microbes have demonstrated significant potential in enhancing soil health by functioning as nutrient mobilizers, plant growth promoters, plant protectors, soil conditioners, and nutrient solubilizers. Various entomopathogenic fungi—such as *Beauveria*, *Metarhizium*, *Paecilomyces*, *Lecanicillium* along with endophytic fungi like *Trichoderma* and *Aspergillus*, have shown promising capabilities in pest control, phosphorus and zinc solubilization, and the transport of essential nutrients such as nitrogen and iron within the soil-plant system. These fungi also promote plant growth by producing phytohormones such as indole acetic acid and enhance the colonization of beneficial microbial communities. Moreover, many entomopathogens are effective degraders of agrochemicals and xenobiotic compounds, highlighting their value in sustainable soil management. Their combined roles in plant protection and nutrient cycling make them crucial agents for restoring soil fertility. However, further research is needed to fully understand and optimize the use of entomopathogenic and endophytic fungi in sustainable agricultural practices.

References

Abd El-Ghany TM, Masmali IA (2016) Fungal biodegradation of organophosphorus insecticides and their impact on soil microbial population. J Plant Pathol Microbiol 7(5):1000349. https://doi.org/10.4172/2157-7471.1000349

Abdel-Mallek AY, Moharram AM, Abdel-Kader MIA, Omar SA (1994) Effect of soil treatment with the organophosphorus insecticide Profenfos on the fungal flora and some microbial activities. Microbiol Res 149:167–171. https://doi.org/10.1016/S0944-5013(11)80114-X

Ahmad MF, Ahmad FA, Alsayegh AA, Zeyaullah M, AlShahrani AM et al (2024) Pesticides impacts on human health and the environment with their mechanisms of action and possible countermeasures. Heliyon 10:e29128. https://doi.org/10.1016/j.heliyon.2024.e29128

Ahsan SM, Injamum-Ul-Hoque M, Das AK, Rahman MM, Mollah MMI et al (2024) Plant–entomopathogenic fungi interaction: recent progress and future prospects on endophytism-mediated growth promotion and biocontrol. Plants 13(10):1420. https://doi.org/10.3390/plants13101420

Aktar MW, Sengupta D, Chowdhury A (2009) Impact of pesticides use in agriculture: their benefits and hazards. Interdiscip Toxicol 2(1):1–12. https://doi.org/10.2478/v10102-009-0001-7

Andrea MM, Peres TB, Luchini LC, Pettinelli A Jr (2000) Impact of long-term pesticide applications on some soil biological parameters. J Environ Sci Health B 35:297–307. https://doi.org/10.1080/03601230009373271

Arinze E, Yubedee AG (2000) Effect of fungicides on *Fusarium* grain rot endoenzyme production in maize. Glob J Appl Sci 6:629–634. https://doi.org/10.4314/gjpas.v6i4.16178

Behie SW, Bidochka MJ (2014) Ubiquity of insect-derived nitrogen transfer to plants by endophytic insect-pathogenic fungi: an additional branch of the soil nitrogen cycle. Appl Environ Microbiol 80:1553–1560. https://doi.org/10.1128/AEM.03338-13

Behie SW, Moreira CC, Sementchoukova I, Barelli L, Zelisko PM et al (2017) Carbon translocation from a plant to an insect-pathogenic endophytic fungus. Nat Commun 8:14245. https://doi.org/10.1038/ncomms14245

Behie SW, Zelisko PM, Bidochka MJ (2012) Endophytic insect-parasitic fungi translocate nitrogen directly from insects to plants. Science 336(6088):1576–1577. https://doi.org/10.1126/science.1222289

Bhat MM, Bashir Z, Hamid B, Nisa M (2025) Impact of the synthetic fertilizer pesticides on soil health and soil microbiology. In: Bashir Z, Kumar R, Nisa M (eds) Organic farming. CRC Press, pp 282–302

Bose S, Kumar PS, Vo DVN (2021) A review on the microbial degradation of chlorpyrifos and its metabolite TCP. Chemosphere 283:131447. https://doi.org/10.1016/j.chemosphere.2021.131447

Carriger JF, Rand GM, Gardinali PR, Perry WB, Tompkins MS et al (2006) Pesticides of potential ecological concern in sediment from South Florida canals: an ecological risk prioritization for aquatic arthropods. Soil Sediment Contam 15:21–45. https://doi.org/10.1080/15320380500363095

Chalam AV, Sasikala C, Ramana CV, Rao PR (1996) Effect of pesticides on hydrogen metabolism of *Rhodobacter sphaeroides* and *Rhodopseudomonas palustris*. FEMS Microbiol Ecol 19(1):1–4. https://doi.org/10.1111/j.1574-6941.1996.tb00192.x

Chalam AV, Sasikala C, Ramana CV, Uma NR, Rao PR (1997) Effect of pesticides on the diazotrophic growth and nitrogenase activity of purple nonsulfur bacteria. Bull Environ Contam Toxicol 58:463–468. https://doi.org/10.1007/s001289900357

Chen S, Liu C, Peng C, Liu H, Hu M et al (2012) Biodegradation of chlorpyrifos and its hydrolysis product 3,5,6-trichloro-2-pyridinol by a new fungal strain *Cladosporium cladosporioides* Hu-01. PLoS One 7(10):e47205. https://doi.org/10.1371/journal.pone.0047205

Chen SK, Edwards CA, Subler S (2001) A microcosm approach for evaluating the effects of the fungicides benomyl and captan on soil ecological processes and plant growth. Appl Soil Ecol 18:69–82. https://doi.org/10.1016/S0929-1393(01)00135-4

Chen X, Wicaksono WA, Berg G, Cernava T (2021) Bacterial communities in the plant phyllosphere harbour distinct responders to a broad-spectrum pesticide. Sci Total Environ 751:141799. https://doi.org/10.1016/j.scitotenv.2020.141799
Chhipa H, Singh H, Deshmukh SK (2024) *Paecilomyces*: an eco-friendly approach to control agricultural pests and pathogens. In: Deshmukh SK, Sridhar KR (eds) Entomopathogenic fungi: prospects and challenges. Springer Nature, Singapore, pp 359–378. https://doi.org/10.1007/978-981-97-5991-0_14
Ciceoi R, Dobrin A, Moţ A, Fătu C, Dinu MM (2021) Effect of entomopathogenic fungi (*Beauveria bassiana*) on soil nutrients content. Sci Pap Ser B Hortic 65(1):722–727
Contreras-Cornejo HA, Macías-Rodríguez L, Cortés-Penagos C, López-Bucio J (2009) *Trichoderma virens*, a plant beneficial fungus, enhances biomass production and promotes lateral root growth through an auxin-dependent mechanism in *Arabidopsis*. Plant Physiol 149(3):1579–1592. https://doi.org/10.1104/pp.108.130369
Dara SK, Dara SSR (2019) Mycorrhiza-like role of *Beauveria bassiana*: a preliminary study. YouTube. https://youtu.be/88k5rWyceVE. Accessed 5 June 2025
de Mandal S, Passari AK (eds) (2021) Recent advancement in microbial biotechnology: agricultural and industrial approach, 1st edn. Elsevier, Waltham. https://doi.org/10.1016/C2019-0-04546-3
Deka D, Rabha J, Jha DK (2018) Application of myconanotechnology in the sustainable management of crop production system. In: Prasad R (ed) Mycoremediation and environmental sustainability. Fungal biology. Springer, Cham. https://doi.org/10.1007/978-3-319-77386-5_11
Demanou J, Monkiédjé A, Njiné T, Foto SM, Nola M et al (2004) Changes in soil chemical properties and microbial activities in response to the fungicide Ridomil gold plus copper. Int J Environ Res Public Health 1:26–34. https://doi.org/10.3390/ijerph2004010026
Di Ciocco CA (1997) Effect of the fungicide captan on *Azospirillum brasilense* Cd in pure culture and associated with *Setaria italica*. Rev Argent Microbiol 29:152–156
Ding M, Dai H, He Y, Liang T, Zhai Z et al (2024) Continuous cropping system altered soil microbial communities and nutrient cycles. Front Microbiol 15:1374550. https://doi.org/10.3389/fmicb.2024.1374550
do Nascimento JM, Netto JAFV, Valadares RV, de Oliveira Mendes G, da Silva IR et al (2021) *Aspergillus niger* as a key to unlock fixed phosphorus in highly weathered soils. Soil Biol Biochem 156:108190. https://doi.org/10.1016/j.soilbio.2021.108190
Draganova S, Donkova R, Georgieva D (2008) Impact of strains of entomopathogenic fungi on some main groups of soil microorganisms. J Plant Prot Res 48(2):169–179. https://doi.org/10.2478/v10045-008-0020-y
Eijsackers H, Beneke P, Maboeta M, Louw JPE, Reinecke AJ (2005) The implications of copper fungicide usage in vineyards for earthworm activity and resulting sustainable soil quality. Ecotoxicol Environ Saf 62(1):99–111. https://doi.org/10.1016/j.ecoenv.2005.02.017
Gadd GM (2010) Metals, minerals and microbes: geomicrobiology and bioremediation. Microbiology 156:609–643. https://doi.org/10.1099/mic.0.037143-0
Gallori E, Casalone E, Colella CM, Daly S, Polsinelli M (1991) 1,8-Naphthalic anhydride antidote enhances the toxic effects of captan and thiram fungicides on *Azospirillum brasilense* cells. Res Microbiol 142(9):1005–1012. https://doi.org/10.1016/0923-2508(91)90011-X
Gianfreda L, Sannino F, Violante A (1995) Pesticide effects on the activity of free, immobilized and soil invertase. Soil Biol Biochem 27:1201–1208. https://doi.org/10.1016/0038-0717(95)00034-C
González-Guzmán A, Sacristán D, Sánchez-Rodríguez AR, Barrón V, Torrent J et al (2020) Soil nutrients effects on the performance of durum wheat inoculated with entomopathogenic fungi. Agronomy 10(4):589. https://doi.org/10.3390/agronomy10040589
Gregory AS, Ritz K, McGrath SP, Quinton JN, Goulding KW et al (2015) A review of the impacts of degradation threats on soil properties in the UK. Soil Use Manag 31(S1):1–15. https://doi.org/10.1111/sum.12212
Hammouda O (1999) Response of the paddy field cyanobacterium *Anabaena doliolum* to carbofuran. Ecotoxicol Environ Saf 44:215–219. https://doi.org/10.1006/eesa.1999.1823

Hossain SM, Anantharaman N (2005) Studies on aerobic biodegradation of DDT using *Phanerochaete chrysosporium*. Indian J Environ Prot 25(5):454

Hou WP, Wang JF, Nan ZB, Christensen MJ, Xia C et al (2020) *Epichloë gansuensis* endophyte-infection alters soil enzymes activity and soil nutrients at different growth stages of *Achnatherum inebrians*. Plant Soil 455(1):227–240. https://doi.org/10.1007/s11104-020-04682-2

Huang S, Gao J, Zhou L, Gao L, Song M et al (2024) Biodegradation of p-chloroaniline by fungus *Isaria fumosorosea* SP535. Res Sq:1–21. https://doi.org/10.21203/rs.3.rs-4840476/v1

Ingram CW, Coyne MS, Williams DW (2005) Effects of commercial diazinon and imidacloprid on microbial urease activity in soil and sod. J Environ Qual 34:1573–1580. https://doi.org/10.2134/jeq2004.0433

Ismail BS, Jokha Y, Omar O (1996) Effects of glufosinate-ammonium on microbial populations and enzyme activities in soils. Microbios 83:185–190

Jalal MAF, Bondarenko S (2025) Agrochemicals: degradation and metabolism. In: Jalal MAF, Bondarenko S (eds) Radiochemical analysis of agrochemicals. Environmental fate and metabolism. Springer, Cham, pp 51–98. https://doi.org/10.1007/978-3-031-80080-1_2

Kassam R, Yadav J, Jaiswal N, Chatterjee M, Hada A et al (2022) Identification and potential utility of *Metarhizium anisopliae* (ITCC9014) for the management of root-knot nematode, *Meloidogyne incognita*. Indian Phytopathol 75(3):875–881. https://doi.org/10.1007/s42360-022-00498-5

Kaur G, Kaur D, Gupta S (2021) The role of microorganisms in remediation of environmental contaminants. In: Prasad R (ed) Environmental pollution and remediation. Environmental and microbial biotechnology. Springer, Singapore, pp 421–450. https://doi.org/10.1007/978-981-15-5499-5_15

Khan MS, Zaidi A, Rizvi PQ (2006) Biotoxic effects of herbicides on growth, nodulation, nitrogenase activity, and seed production in chickpeas. Commun Soil Sci Plant Anal 37:1783–1793. https://doi.org/10.1080/00103620600710645

Kızılkaya R, Akça İ, Aşkın T, Yılmaz R, Olekhov V et al (2012) Effect of soil contamination with azadirachtin on dehydrogenase and catalase activity of soil. Eurasian J Soil Sci 1:98–103

Kumar CS, D'Silva S, Praveena R, Kaprakkaden A, Krishnan LA et al (2024) Zinc solubilization and organic acid production by the entomopathogenic fungus, *Metarhizium pingshaense* sheds light on its key ecological role in the environment. Sci Total Environ 923:171348. https://doi.org/10.1016/j.scitotenv.2024.171348

Lamichhane JR, Dachbrodt-Saaydeh S, Kudsk P, Messéan A (2016) Toward a reduced reliance on conventional pesticides in European agriculture. Plant Dis 100(1):10–24. https://doi.org/10.1094/PDIS-05-15-0574-FE

Leach JE, Triplett LR, Argueso CT, Trivedi P (2017) Communication in the phytobiome. Cell 169: 587–596. https://doi.org/10.1016/j.cell.2017.04.025

Li HY, Wei DQ, Shen M, Zhou ZP (2012) Endophytes and their role in phytoremediation. Fungal Divers 54(1):11–18. https://doi.org/10.1007/s13225-012-0165-x

Liu H, Li J, Carvalhais LC, Percy CD, Verma JP et al (2021) Evidence for the plant recruitment of beneficial microbes to suppress soil borne pathogens. New Phytol 229:2873–2885. https://doi.org/10.1111/nph.17057

Madhuri RJ, Rangaswamy V (2002) Influence of selected insecticides on phosphatase activity in groundnut (*Arachis hypogeae* L.) soils. J Environ Biol 23:393–397

Mandal A, Sarkar B, Mandal S, Vithanage M, Patra AK et al (2020) Impact of agrochemicals on soil health. In: Prasad MNV (ed) Agrochemicals detection, treatment and remediation, pesticides and chemical fertilizers. Butterworth-Heinemann, pp 161–187. https://doi.org/10.1016/B978-0-08-103017-2.00007-6

Martınez-Toledo MV, Salmeron V, Rodelas B, Pozo C, Gonzalez-Lopez J (1998) Effects of the fungicide Captan on some functional groups of soil microflora. Appl Soil Ecol 7(3):245–255. https://doi.org/10.1016/S0929-1393(97)00026-7

Martino E, Perotto S, Parsons R, Gadd GM (2003) Solubilization of insoluble inorganic zinc compounds by ericoid mycorrhizal fungi derived from heavy metal polluted sites. Soil Biol Biochem 35(1):133–141. https://doi.org/10.1016/S0038-0717(02)00247-X

Mayanglambam T, Vig K, Singh DK (2005) Quinalphos persistence and leaching under field conditions and effects of residues on dehydrogenase and alkaline phosphomonoesterases activities in soil. Bull Environ Contam Toxicol 75:1067–1076. https://doi.org/10.1007/s00128-005-0858-x

Mendes GDO, Bahri-Esfahani J, Csetenyi L, Hillier S, George TS et al (2021) Chemical and physical mechanisms of fungal bioweathering of rock phosphate. Geomicrobiol J 38(5): 384–394. https://doi.org/10.1080/01490451.2020.1863525

Mohapatra D, Rath SK, Mohapatra PK (2021) Evaluating a preparation of malathion-tolerant *Aspergillus niger* MRU01 for accelerated removal of four organophosphorus insecticides. J Chem Technol Biotechnol 96(6):1603–1610. https://doi.org/10.1002/jctb.6679

Monkiedje A, Ilori MO, Spiteller M (2002) Soil quality changes resulting from the application of the fungicides mefenoxam and metalaxyl to a sandy loam soil. Soil Biol Biochem 34:1939–1948. https://doi.org/10.1016/S0038-0717(02)00211-0

Moreno-Gavíra A, Diánez F, Sánchez-Montesinos B, Santos M (2021) Biocontrol effects of *Paecilomyces variotii* against fungal plant diseases. J Fungi 7(6):415. https://doi.org/10.3390/jof7060415

Mukherjee A, Bhowmick S, Yadav S, Rashid MM, Chouhan GK et al (2021b) Re-vitalizing of endophytic microbes for soil health management and plant protection. 3 Biotech 11:1–17. https://doi.org/10.1007/s13205-021-02931-4

Mukherjee A, Chouhan GK, Gaurav AK, Jaiswal DK, Verma JP (2021a) Development of indigenous microbial consortium for biocontrol management. In: Verma JP, Macdonald C, Gupta V, Podile A (eds) New and future developments in microbial biotechnology and bioengineering phytomicrobiome for sustainable agriculture. Elsevier, Amsterdam, pp 91–104. https://doi.org/10.1016/B978-0-444-64325-4.00009-2

Mukherjee A, Gaurav AK, Singh S, Chouhan GK, Kumar A et al (2019) Role of potassium (K) solubilising microbes (KSM) in growth and induction of resistance against biotic and abiotic stress in plant: a book review. Clim Chang Environ Sustain 7(2):212–214

Mukherjee A, Singh B, Verma JP (2020) Harnessing chickpea (*Cicer arietinum* L.) seed endophytes for enhancing plant growth attributes and bio-controlling against *Fusarium* sp. Microbiol Res 237:126469. https://doi.org/10.1016/j.micres.2020.126469

Mumtaz MZ, Barry KM, Baker AL, Nichols DS, Ahmad M et al (2019) Production of lactic and acetic acids by *Bacillus* sp. ZM20 and *Bacillus cereus* following exposure to zinc oxide: a possible mechanism for Zn solubilization. Rhizosphere 12:100170. https://doi.org/10.1016/j.rhisph.2019.100170

Murindangabo YT, Kopecký M, Perná K, Konvalina P, Bohatá A et al (2024) Relevance of entomopathogenic fungi in soil–plant systems. Plant Soil 495(1):287–310. https://doi.org/10.1007/s11104-023-06325-8

Muvea AM, Meyhöfer R, Subramanian S, Poehling HM, Ekesi S et al (2014) Colonization of onions by endophytic fungi and their impacts on the biology of Thrips tabaci. PLoS One 9(9): e108242. https://doi.org/10.1371/journal.pone.0108242

Nandy S, Das T, Tudu CK, Pandey DK, Dey A et al (2020) Fungal endophytes: futuristic tool in recent research area of phytoremediation. S Afr J Bot 134:285–295. https://doi.org/10.1016/j.sajb.2020.02.015

Niewiadomska A (2004) Effect of carbendazim, imazetapir and thiram on nitrogenase activity, the number of microorganisms in soil and yield of red clover (*Trifolium pratense* L.). Pol J Environ Stud 13:403–410

Niewiadomska A, Klama J (2005) Pesticide side effect on the symbiotic efficiency and nitrogenase activity of Rhizobiaceae bacteria family. Pol J Microbiol 54:43–48

Nowak A, Nowak J, Klodka D, Przybulewska K, Telesinski A et al (2004) Changes in the microflora and biological activity of the soil during the degradation of isoproturon. J Plant Dis Prot 19:1003–1016

Nowak M, Soboń A, Litwin A, Różalska S (2019) 4-n-Nonylphenol degradation by the genus Metarhizium with cytochrome P450 involvement. Chemosphere 220:324–334. https://doi.org/10.1016/j.chemosphere.2018.12.114

Nweke CO, Ntinugwa C, Obah IF, Ike SC, Eme GE et al (2007) In vitro effects of metals and pesticides on dehydrogenase activity in microbial community of cowpea (*Vigna unguiculata*) rhizoplane. Afr J Biotechnol 6:290–295. http://www.academicjournals.org/AJB

Omar SA, Abdel-Sater MA (2001) Microbial populations and enzyme activities in soil treated with pesticides. Water Air Soil Pollut 127:49–63. https://doi.org/10.1023/A:1005209516845

Omar SA, Abd-Alla MH (1992) Effect of pesticides on growth, respiration and nitrogenase activity of *Azotobacter* and *Azospirillum*. World J Microbiol Biotechnol 8:326–328. https://doi.org/10.1007/BF01201891

Ortega SN, Nitschke M, Mouad AM, Landgraf MD, Rezende MOO et al (2011) Isolation of Brazilian marine fungi capable of growing on DDD pesticide. Biodegradation 22:43–50. https://doi.org/10.1007/s10532-010-9374-8

Parks CG, Hoppin JA, De Roos AJ, Costenbader KH, Alavanja MC et al (2016) Rheumatoid arthritis in agricultural health study spouses: associations with pesticides and other farm exposures. Environ Health Perspect 124(11):1728–1734. https://doi.org/10.1289/EHP129

Patil BC, Sharma RP, Salunkhe DK, Salunkhe K (1972) Evaluation of solanine toxicity. Food Cosmet Toxicol 10(3):395–398. https://doi.org/10.1016/S0015-6264(72)80258-X

Pedrinho A, Mendes LW, de Araujo Pereira AP, Araujo ASF, Vaishnav A et al (2024) Soil microbial diversity plays an important role in resisting and restoring degraded ecosystems. Plant Soil 500(1):325–349. https://doi.org/10.1007/s11104-024-06489-x

Pineda M (2014) Efecto del corte de pelo y el uso de hongos entomopatogenos (*Beauveria bassiana* y *Metarhizium anisopliae*) en el control de garrapata (*Bophilus microplus*) en ganado bovino. Doctoral dissertation, Zamorano: Escuela Agrícola Panamericana, Zamorano, Honduras, p 12

Prasad SM, Zeeshan M, Kumar D (2011) Toxicity of endosulfan on growth, photosynthesis, and nitrogenase activity in two species of *Nostoc* (*Nostoc muscorum* and *Nostoc calcicola*). Toxicol Environ Chem 93:513–525. https://doi.org/10.1080/02772248.2010.542157

Prashar P, Shah S (2016) Impact of fertilizers and pesticides on soil microflora in agriculture. In: Lichtfouse E (ed) Sustainable agriculture reviews, vol 19. Springer, Cham, pp 331–361. https://doi.org/10.1007/978-3-319-26777-7_8

Prieto I, Klimm A, Roldán F, Vetter W, Arbeli Z (2021) Evidence for cometabolic transformation of weathered toxaphene under aerobic conditions using camphor as a co-substrate. J Appl Microbiol 131(1):221–235. https://doi.org/10.1111/jam.14963

Purnomo AS, Kamei I, Kondo R (2008) Degradation of 1,1,1-trichloro-2,2-bis (4-chlorophenyl) ethane (DDT) by brown-rot fungi. J Biosci Bioeng 105(6):614–621. https://doi.org/10.1263/jbb.105.614

Qayyum MA, Wakil W, Arif MJ, Sahi ST, Dunlap CA (2015) Infection of *Helicoverpa armigera* by endophytic *Beauveria bassiana* colonizing tomato plants. Biol Control 90:200–207. https://doi.org/10.1016/j.biocontrol.2015.04.005

Ramteke BD, Thakur KD, Bramhankar SB, Gawande SP, Isokar SS et al (2022) Isolation of *Lecanicillium* spp., its compatibility with different insecticides and bioassay against cotton aphids. Pharm Innov J 11(9):500–503

Raya-Díaz S, Sánchez-Rodríguez AR, Segura-Fernández JM, Del Campillo MDC, Quesada-Moraga E (2017) Entomopathogenic fungi-based mechanisms for improved Fe nutrition in sorghum plants grown on calcareous substrates. PLoS One 12:1–28. https://doi.org/10.1371/journal.pone.0185903

Riah W, Laval K, Laroche-Ajzenberg E, Mougin C, Latour X et al (2014) Effects of pesticides on soil enzymes: a review. Environ Chem Lett 12(2):257–273. https://doi.org/10.1007/s10311-014-0458-2

Rigobelo EC (2024) Filamentous fungi as plant growth promoters. In: Rigobelo EC (ed) Microbial services for cereal crops: reducing costs and environmental impact. Springer Nature Switzerland, Cham, pp 79–93. https://doi.org/10.1007/978-3-031-63149-8_5

Rigobelo EC, Nicodemo D, Babalola OO, Desoignies N (2024) Purpureocillium lilacinum as an agent of nematode control and plant growth-promoting fungi. Agronomy 14(6):1225. https://doi.org/10.3390/agronomy14061225

Różalska S, Pawłowska J, Wrzosek M, Tkaczuk C, Długoński J (2013) Utilization of 4-n-nonylphenol by *Metarhizium* sp. isolates. Acta Biochim Pol 60(4):677

Różalska S, Soboń A, Pawłowska J, Wrzosek M, Długoński J (2015) Biodegradation of nonylphenol by a novel entomopathogenic *Metarhizium robertsii* strain. Bioresour Technol 191:166–172. https://doi.org/10.1016/j.biortech.2015.05.011

Sagar A, Rathore P, Ramteke PW, Ramakrishna W, Reddy MS et al (2021) Plant growth promoting rhizobacteria, arbuscular mycorrhizal fungi and their synergistic interactions to counteract the negative effects of saline soil on agriculture: key macromolecules and mechanisms. Microorganisms 9:1491. https://doi.org/10.3390/microorganisms9071491

Saleem M, Hu J, Jousset A (2019) More than the sum of its parts: microbiome biodiversity as a driver of plant growth and soil health. Annu Rev Ecol Evol Syst 50(1):145–168. https://doi.org/10.1146/annurev-ecolsys-110617-062605

Sánchez-Rodríguez AR, Raya-Díaz S, Zamarreño ÁM, García-Mina JM, del Campillo MC et al (2018) An endophytic *Beauveria bassiana* strain increases spike production in bread and durum wheat plants and effectively controls cotton leafworm (*Spodoptera littoralis*) larvae. Biol Control 116:90–102. https://doi.org/10.1016/j.biocontrol.2017.01.012

Sannino F, Gianfreda L (2001) Pesticide influence on soil enzymatic activities. Chemosphere 45:417–425. https://doi.org/10.1016/S0045-6535(01)00045-5

Senthil Kumar CM, Jacob TK, Devasahayam S, Thomas S, Geethu C (2018) Multifarious plant growth promotion by an entomopathogenic fungus *Lecanicillium psalliotae*. Microbiol Res 207:153–160. https://doi.org/10.1016/j.micres.2017.11.017

Sharma A, Kumar V, Handa N, Bali S, Kaur R et al (2018) Potential of endophytic bacteria in heavy metal and pesticide detoxifcation. In: Egamberdieva D, Ahmad P (eds) Plant microbiome: stress response. Microorganisms for sustainability. Springer Belin, pp 307–336. https://doi.org/10.1007/978-981-10-5514-0_14

Sharma A, Srivastava A, Shukla AK, Srivastava K, Srivastava AK et al (2020) Entomopathogenic fungi: a potential source for biological control of insect pests. In: Solanki MK, Kashyap PL, Kumari B (eds) Phytobiomes: current insights and future vistas. Springer Singapore, Singapore, pp 225–250. https://doi.org/10.1007/978-981-15-3151-4_9

Siewiera P, Bernat P, Różalska S, Długoński J (2015) Estradiol improves tributyltin degradation by the filamentous fungus *Metarhizium robertsii*. Int Biodeterior Biodegradation 104:258–263. https://doi.org/10.1016/j.ibiod.2015.06.014

Singh DK, Kumar S (2008) Nitrate reductase, arginine deaminase, urease and dehydrogenase activities in natural soil (ridges with forest) and in cotton soil after acetamiprid treatments. Chemosphere 71:412–418. https://doi.org/10.1016/j.chemosphere.2007.11.005

Singh G, Wright D (1999) Effects of herbicides on nodulation, symbiotic nitrogen fixation, growth and yield of pea (*Pisum sativum*). J Agric Sci 133:21–30. https://doi.org/10.1017/S0021859699006735

Singh J, Singh DK (2005) Dehydrogenase and phosphomonoesterase activities in groundnut (*Arachis hypogaea* L.) field after diazinon, imidacloprid and lindane treatments. Chemosphere 60:32–42. https://doi.org/10.1016/j.chemosphere.2004.11.096

Sisay K, Legese H (2020) Entomopathogenic fungi as a biological pest management option: a review. Int J Res Stud Agric Sci 6(6):1–10. https://doi.org/10.20431/2454-6224.0606001

Sofo A, Scopa A, Dumontet S, Mazzatura A, Pasquale V (2012) Toxic effects of four sulphonylureas herbicides on soil microbial biomass. J Environ Sci Health B 47:653–659. https://doi.org/10.1080/03601234.2012.669205

Srinivasulu M, Mohiddin GJ, Subramanyam K, Rangaswamy V (2012) Effect of insecticides alone and in combination with fungicides on nitrification and phosphatase activity in two groundnut (*Arachis hypogeae* L.) soils. Environ Geochem Health 34:365–374. https://doi.org/10.1007/s10653-011-9399-x

St Leger RJ, Wang JB (2020) *Metarhizium*: jack of all trades, master of many. Open Biol 10(12): 200307. https://doi.org/10.1098/rsob.200307

Sukul P (2006) Enzymatic activities and microbial biomass in soil as influenced by metalaxyl residues. Soil Biol Biochem 38:320–326. https://doi.org/10.1016/j.soilbio.2005.05.009

Swathy K, Vivekanandhan P, Yuvaraj A, Sarayut P, Kim JS et al (2024) Biodegradation of pesticide in agricultural soil employing entomopathogenic fungi: current state of the art and future perspectives. Heliyon 10(1):e23406. https://doi.org/10.1016/j.heliyon.2023.e23406

Tu CM (1981) Effects of some pesticides on enzyme activities in an organic soil. Bull Environ Contam Toxicol 27:109–114

Taşeli BK (2006) Dehalogenation of lindane by *Penicillium camemberti*. Bull Environ Contam Toxicol 77:882–887. https://doi.org/10.1007/s00128-006-1226-1

Van Zwieten L, Rust J, Kingston T, Merrington G, Morris S (2004) Influence of copper fungicide residues on occurrence of earthworms in avocado orchard soils. Sci Total Environ 329(1 3): 29–41. https://doi.org/10.1016/j.scitotenv.2004.02.014

Vassileva M, Mendes GDO, Deriu MA, Benedetto GD, Flor-Peregrin E et al (2022) Fungi, P-solubilization, and plant nutrition. Microorganisms 10:1716. https://doi.org/10.3390/microorganisms10091716

Vivekanandhan P, Swathy K, Sarayut P, Patcharin K (2025) Enzymatic, cellular breakdown and lysis in treatment of *Beauveria brongniartii* on *Spodoptera litura* (Fabricius, 1775). Sci Rep 15(1):7175. https://doi.org/10.1038/s41598-024-78151-6

Wang Y, Xiang S, Chen R, Chen L, Lan W et al (2025) Enhancing Miscanthus floridulus remediation of soil cadmium using *Beauveria bassiana* FE14: plant growth promotion and microbial interactions. Ecotoxicol Environ Saf 290:117745. https://doi.org/10.1016/j.ecoenv.2025.117745

Wu C, Tang D, Dai J, Tang X, Bao Y et al (2022) Bioremediation of mercury-polluted soil and water by the plant symbiotic fungus *Metarhizium robertsii*. Proc Natl Acad Sci USA 119(47): e2214513119. https://doi.org/10.1073/pnas.2214513119

Xiao P, Mori T, Kamei I, Kiyota H, Takagi K et al (2011) Novel metabolic pathways of organochlorine pesticides dieldrin and aldrin by the white rot fungi of the genus *Phlebia*. Chemosphere 85(2):218–224. https://doi.org/10.1016/j.chemosphere.2011.06.028

Xu F, Li C, Chen Y, Wu J, Bai H et al (2024) Soil microbial community structure and soil fertility jointly regulate soil microbial residue carbon during the conversion from subtropical primary forest to plantations. Geoderma 441:116767. https://doi.org/10.1016/j.geoderma.2023.116767

Zhang Z, Tian Y, Sui L, Lu Y, Cheng K et al (2024) First record of *Aspergillus nomiae* as a broad-spectrum entomopathogenic fungus that provides resistance against phytopathogens and insect pests by colonization of plants. Front Microbiol 14:1284276. https://doi.org/10.3389/fmicb.2023.1284276

Part IV

Biopesticides and Pest Management

Synergistic Strategies: Enhancing Pest Management Through the Compatibility of EPF and Other Biocontrol Agents

13

Ipsita Samal, Tanmaya Kumar Bhoi, Deepak Kumar Mahanta, J. Komal, Bimal Kumar Sahoo, Ansh Raj, Sanjan Kumar Bharti, and Amit Kumar

Abstract

Entomopathogenic fungi (EPF) have emerged as effective biocontrol agents in integrated pest management (IPM) systems, offering a sustainable alternative to chemical pesticides. This chapter provides a comprehensive overview of the compatibility of EPF with other biocontrol agents, including bacteria, nematodes, and insect parasitoids. We explore the mechanisms that govern both synergistic and antagonistic interactions among these agents, emphasizing the importance of environmental conditions, host specificity, and application methods. Through case studies and experimental approaches, we highlight successful combinations that enhance pest control efficacy while minimizing negative interactions. This chapter concludes with insights into future research directions aimed at optimizing the use of EPF in conjunction with other biocontrol strategies, ultimately contributing to more sustainable agricultural practices.

I. Samal (✉) · A. Raj · S. K. Bharti · A. Kumar
ICAR-National Research Centre on Litchi, Muzaffarpur, Bihar, India

T. K. Bhoi
Forest Protection Division, ICFRE – Arid Forest Research Institute (ICFRE-AFRI), Jodhpur, Rajasthan, India

D. K. Mahanta
Forest Entomology Discipline, Forest Protection Division, Indian Council of Forestry Research and Education (ICFRE)-Forest Research Institute (ICFRE-FRI), Dehradun, Uttarakhand, India

J. Komal
Central Silk Board-Basic Tasar Silkworm Seed Organization, Bilaspur, Chattisgarh, India

B. K. Sahoo
Tamil Nadu Agricultural University, Coimbatore, Tamil Nadu, India

S. K. Deshmukh et al. (eds.), *Frontiers in Entomopathogenic Fungi*,
https://doi.org/10.1007/978-981-95-6529-0_13

Keywords

Agricultural practices · Behavioral role · Compatibility · Ecological role · Host specificity · Synergism

1 Introduction

Integrated pest management (IPM) is increasingly favoring biological solutions, including the utilization of pathogens, parasitic wasps, and predatory insects and mites, over synthetic pesticides. *Beauveria bassiana*, *Metarhizium* spp., *Lecanicillium muscarium* (formerly *Verticillium lecanii*), and *Isaria* (formerly *Paecilomyces*) species are prominent entomopathogenic fungi (EPF) utilized in insect pest management (Jeffers and Chong 2021). Various fungi infect numerous arthropod species by penetrating the host's outer layer, proliferating within the organism, and often causing devastating epidemics (epizootics) among pest populations (Baker et al. 2020). EPF are utilized in numerous IPM programs due to their efficacy in pest control and their environmental safety. Nonetheless, the effectiveness of many biocontrol agents exhibits significant variability in field conditions (Lacey 2016). Consequently, the utilization of EPF in conjunction with other natural predators is currently under investigation to enhance pest management and control. When pests emerge in the ecosystem, they may simultaneously be targeted by many agents, including predators, pathogens, and parasitoids.

These attacks may be neutral or may lead to either antagonism or synergism (Francis et al. 2020). The primary objective of biocontrol is to integrate advantageous functions while eliminating detrimental ones. This chapter elucidates EPF in IPM and their potential integration with other natural control methods. We analyze the notions of compatibility and synergism, elucidate the interactions among many ecological, physiological, and molecular processes between EPF and natural enemies, and investigate cases from several global cropping systems (Smagghe et al. 2023). Research, tests, and meta-analyses indicate both the benefits and drawbacks of concomitant use of these medications. We address the fundamentals of formula preparation and use, environmental factors influencing their efficacy, primary challenges in this domain, and critical areas for future investigation. Understanding that EPF synergizes with predators, parasitoids, and biocontrol agents enables pest management systems to effectively and consistently regulate agricultural pests.

2 Compatibility and Synergism: Concepts and Definitions

Biocontrol agents are considered compatible only if they do not compromise each other's efficacy (Nawaz et al. 2022). For this combination to be successful, both the fungus and the predator/parasitoid must minimally disrupt the host's essential processes. Laboratory tests indicate that *B. bassiana* and *M. anisopliae* exert no adverse impact on the lifespan or reproductive capacity of the beneficial ladybird *Coccinella septempunctata*. Ensure that the chemicals employed are suitable to prevent issues with pest management (Sain et al. 2019).

The synergistic effect of two agents employed concurrently for biological control surpasses their impacts (Iqbal et al. 2019). Employing numerous strategies increases the likelihood of significantly reducing insect populations compared to using a single strategy. An additive effect occurs when the combined impact of two medications equals their individual effects; if the combined effect is diminished, the drugs are considered to antagonize each other. Reports indicate that certain natural adversaries may exhibit neutrality toward one another, while others may engage in competition or even cooperation. Researchers are endeavoring to integrate treatments that exhibit interactions equal to or superior to their individual applications.

Various tests are employed to verify the efficacy and safety of the combination of two agents (Table 13.1). Does the fungus provide any challenges for the biocontrol agent? Utilizing EPF-based biopesticides in accordance with the guidelines typically does not adversely affect key predators or parasitoids. It may be beneficial to ascertain whether the agents are more effective at overcoming the bug collectively than they are independently. In the current thorough investigation, most cases (38 out of 49) demonstrated that the simultaneous use of hymenopteran parasitoids and EPF did not adversely impact the efficacy of the parasitoids. Six tests were unsuccessful due to incorrect scheduling or the parasitoid being overrun by fungus. Approximately 26% of the evaluated combinations exhibited no interactions between the medications, whereas only a limited number demonstrated indications of synergy. The simultaneous application of a parasitoid wasp and an entomopathogen resulted in a more significant reduction of *Helicoverpa armigera* due to their synergistic effect. It is undeniable that, although countries are increasingly refraining from interference, the potential for enhanced cooperation exists but is hardly realized. This article will thereafter explore the specifics and prerequisites necessary for the initiation of synergism.

3 Mechanisms of Interaction Between Fungi and Biocontrol Agents

3.1 Ecological and Behavioral Mechanisms

Depending on the environment where several natural enemies coexist, their effects on the hosts may vary. The life stages of insects and the activity periods of predatory insects, parasitoids, and fungi may be complementary to each other (Roca-Couso et al. 2021). An egg parasitoid only attacks the eggs of the insect, but an entomopathogenic fungus (EPF) can attack the pest at all stages, including the larval and adult stages. The host may fare differently in case biocontrol chemicals are used. Many predators and parasitoids avoid diseased hosts and those with fungal infections to avoid competition with conspecifics over food resources. To minimize the risk of being infected, an adult ladybird can avoid eating fungus-infected food

Table 13.1 List of biocontrol agent(s) (combined with EPF) and their target pest(s), along with application method and timing of applications

Biocontrol agent (combined with EPF)	Target pest	Crop/system	EPF (species/ strain)	Application method	Timing/ sequence	Observed outcome	Notes (effectiveness/ recommendation)
Predatory mite *Amblyseius swirskii* (generalist)	Western flower thrips (*Frankliniella occidentalis*)	Greenhouse vegetables (e.g., cucumber)	*Beauveria bassiana* (e.g., strain GHA)	Foliar spray of fungus; release of mites	Simultaneous (concurrent use)	Neutral (no synergy)	No increased combined effect was observed; the two agents simply complement each other (mite provides preventive suppression, fungus serves as a curative for outbreaks). Coapplication is feasible, as neither agent significantly interferes with the other.
Predatory mite *Phytoseiulus persimilis* (spider mite predator)	Two-spotted spider mite (*Tetranychus urticae*)	Vegetables/ ornamentals (greenhouse)	*Beauveria bassiana*	Foliar spray of fungus; release of mites	Mites introduced onto fungus-treated plants	Synergistic	Predator performance was not hindered (mites only spent more time grooming with no loss in predation) and the combined treatment gave greater spider mite control than either alone. The fungus and predator together achieved higher mite mortality,

							indicating a true synergistic effect in pest suppression.
Predatory mirid bug *Macrolophus pygmaeus* (syn. *M. caliginosus*)	Greenhouse whitefly (*Trialeurodes vaporariorum*)	Greenhouse tomato/ vegetable crops	*Beauveria bassiana* (e.g., Naturalis-L)	Foliar spray of fungus; predator release	Simultaneous (fungus applied in presence of predator)	Additive (compatible)	The mirid predator was unaffected by the fungus (no significant impact on its survival or predation). Combined use resulted in whitefly mortality comparable to the sum of individual effects (additive control), so the fungus can be used alongside the predator without reducing overall efficacy.
Predatory ladybird beetle *Cryptolaemus montrouzieri* (mealybug destroyer)	Citrus mealybugs (*Planococcus* spp.) and other scale insects	Orchards (citrus, tropical fruit) or greenhouses	*Beauveria bassiana* (oil formulation, e.g., strain Bb112)	Foliar spray (targeting pest colonies)	Fungus applied to infested plants; beetles released for biological control	Neutral/ additive	The fungus is essentially safe for the lady beetle—studies show >90% adult survival even at high spore concentrations. Therefore, the beetles continue predation unabated alongside the fungus. The two agents target different

(continued)

Table 13.1 (continued)

Biocontrol agent (combined with EPF)	Target pest	Crop/system	EPF (species/strain)	Application method	Timing/sequence	Observed outcome	Notes (effectiveness/recommendation)
							pest stages (fungus infects various stages, predator feeds on eggs/larvae), providing complementary control without interfering with each other.
Parasitoid wasp *Encarsia formosa* (whitefly parasitoid) + EPF Lecanicillium muscarium	Greenhouse whiteflies (*Trialeurodes vaporariorum*)	Greenhouse vegetables (tomato, cucumber)	*Lecanicillium muscarium* (e.g., strain Ve6, product "Mycotal")	Foliar spray of spores; wasp releases (parasitizes nymphs)	Often simultaneous in practice (can be sequential)	Additive (with some host-use competition)	Both the fungus and the wasp can successfully kill whiteflies in combination (total pest mortality similar to using each alone). However, the highly virulent *L. muscarium* often infects and kills many whitefly nymphs before parasitoid emergence, thereby reducing wasp progeny (fungal infection of parasitized hosts limits parasitism to

							~10%). To maximize effectiveness, it's recommended to allow parasitism to occur (e.g., release *Encarsia* first) before applying the fungus, so the agents don't significantly hinder one another.
Parasitoid wasp *Encarsia formosa* + EPF Beauveria bassiana	Greenhouse whiteflies (*Trialeurodes vaporariorum*)	Greenhouse vegetables (tomato, cucumber)	*Beauveria bassiana* (e.g., strain ATCC 74040)	Foliar spray of spores; wasp releases	Simultaneous or fungus, applied after some parasitism	Additive (largely compatible)	*B. bassiana* is less aggressive against whitefly nymphs than *Lecanicillium*, so fewer parasitized hosts are killed by the fungus. This leads to higher parasitoid emergence rates, and the combined treatment maintains high whitefly mortality without strong interference between fungus and wasp. For best results, apply *B. bassiana* after parasitoids have been working for a day or two, ensuring most

(continued)

Table 13.1 (continued)

Biocontrol agent (combined with EPF)	Target pest	Crop/system	EPF (species/strain)	Application method	Timing/sequence	Observed outcome	Notes (effectiveness/recommendation)
							parasitism occurs before fungal kill.
Parasitoid wasp *Aphidius matricariae* (aphid parasitoid)	Green peach aphid (*Myzus persicae*) and other aphids	Vegetables, greenhouse ornamentals	*Beauveria bassiana* (e.g., strain GHA)	Foliar spray (targeting aphid colonies); wasp releases (parasitizing aphids)	Sequential (optimal: parasitoid release $\geq$48 h before fungus spray)	Additive (if properly timed)	Generally, aphid parasitoids are compatible with *B. bassiana* applications. For optimal control, introduce *Aphidius* wasps at least 2 days prior to spraying the fungus. This timing allows parasitoids to form mummies; fungus applied too early can infect hosts and slightly prolong mummy formation or reduce wasp emergence. When timed correctly, both agents together improve aphid suppression (fungus kills aphids that are not parasitized, while wasps develop in

							others) without antagonism.
Parasitoid wasp *Microplitis prodeniae* (larval parasitoid)	Fall armyworm (*Spodoptera frugiperda*) larvae	Field crops (maize and sorghum)	*Beauveria bassiana* (local strain)	Foliar application (spray or dip) of spores; parasitoid wasp releases (attacks larvae)	Simultaneous or staggered (wasp released onto fungus-exposed larvae)	Synergistic	The joint use of *B. bassiana* and *Microplitis* was far more effective than either alone—combined treatment significantly increased larval mortality (early-instar FAW) over single-agent treatment. The parasitoid's activity (stinging and host feeding) helped spread fungal conidia and breach the host cuticle, enhancing fungal infection rates. This clear synergistic interaction suggests that releasing parasitoids alongside a fungal spray can yield improved control of armyworm infestations.
Entomopathogenic nematode (EPN) *Heterorhabditis bacteriophora*	Scarab beetle grubs (e.g., barley chafer	Field crops (cereals, pasture/turf)	*Metarhizium anisopliae*	Soil application (soil drench or granules of fungus)			

(continued)

Table 13.1 (continued)

Biocontrol agent (combined with EPF)	Target pest	Crop/system	EPF (species/ strain)	Application method	Timing/ sequence	Observed outcome	Notes (effectiveness/ recommendation)
	Coptognathus curtipennis)						

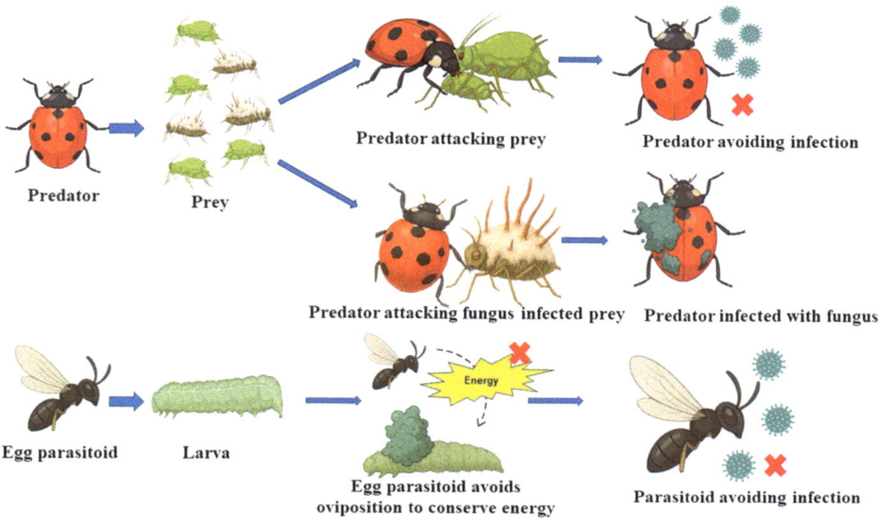

Fig. 13.1 Behavioral responses of natural enemies to fungal-infected hosts in a tritrophic interaction. *Top panel*: A predatory beetle attacks both healthy and fungus-infected aphid prey. While consumption of healthy prey allows the predator to avoid infection, ingestion of infected prey results in fungal transmission to the predator. *Bottom panel*: An egg parasitoid evaluates larval hosts for oviposition. When encountering a fungus-infected larva, the parasitoid avoids oviposition to conserve energy and minimize the risk of infection. These avoidance behaviors highlight the selective pressures imposed by EPF on predator and parasitoid host-selection strategies

(Adeleke et al. 2022). Some parasitoid wasps might sense the signs of a fungal infection in their hosts and avoid ovipositing in them. Avoiding oviposition on an unhealthy host saves energy for the parasitoid, although forgetting to lay eggs on enough hosts can lead to population decrease (Figs. 13.1 and 13.2). Table 13.2 further explains the various EPFs and various other management strategies and their effects on the control of pests.

In some situations, interspecies relationships encourage cooperation. Natural predators can carry a fungus spore to a different location (Peng et al. 2021). Some predators gather EPF conidia and move to other areas where there are pests, and they aid in the spread of the sickness. The introduction of the *Amblyseius swirskii* mites, which fed on the *B. bassiana* conidia on infested crops, led to increased infections of thrips by the fungus compared to plants treated with *B. bassiana* alone. The thrips avoided contact with the colony, which allowed *A. swirskii* to attack and infect them, heavily impacting the Western flower thrips (Palmieri et al. 2022). On the other hand, the lower motile mite (*Neoseiulus cucumeris*) led to a weak number of conidia and lower effectiveness. As a result, the fungus spreads to new places in a scenario where a predator is in search of its prey. Sometimes, a predator attack may make a pest more vulnerable to diseases caused by other sources. However, their infections can be transported to other parts and individuals by parasitoids and predators. The studies have shown that predation makes thrips larvae more vulnerable to infection

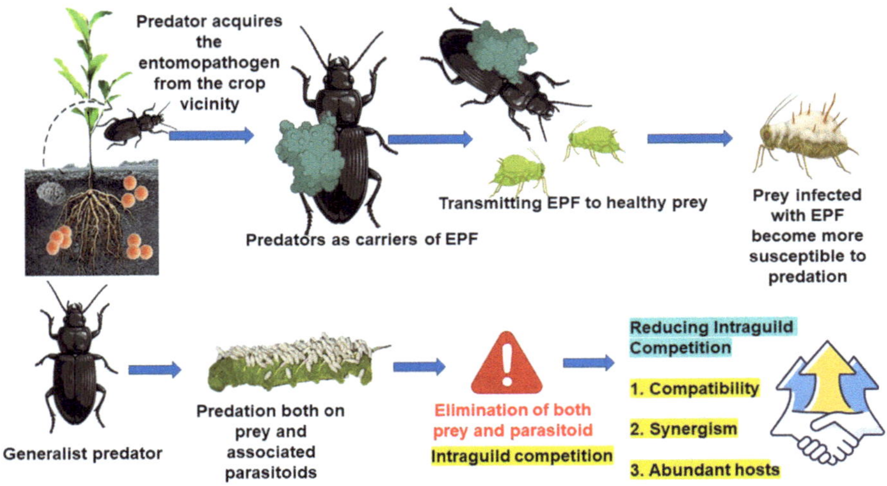

Fig. 13.2 Dual ecological role of generalist predators in entomopathogenic fungal dynamics and intraguild interactions. *Top panel*: Predators acquire entomopathogenic fungus from the crop environment and act as mechanical carriers, transmitting fungal spores to healthy prey. Infection with an entomopathogenic fungus increases prey susceptibility to predation. *Bottom panel*: Generalist predators may consume both herbivorous prey and their associated parasitoids, leading to intraguild predation and the elimination of both natural enemies. Reducing such intraguild competition relies on ecological compatibility, synergism among agents, and the availability of abundant hosts

with *B. bassiana* than their counterparts that have not been exposed to predators (Manzar et al. 2022).

At the same time, some ecological connections can be negative. The IGP may also operate as an antagonist: in case the generalist predator eats both the pest and the infected variant of the pest, it leads to the elimination of both the pest and the disease or the parasitoid (Attwood 2025). A predator that eats an insect with parasites eliminates the entire progeny of the parasites. In the event that these sessions are not monitored, it is possible that the end result of the discussions could be reduced. When there are many healthy prey, the chances of intraguild predation are unlikely because predators can use the alternative sources of food (Michalko et al. 2021). Intraguild predation could be reduced when parasitoids attack insect hosts at an early stage of development and make more resources available. The compatibility takes place when the species use the same habitats differently, and the synergy comes into play when there are vectors and complementary predators. Intraguild conflict should be reduced to decrease antagonism (Cicero et al. 2024).

Table 13.2 List of EPF along with other biocontrol agents, their mechanism of effectiveness, and their impact on insect pests

Entomopathogenic fungus	Other biocontrol agent	Mechanism	Impact	Reference
Metarhizium anisopliae	Chemical insecticide (Spinosad)	Synergistic enhancement of wireworm control via combined pathogenicity	Synergistic pest mortality	Koller et al. (2023)
Beauveria bassiana	Parasitoids (e.g., *Trichogramma pretiosum*)	Delayed EPF application minimizes parasitoid fitness disruption	Additive/ compatible control	Půža and Tarasco (2023)
Metarhizium anisopliae	Entomopathogenic nematodes (EPNs)	Antifungal compounds from nematode symbionts (e.g., *Photorhabdus*) inhibit EPF	Antagonistic under coapplication	Devi (2019)
Beauveria bassiana	EPNs	EPF-induced host stress increases susceptibility to nematode infection	Synergistic mortality in stressed hosts	Půža and Tarasco (2023)
Lecanicillium muscarium	Parasitoids (e.g., *Encarsia formosa*)	Optimal dosing avoids parasitoid developmental interference	Additive effect with proper timing	

3.2 Physiological and Developmental Interactions

The host experiences a lot of internal changes when infected with an EPF, and the host of interest is a preferred host of a parasitoid (Ma et al. 2024) (Fig. 13.3). Fungi have to compete with predators and parasitoids over the nutrients of the host, and changes in the tissue of the host might reduce the likelihood of survival. Sometimes, in case the fungal host dies, the predators might not develop to their full maturity, and this might starve them (Wang et al. 2024). A parasitoid wasp has to accelerate its development during the larval stage since the host of the fungus might reduce the chances of the wasp emerging. *B. bassiana* parasite may also cause the death of the caterpillar, as it has a high rate of mortality (Ghorui et al. 2024).

The researchers have noted that fungus-infected hosts often suppress the development of parasitoids. This degree of obstruction depends upon the type and strain of fungus, its concentration in the host, and the timing between fungal infection and larval parasitism by the parasitoid. Usually, a parasitoid that is infected as soon as it attacks its victim is not likely to survive, but the ones approaching pupation may survive. Therefore, some females do not lay their eggs in the already infected hosts and postpone oviposition until the conidia appear. In turn, violent conflict between people is reduced (Sui et al. 2024).

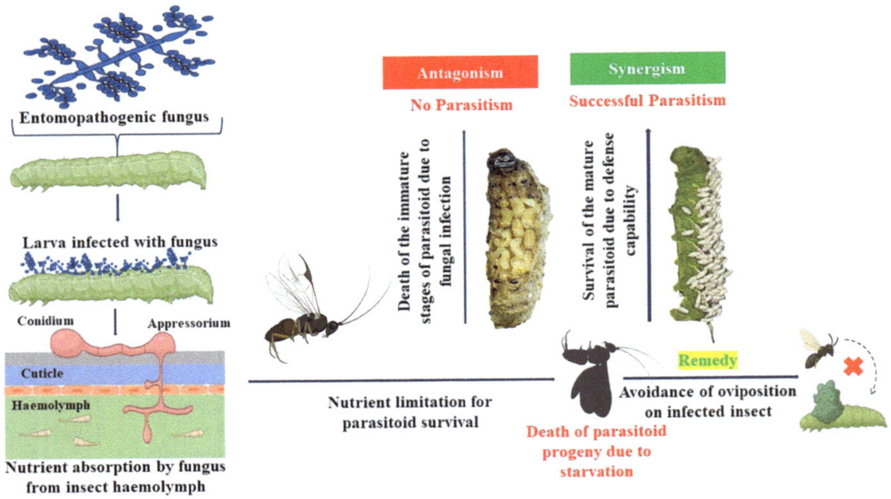

Fig. 13.3 Interaction outcomes between EPF and parasitoids within a shared host. Following fungal infection, the pathogen penetrates the insect cuticle via appressoria and absorbs nutrients from the host hemolymph, creating nutrient limitations for developing parasitoids. This interaction can lead to antagonism, where immature parasitoid stages die due to resource competition and fungal virulence. Alternatively, synergism may occur if mature parasitoids survive due to their defense capabilities. As a behavioral adaptation, parasitoids may avoid oviposition on fungus-infected hosts to prevent progeny starvation and ensure reproductive success

Alternatively, parasitoids could stimulate the fungus growth by influencing the host in certain ways. Once entering the host, the parasitoids use venom and produce various compounds to weaken the host and, therefore, make it more suitable for the advantage of the offspring (Hong et al. 2023). One immunosuppressed person can demonstrate a higher risk of getting another infection. After being infected by *H. hebetor*, the larvae of Galleria *mellonella* were characterized by greater vulnerability to infections by *B. bassiana*. After the attack of the parasitoid, the insect was more vulnerable to the fungal infection because of the changes in defense cells, natural antimicrobials, and microbiota of the gut (de Bekker et al. 2021). The wasp venom suppresses the immune system of the host, and germs multiply as the fungus takes over and kills the host. *Pimpla hypochondriaca* venom together with *B. bassiana* improves larval infections. An example of a parasitoid is that it seems to increase the vulnerability of the host toward the next fungus (Cerdeira and Brigagão 2024).

Besides, pests are vulnerable to many fungal infections, therefore making the work of their predators easier. Certain fungi are able to cause paucimobility and feeding inhibition in insects before fatal infection. A weak pest has limited mobility, making it vulnerable to attack by other organisms. Even though the situation is hard to put into quantitative terms, it could be a sign that a disease erodes resilience and makes a host more susceptible to attacks. As an example, small amounts of fungi (or similar compounds) can suppress the activity of pests and decrease their number to the point when it is easy to kill them by predators (Ma et al. 2024). The combination of the chemicals produces better results than when they are used separately in the control of the pests.

3.3 Molecular and Microbial Interactions

EPF, other biocontrol agents, and insects are connected to the immune system and some signals in the body. Toll, IMD, and melanization are some of the mechanisms that insects use to resist attacks by parasitoids and fungi (Makua 2021). The immune system can produce different results when one is infected with numerous viruses or diseases. When a *Helicoverpa* caterpillar is infected with *B. bassiana*, its Toll pathway is activated, thus triggering its immunological response to the infection (Aarthi et al. 2022). An environment parasitoid can defend its egg against predators by activating its immune system using its venom or viruses. The problem might be present without us noticing, hence the advantage of the fungus. There is an increasing number of scientists now studying the difference in gene expression between hosts in response to parasitoids. Transcriptomic examination of the *Pteromalus puparum* parasitizing *Pieris* caterpillars infected with *B. bassiana* indicates that different immune and developmental genes inside the wasp are modified by the fungus (Devi 2021).

After oviposition, the parasitoids allow their germs to enter the host, which may cause a certain impact on the physiological processes of the host. They can treat fungal infections that are at a microscopic level in organisms. An alteration in the intestinal microbiome occurred following exposure of the patient to *H. hebetor*. In some cases, the bacteria in the bite areas grew faster within the hosts as their immune systems were weakened; this often resulted in a severe infection, which was known as septicemia. Even though the fungus has a reduced impact on the host, it shows that the parasitoid has the ability to not only promote but also hinder the activity of the fungus in the host system. In case the bacteria in the gut cause overgrowth and produce antifungals, the result will be antagonistic; however, in case they just harm the host, it would be considered synergistic (Vivekanandhan et al. 2024).

It is also necessary to determine whether the fungal products in the same formulation cause any negative effects on the other type of biocontrol agents. Most EPF grow destruxins, beauvericin, and enzymes. Many pesticides are produced to kill insects but are nontoxic to plants and other animals; they can have a different effect on insects depending on the species. Some of the compounds produced by insects repel predators or lower oviposition by parasites. Most of the existing commercial EPF-based products are currently tested to make sure that they do not contain extremely harmful compounds to biocontrol insects (Shahbaz et al. 2024). EPF use on ladybird beetles, lacewings, and predatory mites has proved safe. In addition, several methodologies in molecular biology are being considered by the researchers to improve *B. bassiana*, i.e., genetically engineered *B. bassiana* to produce a toxin made by *Bacillus thuringiensis*, which makes the new strain deadlier to the larvae of mosquitoes. It allows the fungus and the Bt toxin to work simultaneously at the molecular level in the same organism. It is not a product of the fusion of multiple bacteria, yet it is one of the examples of advancement in the field of biocontrol by means of molecular usage (Ortiz-Urquiza and Keyhani 2016). The same group involves species that interact in different ways. Some parasites compete over the same habitats, others team up with one another in battles, some compete over nutriment, and others strengthen the immunological defense of the host. The data can be applied to create treatments that kill pests without destroying the beneficial chemicals used in the agricultural sector (Wang et al. 2021).

4 Global Examples of Combined Biocontrol in Multi-Crop Systems

Scientists have conducted test studies involving EPF against the predators, the parasitoids, or biocontrol organisms in greenhouses and fields around the world. This chapter defines some examples and case studies of how integrated assessment can work with alternative methods.

4.1 Greenhouse Vegetables

When there is an existence of whiteflies in the vegetable greenhouse, it has been the norm to eradicate them by using an insecticide. Whiteflies attack tomatoes, cucumbers, and ornamentals in greenhouses everywhere. Many farmers use the parasitic wasp, *Encarsia Formosa*, to control the whitefly nymphs. Ancora and Pfr-97 are fungal species *B. bassiana* and *I. fumosorosea*. They have worked well when applied in combination (Sani et al. 2020). When *I. fumosorosea* sprays are used alongside *E. formosa* wasp releases, the whiteflies are eliminated efficiently in the commercial greenhouse tomato production process without any harm. Since *Encarsia* feeds on nymphs, which are on the underside of the leaves, unlike sprays, which act on the ones on the upper side, it is unlikely that *Encarsia* will attack the ones that have been sprayed with chemicals (van Lenteren et al. 1996).

Field survey demonstrated that *B. bassiana* is not affected by parasitoids, *Encarsia* and *Delphastus*, which do not feed on whiteflies killed by fungal infection and prefer to avoid parasitized ones. The combination of *E. formosa* with *Lecanicillium muscarium* has similar or even better results than the respective applications, and it does not increase the parasitism level. Accordingly, in the event of numerous pests entering the greenhouse, integrated pest management (IPM) uses parasitoids and predators, along with the use of fungal biopesticides, to control the number of pests.

4.2 Field Crops

Caterpillars (Lepidopteran pests) are the most formidable pests in pulse, grain, and cotton fields, susceptible to parasitic organisms and illnesses. Caterpillars (Lepidoptera) are the most damaging pests in pulse, grain, and cotton fields and are vulnerable to parasitic organisms and diseases. Management of *Helicoverpa armigera* in chickpea production has been achieved. It is common to see farmers in various countries control *H. armigera* with the help of *Trichogramma*, *Habrobracon hebetor*, NPV, and *B. bassiana*. Researchers in Iran determined that a combination of an entomopathogenic fungus and a virus in a larval parasitoid would suppress *H. armigera* better than the use of either agent alone. Integrated use of both parasitoids and the virus led to an increase in the mortality of the larvae as compared to the use of either of the agents alone. Basically, the larvae were exposed to parasitoids, but other larvae were subjected to the virus, placing the pest outbreak dangerously near collapse. *Metarhizium anisopliae* is used in Asian environments to suppress the number of *Spodoptera* caterpillars with the help of *Trichogramma chilonis* and/or insects of the genus Bracon (Nozad-Bonab et al. 2021). To prevent attack by parasitoids on the control insects, fungicide ought to be applied immediately after the parasitoids start emerging. Field trials showed that the use of both insecticides reduced the number of bugs very fast. The tomato leafminer (*Tuta absoluta*) is already invading tomato crops in Africa, Europe, and Asia.

The larvae of *Dolichogenidea gelechiidivoris* and *M. anisopliae* were used by African laboratories to fight Tuta (Aigbedion-Atalor et al. 2020). In the study, mice died under both treatments in both laboratory and cage results; the use of one drug at a time also decreased the risk of death. When their methodologies were combined, it was also clear that their strategies surpassed their expectations, especially in the parasitoid. Although the fungus did not have much of a significant effect on increasing the life span of individuals after parasitoid attack, many dead Tuta showed that the approach would work in the control of Tuta in tomato farming. The strategy is becoming popular in business because the use of pesticides is not effective in controlling *T. absoluta* (Aigbedion-Atalor et al. 2021).

4.3 Orchards and Perennial Crops

Many pests can damage tree crops, and the examination of underground insects and the ones living on the tree's canopy is a must. Damage of date palm and coconut plantations is being caused by the red palm weevil (*Rhynchophorus ferrugineus*). The combined research has proven that *B. bassiana* or *M. anisopliae*, in combination with entomopathogenic nematodes (EPNs) that attack the eggs of weevils, could help the weevils emerge in palms and reduce the growth of their population. EPF and EPN integration have resulted in desirable developments (Alwaneen et al. 2024). It was observed that a higher mortality rate of weevil larvae was reached in the case of the combination of *B. bassiana* and *Heterorhabditis*, compared to the effect of each agent separately.

The dual action of *B. bassiana* and nematodes had a greater effect on larvae of early instar than the combination of *Metarhizium* and nematodes (Půža and Tarasco 2023). According to the concept, a fly or an insect infected by both fungi and nematodes or having been infected by non-lethal nematodes is more prone to the fungal infection. It has been reported that the combination of EPF and EPN has been found to be more effective against citrus root weevils and black vine weevils in strawberries than was expected. Several logistical issues are encountered with two living agents, as expounded below; however, these tests prove their fruitfulness against some of the stubborn pests within crops.

4.4 Cotton and Ornamentals

The sap of cotton, ornamental plants, and most crops, especially in Asia, Africa, and other places, is preferred by the solenopsis mealybug (sucking pest). In a recent study by the Egyptian researchers, the ladybird *Harmonia convergens* was combined with the new fungus *Alternaria murispora* to manage pests (El Aalaoui et al. 2025). The use of *A. murispora* and Harmonia almost eliminated the presence of mealybugs on the plants after 5 weeks of their application. Active plants were found in the places where mealybug predators and fungus were used simultaneously (Wainwright 2021). As a result, ladybirds eliminated mealybugs, which allowed fungal infection in the remaining mealybugs. The mealybugs developed into mummified sporulators after several weeks, and their predators had increased (Subramanian et al. 2021). It was shown that similar results can be achieved using the application of an EPF and a common predator, similar to chemical pest methods of control. Science has affirmed that the North American beetles, under the name of Harmonia, together with native fungi, are capable of fighting the Giant African Land Snail in combination (Pant and Dawadi 2025).

There are also a number of situations of interrelating pest management and crop protection in agriculture and chemical control. *Metarhizium* and other microorganisms have been studied as a means to control mosquitoes in rice fields. The two species (*B. bassiana* and *Anisopteromalus calandrae*) were used to control weevils in stored grain (Murugesan et al. 2024), but excessive use of *B. bassiana* seemed to reduce the effectiveness of the parasitoids; hence, low pest control was recorded when compared to using only *A. calandrae*. The diamondback moth's mortality was higher when stinkbugs (*Podisus*) were applied after the application of *B. bassiana* than when *B. bassiana* or stinkbugs were used alone (additive effect). Fungal infections, as well as Bt, have a number of applications. Plants treated with the combination of *B. bassiana* and Bt had a greater mortality rate among the caterpillars as compared to the use of either *B. bassiana* or Bt alone. The alteration of the midgut by the Bt and the fungus could possibly help the fungus to infiltrate fungi in the insect (Mamoon-ur-Rashid et al. 2025). Recent studies and literature reviews suggest that the integration of EPF with parasitoids or predators normally works without any hindrance, but it is hardly successful in pest control. The cases depict that such synergy between crops and regions is possible in a variety of places.

However, there is a need to carefully plan how to make use of their involvement to ensure that success is achieved. The following section describes critical steps that can support effective teamwork in this area.

5 The Manufacturing Processes, Applications, and Environmental Considerations

5.1 Composition and Distribution

The application of prepackaged plans and their deployment plays an important role in the control of biocontrol (Valantin-Morison et al. 2022). EPFs are sold in the form of wettable powders, oils, or granules in which the requisite fungal spores are in place. They can gain quite large volumes of living adults or host material, as in the case of parasitized eggs on cards in *Encarsia*. Generally, it is impossible to combine a number of chemicals in the equipment spray tank. A person can apply the pathogens predators, or parasitoids, to the foliage at the right time. There are times when some drugs are mixed together to achieve some therapeutic end. Some scientists have integrated fungi and entomopathogenic nematodes in a single liquid formulation to eliminate soil pests. Dry granular material that contains fungal spores and nematode infective juveniles has been produced as samples using an inert medium. On the contrary, they have a hard time maintaining their profitability. At the moment, people apply each product separately and, at the same time, take them together whenever needed (Hamrouni et al. 2025).

Its use with beneficial insects is shown in many sources by a wide range of manufacturers of commercial products of EPF. The safety of Metarhizium or other similar species, used as registered formulations and prior to use, is determined according to the recognized guidelines (Sharma et al. 2023). The oil products can have an adverse impact on some small parasitoids when there is an encounter, and hence, it is advisable to use a water-based solution when the parasitoids are in your garden. Care should also be exercised when choosing the adjuvants and the spreaders for the spray, as some of them can negatively impact animals or reduce the effectiveness of the spores. According to recent reports, efficient fungal control lowers the danger to predators (Bava et al. 2022). When Metarhizium is used to control mosquitoes, low doses in water kill small mosquito larvae and mostly leave large predatory mosquito larvae alive, but with reduced exposure. Use the recommended amount of fungus and strive to prevent any contact with the helpful insects in the process of treating your crops (Mahathir et al. 2025).

As it has been confirmed earlier, the transmission of information between the parasites and their hosts is natural. Alternatively, we can create databases of fungal conidia that predators or parasitoids can collect in the course of their search which will not be released until needed. Further research shows that the use of predatory mites could be a reliable way of pest control in the future.

5.2 Formulation, Applications, and Environmental Factors

5.2.1 Formulation and Delivery

The assembly and application of biocontrol agents take some practical skills. EPF fungi are mostly available in the form of wettable powders, oil emulsions, or granules impregnated with spores (Palmieri et al. 2022). Adults of predators and parasitoids may be supplied, and the materials of their host may be stolen, since they are replicated by the use of parasitized eggs on cards in the instance of Encarsia wasps. In this way, it is usually not possible to put all of the ingredients into the spray tank. One of the best applications of coapplication is the application of the fungus to the leaves and simultaneously releasing the predators or applying the fungus to the soil and simultaneously releasing the parasitoids in the upper region (Khandagale et al. 2023).

Several medicines are used during treatment in a number of situations. In some studies fungus has been mixed with entomopathogenic nematodes in a single liquid to treat soil pests. Inert medium containers of dry granular powders containing fungal spores and nematode infective juveniles have been prepared. Nevertheless, these companies do not find it so easy to continue making profits. Currently, one might expect to use each product separately, nonetheless, never forgetting about the necessity to use them in combination (Abd-Elgawad 2025).

5.2.2 Commercial Manufacturers Publish Their Products with Beneficial Insects

Most registered formulations of *B. bassiana*, *Metarhizium*, and other species are usually tested to determine their safety prior to their application as recommended (Behle and Birthisel 2023). Some of the oils can be directly detrimental to small parasitoids, and in gardens containing parasitoids, application of water-based solutions is recommended. Great care should also be taken in the selection of the adjuvants and spreaders used in spraying, since they may be harmful to the animals or less effective in the functioning of the spores. Research indicates that a small amount of fungus or confining it to a small space is likely to decrease the danger to the predators (McDonald-Howard 2023). Since Metarhizium is applied in controlling the mosquitoes, its application in low doses in water prevents the death of small larvae of the mosquitoes and also, to a greater extent, spares the predatory larvae with the long legs, as it reduces their exposure to the chemical. Use fungus in a non-hazardous quantity and attempt to avoid other beneficial bugs on your crops.

The other potential means of transmitting the pathogen is auto-dissemination by natural enemies. Alternatively, we can prepare cultures of fungal conidia, which predators or parasitoids can carry around on their travels, and then release them into the environment. This avenue has been proposed to be reliable in the near future by experimenting with predatory mites in greenhouses (Uzun Yiğit 2023).

5.2.3 Application Timing and Sequencing

Much research indicates that the time of taking medicine affects its absorption. An excellent style of preventing bad things is to have practitioners place their agents at varying times. Many authors say that the introduction of the parasitoid or predator early enough aids it to perform its role, and then the fungus ought to be introduced. It is because of this that the biocontrol insect has a low likelihood of being exposed to numerous pathogens at the beginning of the process (Ahmad et al. 2021). In case parasitoid wasps are the ones who take care of caterpillars, introduce the fungus when the largest part of the larvae have already become adults. According to Koller and others (2023), to maximize the effectiveness of biopesticides, one should apply it just before the introduction of parasitoid insects. When the fungus was used in large amounts or administered rapidly to the hosts, this tended to cause difficulties, as occasionally it would kill the hosts before the parasitoid reached them or would infect the parasitoid instead. This means that a combination is required: the two treatments would be effective when one of them is a fungus that is less harmful or dangerous to the host. Meanwhile, the rapid and extensive introduction of fungi can save the parasitoids from extinction because it transports them to new locations (Elkhateeb et al. 2021).

On other occasions, an alternative approach is embarked upon where fungi are used to reduce insect populations, and the introduction of predators eliminates the remaining ones. Given that the population of the pests is initially high, this technique can result in numerous copies of the fungus, but when a few of the pests escape detection, then the predators come in handy since they kill the pests. The use of medications should not be combined in most cases where their utilization is not safe. Some experiments have demonstrated that in case *B. bassiana* and predatory insects are applied simultaneously, aphid control is achieved (the release of predatory insects and applying *B. bassiana* at the same time did not decrease the number of aphids eaten). Field workers usually inspect the health of the plants and the level of pests and natural enemies, after which they apply the fungus on the plants when it is most active and when the presence of natural enemies is not an issue.

5.2.4 Environmental Factors

EPFs cannot be effective when used together with other biocontrol agents because they are sensitive to environmental conditions. Most EPF require external humidity and moisture to thrive and infect their hosts since their spores germinate on the exterior surface of the host (Ramkumar et al. 2025). Relative humidity should be maintained at its best (68–100%), and the white flies flourish at a temperature of 22–30 °C. In dry conditions, fungi cannot infect the host, yet predators and parasitoids may attack. Hence, the fungus will be on good terms with the plants when it is able to do its duties. Fungi will be well developed in the tropics or in extremely humid air during nights. Since the fungus in the fields is applied in the evening or by spraying water, it is effective. When the absence of moisture makes the fungus inactive, the work of the pest control is assigned to predators and parasitoids, and this state of affairs might be difficult for them (Quesada-Moraga et al. 2024). Simultaneously, strong rain can shift the soil or liquid in the plants and eliminate any spores or tiny bugs that are in the ground.

Occasionally, application of fungicide just before night's dew and before rains makes the plants stronger. The amount of shade in your planting mix also aids in the conservation of moisture, which can increase the outcomes. When temperatures increase, it causes all the organisms involved to accelerate their activities and metabolism. Standard EPFs are most effective when 20–30 °C is the temperature. There are optimal temperatures that offer the best growth and reproduction among predators and parasitoids (Qayyum et al. 2024). Hot weather decreases the life span of some of the parasitoid species, and also renders some fungi ineffective. In extremely cold areas, the growth of the fungus might be slow, and thus, the insects have a longer life; the predator might find it difficult to get sufficient food. They perform best in spring and fall, but when the outside weather is too hot, they lose most of their skills. As per the test results, the success of some treatments was not different at the different temperatures, e.g., the combination of a fungus and a predator proved highly effective at both test temperatures in a trial against mealybugs. In addition, the environment is also likely to be highly affected by higher temperatures (Gostinčar et al. 2022). The combination of fungi in warm regions and subsequent introduction of parasitoids works primarily because it eliminates the possibility of fungi being exposed to heat (by alternating between the two methods with time) (Jaronski 2023).

5.2.5 Light and UV Exposure

The sunlight and UV ray exposure can eliminate a lot of fungal spore spots, and one day of intense midday light could resolve numerous such spots (Khan et al. 2022). That is why fungi are more likely to be observed in the evening or during cloudy skies. Predators and parasitoids are bothered by bright sunlight; however, they are not as susceptible to intense UV radiation as are bacteria. The more there is UV radiation, the less probability of fungal growth, and most of the daytime, pest control is performed by predators. In the absence of light, the fungus would long survive and would further damage the moss due to the altered conditions. UV-protective spore fungicides may maintain the fungus in an active phase and boost the performance of other formulation agents over a more extended period (Zhao et al. 2022).

5.3 Host Density and Distribution

The population of an insect may be larger, and this aspect could worsen the problems of pests. Fungi can find more hosts to spread fast, i.e., fungi can prosper in areas with more pests. Nonetheless, the fewer parasitoid eggs could decrease their level of control of pests during population outbreaks. In some cases, the fungus prevents the initial phase of the epidemic, and then the parasitoids take over and eliminate the survivors by completing the work—an efficient work division (Angon et al. 2023). In the event of a reduction in the number of pests in the environment, fungi might struggle, but some predators and parasitoids may be able to complete their job, and that is another mechanism of complementarity (Perumal et al. 2024). It is good to know where the pests have their habitat: some of the pests hiding among the leaves may be difficult to locate and treat, or the predators may not get access to them easily; on the other hand, pests that are easily accessible may be treated easily.

Having different agents is advantageous since an individual can identify issues that other people could not (Oberemok et al. 2023; Khulbe and Batra 2024).

5.3.1 Crop and Canopy Structure

The way you cluster plants in the garden can enhance or lessen the probability of different species thriving (Qayyum et al. 2021). The canopies of tropical fruit trees and vine crops are dense and humid, such that several pests take refuge and fungi grow readily. There are areas in which conidia application is not optimal that may attract predators. This is beneficial to the parasitoids since they are able to locate their prey. You should therefore make sure that the nutrients in the mixture are suitable for the crop you are cultivating. Applying fungicide underground can inhibit the growth of fungus in the roots, and the use of parasitoids by spraying them through the canopy eliminates borers in the corn stalks (Gonzalez-Guzman et al. 2021a).

5.3.2 Interactions with Other IPM Tactics

In IPM, biocontrol is occasionally augmented with the use of plants with natural defenses, pheromone traps, or low-level plant insecticides. Some things can transform the fungus or the objects that keep it at bay. An excellent case is the mixture of a moderate amount of neem extract along with *B. bassiana* and a predator, which can have a subtle impact on the behavior of insects and the growth of fungi. There needs to be strategies to conserve *B. bassiana* and predators, assuming that the agents are properly employed (many botanicals and mineral oils are frequently tolerated by them). The concept of biorational IPM is based on the premise that none of the methods should impair each other or reduce their performance (Quesada-Moraga et al. 2022).

The impact on the environment and the use of the products should be considered in order to achieve synergy. The use of biological agents in campaigns is also, in many cases, misguided, and as a result, biocontrol is proven to fail in practical tests. Accordingly, the application of parasitoids provided farmers and workers with motivation to develop usable principles: spray fungicides during favorable weather, wait when parasitoids are active, select pesticides that do not kill beneficial insects, and monitor any changes in the moisture content of the air to anticipate the occurrence. These elements can be made to work better by modern methods such as predictive modeling and controlled drug delivery.

6 Difficulties, Constraints, and Considerations

Even though multi-agent biocontrol is attractive, there are some problems that must not be ignored.

6.1 Potential Incompatibilities

At certain times, not all kinds of EPFs and biocontrol agents are well compatible. When fungus attacks the beneficial insects, it might get into their bodies or cause them to become ill enough to die (Rajula et al. 2021). As it was already mentioned, when the conditions are favorable, some fungi (Beauveria and Metarhizium) may attack various insects, both the ones we like and the ones we do not want to have. In some cases, *B. bassiana* can also go after ladybird pupae or lacewing larvae in high concentrations, but this does not occur in the field often (Saldanha et al. 2022).

Laboratory tests have revealed that some insecticides may be toxic to the predators or parasitoids by killing them as well as inhibiting their egg-laying and shortening their life. At that, the strain should be selected, which will affect a limited number of reactions or, simply speaking, inflict fewer harms on the plant we are breeding. Parasitoids have the ability to conceal or kill fungi that accompany their eggs, thus potentially lessening the utility of fungi (Halder et al. 2021).

6.2 Timing and Life-History Trade-Offs

When the events occur at an inappropriate time, a synergistic combination may be turned against itself. In case fungi destroy the plant prior to the infection of the parasitoids, they may kill the hosts or infect them, thus forcing the parasitoids to leave the plant (Erdos et al. 2024). Findings of the semi-field experiment indicated that *B. bassiana* was not very effective in combating grain pests compared to parasitoids on their own. Due to the subsequent research, it was discovered that the application of fungal beta on plants eliminated the infected pests prior to the introduction of wasp parasitoids and made the adult wasps ill, contributing to the decrease of their population. It implies that not all combinations of a host, parasitoid, and fungus can work harmoniously. Without proper use, agents may conflict and leave the outcomes still less effective than those that one agent can produce individually (Mangan et al. 2024).

6.3 Environmental Constraints

To perform properly, fungi require certain conditions, and in case they are not met, e.g., everything is too dry, a predator or a parasitoid must replace their role. In case the mouse cannot solve the problem by itself, then the mouse can run away before it is caught. In case of dry weather where there are not many predators to deal with, relying on *B. bassiana* may make the fungus less useful in pest management. Agriculture is vulnerable because farmers cannot confidently predict the weather. Fungus-insect combinations in spring and fall and other methods in summer or following a drought are most likely to yield the best results (Quesada-Moraga et al. 2024).

6.4 Operational Complexity

To be able to work properly, fungi require certain conditions and, in the case where these conditions are not maintained, e.g., when things are too dry, a predator or a parasitoid is needed to take their place. In case the mouse cannot manage the issue on its own, there is a possibility that the mouse has time to run away before being captured. When the climate is dry and there are not many predators, reliance on *B. bassiana* may lead to the fungus being ineffective in the control of pests. The agricultural industry is vulnerable because farmers cannot accurately predict the weather. Fungus-insect combinations in spring and fall and other methods in summer or after a drought are likely to work the best (Quesada-Moraga et al. 2024).

6.4.1 Regulatory and Compatibility Data

All the biocontrol agents, including fungi, insects, or nematodes, are required to pass a clearance before their use at farms. The mixture is hardly regulated by new policies in case a manufacturer acts according to the instructions of the supplier because it has independent approvals (Sain et al. 2022). Certain microbial pesticides should be tested on their impact on the honeybees and other such beings, but this is not necessary in all forms of beneficial insects that are used in biological pest control. Individuals seek advice in books or turn to other people who have gone through the same problem. The weakness of this method is in the fact that it does not take all the possible mixtures into account. Very unusual combinations of animals are not always determinable as to their compatibility. In other cases, parasitoid specialists test whether a freshly collected fungus will have an impact on the population of their parasitoids prior to their application (Sharma et al. 2023).

6.4.2 Unpredictable Field Results

Controlled experiments do not necessarily yield the same encouraging outcomes in the open setting (Sharma et al. 2023). The method can have an excellent outcome in experiment A and minimal or no outcomes in experiment B. This may occur due to the number of pests, their lifestyle, the different means of applying pesticides, or natural variations. Thus, multi-agent techniques may be considered unreliable by a great number of people in the agriculture industry. Although EPF, along with predator control, appears to be highly effective in the controlled field trials, in the open fields, the fungus might not spread sufficiently to obtain the same effects as those obtained with the use of predators alone. Individuals who study this topic attempt to understand more about these differences by conducting experiments on the release of agents as well as in situations that are more challenging to cooperate (Dearlove 2022).

6.4.3 Risk of Mutual Interference

Although the researchers give so much importance to the fungi that kill the predators and parasitoids, predators have their way of killing the hosts, which act as carriers of the disease. In the case where a predator kills most of the diseased insects, the fungus will be considered ineffective in case it fails to infect more of the insects. Application of the small snail as a pest control is beneficial, but it may prevent natural fungal infections (Bava et al. 2022). By feeding on its prey, a predator reduces the fungus's reproduction opportunity. In case the fungus is unable to decompose the predator due to a drop in its population, there is a likelihood that the pest may be reintroduced. It is uncommon but must be considered when fungal recycling seems probable, as in outbreaks of mass forest pests. As a possible solution, it can be recommended to give each muscle a break when the other one is in action (Sun et al. 2023).

6.4.4 Target Specificity Mismatch

The mixture is most preferred when the active components target the same pests. Failure to apply any agent properly may hurt other living organisms or fail to get rid of other pests. In the event that a predator keeps the populations of local insects at bay and a fungicide is applied to a pest, the predators might become hungry since some of their primary foods are killed with the application of the fungicide. Since the target pests have different spreading patterns, then agents should be assigned in a matched manner. Very many case studies only look at a limited number of pest species, but it is difficult to guarantee that all pests coexist harmoniously in polyculture.

Sharing control of insects with a number of biocontrol agents involves a lot of planning to prevent the possibility of unanticipated results. It can be difficult to manage them when the time is wrong, the environment is not favorable, or when the insects under management are too different. Therefore, despite the numerous valuable findings, the application of such combinations in agricultural protection programs remains uncommon. Moreover, the observation of such challenges has led to the development of new strategies for improved future development (Stiling and Cornelissen 2005). The strategy of biocontrol through fungi and other natural predators is never stagnant. Research and development can help us get superior results through synergetic strategies in some crucial areas.

6.4.5 Increase Field Validation and Long-Term Trials

It is best to have both compounds targeting the same pest or group of pests in order to have combined control. Failure to target any agent properly may either kill useful organisms or leave some more minor pests untouched. In a situation where a fungicidal solution is used to contain a pest and a predator in the neighborhood feeds on insects, the fungicide may reduce the population of insects, and this may lead to the predator starving to death. Distribution of the pest influences the outcome, so there should be a proper pairing of the agents with the appropriate skills. The majority of the case studies select one pest species, and in polyculture, it is difficult to maintain that all the pests remain satisfied with one another (Van Lenteren et al. 2018).

In the case of the application of EPFs in combination with other biocontrol agents, it is important to plan the process carefully in order to prevent any potential difficulties. An attempt to control insects at the wrong time of year, under poor conditions, or with species/strains improperly matched may present additional problems. This implies that, although there exist numerous useful results, the usage of such combinations is not a widespread phenomenon in the areas of crop protection. The knowledge of these concerns, in its turn, has contributed to the development of new methods to enhance the outcomes (De Clercq et al. 2011).

6.4.6 Optimize Timing and Application Protocols

This is to come up with ways or mechanisms that will be able to select the most appropriate times, order, and quantity of multiple treatments to be simultaneously applied. Based on this system, an individual could determine when to use the fungus correctly and minimize the occurrence of the pests and still not kill the parasitoid. In experiments, the researchers would be able to investigate what happens when fungus is applied before parasitoids and vice versa, as well as evaluate the effects of introducing them simultaneously or in different periods. Another alternative is to analyze the current information (such as spore traps and natural enemies) and determine whether the environment is conducive to disease before the employment of products (Gonzalez-Guzman et al. 2021b).

6.4.7 Strain/Species Selection and Breeding

The effectiveness of the enemy strains against the controlled entities can be different. In the long run, researchers can examine a large number of fungi to determine which of them are harmful to the desired pest yet have no effect on its natural predators. Selecting or developing fungi that are less toxic or slow down the hosts could assist the parasite to avoid detection and, nevertheless, keep pests at bay. Moreover, selecting strains of parasitoids or predators that are strongly resistant to fungi, either in their construction or in their behavior, would make them compatible with bees. Individuals may attempt to develop ladybird strains that have reduced chances of being infected by Beauveria. It is also done by genetic modification and selective breeding; scientists are incorporating the features of other biocontrol agents or symbiotic viruses into fungi. New concepts on how to perform a number of actions are provided by the transgenic biocontrol strategies despite the regulations. The investigations on the EPF strains that lead to the diseases of plants and the utilization of natural air enemies could provide an entirely new approach, and the inner features of EPF play a significant role (Fenton et al. 2002).

6.4.8 Mechanistic Studies on Interactions

Studying at the molecular level and physiological level will lead to better effects of mixtures. Genomics, transcriptomics, and metabolomics laboratory experiments can demonstrate the response of insects to pests and insecticides and whether one antagonistic agent can change the outcome due to another. In the event that researchers identify a major immunological cause of the antagonism of the body, then they can strive to develop methods or medications that would be effective regardless of the antagonism. Insects could harbor infections; that can be determined by looking at the microbes in the insects and testing the impact of the parasitoid venom in the gut. It would be helpful to investigate whether the interactions of the plant defense system and the chemo-immunome are harmful or beneficial to each combination of the two (Khachatourians and Qazi 2008).

6.4.9 Innovative Delivery Systems

Future studies over the next couple of years can enhance the introduction of blends to geology. Still, as an example, the fungi and predator nutrition may be microencapsulated, and the gel materials may be used to distribute the fungi gradually and keep the plant moist. We could take advantage of predators by spraying them with a spore spray, or we could use predator species that harbor the required beneficial bacteria. Parasitoids and fungi can be applied directly to regions of a field using drones and new methods, potentially enhancing their efficacy in controlling the particular pests that are present. This reduces the likelihood of infection of the parasitoids by delaying the time they release fungus (Butt et al. 2016).

6.4.10 Multi-Agent and Multi-Tactic Integration

It is understood that in the future the integrated pest management (IPM) may be based on the utilization of numerous organisms and their modes, e.g., predators, parasitoids, fungi, nematodes, and even at times viruses. Figure 13.4 shows the incorporation of EPF usage along with other strategies of pest management and their effect on the control of the pests (Table 13.3). A study of the way various species in a community relate to other species at other trophic levels is hard. It demands the development of the ecological model and the conducting of organized experimentation (Kaçar et al. 2023). Moreover, the application of these biological agents and other sustainable agriculture approaches is greatly valued. Incorporation of fungi in the trap crop and inclusion of parasitoids in the main crop could enhance the efficacy of repellents and attractants. Since each new piece adds some complexity to the system, the analysis should aim at determining ways of making the system less complex and areas where a slight change has significant effects (Kennedy and Lekshmi 2022).

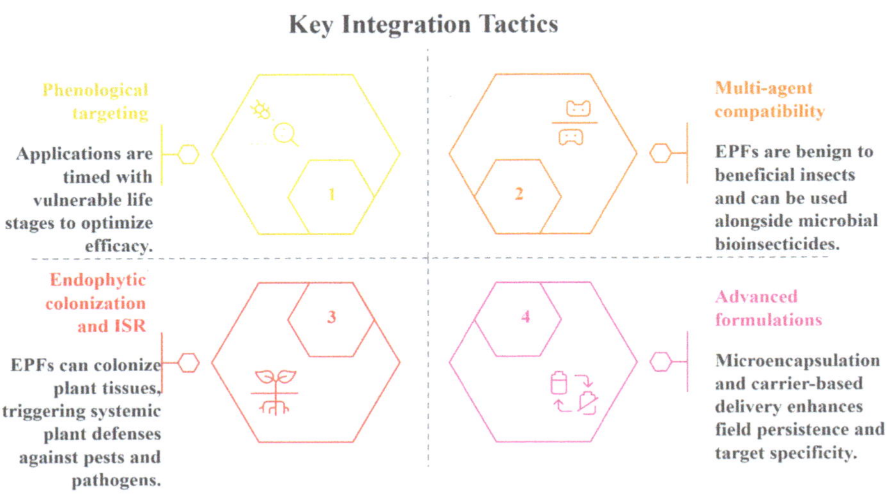

Fig. 13.4 Impact of entomopathogenic fungi with other tactics of pest management

Table 13.3 Impact of entomopathogenic fungi and other pest management tactics on insect pest management in different crops

Targeted insect	Crop	Entomopathogenic fungus	Other tactics	Reference
Weevils (*Curculionidae* spp.)	Nut trees, fruit crops, storage roots, palms	*Beauveria* spp./ *Metarhizium* spp.	Habitat timing (apply when larvae are active in soil/ foliage); "attract-and-kill" baits; combined with BT toxins; use as an endophyte in plants; compatibility with parasitoids/predators	Maniania et al. (2017)
Whitefly (*Bemisia tabaci*)	Vegetables, ornamentals, papaya	*Beauveria bassiana*, *Isaria fumosorosea*	EPF colonization as plant endophyte	Sani et al. (2020)
			Encourages induced systemic resistance (ISR)	
			Compatible with natural enemies	
			Formulated with adjuvants	
Aphids (*Aphis gossypii*, others)	Papaya, general horticulture	*Beauveria*, *Lecanicillium lecanii*, *Penicillium* and *Fusarium* spp.	Enhances ISR	Mukherjee et al. (2020)
			Delivered as a root-colonizing inoculant	
			Integrated with other beneficial microbes	
Locusts/ grasshoppers/ beetles	Pasture, sugarcane	*Metarhizium anisopliae*, *M. robertsii*	Soil drenching with conidia	Zembrzuski (2023)
			Compatible with grazing/burn cycles	
			Microencapsulation for longevity	
Soil-dwelling scarab grubs (*Melolontha* spp.)	Forestry/field crops	*Beauveria brongniartii*	Application via colonized carriers (e.g., barley kernels)	Graf (2024)
			Compatible with parasitoid wasps	
			Long-term soil persistence	

7 Climate Change and New Pest Invasions

Changes in the partnerships between fungi, insects, and other biocontrol agents will be due to climate change. Rising temperatures can reduce the efficiency of fungi and this might allow the spread of certain pests and predators to new regions. Future researchers can explore ways of applying biocontrol in other climatic conditions, either by selecting resistant strains or by timing their interventions. Moreover, the increase in the network of international trade also reveals new pests, such as *Tuta absoluta* and *Phenacoccus solenopsis*, which have emerged within the past few decades. It will be extremely significant to develop the means of controlling the new invasive species. The available knowledge regarding past pest problems might be required in dealing with current pest problems today (Quesada-Moraga et al. 2024). Such principles could be identified by comparing the results of numerous studies; e.g., "In the case of pests that can breed rapidly, a rapid-acting pathogen and a slow-growing parasitoid are most effective to use together" (Skendžić et al. 2021).

7.1 Farmer Training and Implementation

The study findings are very important to use. Investigators should get to know the attitude of farmers toward the multi-agent integrated pest management and determine the issues they encounter when applying it. The use of scientific concepts in practice, the cost analysis, and simple guide preparation are steps to assist farmers (Faria et al. 2023). An application in the hands of a cultivator might suggest, "In the event of applying parasitoid X in week 1, plan to apply fungus Y in week 3, provided weather conditions are A, B, or C." When a relationship between some activity and crop outcomes is observed, it is more likely that scientists notice it and investigate further (Maina et al. 2018).

8 Conclusion and Perspectives

There is the possibility of improving the management of pests with the help of entomopathogenic fungi and other biocontrol agents. Learning about your successes and your failures can increase the output of your future strategies. Our goal is to achieve integrated pest management by having the different natural enemies cooperate as they do in nature to hold the pest insects in check with reduced application of pesticides. To make this successful, individuals and experts working in the fields of microbiology, entomology, agronomy, and climate science should collaborate. Based on our study in the laboratory and field sites, we propose that the integration of these fungi with predators, parasitoids, and other fauna might be the new trend in the control of ecological pests.

References

Aarthi HP, Shashikala M, Ramya AR (2022) Behavioral and physiological response of insects against microbial pest control strategies. In: Naresh RK (ed) Research trends in agriculture sciences, vol 33. AkiNik Publications, New Delhi, pp 3–18

Abd-Elgawad MM (2025) Integrated nematode management strategies: optimization of combined nematicidal and multi-functional inputs. Plants 14(7):1004. https://doi.org/10.3390/plants14071004

Adeleke BS, Ayilara MS, Akinola SA, Babalola OO (2022) Biocontrol mechanisms of endophytic fungi. Egypt J Biol Pest Control 32(1):46. https://doi.org/10.1186/s41938-022-00547-1

Ahmad G, Khan A, Khan AA, Ali A, Mohhamad HI (2021) Biological control: a novel strategy for the control of the plant parasitic nematodes. Antonie Van Leeuwenhoek 114(7):885–912. https://doi.org/10.1007/s10482-021-01577-9

Aigbedion-Atalor PO, Mohamed SA, Hill MP, Zalucki MP, Azrag AG et al (2020) Host stage preference and performance of *Dolichogenidea gelechiidivoris* (Hymenoptera: Braconidae), a candidate for classical biological control of *Tuta absoluta* in Africa. Biol Control 144:104215. https://doi.org/10.1016/j.biocontrol.2020.104215

Aigbedion-Atalor PO, Hill MP, Ayelo PM, Ndlela S, Zalucki MP et al (2021) Can the combined use of the mirid predator *Nesidiocoris tenuis* and a braconid larval endoparasitoid *Dolichogenidea gelechiidivoris* improve the biological control of *Tuta absoluta*? Insects 12(11):1004. https://doi.org/10.3390/insects12111004

Alwaneen WS, Tahir M, Avery PB, Wakil W, Kavallieratos NG et al (2024) Initial evaluation of the entomopathogenic fungi *Beauveria bassiana* and *Metarhizium robertsii*, and the entomopathogenic nematode *Heterorhabditis bacteriophora*, individually and in combination against the noxious *Helicoverpa armigera* (Lepidoptera: Noctuidae). Agronomy 14(7):1395. https://doi.org/10.3390/agronomy14071395

Angon PB, Mondal S, Jahan I, Datto M, Antu UB et al (2023) Integrated pest management (IPM) in agriculture and its role in maintaining ecological balance and biodiversity. Adv Agric 2023(1): 5546373. https://doi.org/10.1155/2023/5546373

Attwood MC (2025) How do parasites and predators choose their victim? A trade-off between quality and vulnerability across antagonistic interactions. Biol Rev Camb Philos Soc 100:2099–2115. https://doi.org/10.1111/brv.70037

Baker BP, Green TA, Loker AJ (2020) Biological control and integrated pest management in organic and conventional systems. Biol Control 140:104095. https://doi.org/10.1016/j.biocontrol.2019.104095

Bava R, Castagna F, Piras C, Musolino V, Lupia C et al (2022) Entomopathogenic fungi for pests and predators control in beekeeping. Vet Sci 9(2):95. https://doi.org/10.3390/vetsci9020095

Behle R, Birthisel T (2023) Formulations of entomopathogens as bioinsecticides. In: Morales-Ramos JA, Rojas MG, Shapiro-Ilan DI (eds) Mass production of beneficial organisms. Academic Press, pp 407–429. https://doi.org/10.1016/B978-0-12-822106-8.00010-5

Butt TM, Coates CJ, Dubovskiy IM, Ratcliffe NA (2016) Entomopathogenic fungi: new insights into host-pathogen interactions. Adv Genet 94:307–364. https://doi.org/10.1016/bs.adgen.2016.01.006

Cerdeira CD, Brigagão MR (2024) Targeting macrophage polarization in infectious diseases: M1/M2 functional profiles, immune signaling, and microbial virulence factors. Immunol Investig 53(7):1030–1091. https://doi.org/10.1080/08820139.2024.2367682

Cicero L, Chavarín-Gómez LE, Pérez-Ascencio D, Barreto-Barriga O, Guevara R et al (2024) Influence of alternative prey on the functional response of a predator in two contexts: with and without intraguild predation. Insects 15(5):315. https://doi.org/10.3390/insects15050315

de Bekker C, Beckerson WC, Elya C (2021) Mechanisms behind the madness: how do zombie-making fungal entomopathogens affect host behavior to increase transmission? MBio 12(5): e01872-21. https://doi.org/10.1128/mbio.01872-21

De Clercq P, Mason PG, Babendreier D (2011) Benefits and risks of exotic biological control agents. BioControl 56:681–698. https://doi.org/10.1007/s10526-011-9372-8

Dearlove EL (2022) Multiple stressor effects in biological pest control: improving efficacy in challenging environments. Doctoral dissertation, University of Warwick, United Kingdom

Devi G (2019) Interaction between entomopathogenic nematodes and entomopathogenic fungi in biocontrol mechanism. J Entomol Zool Stud 7(6):959–964

Devi G (2021) İnsect defense system and immunosuppression strategies of entomopathogenic nematodes—an overview. Int J Environ Agric Biotechnol 6(3):69–77. https://doi.org/10.22161/ijeab.63.7

El Aalaoui M, Rammali S, Kamal FZ, Calin G, Rarinca V, Hritcu LD, Ciobică A, Sbaghi M (2025) Effectiveness of the entomopathogenic fungus *Alternaria murispora* and the predatory ladybird Harmonia convergens in controlling *Phenacoccus solenopsis* (Tinsley). Front Sustain Food Syst 8:1469247. https://doi.org/10.3389/fsufs.2024.1469247

Elkhateeb WA, Mousa KM, ELnahas MO, Daba GM (2021) Fungi against insects and contrariwise as biological control models. Egypt J Biol Pest Control 31:1–9. https://doi.org/10.1186/s41938-020-00360-8

Erdos Z, Studholme DJ, Sharma MD, Chandler D, Bass C, Raymond B (2024) Manipulating multi-level selection in a fungal entomopathogen reveals social conflicts and a method for improving biocontrol traits. PLoS Pathog 20(3):e1011775. https://doi.org/10.1371/journal.ppat.1011775

Faria M, Mascarin GM, Butt T, Lopes RB (2023) On-farm production of microbial entomopathogens for use in agriculture: Brazil as a case study. Neotrop Entomol 52(2):122–133. https://doi.org/10.1007/s13744-023-01033-5

Fenton A, Gwynn RL, Gupta A, Norman R, Fairbairn JP, Hudson PJ (2002) Optimal application strategies for entomopathogenic nematodes: integrating theoretical and empirical approaches. J Appl Ecol 39(3):481–492. https://doi.org/10.1046/j.1365-2664.2002.00727.x

Francis F, Jacquemyn H, Delvigne F, Lievens B (2020) From diverse origins to specific targets: role of microorganisms in indirect pest biological control. Insects 11(8):533. https://doi.org/10.3390/insects11080533

Ghorui M, Chowdhury S, Burla S (2024) The science behind entomopathogenic fungi: mechanisms and applications. In: Deshmukh SK, Sridhar KR (eds) Entomopathogenic fungi: prospects and challenges. Springer Nature Singapore, Singapore, pp 3–35. https://doi.org/10.1007/978-981-97-5991-0_1

González-Guzmán A, Sánchez-Rodríguez AR, Quesada-Moraga E, del Campillo MC, Yousef-Yousef M (2021a) Optimizing wheat seed treatment with entomopathogenic fungi for improving plant growth at early development stages. Span J Agric Res 19(4):e1004. https://doi.org/10.5424/sjar/2021194-17120

Gonzalez-Guzman A, Raya-Diaz S, Sacristán D, Yousef M, Sánchez-Rodríguez AR et al (2021b) Effects of entomopathogenic fungi on durum wheat nutrition and growth in the field. Eur J Agron 128:126282. https://doi.org/10.1016/j.eja.2021.126282

Gostinčar C, Zalar P, Gunde-Cimerman N (2022) No need for speed: slow development of fungi in extreme environments. Fungal Biol Rev 39:1–14. https://doi.org/10.1016/j.fbr.2021.11.002

Graf T (2024) The control of invasive and native scarab beetles with the entomopathognic fungi *Metarhizium brunneum* and *Beauveria brongniartii*. Doctoral dissertation, University of Zurich, Switzerland

Halder J, Divekar PA, Rani AT (2021) Compatibility of entomopathogenic fungi and botanicals against sucking pests of okra: an ecofriendly approach. Egypt J Biol Pest Control 31:1–7. https://doi.org/10.1186/s41938-021-00378-6

Hamrouni R, Regus F, Farnet Da Silva AM, Orsiere T, Boudenne JL, Laffont-Schwob I, Christen P, Dupuy N (2025) Current status and future trends of microbial and nematode-based biopesticides for biocontrol of crop pathogens. Crit Rev Biotechnol 45(2):333–352. https://doi.org/10.1080/07388551.2024.2370370

Hong S, Sun Y, Chen H, Zhao P, Wang C (2023) Fungus–insect interactions beyond bilateral regimes: the importance and strategy to outcompete host ectomicrobiomes by fungal parasites. Curr Opin Microbiol 74:102336. https://doi.org/10.1016/j.mib.2023.102336

Iqbal M, Shaheen FA, Mahmood R, Rafique MK, Bodlah I et al (2019) Synergistic effect of entomopathogenic fungi and bacteria against pulse beetle, *Callosobruchus* chinensis. Pak J Zool 51(5):1685–1691. https://doi.org/10.17582/journal.pjz/2019.51.5.1685.1691

Jaronski ST (2023) Mass production of entomopathogenic fungi—state of the art. In: Morales-Ramos JA, Rojas MG, Shapiro-Ilan DI (eds) Mass production of beneficial organisms,

invertebrates and entomopathogens. Academic Press, pp 357–413. https://doi.org/10.1016/B978-0-12-391453-8.00011-X
Jeffers A, Chong JH (2021) Biological control strategies in integrated pest management (IPM) programs. Clemson University Cooperative, Land-Grant Press Clemson Extension, LGP, 1111, pp 1–9
Kaçar G, Butrón A, Kontogiannatos D, Han P, Peñaflor MF, Farinós GP et al (2023) Recent trends in management strategies for two major maize borers: *Ostrinia nubilalis* and *Sesamia nonagrioides*. J Pest Sci 96(3):879–901. https://doi.org/10.1007/s10340-023-01595-8
Kennedy JS, Lekshmi JK (2022) Holistic pest management strategies in tropical plant species. In: Khan MS (ed) Tropical plant species and technological interventions for improvement. IntechOpen, pp 1–27. https://doi.org/10.5772/intechopen.105104
Khachatourians GG, Qazi SS (2008) Entomopathogenic fungi: biochemistry and molecular biology. In: Brakhage AA, Zipfel PF (eds) Human and animal relationships, The Mycota, vol 6. Springer, Berlin, Heidelberg, pp 33–61. https://doi.org/10.1007/978-3-540-79307-6_3
Khan M, McDonald M, Mundada K, Willcox M (2022) Efficacy of ultraviolet radiations against coronavirus, bacteria, fungi, fungal spores and biofilm. Hygiene 2(3):120–131. https://doi.org/10.3390/hygiene2030010
Khandagale PP, Padsala J, Landge SA, Siddhapara MR, Patel PR (2023) Effective augmentative release of natural enemies and agro-ecosystem management in integrated pest management. Biol Forum 15(12):425–430
Khulbe A, Batra P (2024) Insect-pests and diseases in greenhouse cultivation and their biological control. In: Singh MC, Sharma KK (eds) Protected cultivation. Apple Academic Press, pp 217–254
Koller J, Sutter L, Gonthier J, Collatz J, Norgrove L (2023) Entomopathogens and parasitoids allied in biocontrol: a systematic review. Pathogens 12(7):957. https://doi.org/10.3390/pathogens12070957
Lacey LA (ed) (2016) Microbial control of insect and mite pests: from theory to practice. Academic Press, p 461. https://doi.org/10.1016/C2015-0-00092-2
Ma M, Luo J, Li C, Eleftherianos I, Zhang W, Xu L (2024) A life-and-death struggle: interaction of insects with entomopathogenic fungi across various infection stages. Front Immunol 14: 1329843. https://doi.org/10.3389/fimmu.2023.1329843
Mahathir AMBM, Hadibarata T, Jusoh MNHB, Tan IS, Tan YH (2025) Biological control of insect pests using entomopathogenic enzymes: opportunities and challenges. Biointerface Res Appl Chem 15(2):1–21. https://doi.org/10.33263/BRIAC152.022
Maina UM, Galadima IB, Gambo FM, Zakaria D (2018) A review on the use of entomopathogenic fungi in the management of insect pests of field crops. J Entomol Zool Stud 6(1):27–32
Makua MS (2021) The immune response of *Sirex noctilio* to pathogen infection. Doctoral dissertation, University of Pretoria, South Africa
Mamoon-ur-Rashid M, Baloch MS, Begum HA, Law D, Elsadek MF et al (2025) Supplementary bioactivities of Curcuma longa extracts and parasitoid *Anisopteromalus calandrae* as a postharvest eco-friendly management tactic against maize weevil, *Sitophilus zeamais* (Motschulsky) (coleoptera: Curculionidae). Entomol Res 55(1):e70013. https://doi.org/10.1111/1748-5967.70013
Mangan R, Tinsley MC, Ferrari E, Polanczyk RA, Bussiere LF (2024) Exploiting pathogen defence trade-offs to manage risks of crop pests evolving resistance to biocontrol. bioRxiv. https://doi.org/10.1101/2024.10.19.619212
Maniania NK, Ekesi S, Dolinski C (2017) Entomopathogens routinely used in pest control strategies: orchards in tropical climate. In: Lacey LA (ed) Microbial control of insect and mite pests, from theory to practice. Academic Press, pp 269–282. https://doi.org/10.1016/B978-0-12-803527-6.00018-4
Manzar N, Kashyap AS, Goutam RS, Rajawat MVS, Sharma PK et al (2022) *Trichoderma*: advent of versatile biocontrol agent, its secrets and insights into mechanism of biocontrol potential. Sustainability 14(19):12786. https://doi.org/10.3390/su141912786

McDonald-Howard K (2023) Investigations into the efficacy of the slug and snail biological control *Phasmarhabditis hermaphrodita*. Doctoral dissertation, Liverpool John Moores University, United Kingdom

Michalko R, Uhrinec M, Khum W, Sentenská L (2021) The benefits of intraguild predation for a top predator spider. Ecol Entomol 46(2):283–291. https://doi.org/10.1111/een.12960

Mukherjee A, Debnath P, Ghosh SK, Medda PK (2020) Biological control of papaya aphid (*Aphis gossypii* Glover) using entomopathogenic fungi. Vegetos 33:1–10. https://doi.org/10.1007/s42535-019-00072-x

Murugesan P, Krishanan A, Ramasamy P, Chandrasekaran K, Rajendran J et al (2024) Biorational methods for effective pest control management in stored products for agricultural sustainability. Entomol Res 54(1):e12697. https://doi.org/10.1111/1748-5967.12697

Nawaz A, Razzaq F, Razzaq A, Gogi MD, Fernández-Grandon GM et al (2022) Compatibility and synergistic interactions of fungi, *Metarhizium anisopliae*, and insecticide combinations against the cotton aphid, *Aphis gossypii* Glover (Hemiptera: Aphididae). Sci Rep 12(1):4843. https://doi.org/10.1038/s41598-022-08841-6

Nozad-Bonab Z, Hejazi MJ, Iranipour S, Arzanlou M, Biondi A (2021) Lethal and sublethal effects of synthetic and bio-insecticides on *Trichogramma brassicae* parasitizing *Tuta absoluta*. PLoS One 16(7):e0243334. https://doi.org/10.1371/journal.pone.0243334

Oberemok VV, Gal'chinsky NV, Useinov RZ, Novikov IA, Puzanova YV et al (2023) Four most pathogenic superfamilies of insect pests of suborder sternorrhyncha: invisible superplunderers of plant vitality. Insects 14(5):462. https://doi.org/10.3390/insects14050462

Ortiz-Urquiza A, Keyhani NO (2016) Molecular genetics of *Beauveria bassiana* infection of insects. Adv Genet 94:165–249. https://doi.org/10.1016/bs.adgen.2015.11.003

Palmieri D, Ianiri G, Del Grosso C, Barone G, De Curtis F et al (2022) Advances and perspectives in the use of biocontrol agents against fungal plant diseases. Horticulturae 8(7):577. https://doi.org/10.3390/horticulturae8070577

Pant J, Dawadi P (2025) Effective control of invasive grape mealybugs (*Maconellicoccus hirsutus*) involves cautious release of ladybird beetles in variable temperature parameters. J Appl Entomol 5(1):01–06

Peng Y, Li SJ, Yan J, Tang Y, Cheng JP et al (2021) Research progress on phytopathogenic fungi and their role as biocontrol agents. Front Microbiol 12:670135. https://doi.org/10.3389/fmicb.2021.670135

Perumal V, Kannan S, Pittarate S, Krutmuang P (2024) A review of entomopathogenic fungi as a potential tool for mosquito vector control: a cost-effective and environmentally friendly approach. Entomol Res 54(3):e12717. https://doi.org/10.1111/1748-5967.12717

Půža V, Tarasco E (2023) Interactions between entomopathogenic fungi and entomopathogenic nematodes. Microorganisms 11(1):163. https://doi.org/10.3390/microorganisms11010163

Qayyum MA, Saeed S, Wakil W, Nawaz A, Iqbal N et al (2021) Diversity and correlation of entomopathogenic and associated fungi with soil factors. J King Saud Univ Sci 33(6):101520. https://doi.org/10.1016/j.jksus.2021.101520

Qayyum MA, Saeed S, Wakil W, Sharif U, Taha H et al (2024) Entomopathogenic fungi: prospects and challenges. In: Deshmukh SK, Sridhar KR (eds) Entomopathogenic fungi. Springer, Singapore, pp 57–79. https://doi.org/10.1007/978-981-97-5991-0_3

Quesada-Moraga E, Garrido-Jurado I, Yousef-Yousef M, González-Mas N (2022) Multitrophic interactions of entomopathogenic fungi in BioControl. BioControl 67(5):457–472. https://doi.org/10.1007/s10526-022-10163-5

Quesada-Moraga E, González-Mas N, Yousef-Yousef M, Garrido-Jurado I, Fernández-Bravo M (2024) Key role of environmental competence in successful use of entomopathogenic fungi in microbial pest control. J Pest Sci 97(1):1–15. https://doi.org/10.1007/s10340-023-01622-8

Rajula J, Karthi S, Mumba S, Pittarate S, Thungrabeab M et al (2021) Current status and future prospects of entomopathogenic fungi: a potential source of biopesticides. In: De Mandal S, Passari AK (eds) Agricultural and industrial approach. Recent advancement in microbial biotechnology. Elsevier, pp 71–98. https://doi.org/10.1016/B978-0-12-822098-6.00013-6

Ramkumar, Chaudhary M, Sahu P, Singh K, Kumar R et al (2025) Effect of humidity on pest and disease incidence in crops. In: Singh A, Pandey S, Kumar A (eds) Climate change and biotic factors. Apple Academic Press, pp 3–42

Roca-Couso R, Flores-Félix JD, Rivas R (2021) Mechanisms of action of microbial biocontrol agents against *Botrytis cinerea*. J Fungi 7(12):1045. https://doi.org/10.3390/jof7121045

Sain SK, Monga D, Kumar R, Nagrale DT, Hiremani NS, Kranthi S (2019) Compatibility of entomopathogenic fungi with insecticides and their efficacy for IPM of *Bemisia tabaci* in cotton. J Pestic Sci 44(2):97–105. https://doi.org/10.1584/jpestics.D18-067

Sain SK, Monga D, Kranthi S, Hiremani NS, Nagrale DT et al (2022) Evaluation of the bioefficacy and insecticide compatibility of entomopathogens for management of whitefly (Hemiptera: Aleyrodidae) on upland cotton under laboratory and polyhouse conditions. Neotrop Entomol 51(4):600–612. https://doi.org/10.1007/s13744-022-00964-9

Saldanha MA, Costa EC, Muniz MFB, Quevedo AC (2022) Compatibility levels of fipronil with entomopathogenic fungi. Rev Bras Cienc Agrar 17(2):e1392. https://doi.org/10.5039/agraria.v17i2a1392

Sani I, Ismail SI, Abdullah S, Jalinas J, Jamian S et al (2020) A review of the biology and control of whitefly, *Bemisia tabaci* (Hemiptera: Aleyrodidae), with special reference to biological control using entomopathogenic fungi. Insects 11(9):619. https://doi.org/10.3390/insects11090619

Shahbaz M, Palaniveloo K, Tan YS, Palasuberniam P, Ilyas N et al (2024) Entomopathogenic fungi in crops protection with an emphasis on bioactive metabolites and biological activities. World J Microbiol Biotechnol 40(7):217. https://doi.org/10.1007/s11274-024-04022-x

Sharma A, Sharma S, Yadav PK (2023) Entomopathogenic fungi and their relevance in sustainable agriculture: a review. Cogent Food Agric 9(1):2180857. https://doi.org/10.1080/23311932.2023.2180857

Skendžić S, Zovko M, Pajač Živković I, Lešić V, Lemić D (2021) Effect of climate change on introduced and native agricultural invasive insect pests in Europe. Insects 12(11):985. https://doi.org/10.3390/insects12110985

Smagghe F, Spooner-Hart R, Chen ZH, Donovan-Mak M (2023) Biological control of arthropod pests in protected cropping by employing entomopathogens: efficiency, production and safety. Biol Control 186:105337. https://doi.org/10.1016/j.biocontrol.2023.105337

Stiling P, Cornelissen T (2005) What makes a successful biocontrol agent? A meta-analysis of biological control agent performance. Biol Control 34(3):236–246. https://doi.org/10.1016/j.biocontrol.2005.02.017

Subramanian S, Boopathi T, Nebapure SM, Yele Y, Shankarganesh K (2021) Mealybugs. In: Omkar (ed) Polyphagous pests of crops. Springer Singapore, pp 231–272. https://doi.org/10.1007/978-981-15-8075-8_5

Sui L, Zhu H, Wang D, Zhang Z, Bidochka MJ et al (2024) Tripartite interactions of an endophytic entomopathogenic fungus, Asian corn borer, and host maize under elevated carbon dioxide. Pest Manag Sci 80(9):4575–4584. https://doi.org/10.1002/ps.8163

Sun R, Hong B, Reichelt M, Luck K, Mai DT et al (2023) Metabolism of plant-derived toxins from its insect host increases the success of the entomopathogenic fungus *Beauveria bassiana*. ISME J 17(10):1693–1704. https://doi.org/10.1038/s41396-023-01480-3

Uzun Yiğit A (2023) Auto-dissemination of *Cordyceps fumosorosea* amongst adult females of the two-spotted spider mite. Exp Appl Acarol 91(2):279–290. https://doi.org/10.1007/s10493-023-00845-9

Valantin-Morison M, Lasserre-Joulin F, Martinet V, Meiss H, Messéan A, Meynard JM, Rouabah A (2022) Integrating biocontrol into cropping system design. In: Fauvergue X, Rusch A, Barret M, Bardin M, Jacquin-Joly E et al (eds) Extended biocontrol. Springer Netherlands, Dordrecht, pp 233–244. https://doi.org/10.1007/978-94-024-2150-7_20

van Lenteren JC, van Roermund HJ, Sütterlin S (1996) Biological control of greenhouse whitefly (*Trialeurodes vaporariorum*) with the parasitoid *Encarsia formosa*: how does it work? Biol Control 6(1):1–10. https://doi.org/10.1006/bcon.1996.0001

Van Lenteren JC, Bolckmans K, Köhl J, Ravensberg WJ, Urbaneja A (2018) Biological control using invertebrates and microorganisms: plenty of new opportunities. BioControl 63:39–59. https://doi.org/10.1007/s10526-017-9801-4

Vivekanandhan P, Swathy K, Sarayut P, Patcharin K (2024) Biology, classification, and entomopathogen-based management and their mode of action on *Tuta absoluta* (Meyrick) in Asia. Front Microbiol 15:1429690. https://doi.org/10.3389/fmicb.2024.1429690

Wainwright H (2021) Biological control of pests. In: Cut flowers and foliages. CABI, pp 316–341

Wang H, Peng H, Li W, Cheng P, Gong M (2021) The toxins of *Beauveria bassiana* and the strategies to improve their virulence to insects. Front Microbiol 12:705343. https://doi.org/10.3389/fmicb.2021.705343

Wang X, Yuan Q, He L, Wang Z, Li G et al (2024) Biological and physiological effects in Bemisia tabaci feeding on tomatoes endophytically colonized by *Beauveria bassiana*. Pest Manag Sci 80(8):4085–4097. https://doi.org/10.1002/ps.8121

Zembrzuski D (2023) Connecting the nutritional ecology and physiology of *Melanoplus sanguinipes* to immunology under *Metarhizium robertsii* infection. Doctoral dissertation, Arizona State University, Arizona State University, USA

Zhao HX, Zhang TY, Wang H, Hu CY, Tang YL, Xu B (2022) Occurrence of fungal spores in drinking water: a review of pathogenicity, odor, chlorine resistance and control strategies. Sci Total Environ 853:158626. https://doi.org/10.1016/j.scitotenv.2022.158626

Entomopathogenic Fungi as Biocontrol Agents

14

Abhishek Rana, Jeer Vinayaka, Karthik Ramappa, C. V. Nandeesha, Saurabh Dubey, Imtinungsang Jamir, Ravinder Singh Chandel, Abhishek Katoch, and Pooja Kapoor

Abstract

Entomopathogenic fungi (EPF) play an important role in integrated pest management due to their potential to infect and kill large numbers of insect pests. Important EPF being used are *Beauveria bassiana*, *Metarhizium anisopliae*, *Lecanicillium lecanii*, and *Hirsutella thompsonii*, known for their pathogenicity through spore adhesion, cuticle penetration, and systemic colonization. Host specificity varies among species and strains, enabling targeted control with minimal impact on non-target organisms. EPF have been successfully applied

A. Rana (✉)
Krishi Vigyan Kendra, Gopalganj, Dr. Rajendra Prasad Central Agricultural University, Pusa, Samastipur, Bihar, India

J. Vinayaka
Krishi Vigyan Kendra, Saran, Dr. Rajendra Prasad Central Agricultural University, Pusa, Samastipur, Bihar, India

K. Ramappa · R. S. Chandel
Department of Entomology, CSK Himachal Pradesh Krishi Vishvavidyalaya, Palampur, Himachal Pradesh, India

C. V. Nandeesha
Department of Plant Pathology, Junagadh Agricultural University, Junagadh, Gujarat, India

S. Dubey
Krishi Vigyan Kendra, Madhopur, West Champaran, Dr. Rajendra Prasad Central Agricultural University, Pusa, Samastipur, Bihar, India

I. Jamir
Krishi Vigyan Kendra, Lada, Dr. Rajendra Prasad Central Agricultural University, Pusa, Samastipur, Bihar, India

A. Katoch · P. Kapoor
Flavorite Hydroponics Tomatoes, Shepparton, VIC, Australia

S. K. Deshmukh et al. (eds.), *Frontiers in Entomopathogenic Fungi*,
https://doi.org/10.1007/978-981-95-6529-0_14

in various field conditions against pests in crops, forests, and stored products, either through direct spraying or soil application. The compatibility of EPF with other pest management strategies, such as chemical pesticides, parasitoids, and cultural practices, is vital for integrated pest management (IPM) programs. However, their efficacy can be affected by abiotic factors like temperature, humidity, and UV radiation. Challenges such as limited persistence, slow action compared to chemicals, and difficulties in large-scale production and formulation remain significant hurdles. Recent advancements in biotechnology, formulation technology, and strain improvement offer promising prospects for enhancing fungal virulence and environmental tolerance. Future research should focus on improving field stability, understanding fungal ecology, and integrating EPF with other sustainable approaches to formulate a strong IPM.

Keywords

Compatibility · Field application · Insects · Toxins · Virulence

1 Introduction

Insect pests present a significant global challenge to ecosystem stability and food security. At present, insecticides are the predominant solution for controlling these insect pests. The prolonged applications of these insecticides have led to the development of risks to non-target organisms, insect resistance, and harmful effects on the environment. All these concerns have led to an increased interest in eco-friendly alternatives for pest management strategies (Abrol and Shankar 2015; Islam 2017).

Biological control is one such alternative that is being used worldwide for managing harmful insect pests. Fungal biological control is one such tool, which involves the use of EPF to manage the pest population to prevent crop loss. Fungal entomopathogens are naturally occurring microorganisms that parasitize insects. The word “entomopathogen” is composed of two Greek words, “entomon” (insects) and “genes” (arising in). Therefore, the etymological meaning of “entomopathogenic fungi” is the microorganisms that arise in insect pests. The use of biocontrol agents for managing plant pathogens was first explored in the mid-1920s (Choudhary and Johri 2009). Biocontrol agents offer several advantages, including high efficacy, safety for beneficial insects, minimal environmental residual effects, and the promotion of biodiversity in ecosystems (Shahid et al. 2012; O’Brien 2017). About 750 species have been reported to infect various species of pests (Shapiro-Ilan et al. 2017). These are parasitic microorganisms from various groups, including Oomycetes, Chytridiomycota, Microsporidia, Entomophthoromycota, Basidiomycota, and Ascomycota, with the capability to infect and kill hosts (Litwin et al. 2020). The most important EPF include *Beauveria* Vuill., *Cordyceps* (*Isaria*), *Hirsutella* Pat., *Metarhizium* Sorokin, and *Nomuraea* (Humber 2012). EPF have a wide host range targeting different stages of insects, i.e., from the immature to the adult stage (Mantzoukas and Eliopoulos 2020). All developmental stages of insects, including eggs, larvae, pupae, nymphs, and adults, can be infected by EPF (Bitencourt et al. 2021). These fungi produce spores on the host's surface, grow outside for some time, enter the body, and lead to the death of the host after complete infection (Brivio and Mastore 2020). EPF acquire a unique mode of action as they make their entry inside their hosts through the cuticle, whereas other microorganisms, such as bacteria and viruses, require ingestion to exert their effects (St Leger and Wang 2010). Additionally, EPF can control insect pests by producing various extracellular enzymes that effectively degrade the insect's integument (Ortiz-Urquiza and Keyhani 2013; Mondal et al. 2016). After infecting and killing the host, the fungus continues to generate new spores on the dead body, which then disperse and perpetuate the life cycle on new hosts (Jaber and Ownley 2018). However, a detailed knowledge of EPF is essential for effectively implementing fungi-mediated biological control.

This chapter provides an overview of the occurrence, diversity, and pathogenic mechanisms of EPF, including their modes of action and toxin production. It also explains the compatibility of different species of EPF with other pest management approaches and the practical aspects of field application. Additionally, this chapter discusses the benefits, ecological limitations, and recent advances in EPF-based biological control.

2 Pathogenicity of Entomopathogenic Fungi

EPF exhibit a unique and distinctive mechanism of action, invading the insect body through the cuticle or mouthparts (Ortiz-Urquiza and Keyhani 2013). The cuticle serves as the initial barrier between the fungi and their nutritional source within the hemocoel (Kaczmarek et al. 2020). Insect death results from a combination of factors, including tissue disruption due to hyphal invasion, nutrient depletion, and the production of various mycotoxins within the insect's body (Gul et al. 2014). EPF have developed several mechanisms for detecting host surfaces, adhesion, and penetration (Sinha et al. 2016). After adhesion, hyphal growth is followed by a series of enzymatic activities involving different enzymes (Ortiz-Urquiza and Keyhani 2013; Gabarty et al. 2014). Structural adaptations, such as appressoria formed by fungal hyphae, along with mechanical pressure from the hyphae, enhance the penetration process (Butt et al. 2016). In addition to immune responses, the strategies used by the insects to defend against pathogens include the production of antimicrobial proteins, lipids, and metabolites in the epicuticle; shedding of old cuticle during stage metamorphosis; and raising body temperature to adapt to changing environmental conditions (Wang 2018). Together, these components lead to the development of a defense system in an insect's body before infection. This ongoing battle between EPF and host insects has driven complex evolutionary adaptations on both sides (Lu and St Leger 2016).

2.1 Adhesion and Conidial Germination

Spores adhere to the surface of the host insect through passive movement via air and hydrophobic interactions (Holder and Keyhani 2005). Spores can germinate at various points on the cuticle, although specific sites have been observed on different insects (Ortiz-Urquiza et al. 2016). Holder et al. (2007) observed that desiccated spores of *Beauveria bassiana* have hydrophobin-coated hydrophobic rodlets on the outermost cuticular layer of *Heliothis virescens*. This hydrophobic layer is specific to the conidial stage and absent in the mycelial phase. Dry spores attach to the cuticle through hydrophobic forces created by these rodlets (Bidochka et al. 1995). In *B. bassiana*, *Hyd1* and *Hyd2* are two hydrophobin proteins that are responsible for the proper functioning of surface rodlets, which contribute to surface hydrophobicity and virulence (Cho et al. 2007). Wang and St Leger (2007) identified two adhesion genes, *Mad1* and *Mad2*, in *Metarhizium anisopliae*. *Mad1* genes are involved in the passive adhesion and adsorption process, while *Mad2* is more related to the active adhesion stage (Ortiz-Urquiza et al. 2013).

Upon attachment to the host's surface, spores begin to germinate, and their growth is significantly influenced by different factors like oxygen, temperature, pH, and nutrients. There are other toxins produced by the host cuticle that affect the growth and further spread of spores. Fungi having a wide host range can infect their hosts in the presence of a variety of nonspecific carbon and nitrogen sources (St Leger et al. 1989). However, according to Ortiz-Urquiza and Keyhani (2013), some entomopathogenic fungi need specific compounds for completion of their life cycle and have a limited host range. Napolitano and Juárez (1997) suggested that insect hydrocarbons, such as aliphatic and methyl-branched alkanes, are essential for the growth of *B. bassiana*.

2.2 Cuticle Penetration

The main body of the cuticle, the procuticle, consists of chitin fibrils embedded in a protein matrix, along with quinones and lipids (Xing and Yang 2020). The entomopathogenic fungus secretes a suite of hydrolytic enzymes, including chitinases, proteases, and lipases, which degrade the chitin, protein, and lipid components of the cuticle. Since more than 70% of the insect cuticle is composed of protein, endoproteases play a crucial role in the penetration of the cuticle (Charnley 2003). For lipid digestion of the cuticle, *B. bassiana* requires eight cytochrome P450 genes (CYP52-X1, CYP655-C1, CYP5337-A1, CYP52-G11, CYP539-B5, CYP617-N1, CYP53-A26, and CYP584-Q1), encoding enzymes, as well as four catalases (catB, catP, catC, and catD), aldehyde dehydrogenases (ADH-2, FALDH), long-chain alcohol (3-oxoacyl carrier protein reductase), and a recognized hydrocarbon carrier protein (Acyl-CoA oxidase) to participate (Pedrini et al. 2010, 2013).

The fungus enters the cuticle of the insect by forming a germ tube, which differentiates into an appressorium, a specialized structure that exerts physical force on the cuticle (Ortiz-Urquiza and Keyhani 2013; Gabarty et al. 2014). The appressorium is likely an adaptation that concentrates chemico-physical energy in a small area to facilitate efficient entry. The formation of the appressorium is influenced by the topography of the host surface (Soliman 2020), as well as by cyclic AMP (cAMP) and intracellular second messengers like Ca^{2+} (Ryder and Talbot 2015).

2.3 Invasion of the Hemocoel and Toxin Production

Once fungi penetrate the insect's body, they enter the hemocoel, the open circulatory system of the host. Here, it transforms into a yeast-like phase, producing blastospores that proliferate rapidly in the nutrient-rich hemolymph. In this stage, the growth of the fungus interferes with various functions of hemocytes, such as encapsulation and phagocytosis, that contribute to the insect's immune system.

EPF often produce secondary metabolites, or mycotoxins, that disrupt host physiology. Mitosporic fungi produce a wide variety of mycotoxins that negatively impact insect health (Armendariz et al. 2017). Cytotoxins disrupt cellular structures and facilitate hyphal penetration (Sheppard and Filler 2015). Neuromuscular toxins affect insect health by causing general paralysis, reduced irritability, and sluggishness (Sandhu et al. 2012). Several toxins, such as beauverolides, beauvericin, isarolides, and brassinolide, have been isolated from *B. bassiana*. (Eyal et al. 1994).

2.3.1 Destruxins

Metarhizium anisopliae produces destruxins and cytochalasins in infected hosts. Destruxins can activate calcium channels, leading to depolarization of the lepidopteran muscle membrane, and they also interfere with the function of insect hemocytes (Golo et al. 2014). These toxins were first isolated from *M. anisopliae* (Kodaira 1961). More than 28 distinct destruxins have been isolated, primarily from *Metarhizium* species, each exhibiting various activity levels (Peng et al. 2005). The concentration of destruxins has been linked to virulence and host specificity (Mukherjee and Vilcinskas 2018). Studies on destruxin activity have shown that they cause deregulation of the insect's cellular immune system, preventing nodule formation (Peng et al. 2005) and disrupting phagocytosis (Pal et al. 2007). Their immunosuppressive effects have also been observed in insects due to the production of destruxins (Haron et al. 2020). Although the complete mechanism remains unclear, recent research investigated the effect of destruxins on free Ca^{2+} and hydrogen (H^+) ions in the hemocytes of *Exolontha serrulata*, *Bombyx mori* and *Spodoptera litura* using laser scanning confocal microscopy (LSCM) (Chen et al. 2014). The findings demonstrated that destruxins-A (DTX-A) and destruxins-B (DTX-B) triggered a rapid influx of Ca^{2+} into the hemocytes. However, the channel inhibitors 2-APB (2-aminoethoxydiphenyl borane) and U73122 had no impact on this influx. Similarly, DTX-A (690 μmol/L) and DTX-B (20 μmol/L) induced a notable decrease in intracellular free H^+ (Chen et al. 2014). These results suggest that the primary role of destruxins is to aid the establishment of the pathogen in host tissues by altering its ion channels, which may ultimately lead to insect death following infection.

2.3.2 Beauvericin

Beauvericin is extracted from *Fusarium semitectum*, *Fusarium moniliforme*, *Beauveria bassiana*, *Cordyceps fumosorosea*, and certain plant pathogenic fungi from Basidiomycota (Wang and Xu 2012; Qasim et al. 2018). Beauvericin tends to form complexes with other cations (Bertero et al. 2020). It disrupts cells and causes toxicity in certain insects, like mosquitoes, butterflies, and potato beetles (Luz et al. 2017). Beauvericin induces cell apoptosis by increasing oxidative stress through the generation of reactive oxygen species and by interfering with mitochondrial and hormonal signaling pathways (Qadri et al. 2011). Similar to destruxins, beauvericin also modulates the influx of Ca^{2+} into the cells. The increasing levels of Ca^{2+} induced by beauvericin may lead to cytotoxicity.

2.3.3 Bassianolide

Bassianolide is a harmful metabolite derived from *Lecanicillium lecanii* (Ravindran et al. 2018) and *B. bassiana* strains, often found in the cadavers of *Bombyx mori* pupae (Xu et al. 2009). Chemically, bassianolide is a cyclic polymer of d-α-hydroxyisovaleryl L-N-methylleucinol. Like other depsipeptides, it functions as an ionophore (Kanaoka et al. 1978). When injected into *B. mori* larvae at a dose of 2 μg/larva suspended in water induce atonic symptoms. Recent studies have shown that bassianolide produced by *Lecanicillium muscarium* can impact acetylcholine receptors in insect muscles, leading to a reduction in the production of acetylcholine-E in whiteflies (Ali et al. 2017; Villegas-Mendoza et al. 2019).

2.3.4 Leucinostatins

Leucinostatins are primarily isolated from *Purpureocillium lilacinum*, *Paecilomyces marquandii*, and *Cordyceps farinosa* (Wang et al. 2016). Chemically, they consist of l-threo-β-hydroxyleucine, 2-amino-6-hydroxy-4-methyl-8-oxodecanoic acid (AHMOD), and cis-4-methyl-L-proline (Mori et al. 1982). Leucinostatins possess antimicrobial activity against various fungi and Gram-positive bacteria (Fukushima et al. 1983). This toxin can also affect mice, exhibiting both intraperitoneal and oral toxicity. The pure isolation of leucinostatins from *C. farinosa* extracts has demonstrated notable insecticidal effects against the Colorado potato beetle (Gökce et al. 2006). Additionally, leucinostatins are involved in oxidative phosphorylation within mitochondria and show nematicidal activity (Park et al. 2004) and anti-trypanosomal properties (Ishiyama et al. 2009).

2.4 Sporulation and Transmission

After the host dies, the fungus emerges from the cadaver, producing new conidia on the surface of the corpse. This external sporulation facilitates dispersal to new hosts, completing the infection cycle. The ability to sporulate prolifically under favorable conditions enhances the pathogen's potential for epizootics in insect populations.

3 Major EPF and Their Applications for Insect Pest Control

Their growing use in pest management is driven by their direct toxicity to a broad range of insect pests and the reduced environmental, health, and ecological risks associated with their application (Sabbour et al. 2012). Additionally, many EPF can colonize plants as endophytes, exerting indirect pest control by inducing the production of toxic plant metabolites or altering plant volatile organic compound profiles. These changes influence the tripartite interactions between host plants, insect pests, and their natural enemies (Francis et al. 2022). An estimated 700–800 species of EPF are mainly distributed within Ascomycota (order Hypocreales) (Naranjo-Ortiz and Gabaldón 2019), but only a few stand out for their commercial use, wide host range, ease of culture, and effectiveness. These fungi, which typically belong to Ascomycota, are commonly found in soil or associated with infected insects in natural ecosystems (Chet et al. 1993; Li et al. 2001). Worldwide, many species belonging to Hypocreales have been successfully formulated and commercialized for large-scale pest control (Bamisile et al. 2021) (Table 14.1). Among the significant genera exhibiting entomopathogenic properties, genera that have been mostly intensively investigated for microbial insecticides include *Beauveria*, *Metarhizium*, *Verticillium*, *Paecilomyces*, *Hirsutella*, *Erynia*, *Nomuraea*, *Aspergillus*, *Aschersonia*, *Tolypocladium*, *Entomophthora*, and others.

The application of EPF in field settings necessitates a comprehensive understanding of formulation science, delivery mechanisms, and ecological interactions. The efficacy of EPF in field applications is critically dependent on the formulation, which serves to protect and enhance the viability and virulence of fungal propagules (typically conidia). Key considerations in EPF formulation include the following.

Viability and Stability: EPF conidia are susceptible to environmental stressors such as UV radiation, desiccation, and temperature extremes. Formulations must ensure conidial viability during storage and post-application in the field. Studies indicated that oil-based formulations can offer superior protection against UV radiation compared to aqueous suspensions (Batta 2003).

Table 14.1 Common EPF and their target insect pests

Entomopathogenic fungus	Target pest
Beauveria bassiana	European corn borer, gram pod borer, white grubs, thrips, weevils, whiteflies, aphids and termites
Cephalosporium lecanii	Coffee green bug
Entomophthora virulenta	Mustard aphid
Hirsutellathompsonii	Eriophid mites
Lecanicillium spp.	Whiteflies, thrips and aphids
Mearhizium anisopliae	White grub, termites, beetles and grasshoppers
Verticillium lecanii	Aphids and white flies

Adhesion and Spreading: An effective formulation enhances the potential of EPF to infect their host by successful adhesion of its conidia to the cuticle. Adjuvants, such as surfactants and oils, are often incorporated to improve spreading and adhesion on hydrophobic insect surfaces.

Dispersibility: The formulation should facilitate the uniform dispersal of conidia in the target environment. This is particularly important when applications are to be conducted aerially, where droplet size and distribution influence the probability of insect-fungus contact.

Compatibility: Formulations must be compatible with application equipment and, in some cases, with other pesticides being used in integrated pest management systems.

Common formulation types employed and their scientific rationale include the following.

Wettable Powders (WPs): These are dry formulations that require suspension in water. WPs offer relatively good stability and are suitable for a range of application methods.

Oil-based Formulations: Low humidity conditions are often encountered in the field, and oil-based formulations can enhance conidial persistence and virulence under such conditions. Studies have shown improved efficacy of *Metarhizium* spp. when applied in oil formulations, likely due to increased cuticle penetration and reduced desiccation (Bukhari et al. 2011).

Emulsions: These formulations combine the benefits of oil and water, allowing for efficient application and enhanced conidial stability.

These formulations are being applied directly to target pests under field conditions or endophytically. In endophytic colonization strategies, plant species can be inoculated through various methods, including seed coating and seedling injection (Brownbridge et al. 2012), stem injection (Tefera and Vidal 2009), soil drenching, and foliar spraying (Gurulingappa et al. 2010). Approximately 67.8% of EPF-based formulations available in the market are based exclusively on *B. bassiana* and *M. anisopliae* (De Faria and Wraight 2007).

3.1 *Beauveria bassiana*

A cosmopolitan species, *B. bassiana* infects a wide range of insects, including beetles, moths, and aphids (Fig. 14.1). Its pathogenicity relies on efficient cuticle penetration and the production of beauvericin. Many locally occurring *Beauveria* species have been successfully isolated from a variety of insects worldwide, which are of agricultural importance. This fungus acts as a pathogen against various agriculturally important insect species, causing white muscardine disease (Sandhu and Vikrant 2004; Thakur et al. 2005; Jain et al. 2008). *B. bassiana* has demonstrated virulence against *Leptinotarsa decemlineata*, making it an essential agent in insect

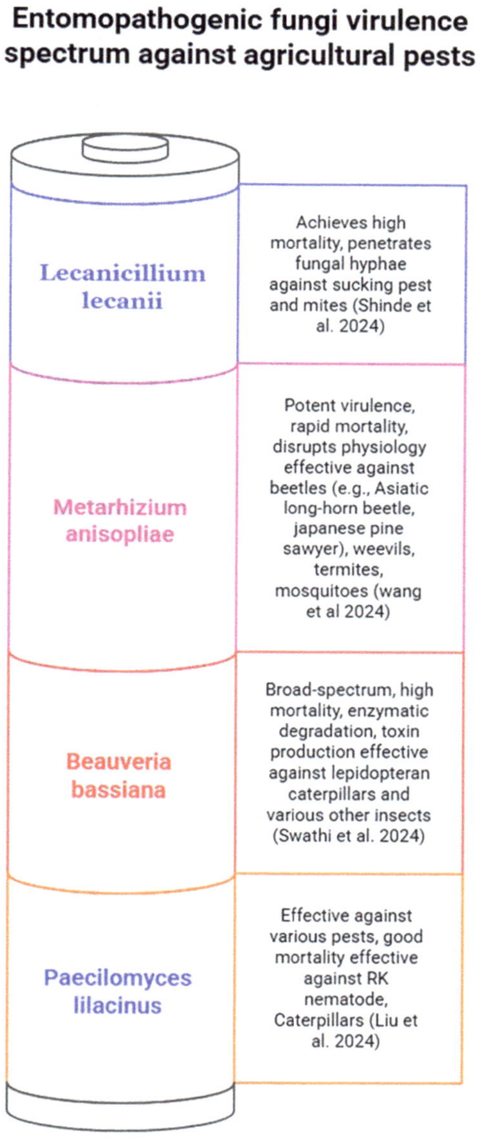

Fig. 14.1 EPF and their virulence against different pests

pest management strategies. Many weevils belonging to Curculionidae are highly susceptible to white muscardine disease produced by *B. bassiana* (Beavers et al. 1983). The Tsetse fly, *Glossina morsitans* (vector of tropical infectious diseases); the sand fly, *Phlebotomus* (transmits *Leishmania*); and bugs belonging to *Triatoma* and *Rhodnius* (vectors of Chagas disease) are pests of medical importance that serve as hosts of this fungus. The Colorado potato beetle, the codling moth, termites, *Helicoverpa armigera*, *Hyblaeapara*, and *Eutectona machaeralis* are some agriculturally important hosts of *B. bassiana* (Thakur and Sandhu 2010).

Furthermore, the different qualities, such as high host specificity, wide host range, and persistence for longer durations in the host population and environment, make this fungus more suitable against different pest species. Many genetically distinct variants of the entomopathogenic fungi, *B. bassiana*, have been reported worldwide, which differ in their ability to produce pathogenesis. *Beauveria bassiana* is also applied against *Ostrinia nubilalis*, *Dendrolimus* spp., and *Nephotettix* spp.

When entomotoxic proteins of *B. bassiana* were tested against two hemipteran insects, *Dysdercus cingulatus* Fab. and *Phenacoccus solenopsis* Tinsely, these were found to be very effective in their management (Tomson et al. 2021). This fungus is effective against a range of pests, including white grubs and thrips. Field trials reported up to 84% pest mortality when combined with other EPF species. *Beauveria bassiana* also shows promise as an endophyte, colonizing plant tissues and providing systemic protection (Bamisile et al. 2021). In greenhouse-grown cucumber, *B. bassiana* reduced immature as well as adult populations of *Frankliniella occidentalis* (Pergande) with its single application (Wu et al. 2016). Similarly, Wu et al. (2013) evaluated a novel strain of *B. bassiana* (SZ-26) under laboratory and greenhouse conditions for the control of onion thrips, *Thrips tabaci* and reported high virulence, with substantial reductions in adult and larval numbers. In another study, injection of *B. bassiana* into naturally infested palm trees led to a reduction of up to 90% in populations of the red palm weevil, *Rhynchophorus ferrugineus* Oliv. (Sewify et al. 2009). Furthermore, the commercial formulation BbWeevil, based on *B. bassiana* conidia, demonstrated effective control of key storage grain beetles, including *Oryzaephilus surinamensis*, *Sitophilus granarius*, and *Tribolium castaneum* (Khashaveh et al. 2011). Additionally, endophytic colonization of cotton plants by *B. bassiana* was shown to suppress the reproduction and feeding behavior of *Aphis gossypii* Glover (Gurulingappa et al. 2010). In Argentina, under field conditions, the highest mortality of *Dichroplus maculipennis* (Blanchard) was observed when *B. bassiana* was applied in combination with canola oil and wheat bran (Pelizza et al. 2019).

In *Solanum melongena* agroecosystems, *B. bassiana* (isolate Bb62) exhibits significant virulence against *Bemisia tabaci*, achieving 25.2% ovicidal mortality and 72.9% second-instar nymphal mortality at 1×10^8 conidia/mL, with a combined application of 0.5% neem enhancing efficacy to 29.5% egg and 97.2% nymphal mortality, reducing LT50 to 2.08 days through conidial cuticle penetration and neem's antifeedant effects (Islam et al. 2010). In *Brassica* and *Triticum* fields of Pakistan, isolate BB-01 demonstrated robust pathogenicity against aphid species, with LC_{50} values of 6.28×10^5 for *Brevicoryne brassicae*, 1.36×10^6 for *Lipaphis erysimi*, 6.76×10^6 for *Schizaphis graminum*, and 1.15×10^6 for *Rhopalosiphum padi* at 1×10^8 spores/mL, achieving 98.4–100% mortality and minimal impact on the predator *Coccinella septempunctata* (Akmal et al. 2013).

Due to the sensitivity of *B. bassiana* spores to ultraviolet (UV) radiation, their viability and effectiveness can be significantly reduced when exposed to direct sunlight. Therefore, for optimal field performance, it is recommended to apply this entomopathogenic fungus during the late afternoon or evening hours when UV intensity is minimal. Additionally, temperatures between 64 and 85 °F (18–29 °C) are ideal for promoting the product's efficacy. Alternatively, application can be effectively made on cloudy or rainy days, as these conditions help maintain the viability of the spores, enhance adhesion to insect cuticles, and promote favorable humidity for germination and infection. It is generally recommended to avoid using fungicides in close succession or with products containing *B. bassiana*, as such practices may adversely affect the viability and efficacy of this entomopathogenic fungus.

3.2 *Lecanicillium lecanii*

Lecanicillium lecanii is a widely distributed entomopathogenic fungus with significant potential in biological pest control. It thrives in warm, humid environments and is especially effective in tropical and subtropical regions, where it can cause large-scale epizootics among pest populations (Nunez et al. 2008) (Fig. 14.1). Due to its adaptability and efficacy, *L. lecanii* has become a valuable component in integrated pest management strategies, particularly in greenhouse agriculture. This fungus is known for its ability to infect and kill a wide range of soft-bodied insect hosts, including aphids, whiteflies, and other pests from Homoptera, Coleoptera, and Lepidoptera. It infects both nymphs and adults by attaching to the leaf undersides through a filamentous mycelium (Kerry et al. 1982) and secretes extracellular enzymes such as chitinase, which help degrade the insect cuticle and enhance colonization efficiency. In the 1970s, it was developed to manage whiteflies and aphid species like the green peach aphid, *Myzus persicae*, in greenhouse chrysanthemum cultivation (Hamlen 1979). *L. lecanii* has been shown to cause 100% mortality of *Aphis gossypii* with an LT_{50} of 2.74 days (Kim et al. 2008). Beyond insect pests, *L. lecanii* has demonstrated nematocidal properties too. It was identified as a major parasite responsible for significant reductions in cereal-cyst nematode populations in monocultures of susceptible crops (Kerry et al. 1982). These capabilities highlight its versatility in pest management. *Verticillium chlamydosporium* also shows promise as a biological control agent against cyst and root-knot nematodes, although its effectiveness varies depending on the isolate.

Overall, EPF like *Lecanicillium lecanii* are emerging as powerful tools in sustainable agriculture. Their ability to produce secondary metabolites that disrupt insect hormonal systems, combined with their adaptability and environmental compatibility, makes them effective alternatives to chemical pesticides in modern pest management systems.

3.3 *Hirsutella thompsonii*

Hirsutella is a genus of asexually reproducing fungi in *Ophiocordycipitaceae*, known for its pathogenicity against a wide range of arthropods, including insects, mites, and nematodes. This genus comprises over 50 species capable of colonizing diverse insect hosts. Among these, *H. thompsonii* has gained significant attention for its biocontrol potential. Several studies have demonstrated that crude filtrates of *H. thompsonii* exhibit strong toxicity against various arthropod groups such as moth, fly, and mosquito larvae, aphids, and mites, often resulting in their death.

In India, *H. thompsonii* has been effectively deployed to control the coconut mite, a major pest in coconut production. Field applications of *H. thompsonii* formulations commercially available under the trade name "Mycohit," a talc-based product with approximately 12% moisture content, have led to significant reductions in mite populations. Efficacy rates in the field have ranged from 85.6% to 97.1%, with concurrent reductions in nut losses and improvements in overall crop quality. Beyond coconut cultivation, *H. thompsonii* has also shown promise in vegetable production systems. Laboratory and greenhouse studies have demonstrated its effectiveness against the red spider mite, a common pest of okra. At a concentration of 4.6×10^8 spores/mL, the fungus achieved mortality rates of up to 92.31% in immature stages and 82.05% in adults, reinforcing its potential as a viable biocontrol agent for vegetable crops. With its broad host range, demonstrated efficacy, and suitability for commercial formulation, *Hirsutella thompsonii* represents a valuable tool in integrated pest management strategies aimed at reducing dependence on chemical pesticides.

3.4 *Nomuraea* spp.

Nomuraea rileyi, a dimorphic hyphomycete, has been reported to cause epizootics in many insect pest species belonging to Lepidoptera, including *Spodoptera litura*, as well as to Coleopteran pest species (Ignoffo 1981). Due to high host specificity and environment-friendly behavior, *N. rileyi* has been reported to infect many insect hosts, such as *Pseudoplusia includens*, *Heliothis zea*, *Plathypena scabra*, *Bombyx mori*, *Trichoplusia ni*, and *Anticarsia gemmatalis*. The hairy caterpillar, *Spilosoma*, was also found to be severely infected by this fungus and hence studied in detail for its myco-biocontrol (Mathew et al. 1998). Similarly, a severe epizootic of *N. rileyi* was noticed against the hedge plant eater, *Junonia orithya* (Rajak et al. 1991).

3.5 *Paecilomyces lilacinus*

Paecilomyces lilacinus is a highly effective entomopathogenic fungus with broad-spectrum activity against numerous economically important insect pests, particularly in greenhouse and field agroecosystems. Under field conditions, *P. lilacinus* has shown remarkable efficacy against the woolly apple aphid, outperforming other fungal biocontrol agents. In greenhouse settings, *P. lilacinus* has been identified as one of the most effective agents for managing a variety of major pests, including *Trialeurodes vaporariorum*, *Tetranychus urticae*, *Aphis gossypii*, and *Frankliniella occidentalis* (Moreno-Gavíra et al. 2020). In Poland, studies on *P. lilacinus* strain Pl have demonstrated high virulence against *Aphis gossypii*, achieving 90–96% mortality of both winged and wingless forms under laboratory conditions and 95–100%

under greenhouse conditions at a concentration of 1×10^6 conidia/mL (Fiedler and Sosnowska 2007). Against *Trialeurodes vaporariorum*, the fungus targets nymphal stages, causing 84% mortality in laboratory assays and 72% mortality in greenhouse trials. In the case of *Frankliniella occidentalis*, soil application of *P. lilacinus* spore suspensions significantly reduces prepupal and pupal populations, leading to a 60% reduction in thrips' density on cucumber leaves after 21 days in greenhouse conditions (Fiedler and Sosnowska 2007). Further evidence of *P. lilacinus* has broad-spectrum efficacy comes from Vietnam, where the isolate PL01 displayed strong pathogenicity against *Plutella xylostella* and *Spodoptera litura*. At a concentration of 1×10^8 spores/mL, LT_{50} values were 2.51 days for *P. xylostella* and 7.09 days for *S. litura*, with 100% mortality of *P. xylostella* observed within 6 days (Nguyen et al. 2017). Collectively, these findings position *P. lilacinum* as a potent and versatile biological control agent with proven efficacy across a diverse array of pests and growing conditions, making it a valuable component of sustainable pest management programs.

3.6 *Metarhizium* spp.

Metarhizium anisopliae is a highly potent entomopathogenic fungus that has been extensively explored for the biological control of economically significant insect pests (Sandhu et al. 1993; Sandhu and Mishra 1994) (Fig. 14.1). Its effectiveness lies in its ability to penetrate the host insect via the cuticle and adapt to a variety of environmental conditions, making it a cornerstone in integrated pest management strategies. This fungus has demonstrated broad-spectrum activity against both soil-dwelling and foliar insect pests. It predominantly affects members of Orthoptera, contributing to its diverse pest control potential.

Field studies in sugarcane have shown that applications of *M. anisopliae* can reduce white grub damage by up to 93% over 2 years. Similarly, talc-based formulations have proven effective in protecting potato tubers from cutworm infestations. Recent studies have highlighted its efficacy against major agricultural pests. In *Zea mays* agroecosystems of China, *M. anisopliae* exhibited potent virulence against *Spodoptera frugiperda*, causing 86% ovicidal mortality, 57.1% neonate mortality, and 77.8% antifeedant activity in second instars at a concentration of 1×10^8 conidia/mL. This effectiveness is attributed to conidial penetration and destruxin-mediated immunosuppression, which disrupts host physiology (Idrees et al. 2023). In Indian cotton (*Gossypium hirsutum*) fields, *M. anisopliae* has been shown to effectively manage multiple pest species (Sahayaraj and Borgio 2010). They reported significant mortality rates of different pests, including the following:

Aphis craccivora (Hemiptera: Aphididae; LC_{50}: 1.84×10^4 spores/mL)
Dysdercus cingulatus (Hemiptera: Pyrrhocoridae; 75% adult mortality)
Helicoverpa armigera (Lepidoptera: Noctuidae; 93.33% instar mortality)
Pericallia ricini (Lepidoptera: Arctiidae; LC_{50}: 1.62×10^4 spores/mL)

Fig. 14.2 Sugarcane white grub *Holotrichia serrata*: (**a**) Healthy grubs. (**b**) Mycelial growth of *M. anisopliae* in soil. Fungus-infected larvae: (**c**) First instar. (**d**) Second instar. (**e**) Third instar. (Source: Vinayaka (2014))

These effects are largely driven by the production of cuticle-degrading enzymes such as glucosidase and trehalase, as well as direct hyphal invasion.

In forest ecosystems, *M. anisopliae* has been identified as a promising myco-biocontrol agent against the teak skeletonizer, *Eutectona machaeralis* (Sandhu et al. 2000). Moreover, *M. anisopliae* has demonstrated control not only over insect pests but also over plant pathogens, such as *Ophiostoma ulmi*, the causative agent of Dutch elm disease (Sharma et al. 2023). Vinayaka (2014) also tested *M. anisopliae* against first, second, and third instars of sugarcane white grub, *Holotrichia serrata*, and found it very effective (Fig. 14.2). Its wide host range, robust infectivity, and environmental adaptability make it a versatile and effective biological control agent in both agricultural and forestry settings.

All these aforementioned EPF are not just laboratory curiosities—they represent critical tools in the global shift toward sustainable agriculture and IPM. Continued research into their formulation, application, and compatibility with other pest control methods will enhance their role in future food systems.

4 Pest Management Strategies

The integration of EPF into agricultural practices aligns with the principles of sustainable farming. Their deployment reduces dependence on synthetic chemical pesticides, thereby lowering agrochemical residues in soil and water bodies. This contributes to the conservation of beneficial insect fauna, pollinators, and other non-target organisms, thus supporting agroecosystem biodiversity. EPF are increasingly being incorporated into pest control strategies because they are compatible with chemical, biological, and cultural controls, providing environmentally friendly alternatives to synthetic pesticides.

4.1 Compatibility with Biocontrol Agents

Field studies show compatibility in some cases; for example, combining *B. bassiana* with *Neoseiulus cucumeris* enhanced thrips control without harming the predator. *Paecilomyces fumosoroseus* paired with *Dicyphus hesperus* reduced whitefly density by 62% in greenhouses, demonstrating additive effects (Alma 2005). *Beauveria bassiana* can be isolated from hibernating coccinellids, indicating potential natural infections, but field applications at lower temperatures may reduce predator mortality. Predators can also vector fungi, as seen with *Coccinella septempunctata* transmitting *Erynianeo aphidis* to aphid populations, initiating epizootics (Roy and Pell 2000). However, high fungal doses or direct exposure, such as 95.8% mortality in *Teretrius nigrescens* adults at 10^9 conidia/mL of *B. bassiana*, highlight the need for careful application.

EPF interactions with parasitoids are influenced by timing, dosage, and host discrimination. *Beauveria bassiana* is compatible with many parasitoids when applied strategically. For example, it did not affect *Encarsia formosa* development in *Trialeurodes vaporariorum* nymphs when sprayed post-parasitism (Van de Veire and Degheele 1996). In greenhouse trials, *B. bassiana* with *Aphidius matricariae* synergistically suppressed *Myzus persicae*. However, *B. bassiana* reduced *Tamarixia radiata* emergence by 85% when applied 3 days post-parasitism but did not affect older parasitoid stages (Chow et al. 2016). *Lecanicillium muscarium* was compatible with *Aphidius colemani* when applied 6 days after parasitism, avoiding negative effects on emergence and sex ratio (Mohammed and Hatcher 2017). Parasitoids like *Aphidius rhopalosiphi* and *E. formosa* can discriminate against *B. bassiana*- or *Aschersonia aleyrodis*-infected hosts, reducing oviposition in treated hosts (Fransen and van Lenteren 1993). Timing is critical; *N. rileyi* inhibited *Microplitis croceipes* development if hosts were infected 1 day post-parasitism, but effects diminished after 4 days (King and Bell 1978). Field experiments showed *B. bassiana* and *M. anisopliae* had minimal impact on *Cephalonomia stephanoderis* when applied days before or after release, with mortalities below 7% (Aristizabal et al. 2016).

4.2 Compatibility with Botanicals

EPF like *B. bassiana*, *M. anisopliae*, and *V. lecanii* are often compatible with botanicals, particularly neem, though outcomes differ by pest and formulation. For *Tetranychus urticae*, *B. bassiana* combined with flufenoxuron exhibited clear synergism, achieving nearly 100% larval mortality, while azadirachtin with *B. bassiana* showed an additive effect (Hernández et al. 2012). Against *Pectinophora gossypiella*, a combination of *B. bassiana*, *M. anisopliae*, and *V. lecanii* with neem extract at high conidial concentrations (1×10^8 CFU/mL) resulted in the highest larval mortality (74.67%) (Farooq et al. 2020). This combination also minimized pupation and adult emergence, indicating sublethal effects critical for long-term pest suppression (Farooq et al. 2020). For *Sitobion avenae*, *B. bassiana* with eucalyptus extract caused the highest mortality (87%) and lowest fecundity (7 nymphs/female), outperforming neem combinations, which showed lower efficacy (54% mortality) due to potential inhibition of fungal growth by neem's secondary metabolites (Ali et al. 2018). Terpenoids of *Eucalyptus* likely enhanced fungal penetration by disrupting aphid physiology (Ali et al. 2018). Neem products are frequently synergistic with EPF across pests like *Spodoptera litura*, *Bemisia tabaci* and *Epilachna dodecastigma* (Reddy and Chowdary 2021). Other botanicals, like pyrethrum and matrine, also show synergism with *M. anisopliae* and *V. muscarium*, respectively, reducing pest survival times and enhancing control (Reddy and Chowdary 2021). Adjuvants, such as surfactants and oils, improve EPF efficacy by enhancing conidial adhesion and environmental persistence. Silicone surfactants increased *B. bassiana* conidial survival and *T. urticae* mortality compared to mineral oil-based adjuvants (Holka and Kowalska 2023).

4.3 Cultural Practices

Crop rotation and intercropping practices provide beneficial microclimates for EPF survival, which prolongs their pest-suppressive activity (Pell et al. 2001). Similarly, reduced tillage practices conserve fungal spores in soil, which ensures long-term pest control (Quesada-Moraga et al. 2023).

4.4 Chemical Insecticides

Strains Ij-102 (*Metarhizium* spp.) and Bb-4511 (*B. bassiana*) can tolerate lower doses of the insecticides spiromesifen and flonicamid, maintaining over 80% conidial viability (Sain et al. 2019). In macadamia orchards, *M. anisopliae* and *B. bassiana* were found compatible with trichlorfon, acephate, indoxacarb, sulfoxaflor, and spinetoram (Khun et al. 2020). Applying EPF 72 h before insecticides like abamectin helps reduce negative interactions, granting the fungal spores time to penetrate the insect cuticle before exposure to chemicals.

5 Factors Influencing Pathogenicity of EPF

The deployment of EPF as biocontrol agents for managing insect pests represents a sustainable alternative to chemical pesticides (Bamisile et al. 2021). However, their practical application is impeded by a suite of challenges and limitations arising from biological, environmental, technical, and regulatory factors. These hurdles collectively restrict the scalability, reliability, and widespread adoption of fungal-based pest control strategies in agriculture (Singh et al. 2020; Wang et al. 2024).

A foremost challenge is the strict dependence of EPF on specific environmental conditions for effective performance. Species such as *B. bassiana* and *M. anisopliae*, commonly employed in biocontrol, require optimal humidity, temperature, and protection from ultraviolet light to germinate and infect their insect targets (Abdul Qayyum et al. 2021; Tandon et al. 2024). High humidity facilitates spore germination and penetration of the insect's outer layer, while extreme heat or cold can impair fungal activity by denaturing critical enzymes or desiccating spores. For example, in arid or semi-arid regions, maintaining adequate moisture levels is often impractical, reducing fungal efficacy (Parveen and Jeyarani 2023). Similarly, UV radiation from sunlight degrades spores exposed on plant surfaces, shortening their persistence and effectiveness in open fields. These environmental constraints limit the geographic and seasonal applicability of fungal biocontrol agents, particularly in climates with unpredictable weather patterns.

The specificity of EPF to insect hosts further complicates their use (Andrade-Hoyos et al. 2025). While this trait minimizes harm to beneficial insects, such as pollinators or natural predators, it also restricts the range of pests a single fungal species can control. For instance, a fungus effective against aphids may have little impact on beetles or caterpillars, necessitating the use of multiple fungal strains to address diverse pest populations within a single cropping system (Murindangabo et al. 2024). This requirement increases both the complexity and cost of pest management programs, as farmers must identify, source, and apply different fungi tailored to each pest. Moreover, some insects exhibit natural defenses, such as grooming behaviors that remove fungal spores from their bodies or immune responses that suppress infection, further reducing the fungi's effectiveness against certain targets (Zhang et al. 2025).

Technical difficulties in mass production, formulation, and storage present additional barriers. Producing large quantities of viable fungal spores demands advanced fermentation processes and stringent quality control to ensure consistency and potency (Grzywacz et al. 2023). Once produced, these spores must be formulated into products, such as oil suspensions, granules, or powders, that maintain their viability during storage and application. Oil-based formulations, for instance, can improve spore adhesion to insects and offer some protection against drying, but they may be costlier or harder to apply evenly across crops (Awan et al. 2021). Storage poses another challenge, as fungal spores often have a limited shelf life and require controlled conditions (e.g., cool temperatures and low humidity) to remain active. These logistical demands can strain resources, particularly in regions with limited infrastructure, and may discourage adoption by small-scale farmers.

Another significant limitation is the relatively slow speed of action exhibited by EPF compared to chemical pesticides (Sharma and Sharma 2021). While synthetic insecticides often kill pests within hours, fungal agents typically take days to weeks to infect and eliminate their hosts. This delay stems from the biological process of infection: spores must germinate, penetrate the insect's outer layer, and proliferate within the host before causing death. During this period, pests may continue feeding and reproducing, leading to ongoing crop damage that undermines the fungi's ability to prevent economic losses. In the scenarios where rapid pest control is critical, such as during outbreaks of highly destructive species, this slower action renders fungal biocontrol less competitive than conventional alternatives (Bamisile et al. 2021).

The potential for insect pests to develop resistance to EPF is an emerging concern that could compromise their long-term utility (Méndez-González et al. 2022). Although less documented than resistance to chemical pesticides, the possibility exists, particularly under repeated exposure to a single fungal strain. Insects may evolve enhanced immune defenses, behavioral adaptations (e.g., avoiding spore-laden areas), or physiological changes that inhibit fungal growth. Such resistance would diminish the effectiveness of fungal agents over time, mirroring the challenges faced by chemical pest control methods and necessitating continuous research to identify or develop new strains (Lu and St Leger 2016).

Regulatory obstacles further hinder the development and commercialization of EPF. A strict regulatory framework enhances the overall costs of microbial biopesticides and delays their commercialization (Sabbahi et al. 2022). Biological control agents must undergo rigorous evaluation to secure approval for agricultural use, a process that requires extensive data on efficacy, safety to humans and wildlife, and environmental impact. These assessments are time-consuming and expensive, often spanning several years and demanding significant financial investment. For smaller companies or research institutions, the regulatory burden can be prohibitive, slowing the pipeline of new fungal products to market. Additionally, differing standards across countries complicate international trade and deployment, limiting the global reach of these biocontrol tools.

The integration of EPF into practical pest management also faces challenges related to farmer perception and knowledge gaps (Chandler et al. 2011). Many agricultural producers are accustomed to the immediate and predictable results of chemical pesticides, whereas fungal agents require a greater understanding of application timing, environmental conditions, and pest life cycles for optimal results. Without adequate training or demonstration of cost-effectiveness, adoption rates may remain low, particularly in resource-limited settings where experimentation with new methods carries a higher risk. Addressing these issues requires a multifaceted approach that combines scientific innovation, improved infrastructure, and education to unlock the full potential of fungal-based pest management.

6 Conclusion and Future Prospects

EPF represent a promising and environmentally sound approach to pest management, offering targeted pathogenicity against a wide range of insect pests. Various fungal species, such as *Beauveria bassiana* and *Metarhizium anisopliae*, have demonstrated effectiveness in biocontrol, provided they are selected based on rigorous criteria, including virulence, environmental adaptability, and host specificity. The success of EPF-based control depends heavily on careful selection of target pests and understanding host-pathogen interactions to ensure precision and minimize non-target effects. Field applications require optimization of formulation and timing, while continuous monitoring and evaluation are essential to assess efficacy under real-world conditions. Importantly, EPF can be integrated with other pest management strategies, showing good compatibility with natural enemies and reduced chemical inputs. Although generally considered safe for the environment and non-target organisms, challenges such as variable field performance, limited shelf life, and regulatory hurdles must be addressed. Overall, EPF serves as a valuable component of IPM, supporting sustainable agriculture with reduced ecological impact.

Despite the obstacles outlined, the future of EPF as biocontrol agents is bright, with emerging research and technologies poised to enhance their efficacy and practicality. Advances in genetics, formulation, integrated management, and education offer pathways to overcome current limitations and position these fungi as a cornerstone of sustainable pest control. Genetic engineering holds transformative potential for improving fungal performance. By modifying fungal strains, scientists could enhance traits such as virulence, environmental resilience, or speed of infection. For example, increasing the production of toxins or enzymes that degrade insect defenses could accelerate pest mortality, while introducing genes for UV or drought tolerance could broaden the fungi's operational range. However, the use of genetically modified fungi must navigate regulatory and public acceptance challenges, requiring thorough safety assessments to ensure ecological balance.

Innovations in formulation and delivery methods are another promising frontier. Techniques like microencapsulation could shield spores from harsh conditions, while precision tools such as drones or targeted sprayers could improve application efficiency, ensuring spores reach pests effectively. These advancements could reduce costs and enhance reliability, making fungal agents more appealing to farmers. Integrating EPF into broader pest management strategies offers a holistic approach to overcoming their limitations. Combining fungi with low-dose chemical pesticides, cultural practices, or other biological agents could enhance control efficacy, delay resistance, and minimize environmental harm. Research into pest and fungal ecology will further refine these strategies, identifying optimal conditions and synergies for deployment. Raising awareness through education and field demonstrations can drive adoption. By showcasing the economic and ecological benefits of fungal biocontrol, stakeholders can build confidence among farmers and policymakers, fostering a shift toward sustainable agriculture. Together, these prospects signal a future where EPF plays a vital role in reducing pesticide reliance and supporting resilient farming systems.

References

Abdul Qayyum M, Bilal H, Ali H, Raza H, Wajid M (2021) Factors affecting the epizootics of entomopathogenic fungi – a review. J Bioresour Manag 8(4):75–85. https://doi.org/10.35691/JBM.1202.0204

Abrol DP, Shankar U (2015) Integrated pest management. In: Gupta SK (ed) Breeding oilseed crops for sustainable production: opportunities and constraints. Academic Press, p 523

Akmal M, Freed S, Malik MN, Gul HT (2013) Efficacy of *Beauveria bassiana* (Deuteromycotina: Hypomycetes) against different aphid species under laboratory conditions. Pak J Zool 45(1): 71–78

Ali S, Zhang C, Wang Z, Wang XM, Wu JH et al (2017) Toxicological and biochemical basis of synergism between the entomopathogenic fungus *Lecanicillium muscarium* and the insecticide matrine against *Bemisia tabaci* (Gennadius). Sci Rep 7:46558. https://doi.org/10.1038/srep46558

Ali S, Farooqi MA, Sajjad A, Ullah MI, Qureshi AK et al (2018) Compatibility of entomopathogenic fungi and botanical extracts against the wheat aphid, *Sitobion avenae* (Fab.) (Hemiptera: Aphididae). Egypt J Biol Pest Control 28:97. https://doi.org/10.1186/s41938-018-0101-9

Alma CR (2005) Combining the fungal agent *Paecilomyces fumosoroseus* with the predatory bug *Dicyphus hesperus* for biological control of whitefly in greenhouse tomato. IOBC/WPRS Bull 28(1):7–10

Andrade-Hoyos P, Aguilar-Marcelino L, Luna-Cruz A, Vallejo-Pérez MR, Molina-Gayosso E et al (2025) Specificity and biocontrol potential of entomopathogenic fungi. In: Deshmukh SK, Sridhar KR (eds) The role of entomopathogenic fungi in agriculture. CRC Press, pp 117–139

Aristizabal LF, Bustillo AE, Arthurs SP (2016) Integrated pest management of coffee berry borer: strategies from Latin America that could be useful for coffee farmers in Hawaii. Insects 7(1):6. https://doi.org/10.3390/insects7010006

Armendariz MCR, de la Torre AH, Fernandez AJG, Weller DG, Montelongo SP et al (2017) Metal toxicity in foods. In: Sachan A, Suzanne H (eds) Food toxicology: current advances and future challenges. Apple Academic Press, pp 291–336

Awan UA, Meng L, Xia S, Raza MF, Zhang Z et al (2021) Isolation, fermentation, and formulation of entomopathogenic fungi virulent against adults of *Diaphorina citri*. Pest Manag Sci 77:4040–4053. https://doi.org/10.1002/ps.6429

Bamisile BS, Akutse KS, Siddiqui JA, Xu Y (2021) Model application of entomopathogenic fungi as alternatives to chemical pesticides: prospects, challenges, and insights for next-generation sustainable agriculture. Front Plant Sci 12:741804. https://doi.org/10.3389/fpls.2021.741804

Batta YA (2003) Production and testing of novel formulations of the entomopathogenic fungus *Metarhizium anisopliae* (Metschinkoff) Sorokin (Deuteromycotina: Hyphomycetes). Crop Prot 22:415–422. https://doi.org/10.1016/S0261-2194(02)00200-4

Beavers JB, McCoy CW, Kaplan DT (1983) Natural enemies of sub-terranean *Diaprepes abbreviatus* (Coleoptera: Curculionidae) larvae in Florida. Environ Entomol 12:840–843. https://doi.org/10.1093/ee/12.3.840

Bertero A, Fossati P, Tedesco DEA, Caloni F (2020) Beauvericin and enniatins: in vitro intestinal effects. Toxins 12(11):686. https://doi.org/10.3390/toxins12110686

Bidochka MJ, St Leger RJ, Joshi L, Roberts DW (1995) The rodlet layer from aerial and submerged conidia of the entomopathogenic fungus *Beauveria bassiana* contains hydrophobin. Mycol Res 99:403–406. https://doi.org/10.1016/S0953-7562(09)80637-0

Bitencourt ROB, Mallet JRS, Mesquita E, Golo PS, Fiorotti J et al (2021) Larvicidal activity, route of interaction and ultrastructural changes in *Aedes aegypti* exposed to entomopathogenic fungi. Acta Trop 213:105732. https://doi.org/10.1016/j.actatropica.2020.105732

Brivio MF, Mastore M (2020) When appearance misleads: the role of the entomopathogen surface in the relationship with its host. Insects 11:1–24. https://doi.org/10.3390/insects11060387

Brownbridge M, Reay SD, Nelson TL, Glare TR (2012) Persistence of *Beauveria bassiana* (Ascomycota: Hypocreales) as an endophyte following inoculation of radiata pine seed and seedlings. Biol Control 61:194–200. https://doi.org/10.1016/j.biocontrol.2012.01.002

Bukhari T, Takken W, Koenraadt CJ (2011) Development of *Metarhizium anisopliae* and *Beauveria bassiana* formulations for control of malaria mosquito larvae. Parasites Vectors 4:23. https://doi.org/10.1186/1756-3305-4-23

Butt TM, Coates CJ, Dubovskiy IM, Ratcliffe NA (2016) Entomopathogenic fungi: new insights into host–pathogen interactions. Adv Genet 94:307–364. https://doi.org/10.1016/bs.adgen.2016.01.006

Chandler D, Bailey AS, Tatchell GM, Davidson G, Greaves J et al (2011) The development, regulation and use of biopesticides for integrated pest management. Philos Trans R Soc B 366: 1987–1998. https://doi.org/10.1098/rstb.2010.0390

Charnley AK (2003) Fungal pathogens of insects: cuticle degrading enzymes and toxins. Adv Bot Res 40:241–321. https://doi.org/10.1016/S0065-2296(05)40006-3

Chen XR, Hu QB, Yu XQ, Ren SX (2014) Effects of destruxins on free calcium and hydrogen ions in insect hemocytes. Insect Sci 21:31–38. https://doi.org/10.1111/1744-7917.12028

Chet T, Schichler H, Haran S, Appenheim AB (1993) Cloned chitinase and their role in biological control of plant pathogenic fungi. In: Proceedings of the international symposium on chitin enzymology, Senigallia, Italy, pp 47–48

Cho EM, Kirkland BH, Holder DJ, Keyhani NO (2007) Phage display cDNA cloning and expression analysis of hydrophobins from the entomopathogenic fungus *Beauveria* (*Cordyceps*) *bassiana*. Microbiology 153:3438–3447. https://doi.org/10.1099/mic.0.2007/008532-0

Choudhary DK, Johri BN (2009) Interactions of *Bacillus* spp. and plants – with special reference to induced systemic resistance (ISR). Microbiol Res 164:493–513. https://doi.org/10.1016/j.micres.2008.08.007

Chow A, Dunlap CA, Jackson MA, Flores D, Patt JM, Sétamou M (2016) Oviposition behavior and survival of Tamarixia radiata (Hymenoptera: Eulophidae), an ectoparasitoid of the Asian citrus psyllid, Diaphorina citri (Hemiptera: Liviidae), on hosts exposed to an entomopathogenic fungus, Isaria fumosorosea (Hypocreales: Cordycipitaceae), under laboratory conditions. J Econ Entomol 109:1–11. https://doi.org/10.1093/jee/tow016

De Faria MR, Wraight SP (2007) Mycoinsecticides and mycoacaricides: a comprehensive list with worldwide coverage and international classification of formulation types. Biol Control 43(3): 237–256. https://doi.org/10.1016/j.biocontrol.2007.08.001

Eyal J, Mabud MA, Fischbein KL, Walter JF, Osborne LS et al (1994) Assessment of *Beauveria bassiana* Nov. EO-1 strain, which produces a red pigment for microbial control. Appl Biochem Biotechnol 44:65–80. https://doi.org/10.1007/BF02921852

Farooq MA, Atta B, Gogi MD, Arif MJ, Arain QA (2020) Compatibility of entomopathogenic fungi and *Azadirachta indica* extract against the cotton pink bollworm, *Pectinophora gossypiella* (Saunders) (Lepidoptera: Gelechiidae) under controlled conditions. Egypt J Biol Pest Control 30:63. https://doi.org/10.1186/s41938-020-00260-x

Fiedler Z, Sosnowska D (2007) Nematophagous fungus *Paecilomyces lilacinus* (Thom) Samson is also a biological agent for control of greenhouse insects and mite pests. BioControl 52:547–558. https://doi.org/10.1007/s10526-006-9052-2

Francis F, Fingu-Mabola JC, Ben Fekih I (2022) Direct and endophytic effects of fungal entomopathogens for sustainable aphid control: a review. Agriculture 12:2081. https://doi.org/10.3390/agriculture12122081

Fransen JJ, van Lenteren JC (1993) Host selection and survival of the parasitoid *Encarsia formosa* on greenhouse whitefly, *Trialeurodes vaporariorum*, in the presence of hosts infected with the fungus *Aschersonia aleyrodis*. Entomol Exp Appl 69:239–249. https://doi.org/10.1111/j.1570-7458.1993.tb01747.x

Fukushima K, Arai T, Mori Y, Tsuboi M, Suzuki M (1983) Studies on peptide antibiotics, leucinostatins. I. Separation, physicochemical properties and biological activities of leucinostatins A and B. J Antibiot 36:1606–1612. https://doi.org/10.7164/antibiotics.36.1606

Gabarty A, Salem HM, Fouda MA, Abas AA, Ibrahim AA (2014) Pathogenicity induced by the entomopathogenic fungi *Beauveria bassiana* and *Metarhizium anisopliae* in *Agrotis ipsilon* (Hufn.). J Radiat Res Appl Sci 7:95–100. https://doi.org/10.1016/j.jrras.2013.12.004

Gökce A, Isaacs R, Whalon ME (2006) Behavioural response of Colorado potato beetle (*Leptinotarsa decemlineata*) larvae to selected plant extracts. Pest Manag Sci 62:1052–1057. https://doi.org/10.1002/ps.1271

Golo PS, Gardner DR, Grilley MM, Takemoto JY, Krasnoff SB et al (2014) Production of destruxins from *Metarhizium* spp. fungi in artificial medium and in endophytically colonized cowpea plants. PLoS One 9:e97064. https://doi.org/10.1371/journal.pone.0104946

Grzywacz D, Moore S, Luke B, Subramanian S, Moore D et al (2023) Mass production of entomopathogens in less industrialized countries. In: Morales-Ramos JA, Rojas MG, Shapiro-Ilan DI (eds) Mass production of beneficial organisms. Academic Press, pp 431–462. https://doi.org/10.1016/B978-0-12-822106-8.00001-4

Gul HT, Saeed S, Khan FZA (2014) Entomopathogenic fungi as effective insect pest management tactic: a review. Appl Sci Bus Econ 1:10–18

Gurulingappa P, Sword GA, Murdoch G (2010) Colonization of crop plants by fungal entomopathogens and their effects on two insect pests when in planta. Biol Control 55:34–41. https://doi.org/10.1016/j.biocontrol.2010.06.011

Hamlen RA (1979) Biological control of insects and mites on European greenhouse crops: research and commercial implementation. Proc Fla State Hortic Soc 92:367–368

Haron E, Ahmed M, Ali S, Abas A, Elshaier M (2020) Evaluate the effects of entomopathogenic fungi isolates on wheat aphid, *Schizaphis graminum* (Rondani) (Hemiptera: Aphididae). Egypt Acad J Biol Sci A Entomol 13:149–159. https://doi.org/10.21608/eajbsa.2020.86254

Hernández MM, Martínez-Villar E, Peace C, Pérez-Moreno I, Marco V (2012) Compatibility of the entomopathogenic fungus *Beauveria bassiana* with flufenoxuron and azadirachtin against *Tetranychus urticae*. Exp Appl Acarol 58:395–405. https://doi.org/10.1007/s10493-012-9594-1

Holder DJ, Keyhani NO (2005) Adhesion of the entomopathogenic fungus *Beauveria* (*Cordyceps*) *bassiana* to substrata. Appl Environ Microbiol 71:5260–5266. https://doi.org/10.1128/AEM.71.9.5260-5266.2005

Holder DJ, Kirkland BH, Lewis MW, Keyhani NO (2007) Surface characteristics of the entomopathogenic fungus *Beauveria bassiana* (*Cordyceps*). Microbiology 153:3448–3457. https://doi.org/10.1099/mic.0.2007/008524-0

Holka M, Kowalska J (2023) The potential of adjuvants used with microbiological control of insect pests with emphasis on organic farming. Agriculture 13:1659. https://doi.org/10.3390/agriculture13091659

Humber RA (2012) Identification of entomopathogenic fungi. In: Lacey LA (ed) Manual of techniques in invertebrate pathology. Elsevier, pp 151–187. https://doi.org/10.1016/b978-0-12-386899-2.00006-3

Idrees A, Afzal A, Qadir ZA, Qadir A, Li J (2023) Virulence of entomopathogenic fungi against fall armyworm, *Spodoptera frugiperda* (Lepidoptera: Noctuidae) under laboratory conditions. Front Physiol 14:1107434. https://doi.org/10.3389/fphys.2023.1107434

Ignoffo CM (1981) The fungus *Nomuraea rileyi* as a microbial insecticide. In: Burges HD (ed) Microbial control of pests and plant diseases. Academic Press, London, pp 513–538

Ishiyama A, Otoguro K, Iwatsuki M, Namatame M, Nishihara A et al (2009) In vitro and in vivo antitrypanosomal activities of three peptide antibiotics: leucinostatin A and B, alamethicin I and tsushimycin. J Antibiot 62:303–308. https://doi.org/10.1038/ja.2009.32

Islam W (2017) Eco-friendly approaches for the management of red flour beetle: *Tribolium castaneum* (Herbst). Sci Lett 5:105–114

Islam MT, Castle SJ, Ren S (2010) Compatibility of the insect pathogenic fungus *Beauveria bassiana* with neem against sweetpotato whitefly, *Bemisia tabaci*, on eggplant. Entomol Exp Appl 134:28–34. https://doi.org/10.1111/j.1570-7458.2009.00933.x

Jaber LR, Ownley BH (2018) Can we use entomopathogenic fungi as endophytes for dual biological control of insect pests and plant pathogens? Biol Control 116:36–45. https://doi.org/10.1016/j.biocontrol.2017.01.018

Jain N, Rana IS, Kanojiya A, Sandhu SS (2008) Characterization of *Beauveria bassiana* strains based on protease and lipase activity and their role in pathogenicity. J Basic Appl Mycol 1–2: 18–22

Kaczmarek A, Bogus MI, Włóka E, Wrońska AK, Krawiel A et al (2020) The interaction between cuticle free fatty acids (FFAs) of the cockroaches *Blattella germanica* and *Blatta orientalis* and hydrolases produced by the entomopathogenic fungus *Conidiobolus coronatus*. PLoS One 15: e0228592. https://doi.org/10.1371/journal.pone.0235785

Kanaoka M, Isogai A, Murakoshi S, Ichinoe M, Suzuki A et al (1978) Bassianolide, a new insecticidal cyclodepsipeptide from *Beauveria bassiana* and *Verticillium lecanii*. Agric Biol Chem 42:629–635. https://doi.org/10.1080/00021369.1978.10863029

Kerry BR, Crump DH, Mullen LA (1982) Studies of the cereal cyst nematode, *Heterodera avenae* under continuous cereals, 1975–1978. II. Fungal parasitism of nematode females and eggs. Ann Appl Biol 100:489–499. https://doi.org/10.1111/j.1744-7348.1982.tb01415.x

Khashaveh A, Ghosta Y, Safaralizadeh MH, Ziaee M (2011) The use of entomopathogenic fungus, *Beauveria bassiana* (Bals.) Vuill. in assays with storage grain beetles. J Agric Sci Technol 13: 35–43

Khun KK, Ash GJ, Stevens MM, Huwer RK, Wilson BA (2020) Compatibility of *Metarhizium anisopliae* and *Beauveria bassiana* with insecticides and fungicides used in macadamia production in Australia. Pest Manag Sci 77:709–718. https://doi.org/10.1002/ps.6065

Kim JJ, Goettel MS, Gillespie DR (2008) Evaluation of *Lecanicillium lecanii* for simultaneous biological control of aphids and powdery mildew in greenhouse cucumbers. Biol Control 47: 235–240. https://doi.org/10.1016/j.biocontrol.2008.08.006

King EG, Bell JV (1978) Interactions between a braconid, *Microplitis croceipes*, and a fungus, *Nomuraea rileyi*, in laboratory-reared bollworm larvae. J Invertebr Pathol 31:337–340. https://doi.org/10.1016/0022-2011(78)90225-2

Kodaira Y (1961) Toxic substances to insects, produced by *Aspergillus ochraceus* and *Oopsra destructor*. Agric Biol Chem 25:261–262. https://doi.org/10.1080/00021369.1961.10857803

Li ZZ, Li CR, Huang B, Meizhen MZ (2001) Discovery and demonstration of the teleomorph of *Beauveria bassiana* (Bals.) Vuill., an important entomogenous fungus. Chin Sci Bull 46:751–753. https://doi.org/10.1007/BF03187215

Litwin A, Nowak M, Różalska S (2020) Entomopathogenic fungi: unconventional applications. Rev Environ Sci Biotechnol 19:23–42. https://doi.org/10.1007/s11157-020-09525-1

Lu HL, St Leger RJ (2016) Insect immunity to entomopathogenic fungi. Adv Genet 94:251–285. https://doi.org/10.1016/bs.adgen.2015.11.002

Luz C, Saladino F, Luciano FB, Mañes J, Meca G (2017) Occurrence, toxicity, bioaccessibility and mitigation strategies of beauvericin, a minor Fusarium mycotoxin. Food Chem Toxicol 107: 430–439. https://doi.org/10.1016/j.fct.2017.07.032

Mantzoukas S, Eliopoulos PA (2020) Endophytic entomopathogenic fungi: a valuable biological control tool against plant pests. Appl Sci 10:3605. https://doi.org/10.3390/app10010360

Mathew SO, Sandhu SS, Rajak RC (1998) Bioactivity of *Nomuraea rileyi* against *Spilosoma obliqua*: effect of dosage, temperature and relative humidity. J Indian Bot Soc 77:23–25

Méndez-González F, Castillo-Minjarez JM, Loera O, Favela-Torres E (2022) Current developments in the resistance, quality, and production of entomopathogenic fungi. World J Microbiol Biotechnol 38:115. https://doi.org/10.1007/s11274-022-03301-9

Mohammed AA, Hatcher PE (2017) Combining entomopathogenic fungi and parasitoids to control the green peach aphid *Myzus persicae*. Biol Control 110:44–55. https://doi.org/10.1016/j.biocontrol.2017.03.012

Mondal S, Baksi S, Koris A, Vatai G (2016) Journey of enzymes in entomopathogenic fungi. Pac Sci Rev A Nat Eng Sci 18:85–99. https://doi.org/10.1016/j.psra.2016.10.001

Moreno-Gavíra A, Huertas V, Diánez F, Sánchez-Montesinos B, Santos M (2020) *Paecilomyces* and its importance in the biological control of agricultural pests and diseases. Plants 9:1746. https://doi.org/10.3390/plants9121746

Mori Y, Tsuboi M, Suzuki M, Fukushima K, Arai T (1982) Structure of leucinostatin A, new peptide antibiotic from *Paecilomyces lilacinus* A-267. J Chem Soc Chem Commun 1982:94–96. https://doi.org/10.1039/C39820000094

Mukherjee K, Vilcinskas A (2018) The entomopathogenic fungus *Metarhizium robertsii* communicates with the insect host *Galleria mellonella* during infection. Virulence 9:402–413. https://doi.org/10.1080/21505594.2017.1405190

Murindangabo YT, Kopecký M, Perná K, Konvalina P, Bohatá A et al (2024) Relevance of entomopathogenic fungi in soil–plant systems. Plant Soil 495:287–310. https://doi.org/10.1007/s11104-023-06325-8

Napolitano R, Juárez MP (1997) Entomopathogenous fungi degrade epicuticular hydrocarbons of *Triatoma infestans*. Arch Biochem Biophys 344:208–214. https://doi.org/10.1006/abbi.1997.0163

Naranjo-Ortiz MA, Gabaldón T (2019) Fungal evolution: diversity, taxonomy and phylogeny of the fungi. Biol Rev Camb Philos Soc 94(6):2101–2137. https://doi.org/10.1111/brv.12550

Nguyen HC, Tran TVA, Nguyen QL, Nguyen NN, Nguyen MK et al (2017) Newly isolated *Paecilomyces lilacinus* and *Paecilomyces javanicus* as novel biocontrol agents for *Plutella xylostella* and *Spodoptera litura*. Not Bot Horti Agrobot Cluj Napoca 45:280–286. https://doi.org/10.15835/nbha45110726

Nunez E, Iannacone J, Gomez HG (2008) Effect of two entomopathogenic fungi in controlling *Aleurodicus cocois* (Curtis, 1846) (Hemiptera: Aleyrodidae). Chil J Agric Res 68:21–30

O'Brien PA (2017) Biological control of plant diseases. Australas Plant Pathol 46:293–304. https://doi.org/10.1007/s13313-017-0481-4

Ortiz-Urquiza A, Keyhani NO (2013) Action on the surface: entomopathogenic fungi versus the insect cuticle. Insects 4:357–374. https://doi.org/10.3390/insects4030357

Ortiz-Urquiza A, Keyhani NO, Quesada-Moraga E (2013) Culture conditions affect virulence and production of insect toxic proteins in the entomopathogenic fungus *Metarhizium anisopliae*. Biocontrol Sci Tech 23:1199–1212. https://doi.org/10.1080/09583157.2013.822474

Ortiz-Urquiza A, Fan Y, Garrett T, Keyhani NO (2016) Growth substrates and caleosin-mediated functions affect conidial virulence in the insect pathogenic fungus *Beauveria bassiana*. Microbiology 162:1913–1921. https://doi.org/10.1099/mic.0.000375

Pal S, St Leger RJ, Wu LP (2007) Fungal peptide destruxin A plays a specific role in suppressing the innate immune response in *Drosophila melanogaster*. J Biol Chem 282:8969–8977. https://doi.org/10.1074/jbc.M605927200

Park JO, Hargreaves JR, McConville EJ, Stirling GR, Ghisalberti EL et al (2004) Production of leucinostatins and nematicidal activity of Australian isolates of *Paecilomyces lilacinus* (Thom) Samson. Lett Appl Microbiol 38:271–276. https://doi.org/10.1111/j.1472-765X.2004.01488.x

Parveen SS, Jeyarani S (2023) Laboratory evaluation of temperature effects on germination, radial growth and sporulation of entomopathogenic fungi and on their pathogenicity to red spider mite, *Tetranychus urticae* Koch. Indian J Agric Res 57:376–382

Pedrini N, Zhang S, Juárez MP, Keyhani NO (2010) Molecular characterization and expression analysis of a suite of cytochrome P450 enzymes implicated in insect hydrocarbon degradation in the entomopathogenic fungus *Beauveria bassiana*. Microbiology 156:2549–2557. https://doi.org/10.1099/mic.0.039735-0

Pedrini N, Ortiz-Urquiza A, Huarte-Bonnet C, Zhang S, Keyhani NO (2013) Targeting of insect epicuticular lipids by the entomopathogenic fungus *Beauveria bassiana*: hydrocarbon oxidation within the context of a host-pathogen interaction. Front Microbiol 4:24. https://doi.org/10.3389/fmicb.2013.00024

Pelizza SA, Mariottini Y, Russo LM, Vianna MF, Scorsetti AC et al (2019) Application of *Beauveria bassiana* using different baits for the control of grasshopper pest *Dichroplus maculipennis* under field cage conditions. J King Saud Univ Sci 31:1511–1515. https://doi.org/10.1016/j.jksus.2019.02.014

Pell J, Eilenberg J, Hajek A, Steinkraus D (2001) Biology, ecology and pest management potential of Entomophthorales. In: Butt TM, Jackson C, Magan N (eds) Fungi as biocontrol agents: progress, problems and potential. CABI, Wallingford, pp 71–154. https://doi.org/10.1079/9780851993560.0071

Peng KC, Huang HS, Tzeng YM, Liu SY (2005) Circular dichroism analysis of destruxins from *Metarhizium anisopliae*. J Biochem Biophys Methods 62:41–50. https://doi.org/10.1016/j.jbbm.2004.08.001

Qadri SM, Kucherenko Y, Lang F (2011) Beauvericin induced erythrocyte cell membrane scrambling. Toxicology 283:24–31. https://doi.org/10.1016/j.tox.2011.01.023
Qasim M, Lin Y, Dash CK, Bamisile BS, Ravindran K et al (2018) Temperature-dependent development of Asian citrus psyllid on various hosts, and mortality by two strains of *Isaria*. Microb Pathog 119:109–118. https://doi.org/10.1016/j.micpath.2018.04.019
Quesada-Moraga E, Garrido-Jurado I, Gonzalez-Mas N, Yousef-Yousef M (2023) Ecosystem services of entomopathogenic ascomycetes. J Invertebr Pathol 201:108015. https://doi.org/10.1016/j.jip.2023.108015
Rajak RC, Sandhu SS, Mukherjee S, Kekre S, Gupta A (1991) Natural outbreak of *Nomuraea rileyi* on *Junonia orithyia*. J Biol Control 5:123–124
Ravindran K, Sivaramakrishnan S, Hussain M, Dash CK, Bamisile BS et al (2018) Investigation and molecular docking studies of bassianolide from *Lecanicillium lecanii* against *Plutella xylostella* (Lepidoptera: Plutellidae). Comp Biochem Physiol C Toxicol Pharmacol 206–207: 65–72. https://doi.org/10.1016/j.cbpc.2018.03.004
Reddy DS, Chowdary NM (2021) Botanical biopesticide combination concept—a viable option for pest management in organic farming. Egypt J Biol Pest Control 31:23. https://doi.org/10.1186/s41938-021-00366-w
Roy HE, Pell JK (2000) Interactions between entomopathogenic fungi and other natural enemies: implications for biological control. Biocontrol Sci Technol 10:737–752. https://doi.org/10.1080/09583150020011708
Ryder LS, Talbot NJ (2015) Regulation of appressorium development in pathogenic fungi. Curr Opin Plant Biol 26:8–13. https://doi.org/10.1016/j.pbi.2015.05.013
Sabbahi R, Hock V, Khalil A, Sanaa S, Belkheir H (2022) A global perspective of entomopathogens as microbial biocontrol agents of insect pests. J Agric Food Res 10:100376. https://doi.org/10.1016/j.jafr.2022.100376
Sabbour MM, Abd El-Rahman A, Abd El-Raheem MA, Ragei M (2012) Pathogenicity of entomopathogenic fungi against olive insect pests under laboratory and field conditions in Egypt. J Appl Sci Res 8:3448–3452
Sahayaraj K, Borgio JF (2010) Virulence of entomopathogenic fungus *Metarhizium anisopliae* (METSCH.) Sorokin on seven insect pests. Indian J Agric Res 44:195–200
Sain SK, Monga D, Kumar R, Nagrale DT, Hiremani NS et al (2019) Compatibility of entomopathogenic fungi with insecticides and their efficacy for IPM of *Bemisia tabaci* in cotton. J Pestic Sci 44:97–105. https://doi.org/10.1584/jpestics.D18-067
Sandhu SS, Mishra M (1994) Larvicidal activity of fungal isolates *Beauveria bassiana*, *Metarhizium anisopliae* and *Aspergillus flavus* against mosquito sp. *Culex pipiens*. In: Proceedings of the National Symposium on Advances in Biological Control of Insect Pests, Muzaffarnagar, India, pp 145–150
Sandhu SS, Vikrant P (2004) Myco-insecticides: control of insect pests. In: Gautam SP, Sandhu SS, Sharma A, Pandey AK (eds) Microbial diversity: opportunities & challenges. Indica Publishers, New Delhi, pp 165–176
Sandhu SS, Rajak RC, Agarwal GP (1993) Studies on prolonged storage of *Beauveria bassiana* conidia: effects of temperature and relative humidity on conidial viability and virulence against chickpea borer *Helicoverpa armigera*. Biocontrol Sci Tech 3:47–53. https://doi.org/10.1080/09583159309355258
Sandhu SS, Rajak RC, Hasija SK (2000) Potential of entomopathogens for the biological management of medically important pests: progress and prospect. Glimpses Plant Sci 2000:110–117
Sandhu SS, Sharma AK, Beniwal V, Goel G, Batra P et al (2012) Myco-biocontrol of insect pests: factors involved, mechanism, and regulation. J Pathog 2012:1–10. https://doi.org/10.1155/2012/126819
Sewify GH, Belal MH, Al-Awash SA (2009) Use of the entomopathogenic fungus *Beauveria bassiana* for the biological control of the red palm weevil, *Rhynchophorus ferrugineus* Olivier. Egypt J Biol Pest Control 19:157–163

Shahid AA, Rao AQ, Bakhsh A, Husnain T (2012) Entomopathogenic fungi as biological controllers: new insights into their virulence and pathogenicity. Arch Biol Sci 64:21–42. https://doi.org/10.2298/ABS1201021S

Shapiro-Ilan D, Hazir S, Glazer I (2017) Basic and applied research: entomopathogenic nematodes. In: Lacey LA (ed) Microbial control of insect and mite pests: from theory to practice. Academic Press, pp 91–105

Sharma R, Sharma P (2021) Fungal entomopathogens: a systematic review. Egypt J Biol Pest Control 31:1–13. https://doi.org/10.1186/s41938-021-00404-7

Sharma A, Sharma S, Yadav PK (2023) Entomopathogenic fungi and their relevance in sustainable agriculture: a review. Cogent Food 9:2180857. https://doi.org/10.1080/23311932.2023.2180857

Sheppard DC, Filler SG (2015) Host cell invasion by medically important fungi. Cold Spring Harb Perspect Med 5:a019687. https://doi.org/10.1101/cshperspect.a019687

Singh S, Singh AP, Chaurasiya DK, Mukherjee S, Mondal G (2020) Entomopathogenic fungi as biocontrol agents in agriculture. Res Today 2:264–266

Sinha KK, Choudhary AK, Kumari P (2016) Entomopathogenic fungi. In: Omkar (ed) Ecofriendly pest management for food security. Academic Press, pp 475–505. https://doi.org/10.1016/B978-0-12-803265-7.00015-4

Soliman N (2020) Toxicological and biochemical effects of *Beauveria bassiana* (Bals.) on peach fruit fly, *Bactrocera zonata* (Saunders) immature stages. J Plant Prot Pathol 11:579–585. https://doi.org/10.21608/jppp.2020.133805

St Leger RJ, Wang C (2010) Genetic engineering of fungal biocontrol agents to achieve greater efficacy against insect pests. Appl Microbiol Biotechnol 85:901–907. https://doi.org/10.1007/s00253-009-2306-z

St Leger RJ, Butt TM, Staples RC, Roberts DW (1989) Synthesis of proteins including a cuticle-degrading protease during differentiation of the entomopathogenic fungus *Metarhizium anisopliae*. Exp Mycol 13:253–262. https://doi.org/10.1016/0147-5975(89)90047-9

Tandon V, Rana A, Kapoor P, Katoch A (2024) Biocontrol potential and specificity of entomopathogenic fungi. In: Deshmukh SK, Sridhar KR (eds) Entomopathogenic fungi: prospects and challenges. Springer Nature, Singapore, pp 127–140. https://doi.org/10.1007/978-981-97-5991-0_6

Tefera T, Vidal S (2009) Effect of inoculation method and plant growth medium on endophytic colonization of sorghum by the entomopathogenic fungus *Beauveria bassiana*. BioControl 54:663–669. https://doi.org/10.1007/s10526-009-9216-y

Thakur R, Sandhu SS (2010) Distribution, occurrence and natural invertebrate hosts of indigenous entomopathogenic fungi of Central India. Indian J Microbiol 50:89–96. https://doi.org/10.1007/s12088-010-0007-z

Thakur R, Rajak RC, Sandhu SS (2005) Biochemical and molecular characteristics of indigenous strains of the entomopathogenic fungus *Beauveria bassiana* of Central India. Biocontrol Sci Tech 15:733–744. https://doi.org/10.1080/09583150500136584

Tomson M, Sahayaraj K, Sayed S, Elarnaouty SA, Petchidurai G (2021) Entomotoxic proteins of *Beauveria bassiana* Bals. (Vuil.) and their virulence against two cotton insect pests. J King Saud Univ Sci 33:101595. https://doi.org/10.1016/j.jksus.2021.101595

Van de Veire M, Degheele D (1996) Compatibility of the entomopathogenic fungus *Beauveria bassiana* with the parasitoid *Encarsia formosa* for control of whitefly on poinsettia. Meded Fac Landbouwkd Toegep Biol Wet Univ Gent 61:423–430

Villegas-Mendoza JM, Mireles-Martínez M, Rodríguez-Castillejos G, Santiago-Adame R, Rosas-García NM (2019) Toxicity evaluation of a methanolic extract from *Beauveria bassiana* against *Spodoptera exigua* and *Spodoptera frugiperda* (Lepidoptera: Noctuidae) larvae. Southwest Entomol 44:715–722

Vinayaka J (2014) Species composition of white grubs in rainfed ecosystem of Dharwad and evaluation of EC formulations of *Metarhizium anisopliae* (Meschinikoff) Sorokin. MSc thesis, University of Agricultural Sciences, Dharwad, Bengaluru, India, p 76

Wang L (2018) Endophytic entomopathogenic fungi enhance the growth of *Phaseolus vulgaris* L. (Fabaceae) and negatively affect the development and reproduction of *Tetranychus urticae* Koch (Acari: Tetranychidae). Microb Pathog 125:385–392. https://doi.org/10.1016/j.micpath.2018.09.044

Wang C, St Leger RJ (2007) The MAD1 adhesin of *Metarhizium anisopliae* links adhesion with blastospore production and virulence to insects, and the MAD2 adhesin enables attachment to plants. Eukaryot Cell 6:808–816. https://doi.org/10.1128/ec.00409-06

Wang Q, Xu L (2012) Beauvericin, a bioactive compound produced by fungi: a short review. Molecules 17:2367–2377. https://doi.org/10.3390/molecules17032367

Wang G, Liu Z, Lin R, Li E, Mao Z et al (2016) Biosynthesis of antibiotic leucinostatins in biocontrol fungus *Purpureocillium lilacinum* and their inhibition on Phytophthora revealed by genome mining. PLoS Pathog 12:e1005685. https://doi.org/10.1371/journal.ppat.1005685

Wang L, Keyhani NO, Xia Y, Xie J (2024) The potential and limitations of entomopathogenic fungi as biocontrol agents for insect pest management. Entomol Gen 44(4):797–811. https://doi.org/10.1127/entomologia/2024/1725

Wu S, Gao Y, Xu X, Zhang Y, Wang J et al (2013) Laboratory and greenhouse evaluation of a new entomopathogenic strain of *Beauveria bassiana* for control of the onion thrips *Thrips tabaci*. Biocontrol Sci Technol 23:794–802. https://doi.org/10.1080/09583157.2013.794896

Wu S, Gao Y, Smagghe G, Xu X, Lei Z (2016) Interactions between the entomopathogenic fungus *Beauveria bassiana* and the predatory mite *Neoseiulus barkeri* and biological control of their shared prey/host *Frankliniella occidentalis*. Biol Control 98:43–51. https://doi.org/10.1016/j.biocontrol.2016.04.001

Xing Y, Yang J (2020) Stiffness distribution in natural insect cuticle reveals an impact resistance strategy. J Biomech 109:109952. https://doi.org/10.1016/j.jbiomech.2020.109952

Xu Y, Orozco R, Wijeratne EM, Espinosa-Artiles P, Gunatilaka AA, Stock SP et al (2009) Biosynthesis of the cyclooligomer depsipeptide bassianolide, an insecticidal virulence factor of *Beauveria bassiana*. Fungal Genet Biol 46:353–364. https://doi.org/10.1016/j.fgb.2009.03.001

Zhang D, Qi H, Zhang F (2025) Parasitism by entomopathogenic fungi and insect host defense strategies. Microorganisms 13:283. https://doi.org/10.3390/microorganisms13020283

Application of Entomopathogenic Fungi as Biocontrol Agents for Insect Pest Management

15

Parthasarathy Seethapathy and Rajadurai Gothandaraman

Abstract

Entomopathogenic fungi (EPF) have emerged as a major biopesticide in the framework of sustainable pest management, providing an eco-friendly alternative to synthetic pesticides. These fungi, naturally found in diverse ecosystems, parasitize and kill insect hosts through direct penetration of the cuticle, followed by internal propagation and host death. Their host specificity, minimal non-target effects, and compatibility with integrated pest management (IPM) approaches support EPF as highly suitable for current farming systems. This chapter provides recent updates in the application of EPF as biocontrol agents, highlighting encapsulated formulations, precision delivery, and integration into IPM systems suitable for agroecological crop protection. It also explores the ecological roles of EPF and addresses field-level challenges to enhance their efficacy and sustainability. Advancements in mass multiplication, formulation, and application methods have helped the development of commercial biopesticides, but challenges such as environmental sensitivity, unpredictable field performance, and short shelf life limit their widespread adoption. Recent innovations in the formulation industry, molecular biology, and delivery systems, including nanoscience and drone-based spraying, are expanding the practical application of EPF in field conditions. Furthermore, their potential roles beyond pest suppression include endophytic and rhizospheric colonization, plant growth promotion, and soil health improvement. This chapter also addresses environmental,

P. Seethapathy (✉)
Department of Plant Pathology, Amrita School of Agricultural Sciences, Amrita Vishwa Vidyapeetham, Coimbatore, India

R. Gothandaraman
Department of Plant Biotechnology, Centre for Plant Molecular Biology and Biotechnology, Tamil Nadu Agricultural University, Coimbatore, India

S. K. Deshmukh et al. (eds.), *Frontiers in Entomopathogenic Fungi*,
https://doi.org/10.1007/978-981-95-6529-0_15

biosafety, and regulatory considerations that influence EPF deployment. Addressing through case studies and field applications, this chapter highlights the practical relevance and emerging research directions intended at optimizing EPF-based biopesticides. Overall, EPF is a promising frontier in the development of biological pest management approaches aligned with global SDGs toward zero hunger, ecological strengthening, and sustainable production.

Keywords

Biocompatibility · Bioformulation · Biopesticides · Endophyte · Mycoinsecticides

1 Introduction

Insect pests are the most limiting factors in crop production, contributing to an estimated 10.80% loss in global-level crop yields during the post-Green Revolution era (Dhaliwal et al. 2015). Recently, annual crop production losses at the global level have been estimated at 18–26%, corresponding to a monetary value of approximately $470 billion (Mantzoukas and Eliopoulos 2020). To alleviate these losses, insecticides or pesticides have been widely employed and have become a critical component in pest management due to their relatively low application effort, high efficacy, and operational convenience (Sharma and Sharma 2021). However, the intensive and prolonged use of chemical insecticides has led to the advance of resistance in approximately 500 pest species, often to one or more classes of insecticides (Kumar and Kalita 2017).

Consequently, entomopathogens that comprise fungi, viruses, protozoa, and bacteria are increasingly recognized as essential biological control agents in integrated pest management (IPM) strategies (Harish et al. 2021; Parthasarathy et al. 2024b). These antagonistic microorganisms function as natural controllers of insect pest populations, serving to suppress outbreaks and reduce reliance on chemical pesticides. Naturally occurring entomopathogens contribute significantly to the ecological balance by maintaining insect densities below economic threshold levels (Roy and Cottrell 2008). The term "entomopathogens" was first introduced by Tanzini to describe microbial agents that can control insect pests effectively without causing economic harm to agricultural and horticultural crops. These fungi were categorized as either facultative or obligate parasites, depending on their reliance on the host for survival and reproduction. The field application of entomopathogenic fungi (EPF) as biocontrol agents is not only effective in managing crop pest populations but also ensures environmental sustainability, reducing harmful effects to humans and non-target animals due to the lack of residual effects. Within agroecosystems, along with natural enemies, predators, parasitoids, and EPF regulate key pest populations and prevent their escalation to economically damaging levels, especially in perishable fruits and vegetables.

EPF are among the earliest organisms employed in the biological control of insect pests. Owing to their eco-friendly nature and biological perseverance, EPF are widely chosen for targeting insect pests across numerous developmental phases

(Tahira et al. 2014). These EPF are recognized for their capacity to infect insects under favorable conditions by directly entering the cuticle of the host and proliferating within the hemolymph, ultimately leading to host death. To date, over 700 fungal species from 90 diverse genera have been identified as EPF (Maina et al. 2018; Bamisile et al. 2021; Karthi et al. 2024). Notable EPF species include *Aspergillus nidulans*, *A. fumigatus*, *Batkoa apiculata*, *Beauveria bassiana*, *Cladosporium cladosporioides*, *Conidiobolus obscurus*, *Cordyceps* spp., *Entomophaga grylli*, *Entomophthora muscae*, *E. planchoniana*, *Metarhizium anisopliae*, *M. flavoviride*, *Cordyceps fumosorosea*, *Nomuraea rileyi*, *Pandora neoaphidis*, *Penicillium citrinum*, and *Trichoderma harzianum*, all of which function as natural antagonists of insect pests. Though some species, such as *B. bassiana*, *M. anisopliae*, *C. fumosorosea*, *Purpureocillium lilacinum*, and *Lecanicillium* spp., are predominantly associated with the phylloplane, most EPF are soil-borne organisms capable of infecting and killing arthropods through cuticular penetration and colonization of the host hemolymph.

The most studied genera of EPF for crop pest management include *Akanthomyces* Lebert, *Beauveria* Vuill., *Cordyceps* Fr., *and Metarhizium* Sorokīn (Ascomycota: Hypocreales). These EPF have been extensively studied due to their omnipresence in nature, host range, high efficacy in pest control, and ease of use in the development of biocontrol formulations for commercial products. They are well characterized for the mass production of free conidia from conidiophores, which facilitates their dissemination by wind and water. EPF are widely distributed across diverse ecosystems and can be found in a variety of geographic and climatic regions, including both cultivated and natural soils (Garrido-Jurado et al. 2015). They are capable of infecting insect pests at all growth stages, from eggs, larvae, and pupae to adults. Although primarily recognized for their role as biological control agents, these EPF have also demonstrated the ability to protect host plants from diseases and mitigate damage caused by pathogenic microorganisms and nematodes (Chen et al. 2015; Parthasarathy 2023a; Saha et al. 2025). The objective of this chapter is to examine the significance of EPF in the context of application in IPM for sustainable agriculture and their field-level commercial developments toward the biological control of insect pests and nematodes to fulfill UN SDGs, especially Zero Hunger SDG-2.

2 Diversity and Key Genera in Biocontrol

EPF primarily belong to Hypocreales, Onygenales (specifically *Ascosphaera*), Entomophthorales, and Neozygitales (Entomophthoromycota) (Georgiev et al. 2013). These fungi target various insect orders, such as Coleoptera, Diptera, Hemiptera, Hymenoptera, Lepidoptera, and Orthoptera (Maina et al. 2018; Mongkolsamrit et al. 2020; Saminathan et al. 2025). Some species in Hypocreales infect diverse host insects, whereas those in the Entomophthorales tend to focus on particular insect species (Bidochka and Hajek 1998). These fungi have the ability to infect a diverse array of insect pests, nematodes, and mite species, such as thrips, aphids, and lepidopterous larvae, which are significant agricultural pests across the globe.

2.1 Major Genera Used in Crop Pest Management

2.1.1 *Beauveria*

Beauveria species (Cordycipitaceae) (White Muscardine Fungi) belong to Ascomycota and exhibit a filamentous morphology (Bamisile et al. 2021) and produce unicellular, small hydrophobic anamorphic conidia. This genus comprises several economically and biologically important species, including *B. amorpha, B. bassiana, B. brongniartii, B. caledonica, B. peruviensis, B. polyrhachicola, B. pseudobassiana, B. rudraprayagi, B. songmingensis, B. subscarabaeidicola, B. staphylinidicola, B. sungii,* and *B. yunnanensis* (Islam et al. 2021). Among them, *B. bassiana* is a cosmopolitan type species that occurs naturally in soil, rhizosphere, or biological endophytic inhabitants associated with many plant species. It is widely recognized for its entomopathogenic activities, being the causative agent of white muscardine sickness in various insect taxa (Rohrlich et al. 2018). *B. bassiana* is a well-known entomopathogenic fungus widely used in biological control programs. It represents the asexual (anamorphic) stage of the fungus, whereas *Cordyceps bassiana* corresponds to its sexual (teleomorphic) form (Parthasarathy 2023a). Numerous *Beauveria* isolates demonstrate high host specificity, and infected hosts exhibit extensive colonization, with the exoskeleton becoming densely coated with white mycelial growth, frequently leading to the formation of anamorphic synnemata (Rohrlich et al. 2018). *B. bassiana* can be readily identified through microscopic analysis, which reveals characteristic conidiogenous cells with globose bases and elongated, denticulate rachides. These structures produce globose conidia measuring less than 3.5 μm in diameter (de Costa Eula Maria et al. 2011). The conidia are typically unicellular, hyaline, and spherical and are often observed in dense aggregations formed by numerous conidiogenous cells (Kary and Alizadeh 2017). *Beauveria* species are capable of causing epizootics and are therefore considered effective biopesticides for the management of various insect pests worldwide, including malaria-vectoring mosquitoes, termites, and whiteflies (Lovett and St Leger 2018). *Beauveria* spores can be formulated into wettable powders or emulsifiable suspensions, both of which possess insecticidal properties suitable for application on infested crops. *B. bassiana* is known to parasitize a broad range of arthropod pests and is generally regarded as a non-selective general biocontrol agent for pests, nematodes, and diseases. Major pests that have demonstrated susceptibility to this *B. bassiana* include the European corn borer, pine caterpillar, and green leafhopper (Mascarin and Jaronski 2016). Also, the study (Mushyakhwo et al. 2025) evaluated nine EPF; among them, *B. bassiana* NCHU-157 induced the highest mortality rates in *Thrips palmi*, causing 35% larval and 98% adult mortality within 7 days post-inoculation.

2.1.2 *Akanthomyces*

Akanthomyces (Cordycipitaceae) (*Verticillium* spp./*Lecanicillium* spp.) includes several important parasitic species such as *A. dipterigenus*, *A. lecanii* (formerly *Verticillium lecanii*), *A. lepidopterorum*, *A. muscarius*, *A. uredinophilus*, and *Verticillium chlamydosporium* (Lin et al. 2016; Shi-Kunne et al. 2018) targeting diptera, hemiptera, and lepidoptera orders. Climatic conditions significantly influence fungal populations; notably, due to its widespread distribution, *A. lecanii* (syn. *V. lecanii*) can cause extensive epizootics in temperate and tropical regions with warm and humid climatic conditions. *Verticillium* species are characterized by the verticillate pattern of branching of their conidiophores from the mycelium (Shi-Kunne et al. 2018). *V. lecanii* is not known to express a sexual reproductive stage (teleomorph) and propagates mostly through non-motile asexual conidia. Upon germination, these asexual conidia give rise to hyphae, which further develop into conidiophores that subsequently produce new conidia within a short period. *Akanthomyces* is likely not restricted to a specific developmental stage, as it can infect both nymphs and adults by its mode of action. It adheres to the lower side of leaves through filamentous mycelia production. *Lecanicillium* species are distributed across diverse environments, including soil, arthropod hosts, and decaying fungal matter (Nicoletti and Becchimanzi 2020).

The effectiveness of *Lecanicillium* spp. as EPF is primarily influenced by host ecological factors, particularly optimal temperature and humidity under field conditions (Gangireddy Eswara Reddy 2021). Key species include *L. anqingense, L. attenuatum, L. lecanii, L. longisporum, L. nodulosum, L. muscarium, L. primulinum, L. renii,* and *L. uredinophilum* (Nicoletti and Becchimanzi 2020). These fungi have demonstrated efficacy against a variety of insect pests, particularly sap-sucking species such as aphids, whiteflies, mealybugs, and scale insects. In a case study, *A. lecanii* (syn. *V. lecanii*) has demonstrated effectiveness as a biocontrol agent against certain insect pests, including *Trialeurodes vaporariorum* (greenhouse whitefly) (Chen et al. 2008).

2.1.3 *Metarhizium*

M. anisopliae (Cordycipitaceae) is a well-characterized, soil-dwelling entomopathogenic fungus with a broad ecological distribution, spanning from arctic regions to tropical environments worldwide (Karthi et al. 2024). Commonly known as the green muscardine fungus, it was first described by Metschnikoff in 1879 as *Entomophthora anisopliae* and was subsequently reclassified under *Metarhizium*. Compared to *Beauveria*, *Metarhizium* exhibits a relatively narrower host range, though it can infect over 200 insect species across seven insect orders, particularly soil-dwelling insects. Susceptible hosts include coleopteran beetles (mostly within Scarabaeidae), root weevils, flies, thrips, and fungus gnats (Yapa et al. 2025). Among EPF, the *M. anisopliae* complex is considered one of the most comprehensively studied at the morpho-molecular level, particularly in its virulence mechanisms and host specificity. This genus comprises many important species: *Metarhizium acridum, M. anisopliae, M. brunneum, M. flavoviride, M. frigidum, M. majus, M. humberi, M. guizhouense, M. novozealandicum, M. pemphigi, M. pingshaense, and M. rileyi*, among which *M. anisopliae* is recognized as a highly potent EPF (Glare et al. 2021; Yapa et al. 2025). A distinguishing feature of this genus is the intricate branching pattern of the conidiophores, which are densely interwoven and exhibit rounded to conical apices. The conidiogenous cells are aligned in a compact hymenial arrangement (Mongkolsamrit et al. 2020).

Metarhizium species produce conidia arranged in chains, which tend to aggregate into conidial columns. These conidia are aseptate, lacking cross walls, and typically exhibit pale to bright green or yellow-green coloration. They are ovoid to cylindrical and are produced in chains, which commonly aggregate into prismatic or cylindrical columns or occasionally form a solid mass of parallel chains (Leadmon et al. 2020). These conidia typically adhere to the upper 20 cm of the soil profile and have demonstrated the ability to remain viable for up to 34 years. As a case study, the teak skeletonizer, *Eutectona machaeralis*, was selected as a test insect to evaluate the pathogenicity of *M. anisopliae*, which demonstrated strong potential as a myco-biocontrol agent against this pest. Another study (Mushyakhwo et al. 2025) found that among the nine EPF evaluated, *Metarhizium lepidiotae* NCHU-9 resulted in 31% larval and 81% adult mortality rates on *T. palmi* 7 days post-inoculation, similar to *B. bassiana* NCHU-157.

M. rileyi (Farlow) Kepler, SA Rehner, and Humber (formerly *Nomuraea rileyi*) (Hypocreales: Clavicipitaceae), *M. atypicola,* and *M. rileyi* are two significant EPF species within this genus (Boucias et al. 2000). Numerous insect species have exhibited epizootic mortality resulting from *M. rileyi* infection, which is attributed to its dimorphic hyphomycetous nature (Ramos et al. 2024). The hyphae are septate, possessing cross walls, and are white in color with a flocculent overgrowth. In a culture, colony growth is typically sparse; however, on the insect host, the fungus proliferates densely, often invading the entire host body. As sporulation progresses, the fungal colony color transitions from green to purple-gray or deep purple. The conidiogenous cells are relatively short with blunt apices (Kothari and Sandhu 2024). *M. rileyi* exhibits a high degree of host specificity and does not pose any adverse effects on the environment, making it a promising antagonistic agent for sustainable insect pest management. Several insect species, including *Spodoptera exigua*, *Spodoptera frugiperda*, *Trichoplusia* spp., *Heliothis zea*, *Plathypena scabra*, and *Bombyx mori*, have been identified as susceptible insect hosts supporting its infection and development (Montecalvo and Navasero 2021; Gebreslasie et al. 2023; Kothari and Sandhu 2024; Yapa et al. 2025).

2.1.4 *Cordyceps*

Cordyceps (*Paecilomyces* spp./*Isaria* spp.), a genus within Cordycipitaceae, includes two notable species: *C. fumosorosea* (syn. *Paecilomyces fumosoroseus/ Isaria fumosorosea*) and *P. lilacinum* (Tkaczuk and Majchrowska-Safaryan 2023). In many species, the conidiophores are synnematous, with well-developed septa and whorls of branches along with conidiogenous cells (phialides) (Nguyen and Doan 2024). Morphologically, the conidiogenous cells are oval to flask-shaped or subglobose, typically possessing a distinct neck. These cells may occur singly or in whorls, while the conidia are single-celled and range in color from hyaline to light-colored (Nguyen and Doan 2024). *Cordyceps* species are recognized as EPF with pathogenic capabilities and have demonstrated significant effectiveness in managing crop pests and plant-parasitic nematodes through infection-induced mortality (Tahira et al. 2014). *C. fumosorosea* is known to cause a disease in whiteflies referred to as “yellow muscardine sickness” and functions as a natural antagonist of many pests (Sandhu et al. 2012). A study says *Cordyceps* spp. exhibit strong epizootic activity against *Trialeurodes* spp. and *Bemisia* spp. under both greenhouse and field conditions (Deshmukh and Sandhu 2023). Associated soil characteristics play a critical role in the efficacy of biocontrol agents; tropical and subtropical agricultural soils have been reported to offer the most favorable conditions for their activity (Deshmukh and Sandhu 2023).

Isaria includes several important EPF that were previously classified under *Paecilomyces* and are now under *Cordyceps*. Prominent species within this genus include *Isaria farinosa* and *I. fumosorosea* (Zimmermann 2008). Recent taxonomic changes, such as *Cordyceps farinosa* (syn. I. *farinosa*) and *C. fumosorosea* (syn. *I. fumosorosea*), are EPF that have shown great potential in the control of a wide range of insect pests (Nguyen and Doan 2024). These fungi exhibit a broad host range, with a primary affinity for lepidopteran insects. During infection, they are known to target and invade the reproductive tissues of their insect hosts (Gao et al. 2017). *I. farinosa* was isolated from various living host insect orders, including Acari, Araneae, Coleoptera, Diptera, Hemiptera, Hymenoptera (specifically Formicidae), Lepidoptera, Thysanoptera, and Neuroptera. *I. fumosorosea* is considered a species complex, with various strains demonstrating efficacy against a range of insect pests, including Acari, Hemiptera, Hymenoptera (specifically Formicidae), and certain dipteran species (Zimmermann 2008). The study by Bugti et al. (2018) identified the *I. fumosorosea* Ifu13a strain as an effective biocontrol agent against various sap-sucking insect pests. It was applied at a concentration of 10^8 conidia/mL, the Ifu13a strain achieved mortality rates ranging from 81% to 100% against *Jacobiasca formosana* (tea jassid), *Aphis gossypii* (cotton aphid), *Bemisia tabaci* (tobacco whitefly), and *Stephanitis nashi* (pear lace bug) (Bugti et al. 2018).

A notable limitation of using *Isaria* species as biocontrol agents is their potential impact on non-target insects. *I. farinosa* has been reported to infect beneficial and non-target organisms such as the silkworm (*Bombyx mori*), parasitoid wasps (*Dibrachys affinis*), honeybees (*Apis mellifera*), and ground beetles (*Broscus cephalotes*) (Zimmermann 2008). However, the development of resistance to *Isaria* species among target pest populations appears to be minimal. A study depicts that continuous application of the fungus over five successive generations of whiteflies did not result in any observable resistance in its biotypes (Gao et al. 2017).

2.1.5 *Hirsutella*

Hirsutella comprises three notable species: *Hirsutella thompsonii*, *H. citriformis*, and *H. gigantea*. Accurate identification and differentiation of *Hirsutella* species from other fungal genera can be challenging due to the high number of species and lack of significant morphological variability within the genus. Anamorphically, it produces synnemata, which are erect and may be cylindrical or slightly tapered, ranging from long and hair-like to short and verrucose, and may be unbranched, sparsely branched, or possess numerous short lateral branches. The conidiogenous cells of *H. thompsonii* are typically subglobose in shape, characterized by a swollen basal region (Mongkolsamrit et al. 2020). Antagonistically, among the members, *H. thompsonii* has demonstrated potential as a biocontrol agent, particularly against small arthropods such as the citrus rust mite, and is also pathogenic to certain species within Lepidoptera and Hemiptera (Pérez-González et al. 2022).

Jampameung et al. (2024) explored six isolates of *H. thompsonii* that affect the coconut mite (*Aceria guerreronis*), parasitizing both eggs and adult forms. Their findings showed that isolates TBRC 6805 and TBRC 6933 of *H. thompsonii* caused the highest mortality rate, reaching 100% by the third day after inoculation. The hemp russet mite (*Aculops cannabicola*) poses a significant threat to indoor and greenhouse hemp (*Cannabis sativa* L.), and unfortunately, there are limited options for effectively managing this pest. But *H. thompsonii* targets and eliminates the hemp russet mite, presenting a promising sustainable approach for controlling this pest (Ricciardi et al. 2024). *L. lecanii* and *H. thompsonii* are effective in managing the broad mite, *Polyphagotarsonemus latus*, on mulberry. Notably, *H. thompsonii* achieves a higher rate of mite mortality than *L. lecanii* (Ramesha and Sannappa 2024).

2.2 Host Range and Specificity

EPF cause lethal infections in insects and plays a significant role in regulating insect populations through natural epizootics. EPF, especially those belonging to Hypocreales, represent a fascinating and often widely studied group of fungi that target insect pests and other arthropods around the world. Even though they play a vital role in ecosystem balancing, there is still a lack of detailed information about the host specificity and geographical distribution of these hypocrean EPF. In response to this need, EntomoFun 1.0 was developed as a freely accessible database that brings together worldwide records of EPF associations within Hypocreales (De Wint et al. 2024). This database features 1791 entries that provide insights into EPF species and their insect host taxa, the countries where they are found, and the life stages of these hosts. EntomoFun 1.0 (https://datadryad.org/dataset/doi:10.5061/dryad.1zcrjdg17) is built upon a foundation of 600 literature sources, along with herbarium specimens from the Royal Botanical Gardens, Kew (De Wint et al. 2024). The representation of records related to the taxonomy of EPF is organized in the following sequence: Order, Family, Genus, and Species. Ophiocordycipitaceae makes up 78%, followed by Cordycipitaceae at 20% and Clavicipitaceae at 2% (De Wint et al. 2024).

The ecological host range of EPF encompasses many species, which the parasite naturally forms symbiotic relationships with in the environment, resulting in viable progeny. These fungi exhibit a broad host range, with over 1800 documented associations between fungal species and various insect hosts. The term "host range" refers to the spectrum of species that support the survival and reproduction of EPF (Parthasarathy et al. 2024a). This range is dynamic and subject to evolutionary changes in nature. In contrast, the physiological host range is determined through controlled laboratory experiments and may include species that are susceptible to infection under experimental conditions but are not necessarily infected in natural conditions. A valid pathogen–host association is established when an insect is naturally or experimentally infected and the pathogen completes its life cycle by producing infectious propagules (Parthasarathy 2023a). If infection does not occur despite exposure, no such association is recognized. The host specificity of EPF varies significantly both among and within genera. The muscardine fungi, *M. anisopliae* and *B. bassiana*, are widely studied EPF and exhibit a broad host range in arthropods and other fungi. Recently, several species from Lepidoptera, Coleoptera, Hymenoptera, Diptera, and Homoptera have been documented as hosts of these fungi (Tahira et al. 2014; Lovett and St Leger 2018; Maina et al. 2018). *M. anisopliae* primarily infects Coleopteran species, including more than 70 species of Scarabaeidae (Mongkolsamrit et al. 2020). Insect-pathogenic fungal strains tend to be host-specific, with rare instances where they may infect both beneficial and pest species. *Zoophthora radicans* has been isolated from over 80 insect species across Lepidoptera, Diptera, Coleoptera, and Homoptera (Sacco and Hajek 2023). With a few exceptions, most strains are highly specific and typically cause greater infection in closely related hosts, while the host range is more restricted in other EPF groups.

In most cases, infection by EPF leads to the death of susceptible insect hosts. On rare occasions, EPF may cause mortality in certain species under specific or conducive conditions. However, insect susceptibility to various EPF can differ under laboratory and field conditions. Additionally, the susceptibility of different developmental stages of insects varies depending on the specific characteristics of the fungal species. Some fungi are capable of producing toxins and enzymes and can cause infections in vertebrates; as such, certain fungal species are also recognized as common allergens (Bidochka and Hajek 1998; Tahira et al. 2014; Leadmon et al. 2020; Sharma et al. 2023). Some hypocrean fungi have been implicated in allergic reactions in humans and livestock. EPF are generally not allergenic to humans, and vertebrates are not considered susceptible to EPF infections. Exceptions exist, such as *Aspergillus flavus*, a known pathogen of vertebrates, which is one of the reasons certain fungi are not used in commercial applications (Bischoff et al. 2009; Sharma and Sharma 2021).

Host switching among fungal biocontrol agents is a critical factor in understanding fungal virulence and survival in relation to both target and non-target hosts (Mascarin and Jaronski 2016; Lovett and St Leger 2018). Such host-switching events can have detrimental effects on biological systems by disrupting the population dynamics of non-target species (Zimmermann 2008). Given the broad host range of EPF, the possibility of "host-jumping" events cannot be ruled out; however, detecting and tracing these events, particularly from an evolutionary perspective, remains challenging (Sharma and Sharma 2021; Sharma et al. 2023). A phylogenetic analysis using mitochondrial DNA sequences from *Cordyceps* species (including truffle-associated and insect-parasitic species) to investigate inter-kingdom host-jumping events (Harish et al. 2022). Further, findings revealed a host shift from cicada nymphs to Hart's truffle, exemplifying an evolutionary transition across kingdoms. Nevertheless, they were unable to resolve the evolutionary pathways underlying major host shifts between insect orders with sufficient clarity (Nguyen and Doan 2024).

2.3 Natural Occurrence and Ecological Adaptation

EPF encompass taxa from various fungal groups; the most prominent and commonly encountered EPF are found within Hypocreales and Entomophthorales. EPF are distributed across a wide range of habitats, including agricultural fields, aquatic environments, forests, pastures, urban areas, and deserts. Evidence suggests that they can also occur in locations devoid of arthropod hosts (Mascarin et al. 2024). The distribution of EPF varies among species; many are cosmopolitan and can be found in diverse geographic regions (Tahira et al. 2014). *B. bassiana* has been reported in the tropical rainforests of Thailand as well as in temperate regions such as Canada (Mascarin and Jaronski 2016). Numerous studies have examined the influence of climatic conditions, geographic location, habitat type, soil characteristics, cropping systems, and biotic interactions on the occurrence and dispersal of insect-pathogenic fungi. The ecological adoption of *B. bassiana, M. anisopliae*, and *Tolypocladium cylindrosporum* was observed in Norway, as well as the occurrence of *C. farinosa* (*I. farinosa*), *B. bassiana*, and *M. anisopliae* in Finland. Furthermore, EPF have also been documented in extreme environments such as Arctic Greenland and Antarctica as a natural occurrence. *Paecilomyces antarctica*, an endemic fungal species, was isolated from the Antarctic springtail (*Cryptopygus antarcticus*) in the Antarctic region as a natural occurrence. Studies have shown that altitude has a minimal effect on the presence of EPF, as evidenced by the detection of *B. bassiana* at elevations as high as 1608 m. Conversely, there are some findings observed that altitude, up to 5200 m, significantly influences the species diversity of insect-associated fungi (Garrido-Jurado et al. 2015; Quesada-Moraga et al. 2022, 2024).

The dominance of specific EPF varies across different habitat types. Members of Hypocreales are more diverse in the upper soil layers and surface litter, whereas *Gibellula*, which primarily infects spiders, is more prevalent in subsurface environments, as observed in Białowieża Forest in Poland (Mendes-Pereira et al. 2023). Entomophthorales are recognized as key pathogens of insect pests in temperate forest ecosystems, but their occurrence is relatively rare in tropical forests. In contrast, *Cordyceps* tends to be more abundant in moist tropical forests (Nguyen and Doan 2024). Additionally, *Beauveria*, *Metarhizium*, and *Cordyceps* (*Isaria*) are frequently associated with soil-dwelling insect hosts. While *B. bassiana* and *M. anisopliae* are present across a wide range of ecosystems, *B. bassiana* is generally more dominant in natural habitats. Its distribution is significantly influenced by agricultural practices such as soil tillage, which can negatively affect its persistence (Quesada-Moraga et al. 2024). Numerous studies have indicated that the diversity of EPF are considerably reduced in cultivated soils compared to forested environments. In forest ecosystems, EPF are typically more abundant in the soil and leaf litter of temperate forests than in tropical forests. Nonetheless, temperate forests support a markedly greater diversity of EPF than agricultural soils (Tkaczuk and Majchrowska-Safaryan 2023).

3 Biology and Mode of Action

3.1 Biological Characteristics

The life stages of EPF consist of an active parasitic phase (from host infection to its death) and a latent saprophytic phase (after host death). Unlike other entomopathogens, such as bacteria and viruses (Harish et al. 2021), which typically enter the insect host via ingestion, EPF initiate infection by directly penetrating the host's external cuticle. The life cycle of EPF begins with the attachment and germination of spores on their targeted host cuticle. The conidia of EPF are generally enveloped by a mucilaginous layer rich in proteins and glucans, which facilitates adhesion to the insect cuticle. Spore germination leads to the formation of an infectious germ tube, which penetrates the host cuticle, facilitating the entry of the fungus into the insect's body (Bidochka and Hajek 1998).

Once inside, fungal spores proliferate and produce toxins and enzymes, resulting in the acquisition of nutrients and host mortality. Under favorable environmental conditions, the fungus emerges from the insect cadaver and liberates its next-generation spores into the surrounding environment. Many species can also produce resting spores under unfavorable environments, initiating new infections when conditions become conducive. The fungus relies on effective transmission strategies for successful infection and dissemination to new hosts. The key factors influencing fungal reproduction, survival, and pathogenicity are the presence of a susceptible host and optimal environmental conditions (Seethapathy 2024).

3.2 Infection Process

Although entomopathogenic infections are a common route of infection for bacteria, viruses, and protozoa, they are relatively different among EPF (Harish et al. 2021; Ma et al. 2024). These EPF typically show a growth pattern that can take on two distinct forms. In the asexual vegetative cycle, spores begin to germinate and grow in a filamentous manner, leading to the production of sympodial conidia. This method is favored particularly in the absence of a specific insect host. During the availability of hosts, EPF shifts into a pathogenic life cycle. Initiating oral infection, processes such as mycosis and colonization occur; however, it remains unclear whether EPF infections become established in the intestinal tract or the buccal cavity (Lai et al. 2024).

Genome sequencing of well-studied EPF, such as *Metarhizium* spp. and *B. bassiana*, has facilitated the identification of genes potentially involved in oral infection pathways. Likewise, metabolomic analyses have predicted and revealed several metabolic pathways associated with oral infections by *B. bassiana*. EPF capable of infecting insects via multiple routes holds significant promise as candidates for optimized use in biological control strategies.

The primary route of infection for EPF is through cuticle penetration; however, several alternative infection routes have also been identified (Islam et al. 2021). The process of pathogenicity begins when conidiospores germinate on the insect cuticle, leading to their penetration through hyphal tubes. These alternative pathways may enhance EPF efficacy against fungal-resistant arthropods that have evolved antifungal defenses in their cuticle. Alternative infection routes were proposed decades ago when non-germinated spores of *M. anisopliae* were recovered from the gut of *Oryctes* larvae, and fungal hyphae were observed localized around the mandibles and esophagus of *Ephestia kuhniella*. Within the organism, the fungus transitions into a yeast-like form, generating hyphal bodies that propagate through a process of budding. Once it has killed the host, it returns to its saprotrophic stage, and this shift is crucial to its ability to cause disease.

The findings revealed insect mortality associated with hyphal development in the digestive tract; however, no conidial germination was observed in the gut, suggesting that infection likely occurred through the insect's mouthparts rather than through gut colonization. A study (Khan and Khan 2023) reported similar findings, wherein *Sitophilus granarius* Linnaeus was fed a mixture of *M. anisopliae* and *B. bassiana* conidia along with diatomaceous earth. To prevent cuticular penetration, the insect cuticles were surface-disinfected. The results indicated that beetle mortality was associated with hyphal development within the digestive tract. Although the findings from these and previous studies suggest that fungal spores exhibit a greater tendency to adhere to the buccal cavity lining than to the intestinal epithelium, the precise mechanisms underlying insect mortality following oral ingestion of conidia remain poorly understood. Consequently, more targeted physiological, molecular, and metagenomic investigations are warranted to elucidate this infection process.

3.2.1 Mode of Action

EPF engage with their insect hosts through a series of stages. Initially, they attach to the insect's exoskeleton using conidia and blastospores. Following this, they germinate and form infectious structures, including appressoria and pegs. Following this, they break through the host's outer layer, enter the body cavity, and begin to multiply. Ultimately, the EPF release spores following the death of the insect. EPF make its way into their hosts by penetrating the cuticle and accessing the hemolymph. Once they enter the hemocoel, they grow by taking in the nutrients found in the hemolymph, especially when the environmental conditions are just right. Once inside the insect host, EPF face a range of robust immune responses. Thereby, EPF have developed a variety of strategies to outsmart and disarm the insect's immune system. These methods involve strategies that alter their cell wall to escape detection, along with disrupting the host's immune response by releasing secondary metabolites and secretions. Mitosporic fungi produce a wide array of fungal toxins that negatively impact insect health (Mantzoukas and Eliopoulos 2020; Sharma and Sharma 2021).

These fungal toxins exhibit diverse effects on various insect tissues. Cytotoxic compounds disrupt cellular structures and facilitate hyphal penetration (Fadiji and Babalola 2020). Neuromuscular toxins compromise insect health, resulting in general paralysis, reduced responsiveness, and sluggish behavior. Several toxins from

B. bassiana, including beauvericin, beauverolides, bassiacridin, beauveriolides, bassianin, oxalic acid, tenellin, oosporein, isarolides, and brassinolide (Tahira et al. 2014; Sharma et al. 2023). Destruxins (DTXs) and cytochalasins are secondary metabolites produced by *M. anisopliae* and *Beauveria felina* within infected hosts. DTXs activate calcium channels, leading to depolarization of lepidopteran muscle membranes, and also impair the function of insect hemocytes (Litwin et al. 2020; Ahsan et al. 2024; Saha et al. 2025). Below are some key toxins produced by EPF.

4 Formulation and Mass Production Technologies

4.1 Solid-State Versus Liquid-State Fermentation

EPF are significantly important and have been widely used for biological control of various notorious insect pests. Conventionally, infected insect cadavers are collected and studied, then introduced into field environments across different seasons. In certain cases, insects are artificially reared in laboratory conditions, infected, and subsequently released into pest populations, either in the form of active individuals or as cadavers covered with sporulating fungi. Culturing these fungi on artificial media remains the most common method for their propagation. Artificial media or broth adequately support the growth of conidial cultures for field colonization. *B. bassiana* develops simply on both solid and liquid media, including Sabouraud and potato dextrose agar and broth. When grown on solid media, it generates aerial conidia that are both morphologically and infectively the same as those found on insect cadavers (Kary and Alizadeh 2017).

Formulations of *B. bassiana* are available in various forms, including wettable powder, talc, and aqueous suspensions. The recommended usage levels typically range from 10^7 to 10^8 conidia/ml, with application rates of 400 to 750 liters per hectare, tailored to meet the specific needs of different crops (Ain et al. 2021; Davari et al. 2021). Using an oil-based formulation of *B. bassiana* on the larval population of *Spodoptera litura* led to a 72% mortality rate, which notably improved crop production in groundnut (Ravi Sankar and Padmaja 2014; Hernández-Velázquez et al. 2000). Additionally, plant-based substrates such as cracked barley, rice, and wheat bran have been extensively employed in large-scale production systems. Millet-based media have also been reported as effective substrates for fungal mass production. Millet serves as a relatively simple and effective substrate, providing essential nutrients for fungal growth in soil-based media without the need for an insect host (Davari et al. 2021). In this method, millet grains are placed in a polyvinyl bag with water containing suitable acids. The mixture is initially heated to 90 °C for 1 h, followed by autoclaving at 121 °C for 30 min. Subsequently, the bag is incubated at 25 °C under a 16:8-h light-to-dark photoperiod for 3 weeks. After incubation, the culture is air-dried until the moisture content drops below 5%. This process yields granules with a 1.1×10^8 conidial concentration of *B. bassiana* per gram of grain, achieving a germination rate of 98.2% at 20 °C after 24 h. In Russia (Harper and Huang 1986), large-scale conidial production is achieved through extensive mycelial cultivation in fermenters. The resulting mycelial mass is subsequently transferred to shallow trays, approximately 1.0 cm deep, where conidia formation occurs after several days (Mongkolsamrit et al. 2020). It is noteworthy that the choice of substrate significantly influences the thermotolerance of the resulting conidia. Several studies have demonstrated that the storage stability and virulence of conidia can be influenced by the amino acid composition of the growth medium. Understanding the surface chemistry of EPF is essential for elucidating their ecological interactions.

Mycelium is also produced through a submerged fermentation process, where fungal biomass is harvested as thin mats, around 3 to 6 mm in thickness, which are subsequently treated with a sugar solution, naturally 10% maltose or sucrose, to act as a desiccation protectant. Although the mycelial mats become non-viable during processing, they are milled into particles approximately 2 mm in size at ambient temperature (Leadmon et al. 2020; Karthi et al. 2024). These particles can then be applied to the field, where exposure to moisture induces conidial production (Islam et al. 2021). The mycelial fragments can generate conidia continuously over several days, thus providing a sustained source of fresh inoculum in the field. For the preparation of these mycelial mats, the biomass is separated from the culture medium and dried at room temperature. Given the limited shelf life of conidia on plant surfaces under sunlight, the formulation of effective mycopesticides remains a significant challenge. Unlike chemical insecticides, organic solvents are unsuitable for fungal formulations due to their potential antiseptic effects on the propagules. Instead, microencapsulation techniques using pre-gelatinized or alginate starch have been identified as promising alternatives for preserving and delivering dried mycelium effectively (Islam et al. 2021; Saminathan et al. 2025).

The surrounding conditions significantly influence the virulence of EPF (Liu et al. 2025). Therefore, it is essential to understand and optimize the substrates used for culturing these fungi. Various substrates such as carrot-based media, rice grains, broken maize grains, wheat bran, softwood sawdust, cottonseed cake, cowpea leaves, and soybean chunks have been recommended for the cultivation and mass multiplication of *B. bassiana* (Skinner et al. 2014). In addition, alternative materials were discussed (Tahira et al. 2014), including vegetable waste, rice husk, cereal grains, pulse seeds, sawdust, and liquid media (coconut water, water from washing wheat or rice, and water from cooked rice), which have been proposed for the commercial-scale production of *B. bassiana*. Wheat grains supported the highest spore yield for *B. bassiana*. Conversely, sorghum was found to be the most effective substrate for spore production in *A. lecanii* (formerly *V. lecanii*) and *C. fumosorosea*. Additionally, substrates such as carrot, bhendi, and jackfruit seeds were observed to promote both sporulation and mycelial growth in all three fungal species tested. Coconut water similarly enhanced fungal growth and spore production. Furthermore, paddy chaffy grains were reported to stimulate the growth of all three *Fusarium* isolates examined (Tahira et al. 2014).

M. rileyi is a well-documented entomopathogenic fungus, known to infect a broad range of insect hosts, particularly polyphagous pest species within *Spodoptera*, *Heliothis*, *Plutella*, *Pseudoplusia*, *Rachiplusia*, and *Trichoplusia*

(Chen et al. 2008). However, the lack of reliable and cost-effective methods for its large-scale cultivation has limited its potential for commercial development and widespread application. Many investigated the use of various agricultural substrates, including rice, sorghum, wheat, and agro-industrial wastes such as discarded raw potatoes and bananas, for the commercial-scale production of *M. rileyi*. Among the tested substrates, rice supported the highest spore yield, with 5.53×10^7 spores per gram recorded on the 11th day post-inoculation. This was followed by discarded raw bananas (4.2×10^7 spores/g) and sorghum (4×10^7 spores/g), indicating their potential suitability for large-scale fungal cultivation (Jaber and Araj 2018).

4.2 Development of Oil, Wettable Powder, and Granular Formulations

Biopesticide formulations are generally categorized into two main types: dry and liquid formulations (Mascarin and Jaronski 2016). Dry formulations intended for direct application include powders, seed dressing powders (DS), granules (G), microgranules (MG), water-dispersible granules (WDG), and wettable powders (WP). Liquid formulations encompass emulsifiable concentrates (EC), suspension concentrates (SC), oil dispersions (OD), suspo-emulsions (SE), and capsule suspensions (CS) (Sharma et al. 2023). Among 171 commercial EPF-based products, the majority are formulated using *B. bassiana* (33.9%), *M. anisopliae* (33.9%), *C. fumosorosea* (5.8%), and *B. brongniartii* (4.1%) (Maina et al. 2018). Notably, both formulated and unformulated *B. bassiana*-based biopesticides effectively target pest species; only approximately half of the referenced studies included field efficacy evaluations (Mascarin and Jaronski 2016). This is a significant consideration, and discrepancies exist between claimed and actual efficacy in both salable and non-commercial strains from *Beauveria*, *Metarhizium*, and *Isaria*, with notable inconsistencies observed particularly among commercial formulations.

Following the isolation of pathogenic fungal strains, the subsequent step involves the identification and application of appropriate formulations for the management of arthropod pests. The effectiveness of EPF is highly influenced by the formulation type. Since the 1960s, a substantial number of EPF-based acaricides and insecticides have been developed and commercialized globally. Commercial biopesticides are most commonly formulated using strains of *Beauveria*, *Metarhizium*, and *Trichoderma* (Cai and Dimopoulos 2025). The EPF products available in the market occur in the form of dry powders or emulsified oils. Additives are frequently included to shield the spores from UV rays, enhance their adherence to foliage, or boost the humidity surrounding them. A photoactive dye, Phloxine B, has been proposed as a potential protector against phototoxicity. For powdered preparations, it is advisable to include moisture absorbers like calcium chloride, magnesium sulfate, white carbon, sodium sulfate, or silica gel (Kozłowska et al. 2019). Using flour as a nutritional additive to support the germination of fungal conidia, along with employing oil as a hygroscopic and lipophilic substance for arid conditions. Oils are incorporated into liquid and powder formulations to enhance product shelf life and improve performance in dry conditions. Adding both mineral and vegetable oils (corn and cottonseed oil) enhances the adhesion of spores to the surface of the insect cuticle (Ambethgar 2018; Karthi et al. 2024).

The design of formulations and application methods plays a crucial role in determining the efficacy, consistency, shelf life, and virulence of commercially available fungal-based biopesticides. For example, sprayable formulations are typically intended to produce immediate effects against plant pests and pathogens. To enhance their effectiveness, adjuvants such as surfactants are often incorporated to improve adhesion to plant surfaces or insect cuticles (Chaud et al. 2021; Friuli et al. 2023). Granular formulations, which are typically applied to soil, are designed to exert antagonistic effects over an extended period. To support sustained activity, such formulations often incorporate stabilizers and nutrient sources. Importantly, all adjuvants used in fungal-based biopesticides, including oils, surfactants, stabilizers, solvents, and hygroscopic agents, must be carefully selected to avoid compromising the efficacy of the biopesticide. In some cases, these additives may exert synergistic effects, enhancing desiccation resistance, spreading efficiency, plant uptake, and overall biocontrol performance (Quesada-Moraga et al. 2022). Wettable powder formulations, which consist of a fine powder that is mixed with water to form a suspension for application, are among the most widely used and readily applicable forms of EPF. Another commonly used type is the suspension concentrate, which contains a stable dispersion of the active ingredient (a.i.) in water and is intended to be further diluted before application. More recently, oil-miscible concentrate formulations have gained increasing attention due to their improved efficacy and application potential.

4.2.1 Wettable Powders

Wettable powder-based formulation of EPF is the most commonly produced and typically comprises 50–80% powder, 15−45% filler, 1−10% dispersing agents, and 3−5% surfactants (Islam et al. 2021). When mixed with water, it can be applied to foliage using standard insecticidal spraying techniques, including ultra-low volume and hydraulic applications. Additionally, it is suitable for soil drenching. Wettable powder-based formulation ensures the effective performance of the biopesticide and is developed using a diverse range of compounds, each selected for specific physicochemical properties aimed at enhancing efficacy and spore viability (Parthasarathy et al. 2024a). Additional compounds, such as formulation additives, are often incorporated to protect fungal spores from UV radiation, improve their adhesion to plant surfaces by reducing wash-off, and enhance local humidity around the spores to facilitate germination under suboptimal environmental conditions. The phloxine B is applied at a concentration of 0.005 g/m^2 and has been identified as a potential agent to mitigate phytotoxic effects. Moisture content plays a critical role in maintaining conidial viability, which directly influences the shelf life of biopesticide formulations during storage. To address this, the incorporation of moisture-absorbing agents such as white carbon, magnesium silicate, calcium chloride, magnesium sulfate, or sodium sulfate has been recommended for use in 10% wettable powder formulations containing conidia (Kozłowska et al. 2019).

4.2.2 Oil Formulations

Natural oils are used as substrates in numerous formulations to enhance the shelf life of fungal products and to improve their efficacy under harsh climatic conditions. These oils facilitate the wetting of waxy, lipophilic, or hydrophobic surfaces commonly found on plant foliage and insect cuticles. The existence and viability of conidia are significantly influenced by the incorporation of oil into the spore powder. Oils may enhance the adhesion of spores to the insect cuticle, facilitate penetration, and disrupt the waxy outer layer, thereby promoting germination. Isoparaffinic hydrocarbon solvents, including mineral oil and paraffin oil, serve as carriers in oil-based formulations. Additionally, other oils such as vegetable oils, methyl oleate, corn oil, and cottonseed oil (functioning as a wetting agent and emulsifier) have also been demonstrated in such applications. According to Saminathan et al. (2025), an oil-emulsion formulation of *M. anisopliae* significantly improved its effectiveness in controlling the red palm weevil. Their research showed that when *Dysdercus cingulatus* was exposed to local isolates of *M. anisopliae*, the mortality rates varied between 70% and 100% at elevated concentrations. A formulation of *M. anisopliae* in oil emulsion has shown improved effectiveness in combating the red palm weevil (*Rhynchophorus ferrugineus*). This formulation enhanced the stability and pathogenicity of the fungus, leading to increased mortality rates and underscoring its potential role in pest management (Lei et al. 2023).

4.2.3 Encapsulation or Granular Formulations

Bio-formulations developed through encapsulation or granulation of fungal propagules derived from liquid cultures are increasingly recognized as effective and environmentally sustainable alternatives to conventional chemical acaricides and insecticides. These formulations have the potential to enhance the efficacy of the fungal agents, improve their persistence under field conditions, and extend shelf life (Immediato et al. 2017; Sarma et al. 2023). Such propagules present a promising advancement toward replacing aerial conidia, which currently remain the predominant form used in fungal biopesticide products. Granular formulations of EPF agents are receiving attention as a cost-effective and practical application of biocontrol agents and bioinoculants in agriculture (Davari et al. 2021). In the management of soil-dwelling arthropod pests, small granules serve as an efficient delivery system for fungal propagules to the pest habitat. Likewise, dry microsclerotia (MS)-based granules of *M. brunneum* have demonstrated good efficacy in controlling sugar beet root maggots compared to aerial conidia (Friuli et al. 2023).

Under moist conditions, the MS can germinate, develop as vegetative mycelium, and subsequently sporulate, releasing large quantities of infective conidia directly at the site of pests. In contrast to MS, blastospores (BLS) are environmentally sensitive, yeast-like fungal cells that have also been evaluated as granular formulations for the control of ticks (Seethapathy 2024). However, a direct comparison between MS- and BLS-based formulations for cattle tick management, including their persistence in soil, had not been conducted. Due to their greater sensitivity to desiccation, storage, and adverse conditions, BLS have been relatively underutilized, particularly in soil-based applications. Recent technological advancements in BLS production, along with the selection of more resilient fungal strains capable of withstanding environmental stressors, are beginning to address these limitations (Mascarin et al. 2024; Seethapathy 2024). Similarly, microcrystalline cellulose (MC)-based granules containing microsclerotia of *M. robertsii* were shown to be effective in reducing larval populations of the same tick species, achieving a relative efficacy of 64.8% during the humid season following soil application in semi-natural pasture conditions (Saminathan et al. 2025).

4.2.4 Enhancing Shelf Life, Viability, and Virulence

Shelf life represents a crucial factor in the development and formulation of fungal-based biopesticides. According to Seethapathy (2024), preservation strategies should be designed to enhance product stability under challenging conditions, including high temperatures typical of tropical climates and extended transport durations, often ranging between 30 and 50 °C. Due to their superior stability during storage and distribution, wettable powder (WP) formulations, followed by water-dispersible granules (WDG), are generally favored. Numerous researchers have emphasized the importance of evaluating the performance of fungal biopesticides under field conditions, as opposed to controlled laboratory environments, to yield more realistic and applicable data on formulation efficacy. However, studies that assess both formulation performance and efficacy in the field after storage remain limited (Mascarin and Jaronski 2016). Strategies for extending the shelf life of fungal-based biopesticides encompass the optimization of nutrient formulations, regulation of oxygen and moisture levels through vacuum techniques and active packaging systems, the incorporation of relative humidity and oxygen scavengers, and the application of advanced drying methods (Mascarin and Jaronski 2016). Given the variability in preservation requirements among fungal strains, it is essential to identify strain-specific optimal storage conditions through targeted screening. Additionally, conidia generated via liquid fermentation are generally more susceptible to viability loss due to desiccation compared to those produced through solid-state fermentation. Microsclerotia-based granules that retain a prolonged shelf life under both cool and ambient storage conditions, while maintaining consistent conidial production capacity, represent ideal candidates for field applications as essential components of biological pest management strategies (Mascarin et al. 2024).

B. bassiana is a widely used entomopathogenic fungus known for its efficacy against a broad spectrum of insect pests. It is in high demand among agricultural producers for its ability to infect and control pests such as aphids, mites, moths, stink bugs, thrips, whiteflies, weevils, lace bugs (tingids), and mealybugs (Mascarin and Jaronski 2016). While *B. bassiana* is commonly recommended for pest management in the cultivation of cereals, fruits, legumes, vegetables, herbs, spices, and ornamental flowers, its application can be extended to other crop systems affected by the same insect pests. Other EPF, including *M. anisopliae* and *I. fumosorosea*, are also commonly incorporated as active ingredients in biopesticide formulations. These fungi provide effective protection for a wide range of insect pests, including aphids, ants, caterpillars, leaf miners, beetles, flies, mealybugs, psyllids, root grubs, locusts, thrips, termites, root weevils, whiteflies, and rootworms (Ambethgar et al. 2019; Leadmon et al. 2020; Gebreslasie et al. 2023). *P. lilacinus* is the primary fungal strain employed for the biological control of plant-parasitic nematodes. It exhibits high efficacy against several nematode groups, including root-knot, reniform, cyst, golden cyst, and citrus nematodes.

Notably, it is particularly effective against genera such as *Belonolaimus*, *Globodera*, *Heterodera*, *Meloidogyne*, *Nacobbus*, *Radopholus*, *Rotylenchulus*, and *Pratylenchus* (Jaber and Araj 2018; Mongkolsamrit et al. 2020; Bamisile et al. 2021; Sharma et al. 2023). Additionally, various EPF, such as *V. lecanii* and *L. muscarium*, along with mycoparasitic fungi like *Ampelomyces quisqualis* (Parthasarathy et al. 2025) and non-toxigenic strains of *Aspergillus flavus*, *Chaetomium globosum*, *Coniothyrium minitans*, and *Ulocladium oudemansii*, are emerging as significant contributors to current trends in the global biopesticide market (Seethapathy et al. 2023; Rajalakshmi et al. 2025). Several commercially significant brands, such as those from BASF (Green Guard, Green Muscle, and TrichoPlus), AgriLife (Biokuprum, Downycare, Powderycare, and Racer), Dudutech (Beauvitech, Lecatech, Mytech and Trichotech), LAM International (Mycostar), and Verdera (Rotstop), are popular regarding the specific EPF strains incorporated in their formulations.

5 Application Strategies in Agricultural Systems

In agroecological crop protection (ACP), EPF are crucial microbial agents used for biological pest control. They are increasingly recognized as a potential alternative to agrochemicals, offering an eco-friendly method for effective pest management. EPF are naturally occurring fungal pathogens that can infect various insect and mite species, playing a vital role in regulating pest populations and preventing damage to crops. Various application techniques have been established to enhance the efficacy of EPF in agricultural environments. These methods not only target pest populations directly but also contribute to improved crop vigor, thereby promoting healthier and more productive farming systems. Collectively, these strategies highlight the adaptability of EPF and underscore their significant potential in advancing IPM practices (Quesada-Moraga et al. 2022; Cai and Dimopoulos 2025).

5.1 Foliar and Soil Application Techniques

Research indicates that *Beauveria brongniartii* is quite effective in managing various insect pests, including *Anoplophora glabripennis* and *Oryctes elegans*. Treated insects experience a decrease in feeding and reproduction, leading to increased mortality rates when compared to control groups. In field studies, researchers have effectively utilized *B. brongniartii* in fiber bands to manage the Asian longhorned beetle (*Anoplophora glabripennis*), resulting in faster mortality rates among treated beetles compared to the controls. *Beauveria caledonica* has proven to be effective in controlling insect pests like *Hylobius abietis* (large pine weevil) and *Cosmopolites sordidus* (banana weevil) (Mascarin et al. 2024). Additionally, it has been identified as a native pathogen of bark beetles, including *Ips typographus*, in regions such as Bulgaria (Barta et al. 2020). According to the findings of Montecalvo and Navasero (2021), *M. rileyi* and *B. felina* have been observed to infect the *S. frugiperda* (fall armyworm) in maize fields. The experimental findings showed that using a talc-based formulation (5×10^{13} spores/ha) of *M. anisopliae* alongside FYM led to a remarkable 93% reduction in white grub damage to sugarcane, while also decreasing grub populations by 77% over 2 years (Saminathan et al. 2025).

A sustainable and eco-friendly approach has been developed to manage key rice insect pests, including *Chilo suppressalis*, *Cnaphalocrocis medinalis*, and *Nilaparvata lugens* (Peng et al. 2021). This strategy aims to minimize the impact on natural enemies by selecting a particular entomopathogenic fungal strain as a biological control agent. In extensive multi-year field studies, *M. anisopliae* CQMa421 demonstrated effectiveness comparable to that of chemical pesticides for pest management over 8 years (Peng et al. 2021). The effectiveness of *M. anisopliae* formulations was highlighted, and it was suggested to use them in combination with compost (5 × 10 spores/gm) to safeguard potato tubers from *Agrotis ipsilon* (cutworm) (Pandey 2013). This study focused on assessing the effectiveness and long-lasting ability of *M. anisopliae* strain ESALQ1604 in a sugarcane field, utilizing microsatellite molecular markers for documentation (Iwanicki et al. 2019). Also examined the diversity of *Metarhizium* species found in insects, soil, and sugarcane roots, comparing areas that were sprayed and those that were not sprayed.

The *M. anisopliae* strain ESALQ1604 was successfully recovered from spittlebugs at 7, 30, and 60 days after application, representing 50–70.5% of all insects killed by *M. anisopliae* during those intervals, respectively. Every haplotype from the spittlebug was linked to one specific subclade of *M. anisopliae*. In the study, *Metarhizium robertsii*, *M. anisopliae*, and *M. brunneum* were identified, with one lineage remaining unassigned taxonomically. Notably, only *M. anisopliae* and *M. brunneum* were successfully isolated from root samples (Iwanicki et al. 2019). This study highlighted the remarkable variety of *Metarhizium* species present in the sugarcane agroecosystem and emphasized the significance of the local *M. anisopliae* population in managing spittlebugs. *I. fumosorosea* is a promising solution for managing *Aleurodicus rugioperculatus* (rugose spiraling whitefly) in oil palm plants. It enables farmers to produce mother cultures cost-effectively, utilizing materials that are readily available in their local environment (Subbanna and Kalidas 2024).

5.2 Seed Treatment and Endophytic Colonization

EPF serve as a natural biocontrol, supports plant growth, and enhances resilience against various crop stresses (González-Guzmán et al. 2021). The study conducted three experiments to explore (i) how varying concentrations of the surfactant Tween 80 (0, 0.5, 1, 5, and 10%) influence the coating of wheat seeds with conidia of

M. brunneum, as well as the viability of both seeds and conidia; (ii) growth performances of wheat seedlings in their early stages after being inoculated with *B. bassiana* or *M. brunneum* through seed coating or soil drenching; and (iii) the impact of soil sterilization and seed disinfection on the chlorophyll concentration in leaves (measured by SPAD) and the colonization by *B. bassiana* or *M. brunneum*. The concentration of Tween 80 significantly enhanced seed coating by up to 127%, leaving the germination of both wheat seeds and fungal conidia unaffected. The length of seedlings from inoculated plants showed a notable increase with *B. bassiana* and *M. brunneum*, achieving growth rates of 67% and 46%, respectively, through seed coating. Seed disinfection played a crucial role in improving wheat SPAD values by 10–18% when using *B. bassiana* or *M. brunneum* compared to the control. This, along with soil sterilization, resulted in the highest rates of endophyte colonization, reaching up to 83.3% with both fungi (González-Guzmán et al. 2021).

Another study assessed the impact of endophyte seed treatment with EPF on the growth rate of aphid populations, their offspring performance, and fecundity, along with its influence on *Vicia faba* when exposed to aphid feeding (Akello and Sikora 2012). Fava beans treated with endophytes showed a notably reduced population of *Acyrthosiphon pisum* compared to the untreated controls. The most significant reduction effects were seen in plants treated with *Gibberella moniliformis*, *Trichoderma asperellum*, and *B. bassiana* isolates. All isolates of *Metarhizium anisopliae* and *Hypocrea lixii* showed minimal effects on the population growth of *A. pisum*. In a similar manner, treating seeds with endophytes negatively impacted the fitness, development, and reproductive success of the offspring. No matter the aphid species, the offspring born to females that were fed on endophyte-treated plants for two subsequent generations had a significantly lower birth rate compared to females fed on control plants. Consequently, all plants treated with endophytes showed a significant reduction in the number of *Aphis fabae* and *A. pisum* nymphs when compared to the untreated controls on the final day. At the same time, treating seeds with endophytes significantly improved seedling survival, achieving rates between 20% and 100%, while the control group showed no survival after 20 days of infestation (Akello and Sikora 2012). This study shows that treating seeds with endophytic EPF benefits *V. faba*, making it more resistant against aphids, making it a useful strategy in IPM systems (Parsa et al. 2018). Further study explored the potential of *B. bassiana* and *M. anisopliae* as seed treatments to establish themselves endophytically in common beans. The endophytic EPF colonization in sterile sand/peat was approximately 40% greater for the fungus treatments and about six times higher for volunteer fungi (Harish et al. 2023), which are the naturally occurring fungal endophytes in samples, compared to sterile vermiculite (Parsa et al. 2018).

The colonization of *B. bassiana* and *M. anisopliae* showed the least variability in sterile vermiculite, while it exhibited the greatest variability in a mixture of sterile soil, sand, and peat. They evaluated soil sterilization effects on endophytic colonization in a distinct experiment that involved six different soils collected from the field. Soil sterilization had the most significant effect on colonization, accounting for 70.8% of the total variance, whereas the fungal isolate used for seed inoculation contributed 8.4% to the variance. In the natural microbial soil, seed inoculations with *B. bassiana* and *M. anisopliae* may not consistently result in reliable levels of endophytic EPF colonization (Parsa et al. 2018).

5.3 Use in Protected Cultivation and Greenhouses

The protected cultivation of vegetables such as cucumber, capsicum, tomato, and gerbera is highly prone to sap feeders and vector-borne pathogens. Economic pests such as *A. gossypii* and *Myzus persicae* (aphids), *Liriomyza trifolii* (serpentine leaf miner), *Polyphagotarsonemus latus* (broad mite), *Franklinella occidentalis* and *Scirtothrips dorsalis* (thrips), *Spodoptera litura* (tobacco caterpillar), *T. vaporariorum* and *B. tabaci* (whitefly), and *Tetranychus ludeni* and *T. urticae* (red spider mite) pose significant threats to crops in protected cultivation (Singh and Joshi 2020). The major insect pests include aphids such as *A. gossypii* and *M. persicae*, along with *T. vaporariorum* and *B. tabaci*. These pests are responsible for vector-borne viral diseases, leading to various forms of damage, including chlorosis, necrosis, wilting, stunting, flower and fruit abortion, leaf distortion, and defoliation (Parthasarathy 2023a). The favorable environmental conditions found in protected cultivation make pest management quite challenging. The primary method involves the use of chemical pesticides; however, the excessive and indiscriminate application of these substances has led to insect pests developing resistance. Moreover, the overuse of chemical insecticides is associated with negative impacts such as harmful residues on crops, the death of non-target organisms, and the emergence of pesticide resistance (Seethapathy et al. 2016).

It is significant in the current situation that insect pests affecting controlled environment agriculture conditions in India and its management strategies using EPF, as well as enhancements in virulence through various genetically modified techniques, are most crucial (Harish et al. 2022). The prolonged and widespread application of insecticides has led to these insect pests developing resistance to them. There is a clear need for effective alternatives to control these pests. Biological control serves as a cornerstone of IPM, significantly contributing to the suppression of arthropod pests. Various IPM programs highlight the use of microbial formulations, which are environmentally friendly and safe for all forms of life, offering significant benefits to farmers and entrepreneurs (Parthasarathy et al. 2024a). EPF effectively reduce aphid and whitefly populations on vegetable and tuber/bulb crops (Seethapathy et al. 2025), making them a strong recommendation for organic production in greenhouse and field conditions and a valuable component of IPM.

5.4 Case Studies on Field-Level Successes

This study aimed to assess how effective the EPF, *I. fumosorosea*, is both on its own and when paired with a lower dose of certain insecticides in managing *A. rugioperculatus* on coconut (Sandeep et al. 2022). The goal was to develop a sustainable approach to IPM. The findings showed that using a combination of *I. fumosorosea* (ICAR-NBAIR Pfu-5) at 5 ml/l along with profenophos 50 EC at 2 ml/l resulted in a significant reduction in the percentage of *A. rugioperculatus* infestation (82.97%), intensity of RSW (80.49%), and the average number of live *A. rugioperculatus* colonies (79.68%). This was closely followed by the combination of *I. fumosorosea* (ICAR-NBAIR Pfu-5) at 5 ml/l and buprofezin 25% SC at 1.25 ml/l, which also significantly decreased the percentage of RSW infestation (79.35%), the intensity of RSW (74.79%), and the mean number of live RSW colonies (74.20%) compared to using *I. fumosorosea* (ICAR-NBAIR Pfu-5) alone and the untreated control. Using a combination of *I. fumosorosea* (5 ml/l) and profenophos 50 EC (2 ml/l), applied as a spray twice with a 15-day interval, proved effective in decreasing the RSW population on coconut (Sandeep et al. 2022). Another study focused on assessing these isolates in their effectiveness against the rose aphid, *M. rosae*, in field conditions (Sayed et al. 2019).

This study examined both indigenous isolates and a commercial strain of *B. bassiana* for their effectiveness in controlling aphids at various infestation levels on roses. Four genetically distinct isolates of *B. bassiana* (isolates 1, 2, 3, and 4) were utilized to determine the LC50 for the rose aphid. The same isolates were employed to manage aphid infestations on rose plants in the field, and their effectiveness was compared to a commercial strain (Naturalis) using two concentrations (2.3×10^6 and 4.6×10^6 conidia/ml) for each isolate. The bioassay results showed that the commercial isolate had a significantly different LC50 value of 6.46×10^4 compared to the other three isolates, which had LC50 values of 1.46×10^5, 1.52×10^5, and 1.71×10^5. Field trials showed that a concentration of 4.6×10^6 conidia/ml, whether from indigenous or commercial strains (Sayed et al. 2019), resulted in the most significant reductions in rose aphid infestation and the highest yield of rose oil.

6 Integration into IPM Programs

6.1 Compatibility with Chemical Pesticides, Parasitoids, and Predators

Application of hypocrealean EPF for biocontrol is an essential strategy in IPM programs that aim to gradually substitute synthetic chemical insecticides with more eco-friendly pest control options (Quesada-Moraga et al. 2022). These EPF are notable among entomopathogens due to their unique method of infection through the arthropod pests. Additionally, they form close relationships with plants, adopting an endophytic lifestyle and demonstrating rhizosphere competence, which allows them to play a significant role in IPM and crop production. Also, using insecticides together with EPF offers a hopeful approach to pest control while reducing negative chemical impacts. The study examined the manner in which five insecticides (Pelizza et al. 2018), two of which are biorational, interact with five fungal entomopathogenic strains in a laboratory setting. The goal was to assess the potential effectiveness of these combinations in targeting *Rachiplusia nu* (diamondback moth). The insecticides were evaluated for compatibility using four doses in an in vitro bioassay, and their lethality against *R. nu* was assessed through inoculations at three different doses. The use of those insecticides in combination with *B. bassiana*, *M. anisopliae*, and *M. robertsii* resulted in high mortality rates in *R. nu* larvae compared to the effects of each agent used individually.

Our observations revealed notable variations in the conidial viability, vegetative growth, and conidia production among the five strains of EPF when subjected to various doses of chemical insecticides. The combination of gamma-cyhalothrin and *B. bassiana* resulted in the highest mortality rate of *R. nu* larvae, showing synergistic effects at 50% and 25% of the maximum field doses. In this study (Sain et al. 2019), they assessed 10 EPF grown in polyhouse environments to determine their effectiveness against whitefly nymphs alongside chemical and botanical insecticides. Ij-102 showed the highest overall biological efficacy index, with Bb-4511 and Ij-089 following closely behind. A study was carried out in a controlled environment to assess how botanical and chemical pesticides influence mycelial growth and spore production, utilizing the poisoned food technique. Pesticides significantly impacted mycelial growth and conidial production, showing reductions between 69.5 and 94.1% and 25.6 and 87.6%, respectively. Ij-089, Ij-102, Ma-1299, and Bb-4511 emerged as the most compatible options among the chemicals and botanicals assessed. In comparison, buprofezin, diafenthiuron, flonicamid, pyriproxyfen, and spiromesifen showed greater compatibility with EPF at half doses than the other chemical pesticides, such as imidacloprid, fipronil, profenophos, and triazophos (Sain et al. 2019).

EPF and natural enemies from arthropods can play a role in reducing insect pest populations, whether acting alone or as part of a group of species, toward ensuring ACP. It is crucial to evaluate the interactions among natural enemies of insects, as they have evolved to operate within a multitrophic interaction. This understanding is essential for effectively utilizing them in pest management strategies. Natural enemies can engage with one another in ways that either enhance their effects together, such as improving the spread of insect pathogens, or work against each other, like through parasitism, infection, predation, and competition. The interaction of EPF with plants involves a complex relationship at various levels, including insect pests, plants, and their natural predators (Quesada-Moraga et al. 2024). There is a need for more studies to explore the multitrophic interactions between insect pests and fungal natural enemies. Overall, multiple studies should focus on the beneficial interactions between natural enemies of arthropods and EPF in managing insect populations. Further investigation is needed to explore the multitrophic interactions within the natural enemy community and EPF in the agroecosystem.

6.2 Synergistic and Additive Effects in IPM

EPF demonstrate strong compatibility based on the specific crop, insect pest, and environmental conditions. They can be utilized alone, rotated, or combined with beneficial insects like pollinators, predators, and parasitoids, as well as agrochemicals like insecticides, herbicides, or other beneficial microflora. Consequently, they can be seamlessly integrated into IPM programs for both organic and conventional farming systems. Numerous biopesticides feature minimal or no re-entry and handling intervals, and they do not require regular oversight from regulatory bodies or retailers. The study by Akbar et al. (2012) found that using sublethal doses of insecticides alongside EPF can create synergistic effects, which ultimately increase insect mortality. This method provides multiple benefits compared to relying solely on insecticides: it enhances pest mortality, lowers the need for insecticide applications, minimizes environmental pollution, and slows down the development of resistance. In 2014, Fang et al. (2014) discovered that a genetically engineered strain of *M. acridum*, which expressed a combination of four insect-specific toxins, produced a synergistic effect that decreased the LT50 for acridids by 48%. In experiments where both engineered strains of *M. acridum* and the wild-type were exposed to non-locust hosts like cockroaches and mosquitoes, none succeeded in causing infection or death.

The significance of host specificity is crucial when it comes to introducing a biopesticide into the environment (Moonjely et al. 2016). Nanotechnology allows for the blending of biopesticides with other nano-based materials, like adjuvants or essential oils, resulting in synergistic combinations that improve pest control effectiveness. This method could potentially lower the chances of pests developing resistance (Bihal et al. 2023). The nanoencapsulation of *B. bassiana*, a widely utilized fungal biopesticide, has demonstrated enhanced stability and extended effectiveness when compared to conventional formulations (Pillay et al. 2024). They studied the eco-friendly production of silver nanoparticles through extracts of *B. bassiana*, emphasizing their potential as effective ingredients in nanocosmetics. Fungal biopesticides are likely to play a significant role in the future, especially when used in combination with other pest control strategies. Strategies for IPM that combine ACP with natural predators, cultural practices, and resistant crop varieties can offer comprehensive and sustainable solutions.

6.3 Commercial Applications

EPF like *B. bassiana*, *L. lecanii*, and *M. anisopliae* are effective against various insect pests, often leading to widespread outbreaks in their hosts. Their beneficial traits make them promising candidates for development as commercial biocontrol agents (Table 15.1). Currently, microbial biopesticides derived from EPF primarily serve specialized markets. Biopesticides presently make up a small segment of the global crop protection market. In the last 5 years (Sabbahi et al. 2022), it was valued at approximately US$3 billion, which is around 5% of the total crop protection market. Biopesticides represent only a small segment of the overall pesticide market, contributing approximately US$3–4 billion to the total of US$61.3 billion. Biopesticides are anticipated to exceed conventional pesticides in growth, with annual increases projected between 10% and 20% (Parthasarathy 2023b). The Annual Biocontrol Industry Meeting (ABIM) in 2022 also shared insights on the

Table 15.1 Commercial products of entomopathogenic fungi against crop pests

EPF	Target pest	Trade name
Beauveria bassiana	Naturalis, Myco-Jaal, Biosoft, Biorin/Kargar, Mycotrol ESO, Bb-Protec, Botanigard ES, Bio-Power, Ostrinol, Nagestra, Conidia, Daman, Racer, Naturalis, Biowonder, and Beauvitech-WP	Grasshoppers, aphids, whiteflies, cutworms, lepidopterans, mealybugs, coleopterans, thrips, leafhoppers, and stemborers
B. brongniartii	Betel, Engerlingspilz, Bas-Eco, and Melocont	European cockchafer beetles, and Coleoptera (Scarabaeidae)
Hirsutella thompsonii	ABTEC Hirsutella, Mycar and No-Mite	Mites
Isaria fumosorosea	Pfr-97 WDG, Nofly™WP	Psyllids, thrips, whitefly, mealybugs, and aphids
Lecanicillium lecanii	Phule Bugicide, Vertelac and Mycotal	Aphids, whiteflies, and thrips
L. muscarium	Mycotal	Aphids, whiteflies, psyllids, scales, and thrips
Metarhizium anisopliae	BioCane, BioGreen, Metarril, Ory-X, Bio-Magic, Novacrid, Pacer, Metarhizium, Achieve, Met52/BIO1020 granular, Bio-Catch-M, Kalichakra, Multiplex, Green Muscle, and Biomet/Ankush	Scarabids, aphids, thrips, whitefly, and mites
M. brunneum	Attracap, Met52 EC	Aphids, beetles, whiteflies, psyllids, scales, and thrips
Purpureocillium lilacinus	Mytech-WP, Bio-Nematon, Paecilo, and MeloCon WG	Plant-parasitic nematodes

biologicals market share in agriculture. The report highlighted that insect pest biocontrol emerged as the most rapidly expanding category of the biocontrol market, boasting a compound annual growth rate (CAGR) of 13.6%.

Biostimulants and biofertilizers are projected to grow at impressive rates of 12.0% and 12.5%, respectively. The biocontrol market is projected to grow significantly, rising from approximately $5 billion to an estimated $15 billion USD by 2029. The leading category in biocontrol today is biochemicals, encompassing pheromones, extracts of plants and plant growth regulators. By 2029, microbial biopesticides are anticipated to have a nearly equal contribution. In the European Union (EU), there are fewer registered biopesticide-active chemicals compared to countries like the United States, Canada, or Australia. This is largely due to the lengthy and complex registration processes in the EU, which mirror those used for chemical pesticide registration. In the US market, there are around 390 registered biopesticide-active ingredients, whereas the EU market features just 68 similar products. While many biopesticides are registered in Africa, only a limited number have been produced locally. Progress is underway, with South Africa and Kenya at the forefront of biopesticide development and application (Parthasarathy 2023b). Out of the 31 products registered in South Africa, 7 are produced locally (Sabbahi et al. 2022). In India, more than 60 products originating from *L. lecanii*, which is still categorized under the species *Verticillium*, are available in both liquid and dry forms. These products are utilized to manage aphids, scales, and a variety of other soft-bodied sucking pests affecting numerous crops.

To meet the growing demands of the market, there is a need to create more efficient systems for mass production, formulation, and delivery. It is essential to carry out additional experiments across various crops to better understand how abiotic and biotic factors influence effectiveness and persistence (Parthasarathy 2023b). Ultimately, it is important to focus more on integrating EPF into IPM and sustainable agriculture programs. Specifically, to develop methods that combine EPF with natural enemies, ensuring they work well together and are effective.

7 Constraints and Challenges in Field Deployment

The process of obtaining commercial approvals for EPF-based biopesticides involves a thorough and often lengthy series of steps. This includes conducting toxicological and environmental assessments, as well as tests to understand their mode of action, host range, and effectiveness (Parthasarathy 2023b). Local manufacturers often find toxicological investigations to be prohibitively expensive and complicated. Regulators may find it challenging to meet these standards. One challenge is finding the right EPF biopesticides while ensuring they meet safety and consistency standards for commercial use. Implementing changes like fast-track registration, priority registration, and zonal authorization could encourage ongoing scientific advancements and enhance patenting in the field of EPF biocontrol. Streamlining the registration processes for biopesticides on a global scale would certainly boost their commercialization across the globe. To date, the Organization for Economic Cooperation and Development (OECD), the International Organization for Biological Control (IOBC), and the European and Mediterranean Plant Protection Organization (EPPO) have taken remarkable steps toward achieving this harmonization (Parthasarathy 2023b), though their success has been slightly constrained (Sabbahi et al. 2022). The ongoing shift toward minimizing chemical pesticide use and streamlining regulations for low-risk substances could pave the way for the widespread production and global application of EPF. The limited number of registered products for EPF of insect pests can be attributed to poor technology transfer. As a result, the agricultural sector, especially in less industrialized countries, has not fully recognized its economic potential. Moreover, for potential EPF candidates, a solid grasp of organism-specific research methods for mass production and applications is often lacking, which turns the mass production of EPF in vivo into a challenging and resource-intensive task. Additionally, a significant challenge in utilizing EPF is making sure they arrive at the correct location regularly and in sufficient numbers to be effective while also maintaining their presence consistently.

By the early 2050s, microbial biopesticides are expected to match the use of conventional insecticides (Sabbahi et al. 2022). However, significant uncertainties regarding acceptance rates, especially in Southeast Asia and Africa, contribute to the variability in these predictions. Research indicates that this gap could stem from the steep costs and often protracted timelines associated with the European active ingredient approval process, along with the intricate nature of registration procedures. Another reason for limiting the number of licensed EPF is that many farmers select pest management technologies based on economic, technological, and practical factors. To encourage producers to adopt EPF, insect management tools need to be easy to implement and assess while also providing cost, demand, and supply efficiency benefits compared to other methods. Some farmers prefer granular formulations that can be easily stored and applied by machines, as opposed to liquids that require costly equipment or cold storage. Conversely, farmers often feel hesitant to adopt biological methods because they lack the necessary knowledge and technical support. To facilitate effective information sharing, training and field demonstrations are essential for bridging the gap between technology developers and end users (Paramasivam et al. 2022). Effective pest control considers the tangible benefits of increased yields and reduced risks alongside external costs while also taking into account the non-financial aspects that affect farmers' choices and well-being. This analysis should form the foundation for government regulations on pest management. Regulations that promote the registration of low-risk materials with enticements, national policies intended to reduce the use of conventional insecticides, and the encouragement of IPM strategies through measures like tax credits for organic or sustainable farming practices could contribute to accelerating the commercialization and accessibility of microbial biopesticides in the market (Parthasarathy 2023b). The commercialization and application of EPF biopesticides that target various significant insect pests are essential for managing agricultural ecosystems and are mandated in the current situation. This approach supports sustainable food production while safeguarding One Health and the environment. While countries such as the USA, China, and those in Europe have embraced these techniques, India and Africa still have room for improvement in this area moving forward.

8 Conclusion

Using EPF is more cost-effective than relying on chemicals, as it leverages the ongoing infection cycle due to its ability to remain in the soil over time. This could serve as a valuable resource for IPM programs. In the coming years, these insect pests are likely to develop some resistance to EPF. To enhance the effectiveness of various entomopathogens, the adoption of innovative methods like consortia development, encapsulated formulations for slow release, and genetic engineering of fungal formulations may be essential. These augmentatives and inundative approach in EPF toward crop pest control show great promise, especially with advancements in pathogenesis and biotechnological innovations in the near future. Based on recent advancements, the EPF formulations are expected to have an extended shelf life, be non-allergenic for humans, and ensure safety for non-target species. In-depth studies focused on the interactions between hosts, environments, and pathogens are essential for developing innovative mycoinsecticide formulations. EPF applications should address the increasing challenges posed by insect pests and enhance both the yield and quality of food crops. This may lead to more sustainable farming practices and improved food security toward SDGs like Zero Hunger SDG-2.

Acknowledgments The corresponding author is grateful for the award of the DBT-Biotechnology Industry Research Assistance Council (BIRAC), New Delhi, Fellowship (E28XZ) for the Development of Bioformulation of Mycoparasitic Fungi by the University Innovation Cluster (UIC) of Tamil Nadu Agricultural University, which is highly appreciated.

References

Ahsan SM, Injamum-Ul-Hoque M, Das AK, Rahman MM, Mollah MM et al (2024) Plant–entomopathogenic fungi interaction: recent progress and future prospects on endophytism-mediated growth promotion and biocontrol. Plants 13. https://doi.org/10.3390/plants13101420

Ain Q, Mohsin AU, Naeem M, Shabbir G (2021) Effect of entomopathogenic fungi, *Beauveria bassiana* and *Metarhizium anisopliae*, on *Thrips tabaci* Lindeman (Thysanoptera: Thripidae) populations in different onion cultivars. Egypt J Biol Pest Control 31. https://doi.org/10.1186/s41938-021-00445-y

Akbar S, Freed S, Hameed A, Gul HT, Akmal M et al (2012) Compatibility of *Metarhizium anisopliae* with different insecticides and fungicides. Afr J Microbiol Res 6(17):3956–3962. https://doi.org/10.5897/AJMR12.417

Akello J, Sikora R (2012) Systemic acropedal influence of endophyte seed treatment on *Acyrthosiphon pisum* and *Aphis fabae* offspring development and reproductive fitness. Biol Control 61:215–221. https://doi.org/10.1016/j.biocontrol.2012.02.007

Ambethgar V (2018) Strategic approaches for applications of entomopathogenic fungi to counter insecticide resistance in agriculturally important insect pests. In: Gehlot P, Singh J (eds) Fungi and their role in sustainable development: current perspectives. Springer, Singapore, pp 221–254. https://doi.org/10.1007/978-981-13-0393-7_13

Ambethgar V, Kollam M, Chinnadurai C, Ramsubhag A, Jayaraman J (2019) Ecology of emerging vector-borne plant viruses and approaches for integrated management in vegetable production systems. Trop Agric 95:81–94

Bamisile BS, Siddiqui JA, Akutse KS, Ramos Aguila LC, Xu Y (2021) General limitations to endophytic entomopathogenic fungi use as plant growth promoters, pests and pathogens biocontrol agents. Plants 10:1–23. https://doi.org/10.3390/plants10102119

Barta M, Takov D, Pilarska D, Doychev D, Horáková MK (2020) Entomopathogenic fungi of the genus Beauveria and their pathogenicity to Ips typographus (Coleoptera: Curculionidae) in the Vitosha National Park, Bulgaria. J For Sci 66:420-435. https://doi.org/10.17221/123/2020-JFS

Bidochka MJ, Hajek AE (1998) A nonpermissive entomophthoralean fungal infection increases activation of insect prophenoloxidase. J Invertebr Pathol 72:231–238. https://doi.org/10.1006/jipa.1998.4782

Bihal R, Al-Khayri JM, Banu AN, Kudesia N, Ahmed FK et al (2023) Entomopathogenic fungi: an eco-friendly synthesis of sustainable nanoparticles and their nanopesticide properties. Microorganisms 11:1617. https://doi.org/10.3390/microorganisms11061617

Bischoff JF, Rehner SA, Humber RA (2009) A multilocus phylogeny of the *Metarhizium anisopliae* lineage. Mycologia 101:512–530. https://doi.org/10.3852/07-202

Boucias DG, Tigano MS, Sosa-Gomez DR, Glare TR, Inglis PW (2000) Genotypic properties of the entomopathogenic fungus *Nomuraea rileyi*. Biol Control 19:124–138. https://doi.org/10.1006/bcon.2000.0857

Bugti GA, Na C, Bin W, Feng LH (2018) Control of plant sap-sucking insects using entomopathogenic fungi *Isaria fumosorosea* strain (Ifu13a). Plant Prot Sci 54:258–264. https://doi.org/10.17221/118/2017-PPS

Cai P, Dimopoulos G (2025) Microbial biopesticides: a one health perspective on benefits and risks. One Health 20:100962. https://doi.org/10.1016/j.onehlt.2024.100962

Chaud M, Souto EB, Zielinska A, Severino P, Batain F et al (2021) Nanopesticides in agriculture: benefits and challenge in agricultural productivity, toxicological risks to human health and environment. Toxics 9:131. https://doi.org/10.3390/toxics9060131

Chen A, Shi Z, Zhang L (2008) The effects of some storage conditions on viability of *Lecanicillium lecanii* conidia to whitefly (Homoptera: *Trialeurodes vaporariorum*). Biocontrol Sci Tech 18: 267–278. https://doi.org/10.1080/09583150801885897

Chen X, Li L, Hu Q, Zhang B, Wu W et al (2015) Expression of dsRNA in recombinant *Isaria fumosorosea* strain targets the TLR7 gene in *Bemisia tabaci*. BMC Biotechnol 15:1–8. https://doi.org/10.1186/s12896-015-0170-8

Davari A, Parker BL, Sullivan CF, Ghalehgolabbehbahani A, Skinner M (2021) Biological control of Western flower thrips, *Frankliniella occidentalis* using a self-sustaining granular fungal treatment. Bull Entomol Res 111:688–693. https://doi.org/10.1017/S0007485321000365

de Costa Eula Maria MB, Costa EM, Pimenta FC, Luz C, Oliveira VD et al (2011) *Beauveria bassiana*: Quercetinase production and genetic diversity. Braz J Microbiol 42:12–21. https://doi.org/10.1590/s1517-83822011000100002

De Wint FC, Nicholson S, Koid QQ, Zahra S, Chestney-Claassen G et al (2024) Introducing a global database of entomopathogenic fungi and their host associations. Sci Data 11:1418. https://doi.org/10.1038/s41597-024-04103-4

Deshmukh L, Sandhu SS (2023) Entomopathogenic fungi: a boon towards organic life support management. In: Singh SK, Kumar D, Shamim M, Sharma R (eds) Applied mycology for agriculture and foods. Apple Academic Press, New York, pp 97–128

Dhaliwal GS, Jindal V, Mohindru B (2015) Crop losses due to insect pests: global and Indian scenario. Indian J Entomol 77:165. https://doi.org/10.5958/0974-8172.2015.00033.4

Fadiji AE, Babalola OO (2020) Exploring the potentialities of beneficial endophytes for improved plant growth. Saudi J Biol Sci 27:3622–3633. https://doi.org/10.1016/j.sjbs.2020.08.002

Fang W, Lu HL, King GF, St Leger RJ (2014) Construction of a hypervirulent and specific mycoinsecticide for locust control. Sci Rep 4:7345. https://doi.org/10.1038/srep07345

Friuli M, Pellegrino R, Lamanna L, Nitti P, Madaghiele M et al (2023) Materials engineering to help pest control: a narrative overview of biopolymer-based entomopathogenic fungi formulations. J Fungi 9. https://doi.org/10.3390/jof9090918

Gangireddy Eswara Reddy S (2021) *Lecanicillium* spp. for the management of Aphids, Whiteflies, Thrips, Scales and Mealy Bugs: review. In: Ranz RER (ed) Arthropods – are they beneficial for mankind? IntechOpen. https://doi.org/10.5772/intechopen.94020

Gao T, Li Y, Ding M, Chai Y, Wang Q (2017) The phosphotransferase system gene ptsI in *Bacillus cereus* regulates expression of sodA2 and contributes to colonization of wheat roots. Res Microbiol 168:524–535. https://doi.org/10.1016/j.resmic.2017.04.003

Garrido-Jurado I, Fernández-Bravo M, Campos C, Quesada-Moraga E (2015) Diversity of entomopathogenic Hypocreales in soil and phylloplanes of five Mediterranean cropping systems. J Invertebr Pathol 130:97–106. https://doi.org/10.1016/j.jip.2015.06.001

Gebreslasie MG, Nishi O, Wasano N, Yasunaga-Aoki C (2023) Varied selectivity of caterpillar-specific *Metarhizium rileyi* and generalist entomopathogenic fungi against last instar larvae and pupae of common cutworm, *Spodoptera litura* (Lepidoptera: Noctuidae). Appl Entomol Zool 58:219–228. https://doi.org/10.1007/s13355-023-00824-x

Georgiev G, Mirchev P, Rossnev B, Petkov P, Georgieva M et al (2013) Potential of *Entomophaga maimaiga* Humber, Shimazu and Soper (Entomophthorales) for suppressing *Lymantria dispar* (Linnaeus) outbreaks in Bulgaria. Proceed Bulg Acad Sci 66. https://doi.org/10.7546/CR-2013-66-7-13101331-14

Glare TR, Scholte Op Reimer Y, Cummings N, Rivas-Franco F, Nelson TL et al (2021) Diversity of the insect pathogenic fungi in the genus *Metarhizium* in New Zealand. New Zeal J Bot 59:440–456. https://doi.org/10.1080/0028825X.2021.1890155

González-Guzmán A, Sánchez-Rodríguez AR, Quesada-Moraga E, del Campillo MC, Yousef-Yousef M (2021) Optimizing wheat seed treatment with entomopathogenic fungi for improving plant growth at early development stages. Span J Agric Res 19:e1004. https://doi.org/10.5424/sjar/2021194-17120

Harish S, Murugan M, Kannan M, Parthasarathy S, Prabhukarthikeyan SR et al (2021) Entomopathogenic viruses. In: Omkar (ed) Microbial approaches for insect pest management. Springer, Singapore, pp 1–57. https://doi.org/10.1007/978-981-16-3595-3_1

Harish S, Parthasarathy S, Prabhukarthikeyan SR, Mahanty A, Teli B et al (2022) Omics: a potential tool to delineate the mechanism of biocontrol agents against plant pathogens. In: Sahu J, Vaishnav A, Singh HB (eds) Plant-microbe interactions. CRC Press, Boca Raton, pp 79–100. https://doi.org/10.1201/9781003171416-5

Harish S, Sendhilvel V, Rajendran L, Parthasarathy S, Raguchander T (2023) Plant associated endophytes as potential agents for the protection of crops from phytopathogens. In: Bastas KK, Kumar A, Sivakumar U (eds) Microbial biocontrol: molecular perspective in plant disease management. Springer, Singapore, pp 89–115. https://doi.org/10.1007/978-981-99-3947-3_4

Harper AM, Huang HC (1986) Evaluation of the Entomophagous fungus *Verticillium lecanii* (Moniliales: Moniliaceae) as a control agent for insects. Environ Entomol 15:281–284. https://doi.org/10.1093/ee/15.2.281

Hernández-Velázquez VM, Berlanga-Padilla AM, Barrientos-Lozano L (2000) Vegetable and mineral oil formulations of *Metarhizium anisopliae* var. *acridum* to control the Central American locust (*Schistocerca piceifrons piceifrons* Walker) (Orthoptera: Acrididae). J Orthoptera Res 9:223. https://doi.org/10.2307/3503650

Immediato D, Iatta R, Camarda A, Giangaspero A, Capelli G et al (2017) Storage of *Beauveria bassiana* conidia suspension: a study exploring the potential effects on conidial viability and virulence against *Dermanyssus gallinae* De Geer, 1778 Acari: Dermanyssidae. Ann Biol Sci 05. https://doi.org/10.21767/2348-1927.1000111

Islam W, Adnan M, Shabbir A, Naveed H, Abubakar YS et al (2021) Insect-fungal-interactions: a detailed review on entomopathogenic fungi pathogenicity to combat insect pests. Microb Pathog 159:105122. https://doi.org/10.1016/j.micpath.2021.105122

Iwanicki NS, Pereira AA, Botelho ABRZ, Rezende JM, Moral RDA et al (2019) Monitoring of the field application of *Metarhizium anisopliae* in Brazil revealed high molecular diversity of *Metarhizium* spp in insects, soil and sugarcane roots. Sci Rep 9:4443. https://doi.org/10.1038/s41598-019-38594-8

Jaber LR, Araj SE (2018) Interactions among endophytic fungal entomopathogens (Ascomycota: Hypocreales), the green peach aphid *Myzus persicae* Sulzer (Homoptera: Aphididae), and the aphid endoparasitoid *Aphidius colemani* Viereck (Hymenoptera: Braconidae). Biol Control 116: 53–61. https://doi.org/10.1016/j.biocontrol.2017.04.005

Jampameung N, Arikit S, Sanguansub S, Buttachon S (2024) Entomopathogenic fungi, *Hirsutella thompsonii* F.E. Fisher safe mycoacaricide for the management of coconut mite, *Aceria guerreronis* Keifer in Thailand. Syst Appl Acarol 29. https://doi.org/10.11158/saa.29.10.1

Karthi S, Vasantha-Srinivasan P, Senthil-Nathan S, Han YS, Shivakumar MS et al (2024) Entomopathogenic fungi promising biocontrol agents for managing lepidopteran pests: review of current knowledge. Biocatal Agric Biotechnol 58:103146. https://doi.org/10.1016/j.bcab.2024.103146

Kary NE, Alizadeh Z (2017) Effects of sub-culturing on genetic and physiological parameters in different *Beauveria bassiana* isolates. J Invertebr Pathol 145:62–67. https://doi.org/10.1016/j.jip.2017.03.008

Khan HAA, Khan T (2023) Efficacy of entomopathogenic fungi against three major stored insect pests, *Rhyzopertha dominica*, *Sitophilus zeamais,* and *Trogoderma granarium.* J Stored Prod Res 104:102188. https://doi.org/10.1016/j.jspr.2023.102188

Kothari M, Sandhu SS (2024) Entomopathogenic perspectives of *Nomuraea rileyi*. In: Deshmukh SK, Sridhar KR (eds) The role of Entomopathogenic fungi in agriculture. CRC Press, Boca Raton, pp 140–161. https://doi.org/10.1201/9781003505228-8

Kozłowska E, Dymarska M, Kostrzewa-Susłow E, Janeczko T (2019) Cascade biotransformation of estrogens by *Isaria fumosorosea* KCh J2. Sci Rep 9:1–8. https://doi.org/10.1038/s41598-019-47225-1

Kumar D, Kalita P (2017) Reducing postharvest losses during storage of grain crops to strengthen food security in developing countries. Foods 6:1–22. https://doi.org/10.3390/foods6010008

Lai Y, Zheng W, Zheng Y, Lu H, Qu S et al (2024) Unveiling a novel entry gate: insect foregut as an alternative infection route for fungal entomopathogens. Innov (Camb) 5:100644. https://doi.org/10.1016/j.xinn.2024.100644

Leadmon CE, Sampson JK, Maust MD, Macias AM, Rehner SA et al (2020) Several *Metarhizium* species produce ergot alkaloids in a condition-specific manner. Appl Environ Microbiol 86:1–13. https://doi.org/10.1128/AEM.00373-20

Lei CJ, RHIR A, Halim NA, Asib N, Zakaria A et al (2023) Bioefficacy of an oil-emulsion formulation of Entomopathogenic fungus, *Metarhizium anisopliae* against adult red palm weevil, *Rhynchophorus ferrugineus*. Insects 14:482. https://doi.org/10.3390/insects14050482

Lin Y, Hussain M, Avery PB, Qasim M, Fang D et al (2016) Volatiles from plants induced by multiple aphid attacks promote conidial performance of *Lecanicillium lecanii*. PLoS One 11:1–14. https://doi.org/10.1371/journal.pone.0151844

Litwin A, Nowak M, Różalska S (2020) Entomopathogenic fungi: unconventional applications. Rev Environ Sci Biotechnol 19:23–42. https://doi.org/10.1007/s11157-020-09525-1

Liu X, Li R, Liu Y, Feng M, Zhou G et al (2025) Roles of Sti1, a homolog to mammalian Hop, in conidiation, environmental adaptation, and virulence of *Beauveria bassiana*. Fungal Biol 129: 101596. https://doi.org/10.1016/j.funbio.2025.101596

Lovett B, St Leger RJ (2018) Genetically engineering better fungal biopesticides. Pest Manag Sci 74:781–789. https://doi.org/10.1002/ps.4734

Ma M, Luo J, Li C, Eleftherianos I, Zhang W et al (2024) A life-and-death struggle: interaction of insects with entomopathogenic fungi across various infection stages. Front Immunol 14. https://doi.org/10.3389/fimmu.2023.1329843

Maina U, Zakaria D, Galadima I, Gambo F (2018) A review on the use of entomopathogenic fungi in the management of insect pests of field crops. J Entomol Zool Stud 6:27–32

Mantzoukas S, Eliopoulos PA (2020) Endophytic entomopathogenic fungi: a valuable biological control tool against plant pests. Appl Sci 10. https://doi.org/10.3390/app10010360

Mascarin GM, Jaronski ST (2016) The production and uses of *Beauveria bassiana* as a microbial insecticide. World J Microbiol Biotechnol 32:1–26. https://doi.org/10.1007/s11274-016-2131-3

Mascarin GM, Golo PS, de Souza Ribeiro-Silva C, Muniz ER, de Oliveira Franco A et al (2024) Advances in submerged liquid fermentation and formulation of entomopathogenic fungi. Appl Microbiol Biotechnol 108. https://doi.org/10.1007/s00253-024-13287-z

Mendes-Pereira T, de Araujo JPM, Kloss TG, Costa-Rezende DH, de Carvalho DS et al (2023) Disentangling the taxonomy, systematics, and life history of the spider-parasitic fungus *Gibellula* (Cordycipitaceae, Hypocreales). J Fungi 9:457. https://doi.org/10.3390/jof9040457

Mongkolsamrit S, Khonsanit A, Thanakitpipattana D, Tasanathai K, Noisripoom W et al (2020) Revisiting *Metarhizium* and the description of new species from Thailand. Stud Mycol 95:171–251. https://doi.org/10.1016/j.simyco.2020.04.001

Montecalvo MP, Navasero MM (2021) *Metarhizium* (=nomuraea) *rileyi* (farlow) samson from *Spodoptera exigua* (hübner) cross infects fall armyworm, *Spodoptera frugiperda* (J.E. Smith) (lepidoptera: Noctuidae) larvae. Philipp J Sci 150:193–199. https://doi.org/10.56899/150.01.16

Moonjely S, Barelli L, Bidochka MJ (2016) Insect pathogenic fungi as endophytes. In: St Leger R (ed) Advances in genetics. Academic, New York, pp 107–135

Mushyakhwo K, Maxwell LA, Nai YS, Srinivasan R, Hwang SY (2025) *Beauveria bassiana*-based management of *Thrips palmi* in greenhouse. BioControl 70:333–343. https://doi.org/10.1007/s10526-025-10308-2

Nguyen MT, Doan TTP (2024) On the *Cordyceps* research in Vietnam. In: Deshmukh SK, Sridhar KR (eds) *Cordyceps* and allied species. Springer, Singapore, pp 99–115

Nicoletti R, Becchimanzi A (2020) Endophytism of *Lecanicillium* and *Akanthomyces*. Agriculture 10:205. https://doi.org/10.3390/agriculture10060205

Pandey A (2013) Field evaluation of *Beauveria bassiana* and *Metarhizium anisopliae* against the Cutworm, *Agrotis ipsilon* (Hufnagel) damaging potato in Uttarakhand Hills. J Biol Control 27:293–297

Paramasivam S, Henry P, Seethapathy P, Rajamohan T (2022) A strategic model for empowering farmers by improving livelihood security through organic farming practices in Tamil Nadu, India. J Agric Sci Sri Lanka 17:471. https://doi.org/10.4038/jas.v17i3.9926

Parsa S, Ortiz V, Gómez-Jiménez MI, Kramer M, Vega FE (2018) Root environment is a key determinant of fungal entomopathogen endophytism following seed treatment in the common bean, *Phaseolus vulgaris*. Biol Control 116:74–81. https://doi.org/10.1016/j.biocontrol.2016.09.001

Parthasarathy S (2023a) Fundamentals of plant pathology. CRC Press, London, p 428. https://doi.org/10.1201/9781032711881

Parthasarathy S (2023b) Biological control of plant diseases: principles and practices, 1st edn. Kalyani Publishers, New Delhi

Parthasarathy S, Lakshmidevi P, Satya VK, Gopalakrishnan C (2024a) Plant pathology and disease management principles and practices. CRC Press, London, p 456. https://doi.org/10.1201/9781032711973

Parthasarathy S, Lakshmidevi P, Yashodha P, Gopalakrishnan C (2024b) Pests and diseases in fruit crops. CRC Press, London. pp. 272. doi:https://doi.org/10.1201/9781003504146

Parthasarathy S, Raguchander T, Muthamilan M, Harish S (2025) Techniques to prepare formulation of mycoparasitic biofungicide for use against powdery mildew disease of plants. Indian Patent No. 561865

Pelizza SA, Schalamuk S, Simón MR, Stenglein SA, Pacheco-Marino SG et al (2018) Compatibility of chemical insecticides and entomopathogenic fungi for control of soybean defoliating pest, *Rachiplusia nu*. Rev Argent Microbiol 50:189–201. https://doi.org/10.1016/j.ram.2017.06.002

Peng G, Xie J, Guo R, Keyhani NO, Zeng D et al (2021) Long-term field evaluation and large-scale application of a *Metarhizium anisopliae* strain for controlling major rice pests. J Pest Sci 94: 969–980. https://doi.org/10.1007/s10340-020-01313-8

Pérez-González O, Gomez-Flores R, Tamez-Guerra P (2022) Insight into biological control potential of *Hirsutella citriformis* against Asian citrus psyllid as a vector of citrus Huanglongbing disease in America. J Fungi 8:573. https://doi.org/10.3390/jof8060573

Pillay B, Amobonye A, Bhagwat P, Pillai S (2024) *Beauveria bassiana* facilitated silver nanoparticles in cosmetic formulation: antimicrobial, antioxidant, and photocatalytic activity. Nano-Struct Nano-Objects 39:101300. https://doi.org/10.1016/j.nanoso.2024.101300

Quesada-Moraga E, Garrido-Jurado I, Yousef-Yousef M, González-Mas N (2022) Multitrophic interactions of entomopathogenic fungi in BioControl. BioControl 67:457–472. https://doi.org/10.1007/s10526-022-10163-5

Quesada-Moraga E, González-Mas N, Yousef-Yousef M, Garrido-Jurado I, Fernández-Bravo M (2024) Key role of environmental competence in successful use of entomopathogenic fungi in microbial pest control. J Pest Sci 97:1–15. https://doi.org/10.1007/s10340-023-01622-8

Rajalakshmi J, Harish S, Rajendran L, Parthasarathy S, Saravanakumari K et al (2025) Crude antibiotics and antifungal metabolites from *Chaetomium globosum* Cg6 suppress *Pythium aphanidermatum*, causal agent of rhizome rot of turmeric. Curr Microbiol 82:105. https://doi.org/10.1007/s00284-025-04076-6

Ramesha BS, Sannappa B (2024) In vitro evaluation of Entomopathogenic fungi for the management of broad mite, *Polyphagotarsonemus latus* occurrence on mulberry. Int J Plant Soil Sci 36: 697–702. https://doi.org/10.9734/ijpss/2024/v36i74781

Ramos Y, Pineda-Guillermo S, Tamez-Guerra P, Orozco-Flores AA, Figueroa de la Rosa JI et al (2024) Natural prevalence, molecular characteristics, and biological activity of *Metarhizium rileyi* (Farlow) isolated from *Spodoptera frugiperda* (JE Smith) Larvae in Mexico. J Fungi 10: 416. https://doi.org/10.3390/jof10060416

Ravi Sankar U, Padmaja V (2014) Preparation and use of oil formulations of *Beauveria bassiana* and *Metarhizium anisopliae* against *Spodoptera litura* larvae. African J Microbiol Res 8:1638–1644. https://doi.org/10.5897/AJMR2013.6593

Ricciardi MM, Smith H, Viloria Z, Dixon E, Szarka D et al (2024) First report of *Hirsutella thompsonii* as an entomopathogenic fungus of hemp russet mite (*Aculops cannabicola*) in Kentucky. Plant Health Progress 25:92–94. https://doi.org/10.1094/PHP-04-23-0035-BR

Rohrlich C, Merle I, Mze Hassani IM, Verger M, Zuin M et al (2018) Variation in physiological host range in three strains of two species of the entomopathogenic fungus *Beauveria*. PLoS One 13:1–15. https://doi.org/10.1371/journal.pone.0199199

Roy HE, Cottrell TE (2008) Forgotten natural enemies: interactions between coccinellids and insect-parasitic fungi. Eur J Entomol 105:391–398. https://doi.org/10.14411/eje.2008.049

Sabbahi R, Hock V, Azzaoui K, Saoiabi S, Hammouti B (2022) A global perspective of entomopathogens as microbial biocontrol agents of insect pests. J Agric Food Res 10:100376. https://doi.org/10.1016/j.jafr.2022.100376

Sacco NE, Hajek AE (2023) Diversity and breadth of host specificity among arthropod pathogens in the Entomophthoromycotina. Microorganisms 11:1658. https://doi.org/10.3390/microorganisms11071658

Saha N, Sharma A, Bora P (2025) Expanding the functional landscape of microbial entomopathogens in agriculture beyond pest management. Folia Microbiol (Praha):343–357. https://doi.org/10.1007/s12223-025-01251-x

Sain SK, Monga D, Kumar R, Nagrale DT, Hiremani NS et al (2019) Compatibility of entomopathogenic fungi with insecticides and their efficacy for IPM of *Bemisia tabaci* in cotton. J Pestic Sci 44:97–105. https://doi.org/10.1584/jpestics.D18-067

Saminathan N, Subramanian J, Sankaran Pagalahalli S, Theerthagiri A, Mariappan P (2025) Entomopathogenic fungi: translating research into field applications for crop protection. Arthropod Plant Interact 19:1–27. https://doi.org/10.1007/s11829-024-10110-4

Sandeep A, Selvaraj K, Kalleshwaraswamy CM, Hanumanthaswamy BC, Mallikarjuna HB (2022) Field efficacy of *Isaria fumosorosea* alone and in combination with insecticides against *Aleurodicus rugioperculatus* on coconut. Egypt J Biol Pest Control 32:106. https://doi.org/10.1186/s41938-022-00600-z

Sandhu SS, Sharma AK, Beniwal V, Goel G, Batra P et al (2012) Myco-biocontrol of insect pests: factors involved, mechanism, and regulation. J Pathog 12:1–10 https://doi.org/10.1155/2012/126819

Sarma BD, Puzari KC, Dutta P, Pandey AK (2023) An alginate-based encapsulation enhances shelf life and bioactivity of the entomopathogenic fungus, *Metarhizium anisopliae*. Egypt J Biol Pest Control 33:69. https://doi.org/10.1186/s41938-023-00714-y

Sayed SM, Ali EF, Al-Otaibi SS (2019) Efficacy of indigenous entomopathogenic fungus, *Beauveria bassiana* (Balsamo) Vuillemin, isolates against the rose aphid, *Macrosiphum rosae* L. (Hemiptera: Aphididae) in rose production. Egypt J Biol Pest Control 29:19. https://doi.org/10.1186/s41938-019-0123-y

Seethapathy P (2024) Mass production of *Beauveria bassiana*. In: Deshmukh SK, Sridhar KR (eds) The role of entomopathogenic fungi in agriculture. CRC Press, Boca Raton, pp 298–320. https://doi.org/10.1201/9781003505228-9

Seethapathy P, Gurudevan T, Subramanian KS, Kuppusamy P (2016) Bacterial antagonists and hexanal-induced systemic resistance of mango fruits against *Lasiodiplodia theobromae* causing stem-end rot. J Plant Interact 11:158–166. https://doi.org/10.1080/17429145.2016.1252068

Seethapathy P, Sankarasubramanian H, Lingan R, Thiruvengadam R (2023) *Chaetomium* sp.: an insight into its antagonistic mechanisms, mass multiplication, and production cost analysis. In: Amaresan N, Dharumadurai D, Babalola OO (eds) agricultural microbiology based entrepreneurship. Making money from microbes, vol 39. Springer, Singapore, pp 267–288. https://doi.org/10.1007/978-981-19-5747-5_17

Seethapathy P, Latha TKS, Warrier V, Sivagnanapazham K (2025) African cassava mosaic virus. In: Amaresan N, Kumar K (eds) Compendium of phytopathogenic microbes in agro-ecology. Springer, Cham, pp 17–50. https://doi.org/10.1007/978-3-031-81884-4_2

Sharma R, Sharma P (2021) Fungal entomopathogens: a systematic review. Egypt J Biol Pest Control 31. https://doi.org/10.1186/s41938-021-00404-7

Sharma A, Sharma S, Yadav PK (2023) Entomopathogenic fungi and their relevance in sustainable agriculture: a review. Cogent Food Agric 9:1–21. https://doi.org/10.1080/23311932.2023.2180857

Shi-Kunne X, Faino L, van den Berg GCM, Thomma BP, Seidl MF (2018) Evolution within the fungal genus *Verticillium* is characterized by chromosomal rearrangement and gene loss. Environ Microbiol 20:1362–1373. https://doi.org/10.1111/1462-2920.14037

Singh H, Joshi N (2020) Management of the aphid, *Myzus persicae* (Sulzer) and the whitefly, *Bemisia tabaci* (Gennadius), using biorational on capsicum under protected cultivation in India. Egypt J Biol Pest Control 30:67. https://doi.org/10.1186/s41938-020-00266-5

Skinner M, Parker BL, Kim JS (2014) Role of Entomopathogenic fungi in integrated pest management. In: Abrol DP (ed) Integrated pest management, current concepts and ecological perspective. Elsevier, pp 169–191

Subbanna ARNS, Kalidas P (2024) Mass production of *Isaria fumosorosea* pfu5 for community approach management of rugose spiraling whitefly in oil palm. J Microbiol Methods 225: 107023. https://doi.org/10.1016/j.mimet.2024.107023

Tahira HG, Saeed S, Khan FZA (2014) Entomopathogenic fungi as effective insect pest management tactic: a review. Appl Sci Bus Econ 1:10–18

Tkaczuk C, Majchrowska-Safaryan A (2023) Temperature requirements for the colony growth and conidial germination of selected isolates of entomopathogenic fungi of the *Cordyceps* and *Paecilomyces* Genera. Agriculture 13:1989. https://doi.org/10.3390/agriculture13101989

Yapa AT, Thambugala KM, Samarakoon MC, de Silva N (2025) *Metarhizium* species as bioinsecticides: potential, progress, applications & future perspectives. New Zeal J Bot 63: 439–461. https://doi.org/10.1080/0028825X.2024.2325006

Zimmermann G (2008) The entomopathogenic fungi *Isaria farinosa* (formerly *Paecilomyces farinosus*) and the *Isaria fumosorosea* species complex (formerly *Paecilomyces fumosoroseus*): biology, ecology and use in biological control. Biocontrol Sci Tech 18:865–901. https://doi.org/10.1080/09583150802471812

Endophytic Entomopathogenic Fungi for the Management of Insect Pests

16

Joginder Pal, Suman Sanjta, Chetna Mahajan, Rishi Mahajan, Praneet Chauhan, Shabnam Katoch, Shikha Sharma, Prakriti, Shalini Chandel, and Vanshdeep Dhanda

Abstract

Endophytic entomopathogenic fungi (EEPF) have emerged as promising biological control agents with dual functionality: suppressing insect pests while enhancing plant health and resilience. These fungi naturally colonize plant tissues and contribute to pest suppression through systemic infection and secondary metabolite production, offering a sustainable alternative to conventional chemical pesticides and further strengthening their role in integrated pest management (IPM) systems. Recent advancements in molecular biology, strain development, and precision delivery technologies have enhanced the efficacy, stability, and field applicability of EEPF. Furthermore, novel strains with improved virulence, broader host range, and environmental resilience are being explored to address limitations under variable agroecological conditions. Despite their potential, several challenges hinder the large-scale commercialization and field adoption

J. Pal (✉) · C. Mahajan · S. Katoch · S. Sharma · Prakriti
Department of Plant Pathology, Chaudhary Sarwan Kumar Himachal Pradesh Krishi Vishvavidyalaya, Palampur, Himachal Pradesh, India

S. Sanjta · V. Dhanda
Department of Entomology, Chaudhary Sarwan Kumar Himachal Pradesh Krishi Vishvavidyalaya, Palampur, Himachal Pradesh, India

R. Mahajan
Microbial Ecology and Applied Microbial Sciences Research Group, Department of Microbiology, College of Basic Sciences, Chaudhary Sarwan Kumar Himachal Pradesh Agricultural University, Palampur, Himachal Pradesh, India

P. Chauhan
Department of Plant Pathology, Dr Khem Singh Gill Akal College of Agriculture, Eternal University, BaruSahib, Sirmaur, Himachal Pradesh, India

S. Chandel
Department of Botany, Sri Sai University, Palampur, Kangra, Himachal Pradesh, India

S. K. Deshmukh et al. (eds.), *Frontiers in Entomopathogenic Fungi*,
https://doi.org/10.1007/978-981-95-6529-0_16

of EEPF. A multidisciplinary approach, integrating genomics, metagenomics, ecological modeling, and risk assessment frameworks, is essential to overcome these barriers. With focused research, technological innovation, and supportive policy frameworks, EEPF can become a cornerstone of climate-resilient and environmentally sustainable agriculture.

Keywords

Biocontrol · Biostimulation · Mycoparasitism · Pathogens · Pest management · Plant defense · Sustainable agriculture

1 Introduction

Fungi have historically played a central role in the biological control of insect pests, but the discovery and use of endophytic entomopathogenic fungi (EEPF) have added a novel dimension to eco-friendly management (Panwar and Szczepaniec 2024). The term endophyte, introduced by de Bary in 1866, refers to microorganisms that colonize plant tissues internally without causing visible disease symptoms. Many entomopathogenic fungi (EPF) exhibit endophytic behavior (Fig. 16.1), thereby offering dual benefits: direct insecticidal action and plant protection (Vega et al. 2012; Maina et al. 2018; Sharma et al. 2023). EEPF possess the ability to colonize plant tissues systemically, enhancing host resistance against insect pests and pathogens while simultaneously promoting plant growth and health (Samal et al. 2023a, b). These fungi display their entomopathogenic behavior by piercing the insect cuticle, proliferating inside the hemocoel, and causing mortality, contributing to epizootics among populations of pests (Mantzoukas and Eliopoulos 2020). Nearly

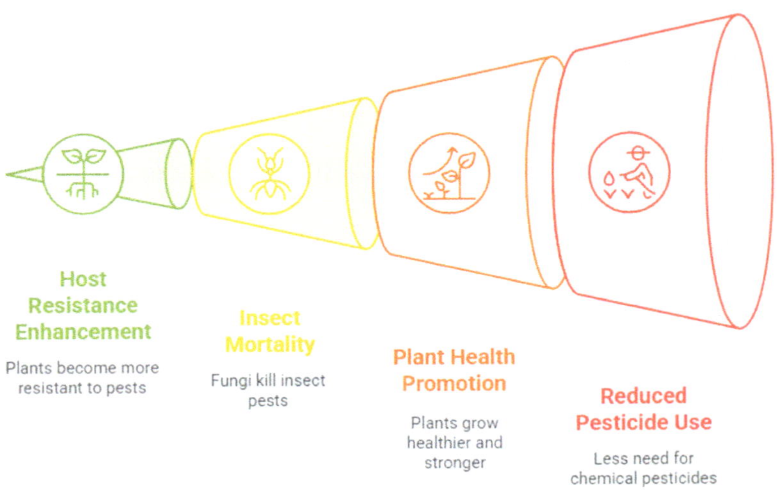

Fig. 16.1 Multifacted role of endophytic entomopathogenic fungi in pest management

700 species of fungi from approximately 90 genera have been identified as insect-pathogenic, with *Beauveria*, *Metarhizium*, *Hirsutella*, *Lecanicillium*, *Verticillium*, and *Paecilomyces* being the most frequently utilized genera (Khachatourians and Qazi 2008; Irsad et al. 2023).

The nonpathogenic nature of endophytic fungi (EF) within plants is attributed to a dynamic equilibrium between host defense mechanisms and fungal virulence. While the host initially perceives these fungi as threats, successful endophytes usually conquer plant immune responses and coexist without inflicting damage (Brader et al. 2017). Notably, many EEPF produce a diverse array of bioactive compounds, such as phenols, terpenoids, flavanols, and quinones, that exhibit insecticidal, antifungal, antioxidant, and antimicrobial properties (Gouda et al. 2016; Hashem et al. 2023).

Some of the best-known EPF, including *Beauveria bassiana* and *Metarhizium anisopliae*, have been shown to colonize plants endophytically while maintaining their insect-pathogenic capabilities (Vega 2018; Ahsan et al. 2024). This dual functionality positions them as potent tools for sustainable agriculture. Unlike topical applications of fungal spores, EEPF operate from within the plant tissues, enabling prolonged persistence and reduced application frequency (Quesada-Moraga 2020). Their systemic colonization facilitates direct contact with insect pests, which enhances control efficiency and cost-effectiveness.

The evolution of endophytic insect-pathogenic fungi (EIPF) reflects complex multi-kingdom interactions between plants, fungi, and insects. EF fungi can cause infestation leading to death of insect pests via antibiosis, antifeedant activity, and enzymatic degradation, while also forming mutualistic associations with their host plants (Samal et al. 2023a, b; Aouali 2024). Such interactions have given rise to both generalist and specialist EIPF species adapted to diverse ecological niches. Despite their promise, several limitations restrict their field utility, including low efficacy under adverse environmental conditions, competition from saprophytic fungi, and extreme sensitivity to desiccation and UV exposure (Ghorui et al. 2024).

Considerable economic losses due to insect pests are reported around the world, which are attributed to insect pests. As claimed by the Food and Agriculture Organization (FAO 2019), crop losses from invasive insect species amount to approximately \$70 billion annually, with a 40% yield reduction. Chemical insecticides remain the most widely used control method, yet their continued use poses major challenges: increasing pest resistance, toxicity to non-target organisms, ecological imbalance, and human health hazards (Khan et al. 2023). These issues have necessitated the search for safer, biologically based alternatives.

By colonizing internal plant tissues and producing insecticidal or growth-promoting secondary metabolites, they create a protective barrier against herbivorous insects. Some strains stimulate plant defense responses, while others directly inhibit pests through toxic or antifeedant compounds (Skinner et al. 2014). Additionally, these endophytes offer ecological benefits by improving plant vigor, nutrient acquisition, and drought resistance, thus reducing reliance on synthetic fertilizers and insecticides.

EPF as biological control agents are especially attractive due to their host specificity, environmental safety, and minimal non-target effects. Nevertheless, only a few strains have been commercialized. Challenges include short shelf life, variability in virulence, and difficulties in large-scale production and storage (Thirumeni et al. 2024). Innovations in formulation techniques and nutrient media have improved fungal viability and longevity, opening new possibilities for field deployment. This paradigm shift from pest eradication via chemical inputs to holistic, ecosystem-based management has emphasized the integration of microbial agents in crop protection. EPF, particularly those with endophytic capabilities, are central to this transition. The integration of EF in integrated pest management (IPM) systems aligns with sustainability goals by reducing pesticide use, enhancing soil health, and preserving biodiversity (Sharma et al. 2020; Sharma and Sharma 2021).

This chapter unravels the role of EEPF in insect pest management. It provides a comprehensive overview of their direct and indirect effects on pests and plant health, highlights recent advances in strain identification and metabolite profiling, and discusses their utility in IPM strategies. Finally, this chapter addresses current challenges such as inconsistent field performance, environmental variability, and non-target effects, along with strategies for overcoming these limitations.

2 Pest Control Mode of Action

The host range of EF includes a vast range of insect orders, including Lepidoptera, Hymenoptera, Diptera, Coleoptera, and Hemiptera, by bypassing the digestive tract and directly penetrating the insect cuticle. This unique mode of infection distinguishes them from bacterial and viral pathogens, which primarily rely on gastrointestinal entry (Litwin et al. 2020; Parveen and Rashtrapal 2024). The infection process begins with the random and passive attachment of fungal conidia (spores) to the insect's outer cuticle, facilitated by wind, water, or physical contact. Upon contact, the conidia adhere to the cuticle surface and germinate, forming germ tubes that differentiate into specialized infection structures called appressoria. These structures are influenced by the physical topography of the underlying surface and modulated by intracellular elicitors like calcium ions (Ca^2) and cyclic AMP (Perlin et al. 2025). Through a penetration peg, the germ tubes exert mechanical pressure and secrete a suite of cuticle-degrading enzymes, including proteases, chitinases, and lipases, that break down the insect's protective barriers (Shree et al. 2024). Fungal spores may target softer or thinner cuticle regions, such as sensory pores or wounds, to gain entry (Sharma et al. 2024).

Following successful penetration, the fungus enters the hemocoel (insect body cavity) and proliferates by forming hyphal bodies or blastospores. During this internal colonization, the fungus exploits host nutrients, secretes toxic metabolites, disrupts physiological processes, blocks internal vessels, and ultimately leads to the host's death (Parulekar-Berde et al. 2021). Upon completion of the internal expansion, the fungus resumes the mycelial stage and re-emerges through the insect cuticle. Remarkably, genes responsible for initial penetration are reactivated to facilitate cuticle exit amidst sporulation (Shin et al. 2020). The degradation process is highly enzyme-driven. Both insects and entomopathogenic fungi produce various

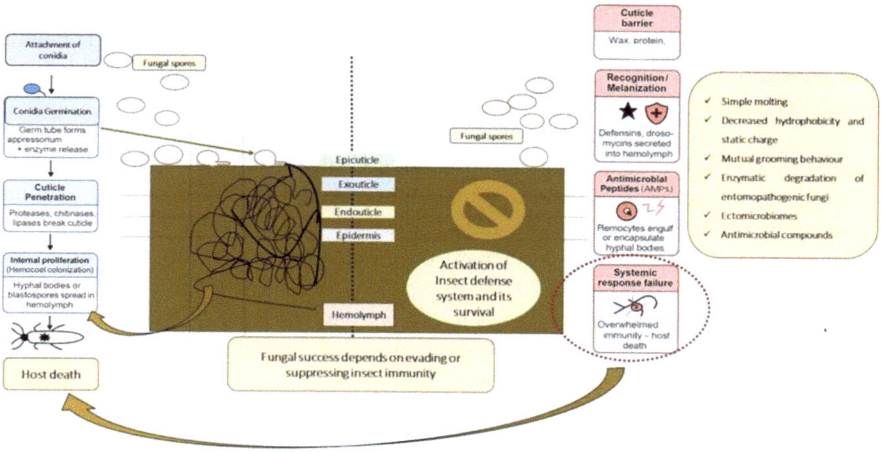

Fig. 16.2 Schematic illustration of the interaction between insects and EEPF

cuticle-degrading enzymes, including trypsin, esterase, chymotrypsin, collagenase, and chymoelastase (Prajapati et al. 2023). Specifically, cuticle endoproteases play a vital role in breaching the host defenses. Proteolytic enzymes act swiftly, while enzymes like N-acetyl glucosaminidase are produced more gradually during infection (Singh et al. 2017). The germination and fungal development are influenced by environmental factors such as nutrient availability, oxygen, pH, water, temperature, and host-produced inhibitory compounds. Throughout this invasion, the insect activates a range of immune defenses, such as the barrier of the cuticle composed of wax and proteins, which acts as a physical barrier, and innate immune responses such as melanization, secretion of antimicrobial peptides (AMPs), and hemocyte activity, which may engulf or encapsulate invading fungal cells.

In some cases, the insect mounts behavioral defenses like grooming, molting, and enzymatic degradation of fungal cells. Ectomicrobiomes and antimicrobial compounds also contribute to resistance mechanisms (Fig. 16.2).

Despite these modes of defense, the success of fungi is largely dependent on their ability to evade or suppress host immunity. If the fungal pathogen overcomes the insect's systemic immune responses, this leads to immune failure and eventual death of the host. Ultimately, the output of these interactions is influenced by two major factors, namely, fungal virulence and the robustness of the insect's immune system. Such mechanisms are effectively exploited in biocontrol strategies for pest management, offering sustainable solutions for reducing crop losses and improving agricultural productivity (Petrisor and Stoian 2017). It has been shown that the fungus shows prompt multiplication after entering the hemocoel. The hyphae extend and branch to form a dense mycelial network that conquers and colonizes the insect's internal tissues. Cellular disintegration is often observed even before hyphal penetration, suggesting that cytotoxic compounds play a role in early tissue damage. During this phase, the fungus produces a variety of secondary metabolites, such as destruxins and beauvericin, that contribute significantly to host death by disrupting cellular function (Gillespie and Claydon 1989). These mycotoxins interfere with host physiology by causing paralysis, reduced responsiveness, sluggish movement, ion channel disruption, apoptosis, and suppression of immune responses.

Insect death generally occurs within a span of a few days to a week after infection, depending on environmental conditions and the virulence of the fungal strain. Once the host dies, the fungus reverts to its saprophytic phase and begins external sporulation. The fungus breaches the cuticle and produces new conidia on the cadaver surface, which are then released into the environment, thus leading to the next infection cycle (Butt et al. 2016).

In Entomophthorales, spores are actively discharged under hydrostatic pressure (except in *Massospora* spp.) and are subsequently dispersed by wind or insect contact (Chaurasiya et al. 2021).

The pathogenic outcome is also influenced by several intrinsic and extrinsic factors, including host species, fungal strain, incubation period, fungal growth rate, and the insect's physiological state. Stress conditions such as poor nutrition, pesticide exposure, or unfavorable weather can weaken insect immunity and increase susceptibility to infection (Ignoffo et al. 1977; Mantzoukas et al. 2022). Moreover, fungal strains vary in their adaptability to abiotic stress. For instance, species adapted to tropical climates tend to have higher thermotolerance than those from temperate regions (Inglis et al. 2001). Mantzoukas et al. (2020) also reported that fungal infections can lead to a wide range of host effects, including behavioral changes, morphological symptoms, and disruptions in developmental and metabolic pathways, which collectively contribute to the eventual death of the insect.

The principal application of EPF lies in their use as biocontrol agents within IPM systems. Their efficacy stems from a combination of mycelial growth, enzymatic degradation of host tissues, and the production of insecticidal toxins that act synergistically to incapacitate or kill insect pests.

Depending on their life strategy and host interaction, EPF are broadly categorized into four types:

Facultative pathogens, which show multiplication at both the sites, i.e., within and outside the host (e.g., *Beauveria bassiana*)

Obligate pathogens, which require a host organism to complete their life cycle (e.g., *Microsporidia daphniae*)

Opportunistic pathogens, which cause disease only in immunocompromised hosts (e.g., *Aspergillus flavus*)

Latent pathogens, which remain inactive under normal conditions but become virulent under congenial periods (Mantzoukas and Eliopoulos 2020)

These diverse biological strategies highlight the versatility of EPF, not only as direct insect pathogens but also as potential endophytic agents that can colonize plant tissues, improve systemic resistance, and contribute to sustainable pest management. Their dual role in targeting insect pests and enhancing plant health underscores their value in ecologically sound agriculture and their growing relevance in integrated crop protection frameworks.

3 Mechanisms of Action

An increasing number of studies have confirmed that EPF acting as endophytes can protect host plants from diseases or mitigate the adverse effects of biotic stressors (Fig. 16.3). However, the detailed mechanisms by which these endophytic fungi confer such protection remain incompletely understood (Ahsan et al. 2024). Since endophytic fungi colonize internal plant tissues, they are shielded from environmental fluctuations and stress, thereby representing a promising class of biological control agents against pathogens and pests. This unique ecological niche suggests that endophytic fungi may offer superior performance compared to other biocontrol agents in managing plant health (Latz et al. 2018).

A variety of mechanisms, including mycoparasitism, resource competition, and antibiosis via the production of fungal secondary metabolites, allow EEPF to exert direct antagonistic effects on plant pathogens. These mechanisms also facilitate indirect interactions through systemic responses within the common host plant (Bamisile et al. 2023). Such interactions are crucial in activating plant defense pathways, inducing systemic resistance, enhancing the production of secondary metabolites, and supporting overall plant growth and development.

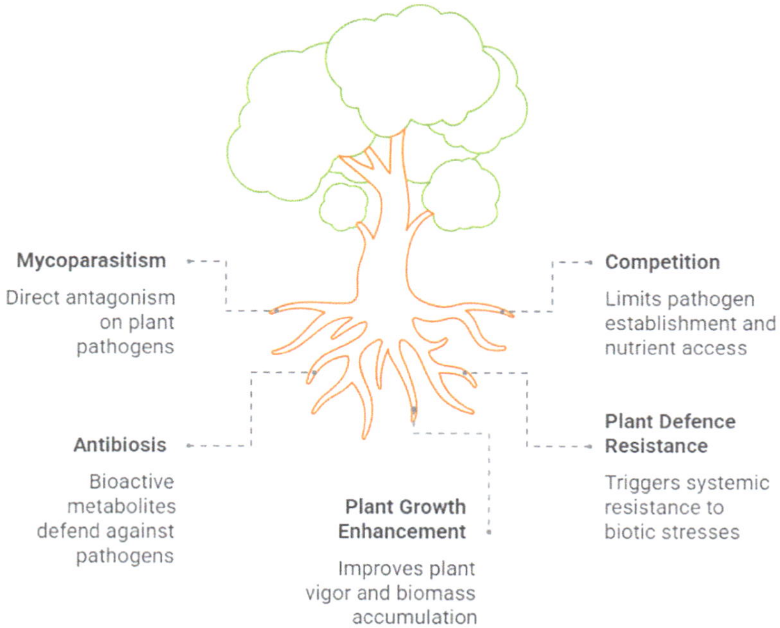

Fig. 16.3 Mechanisms of endophytic fungi in conferring biotic stress tolerance

3.1 Mycoparasitism

Mycoparasitism is a common antagonistic interaction in which a mycoparasitic fungus establishes direct physical contact with the hyphae of a host fungus, often coiling around them, penetrating, absorbing nutrients, and ultimately lysing the host cells. Jeffries (1995) defines mycoparasitism as a specific interaction between a fungal parasite and its fungal host.

This complex physiological process involves the secretion of hydrolytic enzymes such as cellulases, chitinases, and proteases, as well as the production of antifungal secondary metabolites (Mukherjee et al. 2022; Panicker and Sayyed 2022). Among the most studied mycoparasites with biocontrol potential against phytopathogenic fungi are genera such as *Aspergillus*, *Curvularia*, *Fusarium*, and *Trichoderma*.

Endophytic fungi isolated from rice leaves, namely *Curvularia chiangmaiensis*, *Fusarium solani*, and *Trichoderma asperellum*, have shown significant inhibitory effects against the causative agent of rice blast (*Pyricularia oryzae*). These fungi produce hydrolytic enzymes, including cellulases and chitinases, at concentrations ranging between 1.83–4.09 μg/mL and 1.43151 μg/mL, respectively (Xu et al. 2019). Putri et al. (2022) further observed aberrant and stunted growth at the hyphal tips of the pathogen upon interaction with endophytic fungi. Additionally, *Trichoderma* sp., *T. harzianum*, *T. pleuroti*, and two strains of *T. longibrachiatum* isolated from medicinal plants demonstrated pronounced antagonistic activity against *Sclerotinia sclerotiorum* and *Sclerotium rolfsii*.

These fungi exhibited aggressive colonization through hyphal coiling, widespread branching, and hyphal invasion of the pathogens (Li et al. 2024).

3.2 Competition

Endophytic entomopathogenic fungi (EEPF) contribute to plant defense by competing with phytopathogens for space, nutrients, and ecological niches within plant tissues. Rather than acting in isolation, competitive mechanisms often operate synergistically with other antagonistic processes such as mycoparasitism and antibiosis.

Endophytic fungi (EF) can colonize plant tissues either locally or systemically, spreading between or within plant cells. By occupying potential infection sites and scavenging essential nutrients, they delay or prevent pathogen establishment (Grabka et al. 2022). A major advantage of EF over many phytopathogens lies in their ability to enhance nutrient bioavailability, particularly for macronutrients such as nitrogen and phosphorus.

One of the crucial mechanisms through which EF outcompete pathogens is iron sequestration via the release of siderophores, low molecular weight compounds with a high affinity for ferric iron (Fe^3). By limiting the availability of iron to pathogens, siderophores restrict their growth and virulence. For instance, *Trichoderma asperellum* S11, *T. longibrachiatum* S12, and *T. atroviride* PHYTAT7 produce siderophores that inhibit *Rhizoctonia solani* through iron deprivation (de Almeida

Lopes et al. 2018; Rabbee et al. 2019; Grabka et al. 2022). Moreover, in grapevine plants, endophytic colonization by *Beauveria bassiana* significantly reduced the incidence and severity of downy mildew, even before the pathogen *Plasmopara viticola* was introduced, indicating pre-emptive protection due to niche occupation and metabolic interference (Rondot and Reineke 2019).

3.3 Antibiosis

Antibiosis is a critical mechanism by which EPF defend host plants against pathogens and insect pests through the production of bioactive secondary metabolites (Sharma and Sharma 2021; Aravinthraju et al. 2024). These metabolites exhibit a plethora of biological activities, including insecticidal, cytotoxic, antifungal, antibacterial, and antiviral effects (Gibson et al. 2014; Rondot and Reineke 2019; Sharma and Sharma 2021).

Endophytic fungi can induce multiple compounds, such as alkaloids, flavonoids, peptides, quinones, phenols, terpenoids, steroids, polyketides, and other specialized metabolites (Lugtenberg et al. 2016; Wani et al. 2017). Specific strains can be inoculated into host plants to investigate the role of particular metabolites in biocontrol, offering targeted insights into antifungal or insecticidal activity. For instance, 1198 strains of *B. bassiana* produce beauvericin in broth culture, a metabolite reported to suppress damping-off diseases caused by *Pythium myriotylum* and *Rhizoctonia solani* in cotton and tomato plants (Clark 2006; Yin et al. 2022; Adhikari et al. 2024). These metabolites can also modulate host plant metabolism, redirecting metabolic fluxes to promote increased synthesis of protective compounds.

An endophytic fungus, *Mortierella alpina* CS10E4, associated with and isolated from *Crocus sativus* L., provokes the synthesis of apocarotenoids and arachidonic acid in the host, enhancing resistance against corm rot disease (Clark 2006; Wani et al. 2017). In bean plants, endophytic colonization by EF, i.e., *Hypocrea lixii* F3ST1 and *B. bassiana* G1LU3, has been shown to alter the production of plant volatiles such as terpenes, inducing systemic resistance and strengthening the plant's defense against insect pests like *Spodoptera frugiperda* and *Liriomyza huidobrensis* (Chebet et al. 2021). Furthermore, EF also emits volatile organic compounds (VOCs) with significant antimicrobial and insect-repellent properties. These VOCs are typically small, hydrophobic molecules with high vapor pressure and originate from primary and secondary metabolic pathways (Karslı and Şahin 2021).

3.4 Induction of Plant Defense Resistance

Induction of plant defense resistance is one of the primary ways endophytic EEPF enhances the host plant's resilience to biotic stresses. Upon colonization, EPF can trigger induced systemic resistance (ISR) or systemic acquired resistance (SAR), two major defense strategies activated by beneficial or pathogenic microbes, respectively (Maithani et al. 2021; Yu et al. 2022).

Induction of defense pathways due to the association of EEPF has been demonstrated in many studies. For example, *Phomopsis liquidambaris* B3 enhances resistance to *Fusarium oxysporum* by upregulating SA-mediated PR protein expression (Zhu et al. 2021). Similarly, *Trichoderma longibrachiatum* H9 activates both JA-ET and SA signaling pathways in *Cucumis sativus* against *Botrytis cinerea* to induce defense (Pacheco-Trejo et al. 2022). Biological management of blackleg disease in potato (*Pectobacterium carotovora* subsp. *atrosepticum* PHY7) using EF (*Epicoccum nigrum* ASU11) and arbuscular mycorrhizal fungi (AMF) has shown effective activation of plant systemic resistance. This involved the reduction of oxidative stress markers like reactive oxygen species (ROS) and MDA and the enhancement of antioxidant enzymes such as SOD, GPX, CAT, and APX (Bagy et al. 2019).

Fungal elicitors, including chitin, β-glucans, cell wall-degrading enzymes, and proteins, are known to activate defense responses in plants (Guo and Cheng 2022; Kongala and Kondreddy 2023). These elicitors can be delivered in the form of crude mycelial extracts, live bioagents, or oligosaccharides and can trigger robust resistance responses (Sák et al. 2021). Thc6, a zinc finger protein elicitor of *Trichoderma harzianum*, induces ISR against *Curvularia lunata*, the incitant of Curvularia leaf spot in maize (Saravanakumar et al. 2016). Oligosaccharides derived from *Fusarium oxysporum* Dzf17 enhanced resistance in *Dioscorea zingiberensis* by stimulating the activity of PAL, POD, and PPO enzymes (Li et al. 2014). In cotton, root inoculation with *B. bassiana* substantially decreased the bacterial blight intensity, displaying high efficacy comparable to that of 2,6-dichloroisonicotinic acid, a known ISR inducer molecule (Griffin et al. 2006; Ownley et al. 2008).

Colonization by fungal endophytes also enhances the accumulation of phytoalexins, antimicrobial compounds such as terpenoids that play a key role in resistance. For example, *T. harzianum*-inoculated pepper plants exhibited a sevenfold increase in capsidiol, a terpenoid phytoalexin, which delayed lesion development by *Phytophthora capsici* (Ahmed et al. 2000).

Another major response to mitigate the stress includes the generation of reactive oxygen species (ROS) such as hydroxyl radicals (OH), hydrogen peroxide (HO), and superoxide radicals (O). Although these molecules serve as signaling components, their excessive accumulation can damage tissues (Torres et al. 2006). EF augment the plant's ROS scavenging competency via the upregulation of enzymes, namely, peroxidase (POX), superoxide dismutase (SOD), catalase (CAT), and ascorbate peroxidase (APX), which are antioxidative in nature. For example, *Aspergillus jodhpurensis* improved resistance to *R. solani* in tomato by modulating iron uptake and boosting HO and O levels, alongside activating antioxidant responses (Daroodi et al. 2021). Similarly, *T. harzianum* and *T. virens* induced systemic immunity in Luffa plants against downy mildew through upregulation of PAL, CAT, PPO, SOD3, and POD enzymes (Rauf et al. 2022). Additionally, EF-treated seedlings of pearl millet showed increased callose deposition and lignification in response to *Sclerospora graminicola*, as visualized through aniline blue staining (Siddaiah et al. 2017).

3.5 Enhancement of Plant Growth and Kairomones

Despite the remarkable potential of EEPF in managing pathogens and pests, they are further recognized for their significant contribution toward the enhancement of plant growth. An increasing number of studies have concluded that EF can improve plant vigor, biomass accumulation, and physiological resilience.

For instance, *Beauveria bassiana*-colonized squash plants not only displayed reduced disease intensity against *Zucchini yellow mosaic virus* (ZYMV) but also displayed enhanced growth parameters, including increased shoot length, root mass, and overall biomass compared to uninoculated controls (Jaber and Enkerli 2016, 2017). The mechanisms underlying this growth promotion may involve the production of phytohormones (e.g., auxins, gibberellins, and cytokinins) and siderophores, which are low molecular weight iron-chelating compounds that improve nutrient acquisition (Khan et al. 2012). Inoculation with *Metarhizium anisopliae* has been reported to stimulate phytohormone biosynthesis in soybean, contributing to improved root and shoot development. Enhanced iron uptake and modulation of hormone signaling pathways likely mediate these growth-promoting effects. Additionally, EF may indirectly contribute to insect pest deterrence by altering kairomone production in host plants. Kairomones are chemical cues emitted by plants that are perceived and exploited by insects to locate suitable hosts; while beneficial for the receiver (insect), they are disadvantageous to the emitter (plant) (Price et al. 2011).

This dual functionality, growth promotion and pest deterrence, highlights the ecological versatility of entomopathogenic fungal endophytes and underscores their value in integrated plant health strategies.

4 Biocontrol Agents

4.1 Bioefficacy Against Insect Pests

Singkaravanit et al. (2010) defined EPF as a group of fungi that kill an insect by attacking and infecting its host. They act as natural enemies to insects and arachnids and contribute significantly toward the maintenance of insect populations (Eilenberg et al. 2001). A detailed tabulation of endophytic fungi as biocontrol agents against various insect pests has been presented in Table 16.1.

4.2 Biocontrol Agents Against Plant Pathogenic Fungi

Plant pathogenic fungi represent a significant constraint to global crop productivity. The majority of phytopathogens belong to the Ascomycota (e.g., *Fusarium*, *Colletotrichum*, *Botrytis*) and Zygomycota (e.g., *Rhizopus*). Biocontrol strategies involving endophytic entomopathogenic fungi (EEPF) are gaining attention due to their capacity to colonize plant tissues internally and provide long-term protection. Unlike conventional pesticides or surface-colonizing microbes, EEPF remain within

Table 16.1 EEPF as biocontrol agents against various insect pests

Endophytic entomopathogenic fungus	Target pest	Crop	Reference
Acremonium strictum	*Helicoverpa armigera* (cotton bollworm)	Broad bean and tomato	Jallow et al. (2004); Jaber and Vidal (2010)
Beauveria bassiana	*Aphis gossypii* (cotton Aphid)	Cotton and wheat	Gurulingappa et al. (2010)
Beauveria bassiana	*Iraella luteipes* (poppy stem gall wasp)	Opium poppy	Quesada-Moraga et al. (2009)
	Cosmopolites sordidus (banana weevil)	Banana	Akello et al. (2008)
	Sesamia calamistis (Pink stem borers)	Maize	Cherry et al. (2004)
	Myzus persicae (green peach aphid)	Cabbage	Michereff Filho et al. (2011)
	Hypothenemus hampei (coffee borer beetle)	Coffee	Rehner et al. (2006)
	Leptinostarsa decemlineata (Colorado potato beetle)	Potato	Wraight et al. (2007)
	Tuta absoluta (tomato pinworm)	Tomato	Klieber and Reineke (2016)
	Helicoverpa armigera (tomato shoot and fruit borer)	Tomato	Qayyum et al. (2015)
	Liriomyza huidobrensis	Broad bean and common bean	Akutse et al. (2013); Gathage et al. (2016)
Clanostachys rosae	*Thrips tabaci* (onion thrips)	Onion	Muvea et al. (2014)
Clanostachys rosae	*Hyphothenemus hampei*	Coffee	Vega et al. (2008)
Epichloe gansuensis	*Rhopalosiphum padii*	*Achnatherum inebrains*	He et al. (2022)
Hypocrea lixii	*Acyrthosiphon pisum* (pea aphid)	Broad bean	Akutse et al. (2013)
Isaria fumosorosae	*Myzus persicae* *Diaphorina citri*	Lemon and pepper	Mantzoukas et al. (2019)
Lecanicillium longisporum	*Cucumber aphid*	Cucumber	Kim et al. (2010)
Lecanicillium muscarium	*Plutella xylostella*	Cauliflower	Kuchár (2019)
Metarhizium anisopliae	*Lipaphis erysimi* (Mustard aphid), Cabbage aphid (*Aphis gossypii* and *Aphis cracivora*)	Canola and cabbage	Ujjan and Shahzad (2012)
Metarhizium brunneum	*Bemisia tabaci* (Sweet potato whitefly)	Cauliflower	Jaber et al. (2018)

(continued)

Table 16.1 (continued)

Endophytic entomopathogenic fungus	Target pest	Crop	Reference
Metarhizium brunneum	*Aphis glycines* *A. gossypii*	Soybean	Clifton et al. (2018)
	Myzus persicae (green peach aphid)	Sweet pepper	Jaber and Araj (2018)
	Spodoptera littura (cotton leafworm)	Melon	Miranda-Fuentes et al. (2021)
Metarhizium brunneum	*Delia radicum* L. (cabbage root fly)	White cabbage	Cotes et al. (2015)
Metarhizium robertsii	*Agrotis ipsilon* (black cutworm)	Maize	Ahmad et al. (2020)
Metarhizium robertsii	*Sesamia nonagrioides* (Mediterranean corn stalk borer)	Sweet sorghum	Mantzoukas et al. (2015)
Paecilomyces fumosoroseus	*Lipaphis erysimi* (mustard aphid)	Cabbage	Ujjan and Shahzad (2012)
Piriformospora indica	*Lissorhoptrus oryzophilus*	*Oryza sativa*	Cosme et al. (2016)
Purpureocillium lilacinum	*Aphis gossypii* (cotton aphid) and *Helicoverpa zea* (cotton bollworm)	Cotton	Lopez and Sword (2015)
Trichoderma asperellum	*Trialeurodes vaporarium* (green house whitefly)	Green beans	Paradza et al. (2021)
Trichoderma asperellum M2RT4	*Tuta absoluta*	*Tuta absoluta*	Agbessenou et al. (2022)

plant tissues, shielded from environmental stress and external microbial competition (Latz et al. 2018). Endophytic entomopathogenic fungi as biocontrol agents against various phytopathogenic fungi are given in Table 16.2.

4.3 Bioefficacy Against Plant Pathogenic Bacteria

Although research on the potential of fungal endophytes against phytopathogenic bacteria is relatively less explored compared to their antifungal or insecticidal uses, evidence suggests that certain endophytes also possess antibacterial properties. These effects are typically mediated by the production of secondary metabolites such as phenolics, alkaloids, and polyketides, which interfere with bacterial quorum sensing, biofilm formation, or cell wall integrity.

B. bassiana as well as *Trichoderma* spp. are reported to have restricted the growth of *Erwinia amylovora*, *Pseudomonas syringae*, and *Xanthomonas* spp. In addition to direct inhibition, fungal endophytes may also promote plant health and reduce disease incidence through competition for colonization sites, enhancement of host immunity, and modulation of plant defense signaling pathways. A detailed tabulation of endophytic fungi as biocontrol agents against phytopathogenic bacteria has been presented in Table 16.3.

Table 16.2 EEPF as biocontrol agents against phytopathogenic fungi

Endophytic entomopathogenic fungus	Targeted pathogen	Disease/host crop	Reference
Acrophialophora jodhpurensis	*Rhizoctonia solani*	Tomato	Daroodi et al. (2021)
Beauveria bassiana	*Pythium myriotylum* *Rhizoctonia solani*	Damping off of tomato, root rot of cotton	Ownley et al. (2008)
	Fusarium oxysporum f.sp. *cepae*	Basal rot of onion	Flori and Roberti (1993)
	Plasmopara viticola	Downy mildew of grapevine	Jaber (2015)
Cadophora orchidicola	*Fusarium oxysporum* and *Sclerotium rolfsii*	*Kalimeris indica*	Wang et al. (2019)
Chaetomium globosum 12XP1-2-3	*Fusarium pseudograminearum*	*Fusarium* crown rot disease in wheat	Feng et al. (2023)
Diatrype palmicola	*Athelia rolfsii*	*Polyscias fruticosa*	Tanapichatsakul et al. (2020)
Diaporthe sp.	*Colletotrichum acutatum*	*Olea europaea*	Landum et al. (2016)
Epiccocum nigrum	*F. solani*	*Euphorbia milii*	Ali et al. (2023)
Fusarium solani	*Colletotrichum gloeosporioides*	*Sophora tonkinensis*	Yao et al. (2017)
Fusarium subglutinans	*Botrytis cinerea*	Thymus	Mirzaei and Masumi (2022)
Lasiodiplodia pseudotheobromae J-10	*Exserohilum turcicum*, *Pestalotiopsis theae*, and *Colletotrichum capsica*	*Sarcandra glabr*	Luo et al. (2023)
Lecanicillium longisporum	*Sphaerotheca fuliginea*	Powdery mildew of cucumber	Kim et al. (2010)
Metarhizium robertsii	*Fusarium solani* f.sp. *phaseoli*	Root rot of bean	Sasan and Bidochka (2013)
Penicillium citrinum DBR-9	*Bipolaris maydis* and *Alternaria citri*	*Stephania kwangsiensis*	Luo et al. (2019)
Penicillium olsonii ML37	*Fusarium graminearum*	Wheat	Rojas et al. (2022)
Penicillium simplicissimum	*Verticillium dahliae*	*Gossypium arboreum*	Yuan et al. (2017)
Phomopsis liquidambaris B3	*Fusarium oxysporum* *Fusarium proliferatum*	Peanut root rot and rice spikelet disease	Zhu et al. (2021)
Rhizopycnis vagnum	*Fusarium oxysporum*, *Sclerotium rolfsii*	Ginger	Anisha et al. (2018)
Talaromyces assiutensis JTY2	*Alternaria brassicicola*, *Bipolaris oryzae*	*Ceriops tagal*	Li et al. (2022)
Trichoderma asperellum	*Pyricularia oryzae*	Rice blast	Putri et al. (2022)

(continued)

Table 16.2 (continued)

Endophytic entomopathogenic fungus	Targeted pathogen	Disease/host crop	Reference
T. harzianum ThM9	*Pseudoperonospora cubensis*	*Luffa cylindrica*	Rauf et al. (2022)
Trichoderma virens	*Rhizoctonia solani*	Rice	Safari Motlagh et al. (2022)

Table 16.3 Endophytic fungi as biocontrol agents against phytopathogenic bacteria

Endophytic entomopathogenic fungus	Targeted pathogen	Disease/crop	Reference
Amphirosellinia nigrospora JS-1675	*Xanthomonas oryzae* pv. *Oryzae*	*Pteris cretica*	Nguyen et al. (2019)
Aspergillus fumigatus	*Xanthomonas vesicatoria*	*Solanum lycopersicum*	Rashid (2021)
Beauveria bassiana 11–98	*Xanthomonas axonopodis* pv. *malvacearum*	Bacterial blight of cotton	Ownley et al. (2008)
Chaetosphaeronema sp.	*Streptomyces scabies*	Potato scab	Mirzaei and Masumi (2022)
Epicoccum nigrum ASU11	*Pectobacterium carotovora* subsp. *atrosepticum* PHY7	Potato blackleg disease	Bagy et al. (2019)
Epicoccum nigrum ASU11	*Pectobacterium carotovora* subsp. *atrosepticum PHY7*	Potato	Bagy et al. (2019)
Fusarium sambucinum TE-6L	*Ralstonia solanacearum*	*Nicotiana tabacum*	Zhang et al. (2019)
Nigrospora sp. XCE-7	*Pseudomonas syringae*	Halo blight/ *Coffea arabica*	Lu et al. (2022)
Piriformospora indica	*Ralstonia solanacearum*	Bacterial wilt in tomato	Athira and Anith (2020)

4.4 Bioefficacy Against Plant Viruses

Direct antagonism of plant viruses by fungal endophytes is not yet well studied and extensively explored. However, some EPF may indirectly influence viral diseases by strengthening host plant defenses or suppressing insect vectors responsible for virus transmission. For instance, colonization of plants by *Beauveria bassiana* has been shown to reduce populations of aphids and whiteflies, both of which are known vectors for many devastating plant viruses.

Due to the potential of EPF to induce or modulate systemic resistance (ISR) or jasmonic acid signaling pathways, they are gaining attention for further implications in viral disease management. These strategies employed by EPF ultimately lead to enhanced tolerance or reduced viral replication. Although experimental validation on this aspect is limited, such mechanisms could be exploited to mitigate viral diseases in vegetable crops, i.e., tomato and cucumber, as well as in tobacco. A detailed tabulation of endophytic fungi as biocontrol agents against plant viruses has been presented in Table 16.4.

Table 16.4 EEPF as biocontrol agents against plant viruses

Endophytic entomopathogenic fungus	Targeted pathogen	Host crop	Reference
Aspergillus tubingensis FJBJ11	TMV	*Brucea javanica*	Tan et al. (2015)
Aspergillus versicolor	Tobacco mosaic virus	*Nicotiana tabacum*	Yang et al. (2023a, b)
Hypocrea lixii F3ST1	Iris yellow spot virus	*Allium cepa* and Napier grass	Kiarie et al. (2020)
Phialocephala fortinii J2PC4	Southern rice black-streaked dwarf virus	Rice	Zhang et al. (2022)
Piriformospora indica	Blackeye cowpea mosaic virus (BlCMV)	*Vigna unguiculata* var. *sesquipedalis*	Chandran et al. (2021)
Piriformospora indica	Cymbidium mosaic virus	Dendrobium orchids	Safeer and Thara (2022)
Trichoderma asperellum SKT-1	Cucumber mosaic virus (CMV-Y)	*Arabidopsis*	Elsharkawy et al. (2013)
Trichoderma harzianum	Sugarcane mosaic virus (SCMV)	Maize and onion	Kiarie et al. (2020)
Trichoderma harzianum T-22 (T22)	Cucumber mosaic virus (CMV)	*Solanum lycopersicum* var. *cerasiforme*	Vitti et al. (2016)

4.5 Bioefficacy Against Other Plant Pathogens

Plant-parasitic nematodes (PPN), including *Meloidogyne*, *Heterodera*, and *Pratylenchus* species, are major contributors to crop losses worldwide. Recent research has highlighted the role of endophytic fungi in suppressing nematode activity through both direct and indirect mechanisms. These include the secretion of nematicidal compounds, parasitism of eggs and juveniles, competition for nutrients and niches, and activation of host defense response cascades (Khan et al. 2023). A detailed literature review on the use of endophytic fungi as BCAs against plant-parasitic nematodes (PPNs) has been carried out and is presented in Table 16.5.

5 Application as Biostimulants

EEPF represent a novel and sustainable method of protecting plants and promoting their growth. These fungi, traditionally known for their pathogenicity against insects, have demonstrated endophytic colonization capabilities within plant tissues, offering dual benefits of plant biostimulation and pest suppression. Inoculation methods for introducing EPF into plants include foliar spraying, stem injection, seed treatment, and soil irrigation (Quesada-Moraga et al. 2006). Among these, foliar application of spore suspensions is particularly effective in controlling insect pests that inhabit the phylloplane. Endophytic colonization by EPF such as *M. brunneum* and *B. bassiana*

Table 16.5 EEPF as biocontrol agents against plant-parasitic nematodes

Endophytic entomopathogenic fungus	Targeted pathogen	Host crop	Reference
Annulohypoxylon sp. FPYF3050	*Bursaphelenchus xylophilus* (pine wood nematode)	Pine seedlings	Li et al. (2020)
Aspergillus welwitschiae AW2017	*Meloidogyne graminicola*	Rice	Liu et al. (2019)
Fusarium oxysporum V5w2	*Radopholus similis* (burrowing nematode)	Banana	Kisaakye et al. (2022)
Fusarium strain Fu234 and *Chaetomium strain* Ch1001	*Meloidogyne incognita*	Cucumber	Yan et al. (2011)
Penicillium brefeldianum	*Meloidogyne incognita*	Melon	Miao et al. (2019)
Penicillium chrysogenum strain Snef1216	*M. incognita*	Cucumber	Sikandar et al. (2019)
Penicillium citrinum MN518391	*Meloidogyne javanica*	Taify roses	Baazeem et al. (2022)
Purpureocillium lilacinum, *Pochonia chlamydosporia* and *Piriformospora indica*	*Heterodera glycines* (soybean cyst nematode)	Soybean	Silva et al. (2022)
Trichoderma harzianum	*Meloidogyne incognita*	Tomato	Yan et al. (2021)

has been registered in various crops, including tomato, alfalfa, sweet pepper, and melon (Jaber and Araj 2018).

The success of colonization is influenced by plant species, environmental conditions, and the chosen inoculation technique. The use of techniques such as conventional and molecular enhances our understanding of plant-fungus symbioses and supports the refinement of inoculation strategies. However, commercialization, regulatory approval, and production scalability remain critical challenges (Silva et al. 2024).

Insect pests pose a significant threat to crop productivity, economic stability, and global food security. EPF offer an eco-friendly alternative to synthetic insecticides, functioning as mycopesticides with unique infection mechanisms across insect orders. Recent studies reveal that, in addition to their insecticidal activity, EPF also suppress phytopathogens and promote plant growth via rhizospheric colonization. Environmental and biological factors significantly affect EEPF efficacy, and recent genomic research has identified genes that facilitate host adaptation. Advances in genetic engineering, combined with enhanced understanding of fungal virulence, stress resilience, and formulation technology, are expected to improve the performance and adoption of fungal biopesticides.

The endophytes form stable symbiotic relationships within plant tissues, thereby enhancing host resilience to biotic and abiotic stresses (Kumar et al. 2017). EEPF found across fungal divisions such as Zygomycota and Ascomycota are widely distributed in terrestrial ecosystems and are a core component of microbial biocontrol efforts in both field and orchard settings. Infection requires successful contact between conidia and a susceptible host under favorable conditions. Advances in characterization techniques, including morphometrics, extracellular enzyme profiling, pathogenicity assays, and genetic tools, have significantly improved our taxonomic and functional understanding of EPF.

Globally, more than 80 companies have developed over 170 mycoinsecticides and mycoacaricides over the past 40 years (Faria and Wraight 2007). The commercial success of these products hinges on the development of robust formulations that ensure pathogen viability, shelf life, and application efficiency. Commonly used and adapted formulations, namely, granules, oil-based suspensions, and wettable powders, have been developed initially. However, the novel and precision delivery technologies, e.g., microencapsulation and nano-based formulations, are enhancing fungal spore stability under environmental stresses like UV exposure and desiccation (Hazra and Purkait 2019). Moreover, precision agriculture tools and drone-based applications have further increased the precision and effectiveness of EPF deployment in crop fields (Shaari et al. 2024). Continued innovation in formulation and application strategies will be key to maximizing the potential of EPF as viable biopesticides in sustainable agriculture.

6 Compatibilities with Other Management Practices

The frequent and overuse of chemical insecticides employed for insect pest control has led to significant environmental risks and the development of pesticide resistance in target insect pests. These challenges highlight the need for sustainable, eco-friendly alternatives for effective pest suppression and management. In recent years, EPF have been seen as a promising biological control agent due to their natural ability to infect and kill a vast range of insect pests (Sahayaraj 2014). When functioning as endophytes, EPF can play a pivotal role in integrated pest management (IPM) programs across diverse agricultural systems.

EEPF can be utilized individually or in combination with other pest control methods, including chemical insecticides, botanical pesticides, and other biological agents, depending on crop type, pest species, and environmental factors. Incorporating EPF into IPM strategies alongside predators, parasitoids, and botanicals enhances the environmental safety and sustainability (Jaber et al. 2018; González-Mas et al. 2019). Moreover, ensuring compatibility among all components of pest management is crucial for maximizing synergistic effects. Several studies indicate that EPF are highly compatible with natural enemies, including microbial antagonists, parasitoids, and predators (Butt et al. 2016), and their combined use often results in superior pest suppression without negatively affecting other biological control agents.

6.1 Integration with Arthropods

Combined use of EPF along with entomophagous arthropods utilizes the dual approach of fungal infection mechanisms and parasitic or predatory activity of arthropods to achieve effective and enhanced pest suppression. This dual-action strategy helps reduce the number of pest individuals that might otherwise escape treatment by a single control method. The combined application of biological agents often results in additive or even synergistic effects in managing insect pest populations (Ren et al. 2010).

Numerous studies have been performed to assess the compatibility and efficacy of entomopathogenic fungi against predators, parasitoids, and entomopathogenic nematodes. For example, *Chrysoperla carnea* larvae showed a preference for aphids reared on plants colonized by *Beauveria bassiana*, thus showing promising potential for predator-prey interactions in EPF-colonized systems (González-Mas et al. 2019). These findings support the idea that EPF can be effectively combined with natural enemies to enhance IPM approaches.

6.2 Integration with Pesticides

Different types of interaction are reported for EPF in the context of chemicals. EPF are reported to interact with insecticides in varied manners, namely, synergistic, antagonistic, or neutral. Compatibility between EPF and insecticides is a key factor in ensuring successful co-application, as some insecticides may inhibit fungal growth, sporulation, or germination. Therefore, prior assessment of compatibility is critical before incorporating EPF into agrochemical-based pest management programs (Aravinthraju et al. 2024).

Informed selection of specific insecticides, especially reduced-risk compounds, can enhance the effectiveness of EPF. This synergy leads to better pest management, a decrease in the number of chemical treatments, and a reduction in the development of insecticide resistance (Ambethgar 2009). Sharma et al. (2023) reported that locusts and grasshoppers are controlled with higher efficacy when insect growth regulators are combined with oil-based formulations of Metarhizium anisopliae. They further concluded that both treatments, when applied singly, did not show effective control of the pests. Experimental studies have demonstrated variable effects of insecticides on EPF. For instance, imidacloprid exhibited the least inhibition of *M. anisopliae* vegetative growth (11.1%), followed by deltamethrin and cypermethrin (36.7%), thiodicarb (53.5%), and chlorpyrifos (69.2%) (Rachappa et al. 2007). Imidacloprid was found to be compatible with *B. bassiana* and *M. anisopliae*, especially at lower field doses, supporting its use in integrated applications (Abidin et al. 2017).

Fungicides typically exert a more significant adverse effect on EPF than other agrochemicals. Many fungicides used to control phytopathogens can restrict the growth and activity of entomopathogenic fungi. However, through genetic engineering and strain selection, efforts have been made to develop fungicide-tolerant EPF strains (Aravinthraju et al. 2024). For example, although mancozeb significantly inhibited *B. bassiana* growth in vitro, its inhibitory effect in the field was comparatively lower, allowing the fungus to persist and remain effective against pests such as *Leptinotarsa decemlineata* (Dara 2017; Pakdaman Sardrood and Mohammadi Goltapeh 2018). Metalaxyl showed minimal impact in both lab and field conditions, indicating potential compatibility.

Besides insecticides and fungicides, acaricides play a crucial role in pest management, especially in livestock and horticultural systems. Despite limited studies, recent findings suggest that *Metarhizium* spp. microsclerotia and conidia are compatible with certain synthetic acaricides. For instance, the co-application of deltamethrin and *Metarhizium pingshaense* LCM S09 significantly enhanced tick control, particularly in acaricide-susceptible populations (da Silva Carneiro et al. 2022). These results indicate promising avenues for incorporating EPF into broader pesticide-based strategies.

Even though parasitoids have typically demonstrated tolerance to EPF colonization, the effects on predatory arthropods are still unclear and require more explorative investigations (Akutse et al. 2013). A better understanding of tri-trophic interactions among plants, EPF, and insect natural enemies is essential for optimizing such integrated strategies.

6.3 Synergistic Potential in Conjunction with Herbicides

EPF have shown huge potential as endophytes in IPM. However, their sensitivity to herbicides presents a critical limitation. Many herbicides, through repeated application, accumulate in soil ecosystems and may adversely affect the viability and biological activity of these fungi. The fungal characteristics, namely, growth, sporulation, as well as germination, vary depending on the herbicide used and fungal strain (Celar and Kos 2012). For instance, in developmental assays with *Beauveria bassiana*, amidosulfuron and dicamba exhibited relatively lower fungistatic effects compared to other herbicides (Kos and Celar 2013). They confirmed that at full recommended field doses (100%), all tested herbicides significantly inhibited conidial germination.

6.4 Synergistic Integration with Botanical Compounds

Application of EPF as endophytes in combination with plant-based bioactive compounds offers a sustainable and effective strategy for insect pest suppression. A remarkable example is the use of *Beauveria brongniartii* for the management of *Melolontha melolontha* (European cockchafer beetle), which is a major pest in pastures, orchards, and forest margins of Central Europe (Keller et al. 1997). Moreover, the synergistic application of *B. bassiana* with neem has shown a significant decrease in root aphid populations in rice and honeysuckle crops, achieving up to a 62% decrease in aphid density in celery fields in California (Dara 2015).

When EPF start infection, it typically begins with adhesion of conidia to insect cuticle, followed by germination and enzymatic degradation of the cuticle. Systemic colonization through entry into hemolymph results in host mortality due to nutrient competition, toxin production, and disruption of physiological functions (Inglis et al. 2001). The result of infection is dependent upon multiple variables, including fungal virulence, host susceptibility, and environmental conditions. For instance, *B. bassiana* is known to elicit a strong immunological response in insect hosts, eventually leading to death (Shang et al. 2022).

Recent studies have evidenced that the efficacy of EPF is greatly enhanced when applied in combination with botanical insecticides. The combinations of *B. bassiana* with *Chromolaena odorata* leaf powder or *Azadirachta indica* extracts significantly increased mortality in *Spodoptera litura*, although sometimes at the cost of reduced fungal spore viability (Afandhi et al. 2020). Similarly, combined applications of azadirachtin and EPF (e.g., *B. bassiana, P. fumosoroseus*) have been more effective than individual treatments against *Bemisia tabaci*, with up to 27.6% and 20.5% higher nymphal mortality compared to solo applications. In the case of *P. fumosoroseus* combined with azadirachtin, mortality in *Bemisia argentifolii* nymphs reached nearly 90% (Islam et al. 2010).

Tondexir, a hot pepper extract formulation in mineral oil and Palizin (coconut-based insecticidal soap), has been tested alongside *B. bassiana*, *Lecanicillium lecanii*, and *Metarhizium anisopliae* for managing *Galleria mellonella*. Notable differences in fungal germination and efficacy were observed, suggesting the potential for selective combinations in integrated pest strategies (Sohrabi et al. 2019). With the growing interest in phyto-insecticides, their integration with endophytic EPF presents an ecologically viable and potentially synergistic approach to insect pest management.

6.5 Integration with Other Management Practices

The diverse functionalities of EPF, especially their capacity to colonize plants endophytically and simultaneously function as insect pathogens, offer a promising pathway for IPM. A growing body of literature supports the compatibility of EPF with a wide array of BCAs, including natural enemies such as predators (Canassa et al. 2019) and parasitoids (Gathage et al. 2016; Jaber and Araj 2018). The compatibility of EPF with BCAs broadens the base of biological interventions across different trophic levels due to enhanced effectiveness.

Research studies have highlighted the potential of pest and pathogen suppression via application of microbial consortia composed of endophytic fungi and other beneficial microorganisms, i.e., plant growth-promoting rhizobacteria (PGPR). These consortia aim to broaden the spectrum of pest targets and introduce multiple, synergistic modes of action. For instance, a notable study demonstrated that sheath blight of rice (*Rhizoctonia solani*) and rice leaf folder (*Cnaphalocrocis medinalis* Guen.) can be effectively controlled using a consortium that included the B2 strain of *B. bassiana* along with two different strains of *Pseudomonas fluorescens* (Karthiba et al. 2010).

Combined use of EPF along with cultural and biological practices has significantly enhanced pest suppression. Implementation of EPF in combination with practices such as crop rotation, cover cropping, and intercropping, which improve soil structure, moisture retention, and microbial diversity conditions that favor the proliferation and efficacy of EPF (Hokkanen and Menzler-Hokkanen 2024). Moreover, synergistic interactions with soil microbiota such as *Bacillus subtilis* and arbuscular mycorrhizal fungi (AMF), which improve root colonization, nutrient uptake, and stress tolerance, further promote plant health and resilience (Islam et al. 2021).

A crucial mechanism by which EPF provide protection is its capacity to trigger induced systemic resistance (ISR) in host plants. For instance, pre-inoculation with *Lecanicillium* spp. has been reported to impart systemic protection against foliar as well as soilborne pathogens like *Pythium ultimum* and *Sphaerotheca fuliginea* in cucumber (Benhamou and Brodeur 2001; Hirano et al. 2008).

To lessen the dependency on chemical measures, EPF are being investigated for their potential in mitigating post-harvest pest problems. A recent study has shown that stored grains treated with EPF experience less pest damage and reduced weight loss. However, abiotic factors such as grain type significantly influence EPF efficacy and remain underexplored (Mantzoukas and Grammatikopoulos 2020). The combination of EPF with inert carriers like diatomaceous earth has yielded promising results against major coleopteran pests, including *Sitophilus oryzae*, *Rhyzopertha dominica*, *Sitophilus granarius*, and *Tribolium castaneum* (Mantzoukas et al. 2022). Furthermore, the co-application of *B. bassiana* with plant-derived products such as *Chromolaena odorata* leaf powder, neem, chinaberry, lantana, and Mexican sunflower extracts has shown dose-dependent insecticidal activity against pests like *Spodoptera litura* and stored grain beetles (Kalita et al. 2018).

The availability of this growing evidence highlights the integrative and broadened potential of EPF in sustainable pest management strategies. Thus, making EPF valuable components of both pre- and post-harvest agricultural systems.

7 Benefits and Safety Concerns of Biopesticides

Numerous EPF are known for their ability to naturally inhabit plant tissues as endophytes, whereas others can be introduced into plants through seed treatments, foliar sprays, soil drenches, or root inoculations (Vega et al. 2008). The endophytic nature of these fungi has redefined their role in agriculture, expanding their function beyond classical, transient bioinsecticidal agents to more persistent and symbiotic plant partners (Quesada-Moraga et al. 2020). Compared to conventional chemical insecticides, EPF provides a suite of advantages, including high insecticidal efficacy, compatibility with organic farming practices, and minimal ecological disruption (Fig. 16.4) (Litwin et al. 2020).

EPF are well known as a natural regulator of insect populations, wherein they play an important role in suppressing pest outbreaks and further maintaining ecological balance. They contribute toward cycling of soil nutrients by performing

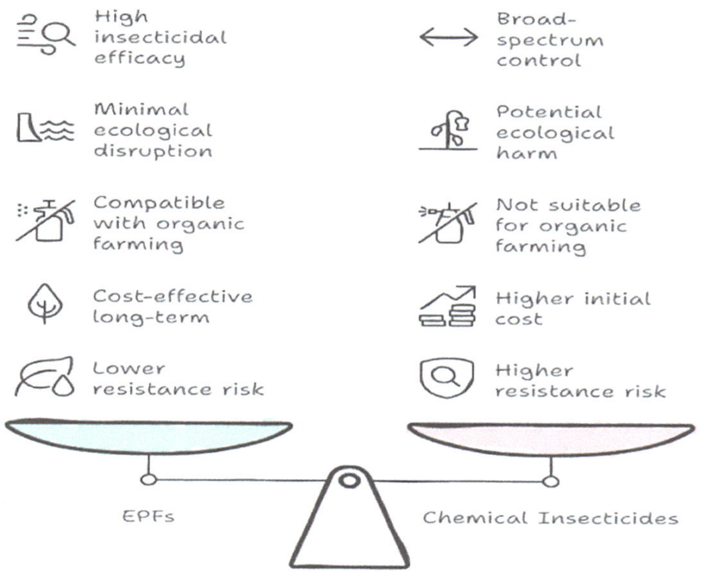

Fig. 16.4 Comparison of the benefits of EPF over chemical insecticides

decomposition of insect cadavers, thereby releasing essential nutrients such as nitrogen, which supports plant growth (Sharma and Sharma 2021). EPF-based biopesticides are generally considered safe for humans, non-target organisms, and the environment, making them promising components of IPM strategies (Sandhu et al. 2012). Moreover, their ability to colonize plants endophytically does not hinder plant development and may, in many cases, stimulate growth (Vega 2018). Given their capacity for self-replication under favorable conditions, EPF are cost-effective over the long term and pose a lower risk of insect resistance development. Hence, their application can significantly reduce reliance on synthetic pesticides (Islam et al. 2022; Sharma et al. 2023).

Successful endophytic colonization by EPF can enhance plant growth, confer protection against insect pests, induce systemic resistance, suppress phytopathogens and nematodes, and mitigate the impacts of abiotic stresses (Vega et al. 2009; Ownley et al. 2010; Sasan and Bidochka 2012; Vega 2018). Their ability to secrete unique secondary metabolites often surpasses the capabilities of chemical pesticides. Additionally, EPF show potential in pharmaceutical fields, with demonstrated antimicrobial, antitumor, antidiabetic, and immunosuppressive properties (Gouda et al. 2016; Yadav 2018). Research has also shown that these microbes enhance soil structure, promote beneficial microbiota, improve water intake capacity as well as nutrition of plants, and act as probiotics to combat pathogens (Sasan and Bidochka 2012; Behie et al. 2015; Jaber and Alananbeh 2018).

Beyond reducing pest populations, endophytic EPF can also play a role in impairing insect reproduction and development. Their colonization adversely affects insect feeding behavior and physiology. For instance, the reproduction of cotton aphids was reduced by 61% and 57% on cotton leaves colonized by *Beauveria bassiana* and *Lecanicillium lecanii*, respectively, compared to uncolonized controls. For example, the weight gain of *Chortoicetes terminifera* (Walker) was reduced up to 60% when they fed on wheat (*Triticum aestivum* L.) colonized by *B. bassiana* (Gurulingappa et al. 2010).

Certain EPF greatly alter the signaling pathways of plant hormones and nutrient acquisition, leading to improved plant growth (Bamisile et al. 2021). Some root-colonizing EPF can improve the bioavailability of soil nutrients, hence increasing their uptake. *Metarhizium brunneum*, *B. bassiana*, and *Isaria farinosa* have been shown to improve iron availability in sorghum, with *M. brunneum* being the most effective by acidifying calcareous soils through organic acid release (Raya-Díaz et al. 2017). Both *M. brunneum* and *B. bassiana* effectively reduced iron chlorosis in sorghum grown under alkaline conditions. In addition to insect control, endophytic EPF also suppress plant pathogens. For instance, seed treatments with *B. bassiana* protected tomatoes and cotton against *Rhizoctonia solani* and *Pythium myriotylum* (Ownley et al. 2008).

Climate change intensifies challenges in agricultural systems via various processes, including alterations in pest distributions and pathogen behavior. EPF offer a sustainable means to address these issues by offering readily adaptive, eco-friendly pest and disease control. Their strain-specific adaptability, combined with modern biotechnological tools, allows customized solutions toward emerging pest threats, aligning with the goals of climate-resilient agriculture. Genomic tools, advanced formulations, and biotechnological innovations are enhancing EPF usefulness in climate-stressed environments (Zanzana et al. 2024).

Despite their promising potential, concerns regarding the safety of EPF and their secondary metabolites persist. Potential risks to non-target organisms, including predators, parasitoids, arthropods, pollinators, and vertebrates such as fish, birds, and amphibians, have been evaluated in various studies. Nevertheless, common EPF-based mycoinsecticides are generally classified as low-risk biopesticides (Zimmermann 2007a, b). Ongoing research continues to focus on enhancing safety profiles and maximizing efficacy through strain selection, formulation refinement, and application strategies.

8 Omics Technologies to Comprehend Interactions

Primarily, to evaluate the antagonistic potential of fungal endophyte, co-cultivation techniques, such as in vitro dual-plate assays, were extensively employed. However, the mechanisms by which endophytic fungi influence and modulate the physiology of the host, as well as volatile organic compounds (VOCs), remain insufficiently understood. Existing knowledge is sparse and often inconsistent across varying environmental conditions (Fontana et al. 2009; Jaber 2015).

Multi-omics approaches, namely, genomics, metabolomics, proteomics, and transcriptomics, offer valuable tools which can be extensively utilized to unravel

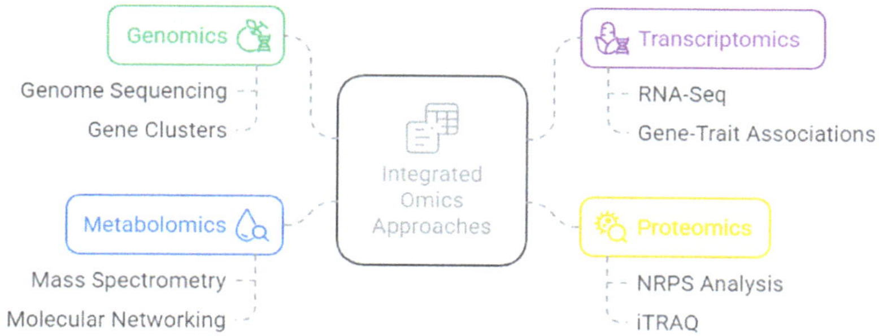

Fig. 16.5 Integrated omics approaches in EEPF research

the intricate interactions among EEPF, host plants, and insect pests (Fig. 16.5). These multi-omics approaches provide a detailed understanding of the biosynthetic potential of EEPF, particularly regarding their secondary metabolite production (Crandall et al. 2020).

Our knowledge and understanding of molecular biology and next-generation sequencing have clarified our basis on EPF biology. Genome sequencing of important EPF species, namely, *Metarhizium anisopliae* and *Beauveria bassiana*, has unraveled many genes encoding virulence factors, including hydrolytic enzymes, secondary metabolite biosynthesis pathways (e.g., destruxins), and host-specificity regulators. For instance, in *Metarhizium*, destruxin biosynthesis has been elucidated, highlighting genes crucial for toxin production and fungal virulence. Studies conducted on transcriptomics have also shown dynamic patterns of gene expression during host infection, particularly involving MAPK and cAMP signaling pathways, which regulate fungal development and pathogenicity (Samal et al. 2023a, b).

With high-throughput genome sequencing and tools such as BLAST, fungal identification has been rendered cost-effective and speedy, enabling effective in-silico comparisons of sequence similarity (Altschul et al. 1990). Gene clusters in EEPF are often organized into operons under a single promoter, facilitating the coordinated regulation of secondary metabolism. These clusters are identified through computational tools that predict biosynthetic gene clusters (BGCs), aiding in the discovery of novel metabolites. For example, genome annotations for *M. acridum*, *M. robertsii* (Gao et al. 2011), *B. bassiana* (Xiao et al. 2012), and *M. anisopliae* (Pattemore et al. 2014) have revealed a wealth of biosynthetic potential for secondary metabolites.

Metagenomics offers rapid identification along with detailed genetic information of unculturable organisms, beyond individual species. Amplicon sequencing and whole-genome shotgun sequencing help characterize microbial communities, with Internal Transcribed Spacer (ITS) regions of nuclear rDNA serving as effective molecular markers for fungal identification (Schoch et al. 2012).

Transcriptome analyses are nowadays conducted through RNA-Seq, instead of microarrays, which provides deeper insights into gene function and expression (Wang et al. 2009; Yuan et al. 2019). Transcriptomics has revealed two main avenues for secondary metabolite discovery: (i) co-regulated clusters of genes that are involved in metabolite biosynthesis; (ii) gene-trait associations linking gene expression to metabolite production or functional phenotypes (Cheng et al. 2020). Such studies have also shown lifestyle shifts in endophytes from biotrophic to necrotrophic under stress conditions such as carbohydrate limitation (Ribeiro et al. 2020). Inoculation studies using *Arabidopsis thaliana* demonstrated transcriptional reprogramming of defense responses, including strain-specific modulation of pathogenesis-related (PR) genes and phytohormonal signaling (JA, SA). Colonization by *B. bassiana* did not change JA or glucosinolate levels in Brassicas; however, EEPF enhanced biomass and sesquiterpenoid accumulation in *Atractylodes lancea* by modulating photosynthesis and central metabolism (Yuan et al. 2018).

Secondary metabolites research has also been explored due to the availability of proteomics tools. Techniques such as NRPS, PKS, DMATS gene cluster analysis, and 2D-PAGE coupled with LC-MS/MS facilitate the identification of post-translational protein modifications and metabolic biomarkers. For low-abundance proteins, methods like iTRAQ are employed to assess proteomic changes induced by endophytic colonization (Wiese et al. 2007). Mass spectrometry-based metabolomics enables rapid, large-scale identification of fungal metabolites without the need for prior purification. Molecular networking platforms like GNPS (Wang et al. 2016) streamline MS/MS data interpretation and dereplication of natural product libraries, accelerating metabolite discovery.

Gene-editing tools such as RNA interference (RNAi) are being explored to engineer recombinant EPF strains with increased virulence (Chen et al. 2015). Despite their efficacy in lab and greenhouse conditions, challenges in dsRNA/siRNA delivery limit field applications (Bamisile et al. 2021). Nonetheless, genetically engineered EEPF offer potential as biocontrol agents, although care must be taken to prevent unintended effects on non-target organisms such as pollinators and beneficial arthropods (Wink 2008; Fadiji and Babalola 2020). The synergistic application of EPF like *B. bassiana* with chemical insecticides has shown promise in reducing pesticide usage and overcoming resistance (Al-Ani et al. 2021). As the study of endophyte-plant-insect tripartite interactions moves beyond culture-dependent assays, culture-independent tools such as GFP-based confocal microscopy and molecular barcoding (e.g., ITS) are gaining importance (Sasan and Bidochka 2012; Behie et al. 2015).

Modern molecular tools, i.e., PCR, RT-PCR, q-PCR, high-throughput sequencing, and DNA microarrays (Chen et al. 2015; Wang et al. 2020), now support comprehensive genome and transcriptome analyses. Technologies such as LC-MS/MS, GC-MS/MS, UPLC-MS/MS, NMR, FTIR, and HS-SPME are frequently used to analyze fungal-derived phytohormones, enzymes, and secondary metabolites (Elbanhawy et al. 2019; Al-Ani et al. 2021). High-throughput sequencing platforms like Illumina and 454 pyrosequencing have advanced fungal microbiome research (Hardoim et al. 2015), while metagenomics facilitates the identification of genes linked to plant growth promotion and host colonization (Akinola and Babalola 2020).

9 Constraints as Bioagents

Despite their ecological potential, the practical use of EEPF as BCAs is constrained due to several technical, environmental, and regulatory challenges (Fig. 16.6).

The major limitation is due to the slower insecticidal action of EPF compared to chemical pesticides. A higher inoculum potential is needed by EPF and it generally takes 25 days for EPF to induce host mortality, during which pests may continue to damage crops (Parveen and Rashtrapal 2024). Additionally, many strains show a narrow host range, increasing the risk of secondary pest outbreaks. The effectiveness of EPF also depends heavily on environmental conditions, especially temperature, humidity, and UV radiation. For instance, temperatures exceeding 30 °C reduce the virulence of key species such as *Beauveria bassiana* and *Metarhizium anisopliae*, while prolonged low temperatures impede fungal growth. High humidity is essential for conidial germination and successful infection, but maintaining such conditions in field environments remains a significant hurdle (Islam et al. 2021).

Their efficacy is greatly hampered due to the short persistence of fungal inoculum in the field and further susceptibility to desiccation and UV degradation. Their performance in inconsistent or suboptimal field conditions often falls shorter than desired, even when more virulent strains are applied. This variability necessitates frequent optimization of formulations and delivery methods to ensure consistent pest suppression across different agroecosystems (Singh et al. 2017).

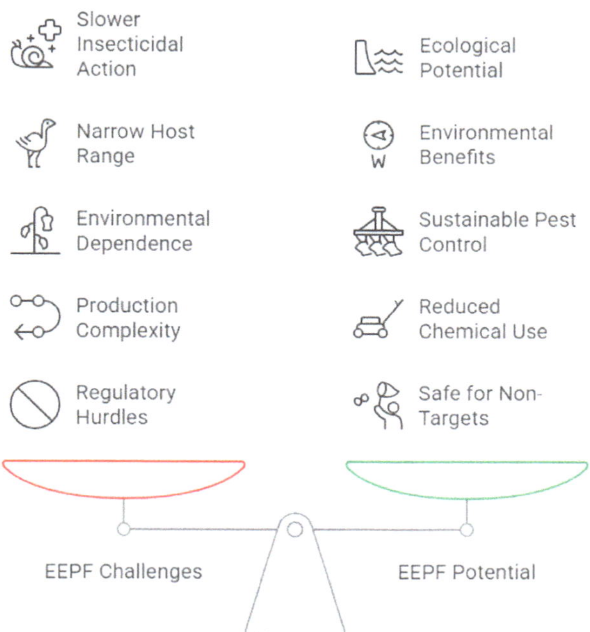

Fig. 16.6 Balancing challenges and potential of EEPF in pest control

Another major constraint lies in the technical and economic complexity of large-scale production. Solid-state fermentation (SSF), commonly used for mass production, is labor-intensive and offers inconsistent yield and quality. Although liquid-state fermentation (LSF) is more scalable, it demands expensive nutrient media and bioreactors, enhancing overall product economics (Formenti et al. 2014). The commercial viability of EPF is further hampered by high formulation costs, limited shelf life, and cold-chain requirements.

Concerns also lie in respect of ecological safety as well as non-target impacts. While EPF are generally regarded as environmentally benign, some species have broad host ranges that could inadvertently affect beneficial arthropods, such as pollinators, predators, decomposers, and parasitoids. Laboratory studies have demonstrated that increased concentrations of *Metarhizium anisopliae* may alter the foraging behavior of honeybees, though such effects are often less pronounced under field conditions. Introducing non-native EPF strains may also disrupt indigenous microbial communities, potentially outcompeting native fungi or altering soil microbial diversity (Devos et al. 2019). Therefore, detailed environmental risk assessments and development of host-specific formulations are crucial to minimize non-target effects. Furthermore, the identification and isolation of endophytic EPF remain technically challenging. Many endophytic strains are unculturable or poorly characterized, making it difficult to assess community structure and functional diversity (Fadiji and Babalola 2020). Their distribution is influenced by many factors, namely, anthropogenic disturbances, geographical location, and vegetation type, further complicating their standardization and deployment in strategies targeted toward pest management (Dong et al. 2016).

10 Conclusion and Future Prospects

EEPF represent a promising frontier in sustainable crop protection, combining insect pest control with additional plant health benefits. These fungi are increasingly recognized as dual-function agents capable of suppressing both insect pests and plant pathogens while simultaneously promoting plant growth and enhancing abiotic stress tolerance. Their integration into IPM programs offers a holistic approach to crop protection, aligning with environmentally sustainable agricultural practices. The potential EEPF application has abundantly increased due to advances in genomics, biotechnology, and formulation science. Precision delivery systems such as microencapsulation and nanotechnology are emerging as pivotal tools to enhance the field efficacy, shelf-life stability, and targeted delivery of these fungi. Similarly, strain development focused on increased virulence, stress tolerance, broader host range, and improved sporulation characteristics remains a key area of ongoing research.

Despite the promising and potential developments, many obstacles still persist. The inconsistent performance of EEPF in field conditions, limited understanding of their multitrophic interactions with host plants and non-target organisms, and inadequate delivery technologies hinder their large-scale commercialization. Moreover, there is a considerable knowledge gap pertaining to the ecological dynamics and persistence of EEPF in various agroecosystems. Comprehensive field trials and ecological assessments are essential to validate their long-term efficacy and environmental safety, while EEPF can influence multitrophic interactions favorably, their impact on the natural enemies of pests must be carefully assessed. Looking ahead, future research should adopt a multidisciplinary and systems-level approach, incorporating tools such as metagenomics, transcriptomics, and synthetic biology to unlock the multiplex interactions among EEPF, host plants, pests, and the surrounding microbial communities. Exploration of untapped fungal diversity from diverse and extreme ecosystems such as arid zones, high-altitude environments, and forests can yield novel strains with unique bioactive properties. Furthermore, the synergy between EEPF and other sustainable IPM components, including resistant cultivars, cultural practices, and natural enemies, can enhance pest suppression while decreasing dependency on synthetic chemicals.

Advances in molecular biology, including genomics and transcriptomics, are paving the way for the identification and development of more efficient fungal strains with enhanced colonization, virulence, and metabolite production capabilities. Furthermore, innovations in formulation and delivery methods, such as seed coatings, root dips, and nano-based carriers, are making it increasingly feasible to apply EPF in diverse cropping systems. The potential for crop-specific and region-adapted EPF strains also opens avenues for customized biocontrol solutions, particularly in high-value or resource-constrained agricultural settings. While challenges such as inconsistent colonization and limited field validation remain, ongoing research and growing regulatory support for biopesticides are accelerating the commercial development of EPF-based products. Overall, entomopathogenic fungi as endophytes represent a forward-looking, environmentally friendly approach toward insect pest management that aligns with the global focus on sustainable and resilient agricultural practices. Future research focusing on strain optimization, delivery methods, and field validation will enhance their practical application in climate-resilient agriculture.

Acknowledgments The authors are grateful to CSK Himachal Pradesh Agriculture University, Palampur, Himachal Pradesh, for granting access to the resources and infrastructure that supported the writing of this chapter. We would also like to thank the editors of this volume for their thoughtful guidance and editorial support.

References

Abidin AF, Ekowati N, Ratnaningtyas NI (2017) Insecticide compatibility to the entomopathogenic fungi *Beauveria bassiana* and *Metarhizium anisopliae*. Scr Biol 4:273–279. https://doi.org/10.20884/1.SB.2017.4.4.695

Adhikari A, Oli D, Poudel P, Magar PB, Pandit S et al (2024) *Pythium*-induced damping-off disease and its management in tomato (*Solanum lycopersicum* L.): a review. Agric Sci Technol 16:1

Afandhi A, Pratiwi VR, Hadi MS, Setiawan Y, Puspitarini RD (2020) Suitable combination between *Beauveria bassiana* (Balsamo) Vuillemin and four plant leaf extracts to control *Spodoptera litura* (Fabricius). AGRIVITA J Agric Sci 42:341–349. https://doi.org/10.17503/agrivita.v42i2.2678

Agbessenou A, Akutse KS, Yusuf AA, Khamis FM (2022) The endophyte *Trichoderma asperellum* M2RT4 induces the systemic release of methyl salicylate and (Z)-jasmone in tomato plant affecting host location and herbivory of *Tuta absoluta*. Front Plant Sci 13:860309. https://doi.org/10.3389/fpls.2022.860309

Ahmad I, Jiménez-Gasco MM, Luthe DS, Shakeel SN, Barbercheck ME (2020) Endophytic *Metarhizium robertsii* promotes maize growth, suppresses insect growth, and alters plant defense gene expression. Biol Control 144:104167. https://doi.org/10.1016/j.biocontrol.2019.104167

Ahmed AS, Sánchez CP, Candela ME (2000) Evaluation of induction of systemic resistance in pepper plants (*Capsicum annuum*) to *Phytophthora capsici* using *Trichoderma harzianum* and its relation with capsidiol accumulation. Eur J Plant Pathol 106:817–824. https://doi.org/10.1023/A:1008780022925

Ahsan SM, Injamum-Ul-Hoque M, Das AK, Rahman MM, Mollah MMI et al (2024) Plant–entomopathogenic fungi interaction: recent progress and future prospects on endophytism-mediated growth promotion and biocontrol. Plants 13:1420. https://doi.org/10.3390/plants13101420

Akello J, Dubois T, Coyne D, Kyamanywa S (2008) Endophytic *Beauveria bassiana* in banana (*Musa* spp.) reduces banana weevil (*Cosmopolites sordidus*) fitness and damage. Crop Prot 27: 1437–1441. https://doi.org/10.1016/j.cropro.2008.07.003

Akinola SA, Babalola OO (2020) Endophyte-enhanced plant growth and biocontrol: mechanisms and prospects. Plant Soil Environ 66:421–430. https://doi.org/10.17221/118/2020-pse

Akutse KS, Maniania NK, Fiaboe KKM, Van den Berg J, Ekesi S (2013) Endophytic colonization of *Vicia faba* and *Phaseolus vulgaris* (Fabaceae) by fungal pathogens and their effects on the life-history parameters of *Liriomyza huidobrensis* (Diptera: Agromyzidae). Fungal Ecol 6:293–301. https://doi.org/10.1016/j.funeco.2013.01.003

Al-Ani LKT, Aguilar-Marcelino L, Salazar-Vidal VE, Becerra AG, Raza W (2021) Role of useful fungi in agriculture sustainability. In: Yadav AN (ed) Recent trends in mycological research, Fungal biology. Springer, Cham, pp 1–44. https://doi.org/10.1007/978-3-030-60659-6_1

Ali SA, Abdelmoaty HS, Ramadan HH, Salman YB (2023) Chemical identification of certain antifungal extracts from *Epicoccum nigrum* against phytopathogenic *Fusarium solani*. Res Sq:1–21. https://doi.org/10.21203/rs.3.rs-2500578/v1

Altschul SF, Gish W, Miller W, Myers EW, Lipman DJ (1990) Basic local alignment search tool. J Mol Biol 215:403–410. https://doi.org/10.1016/S0022-2836(05)80360-2

Ambethgar V (2009) Potential of entomopathogenic fungi in insecticide resistance management (IRM): a review. J Biopest 2:177–193

Anisha C, Jishma P, Bilzamol VS, Radhakrishnan EK (2018) Effect of ginger endophyte *Rhizopycnis vagum* on rhizome bud formation and protection from phytopathogens. Biocatal Agric Biotechnol 14:116–119. https://doi.org/10.1016/j.bcab.2018.02.015

Aouali S (2024) Endophytic entomopathogenic fungi: biology and applications. In: Deshmukh SK, Sridhar KR (eds) Entomopathogenic fungi: prospects and challenges. Springer, Singapore, pp 141–188. https://doi.org/10.1007/978-981-97-5991-0_7

Aravinthraju K, Shanthi M, Murugan M, Srinivasan R, Maxwell LA et al (2024) Endophytic entomopathogenic fungi: their role in enhancing plant resistance, managing insect pests, and synergy with management routines. J Fungi 10:865. https://doi.org/10.3390/jof10120865

Athira S, Anith KN (2020) Plant growth promotion and suppression of bacterial wilt incidence in tomato by rhizobacteria, bacterial endophytes and the root endophytic fungus *Piriformospora indica*. Indian Phytopathol 73:629–642. https://doi.org/10.1007/s42360-020-00283-2

Baazeem A, Alorabi M, Darwesh H, Alotaibi SS, El-Deen AN et al (2022) Biological control of root-knot nematode (*Meloidogyne javanica*) by potential antagonism of endophytic fungi isolated from Taify roses. J King Saud Univ Sci 34:102329. https://doi.org/10.1016/j.jksus.2022.102329

Bagy HMK, Hassan EA, Nafady NA, Dawood MF (2019) Efficacy of arbuscular mycorrhizal fungi and endophytic strain *Epicoccum nigrum* ASU11 as biocontrol agents against blackleg disease of potato caused by *Pectobacterium carotovora* subsp. *atrosepticum* PHY7. Biol Control 134: 103–113. https://doi.org/10.1016/j.biocontrol.2019.03.005

Bamisile BS, Akutse KS, Siddiqui JA, Xu Y (2021) Model application of entomopathogenic fungi as alternatives to chemical pesticides: prospects, challenges, and insights for next-generation sustainable agriculture. Front Plant Sci 12:741804. https://doi.org/10.3389/fpls.2021.741804

Bamisile BS, Afolabi OG, Siddiqui JA, Xu Y (2023) Endophytic insect pathogenic fungi–host plant–herbivore mutualism: elucidating the mechanisms involved in the tripartite interactions. World J Microbiol Biotechnol 39:326. https://doi.org/10.1007/s11274-023-03780-4

Behie SW, Jones SJ, Bidochka MJ (2015) Plant tissue localization of the endophytic insect pathogenic fungi *Metarhizium* and *Beauveria*. Fungal Ecol 13:112–119. https://doi.org/10.1016/j.funeco.2014.08.001

Benhamou N, Brodeur J (2001) Pre-inoculation of Ri T-DNA transformed cucumber roots with the mycoparasite, *Verticillium lecanii*, induces host defense reactions against *Pythium ultimum* infection. Physiol Mol Plant Pathol 58(3):133–146. https://doi.org/10.1006/pmpp.2001.0322

Brader G, Compant S, Vescio K, Mitter B, Trognitz F et al (2017) Ecology and genomic insights into plant-pathogenic and plant-nonpathogenic endophytes. Annu Rev Phytopathol 55:61–83. https://doi.org/10.1146/annurev-phyto-080516-035641

Butt TM, Coates CJ, Dubovskiy IM, Ratcliffe NA (2016) Entomopathogenic fungi: new insights into host–pathogen interactions. Adv Genet 94:307–364. https://doi.org/10.1016/bs.adgen.2016.01.006

Canassa F, Tall S, Moral RA, de Lara IA, Delalibera I Jr et al (2019) Effects of bean seed treatment by the entomopathogenic fungi *Metarhizium robertsii* and *Beauveria bassiana* on plant growth, spider mite populations and behavior of predatory mites. Biol Control 132:199–208. https://doi.org/10.1016/j.biocontrol.2019.02.003

Carneiro ADS, Mesquita E, Meirelles LN, Bittencourt VREP, Golo PS (2022) Compatibility of different *Metarhizium* spp. propagules with synthetic acaricides for controlling *Rhipicephalus microplus*. Rev Bras Parasitol Vet 31:e018221. https://doi.org/10.1590/S1984-29612022018

Celar FA, Kos K (2012) Compatibility of selected herbicides with entomopathogenic fungus *Beauveria bassiana* (Bals.) Vuill. Acta Agric Slov 99(1):57–63. https://doi.org/10.14720/aas.2012.99.1.14520

Chandran K, Sreeja SJ, Johnson JM (2021) Beneficial root endophytic fungus *Piriformospora indica* inhibits the infection of *Blackeye cowpea mosaic virus* in yard long bean with enhanced growth promotion. J Trop Agric 59(1):1–7

Chaurasiya DK, Singh S, Kumar M, Singh AP, Sahni S (2021) Bio-control of insect pests through entomopathogenic fungi: a novel approach. Curr Res Innov Plant Pathol 14:77–99

Chebet ON, Omosa LK, Subramanian S, Nchiozem-Ngnitedem VA, Mmari JO et al (2021) Mechanism of action of endophytic fungi *Hypocrea lixii* and *Beauveria bassiana* in *Phaseolus vulgaris* as biopesticides against pea leafminer and fall armyworm. Molecules 26(18):5694. https://doi.org/10.3390/molecules26185694

Chen Y, Gao Q, Huang M, Liu Y, Liu Z et al (2015) Characterization of RNA silencing components in the plant pathogenic fungus *Fusarium graminearum*. Sci Rep 5:srep12500. https://doi.org/10.1038/srep12500

Cheng JT, Cao F, Chen XA, Li YQ, Mao XM (2020) Genomic and transcriptomic survey of an endophytic fungus *Calcarisporium arbuscula* NRRL 3705 and potential overview of its secondary metabolites. BMC Genomics 21:1–13. https://doi.org/10.1186/s12864-020-06813-6

Cherry AJ, Banito A, Djegui D, Lomer C (2004) Suppression of the stem-borer *Sesamia calamistis* (Lepidoptera; Noctuidae) in maize following seed dressing, topical application and stem injection with African isolates of *Beauveria bassiana*. Int J Pest Manag 50(1):67–73. https://doi.org/10.1080/09670870310001637426

Clark MM (2006) Biological control methods for damping-off of tomato seedlings caused by *Pythium myriotylum*. Master's thesis, University of Tennessee. https://trace.tennessee.edu/utk_gradthes/1527

Clifton EH, Tylka GL, Gassmann AJ, Hodgson EW (2018) Interactions of effects of host plant resistance and seed treatments on soybean aphid (*Aphis glycines* Matsumura) and soybean cyst nematode (*Heterodera glycines* Ichinohe). Pest Manag Sci 74(4):992–1000. https://doi.org/10.1002/ps.4800

Cosme M, Lu J, Erb M, Stout MJ, Franken P et al (2016) A fungal endophyte helps plants to tolerate root herbivory through changes in gibberellin and jasmonate signaling. New Phytol 211(3): 1065–1076. https://doi.org/10.1111/nph.13957

Cotes B, Rännbäck LM, Björkman M, Norli HR, Meyling NV et al (2015) Habitat selection of a parasitoid mediated by volatiles informing on host and intraguild predator densities. Oecologia 179:151–162. https://doi.org/10.1007/s00442-015-3326-2

Crandall SG, Gold KM, Jiménez-Gasco MDM, Filgueiras CC, Willett DS (2020) A multi-omics approach to solving problems in plant disease ecology. PLoS One 15(9):e0237975. https://doi.org/10.1371/journal.pone.0237975

Dara S (2015) Reporting the occurrence of rice root aphid and honeysuckle aphid and their management in organic celery. UCANR eJ Strawb Veg 21. http://ucanr.edu/blogs/blogcore/postdetail.cfm?postnum=18740

Dara SK (2017) Compatibility of the entomopathogenic fungus with some fungicides used in California strawberry. Open Plant Sci J 10:29–34. https://doi.org/10.2174/1874294701710010029

Daroodi Z, Taheri P, Tarighi S (2021) Direct antagonistic activity and tomato resistance induction of the endophytic fungus *Acrophialophora jodhpurensis* against *Rhizoctonia solani*. Biol Control 160:104696. https://doi.org/10.1016/j.biocontrol.2021.104696

de Almeida Lopes KB, Carpentieri-Pipolo V, Fira D, Balatti PA, López SMY et al (2018) Screening of bacterial endophytes as potential biocontrol agents against soybean diseases. J Appl Microbiol 125(5):1466–1481. https://doi.org/10.1111/jam.14041

Devos Y, Craig W, Devlin RH, Ippolito A, Leggatt RA et al (2019) Using problem formulation for fit-for-purpose pre-market environmental risk assessments of regulated stressors. EFSA J 17: e170708. https://doi.org/10.2903/j.efsa.2019.e170708

Dong T, Zhang B, Jiang Y, Hu Q (2016) Isolation and classification of fungal whitefly entomopathogens from soils of Qinghai-Tibet plateau and Gansu Corridor in China. PLoS One 11(5):e0156087. https://doi.org/10.1371/journal.pone.0156087

Eilenberg J, Hajek A, Lomer C (2001) Suggestions for unifying the terminology in biological control. BioControl 46(4):387–400. https://doi.org/10.1023/A:1014193329979

Elbanhawy AA, Elsherbiny EA, Abd El-Mageed AE, Abdel-Fattah GM (2019) Potential of fungal metabolites as a biocontrol agent against cotton aphid, *Aphis gossypii* Glover and the possible mechanisms of action. Pestic Biochem Physiol 159:34–40. https://doi.org/10.1016/j.pestbp.2019.05.013

Elsharkawy MM, Shimizu M, Takahashi H, Ozaki K, Hyakumachi M (2013) Induction of systemic resistance against *Cucumber mosaic virus* in *Arabidopsis thaliana* by *Trichoderma asperellum* SKT-1. Plant Pathol J 29(2):193. https://doi.org/10.5423/PPJ.SI.07.2012.01

Fadiji AE, Babalola OO (2020) Exploring the potentialities of beneficial endophytes for improved plant growth. Saudi J Biol Sci 27:3622–3633. https://doi.org/10.1016/j.sjbs.2020.08.002

Faria M, Wraight SP (2007) Mycoinsecticides and mycoacaricides: a comprehensive list with worldwide coverage and international classification of formulation types. Biol Control 43: 237–256

Feng C, Xu F, Li L, Zhang J, Wang J et al (2023) Biological control of *Fusarium* crown rot of wheat with *Chaetomium globosum* 12XP1-2-3 and its effects on rhizosphere microorganisms. Front Microbiol 14:1133025. https://doi.org/10.3389/fmicb.2023.1133025

Flori P, Roberti R (1993) Treatment of onion bulbs with antagonistic fungi for the control of *Fusarium oxysporum* f. sp. *cepae*. Difesa delle Piante 16(4):5–12

Fontana A, Reichelt M, Hempel S, Gershenzon J, Unsicker SB (2009) The effects of arbuscular mycorrhizal fungi on direct and indirect defense metabolites of *Plantago lanceolata* L. J Chem Ecol 35:833–843. https://doi.org/10.1007/s10886-009-9654-0

Food and Agriculture Organization of the United Nations (FAO) (2019). Climate change fans the spread of pests and threatens plants and crops. https://www.fao.org/newsroom/detail/Climate-change-fans-spread-of-pestsand-threatens-plants-and-crops-new-FAO-study/en

Formenti LR, Nørregaard A, Bolic A, Quintanilla Hernandez DA, Hagemann T et al (2014) Challenges in industrial fermentation technology research. Biotechnol J 9(6):727–738. https://doi.org/10.1002/biot.201300236

Gao Q, Jin K, Ying SH, Zhang Y, Xiao G et al (2011) Genome sequencing and comparative transcriptomics of the model entomopathogenic fungi *Metarhizium anisopliae* and *M. acridum*. PLoS Genet 7(1):e1001264. https://doi.org/10.1371/journal.pgen.1001264

Gathage JW, Lagat ZO, Fiaboe KKM, Akutse KS, Ekesi S et al (2016) Prospects of fungal endophytes in the control of *Liriomyza* leafminer flies in common bean *Phaseolus vulgaris* under field conditions. BioControl 61:741–753. https://doi.org/10.1007/s10526-016-9761-0

Ghorui M, Chowdhury S, Burla S (2024) The science behind entomopathogenic fungi: mechanisms and applications. In: Deshmukh SK, Sridhar KR (eds) Entomopathogenic fungi: prospects and challenges. Springer Nature, Singapore, pp 3–35. https://doi.org/10.1007/978-981-97-5991-0_1

Gibson DM, Donzelli BG, Krasnoff SB, Keyhani NO (2014) Discovering the secondary metabolite potential encoded within entomopathogenic fungi. Nat Prod Rep 31(10):1287–1305. https://doi.org/10.1039/C4NP00054D

Gillespie AT, Claydon N (1989) The use of entomogenous fungi for pest control and the role of toxins in pathogenesis. Pestic Sci 27(2):203–215. https://doi.org/10.1002/ps.2780270210

González-Mas N, Cuenca-Medina M, Gutiérrez-Sánchez F, Quesada-Moraga E (2019) Bottom-up effects of endophytic *Beauveria bassiana* on multitrophic interactions between the cotton aphid, *Aphis gossypii*, and its natural enemies in melon. J Pest Sci 92:1271–1281. https://doi.org/10.1007/s10340-019-01098-5

Gouda S, Das S, Sen S, Patra JK (2016) Endophytes: a treasure house of bioactive compounds of medicinal importance. Front Microbiol 7:1538. https://doi.org/10.3389/fmicb.2016.01538

Grabka R, d'Entremont TW, Adams SJ, Walker AK, Tanney JB et al (2022) Fungal endophytes and their role in agricultural plant protection against pests and pathogens. Plants 11:384. https://doi.org/10.3390/plants11030384

Griffin MR, Ownley BH, Klingeman WE, Pereira RM (2006) Evidence of induced systemic resistance with *Beauveria bassiana* against *Xanthomonas* in cotton. Phytopathology 96(6):S42

Guo J, Cheng Y (2022) Advances in fungal elicitor-triggered plant immunity. Int J Mol Sci 23: 12003. https://doi.org/10.3390/ijms231912003

Gurulingappa P, Sword GA, Murdoch G, McGee PA (2010) Colonization of crop plants by fungal entomopathogens and their effects on two insect pests when in planta. Biol Control 55(1):34–41. https://doi.org/10.1016/j.biocontrol.2010.06.011

Hardoim PR, Van Overbeek LS, Berg G, Pirttilä AM, Compant S et al (2015) The hidden world within plants: ecological and evolutionary considerations for defining functioning of microbial endophytes. Microbiol Mol Biol Rev 79(3):293–320. https://doi.org/10.1128/mmbr.00050-14

Hashem AH, Al-Askar AA, Abd Elgawad H, Abdelaziz AM (2023) Bacterial endophytes from *Moringa oleifera* leaves as a promising source for bioactive compounds. Separations 10:395. https://doi.org/10.3390/separations10070395

Hazra DK, Purkait A (2019) Role of pesticide formulations for sustainable crop protection and environment management: a review. J Pharmacogn Phytochem 8(2):686–693

He Y, Chen T, Zhang H, White JF, Li C (2022) Fungal endophytes help grasses to tolerate sap-sucking herbivores through a hormone-signaling system. J Plant Growth Regul 41:2122–2137. https://doi.org/10.1007/s00344-021-10430-2

Hirano E, Koike M, Aiuchi D, Tani M (2008) Pre-inoculation of cucumber roots with *Verticillium lecanii* (*Lecanicillium muscarium*) induces resistance to powdery mildew. Res Bull Obihiro Univ 29:82–94. http://purl.org/coar/resource_type/c_6501

Hokkanen HM, Menzler-Hokkanen I (2024) Management of agricultural soils to support ecosystem services for pest control. In: Wang J, Liu H, Menzler-Hokkanen I, Jiang H (eds) The concept of ecostacking: techniques and applications. CABI, GB, pp 94–113. https://doi.org/10.1079/9781789248715.0008

Ignoffo CM, Hostetter DL, Sikorowski PP, Sutter G, Brooks WM (1977) Inactivation of representative species of entomopathogenic viruses, a bacterium, fungus, and protozoan by an ultraviolet light source. Environ Entomol 6(3):411–415. https://doi.org/10.1093/ee/6.3.411

Inglis GD, Goettel MS, Butt TM, Strasser H (2001) Use of hyphomycetous fungi for managing insect pests. In: Butt TM, Jackson C, Magan N (eds) Fungi as biocontrol agents: progress, problems and potential. CABI, Wallingford, pp 23–69. https://doi.org/10.1079/9780851993560.0023

Irsad, Shahid M, Haq E, Mohamed A, Rizvi PQ et al (2023) Entomopathogen-based biopesticides: insights into unraveling their potential in insect pest management. Front Microbiol 14:1208237. https://doi.org/10.3389/fmicb.2023.1208237

Islam MT, Castle SJ, Ren S (2010) Compatibility of the insect pathogenic fungus *Beauveria bassiana* with neem against sweetpotato whitefly, *Bemisia tabaci*, on eggplant. Entomol Exp Appl 134(1):28–34. https://doi.org/10.1111/j.1570-7458.2009.00933.x

Islam W, Adnan M, Shabbir A, Naveed H, Abubakar YS et al (2021) Insect-fungal-interactions: a detailed review on entomopathogenic fungi pathogenicity to combat insect pests. Microb Pathog 159:105122. https://doi.org/10.1016/j.micpath.2021.105122

Islam MS, Subbiah VK, Siddiquee S (2022) Field efficacy of proteolytic entomopathogenic fungi against *Ceratovacuna lanigera* Zehntner. Horticulturae 8(9):808. https://doi.org/10.3390/horticulturae8090808

Jaber LR (2015) Grapevine leaf tissue colonization by the fungal entomopathogen *Beauveria bassiana* s.l. and its effect against downy mildew. BioControl 60:103–112. https://doi.org/10.1007/s10526-014-9618-3

Jaber LR, Alananbeh KM (2018) Fungal entomopathogens as endophytes reduce several species of *Fusarium* causing crown and root rot in sweet pepper (*Capsicum annuum* L.). Biol Control 126:117–126. https://doi.org/10.1016/j.biocontrol.2018.08.007

Jaber LR, Araj SE (2018) Interactions among endophytic fungal entomopathogens (Ascomycota: Hypocreales), the green peach aphid *Myzus persicae* Sulzer (Homoptera: Aphididae), and the aphid endoparasitoid *Aphidius colemani* Viereck (Hymenoptera: Braconidae). Biol Control 116:53–61. https://doi.org/10.1016/j.biocontrol.2017.04.005

Jaber LR, Enkerli J (2016) Effect of seed treatment duration on growth and colonization of Vicia faba by endophytic *Beauveria bassiana* and *Metarhizium brunneum*. Biol Control 103:187–195. https://doi.org/10.1016/j.biocontrol.2016.09.008

Jaber LR, Enkerli J (2017) Fungal entomopathogens as endophytes: can they promote plant growth? Biocontrol Sci Tech 27(1):28–41. https://doi.org/10.1080/09583157.2016.1243227

Jaber LR, Vidal S (2010) Fungal endophyte negative effects on herbivory are enhanced on intact plants and maintained in a subsequent generation. Ecol Entomol 35(1):25–36. https://doi.org/10.1111/j.1365-2311.2009.01152.x

Jaber LR, Araj SE, Qasem JR (2018) Compatibility of endophytic fungal entomopathogens with plant extracts for the management of sweetpotato whitefly *Bemisia tabaci* Gennadius (Homoptera: Aleyrodidae). Biol Control 117:164–171. https://doi.org/10.1016/j.biocontrol.2017.11.009

Jallow MF, Cunningham JP, Zalucki MP (2004) Intra-specific variation for host plant use in *Helicoverpa armigera* (Hübner) (Lepidoptera: Noctuidae): implications for management. Crop Prot 23(10):955–964. https://doi.org/10.1016/j.cropro.2004.02.008

Jeffries P (1995) Biology and ecology of mycoparasitism. Can J Bot 73(S1):S1284–S1290. https://doi.org/10.1139/b95-389

Kalita S, Hazarika LK, Das P, Das K (2018) Efficacy of *Chromolaena odorata* leaf powder and *Beauveria bassiana* against stored grain pests. Indian J Entomol 80(3):1005–1010. https://doi.org/10.5958/0974-8172.2018.00152.9

Karslı A, Şahin YS (2021) The role of fungal volatile organic compounds (FVOCs) in biological control. Turk J Biol Control 12:79–92. https://doi.org/10.31019/tbmd.818701

Karthiba L, Saveetha K, Suresh S, Raguchander T, Saravanakumar D et al (2010) PGPR and entomopathogenic fungus bioformulation for the synchronous management of leaf folder pest and sheath blight disease of rice. Pest Manag Sci 66:555–564. https://doi.org/10.1002/ps.1907

Keller S, Schweizer C, Keller E, Brenner H (1997) Control of white grubs (*Melolontha melolontha* L.) by treating adults with the fungus *Beauveria brongniartii*. Biocontrol Sci Technol 7(1):105–116. https://doi.org/10.1080/09583159731090

Khachatourians GG, Qazi SS (2008) Entomopathogenic fungi: biochemistry and molecular biology. In: Brakhage AA, Zipfel PF (eds) Human and animal relationships, The Mycota, vol 6. Springer, Berlin, Heidelberg, pp 33–61. https://doi.org/10.1007/978-3-540-79307-6_3

Khan AL, Hamayun M, Khan SA, Kang SM, Shinwari ZK et al (2012) Pure culture of *Metarhizium anisopliae* LHL07 reprograms soybean to higher growth and mitigates salt stress. World J Microbiol Biotechnol 28:1483–1491. https://doi.org/10.1007/s11274-011-0950-9

Khan BA, Nadeem MA, Nawaz H, Amin MM, Abbasi GH et al (2023) Pesticides: impacts on agriculture productivity, environment, and management strategies. In: Aftab T (ed) Emerging contaminants and plants: interactions, adaptations and remediation technologies. Springer International, Cham, pp 109–134. https://doi.org/10.1007/978-3-031-22269-6_5

Kiarie S, Nyasani JO, Gohole LS, Maniania NK, Subramanian S (2020) Impact of fungal endophyte colonization of maize (*Zea mays* L.) on induced resistance to thrips- and aphid-transmitted viruses. Plants 9(4):416. https://doi.org/10.3390/plants9040416

Kim JJ, Goettel MS, Gillespie DR (2010) Evaluation of *Lecanicillium longisporum* (Vertalec®) against the cotton aphid, *Aphis gossypii*, and cucumber powdery mildew in greenhouse. Crop Prot 29(6):540–544. https://doi.org/10.1016/j.cropro.2009.12.011

Kisaakye J, Fourie H, Haukeland S, Kisitu J, Nakimera S et al (2022) Endophytic non-pathogenic *Fusarium oxysporum*-derived dual benefit for nematode management and improved banana (*Musa* spp.) productivity. Agriculture 12(2):125. https://doi.org/10.3390/agriculture12020125

Klieber J, Reineke A (2016) The entomopathogen *Beauveria bassiana* has epiphytic and endophytic activity against the tomato leaf miner *Tuta absoluta*. J Appl Entomol 140(8):580–588. https://doi.org/10.1111/jen.12287

Kongala SI, Kondreddy A (2023) A review on plant and pathogen derived carbohydrates, oligosaccharides and their role in plant's immunity. Carbohydr Polym Technol Appl 6: 100330. https://doi.org/10.1016/j.carpta.2023.100330

Kos K, Celar FA (2013) Sensitivity of the entomopathogenic fungus *Beauveria bassiana* (Bals.-Criv.) Vuill. to selected herbicides. Pest Manag Sci 69(6):717–721. https://doi.org/10.1002/ps.3427

Kuchár M (2019) Endophytic fungi associated with cabbage in New Zealand and their potential for biological control. PhD thesis, Lincoln University, CA, USA

Kumar S, Devi EL, Sharma SK, Ansari MA, Phurailatpam S et al (2017) Rice breeding strategies of North Eastern India for resilience to biotic and abiotic stresses: a review. Oryza 54:1–12

Landum MC, do Rosário Félix M, Alho J, Garcia R, Cabrita MJ et al (2016) Antagonistic activity of fungi of *Olea europaea* L. against *Colletotrichum acutatum*. Microbiol Res 183:100–108. https://doi.org/10.1016/j.micres.2015.12.001

Latz MA, Jensen B, Collinge DB, Jørgensen HJ (2018) Endophytic fungi as biocontrol agents: elucidating mechanisms in disease suppression. Plant Ecol Divers 11(5–6):555–567. https://doi.org/10.1080/17550874.2018.1534146

Li P, Luo H, Meng J, Sun W, Wang X et al (2014) Effects of oligosaccharides from endophytic *Fusarium oxysporum* Dzf17 on activities of defense-related enzymes in *Dioscorea zingiberensis* suspension cell and seedling cultures. Electron J Biotechnol 17:156–161. https://doi.org/10.1016/j.ejbt.2014.04.012

Li H, Dou G, Gao M, Ren F, Li R et al (2020) *Annulohypoxylon* sp. FPYF3050 produces volatile organic compounds against the pine wood nematode *Bursaphelenchus xylophilus*. Nematology 22(3):245–255

Li YL, Yi JL, Cai J, Zhou XM, Chen L et al (2022) Two new bioactive secondary metabolites from the endophytic fungus *Talaromyces assiutensis* JTY2. Nat Prod Res 36(14):3695–3700. https://doi.org/10.1080/14786419.2021.1881961

Li E, Zhu N, Zhang S, Xu B, Liu L (2024) Efficacy of *Trichoderma longibrachiatum* SC5 fermentation filtrate in inhibiting the *Sclerotinia sclerotiorum* growth and development in sunflower. Int J Mol Sci 26(1):201. https://doi.org/10.3390/ijms26010201

Litwin A, Nowak M, Różalska S (2020) Entomopathogenic fungi: unconventional applications. Rev Environ Sci Biotechnol 19(1):23–42. https://doi.org/10.1007/s11157-020-09525-1

Liu Y, Ding Y, Peng D, Liu SM, Kong LA et al (2019) Evaluation of the biocontrol potential of *Aspergillus welwitschiae* against the root-knot nematode *Meloidogyne graminicola* in rice (*Oryza sativa* L.). J Integr Agric 18:2561–2570. https://doi.org/10.1016/S2095-3119(19)62610-9

Lopez DC, Sword GA (2015) The endophytic fungal entomopathogens *Beauveria bassiana* and *Purpureocillium lilacinum* enhance growth of cultivated cotton (*Gossypium hirsutum*) and negatively affect survival of the cotton bollworm (*Helicoverpa zea*). Biol Control 89:53–60. https://doi.org/10.1016/j.biocontrol.2015.03.010

Lu L, Karunarathna SC, Hyde KD, Suwannarach N, Elgorban AM et al (2022) Endophytic fungi associated with coffee leaves in China exhibited in vitro antagonism against fungal and bacterial pathogens. J Fungi 8(7):698. https://doi.org/10.3390/jof8070698

Lugtenberg BJJ, Caradus JR, Johnson LJ (2016) Fungal endophytes for sustainable crop production. FEMS Microbiol Ecol 92(12):1–17. https://doi.org/10.1093/femsec/fiw194

Luo H, Qing Z, Deng Y, Deng Z, Tang XA et al (2019) Two polyketides produced by endophytic *Penicillium citrinum* DBR-9 from medicinal plant *Stephania kwangsiensis* and their antifungal activity against plant pathogenic fungi. Nat Prod Commun 14(5):1–6. https://doi.org/10.1177/1934578X19846795

Luo H, Meng S, Deng Y, Deng Z, Shi H (2023) In vitro antifungal activity of lasiodiplodin, isolated from endophytic fungus *Lasiodiplodia pseudotheobromae* J-10 associated with *Sarcandra glabra* and optimization of culture conditions for lasiodiplodin production. Arch Microbiol 205(4):140. https://doi.org/10.1007/s00203-023-03440-z

Maina UM, Galadima IB, Gambo FM, Zakaria DJ (2018) A review on the use of entomopathogenic fungi in the management of insect pests of field crops. J Entomol Zool Stud 6(1):27–31

Maithani D, Singh H, Sharma A (2021) Stress alleviation in plants using SAR and ISR: current views on stress signaling network. In: Sharma A (ed) Microbes and signaling biomolecules against plant stress: strategies of plant-microbe relationships for better survival, Rhizosphere biology. Springer, Singapore, pp 7–36. https://doi.org/10.1007/978-981-15-7094-0_2

Mantzoukas S, Eliopoulos PA (2020) Endophytic entomopathogenic fungi: a valuable biological control tool against plant pests. Appl Sci 10(1):360. https://doi.org/10.3390/app10010360

Mantzoukas S, Grammatikopoulos G (2020) The effect of three entomopathogenic endophytes of the sweet sorghum on the growth and feeding performance of its pest, Sesamia nonagrioides larvae, and their efficacy under field conditions. Crop Prot. 127:104952

Mantzoukas S, Chondrogiannis C, Grammatikopoulos G (2015) Effects of three endophytic entomopathogens on sweet sorghum and on the larvae of the stalk borer *Sesamia nonagrioides*. Entomol Exp Appl 154(1):78–87. https://doi.org/10.1111/eea.12262

Mantzoukas S, Zikou A, Triantafillou V, Lagogiannis I, Eliopoulos PA (2019) Interactions between *Beauveria bassiana* and *Isaria fumosorosea* and their hosts *Sitophilus granarius* (L.) and *Sitophilus oryzae* (L.) (Coleoptera: Curculionidae). Insects 10(10):362. https://doi.org/10.3390/insects10100362

Mantzoukas S, Kitsiou F, Natsiopoulos D, Eliopoulos PA (2022) Entomopathogenic fungi: interactions and applications. Encyclopedia 2(2):646–656. https://doi.org/10.3390/encyclopedia2020044

Miao GP, Han J, Zhang KG, Wang SC, Wang CR (2019) Protection of melon against *Fusarium* wilt-root knot nematode complex by endophytic fungi *Penicillium brefeldianum* HS-1. Symbiosis 77:83–89. https://doi.org/10.1007/s13199-018-0565-0

Michereff Filho M, Oliveira SO, De Liz RS, Faria M (2011) Cage and field assessments of *Beauveria bassiana*-based mycoinsecticides for *Myzus persicae* Sulzer (Hemiptera: Aphididae) control in cabbage. Neotrop Entomol 40:470. https://doi.org/10.1590/S1519-566X2011000400010

Miranda-Fuentes P, Yousef-Yousef M, Valverde-García P, Rodríguez-Gómez IM, Garrido-Jurado I et al (2021) Entomopathogenic fungal endophyte-mediated tritrophic interactions between *Spodoptera littoralis* and its parasitoid *Hyposoter didymator*. J Pest Sci 94:933–945. https://doi.org/10.1007/s10340-020-01306-7

Mirzaei S, Masumi S (2022) The antimicrobial activity of endophytic fungi isolated from *Thymus* spp. J Med Plants By-prod 12:117–124. https://doi.org/10.22092/jmpb.2022.342405.1192

Mukherjee PK, Mendoza-Mendoza A, Zeilinger S, Horwitz BA (2022) Mycoparasitism as a mechanism of *Trichoderma*-mediated suppression of plant diseases. Fungal Biol Rev 39:15–33. https://doi.org/10.1016/j.fbr.2021.11.004

Muvea AM, Meyhöfer R, Subramanian S, Poehling HM, Ekesi S et al (2014) Colonization of onions by endophytic fungi and their impacts on the biology of *Thrips tabaci*. PLoS One 9(9): e108242. https://doi.org/10.1371/journal.pone.0108242

Nguyen HT, Kim S, Yu NH, Park AR, Yoon H et al (2019) Antimicrobial activities of an oxygenated cyclohexanone derivative isolated from *Amphirosellinia nigrospora* JS-1675 against various plant pathogenic bacteria and fungi. J Appl Microbiol 126(3):894–904. https://doi.org/10.1111/jam.14138

Ownley BH, Griffin MR, Klingeman WE, Gwinn KD, Moulton JK et al (2008) *Beauveria bassiana*: endophytic colonization and plant disease control. J Invertebr Pathol 98(3):267–270. https://doi.org/10.1016/j.jip.2008.01.010

Ownley BH, Gwinn KD, Vega FE (2010) Endophytic fungal entomopathogens with activity against plant pathogens: ecology and evolution. BioControl 55:113–128. https://doi.org/10.1007/s10526-009-9241-x

Pacheco-Trejo J, Aquino-Torres E, Reyes-Santamaría MI, Islas-Pelcastre M, Pérez-Ríos SR et al (2022) Plant defensive responses triggered by *Trichoderma* spp. as tools to face stressful conditions. Horticulturae 8:1181. https://doi.org/10.3390/horticulturae8121181

Pakdaman Sardrood B, Mohammadi Goltapeh E (2018) Effect of agricultural chemicals and organic amendments on biological control fungi. In: Lichtfouse E (ed) Sustainable agriculture reviews, vol 31. Springer, Cham, pp 217–359. https://doi.org/10.1007/978-3-319-94232-2_5

Panicker S, Sayyed RZ (2022) Hydrolytic enzymes from PGPR against plant fungal pathogens. In: Sayyed R, Singh A, Ilyas N (eds) Antifungal metabolites of rhizobacteria for sustainable agriculture, Fungal biology. Springer, Cham, pp 211–238. https://doi.org/10.1007/978-3-031-04805-0_10

Panwar N, Szczepaniec A (2024) Endophytic entomopathogenic fungi as biological control agents of insect pests. Pest Manag Sci 80:6033–6040. https://doi.org/10.1002/ps.8322

Paradza VM, Khamis FM, Yusuf AA, Subramanian S, Ekesi S et al (2021) Endophytic colonisation of *Solanum lycopersicum* and *Phaseolus vulgaris* by fungal endophytes promotes seedling growth and hampers the reproductive traits, development, and survival of the greenhouse whitefly, *Trialeurodes vaporariorum*. Front Plant Sci 12:771534. https://doi.org/10.3389/fpls.2021.771534

Parulekar-Berde CV, Joshi SA, Berde VB (2021) Fungal communities as biological control agents for different phytopathogenic organisms. In: Yadav AN (ed) Recent trends in mycological research: volume 1: agricultural and medical perspective, fungal biology. Springer, Cham, pp 189–201. https://doi.org/10.1007/978-3-030-60659-6_8

Parveen SS, Rashtrapal PS (2024) Integrated pest management strategies using endophytic entomopathogenic fungi. Plant Sci Today 11:568–574. https://doi.org/10.14719/pst.2740

Pattemore JA, Hane JK, Williams AH, Wilson BA, Stodart BJ et al (2014) The genome sequence of the biocontrol fungus *Metarhizium anisopliae* and comparative genomics of *Metarhizium* species. BMC Genomics 15:660. https://doi.org/10.1186/1471-2164-15-660

Perlin MH, Poulin R, de Bekker C (2025) Invasion of the four kingdoms: the parasite journey across plant and non-plant hosts. Biol Rev 100:936–968. https://doi.org/10.1111/brv.13169

Petrisor C, Stoian G (2017) The role of hydrolytic enzymes produced by entomopathogenic fungi in pathogenesis of insects: mini review. Rom J Plant Prot 10:66–72

Prajapati DR, Patel NB, Gohel NM (2023) A review on entomopathogenic facet of *Fusarium verticillioides*. Environ Ecol 41(4):2365–2374. https://doi.org/10.60151/envec/FYZN6327

Price PW, Denno RF, Eubanks MD, Finke DL, Kaplan I (2011) Insect ecology: behavior, populations and communities. Cambridge University Press, Cambridge

Putri ND, Sulistyowati L, Aini LQ, Muhibuddin A, Trianti I (2022) Screening of endophytic fungi as potential antagonistic agents of *Pyricularia oryza*e and evaluation of their ability in producing hydrolytic enzymes. Biodiversitas J Biol Divers 23:2. https://doi.org/10.13057/biodiv/d230248

Qayyum MA, Wakil W, Arif MJ, Sahi ST, Dunlap CA (2015) Infection of *Helicoverpa armigera* by endophytic *Beauveria bassiana* colonizing tomato plants. Biol Control 90:200–207. https://doi.org/10.1016/j.biocontrol.2015.04.005

Quesada-Moraga E (2020) Entomopathogenic fungi as endophytes: their broader contribution to IPM and crop production. Biocontrol Sci Tech 30(9):864–877. https://doi.org/10.1080/09583157.2020.1771279

Quesada-Moraga E, Ruiz-García A, Santiago-Álvarez C (2006) Laboratory evaluation of entomopathogenic fungi *Beauveria bassiana* and *Metarhizium anisopliae* against puparia and adults of *Ceratitis capitata* (Diptera: Tephritidae). J Econ Entomol 99(6):1955–1966. https://doi.org/10.1093/jee/99.6.1955

Quesada-Moraga E, Munoz-Ledesma FJ, Santiago-Alvarez C (2009) Systemic protection of *Papaver somniferum* L. against *Iraella luteipes* (Hymenoptera: Cynipidae) by an endophytic strain of *Beauveria bassiana* (Ascomycota: Hypocreales). Environ Entomol 38(3):723–730. https://doi.org/10.1603/022.038.0324

Quesada-Moraga E, Yousef M, Garrido-Jurado I (2020) Advances in the use of entomopathogenic fungi as biopesticides in suppressing crop pests. In: Birch N, Glare T (eds) Biopesticides for sustainable agriculture. Burleigh Dodds Science, Cambridge

Rabbee MF, Ali MS, Choi J, Hwang BS, Jeong SC et al (2019) *Bacillus velezensis*: a valuable member of bioactive molecules within plant microbiomes. Molecules 24(6):1046. https://doi.org/10.3390/molecules24061046

Rachappa V, Lingappa S, Patil RK (2007) Effect of agrochemicals on growth and sporulation of *Metarhizium anisopliae* (Metschnikoff) Sorokin. Karnataka J Agric Sci 20(2):410–413

Rashid TS (2021) Bioactive metabolites from tomato endophytic fungi with antibacterial activity against tomato bacterial spot disease. Rhizosphere 17:100292. https://doi.org/10.1016/j.rhisph.2020.100292

Rauf M, Ur-Rahman A, Arif M, Gul H, Ud-Din A et al (2022) Immunomodulatory molecular mechanisms of *Luffa cylindrica* for downy mildews resistance induced by growth-promoting endophytic fungi. J Fungi 8(7):689. https://doi.org/10.3390/jof8070689

Raya-Díaz S, Quesada-Moraga E, Barrón V, Del Campillo MC, Sánchez-Rodríguez AR (2017) Redefining the dose of the entomopathogenic fungus *Metarhizium brunneum* (Ascomycota, Hypocreales) to increase Fe bioavailability and promote plant growth in calcareous and sandy soils. Plant Soil 418:387–404. https://doi.org/10.1007/s11104-017-3303-0

Rehner SA, Posada F, Buckley EP, Infante F, Castillo A et al (2006) Phylogenetic origins of African and Neotropical *Beauveria bassiana* sl pathogens of the coffee berry borer, *Hypothenemus hampei*. J Invertebr Pathol 93(1):11–21. https://doi.org/10.1016/j.jip.2006.04.005

Ren SX, Ali S, Huang Z, Wu JH (2010) *Lecanicillium muscarium* as microbial insecticide against whitefly and its interaction with other natural enemies. In: Méndez-Vilas A (ed) Current research, technology and education topics in applied microbiology and microbial biotechnology. Formatex Badajoz, pp 339–348

Ribeiro THC, Fernandes-Brum CN, de Souza CR, Dias FAN, Almeida-Junior OD et al (2020) Transcriptome analyses suggest that changes in fungal endophyte lifestyle could be involved in grapevine bud necrosis. Sci Rep 10(1):9514. https://doi.org/10.1038/s41598-020-66500-0

Rojas EC, Jensen B, Jørgensen HJ, Latz MA, Esteban P et al (2022) The fungal endophyte *Penicillium olsonii* ML37 reduces *Fusarium* head blight by local induced resistance in wheat spikes. J Fungi 8(4):345. https://doi.org/10.3390/jof8040345

Rondot Y, Reineke A (2019) Endophytic *Beauveria bassiana* activates expression of defence genes in grapevine and prevents infections by grapevine downy mildew *Plasmopara viticola*. Plant Pathol 68(9):1719–1173. https://doi.org/10.1111/ppa.13089

Safari Motlagh MR, Jahangiri B, Kulus D, Tymoszuk A, Kaviani B (2022) Endophytic fungi as potential biocontrol agents against *Rhizoctonia solani* JG Kühn, the causal agent of rice sheath blight disease. Biology 11(9):128. https://doi.org/10.3390/biology11091282

Safeer MM, Thara SS (2022) Integrated management of *Cymbidium mosaic* disease in commercial *Dendrobium* orchids using root endophytic fungi *Piriformospora indica*. Cogent Food Agric 8(1):2139848. https://doi.org/10.1080/23311932.2022.2139848

Sahayaraj K (ed) (2014) Basic and applied aspects of biopesticides. Springer, New Delhi. https://doi.org/10.1007/978-81-322-1877-7

Sák M, Dokupilová I, Kaňuková Š, Mrkvová M, Mihálik D et al (2021) Biotic and abiotic elicitors of stilbenes production in *Vitis vinifera* L. cell culture. Plants 10:490. https://doi.org/10.3390/plants10030490

Samal I, Bhoi TK, Majhi PK, Murmu S, Pradhan AK et al (2023a) Combatting insect-mediated biotic stress through plant-associated endophytic entomopathogenic fungi in horticultural crops. Front Plant Sci 13:1098673. https://doi.org/10.3389/fpls.2022.1098673

Samal S, Mohanty RP, Mohanty PS, Giri MK, Pati S et al (2023b) Implications of biosensors and nanobiosensors for the eco-friendly detection of public health and agro-based insecticides: a comprehensive review. Heliyon 9(5):e15848. https://doi.org/10.1016/j.heliyon.2023.e15848

Sandhu SS, Sharma AK, Beniwal V, Goel G, Batra P et al (2012) Myco-biocontrol of insect pests: factors involved, mechanism, and regulation. J Pathog 2012(1):12689. https://doi.org/10.1155/2012/126819

Saravanakumar K, Fan L, Fu K, Yu C, Wang M et al (2016) Cellulase from *Trichoderma harzianum* interacts with roots and triggers induced systemic resistance to foliar disease in maize. Sci Rep 6:35543. https://doi.org/10.1038/srep35543

Sasan RK, Bidochka MJ (2012) The insect-pathogenic fungus *Metarhizium robertsii* (Clavicipitaceae) is also an endophyte that stimulates plant root development. Am J Bot 99: 101–107. https://doi.org/10.3732/ajb.1100136

Sasan RK, Bidochka MJ (2013) Antagonism of the endophytic insect pathogenic fungus *Metarhizium robertsii* against the bean plant pathogen *Fusarium solani* f. sp. *phaseoli*. Can J Plant Pathol 35(3):288–293. https://doi.org/10.1080/07060661.2013.823114

Schoch CL, Seifert KA, Huhndorf S, Robert V, Spouge JL et al (2012) Nuclear ribosomal internal transcribed spacer (ITS) region as a universal DNA barcode marker for fungi. Proc Natl Acad Sci USA 109(16):6241–6246. https://doi.org/10.1073/pnas.1117018109

Shaari SA, Radzi NAM, Shaaya SA, Azmi KHM, Azhar NA et al (2024) Recent trends in precision agriculture: applications & challenges in precision farming. Int Trans J Eng Manag Appl Sci Technol 15(3):1–10. https://doi.org/10.14456/ITJEMAST.2024.15

Shang J, Tang G, Lu M, Wang C (2022) Host and environmental sensing by entomopathogenic fungi to infect hosts. Curr Clin Microbiol Rep 9(4):69–74. https://doi.org/10.1007/s40588-022-00185-z

Sharma R, Sharma P (2021) Fungal entomopathogens: a systematic review. Egypt J Biol Pest Control 31:1–13. https://doi.org/10.1186/s41938-021-00404-7

Sharma A, Srivastava A, Shukla AK, Srivastava K, Srivastava AK et al (2020) Entomopathogenic fungi: a potential source for biological control of insect pests. In: Solanki M, Kashyap P, Kumari B (eds) Phytobiomes: current insights and future vistas. Springer, Singapore, pp 225–250. https://doi.org/10.1007/978-981-15-3151-4_9

Sharma A, Sharma S, Yadav PK (2023) Entomopathogenic fungi and their relevance in sustainable agriculture: a review. Cogent Food Agric 9(1):2180857. https://doi.org/10.1080/23311932.2023.2180857

Sharma N, Kaur DP, Thakur D (2024) Various insect pathogenic fungi and their mode of action. In: Ghoneim K (ed) Advances in agricultural entomology. AkiNik Publications, New Delhi, pp 17–23

Shin TY, Lee MR, Park SE, Lee SJ, Kim WJ et al (2020) Pathogenesis-related genes of entomopathogenic fungi. Arch Insect Biochem Physiol 105(4):e21747. https://doi.org/10.1002/arch.21747

Shree A, Pal S, Verma PK (2024) Structural diversification of fungal cell wall in response to stress signaling and remodeling during fungal pathogenesis. Physiol Mol Biol Plants 30(5):733–747. https://doi.org/10.1007/s12298-024-01453-6

Siddaiah CN, Satyanarayana NR, Mudili V, Gupta VK, Gurunathan S et al (2017) Elicitation of resistance and associated defense responses in *Trichoderma hamatum* induced protection against pearl millet downy mildew pathogen. Sci Rep 7:43991. https://doi.org/10.1038/srep43991

Sikandar A, Zhang MY, Zhu XF, Wang YY, Ahmed M et al (2019) Efficacy of *Penicillium chrysogenum* strain SNEF1216 against root-knot nematodes (*Meloidogyne incognita*) in cucumber (*Cucumis sativus* L.) under greenhouse conditions. Appl Ecol Environ Res 17(5): 12451–12464. https://doi.org/10.15666/aeer/1705_1245112464

Silva DM, de Souza VHM, Moral RDA, Delalibera Júnior I, Mascarin GM (2022) Production of *Purpureocillium lilacinum* and *Pochonia chlamydosporia* by submerged liquid fermentation and bioactivity against *Tetranychus urticae* and *Heterodera glycines* through seed inoculation. J Fungi 8(5):511. https://doi.org/10.3390/jof8050511

Silva MDGC, Medeiros AO, Converti A, Almeida FCG, Sarubbo LA (2024) Biosurfactants: promising biomolecules for agricultural applications. Sustainability 16:449. https://doi.org/10.3390/su16010449

Singh D, Raina TK, Singh J (2017) Entomopathogenic fungi: an effective biocontrol agent for management of insect populations naturally. J Pharm Sci Res 9:833

Singkaravanit S, Kinoshita H, Ihara F, Nihira T (2010) Cloning and functional analysis of the second geranylgeranyl diphosphate synthase gene influencing helvolic acid biosynthesis in *Metarhizium anisopliae*. Appl Microbiol Biotechnol 87:1077–1088. https://doi.org/10.1007/s00253-010-2556-9

Skinner M, Parker BL, Kim JS (2014) Role of entomopathogenic fungi in integrated pest management. In: Abrol DP (ed) Integrated pest management. Elsevier, pp 169–191. https://doi.org/10.1016/B978-0-12-398529-3.00011-7

Sohrabi F, Jamali F, Morammazi S, Saber M, Kamita SG (2019) Evaluation of the compatibility of entomopathogenic fungi and two botanical insecticides tondexir and palizin for controlling *Galleria mellonella* (Lepidoptera: Pyralidae). Crop Prot 117:20–25. https://doi.org/10.1016/j.cropro.2018.11.012

Tan QW, Gao FL, Wang FR, Chen QJ (2015) Anti-TMV activity of malformin A1, a cyclic penta-peptide produced by an endophytic fungus *Aspergillus tubingensis* FJBJ11. Int J Mol Sci 16: 5750–5761. https://doi.org/10.3390/ijms16035750

Tanapichatsakul C, Pansanit A, Monggoot S, Brooks S, Prachya S et al (2020) Antifungal activity of 8-methoxynaphthalen-1-ol isolated from the endophytic fungus *Diatrype palmicola* MFLUCC 17-0313 against the plant pathogenic fungus *Athelia rolfsii* on tomatoes. PeerJ 8: e9103. https://doi.org/10.7717/peerj.9103

Thirumeni MA, Yashkamal K, Mohamed RR, Shivakumar MS (2024) Entomopathogenic fungi as biopesticides for sustainable agriculture. In: Deshmukh SK, Sridhar KR (eds) Entomopathogenic fungi: prospects and challenges. Springer Nature, Singapore, pp 81–100. https://doi.org/10.1007/978-981-97-5991-0_4

Torres MA, Jones JD, Dangl JL (2006) Reactive oxygen species signaling in response to pathogens. Plant Physiol 141:373–378. https://doi.org/10.1104/pp.106.079467

Ujjan AA, Shahzad S (2012) Use of entomopathogenic fungi for the control of mustard aphid (*Lipaphis erysimi*) on canola (*Brassica napus* L.). Pak J Bot 44:2081–2086

Vega FE (2018) The use of fungal entomopathogens as endophytes in biological control: a review. Mycologia 110:4–30. https://doi.org/10.1080/00275514.2017.1418578

Vega FE, Posada F, Aime MC, Pava-Ripoll M, Infante F et al (2008) Entomopathogenic fungal endophytes. Biol Control 46:72–82. https://doi.org/10.1016/j.biocontrol.2008.01.008

Vega FE, Goettel MS, Blackwell M, Chandler D, Jackson MA et al (2009) Fungal entomopathogens: new insights on their ecology. Fungal Ecol 2:149–159. https://doi.org/10.1016/j.funeco.2009.05.001

Vega FE, Meyling NV, Luangsa-ard JJ, Blackwell M (2012) Fungal entomopathogens. In: Vega FE, Kaya HK (eds) Insect pathology, 2nd edn. Academic Press, San Diego, pp 171–220

Vitti A, Pellegrini E, Nali C, Lovelli S, Sofo A et al (2016) *Trichoderma harzianum* T-22 induces systemic resistance in tomato infected by *Cucumber mosaic virus*. Front Plant Sci 7:1520. https://doi.org/10.3389/fpls.2016.01520

Wang Z, Gerstein M, Snyder M (2009) RNA-Seq: a revolutionary tool for transcriptomics. Nat Rev Genet 10:57–63. https://doi.org/10.1038/nrg2484

Wang M, Carver JJ, Phelan VV, Sanchez LM, Garg N et al (2016) Sharing and community curation of mass spectrometry data with Global Natural Products Social Molecular Networking. Nat Biotechnol 34:828–837. https://doi.org/10.1038/nbt.3597

Wang L, Shen J, Xu L, Gao J, Zhang C et al (2019) A metabolite of endophytic fungus *Cadophora orchidicola* from *Kalimeris indica* serves as a potential fungicide and TLR4 agonist. J Appl Microbiol 126:1383–1390. https://doi.org/10.1111/jam.14239

Wang Z, Wang H, Peng Y, Chen F, Zhao L et al (2020) A liquid chromatography-tandem mass spectrometry (LC-MS/MS)-based assay to profile 20 plasma steroids in endocrine disorders. Clin Chem Lab Med 58:1477–1487. https://doi.org/10.1515/cclm-2019-0869

Wani ZA, Kumar A, Sultan P, Bindu K, Riyaz-Ul-Hassan S et al (2017) *Mortierella alpina* CS10E4, an oleaginous fungal endophyte of *Crocus sativus* L. enhances apocarotenoid biosynthesis and stress tolerance in the host plant. Sci Rep 7:8598. https://doi.org/10.1038/s41598-017-08974-z

Wiese S, Reidegeld KA, Meyer HE, Warscheid B (2007) Protein labeling by iTRAQ: a new tool for quantitative mass spectrometry in proteome research. Proteomics 7:340–350. https://doi.org/10.1002/pmic.200600422

Wink M (2008) Plant secondary metabolism: diversity, function and its evolution. Nat Prod Commun 3:1205–1216. https://doi.org/10.1177/1934578X0800300801

Wraight SP, Sporleder M, Poprawski TJ, Lacey LA (2007) Application and evaluation of entomopathogens in potato. In: Lacey LA, Kaya HK (eds) Field manual of techniques in invertebrate pathology. Springer, Dordrecht, pp 329–359. https://doi.org/10.1007/978-1-4020-5933-9_15

Xiao G, Ying SH, Zheng P, Wang ZL, Zhang S et al (2012) Genomic perspectives on the evolution of fungal entomopathogenicity in *Beauveria bassiana*. Sci Rep 2:483. https://doi.org/10.1038/srep00483

Xu T, Cao L, Zeng J, Franco CM, Yang Y et al (2019) The antifungal action mode of the rice endophyte *Streptomyces hygroscopicus* OsiSh-2 as a potential biocontrol agent against the rice blast pathogen. Pestic Biochem Physiol 160:58–69. https://doi.org/10.1016/j.pestbp.2019.06.015

Yadav AN (2018) Biodiversity and biotechnological applications of host-specific endophytic fungi for sustainable agriculture and allied sectors. Acta Sci Microbiol 1:1–5. https://doi.org/10.31080/ASMI.2018.01.0044

Yan XN, Sikora RA, Zheng JW (2011) Potential use of cucumber (*Cucumis sativus* L.) endophytic fungi as seed treatment agents against root-knot nematode *Meloidogyne incognita*. J Zhejiang Univ Sci B 12:219–225. https://doi.org/10.1631/jzus.B1000165

Yan Y, Mao Q, Wang Y, Zhao J, Fu Y et al (2021) *Trichoderma harzianum* induces resistance to root-knot nematodes by increasing secondary metabolite synthesis and defense-related enzyme activity in *Solanum lycopersicum* L. Biol Control 158:104609. https://doi.org/10.1016/j.biocontrol.2021.104609

Yang EF, Karunarathna SC, Dai DQ, Priyashantha AKH, Promputtha I et al (2023a) Taxonomy and phylogeny of endophytic fungi (Chaetomiaceae) associated with healthy leaves of *Mangifera indica* in Yunnan, China. Diversity 15:1094. https://doi.org/10.3390/d15101094

Yang EF, Karunarathna SC, Tibpromma S, Stephenson SL, Promputtha I et al (2023b) Endophytic fungi associated with mango show in vitro antagonism against bacterial and fungal pathogens. Agronomy 13:169. https://doi.org/10.3390/agronomy13010169

Yao YQ, Lan F, Qiao YM, Wei JG, Huang RS et al (2017) Endophytic fungi harbored in the root of *Sophora tonkinensis* Gapnep: diversity and biocontrol potential against phytopathogens. Microbiol Open 6:e00437. https://doi.org/10.1002/mbo3.437

Yin M, Xiao D, Wang C, Zhang L, Dun B et al (2022) The regulation of BbLaeA on the production of beauvericin and bassiatin in *Beauveria bassiana*. World J Microbiol Biotechnol 38:1. https://doi.org/10.1007/s11274-021-03162-8

Yu Y, Gui Y, Li Z, Jiang C, Guo J, Niu D (2022) Induced systemic resistance for improving plant immunity by beneficial microbes. Plants 11:386. https://doi.org/10.3390/plants11030386

Yuan Y, Feng H, Wang L, Li Z, Shi Y et al (2017) Potential of endophytic fungi isolated from cotton roots for biological control against verticillium wilt disease. PLoS One 12:e0170557. https://doi.org/10.1371/journal.pone.0170557

Yuan HG, Wu SY, Lei ZR, Rondon SI, Gao YL (2018) Sub-lethal effects of *Beauveria bassiana* (Balsamo) on field populations of the potato tuber worm *Phthorimaea operculella* Zeller in China. J Integr Agric 17(4):911–918. https://doi.org/10.1016/S2095-3119(17)61898-7

Yuan J, Zhang W, Sun K, Tang MJ, Chen PX et al (2019) Comparative transcriptomics and proteomics of Atractylodes lancea in response to endophytic fungus *Gilmaniella* sp. AL12 reveals regulation in plant metabolism. Front Microbiol 10:1208. https://doi.org/10.3389/fmicb.2019.01208

Zanzana K, Dannon EA, Sinzogan AA, Toffa JM (2024) Fall armyworm management in a changing climate: an overview of climate-responsive integrated pest management (IPM) strategies for long-term control. Egypt J Biol Pest Control 34:54. https://doi.org/10.1186/s41938-024-00814-3

Zhang P, Yuan XL, Du YM, Zhang HB, Shen GM et al (2019) Angularly prenylated indole alkaloids with antimicrobial and insecticidal activities from an endophytic fungus *Fusarium sambucinum* TE-6L. J Agric Food Chem 67:11994–12001. https://doi.org/10.1021/acs.jafc.9b05827

Zhang Y, Qin B, Narisawa K, Nong Q, Qin L et al (2022) The dark septate endophyte, *Phialocephala fortinii* J2PC4, mitigates southern rice black-streaked dwarf disease and impacts the mortality of white back planthopper. Biol Control 170:104911. https://doi.org/10.1016/j.biocontrol.2022.104911

Zhu Q, Tang MJ, Yang Y, Sun K, Tian LS et al (2021) Endophytic fungus *Phomopsis liquidambaris* B3 induces rice resistance to RSRD caused by *Fusarium proliferatum* and promotes plant growth. J Sci Food Agric 101:4059–4075. https://doi.org/10.1002/jsfa.11042

Zimmermann G (2007a) Review on safety of the entomopathogenic fungus *Metarhizium anisopliae*. Biocontrol Sci Tech 17:879–920. https://doi.org/10.1080/09583150701593963

Zimmermann G (2007b) Review on safety of the entomopathogenic fungi *Beauveria bassiana* and *Beauveria brongniartii*. Biocontrol Sci Tech 17:553–596. https://doi.org/10.1080/09583150701309006

Part V

Biosafety and Future Prospects

Biosafety Assessment and Residual Pathogenic Effects of Entomopathogenic Fungi

17

Maunata Ghorui, Bidisha Mukherjee, Shouvik Chowdhury, Prakash Balu, and Sashidhar Burla

Abstract

Entomopathogenic fungi (EPF) have emerged as promising biological alternatives to chemical pesticides for integrated pest management (IPM). However, their widespread application necessitates a thorough understanding of their safety and residual pathogenic effects. This review critically evaluates the current knowledge on EPF toxicity to non-target organisms, including pollinators, beneficial arthropods, vertebrates, and humans. The implications of secondary metabolite production, endophytic behavior, and environmental persistence are discussed in relation to ecological and human health safety. Regulatory frameworks from major regions, including the USA, EU, Canada, Australia, Brazil, India, China, and Japan, are compared to highlight global standards in EPF risk assessment. Even with strict rules in place, there are still important areas we do not fully understand, especially about how long these substances last, which plants they affect, how toxic the metabolites are, and the risks to humans. The review underscores the urgent need for harmonized safety protocols, robust field data, and genomic tools to enhance risk evaluation and ensure the responsible deployment of EPF in sustainable agriculture.

M. Ghorui · S. Chowdhury
Symbiotic Sciences Pvt Ltd, Gurgaon, Haryana, India

B. Mukherjee
Shaheed Rajguru College of Applied Sciences for Women, University of Delhi, Delhi, India

P. Balu
Department of Biotechnology, School of Life Sciences, Vels Institute of Science, Technology and Advanced Studies (VISTAS), Chennai, Tamil Nadu, India

S. Burla (✉)
ATGC Biotech Pvt. Ltd., Hyderabad, India

S. K. Deshmukh et al. (eds.), *Frontiers in Entomopathogenic Fungi*,
https://doi.org/10.1007/978-981-95-6529-0_17

Keywords

Health risks · Horizontal transfer · Non-target toxicity · Persistence · Regulatory frameworks

1 Introduction

Pests are responsible for the loss of 20% to 40% of global crop production annually. Invasive insect species alone result in an estimated economic cost of approximately $70 billion worldwide (Gula 2023). Crop protection makes up just 5–8% of total farming costs, yet its effectiveness varies by pest type. Managing pests leads to up to 32% yield enhancements. Protection is generally more effective in cash crops, preventing 53–68% of losses, versus 43–50% in food crops. With climate change and global trade accelerating, pest impacts are expected to rise—yield losses may increase by 10–25% for every 1 °C rise in temperature (Gordon 2023).

Since pests cause significant damage to agriculture, farmers often use pesticides as a quick and effective method to protect their crops and reduce losses. In the USA, nearly 1.3 billion USD of pesticides are used annually to combat weeds, insects, fungi, and other pests threatening crops (TraceXtech Technologies 2024). Pesticides, including herbicides, insecticides, and fungicides, are formulated to target pests, offering immediate relief from infestations that can otherwise devastate harvests. However, the benefits of pesticides come with considerable environmental costs. These chemicals do not always remain confined to the area where they are applied. Through runoff and leaching, pesticides can contaminate nearby rivers, lakes, and groundwater. Aquatic organisms, including fish, amphibians, and insect larvae, are particularly vulnerable to pesticide exposure, which can lead to population declines and biodiversity loss. Furthermore, pesticides often affect non-target organisms, such as pollinators and natural pest predators and parasitosis, disrupting ecological balance. Many synthetic pesticides are persistent, leading to bioaccumulation—a process where toxic substances build up in organisms over time and transfer up the food chain, impacting predators and top consumers. The case of DDT is a powerful reminder of these dangers: the pesticide weakened eggshells in birds of prey like bald eagles, pushing them toward near extinction before its ban (US EPA 2025). Similarly, over 120 countries, under the Stockholm Convention, agreed to phase out the toxic pesticide endosulfan, an organochlorine-based insecticide, by 2012, permitting limited use on specific crop-pest combinations, such as cotton for bollworm control, until 2017 (Hogue 2011).

The Directorate of Plant Protection, Quarantine and Storage, India, has circulated a document titled 'List of Banned, Refused Registration, and Restricted Pesticides in India (as of 31.03.2024)', which outlines key regulatory measures taken by the Indian government to ensure pesticide safety. A total of 49 pesticides have been completely banned for manufacture, import, and use due to their high toxicity and environmental persistence; these include chemicals such as Endosulfan, Aldicarb, Carbaryl, Paraquat, Phorate, and Triazophos. Additionally, some pesticide formulations like Dichlorvos and Captafol 80% powder are banned for domestic use but still permitted for export. The list also includes eight pesticides that have been withdrawn from use until the required safety data is submitted and approved, such as Simazine and Warfarin. Furthermore, 18 pesticides have been refused registration altogether, including 2,4,5-T, lead arsenate, and azinphos methyl, meaning they cannot be produced, sold, or used in the country. Several pesticides are also restricted in their usage; for instance, monocrotophos is banned on vegetables, DDT is limited to public health applications, and aluminum phosphide use is restricted to trained experts under government supervision. These actions reflect India's commitment to safeguarding human health and the environment and aligning with global standards for pesticide management (Directorate of Plant Protection, Quarantine and Storage 2024).

To mitigate these impacts, sustainable pest management approaches are increasingly being promoted, with a focus on reducing chemical inputs and restoring ecological balance (Stellarix 2024). One promising solution is the use of EPF—beneficial microorganisms that naturally infect and kill insect pests (Ghorui et al. 2024). EPF, such as *Beauveria bassiana* and *Metarhizium anisopliae*, are environmentally friendly biocontrol agents that offer targeted pest suppression without harming beneficial insects or the surrounding ecosystem. These fungi can be applied to crops much like conventional pesticides, but with significantly lower ecological risks.

Incorporating EPF into IPM strategies can help reduce dependency on chemical pesticides, making agriculture more resilient, sustainable, and aligned with environmental conservation goals. Various companies produce EPN-based products targeting pests like whiteflies, aphids, and thrips, among others. Bioworks Inc., USA, offers PRINCIPLE WP, which contains *B. bassiana* Strain BW149, effective against thrips, aphids, whiteflies, mealybugs, and other pests (Digital 2025). Koppert Biological Systems (Netherlands) markets Mycotal, which is based on *Akanthomyces muscarius* Ve6 strain (formerly known as *Lecanicillium lecanii*), targeting pests such as white fly, thrips, and aphids (Mycotal 2025). Lallemand Plant Care (USA) provides LALGUARD M52 OD (formerly Met52), which includes *Metarhizium brunneum* strain F52 (formerly *M. anisopliae*) and is used for controlling whiteflies, thrips, mites, and aphids (Lallemand Plant Care 2025). BASF (Germany) produces Velifer, which contains *B. bassiana* strain PPRI 5339 and is aimed at managing whiteflies and thrips (BioProtection Portal 2023).

EPF are soil-dwelling organisms that infect and kill insect pests by breaching their outer cuticle and invading their internal tissues, where they grow and consume the host (Dara 2017). These fungi belong to several divisions, including Ascomycota, Zygomycota, Basidiomycota, Glomeromycota, and Chytridiomycota (Mora et al. 2018). EPF play a significant role in regulating insect pest populations within soil ecosystems, employing multiple mechanisms to infect and kill pests, including starvation, toxin production, and enzymatic degradation of the insect cuticle (Dara 2017). These fungi secrete hydrolytic enzymes such as chitinases, proteases, and lipases, which help breach the cuticle—the insect's first line of defense (Vega et al. 2008). Spores or conidia adhere to the insect surface through

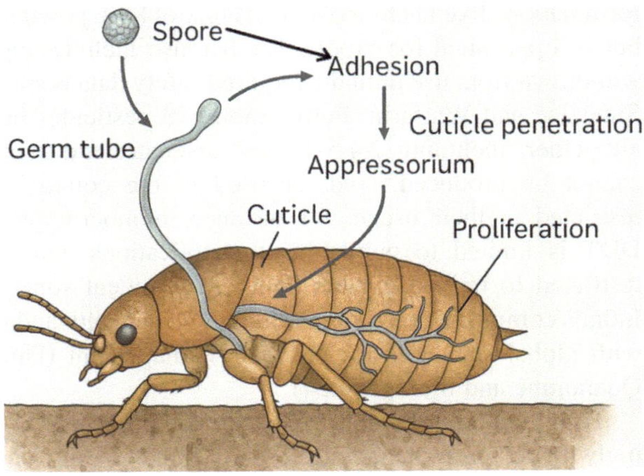

Fig. 17.1 Infection of entomopathogenic fungus on insects

the interaction of surface proteins and host cuticle lipids, sometimes aided by sticky secretions or hydrophobic proteins like hydrophobins (Holder and Keyhani 2005). The overview of infection mechanisms of EPF on insects was illustrated in Fig. 17.1.

Following adhesion, conidia germinate and may form specialized structures like appressoria to physically and enzymatically penetrate the host (Téllez-Jurado et al. 2009). Fungal entry is typically through the cuticle but can also occur via spiracles or sensory organs (St Leger 1991). Once inside the hemocoel, fungal hyphae proliferate, disrupting internal organs and leading to host death (Locke 2001). Virulence is strongly associated with enzyme production, such as PR1 proteases in *M. anisopliae* and chitinases in *B. bassiana* (St Leger 1991; Fan et al. 2007). These traits highlight EPF's potential as biological control agents.

The presence of EPF in natural ecosystems has been widely recognized and documented across the globe (Zimmermann 2007). Over the years, several EPF species have emerged as eco-friendly alternatives to synthetic chemical insecticides, many of which have been phased out due to their environmental hazards, health risks, and the development of resistance among target pest populations (Butt et al. 2001). Some EPF species that have shown commercial potential or are already in use include *M. anisopliae*, *B. bassiana*, *Beauveria brongniartii*, *Lecanicillium* spp. (previously known as *Verticillium* spp.), and *Isaria fumosorosea* (formerly, *Paecilomyces fumosoroseus*). A range of registered biopesticide products based on these fungi has been catalogued, reflecting their practical application in sustainable pest management (de Faria and Wraight 2007). These developments underscore the growing significance of EPF as integral components of IPM strategies, offering sustainable, biologically based solutions for long-term pest management.

The prime objective in testing the applicability and efficacy of chemicals and/or microbiological agents used for plant protection and environmental safety is crucial (Brühl and Zaller 2019; Scheepmaker et al. 2019). Regulatory frameworks aim to distinguish products with acceptable risk profiles, particularly regarding their effects on non-target organisms (Gwynn 2017). In the EU, this is assessed using a tiered system that begins with basic laboratory toxicity tests. If risks appear significant, more advanced tests that simulate real-world conditions are required (Schäfer et al. 2019). Historically, assessment protocols did not differentiate between chemical and microbial agents, overlooking key differences such as specificity, persistence, and the ability of microbes to reproduce (Chandler et al. 2008; OECD 2019). This mismatch often led to unmet data requirements for microbial products, delaying their registration and increasing development costs (Köhl et al. 2019). To address this, separate assessment frameworks have been recommended (Sundh and Goettel 2012). While general guidance on microbial safety has been issued by the OECD (2014), specific protocols, especially for non-target organism testing, are still lacking (Arora et al. 2016). Consequently, there remains a pressing need for the development and harmonization of tailored regulatory protocols that accurately reflect the unique characteristics of microbial agents, thereby facilitating their safe and efficient integration into sustainable agricultural practices.

Biosafety and risk assessment are essential for the responsible use of EPF in pest management. As living organisms, EPF may interact with non-target species or ecosystems in unpredictable ways. Assessing their safety ensures they do not pose risks to human health, beneficial organisms, or the environment. This is particularly important since EPF products differ significantly from chemical pesticides in behavior, persistence, and by-products. Therefore, a thorough and scientifically grounded biosafety and risk assessment framework is crucial to support the safe deployment of EPF, ensuring their benefits in pest management are realized without compromising ecological integrity and/or public health.

This review will critically evaluate the safety profile of EPF, encompassing their effects on non-target organisms, environmental persistence, and risks associated with secondary metabolites, while identifying key knowledge gaps in host specificity, long-term ecological impacts, and the need for harmonized global risk assessment protocols.

2 Safety Assessment

The environmental and animal safety of biological agents has been a recurring concern. Many biochemical compounds produced by fungal endophytes are known to cause adverse effects in livestock. The environmental biosafety of the secondary metabolites and mycotoxins of *Isaria*-based mycopesticides has been the main source of concern (Weng et al. 2019). *B. bassiana, B. brongniartii*, and *M. anisopliae* have all been extensively reviewed for their negative effects on non-target organisms, such as earthworms, silkworms, bees, other pollinators, predators, and parasitoids (Goettel et al. 1989; Danfa and Van Der Valk 1999; Hokkanen and Hajek 2003; Zimmermann 2007). Hence, continuous evaluation of these biological agents, with a focus on their metabolites and impacts on non-target species, remains essential to ensure their safe integration into sustainable pest management practices (Table 17.1).

Table 17.1 Toxicity of entomopathogenic fungi to non-target organisms

Category	Finding	Species affected	Remark	Reference
Pollinators	*Metarhizium*, *Beauveria*, and *Clonostachys* reduced Varroa mites but also reduced adult bee weight. Immune gene expression was triggered.	*Apis mellifera* (honeybee)	EPF may aid mite control but could stress bee physiology	Hamiduzzaman et al. (2012)
	Exposure to *B. bassiana* altered bee recognition, leading to social exclusion.	*Tetragonisca angustula*	Rejected bees at colony entrance, impaired foraging	Almeida et al. (2022)
	Various fungi caused bee mortality; oral exposure more hazardous than contact.	*Scaptotrigona depilis*, *T. angustula*, *A. mellifera*, and *Bombus terrestris*	Strain- and species-specific susceptibility	Leite et al. (2022)
	BoverilWP and MetarrilWP were low to moderately toxic; OctaneSC highly toxic.	*Melipona quadrifasciata*, *P. droryana* and *S. bipunctata*	Risk levels varied by product and bee species	Faita et al. (2023)
Beneficial arthropods	*M. brunneum* showed species-specific toxicity.	*Orius majusculus* (high mortality), *Dalotia coriaria*, *Aphidoletes aphidimyza* (moderate); *Gaeolaelaps aculeifer* (safe)	Importance of species-specific evaluation	De Azevedo et al. (2019)
	Exposure to *Lecanicillium muscarium* and *B. bassiana* reduced aphid longevity and lacewing emergence.	*Aphis gossypii*, *Chrysopa carnea*	Sublethal and transgenerational effects noted	Pavlyushin (2020)
	Early larval and pupal stages of ladybird beetles highly sensitive.	*Coccinella undecimpunctata* and *Hippodamia variegata*	Spray/contact methods more hazardous	Sayed et al. (2021)
	B. bassiana ATCC 74040 showed slight but measurable impacts.	*Chrysoperla lucasina*	Minor sublethal effects; strain mostly compatible	Morda et al. (2024)

(continued)

Table 17.1 (continued)

Category	Finding	Species affected	Remark	Reference
Vertebrates	No toxicity/pathogenicity from *M. anisopliae* via injection, inhalation, oral exposure.	White rats	Early safety confirmation	Schaerffenberg (1968)
	No histological abnormalities; no weight loss or behavior change.	White mice and Guinea pigs	28-day oral exposure	Latch (1976)
	No ocular irritation or tissue infection from spores; fungi cleared by day 21.	Rats, mice, and Guinea pigs	Non-pathogenic	Shadduck et al. (1982)
	M. anisopliae cleared subcutaneously after ~1 month; spores excreted.	Mice and Guinea pigs	Subcutaneous and ingestion routes	El-Kadi et al. (1983)
	B. bassiana detectable for 3 days post-injection; no pathogenicity observed.	Mice	Short-lived presence	Semalulu et al. (1992)
	B. bassiana and *M. anisopliae* considered non-toxic across vertebrate studies.	Rats, birds, and other mammals		Goettel and Jaronski (1997) and Toriello et al. (2005)
	No histological effects in birds consuming spores/infected insects.	Sparrows, pheasants, and hens	No differences in body weight or survival	Althouse et al. (1997), Johnson et al. (2002), and Zimmermann (2007)
	No clinical symptoms or inflammation; minimal germination observed.	CD-1 mice	High-dose oral exposure (10^8 conidia)	Toriello et al. (2005)
	No dermal toxicity of *Isaria fumosorosea* observed over 14 days.	Rats	Confirmed dermal safety	Brunner-Mendoza et al. (2017)

(continued)

Table 17.1 (continued)

Category	Finding	Species affected	Remark	Reference
	Pulmonary exposure to two *M. anisopliae* strains caused high toxicity.	Mice	Strain-specific risk	Goettel et al. (1989)
	Weak sensitization, no acute toxicity, or irritation of *Isaria fumosorosea*.	Rats, rabbits		Weng et al. (2019)
	Occasional human infections (e.g., keratitis) linked to *B. bassiana*.	Humans	Rare but documented	Sachs et al. (1985), Low et al. (1997), and Kisla et al. (2000)

2.1 Toxicity to Non-Target Invertebrates

While EPF are widely regarded as eco-friendly alternatives to chemical pesticides, their environmental safety, especially concerning non-target organisms, requires thorough evaluation. Among the most sensitive and ecologically important non-target groups are pollinators and beneficial arthropods, which play crucial roles in agroecosystem stability and productivity. Therefore, it is imperative to conduct comprehensive toxicity assessments to evaluate the potential risks EPF pose to these vital non-target species, ensuring that their use does not disrupt ecological balance or threaten ecosystem services.

2.1.1 Effects on Pollinators

Pollinators such as honeybees (*Apis mellifera*) and bumblebees (*Bombus* spp.) are essential for the pollination of many food crops. Although EPF generally has a high degree of host specificity, some strains may affect pollinators through direct contact or exposure to fungal spores on treated plants or in the environment. EPF can influence pollinators in both beneficial and harmful ways. Three isolates of *M. anisopliae*, *B. bassiana*, and *Clonostachys rosea* effectively reduced *Varroa destructor* mites, a major parasite affecting honeybee brood. However, direct exposure of brood to *M. anisopliae* or *B. bassiana* reduced adult bee weight, indicating possible harm. Notably, infected mites triggered immune gene expression in bees, countering mite-induced suppression. While EPF can help control *Varroa*, their effects on bee health require careful evaluation (Hamiduzzaman et al. 2012). Exposure of *B. bassiana* to *Tetragonisca angustula*, a native stingless bee, affected their key social behaviors. Bees exposed to fungal conidia were chemically altered, leading nest guards to reject them at the colony entrance. This behavioral response may help prevent the spread of pathogens within the colony but could also reduce foraging efficiency.

These findings suggest EPF can impact bee behavior, highlighting the need for cautious application in agroecosystems (Almeida et al. 2022). Fungal biopesticides are considered low-risk options for pest control, but their impact on pollinators needs evaluation. This study assessed the effects of five concentrations of *B. bassiana*, *M. anisopliae*, and *Cordyceps fumosorosea* on four bee species—*Scaptotrigona depilis*, *Tetragonisca angustula*, *Apis mellifera*, and *Bombus terrestris*—through direct contact and ingestion. All fungi caused varying levels of mortality, with *S. depilis* and *B. terrestris* being more sensitive to *B. bassiana*, and *B. terrestris* and *A. mellifera* showing higher susceptibility to *M. anisopliae* orally. Results highlight potential risks to bees at the individual level under lab conditions (Leite et al. 2022). Stingless bees are key pollinators with environmental and agricultural importance but face risks from insecticide exposure. This study evaluated the impact of three fungal bioinsecticides—*B. bassiana* (Boveril WP), *M. anisopliae* (Metarril WP), and *C. fumosorosea* (Octane SC)—on *Melipona quadrifasciata*, *Plebeia droryana*, and *Scaptotrigona bipunctata*. Mortality varied among species, with *M. quadrifasciata* being most sensitive. While Boveril WP and Metarril WP were considered harmless to moderately harmful, Octane SC was highly toxic. Results highlight that EPF selectivity varies, and careful application is essential to protect pollinators in sustainable farming (Faita et al. 2023).

Laboratory assays often show low acute toxicity, but chronic effects and sublethal impacts such as reduced foraging efficiency or compromised immune responses need further investigation under field conditions. As such, strain selection and formulation refinement are crucial for minimizing unintended effects on pollinator health. Understanding species-specific sensitivity, chronic effects, and behavioral changes is essential to ensuring that EPF applications do not harm essential pollinator populations. Further research on field conditions and the development of selective strains will be key to optimizing the use of EPF in a way that promotes both effective pest control and pollinator conservation.

2.1.2 Impact on Beneficial Arthropods

EPF, though effective in managing insect pests, may also negatively affect beneficial arthropods. Certain fungal species can infect helpful insects such as lacewings and other natural predators, potentially causing their death and disturbing ecological balance, which can undermine integrated biological control efforts. The study assessed the non-target effects of the microbial biological control agent *M. brunneum* applied in soil on four beneficial arthropods: the predatory mite *Gaeolaelaps aculeifer* (Canestrini), the rove beetle *Dalotia coriaria* (Kraatz), the gall midge *Aphidoletes aphidimyza* (Rondani), and the predatory insect *Orius majusculus* (Reuter). *Gaeolaelaps aculeifer* showed no adverse effects, while *Orius majusculus* experienced high mortality and significantly reduced longevity and fecundity. *Dalotia coriaria* and *Aphidoletes aphidimyza* showed intermediate negative responses, including increased mortality and reduced reproductive output. These findings emphasize the importance of evaluating species-specific risks when using entomopathogenic fungi in integrated pest management to prevent unintended harm to beneficial arthropods (De Azevedo et al. 2019).

Studies have shown that EPF, such as *Lecanicillium muscarium* and *Beauveria bassiana*, can exert negative impacts on beneficial arthropods. In aphids (*Aphis gossypii*), commonly used as hosts in biological control systems, exposure to fungal conidia reduced longevity, fertility, and progeny production, with transgenerational effects observed up to the fifth generation. Similarly, lacewings (*Chrysopa carnea*), important predators in agroecosystems, exhibited reduced adult emergence and toxigenic effects in their progeny after exposure to fungal conidia. These effects occurred without visible fungal infection, suggesting the role of fungal metabolites. Such findings highlight the potential for unintended non-target effects and underscore the need for careful integration of EPF in multi-agent biological control strategies (Pavlyushin 2020). The study assessed the impact of an indigenous *Beauveria bassiana* isolate on various life stages of two beneficial predatory beetles, *Coccinella undecimpunctata* and *Hippodamia variegata*. Results revealed that the early developmental stages, particularly the first larval instars of both species and the pupal stage of *C. undecimpunctata*, were negatively affected, with significantly higher mortality observed, especially under the spray and contact methods. These findings suggest that *B. bassiana*, though generally considered safe, can pose risks to vulnerable life stages of beneficial arthropods like predatory ladybird beetles, which are important for biological pest control (Sayed et al. 2021).

The study evaluated the effects of *B. bassiana* strain ATCC 74040 on the predatory insect *Chrysoperla lucasina*, both in laboratory-reared and wild populations. While the results generally indicated compatibility between the fungus and the predator, some slight negative effects were observed on survival, development, or reproduction. Thus, it is important to evaluate case-by-case safety issues prior to field application, even if this *B. bassiana* strain is generally not harmful. It may nevertheless have slight negative effects on beneficial predatory insects (Morda et al. 2024). The unintended consequences on natural predators and other non-target beneficial species, such as reduced survival, fecundity, or developmental delays, underscore the need for species-specific assessments. As such, integrating EPF into integrated pest management strategies requires a nuanced approach to balance pest control efficacy with the preservation of ecosystem services provided by beneficial arthropods. Further research is crucial to fine-tune EPF formulations and application techniques to minimize risks to non-target organisms.

2.2 Toxicity to Vertebrates

Early mammalian safety evaluations of *M. anisopliae* in adult white rats through injection, inhalation, and oral administration revealed no signs of toxicity or pathogenicity (Schaerffenberg 1968). No tissue abnormalities in white mice and guinea pigs fed *M. anisopliae* conidia over 28 days, with no weight loss or abnormal behavior detected during histological analysis (Latch 1976). Research data suggested by demonstrating that rats, mice, and guinea pigs exposed to *M. anisopliae* spores either by injection or exposure did not experience ocular irritation or spore germination in their tissues (Shadduck et al. 1982). Preliminary safety tests of *M. anisopliae* showed no signs of illness, death, or ocular irritation in exposed animals. Tissue reactions were limited to injection sites, with no histological evidence of spore germination. Although fungi were temporarily recovered from certain organs, they were absent by day 21. The results indicated no pathogenicity of *M. anisopliae* in mammals (Shadduck et al. 1982).

Further investigations confirmed that *M. anisopliae* remains non-pathogenic and non-toxic following ingestion, inhalation, and subcutaneous exposure in white mice and guinea pigs, with viable spores being excreted and persisting subcutaneously for about a month (El-Kadi et al. 1983). Mice given an intramuscular injection of *B. bassiana* showed that the fungus was only present in their bodies for 3 days (Semalulu et al. 1992). *Beauveria bassiana* was found to be non-toxic in toxicity tests conducted on rats and other vertebrates (Goettel and Jaronski 1997). Additional studies have consistently reported similar non-toxic outcomes across different animal models (Toriello et al. 2005). Similarly, many bird species have undergone toxicity tests with different strains of EPF, where birds were either given fungal spores blended into their food or insects infected with EPF. For instance, ring-necked pheasants (*Phasianus colchicus Linnaeus*) were exposed to *B. bassiana* conidia, whereas American sparrowhawks (*Falco sparverius Linnaeus*) were fed *B. bassiana* spores (Althouse et al. 1997; Johnson et al. 2002), and hens ate white grubs contaminated with *B. brongniartii*. According to these studies, the birds exposed to EPF did not exhibit any histological abnormalities. Additionally, survival, body weight, and growth indicators did not significantly differ between the treatment and control groups (Zimmermann 2007). The oral toxicity and pathogenicity of *M. anisopliae* var. *anisopliae* were assessed in CD-1 mice. Mice received a single intragastric dose (10^8 conidia/animal) of either viable or nonviable conidia and were monitored through clinical, mycological, and histological assessments. No clinical symptoms, weight loss, or inflammatory reactions were observed. Minimal fungal germination was detected in isolated cases among mice treated with viable conidia. The study concluded that EH-479/2 is nonpathogenic and nontoxic under these conditions (Toriello et al. 2005). An acute cutaneous toxicity test was performed on *Isaria fumosorosea.* Over a 14-day observation period, no clinical signs of disease, skin inflammation, or abnormal weight changes were observed, confirming the dermal safety of this fungal strain (Brunner-Mendoza et al. 2017).

However, contrasting results were reported by Mycotech, a biopesticide company, where two strains of *M. anisopliae* (*var. anisopliae* and *var. acridum*) showed high toxicity in mice following pulmonary exposure (Goettel et al. 1989). Similarly, *I. fumosorosea* was classified as a weak sensitizer, with minimal acute oral, dermal, and inhalation toxicities reported in rats; no eye irritation or dermal sensitization was observed on treated rabbit skin (Weng et al. 2019). On rare occasions, *B. bassiana* has been linked to illnesses in humans, including mycotic keratitis (Sachs et al. 1985; Low et al. 1997; Kisla et al. 2000). Biopesticides based on *Metarhizium* pose a low risk to aquatic organisms. Most EPF, including *M. anisopliae*, *B. bassiana*, and *I. fumosorosea*, are generally non-toxic and non-pathogenic to vertebrates under typical exposure conditions. However, certain strains may show toxicity in specific exposure methods or species. While the overall risk to vertebrates is low, there are reports of toxicity under specific conditions, such as pulmonary exposure or individual sensitivity. Further research is needed to ensure the safe application of EPF-based biopesticides.

2.3 Occupational and Human Health Risks

There is currently no proof that EPF in air or water has harmed human health (Weng et al. 2019). A rare case of human deep tissue infection caused by a *Beauveria* species highlights possible health risks from EPF. While *B. bassiana* is classified as biosafety level 1 and widely used in biocontrol, this clinical strain showed high virulence in insect bioassays. Though it did not grow at 37 °C in vitro, which has caused infections in immunocompromised individuals, it may still pose a risk under certain conditions. Strain-specific differences in virulence suggest that not all EPF strains are equally safe for humans (Henke et al. 2002). While EPF generally pose no health risks, rare infections, such as with *B. bassiana* in immunocompromised individuals, suggest certain strains may pose risks. Strain selection is crucial for ensuring safety.

2.4 Risk Assessment Implications of Secondary Metabolites

Degradation and Stability Secondary metabolites (SMs) of EPF degrade rapidly in submerged media, insect hemolymph (both live and post-mortem), and crops (not detected or below detection limits). They are only stable under specific preserved conditions, which are irrelevant for environmental risk assessment. In natural settings, temperature and UV light cause rapid degradation (OECD 2018).

Endophytic Concerns Some EPF (e.g., *Beauveria*, *Metarhizium*) can occur as endophytes and may theoretically produce SMs within plants. Though this raises potential food chain concerns (Vega 2018), current evidence does not confirm complete fungal life cycles inside plants or arthropod death due to endophytically produced SMs (Gurulingappa et al. 2011). Endophytism may enhance plant productivity via nitrogen translocation (Behie and Bidochka 2013).

Field-Level SM Impact Using *Plutella xylostella* LC_{50} of 17 μg/mL for destruxin E, field application requires 17 mg/L in spray solution. However, destruxin E levels in Microbial Pest Control Agent (MPCA) formulations (97 ± 146 mg/L) diluted for field use result in concentrations ~175 times lower than LC_{50}, indicating negligible risk to target insects (OECD 2018).

Off-Field Risk Drift to off-field areas results in further dilution, especially in vegetated zones. While NTOs (non-target organisms) closely related to targets may be sensitive, the actual exposure remains low. Risk assessment includes consideration of assessment factors and potential additive or synergistic effects of multiple destruxins (e.g., from *Metarhizium* spp.) (European Commission 2002).

While secondary metabolites (SMs) from EPF degrade rapidly in natural conditions and are not a significant environmental risk, the potential impact of endophytic production of SMs and off-field drift should be considered in risk assessments. These factors, along with potential synergistic effects, require careful evaluation to ensure safe use.

3 Residual Pathogenic Effects

EPF are valued for their minimal residual effects on animal health and the environment (Sharma et al. 2023). The effectiveness of these fungal species as biocontrol agents relies not only on their ability to selectively and efficiently eliminate insect pests but also on minimizing or avoiding harmful impacts on beneficial non-target insects in the field (Rizwan et al. 2021). Ongoing evaluation of their residual effects and careful management of their application are essential to ensure minimal environmental and health impacts (Table 17.2).

Table 17.2 Residual effects of entomopathogenic fungi

Category	Key Point	Reference
Crops	*B. bassiana* and *M. anisopliae* have persistent effects on maize through endophytic activity, with residual effects promoting pest resistance and altering plant nutrient composition	Liu et al. (2022) and Krell et al. (2018)
Non-target organisms	EPF can affect non-target insects, including beneficial predators like *Dicyphus tamaninii* and *Chrysoperla carnea*. They may also impact bees and other beneficial species	Broza et al. (2001), James and Lighthart (1994), Ginsberg et al. (2002), and Leite et al. (2022)
	M. brunneum caused high mortality in predatory bugs like *Orius majusculus* and reduced reproductive success in other beneficial arthropods	De Azevedo et al. (2019)
Soil	Soil characteristics influence EPF persistence. Favorable conditions, such as moisture-rich orchard soils, support EPF survival. Soil also protects EPF from UV and temperature fluctuations	Majchrowska-Safaryan and Tkaczuk (2021) and Qayyum et al. (2021)
Water	EPF spores have minimal mobility in water and are unlikely to contaminate groundwater. They can survive in water bodies but have limited proliferation in air over time	Zimmermann (2007) and Shah and Pell (2003)
Air	EPF spores have limited ability to survive or proliferate in the air for extended periods. They are vulnerable to UV radiation and other environmental stressors	Shah and Pell (2003) and Milner et al. (2002)

3.1 Persistence on Crop Surfaces

A study examined the growth-promoting effects of *B. bassiana* and *M. anisopliae* in maize, focusing on whether their action is primarily rhizospheric or endophytic. Fungal populations in hydroponic solution declined rapidly, with less than 10% (*M. anisopliae*) and 1% (*B. bassiana*) remaining after 1 week, and both were nearly undetectable by day 28. In contrast, conidia remained viable in plant-free controls, suggesting that the decline was due to endophytic colonization. This highlights that the residual effects of these fungi on crops are mainly due to their endophytic activity rather than rhizospheric presence (Liu et al. 2022).

EPF can induce lasting changes in plant nutrient composition and defensive metabolites, creating a less favorable environment for herbivores. Such residual pathogenic effects—through sustained production of deterrent compounds—reduce the plant's suitability as a food source and enhance its ability to repel pests over time (Krell et al. 2018). The persistence of EPF on crop surfaces, primarily through endophytic activity, plays a crucial role in enhancing plant defenses and reducing pest suitability, offering sustained pest management benefits with minimal environmental impact.

3.2 Risk of Exposure to Non-Target Insects

The use of EPF in pest control carries the potential risk of adversely affecting non-target insects, including beneficial natural enemies, thereby raising concerns about unintended ecological impacts. *M. anisopliae* caused mortality in *Dicyphus tamaninii* (10%), a predatory mirid, and *Chrysoperla carnea* (4%), a green lacewing (Broza et al. 2001). These findings highlight the risk of EPF affecting non-target insects, as different fungal species show varying pathogenicity. Previous studies confirm that EPF, like *B. bassiana*, can impact non-target species such as collembolans (springtails) and coccinellids (ladybugs), with some fungi causing significant mortality and others having minimal effects (James and Lighthart 1994; Dromph and Vestergaard 2002). *M. anisopliae* showed significant mortality in *Hippodamia convergens* (ladybird beetles) and *Acheta domesticus* (house crickets), with marginal effects on *Oncopeltus fasciatus* (milkweed bugs). Infected insects transmitted the fungus to untreated beetles and crickets, highlighting the risk of non-target effects. While *M. anisopliae* is widespread in North America, field trials are needed to confirm these lab results and assess methods to minimize non-target exposure (Ginsberg et al. 2002).

Beauveria bassiana, *Metarhizium anisopliae*, and *Cordyceps fumosorosea* pose a potential risk to non-target bees. *B. bassiana* was more likely to affect *Bombus terrestris* and *Scaptotrigona depilis*, but *M. anisopliae* was more likely to affect *Apis mellifera* and *B. terrestris* through oral exposure. Mortality did not always increase with higher fungal concentrations, suggesting the risk may not always increase with dosage. While this study indicates that EPFs can impact bee survival, further field studies are needed to fully assess the ecological risks to non-target organisms like bees (Leite et al. 2022). *M. brunneum* (BIPESCO 5/F52), though effective for pest control, may affect non-target beneficial arthropods. Soil applications led to high mortality in the predatory bug *Orius majusculus* (96%) and moderate mortality in the rove beetle *Dalotia coriaria* (7.3%), with reduced longevity and fecundity in both. The gall midge *Aphidoletes aphidimyza* showed increased larval mortality (60%) but unaffected fecundity, while the predatory mite *Gaeolaelaps aculeifer* remained unaffected. These findings highlight the need for multi-parameter and multi-species assessments to gauge non-target risks of biocontrol agents (De Azevedo et al. 2019). Bioassays on non-target insects (honeybees and cockroaches) with local Metarhizium strains showed Less than 5% mortality was recorded, and no mycosis was observed, indicating minimal non-target effects (Bilgo et al. 2018).

3.3 Environmental Fate and Persistence

The effectiveness of EPF under field conditions is greatly influenced by environmental parameters such as temperature, humidity, and UV radiation. Cyclobutane pyrimidine dimers and other photoproducts are DNA damages brought on by UV radiation that can impair the viability and function of fungal cells. Even though filamentous fungi have photolyase enzymes that can repair DNA in the presence of visible light, their effectiveness decreases in the presence of bright light (Yasui et al. 1994; de Laat et al. 1999; Sancar 2003). The ideal temperature range for fungal germination and growth is usually 23–28 °C (Jaronski 2009). Faster conidial germination is also facilitated by high humidity. *B. bassiana* took 72 h at 90% relative humidity to germinate, but just 20 h at 25 °C and 95.5% RH (Luz and Fargues 1997).

The biodiversity or uneven localization of EPF in soils due to climatic and geographic factors significantly impacts fungal endophytes and EPF use. Fungal spore mobility and dissemination in soil are influenced by soil moisture content, soil type, soil organisms, and plant roots (Cabrera-Mora et al. 2019). The possibility that mycopesticide formulations and EPF will float into water bodies or the atmosphere is a common concern when using them in biological control. The downward movement of both wet and dry spores of *M. anisopliae* concluded that there is little chance of the fungus contaminating groundwater (Zimmermann 2007). Their longevity is aided by the production of resting structures or by adopting a saprophytic lifestyle, allowing them to survive extended periods in the absence of hosts. EPF is known to exist in a variety of habitats, including aquatic and terrestrial ecosystems, as well as arctic and tropical climates (Milner et al. 2002). This is mostly because the spores have a limited capacity to survive or proliferate under air conditions over an extended period (Shah and Pell 2003).

The persistence and dispersal of EPF are influenced by a variety of biotic and abiotic factors. Fungal development and infection processes are influenced by biotic factors, such as the characteristics of the insect host and plant-mediated effects (Cory and Ericsson 2009). Abiotic factors such as temperature (Bayissa et al. 2017), solar radiation (Fernández-Marín et al. 2006), humidity and rainfall (Inglis et al. 2001), and soil properties (Quesada Moraga et al. 2007) directly affect the viability and infectivity of fungal spores. These factors collectively determine the ability of the fungus to persist, spread, and cause epizootics under field conditions. Soil serves as both a refuge and a challenge for EPF. It offers protection from harmful UV radiation and buffers against fluctuations in temperature and moisture (Inglis et al. 2001; Rangel et al. 2005), thereby supporting the long-term survival of EPF. Additionally, soil hosts a wide range of potential insect hosts, often in high densities, which promotes the persistence and evolutionary adaptation of EPF to their environment (Humber 2008).

However, soil also contains antimicrobial compounds produced by resident microbes, which can inhibit fungal infectivity. For instance, higher levels of soil fungi stasis have been linked to reduced virulence of *B. bassiana* against the Colorado potato beetle (Groden and Lockwood 1991). Despite these antagonistic pressures, many EPF species persist by producing secondary metabolites within their insect hosts, aiding in competition against opportunistic microbes during the saprotrophic phase (Strasser et al. 2000). These adaptations highlight the ecological resilience and persistence of EPF within the complex soil environment. A large-scale comparative study assessed EPF occurrence in leaf litter and underlying soil across three forest types. *Beauveria*, *Cordyceps*, *Metarhizium*, and *Lecanicillium* spp. were detected in both habitats, with *Beauveria* being the most frequent. Even though leaf litter typically had greater CFUs, the steady recovery of EPF from soil layers shows how persistent they are in forest soils. *Lecanicillium* was the second most abundant genus in Polish forests (Majchrowska-Safaryan and Tkaczuk 2021).

Studies from southern Punjab, based on soil and insect cadaver samples from both natural and cultivated areas, indicate that soil characteristics greatly influence EPF presence and diversity. EPF were consistently isolated from orchard soils, suggesting these less-disturbed and moisture-rich environments support their persistence. Among 225 insect cadavers collected, EPF were most frequently associated with Coleoptera and Lepidoptera. Although *Aspergillus niger* and *Fusarium oxysporum* were the most common fungal isolates overall, the findings underscore that orchard soils and specific insect hosts are more favorable for isolating persistent EPF populations (Qayyum et al. 2021).

4 Genetic Stability and Horizontal Transmission

Metarhizium robertsii gained entomopathogenic traits through horizontal gene transfer (HGT), acquiring 18 genes essential for infecting insects. These genes, especially those involved in cuticle degradation and lipid utilization, enabled the fungus to expand its host range. This highlights how HGT enhances EPF adaptability, potentially impacting gene stability and altering ecological interactions in the environment. Insect cadavers can spontaneously produce entomopathogenic fungal spores that can travel up to 1000–1500 times their own size or, in the case of *Entomophthora muscae*, roughly 5–10 mm. These spores can then spread out through the air and land on adjacent plant surfaces (Małagocka et al. 2015; Ebani and Mancianti 2021; Naundrup et al. 2022). While not immediately infectious, resting spores produced by many fungi in Hypocreales and Entomophthorales can survive in the environment for lengthy periods without a host (Jaronski 2013). These dormant spores either germinate to release infectious spores or directly infect new hosts when the right conditions are met.

4.1 Regulatory Guidelines and Risk Assessment Frameworks

Regulatory approval of microbial pesticides, including EPF, typically demands comprehensive data on strain identity, purity, and production methods, along with detailed safety evaluations. Key jurisdictions such as the USA, EU, Canada, and Australia have established rigorous guidelines addressing environmental behavior, non-target effects, exposure risks, metabolite production, persistence, and host specificity. Table 17.3 below outlines country-specific requirements across these categories, referencing applicable regulatory frameworks.

4.2 United States Environmental Protection Agency

Environmental Protection Agency (EPF) as microbial pesticides fall under the biopesticide category regulated by the EPA. The relevant guidelines are outlined in 40 CFR Part 158 Subpart V, which addresses microbial pesticide data requirements (EPA 2002).

Environmental Fate All EPF uses must include data on viability and persistence in soil and water. If Tier I (initial non-target toxicity) tests raise concerns, Tier II studies such as soil persistence, mobility, and photolysis may be required.

Toxicity Testing Tier I acute toxicity/pathogenicity tests are generally required for outdoor use on non-target organisms—such as birds, fish, aquatic invertebrates, bees, and beneficial arthropods. Higher-tier tests (Tier II–IV) are only triggered if adverse effects are noted. Protocols are detailed under EPA's OPPTS (*Office of Prevention, Pesticides, and Toxic Substances*) 885 series (EPA 2025).

Exposure Scenarios EPA requires evaluation of exposure under typical outdoor use, including spray drift and runoff risks. Standard exposure models and spray-drift studies (e.g., nozzle characterization) are mandated when outdoor applications are proposed.

Metabolites and Toxins Any known or potential secondary metabolites or toxins produced by the EPF must be identified and assessed. If the fungal strain is associated with toxin-producing species (e.g., destruxins from *Metarhizium* or beauvericin from *Beauveria*), sensitive analytical methods and toxicological data must be submitted.

Table 17.3 Regulatory frameworks of entomopathogenic fungi

Country	USA (EPA)	EU (EFSA)	Canada (PMRA)	Australia (APVMA)
Environmental fate data	Required under 40 CFR Part 158 Subpart V. Tier II triggered by Tier I outcomes. Must assess soil and water persistence, viability	Reg. 283/2013 Part 7 requires persistence/ multiplication studies in soil and water. PECs modeled using FOCUS	Tiered: Tier II/III triggered by Tier I outcomes. Persistence in soil/water assessed via microcosms	Tailored requirements. Persistence and movement assessed via lab and field studies as needed
Non-target toxicity testing	Tier I test required for birds, fish, aquatic invertebrates, bees, earthworms and others. Higher-tier tests if needed	Required for birds, bees, soil organisms and aquatic organisms. Based on OECD guidelines	Tier I: Acute tests for birds, fish, bees, earthworms. Higher-tier if effects noted	Required for birds, fish, aquatic invertebrates, bees. Based on same taxa as chemical pesticides
Exposure scenarios	Outdoor use requires drift/ runoff exposure modeling. Subpart L mandates spray-drift studies	PECs based on intended use. Standard FOCUS exposure models used	Exposure calculated for spray/runoff under agricultural/ forestry scenarios	Use-based exposure modeling; typically aligned with proposed agricultural uses
Metabolites/ toxins	Must assess production of toxins (e.g., destruxins, beauvericin); toxicological data required if present	Must identify and quantify toxic metabolites (e.g., destruxins, cyclosporins). Residue definitions include toxins	Metabolite analysis mandatory. Sensitive tests for known toxins (e.g., destruxins) required	Identification and toxicology data required for any secondary metabolites.
Persistence evaluation	Tier II fate testing to assess viability/ survival. Field decline data if needed	Persistence/ multiplication assessed through lab and field studies. Required for non-indigenous EPF	Tier II includes studies on viability decline. Additional field data if persistence >30 days	Data required if persistence beyond typical microbial decay. Decline demonstrated in lab/field.
Host range evaluation	Must evaluate impacts on beneficial and related insects. Host range inferred from non-target tests	Host specificity inferred from non-target arthropod tests and insect order susceptibility data	Host specificity inferred from effects on ecologically/ taxonomically related species	Host spectrum inferred from effects on beneficials and related pests

(continued)

Table 17.3 (continued)

Country	USA (EPA)	EU (EFSA)	Canada (PMRA)	Australia (APVMA)
Reference(s)	40 CFR 158.2150 ǀ US EPA 2002; Series 885—Microbial Pesticide Test Guidelines ǀ US EPA 2025, EPA 2011	Regulation—1107/2009—EN—EUR-LEX 2009; European Commission and Barroso 2013; EFSA 2025	Submission Management and Information Division, PMRA 2001	*Guideline for the Regulation of Biological Agricultural Products*, 2022

Persistence EPA assesses how long the EPF remains viable and whether it replicates in natural media like soil and water. Assessments of environmental risk factors should be supported by field persistence data.

Host Range Host specificity is indirectly evaluated through non-target arthropod testing. A broader-than-expected host range in lab tests may prompt higher-tier specificity studies to ensure minimal impact on non-target and beneficial species.

4.3 European Union—European Food Safety Authority

EPF, categorized as microbial active substances, are regulated under Regulation (EC) No 1107/2009 and its implementing acts (Regulations 283/2013 and 284/2013). Applicants must provide comprehensive data on the microorganism's identity, strain, purity, and biological characteristics.

Environmental Fate According to Annex II of Regulation 283/2013, studies must assess EPF persistence and multiplication in soil and water, including photolysis potential. If the organism is native to the environment, field studies may be needed to confirm natural decline. EFSA guidance and OECD protocols are used to interpret these findings (EFSA 2025).

Toxicity Testing Regulation 283/2013 requires acute and chronic toxicity/pathogenicity testing on a wide range of non-target organisms, like birds, aquatic life, bees, soil invertebrates, and soil microbes. These tests must follow OECD guidelines tailored for microbial agents.

Exposure Scenarios Risk assessments use standard FOCUS models to estimate environmental exposure concentrations (PECs) based on application rates and use patterns. These PECs, combined with toxicity data, help evaluate risks to non-target species.

Metabolites and Toxins EPF capable of producing bioactive metabolites (e.g., destruxins, beauvericin, cyclosporins) must be assessed for their presence in the product and environment. The residue definition includes the microorganism and any pertinent metabolites for regulatory purposes.

Persistence EU regulations require data on the microorganism's ability to multiply and persist in the environment. If lab data suggest prolonged survival, field studies may be requested to confirm that the organism will not spread uncontrollably in non-target habitats.

Host Range While a dedicated host range test is not specified, data on infectivity across insect groups and effects on beneficial arthropods (pollinators, predators, parasitoids) are used to assess specificity and ecological safety.

4.4 Canada—Pest Management Regulatory Agency

EPF are classified as Microbial Pest Control Agents (MPCAs) under Health Canada's Pest Management Regulatory Agency (PMRA) Directive DIR2001–02, which aligns closely with US EPA standards (PMRA 2001).

Environmental Fate Under Part 8 of the Directive, fate studies are tiered based on toxicity findings. If Tier I toxicity data show potential risks, Tier II (lab-based) or Tier III (field-based) tests assess survival, reproduction, and persistence under simulated environmental conditions (soil and water). Non-native or genetically modified strains face stricter scrutiny.

Toxicity Testing A tiered system is used. Acute toxicity/pathogenicity testing on fish, birds, mammals, aquatic invertebrates, bees, and earthworms is required by Tier I. If risks are indicated, Tier II introduces chronic tests, and higher-tier (Tier III/IV) field studies may follow. Protocols cover oral/inhalation studies for birds, 28-day fish tests, and bee contact/oral toxicity.

Exposure Scenarios Risk assessments consider typical use patterns and include modeling of spray drift, runoff, and exposure across soil, water, and air. Applicants must provide realistic environmental concentration estimates based on application rates.

Metabolites and Toxins Applicants must disclose and test for any toxic metabolites. If the EPF is related to known mycotoxin producers (e.g., fungi producing fumonisins or ophiobolins), sensitive assays must prove such toxins are absent or within acceptable levels.

Persistence Persistence data (soil and plant) assess whether the EPF can survive and multiply in the environment. A long half-life (e.g., >30 days) or ongoing reproduction triggers further field studies to ensure environmental safety.

Host Range Testing must include non-target arthropods that are ecologically or taxonomically close to the target pest. This data supports defining the EPF's host specificity. If Tier I tests show broad effects, additional higher-tier specificity testing is required.

4.5 Australia—Australian Pesticides and Veterinary Medicines Authority

In Australia, EPF are regulated as microbial agricultural products under the Australian Pesticides and Veterinary Medicines Authority (APVMA) Guideline for the Regulation of Biological Agricultural Products (Sect. 3.2). EPF are considered microbial pesticides and subject to a tailored risk assessment approach that aligns with, but is distinct from, chemical pesticide evaluations (APVMA 2022).

Environmental Fate Environmental fate data focus on persistence and movement in soil and water. If lab results suggest the EPF may persist, field or glasshouse dissipation studies are required to track decline under realistic use conditions.

Toxicity Testing Toxicity assessments are required for birds, fish, aquatic invertebrates, bees, and beneficial insects, particularly for outdoor applications. Section 4.4 of the guideline provides ecotoxicology test recommendations. If early tests show no adverse effects, additional testing may be waived.

Exposure Scenarios Applicants must provide exposure estimates under worst-case scenarios, especially for spray applications. APVMA requires spray drift analysis (as per Item 40 of their guideline) and exposure assessments for bystanders and aquatic systems in public health or non-agricultural uses.

Metabolites and Toxins Any secondary metabolites or toxins must be identified and characterized. If the fungal species is associated with known toxins, applicants must provide toxicological data or proof of absence.

Persistence Evaluation APVMA evaluates both shelf life (product stability) and environmental persistence. Though microbial products often degrade faster than chemicals, any potential for EPF reproduction and persistence in the environment must be documented and appropriately managed.

Host Range Applicants must submit host specificity data, usually drawn from efficacy studies and non-target insect testing. If the EPF affects non-target insects (e.g., native beetles, parasitoids), labeling restrictions or further testing may be required.

4.6 Brazil

Regulating Bodies Ministério da Agricultura, Pecuária e Abastecimento–Ministry of Agriculture (MAPA), Instituto Brasileiro do Meio Ambiente e dos Recursos Naturais Renováveis (IBAMA)—Brazilian Institute of Environment and Renewable Natural Resources) and Agência Nacional de Vigilância Sanitária (ANVISA)—National Health Surveillance Agency (MAPA 2025; IBAMA 2025; ANVISA 2025).

EPF Category Biological Pesticides (a subset of microbial pesticides).

Key Requirements Strain identity, non-target toxicity (birds, bees, fish, beneficials), environmental fate (soil persistence), and known mycotoxins.

Regulatory Basis National guidelines mirror OECD (*Organisation for Economic Co-operation and Development*) and FAO (*Food and Agriculture Organization*) principles.

4.7 India

Regulating Body Central Insecticides Board and Registration Committee (CIBRC) under the Insecticides Act (1968) (CIBRC 2025).

EPF Category Microbial Biopesticide

Key Requirements Physicochemical characterization, pathogenicity/toxicity (target and non-target insects, mammals, fish, bees, and soil microbes), toxin analysis (demonstrate absence of known toxic metabolites).

Guidance Document Annex D of Biopesticide Guidelines (2011)

4.8 China

Regulating Body Institute for the Control of Agrochemicals, Ministry of Agriculture (ICAMA) (ICAMA 2025).

EPF Category Microbial Biopesticide

Key Requirements Strain ID, viability, growth conditions, non-target safety (birds, fish, mammals, bees), and environmental fate (soil and water).

Approach Harmonized with OECD and EPA standards.

4.9 Japan

Regulating Body Ministry of Agriculture, Forestry and Fisheries (MAFF) (MAFF 2014).

EPF Category Microbial Pesticide.

Key Requirements Acute oral/contact toxicity, pathogenicity in mammals, birds, and bees; soil and water persistence; host-range testing (minimum two insect orders).

Approach Based on OECD guidelines; covered in MAFF's Plant Protection Guidebook.

4.10 Other Countries (South Africa, New Zealand, ASEAN)

EPF Category Generally regulated as Microbial Biopesticides

Key Requirements Identity, non-target toxicity (bees, aquatic organisms, natural enemies), environmental fate (soil/water), and metabolite/toxin assessment.

Approach Aligned with OECD consensus; data from the USA/EU accepted in many cases.

5 Knowledge Gaps and Future Research Directions

Despite the growing application of EPF as environmentally friendly alternatives to chemical pesticides, several critical knowledge gaps remain regarding their safety assessment and potential residual pathogenic effects. One major area of concern is the long-term persistence of EPF spores in different environmental compartments such as soil, water, and plant surfaces, as current data are insufficient to predict their ecological longevity or potential accumulation. There is also a notable lack of long-term field safety data, which limits our ability to understand the real-world ecological impacts of repeated EPF applications over time. Additionally, there is limited understanding of their effects on non-target organisms, including beneficial insects like pollinators, soil fauna, aquatic invertebrates, and vertebrates such as birds and amphibians.

Human health risk assessments require more comprehensive data, particularly concerning exposure through dermal contact, inhalation, or ingestion—especially among agricultural workers and consumers. Another emerging issue is the lack of studies exploring the genetic stability of EPF strains and the possibility of horizontal gene transfer, which could alter their host range or pathogenicity. Furthermore, the toxicological profiles of secondary metabolites produced by EPF are not fully understood, raising concerns about unintended impacts on the environment and non-target species. There is also a significant gap in assessing fungal residues on edible crops, with no standardized protocols currently available to evaluate the viability or safety of these residues at harvest.

To address these issues, future research should prioritize the development of harmonized and standardized testing protocols for EPF safety and efficacy, both in laboratory and field settings. Genomic approaches hold great promise for identifying markers of virulence and safety, enabling more targeted screening and monitoring of EPF strains. Additionally, the development of safer formulations and delivery systems can help minimize non-target exposure and enhance field performance. Establishing long-term field monitoring systems and investigating interactions between EPF and native microbiota are also essential to ensure the responsible and sustainable use of these biological control agents.

6 Conclusion

EPF represent a valuable tool in sustainable pest management strategies. However, ensuring their environmental and human safety requires a comprehensive understanding of their ecological interactions, genetic stability, and toxicological profiles. While global regulatory frameworks provide a structured approach for EPF assessment, notable gaps persist in field-based safety data, secondary metabolite risk evaluation, and long-term persistence in various ecosystems. Addressing these challenges through standardized protocols, molecular characterization, and long-term monitoring will be essential to harness the full potential of EPF while safeguarding environmental and public health. A balanced, science-based regulatory approach—integrating risk-benefit analyses and international cooperation—will be key to fostering responsible innovation in biopesticide development.

References

Almeida FCR, Magalhães DM, Favaris AP, Rodríguez J, Azevedo KEX et al (2022) Side effects of a fungus-based biopesticide on stingless bee guarding behaviour. Chemosphere 287:132147. https://doi.org/10.1016/j.chemosphere.2021.132147

Althouse CM, Petersen BE, McEwen LC (1997) Effects on young American kestrels (Falco sparverius) exposed to *Beauveria bassiana* bioinsecticide. Bull Environ Contam Toxicol 59(4):507–512. https://doi.org/10.1007/s001289900507

ANVISA (2025) Agência Nacional De Vigilância Sanitária – Anvisa. Retrieved April 29, 2025, from https://www.gov.br/anvisa/en

APVMA (2022) Guideline for the regulation of biological agricultural products. Australian pesticides and veterinary medicines authority. Retrieved April 29, 2025, from https://www.apvma.gov.au/registrations-and-permits/data-requirements/agricultural-data-guidelines/biological#data-requirements-for-groups-1-and-2-biologically-derived-chemicals-

Arora NK, Verma M, Prakash J, Mishra J (2016) Regulation of biopesticides: global concerns and policies. In: Arora N, Mehnaz S, Balestrini R (eds) Bioformulations: for sustainable agriculture. Springer, New Delhi, pp 283–299. https://doi.org/10.1007/978-81-322-2779-3_16

Bayissa W, Ekesi S, Mohamed SA, Kaaya GP, Wagacha JM et al (2017) Selection of fungal isolates for virulence against three aphid pest species of crucifers and okra. J Pest Sci 90:355–368. https://doi.org/10.1007/s10340-016-0781-4

Behie SW, Bidochka MJ (2013) Ubiquity of insect-derived nitrogen transfer to plants by endophytic insect-pathogenic fungi: an additional branch of the soil nitrogen cycle. Appl Environ Microbiol 80(5):1553–1560. https://doi.org/10.1128/aem.03338-13

Bilgo E, Lovett B, St. Leger RJ, Sanon A, Dabiré RK et al (2018) Native entomopathogenic *Metarhizium* spp. from Burkina Faso and their virulence against the malaria vector *Anopheles coluzzii* and non-target insects. Parasit Vectors 11:1–6. https://doi.org/10.1186/s13071-018-2796-6

BioProtection Portal (2023) Product details – BioProtection Portal. https://bioprotectionportal.com/productdetails/velifer-86473235/?country=pt&utm _source=google&utm_medium=cpc&gad_source=1&gad_campaignid=22470551435&gbraid=0AAAAApzdjevm0FkyNA5LzajX88n3rDG8f&gclid=CjwKCAjwz_bABhAGEiwAm-P8YYgucRYTIU1iK54GOw_ Om8T2I6AUFp9eOw4FWfHRNVbmdy3lteh5MBoCUNUQAvD_BwE

Broza M, Pereira RM, Stimac JL (2001) The nonsusceptibility of soil Collembola to insect pathogens and their potential as scavengers of microbial pesticides. Pedobiologia 45(6):523–534. https://doi.org/10.1078/0031-4056-00104

Brühl CA, Zaller JG (2019) Biodiversity decline as a consequence of an inappropriate environmental risk assessment of pesticides. Front Environ Sci 7:177. https://doi.org/10.3389/fenvs.2019.00177

Brunner-Mendoza C, Navarro-Barranco H, León-Mancilla B, Pérez-Torres A, Toriello C (2017) Biosafety of an entomopathogenic fungus *Isaria fumosorosea* in an acute dermal test in rabbits. Cutan Ocul Toxicol 36(1):12–18. https://doi.org/10.3109/15569527.2016.1156122

Butt TM, Jackson C, Magan N (eds) (2001) Fungi as biocontrol agents: progress, problems and potential. CABI publishing. https://doi.org/10.1079/9780851993560.0000

Cabrera-Mora JA, Guzmán-Franco AW, Santillán-Galicia MT, Tamayo-Mejía F (2019) Niche separation of species of entomopathogenic fungi within the genera *Metarhizium* and *Beauveria* in different cropping systems in Mexico. Fungal Ecol 39:349–355. https://doi.org/10.1016/j.funeco.2019.02.008

Chandler D, Davidson G, Grant WP, Greaves J, Tatchell GM (2008) Microbial biopesticides for integrated crop management: an assessment of environmental and regulatory sustainability. Trends Food Sci Technol 19(5):275–283. https://doi.org/10.1016/j.tifs.2007.12.009

CIBRC (2025) Directorate of plant protection, quarantine and storage. Retrieved April 29, 2025, from https://ppqs.gov.in/divisions/central-insecticides-board-registration-committee

Cory JS, Ericsson JD (2009) Fungal entomopathogens in a tritrophic context. In: Roy HE, Vega FE, Chandler D, Goettel MS, Pell J, Wajnberg E (eds) The ecology of fungal Entomopathogens. Springer, Dordrecht, pp 75–88. https://doi.org/10.1007/978-90-481-3966-8_6

Danfa A, Van der Valk HC (1999) Laboratory testing of *Metarhizium* spp. and *Beauveria bassiana* on Sahelian non-target arthropods. Biocontrol Sci Technol 9(2):187–198. https://doi.org/10.1080/09583159929776

Dara SK (2017) Entomopathogenic microorganisms: modes of action and role in IPM. UCANR eJ Entomol Biol:1–7. https://www.researchgate.net/publication/317097773

De Azevedo AG, Eilenberg J, Steinwender BM, Sigsgaard L (2019) Non-target effects of *Metarhizium brunneum* (BIPESCO 5/F 52) in soil show that this fungus varies between being compatible with, or moderately harmful to, four predatory arthropods. Biol Control 131:18–24. https://doi.org/10.1016/j.biocontrol.2019.01.002

de Faria MR, Wraight SP (2007) Mycoinsecticides and mycoacaricides: a comprehensive list with worldwide coverage and international classification of formulation types. Biol Control 43(3):237–256. https://doi.org/10.1016/j.biocontrol.2007.08.001

de Laat WL, Jaspers NG, Hoeijmakers JH (1999) Molecular mechanism of nucleotide excision repair. Genes Dev 13(7):768–785. https://doi.org/10.1101/gad.13.7.768

Digital A (2025) PRINCIPLE WP – BioWorks. BioWorks. Retrieved May 9, 2025, from https://bioworksinc.com/products/principle/

Directorate of Plant Protection, Quarantine and Storage (2024) List of banned, refused registration, and restricted pesticides in India (as of 31.03.2024). In Directorate of plant protection, quarantine and storage. Retrieved May 9, 2025, from https://ppqs.gov.in/sites/default/files/list_of_pesticides_which_are_banned_refused_registration_and_restricted_in_use_0.pdf

Dromph KM, Vestergaard S (2002) Pathogenicity and attractiveness of entomopathogenic hyphomycete fungi to collembolans. Appl Soil Ecol 21(3):197–210. https://doi.org/10.1016/s0929-1393(02)00092-6

Ebani VV, Mancianti F (2021) Entomopathogenic fungi and bacteria in a veterinary perspective. Biology 10(6):479. https://doi.org/10.3390/biology10060479

EFSA (2025) European Food Safety Authority guidance and other assessment methodology documents. Retrieved April 29, 2025, from https://www.efsa.europa.eu/en/methodology/guidance

El-Kadi MK, Xara LS, De Matos PF, Da Rocha JV, De Oliveira DP (1983) Effects of the entomopathogen *Metarhizium anisopliae* on Guinea pigs and mice. Environ Entomol 12(1):37–42. https://doi.org/10.1093/ee/12.1.37

EN-EUR-LEX (2009) Retrieved April 29, 2025, from https://eurlex.europa.eu/eli/reg/2009/1107/oj/eng

EPA (2002) 40 CFR 158.2150 – Microbial pesticides nontarget organisms and environmental fate data requirements table. Code of Federal Regulations. Retrieved April 29, 2025, from https://www.ecfr.gov/current/title-40/part-158/section-158.2150

EPA (2011) *Metarhizium anisopliae strain* F52 (029056) Biopesticide Fact Sheet

EPA (2025) Series 885 – Microbial pesticide test guidelines. https://www.epa.gov/test-guidelines-pesticides-and-toxic-substances/series-885-microbial-pesticide-test-guidelines

European Commission (2002) Guidance document on terrestrial ecotoxicology under council directive 91/414/EEC. In Working Document. https://food.ec.europa.eu/system/files/2016-10/pesticides_ppp_app-proc_guide_ecotox_terrestrial.pdf

European Commission, Barroso JM (2013) Commission Regulation (EU) no 283/2013. In: Off J Eur Union

Faita MR, Pereira E, Poltronieri AS (2023) Effect of entomopathogenic fungus-formulated bioinsecticides on stingless bees (Hymenoptera: Apidae: Meliponini) in the laboratory. J Apic Res 63(5):924–931. https://doi.org/10.1080/00218839.2023.2230717

Fan Y, Fang W, Guo S, Pei X, Zhang Y et al (2007) Increased insect virulence in *Beauveria bassiana* strains overexpressing an engineered chitinase. Appl Environ Microbiol 73(1):295–302. https://doi.org/10.1128/aem.01974-06

Fernández-Marín H, Zimmerman JK, Rehner SA, Wcislo WT (2006) Active use of the metapleural glands by ants in controlling fungal infection. Proc Soc Biol Sci 273:1689–1695. https://doi.org/10.1098/rspb.2006.3492

Ghorui M, Chowdhury S, Burla S (2024) The science behind entomopathogenic fungi: mechanisms and applications. In: Deshmukh SK, Sridhar KR (eds) Entomopathogenic fungi. Springer, Singapore, pp 3–35. https://doi.org/10.1007/978-981-97-5991-0_1

Ginsberg HS, Lebrun RA, Heyer K, Zhioua E (2002) Potential nontarget effects of *metarhizium anisopliae* (Deuteromycetes) used for biological control of ticks (Acari: Ixodidae). Environ Entomol 31(6):1191–1196. https://doi.org/10.1603/0046-225x-31.6.1191

Goettel MS, Jaronski ST (1997) Safety and registration of microbial agents for control of grasshoppers and locusts. Mem Entomol Soc Can 129(S171):83–99. https://doi.org/10.4039/entm129171083-1

Goettel MS, Poprawski TJ, Vandenerg JD, Li Z, Roberts DW (1989) Safety to nontarget invertebrates of fungal biocontrol agents. In: Laird M, Lacey LA, Davidson EW (eds) Safety of microbial insecticides. CRC Press, Boca Raton, pp 209–231. https://agris.fao.org/agris-search/search.do?request_locale=es&recordID=US9105333

Gordon A (2023) Yield losses due to pests. AGRIVI. Retrieved April 21, 2025, from https://www.agrivi.com/blog/yield-losses-due-to-pests/

Groden E, Lockwood J (1991) Effects of soil fungistasis on *Beauveria bassiana* and its relationship to disease incidence in the Colorado potato beetle, *Leptinotarsa decemlineata*, in Michigan and Rhode Island soils. J Invertebr Pathol 57(1):7–16. https://doi.org/10.1016/0022-2011(91)90035-o

Gula LT (2023) Researchers helping protect crops from pests. USDA, National Institute of Food and Agriculture. Retrieved April 21, 2025, from https://www.nifa.usda.gov/about-nifa/blogs/researchers-helping-protect-crops-pests

Gurulingappa P, McGee PA, Sword G (2011) Endophytic *Lecanicillium lecanii* and *Beauveria bassiana* reduce the survival and fecundity of Aphis gossypii following contact with conidia and secondary metabolites. Crop Prot 30(3):349–353. https://doi.org/10.1016/j.cropro.2010.11.017

Gwynn R (2017) Registration of microbial control agents (a comparative overview). In: Lawrence A, Lacey LA (eds) Microbial control of insect and mite pests. From theory to practice. Elsevier eBooks, pp 445–450. https://doi.org/10.1016/b978-0-12-803527-6.00030-5

Hamiduzzaman MM, Sinia A, Guzman-Novoa E, Goodwin PH (2012) Entomopathogenic fungi as potential biocontrol agents of the ecto-parasitic mite, Varroa destructor, and their effect on the immune response of honeybees (*Apis mellifera* L.). J Invertebr Pathol 11(3):237–243. https://doi.org/10.1016/j.jip.2012.09.001

Henke MO, De Hoog GS, Gross U, Zimmermann G, Kraemer D et al (2002) Human deep tissue infection with an entomopathogenic *Beauveria* species. J Clin Microbiol 40(7):2698–2702. https://doi.org/10.1128/jcm.40.7.2698-2702.2002

Hogue C (2011) Endosulfan banned worldwide. In Chemical & engineering news C&en Retrieved May 9, 2025, from https://cen.acs.org/articles/89/i19/Endosulfan-Banned-Worldwide.html

Hokkanen HM, Hajek A (eds) (2003) Environmental impacts of microbial insecticides: need and methods for risk assessment. Springer, Singapore. https://doi.org/10.1007/978-94-017-1441-9

Holder DJ, Keyhani NO (2005) Adhesion of the Entomopathogenic fungus *Beauveria* (*Cordyceps*) *bassiana* to Substrata. Appl Environ Microbiol 71(9):5260–5266. https://doi.org/10.1128/aem.71.9.5260-5266.2005

Humber RA (2008) Evolution of entomopathogenicity in fungi. J Invertebr Pathol 98(3):262–266. https://doi.org/10.1016/j.jip.2008.02.017

IBAMA - Instituto Brasileiro do Meio Ambiente e dos Recursos Naturais Renováveis (2025) Retrieved April 29, 2025, from https://www.ibama.gov.br/index.php

ICAMA (2025) Institute for the control of agrochemicals, ministry of agriculture. Retrieved April 29, 2025, from http://www.chinapesticide.org.cn/eng/aboutus

Inglis GD, Goettel MS, Butt TM, Strasser HE (2001) Use of hyphomycetous fungi for managing insect pests. In Butt TM, Jackson C, Magan N, Fungi as biocontrol agents: progress, problems and potential 2001, pp. 23–69 Wallingford: CABI publishing. https://doi.org/10.1079/9780851993560.0023

James RR, Lighthart B (1994) Susceptibility of the convergent lady beetle (Coleoptera: Coccinellidae) to four entomogenous fungi. Environ Entomol 23(1):190–192. https://doi.org/10.1093/ee/23.1.190

Jaronski ST (2009) Ecological factors in the inundative use of fungal entomopathogens. BioControl 55(1):159–185. https://doi.org/10.1007/s10526-009-9248-3

Jaronski ST (2013) Mass production of entomopathogenic fungi—state of the art. In: Morales-Ramos JA, Rojas MG, Shapiro-Ilan DI (eds) Mass production of beneficial organisms, invertebrates and entomopathogens. Elsevier, pp 357–413. https://doi.org/10.1016/b978-0-12-391453-8.00011-x

Johnson DL, Smits JE, Jaronski ST, Weaver DK (2002) Assessment of health and growth of ring-necked pheasants following consumption of infected insects or conidia of entomopathogenic fungi, *Metarhizium anisopliae* var. *acridum* and *Beauveria bassiana*, from Madagascar and North America. J Toxicol Environ Health Part A 65(24):2145–2162. https://doi.org/10.1080/00984100290071847

Kisla TA, Cu-Unjieng A, Sigler L, Sugar J (2000) Medical management of *Beauveria bassiana* keratitis. Cornea 19(3):405–406. https://doi.org/10.1097/00003226-200005000-00031

Köhl J, Booij K, Kolnaar R, Ravensberg WJ (2019) Ecological arguments to reconsider data requirements regarding the environmental fate of microbial biocontrol agents in the registration procedure in the European Union. Biocontrol 64(5):469–487. https://doi.org/10.1007/s10526-019-09964-y

Krell V, Unger S, Jakobs-Schoenwandt D, Patel AV (2018) Endophytic *Metarhizium brunneum* mitigates nutrient deficits in potato and improves plant productivity and vitality. Fungal Ecol 34: 43–49. https://doi.org/10.1016/j.funeco.2018.04.002

Lallemand Plant Care (2025) *LALGUARD M52 OD* | Lallemand plant care. Retrieved May 9, 2025, from https://www.lallemandplantcare.com/en/usa/products/product-details/lalguard-m52-od/

Latch GCM (1976) Studies on the susceptibility of *Oryctes rhinoceros* to some entomogenous fungi. Entomophaga 21(1):31–38. https://doi.org/10.1007/bf02372013

Leite MO, Alves DA, Lecocq A, Malaquias JB, Delalibera JI et al (2022) Laboratory risk assessment of three entomopathogenic fungi used for pest control toward social bee pollinators. Microorganisms 10(9):1800. https://doi.org/10.3390/microorganisms10091800

Liu Y, Yang Y, Wang B (2022) Entomopathogenic fungi *Beauveria bassiana* and *Metarhizium anisopliae* play roles of maize (*Zea mays*) growth promoter. Sci Rep 12(1):15706. https://doi.org/10.1038/s41598-022-19899-7

Locke M (2001) The Wigglesworth lecture: insects for studying fundamental problems in biology. J Insect Physiol 47(4–5):495–507. https://doi.org/10.1016/s0022-1910(00)00123-2

Low CD, Badenoch PR, Coster DJ (1997) *Beauveria bassiana* keratitis cured by deep lamellar dissection. Cornea 16(6):698–699. https://doi.org/10.1097/00003226-199711000-00018

Luz C, Fargues J (1997) Temperature and moisture requirements for conidial germination of an isolate of *Beauveria bassiana*, pathogenic to *Rhodnius prolixus*. Mycopathologia 138(3): 117–125. https://doi.org/10.1023/a:1006803812504

MAFF (2014) Ministry of Agriculture, Forestry and Fisheries of Japan. Droit Francais | European Encyclopédie Du droit. https://droitfrancais.lawlegal.eu/ministry-of-agriculture-forestry-and-fisheries-of-japan/

Majchrowska-Safaryan A, Tkaczuk C (2021) Abundance of entomopathogenic fungi in leaf litter and soil layers in forested habitats in Poland. Insects 12(2):134. https://doi.org/10.3390/insects12020134

Małagocka J, Grell MN, Lange L, Eilenberg J, Jensen AB (2015) Transcriptome of an entomophthoralean fungus (Pandora formicae) shows molecular machinery adjusted for successful host exploitation and transmission. J Invertebr Pathol 128:47–56. https://doi.org/10.1016/j.jip.2015.05.001

MAPA (2025) Ministério Da Agricultura E Pecuária. Retrieved April 29, 2025, from https://www.gov.br/agricultura/pt-br

Milner RJ, Lim RP, Hunter DM (2002) Risks to the aquatic ecosystem from the application of *Metarhizium anisopliae* for locust control in Australia. Pest Manag Sci 58(7):718–723. https://doi.org/10.1002/ps.517

Mora MA, Castilho AM, Fraga ME (2018) Classification and infection mechanism of entomopathogenic fungi. Arq Inst Biol 84:e0552015. https://doi.org/10.1590/1808-1657000552015

Morda W, Nuvoli MT, Ruiu L (2024) Safety of the entomopathogenic fungus Beauveria bassiana for wild and laboratory-reared *Chrysoperla lucasina* strains. Insects 15(8):576. https://doi.org/10.3390/insects15080576

Mycotal (2025) Koppert Global. Retrieved May 9, 2025, from https://www.koppert.com/mycotal/

Naundrup A, Bohman B, Kwadha CA, Jensen AB, Becher PG et al (2022) Pathogenic fungus uses volatiles to entice male flies into fatal matings with infected female cadavers. ISME J 16(10): 2388–2397. https://doi.org/10.1038/s41396-022-01284-x

OECD (2014) Guidance to the environmental safety evaluation of microbial biocontrol agents. In: Series on pesticides and on biocides. https://doi.org/10.1787/9789264221659-en

OECD (2018) Working document on the risk assessment of secondary metabolites of microbial biocontrol agents. In: Series on pesticides (No. 98)
OECD (2019) Report of the 9th biopesticides expert group seminar on test methods for micro-organisms. In: Series on pesticides (No. 100)
Pavlyushin V (2020) Pathogenic post-effect of entomopathogenic fungi on phytophagous pests and entomophagous biocontrol agents. BIO Web Conf 21:00020. https://doi.org/10.1051/bioconf/20202100020
PMRA (2001) Submission management and information division, pest management regulatory agency. Regulatory directive DIR2001-02. In: Guidelines for the registration of microbial pest control agents and products. https://www.canada.ca/content/dam/hc-sc/migration/hc-sc/cps-spc/alt_formats/pacrb-dgapcr/pdf/pubs/pest/pol-guide/dir/dir2001-02-eng.pdf
Qayyum MA, Saeed S, Wakil W, Nawaz A, Iqbal N et al (2021) Diversity and correlation of entomopathogenic and associated fungi with soil factors. J King Saud Univ Sci 33(6):101520. https://doi.org/10.1016/j.jksus.2021.101520
Quesada Moraga E, Navas-Cortés JA, Maranhao EAA, Ortiz-Urquiza A, Santiago-Álvarez C (2007) Factors affecting the occurrence and distribution of entomopathogenic fungi in natural and cultivated soils. Mycol Res 111:947–966. https://doi.org/10.1016/j.mycres.2007.06.006
Rangel DE, Braga GU, Anderson AJ, Roberts DW (2005) Variability in conidial thermotolerance of Metarhizium anisopliae isolates from different geographic origins. J Invertebr Pathol 88(2): 116–125. https://doi.org/10.1016/j.jip.2004.11.007
Rizwan M, Atta B, Arshad M, Khan RR, Dageri A et al (2021) Nondetrimental impact of two concomitant entomopathogenic fungi on life history parameters of a generalist predator, *Coccinella septempunctata* (Coleoptera: Coccinellidae). Sci Rep 11(1):20699. https://doi.org/10.1038/s41598-021-00037-8
Sachs SW, Baum J, Mies C (1985) *Beauvaria bassiana* keratitis. Br J Ophthalmol 69(7):548–550. https://doi.org/10.1136/bjo.69.7.548
Sancar A (2003) Structure and function of DNA photolyase and cryptochrome blue-light photoreceptors. Chem Rev 103(6):2203–2238. https://doi.org/10.1021/cr0204348
Sayed S, Elarnaouty SA, AlOtaibi S, Salah M (2021) Pathogenicity and side effect of indigenous Beauveria bassiana on *Coccinella undecimpunctata* and *Hippodamia variegata* (Coleoptera: Coccinellidae). Insects 12(1):42. https://doi.org/10.3390/insects12010042
Schaerffenberg B (1968) Untersuchungen über die Wirkung der Insektentötenden Pilze *Beauveria bassiana* (Bals. Vuill.) und *Metarrhizium anisopliae* (Metsch.) Sorok. auf Warmblütler. Entomophaga 13(2):175–182. https://doi.org/10.1007/bf02371788
Schäfer RB, Liess M, Altenburger R, Filser J, Hollert H et al (2019) Future pesticide risk assessment: narrowing the gap between intention and reality. Environ Sci Eur 31:1–5. https://doi.org/10.1186/s12302-019-0203-3
Scheepmaker JW, Busschers M, Sundh I, Eilenberg J, Butt TM (2019) Sense and nonsense of the secondary metabolites data requirements in the EU for beneficial microbial control agents. Biol Control 136:104005. https://doi.org/10.1016/j.biocontrol.2019.104005
Semalulu SS, MacPherson JM, Schiefer HB, Khachatourians GG (1992) Pathogenicity of *Beauveria bassiana* in mice. J Vet Med B 39(1–10):81–90. https://doi.org/10.1111/j.1439-0450.1992.tb01141.x
Shadduck JA, Roberts DW, Lause S (1982) Mammalian safety tests of *Metarhizium anosipliae*: preliminary results. Environ Entomol 11(1):189–192. https://doi.org/10.1093/ee/11.1.189
Shah PA, Pell JK (2003) Entomopathogenic fungi as biological control agents. Appl Microbiol Biotechnol 61(5–6):413–423. https://doi.org/10.1007/s00253-003-1240-8
Sharma A, Sharma S, Yadav PK (2023) Entomopathogenic fungi and their relevance in sustainable agriculture: a review. Cogent Food Agric 9(1). https://doi.org/10.1080/23311932.2023.2180857
St Leger RJ (1991) Integument as a barrier to microbial infections. In: Binnington K, Retnakaran A (eds) Physiology of the insect epidermis. Commonwealth Scientific and Industrial Research

Organization, Melbourne, pp 286–308. https://www.scirp.org/reference/referencespapers?referenceid=1867072

Stellarix (2024) Comprehensive guide to sustainable pest management | Stellarix. Retrieved April 21, 2025, from https://stellarix.com/insights/blogs/sustainable-pest-control-in-farming/#:~:text=Sustainable%20pest%20management%20is%20not,safe%20for%20the%20next%20generations

Strasser H, Abendstein D, Stuppner H, Butt TM (2000) Monitoring the distribution of secondary metabolites produced by the entomogenous fungus *Beauveria brongniartii* with particular reference to oosporein. Mycol Res 104(10):1227–1233. https://doi.org/10.1017/s0953756200002963

Sundh I, Goettel MS (2012) Regulating biocontrol agents: a historical perspective and a critical examination comparing microbial and macrobial agents. Biocontrol 58(5):575–593. https://doi.org/10.1007/s10526-012-9498-3

Téllez-Jurado A, Ramírez MGC, Flores YM, Asaff Torres AA, Arana-Cuenca A (2009) Mecanismos de acción y respuesta en la relación de hongos entomopatógenos e insectos. Rev Mex Mic 30:73–80. http://repositoriodigital.academica.mx/jspui/handle/987654321/82800

Toriello C, Pérez-Torres A, Burciaga-Díaz A, Navarro-Barranco H, Pérez-Mejía A et al (2005) Lack of acute pathogenicity and toxicity in mice of an isolate of *Metarhizium anisopliae* var. *anisopliae* from spittlebugs. Ecotoxicol Environ Saf 65(2):278–287. https://doi.org/10.1016/j.ecoenv.2005.07.016

TraceXtech Technologies (2024) Integrated pest management for sustainable agriculture: enhance crop health. Blockchain for food safety, traceability and supplychain transparency. Retrieved April 21, 2025, from https://tracextech.com/integrated-pest-management-for-sustainable-agriculture/

US EPA (2025) US EPA. The case of DDT: Revisiting the impairment | Retrieved April 21, 2025, from https://www.epa.gov/caddis/case-ddt-revisiting-impairment

Vega N (2018) The use of fungal entomopathogens as endophytes in biological control: a review. Mycologia 110(1):4–30. https://doi.org/10.1080/00275514.2017.1418578

Vega FE, Posada F, Aime MC, Pava-Ripoll M, Infante F et al (2008) Entomopathogenic fungal endophytes. Biol Control 46(1):72–82. https://doi.org/10.1016/j.biocontrol.2008.01.008

Weng Q, Zhang X, Chen W, Hu Q (2019) Secondary metabolites and the risks of Isaria fumosorosea and *Isaria farinosa*. Molecules 24(4):664. https://doi.org/10.3390/molecules24040664

Yasui A, Eker AP, Yasuhira S, Yajima H, Kobayashi T et al (1994) A new class of DNA photolyases present in various organisms including aplacental mammals. EMBO J 13(24):6143–6151. https://doi.org/10.1002/j.1460-2075.1994.tb06961.x

Zimmermann G (2007) Review on safety of the entomopathogenic fungi Beauveria bassiana and *Beauveria brongniartii*. Biocontrol Sci Tech 17(6):553–596. https://doi.org/10.1080/09583150701309006

Innovative Applications and Future Potential of Entomopathogenic Fungi

18

Parthasarathy Seethapathy and Atham Bhavana Sruthi

Abstract

Entomopathogenic fungi (EPF) have been recognized as effective biological control agents against a wide range of crop and household pests. Traditionally utilized for their insecticidal activity, recent research works have significantly extended the scope of EPF applications beyond pest management. This chapter explores the novel and emerging uses of EPF and evaluates their future potential in diverse fields. Apart from their wider application in pest management, EPF exhibit multifaceted purposes, including endophytic and rhizospheric colonization, induction of systemic resistance in plants, and enhancement of plant growth and nutrient uptake. Advances in formulations such as nanoencapsulation, oil-based carriers, and smart delivery systems have improved their stability, shelf life, and field efficacy. Further, developments in genetic modifications and molecular tools have flagged the possibility for strain improvement, enhanced virulence, and increased adaptability to abiotic stress. The integration of EPF with digital agriculture platforms, drones, and precision application tools is transforming their deployment under field conditions. In addition, EPFs are being explored for applications in public health for vector control, bioremediation of contaminated soils, and as biofactories for secondary metabolites with pharmaceutical values. Despite these advancements, challenges remain in large-scale commercialization, regulatory support, and performance consistency under

P. Seethapathy (✉)
Department of Plant Pathology, Amrita School of Agricultural Sciences, Amrita Vishwa Vidyapeetham, Coimbatore, India

A. B. Sruthi
Department of Agricultural Entomology, College of Agriculture, Kerala Agricultural University, Thrissur, India

S. K. Deshmukh et al. (eds.), *Frontiers in Entomopathogenic Fungi*,
https://doi.org/10.1007/978-981-95-6529-0_18

variable environments. This chapter provides an overview of the current innovations, highlights research openings, and future directions for the multi-functional exploitation of EPF.

Keywords

Beauveria · Bioformulation · Biopesticides · *Metarhizium* · Nanoencapsulation · SDG-2

1 Introduction

Pests and diseases are responsible for an estimated 20–40% of global food crop losses, with plant diseases alone accounting for about 14.1% of this damage, equating to nearly $470 billion in annual agricultural trade losses (Savary et al. 2019; Cai and Dimopoulos 2025). There are approximately 67,000 pest species that threaten crop plants, including 9000 insects and mites. Therefore, insect pest control measures are essential for ensuring crop yields (Cai and Dimopoulos 2025). Absolute dependence on chemical insecticides leads to critical challenges in pest management, including pest resistance, resurgence of insect populations, chemical residues in the food supply chain, ecological degradation, and health risks to humans and animals. These concerns have intensified global efforts to develop safer, sustainable pest control alternatives that ensure effective crop protection and sustainable productivity. Biological control agents, particularly microbial biopesticides, have emerged as eco-friendly and efficient tools to address these challenges, offering specific actions with minimal environmental impact (Parthasarathy et al. 2024). Biopesticides have demonstrated efficacy in managing pests and plant pathogens and reducing negative impacts on crops, ecosystems, and human well-being. The use of hypocrealean entomopathogenic fungi (EPF) or insect-pathogenic species in biocontrol initiatives of crop pests represents key players for advancing integrated pest management (IPM) (Skinner et al. 2014). This environmentally sustainable IPM approach aims to gradually substitute synthetic chemical pesticides with pest control methods. Consequently, they present considerable advantages for advancing One Health and aiding in the achievement of various United Nations Sustainable Development Goals (SDGs), primarily Zero Hunger (SDG 2).

Among microbial biocontrol agents, EPF have gained significant consideration, representing nearly half of all registered worldwide biopesticide products (Mascarin et al. 2024). EPF are a diverse group of hypocreales fungi that naturally infect insect pests and nematodes by adhering to and piercing the pest cuticle, multiplying within the host, and eventually causing death (Mantzoukas and Eliopoulos 2020; Seethapathy et al. 2025). Following infection, these fungi continue to utilize the insect cadaver, acting as parasites, facultative pathogens, or saprobes. Their origin in nature and minimal environmental side effects make EPF especially promising for sustainable pest control. Notably, EPF have long played a central role in IPM strategies due to their specificity, efficacy, and ecological safety (Sharma et al.

2023). To date, over 800 species of hypocrealean EPF spanning more than 100 genera, including *Beauveria*, *Metarhizium*, *Isaria*, and *Lecanicillium*, have been identified, with many commercial formulations based on their conidia (Subramaniam et al. 2021; Quesada-Moraga et al. 2024). Despite their success, critical questions remain: How can EPF be optimized for broader pest targets and harsher field conditions? What innovations can improve formulation, application, and shelf life? Can entomopathogenic fungi be expanded into new frontiers such as plant growth promotion, bioremediation, or applications in veterinary and medical entomology? Addressing these knowledge and application gaps is vital for advancing the full potential of EPF as multifunctional agents in pest management and beyond. This chapter explores the latest innovations, emerging applications, and future prospects of EPF, aiming to highlight both their current contributions and the untapped opportunities that lie ahead.

2 Scope and Significance

The traditional roles of EPF have primarily focused on their use as biocontrol agents against agricultural pests, leveraging hypocrealean species like *Beauveria bassiana* (Balsamo) Vuillemin and *Metarhizium anisopliae* (Mechnikov) Sorokin for targeted insect management. Concurrently, hypocrealean fungi, including *B. bassiana*, *Cordyceps* spp., *Hirsutella thompsonii* Fisher, *Hypocrella* spp., *Isaria fumosorosea* (Wize) Kepler, B. Shrestha & Spatafora, *Lecanicillium lecanii* (Zimm.), *Metarhizium acridum* (Driver & Milner) J.F. Bisch., Rehner & Humber, *M. anisopliae*, *M. brunneum* Petch, *Ophiocordyceps* spp., *Paecilomyces lilacinus* (Thom) Samson, and *Zoophthora radicans* (Brefeld) Batko, have been commercialized globally as biopesticides (Subramaniam et al. 2021). All these EPF belong to various taxonomic classes of fungi and chromista, including Ascomycota, Basidiomycota, Chytridiomycota, Entomophthoromycota, Microsporidia, and Oomycetes, constituting over a hundred genera and 700 species (Litwin et al. 2020) and some species are reported in Blastocladiomycota and Kickxellomycotina (Mora et al. 2017).

The life cycle of EPF starts upon the release of spores on the host tissue surface. These spores activate and germinate on the host surface, penetrate the surface cuticle, inhabit the inner tissue layers, and secrete toxins and enzymes, ultimately leading to the death of insects. Thereby, its emerging applications are expanding their potential far beyond crop pest control. Recent research works highlight the ability of EPF to promote plant growth through rhizosphere or endophytic colonization, enhanced nutrient uptake, and induced systemic resistance against pests and pathogens. Additionally, EPF are being explored for their role in bioremediation, breaking down pollutants like pesticides and heavy metals in contaminated soils (Saha et al. 2025). Innovations in nanotechnology have enabled the development of fungal-based bio-nanoparticles for sustainable agriculture and health care sectors.

There is also growing demand for their secondary metabolites, which exhibit antimicrobial, antioxidant, and anticancer properties, opening them up for pharmaceutical applications. Novel applications in vector control, such as targeting malaria-transmitting mosquitoes, demonstrate their public health potential (Quesada-Moraga et al. 2024). EPF are being investigated for their potential in managing vector-borne diseases by targeting mosquito and tick larvae. As climate change and pesticide resistance threaten global food security, EPF offer an eco-friendly alternative with multifaceted roles as endophytes, rhizospheric colonizers, plant pathogen antagonists, plant growth promoters, and foundations of economically demanded bioactive compounds (Bamisile et al. 2021; Quesada-Moraga et al. 2024).

Despite over 800 known entomopathogenic species, only a fraction has been industrialized, with anamorphic species *B. bassiana* and *M. anisopliae* dominating 33.9% each of the 171 commercial formulations, followed by *P. fumosorosea* (5.8%) and *B. brongniartii* (4.1%) (Sharma and Sharma 2021; Seethapathy 2024). Their ability to degrade agrochemicals and heavy metals positions them as key players in bioremediation, addressing soil pollution challenges. Their narrow host specificity enhances precision in pest control, positioning EPF as key players in sustainable agriculture, with ongoing innovations expanding their applicability against diverse arthropod pests in agroecosystems worldwide. Innovations in nano-formulations and encapsulation techniques have improved fungal spore viability and delivery and created compatible IPM packages for reducing ecological impact.

Additionally, EPF-derived secondary metabolite-based formulations exhibit promising antimicrobial, anticancer, and immunomodulatory properties, expanding their pharmaceutical demand (Cai and Dimopoulos 2025). With rising pesticide resistance and climate change threats, EPF offers sustainable alternatives, aligning with global food security (SDG 2) and One Health initiatives by WHO and the National One Health Mission. Future research should focus on genetic modifications of strains, enhancements in formulation, and techniques for large-scale production to maximize their potential in agriculture, environmental sustainability, and medicine, indicating a fundamental transition from conventional pest management to comprehensive agroecological management. Thus, innovation in EPF applications bridges ecological and agricultural gaps and paves the way for groundbreaking contributions to medicine and environmental conservation.

3 Agricultural Applications

3.1 As Endophytes for Plant Growth Promotion

EPF, traditionally known for their insecticidal properties, are increasingly recognized as beneficial endophytes capable of enhancing plant growth and fitness. *Metarhizium* and *Beauveria* exhibit versatile characteristics, functioning as both entomopathogens and endophytes (Ahsan et al. 2024). They have an impressive ability to integrate into various plant species, facilitating growth and providing defense against pests. When colonizing plant tissues internally, *B. bassiana* and

Metarhizium spp. establish symbiotic relationships that promote host plant vigor through multiple mechanisms like spore adhesion, cuticle penetration, toxin secretion, and immune system evasion (Quesada-Moraga et al. 2024). Also, EPF demonstrate remarkable adaptability as endophytes as endophytic insect pathogenic fungi (EIPF), colonizing internal plant tissues without causing disease symptoms while providing multifaceted benefits to host plants (Harish et al. 2023). Notable hypocrealean fungi consist of *Beauveria*, *Cordyceps* (*Paecilomyces*), *Hirsutella*, *Isaria*, *Metarhizium*, *Nomuraea*, and *Verticillium*/*Lecanicillium* (Subramaniam et al. 2021), recognized as EIPF. Most of the research confirms that *B. bassiana* and *Metarhizium* spp. successfully establish endophytic relationships in many crops, with inoculation methods such as stem injection and foliar application proving effective against key pests like corn stem borer and fall armyworm in maize.

Beyond these well-studied genera, including diverse ascomycetes like *Lecanicillium*, *Acremonium*, *Clonostachys*, and *Purpureocillium*, naturally colonize various crops; for example, *Metarhizium brunneum*-based seed treatments demonstrate efficacy against wireworms in maize. These EIPF exhibit remarkable ecological plasticity, transitioning between roles as soil-dwelling insect pathogens and plant-beneficial endophytes. Many EPF, including *B. bassiana* and *Metarhizium* sp., are closely related to crop plants through their inhabitation of the rhizosphere or establishment as endophytes in root tissues (Ahsan et al. 2024). These features enhance the ability of EIPF to remain in the rhizosphere system to defend against stresses as biological control mediators. The process by which EIPF colonize the interior tissues of plants is influenced by the host mechanisms. This could be interpreted as a protective mechanism developed by plants to deter herbivorous insects.

Crop plants host a wide range of EIPF, including both entomopathogenic and non-entomopathogenic microbes (Harish et al. 2023). This endophytic relationship of EPF generally reduces the development and activity of polyphagous (e.g., fall armyworm) and sucking insect pests (e.g., whitefly, thrips, and hoppers), while exerting less impact on monophagous insect pests. The alteration of monophagous insect pests to bioactive secondary metabolites or volatiles produced by EPF sheds light on this phenomenon. Herbivores and microbes attacking plants result in modifications in the genetics, physiology, and biochemistry of the crop plants, as demonstrated by changes in gene expression and transcriptional reprogramming. The processes of reprogramming govern the production of ACC deaminase, IAA, nitrogen uptake, phosphate solubilization, zinc solubilization, and siderophore production in EIPF. It has been revealed that *M. robertsii* and *B. bassiana* (but not *L. lecani*) deliver plants with nitrogen bases that are expressed during the parasitization of insect pests (Bidochka and Hajek 1998). Additionally, EIPF modulate phytohormone levels, particularly by producing auxins, cytokinin, and gibberellins, which stimulate root development and shoot growth. Their endophytic presence also induces systemic resistance, priming plants against biotic stressors such as pathogenic fungi, bacteria, nematodes, and herbivorous insects (Seethapathy et al. 2025).

Beyond biotic stress mitigation, EPF enhance plant innate tolerance to abiotic stresses by activating antioxidant defenses and osmoprotectant synthesis. These modifications harmonize the synthesis of both volatile and non-volatile compounds and trigger molecular pathways for defense systems in the plant, simultaneously generating signals for insects. Microbes significantly impact insect suitability and behavior by tritrophic communications, which may influence insect activity and fungal pathogenicity (Ahsan et al. 2024). Recent studies highlight their role in improving soil health by fostering beneficial microbial communities, further supporting sustainable agriculture. The dual functionality of EIPF as both biocontrol agents and plant growth promoters positions them as valuable tools in IPM and climate-resilient farming (Quesada-Moraga et al. 2022, 2024). Their ability to systemically colonize plants through multiple inoculation methods for maintaining entomopathogenic activity of EIPF as promising dual-purpose agents for IPM and crop yield enhancement in sustainable farming systems. Future investigation should emphasize optimizing endophytic colonization, strain-specific interactions, and field applications to maximize their potential in eco-friendly agricultural systems.

3.2 Role in Induced Systemic Resistance and Plant Health

EPF are involved in induced systemic resistance (ISR) in plants by triggering jasmonic acid responses against biotic stressors, enhancing their defense mechanisms without direct pathogen antagonism. EPF such as *B. bassiana* and *Metarhizium* spp., colonize the rhizosphere or endophytically inhabit plant tissues, triggering a cascade of defense-related responses. These fungal MAMPs react with PRR and stimulate the upregulation of pathogenesis-related (PR) proteins, phytohormones such as jasmonic acid (JA) and salicylic acid (SA), and plant secondary metabolites like phenolics and flavonoids, priming the plant for rapid pathogen recognition and defensive response by PTI reaction (Seethapathy et al. 2016). EPF-mediated ISR also enhances antioxidant enzyme activity (e.g., peroxidase, polyphenol oxidase, and catalase) and strengthens physical barriers through lignin and callose deposition.

Studies demonstrate that EPF-treated plants exhibit reduced susceptibility to fungal, bacterial, and viral infections, as well as nematode infestations, indicating a broad spectrum of ISR (Mantzoukas and Eliopoulos 2020; Sharma and Sharma 2021; Ahsan et al. 2024). Furthermore, EPF synergize with beneficial soil microbes, amplifying ISR through microbiome modulation (Qadri et al. 2020). Unlike chemical systemic inducers, EPF provides sustainable, long-term protection while promoting crop growth. Further research works are focusing on elucidating molecular signaling pathways, optimizing EPF-plant interactions, and developing commercial formulations for diverse field applications. Harnessing EPF-induced ISR offers an eco-friendly strategy for integrated pest and disease management, reducing reliance on synthetic agrochemicals and contributing to sustainable agriculture.

3.3 Dual Role in Pest Control and Biofertilization

EPF exhibit a dual role in agriculture by serving as effective biocontrol agents and biofertilizers or biostimulants, enhancing sustainable crop production. As biocontrol agents, EPF such as *B. bassiana* and *M. anisopliae* kill insect pests through multisite action for a longer time in the region/field (Shanmugam and Seethapathy 2017). Concurrently, certain EPF strains contribute to soil fertility by solubilizing phosphorus, fixing nitrogen, and producing phytohormones, thereby promoting plant growth and yield (Harish et al. 2019). *Metarhizium* spp. form symbiotic associations with root systems, improving nutrient intake levels and stress tolerance (Dutta et al. 2014; Bamisile et al. 2021; Nimisha et al. 2024). The primary approach for the management of soil pest propagules, such as larvae, grubs, and pupae, revolves around the use of chemical pesticides. However, the increasing concerns related to safety, environmental contamination, and the insufficient efficacy of proposed insecticides indicate the pressing need for the development of suitable IPM strategies for these pests.

The review work (Sivasabari et al. 2023) proposed the use of EPF enriched with vermicompost in soil as a suitable approach for sustainable soil pest management and mitigation. This compost acts as a substrate and buffering agent for EPF within the soil over an extended duration, facilitating the gradual release of EPF propagules aimed at target pests. This dual-functionality application mitigates pest-associated yield losses and enhances soil health, reducing the need for synthetic fertilizers. Additionally, EPF- and compost-mediated enzymes involved in ISR in plant systems further bolster crop resilience against recurrent biotic and abiotic stresses (Sivasabari et al. 2023). The incorporation of EPF into IPM and sustainable farming systems promotes their innovative applications, offering eco-friendly solutions to global agricultural challenges. Further research should focus on optimizing EPF formulations, improving field efficacy, and exploring novel strains with enhanced multifunctional traits, ensuring their broader adoption in climate-smart agriculture. Thus, EPF represents a promising, multifaceted product for advancing food security and environmental sustainability.

3.4 Management of Emerging and Invasive Pests

EPF have emerged as a sustainable and effective biocontrol strategy for managing emerging and invasive insect pests of crop plants worldwide. These fungi, including genera such as *Beauveria*, *Metarhizium*, and *Hirsutella*, infect pests through cuticular penetration, causing mortality and minimizing environmental impact. Their host specificity and ability to persist in agroecosystems make them ideal for IPM programs, reducing dependence on agrochemicals. Recent advances in EPF formulations, such as microencapsulation and oil-based dispersions, have enhanced their field efficacy and stability under varying climatic conditions (Parthasarathy et al. 2024). EPF also exhibit synergistic effects with other biocontrol agents and selective insecticides, broadening their applicability against resistant pest populations. Additionally, genetic and molecular research is uncovering novel virulence factors and stress tolerance traits, paving the way for engineered fungal strains with improved performance. Despite challenges in mass production, storage, and farmer adoption, EPF hold significant promise for combating invasive pests like *Diaphorina citri* (Asian citrus psyllid) and *Spodoptera frugiperda* (fall armyworm) (Akutse et al. 2014; Litwin et al. 2020; Sharma et al. 2023; Saminathan et al. 2025).

4 Formulation and Delivery Systems

4.1 Mass Production, Storage, and Shelf Life

The large-scale production of EPF is essential for enabling affordable and widespread commercialization, as well as global field application. The commercial viability and field efficacy of EPF depend critically on optimized mass-production protocols and advanced preservation techniques that maintain virulence and viability. A sustained research approach that includes the screening of fungal strains, optimization of media (Table 18.1), and appropriate formulation techniques (Mascarin et al. 2024) is essential for realizing the potential of these spore and mycelial propagules. This should be combined with insights into the molecular methods related to the development and loading constancy to refine the advancement of active biocontrol agents targeting arthropod pests and disease-transmitting vectors. Among the multiple mass-production methods investigated primarily by industry (Mascarin et al. 2024), submerged fluid fermentation is a strong and adaptable process that facilitates the development of different types of propagules for a wide range of insect pest control applications (Seethapathy et al. 2023). A variety of hypocrealean EPF can be readily cultured on artificial substrates through the production of single- or bi-celled forms, such as blastospores and submerged asexual conidia, as well as multicellular forms like hyphae and sclerotial bodies (Shanmugam and Seethapathy 2017; Mascarin et al. 2024) (Table 18.1). Recently, certain EPF may develop resting structures as chlamydospores from the mycelium under unfavorable environments; these resting and resistant structures have largely been ignored in previous studies (Mascarin et al. 2024).

B. bassiana and *P. fumosoroseus* have been cultivated using a range of compact nutritional wastes like cereal grains, vegetable scraps, and cereal husks. *B. bassiana* demonstrated ideal progress and reached its maximum spore production levels on a wheat or rice substrate. Solid kitchen wastes such as beetroot, beans, carrots, melons, jackfruit seeds, and okra demonstrated enhanced growth and sporulation of EPF strains in comparison to cereal substrates (Cai and Dimopoulos 2025). Industrial-scale production typically employs a diphasic system, where initial liquid fermentation (using cost-effective media containing sucrose/glucose and waste brewers' yeast) generates high yields of blastospores or mycelia, followed by solid-substrate cultivation (e.g., rice, barley) to induce prolific conidiation. Maintaining an optimal carbon-to-nitrogen ratio (e.g., 5:1 for *B. bassiana*) is essential for maximizing sporulation, and rigorous sterilization protocols prevent microbial contamination.

Table 18.1 Culture media for EPF

Culture medium	Entomopathogenic fungus
Synthetic/semisynthetic	
Asthana and Hawker's media	*Aschersonia alerodis*
Carrot malt agar	*Nomuraea rileyi*
Czapek-Dox agar (CDA)	*Beauveria bassiana*
Emerson's YpSs agar	*Metarhizium anisopliae*
Oatmeal agar	*Paecilomyces fumosoroseus*
Potato dextrose agar (PDA)	*B. bassiana, Verticillum lecanii, Aschersonia* sp., *Purpureocillium lilacinum*
Sabouraud's dextrose agar (SDA)	*Aschersonia* sp.
Sabouraud's dextrose agar with yeast extract (SDAY)	*B. bassiana, B. brongniartii, V. lecanii*
Sabouraud's maltose agar with yeast extract (SMAY)	*N. rileyi, Zoophthora radicans, Paecilomyces farinosus*
Yeast peptone dextrose agar	*Lecanicillium lecanii*
Barley	*Isaria javanica*
Cassava chips + rice bran + fish meal	*M. anisopliae*
Chopped potato and rice	*B. bassiana*
Coconut water	*M. anisopliae*
Cooked rice	*B. bassiana, M. anisopliae*
Crushed maize	*B. brongniartii*
Crushed sorghum grains with yeast extract	*N. rileyi*
Flaked barley	*M. robertsii, M. pinghaense*
Green gram	*M. anisopliae*
Oats	*M. anisopliae*
Rice and finger millet	*N. rileyi*
Rice husk	*B. bassiana, M. anisopliae*
Sorghum grains	*Verticillium lecanii, Paecilomyces farinosus (I. fumosorosea), B. bassiana*
Sugarcane bagasse	*M. anisopliae*
Wheat	*L. lecanii, L. longisoprum*
Wheat bran	*B. bassiana*

However, repeated subculturing can weaken virulence, necessitating periodic host passage through insect models to re-establish pathogenicity. Post-production, innovative formulation technologies, including microencapsulation, oil-dispersible concentrates, isotonic liquids, and wettable powders, enhance shelf life by protecting spores from environmental stressors (Seethapathy 2024). For short-term storage (months), conidia remain viable at 4 °C in dark conditions, as demonstrated by *B. bassiana* retaining acaricidal efficacy against *Dermanyssus gallinae* under such conditions (Immediato et al. 2017). Long-term preservation (>10 years) employs cryopreservation (−80 °C), lyophilization, or silica gel storage at −20 °C, with glycerin or mineral oil acting as cryoprotectants.

The most captivating future opportunity regarding EPF is in the potential to develop strategies for enhancing strain (Harish et al. 2022). Considerations may encompass enhanced lethality (lower LD_{50}), the ability to initiate disease under low humidity, improved practical usability duration, environmental elasticity such as temperature adaptability and UV resistance, increased mortality rate (lower LT_{50}), better sporulation during large-scale production, and host range expansion (Sharma and Sharma 2021). Starch pellets serve as an ideal delivery system for EPF, particularly *B. bassiana* blastospores. Starch pellets have been shown to enhance the perseverance and attachment of fungal spores on target insect pests, resulting in improved efficacy and increased pest mortality rates (Friuli et al. 2023).

Emerging techniques like cellulose filter paper embedding and microbead encapsulation show promise for strain integrity maintenance. Crucially, storage conditions must balance moisture control (to prevent metabolic activation) with protection from oxidative damage, while periodic viability assays ensure genetic stability. These protocols are underpinned by rigorous quality control measures, including molecular markers to track strain identity and bioassays to confirm virulence retention (Tahira et al. 2014; Shin et al. 2023). Future advancements in formulation and fermentation optimization, cryoprotectant formulations, and predictive modeling of storage degradation kinetics will further enhance the commercial feasibility of EPF as reliable biocontrol agents. The integration of these production and preservation strategies ensures that EPF meets the demands of both large-scale agricultural applications and long-term biodiversity conservation efforts.

4.2 Formulations and Their Application

EPF bioformulations are prepared as encapsulated, liquid metabolites and solid- and cell-free culture supernatants to enhance the efficacy, stability, and field applicability as potent microbial biopesticides. It involves the integration of active fungal ingredients (such as blastospores, conidia, or chlamydospores) with inert carriers, adjuvants, and protective additives to improve physical properties, handling, and delivery (Parthasarathy 2023). The primary objectives of formulation include ensuring microbial viability during storage, maintaining optimal pH, enhancing environmental persistence, reducing the required dosage of active ingredients, and extending shelf life. A well-optimized bioformulation not only improves spore production and storage stability but also enhances application efficiency, post-application survival, and targeted delivery to pest-infested zones. According to Maina et al. (2018), the widely used EPF formulations include technical concentrates with fungal-colonized substrates (26.3%), wettable solid powders (20.5%), and oil-based dispersions (15.2%). Further, the active components utilized in formulations of EPF can be classified as either Technical Material (TC) or Technical Concentrate (TK).

Other formulations, such as granules, baits, peat-based powders, oil-miscible flowable concentrates, freeze-dried powders, water-dispersible granules, and ultra-low volume (ULV) suspensions, are less common but serve niche applications. Formulations based on biopolymers demonstrate significant potential in improving the efficacy of EPF as bioinsecticides. Chitosan and starch with various biopolymers have been employed to enhance spore delivery, tenacity, and adhesion to intended insect targets. These biopolymer-based formulations provide benefits including precise targeting, environmental friendliness, and sustainable practices (Friuli et al. 2023). The scalability of EPF mass production for formulations is a critical issue. This context presents both opportunities and challenges regarding scalability. Obstacles may arise from the intricacy of treating biomaterials, including inconsistency in raw materials, processes, and quality maintenance challenges. Additionally, costs and regulatory compliance restrictions can complicate matters, as meeting safety standards is often time-consuming and expensive (Parthasarathy 2023). Technical reasons also play a role, as upholding product quality and consistency can be challenging. Furthermore, there is the issue of contamination and the confidence of end-users concerning formulation quality. Major types of EPF formulations and their applications are as follows.

4.2.1 Wettable Powders

Wettable powder (WP) formulations consist of finely milled EPF spores blended with inactive carriers and surfactant molecules to facilitate uniform dispersion in water. Wettable powder formulations consist of fungal spores mixed with fine particulate carriers such as talc, kaolin, or clay (Parthasarathy 2023). They are particularly effective in arid environments and can be applied directly to foliage or soil. These formulations are cost-effective, exhibit long shelf lives, and are widely used in agricultural systems. However, they require continuous agitation during spraying to prevent sedimentation. Notable EPF-based wettable powders include those of *P. fumosoroseus*, *B. bassiana*, and *N. rileyi*. The maximum nymphal mortality of *Maconellicoccus hirsutus* was achieved in EPF combination formulations used, i.e., 57.64% with *B. bassiana* (6 g/l) + *L. lecanii* (6 g/l) using WP with 2×10^8 cfu/g each (Gopal et al. 2021). A liquid formulation of *L. lecanii* at (1×10^7 spores/ml) demonstrated high efficacy against the tea thrips (*Scirtothrips bispinosus*). *L. lecanii* at 1500 g (1×10^7 conidia/ml) mixed in 400 L of water was effective against the *S. bispinosus* (Subramaniam et al. 2021).

4.2.2 Liquid Concentrates

Liquid Concentrates (LC) contain high concentrations of fungal spores suspended in water- or oil-based solutions, which are diluted before field application. These formulations are compatible with conventional spray equipment, ensuring ease of use (Parthasarathy 2023). However, they generally have a shorter shelf life compared to dry formulations. Liquid formulation of *V. lecanii* against red spider mite (*Tetranychus cinnabarinus*) of bhendi showed 79.66% to 82.78% mortality at 1% concentration (Chavan et al. 2008). Numerous formulation strategies, including film coating, bio-nano-encapsulation, and liquid isotonic formulation, were developed (Lohse et al. 2015). The formation of blastospores can be transitioned to robust submerged asexual conidia by applying osmotic pressure during their growth phases under fermenter conditions. *B. bassiana* developed from beads and thrived on seeds coated with commercial fungicides. A mixture was prepared with Triton X-114 at

0.1%, molasses at 1%, titanium dioxide at 1%, and a concentration of 10^6 spores/mL, which was then applied to the tips of the leaves. After 14 days, they identified the EIPF through both PCR and microscopic examination of the leaves (Lohse et al. 2015).

4.2.3 Granular Formulations

Granular formulations (GR) incorporate fungal spores into absorbent substrates (e.g., millet, wheat bran, or clay), making them ideal for soil-dwelling pests. These formulations provide enhanced protection against environmental degradation and sustain fungal viability over extended periods (Parthasarathy 2023). The article suggested millet-based *B. bassiana* granules with a conidial concentration of 1.1×10^8 spores/g and a 98.2% germination rate after 24 h at 20 °C (Skinner et al. 2014). A solid culture based on millet granules of *B. bassiana* was tested; the granules performed against brown planthoppers (Lee et al. 2015), *Nilaparvata lugens*, by putting them in water around rice plants in the lab and then infesting brown planthoppers on the plants. In a bioassay using a spray to select the most virulent of four *B. bassiana* isolates, ERL836 was the most virulent and did the least harm to rice plants. When the ERL836 granules were dropped on water, they worked similarly to when the conidia were sprayed. In the lab, millet granules that had been colonized by ERL836 had been placed in water, and brown planthoppers were added at the same time. The granular treatment worked 87.7% of the time, 15 days after it was applied (Lee et al. 2015).

A sustainable EPF treatment has been done to maintain the inoculum level in pot soil while also lowering thrips populations. The oven-desiccated millet mixed with a commercial formulation of the *B. bassiana* strain GHA (BotaniGard) in pot soil leads to an increase in both spore production and the fungal ability to persist in soil (Davari et al. 2021). In the pots where millet was applied, the concentration of spores increased by 3 to 4 times after 30 days when compared to the spore yields observed at 10 days. In the marigold pots that received the GHA mix combined with millet, the number of *Frankliniella occidentalis* adults (Western flower thrips) was notably reduced compared to the untreated controls. Specifically, there were 12% and 10% of *F. occidentalis* adults in the treated pots, while the untreated pots had 70% in sterile and 68% in non-sterile soils (Davari et al. 2021). The addition of millet to the pot soil mixture improved the effectiveness of the EPF treatments by offering a nutrient-rich environment for the fungus to survive for a longer period. These approaches are quite affordable and straightforward for farmers to implement in greenhouses, especially since *B. bassiana* granule formulations are not readily obtainable in the global market.

4.2.4 Bait Formulations

Bait formulations combine EPF with attractive substrates (e.g., cellulose, molasses, or protein-based matrices) to lure target pests, increasing fungal contact and infection rates. This approach is particularly useful for social insects like termites, which exhibit behavioral avoidance of fungal pathogens. The effectiveness of various application methods of *B. bassiana* mycelia was studied (Bextine and Thorvilson

2002), which were alginate pellet encapsulated formulations in tackling *S. invicta* in fields with heavy infestations. Placing pellets right into *S. invicta* dunes led to a noticeable decrease in activity levels. Annotations of ants promptly eliminating pellets from mounds after they were placed inspired the creation of a bait system aimed at promoting the retention of these pellets within the insect mounds. Broadcasting and individual mound applications using *B. bassiana* pellets coated with groundnut oil effectively lowered the *S. invicta* populations in areas with heavy infestations. The direct placement of alginate pellets into *S. invicta* mounds led to a noticeable decrease in activity levels.

Observations of ants quickly removing pellets from their mounds after they were placed prompted the creation of a bait system aimed at promoting the retention of these pellets within the mounds. The use of broadcasting and embattled mound treatments with *B. bassiana* pellets, which are coated in groundnut oil, has shown a decrease in the activity levels of *S. invicta* populations (Bextine and Thorvilson 2002). Using poisoned baits proves to be an effective method for managing insect pests (Pelizza et al. 2019). Incorporating attractants and phagostimulants may enhance bait formulations, potentially leading to more effective treatments for managing grasshopper pests. In laboratory bioassays and field cage circumstances, the mixture of *B. bassiana* conidia with wheat bran and mustard oil showed improved performance, leading to increased mortality of spotted spur-throated grasshoppers (*Dichroplus maculipennis*) (Pelizza et al. 2019).

4.2.5 Encapsulated Formulations

Encapsulation involves embedding fungal spores within protective matrices (e.g., sodium alginate, chitosan, starch, or lipid-based coatings) to enhance stability and controlled release. Alginate is a frequently used material for the production and encapsulation of granules and nano- or microsphere formulations by using *M. anisopliae* and *B. bassiana* conidia. The sodium alginate-encapsulated *M. anisopliae* beads are shown to improve fungal persistence under field conditions (Sarma et al. 2023). Even microspheres containing *B. bassiana* spores have also been developed using starch and chitosan. Chitosan-based formulations offer an environmentally friendly, biodegradable, and non-toxic option, presenting a compelling alternative to synthetic pesticides (Friuli et al. 2023). Innovative encapsulation methods like coacervation, electrostatic encapsulation, liposomal encapsulation, freeze-drying formulation, fluidized bed coating, and spray drying have been investigated to enhance the development of alginate- or carbon-based formulations. Encapsulation techniques provide an accurate and uniform particle size, appearance, and encapsulation competence, leading to effective formulations that exhibit enhanced uniformity in storage, field application, and targeted performance. Formulations based on alginate have been developed as a potential approach to enhance the efficacy of EPF used as bioinsecticides (Bextine and Thorvilson 2002). Formulations of alginate EPF exhibit minimal toxicity to nontarget creatures, ensuring safety for human and animal health and leaving no adverse effects in the environment.

A variety of formulations utilizing alginate have been developed to deliver EPF, including emollients and droplets, aimed at managing different insect pest populations. A study (Sarma et al. 2023) revealed that alginate gel formulations of *M. anisopliae* spores enhanced the tenacity and viability of the asexual conidia over time. The findings from their study indicate that using sodium alginate beads, followed by freeze-drying, effectively preserves the viability of *M. anisopliae*. The results show that freeze-dried beads containing *M. anisopliae* are effective against white grubs, suggesting that these beads could serve as a viable substitute for synthetic pesticides for managing soil-dwelling insect pests, larvae, and grubs (Sarma et al. 2023). This study also supports future developments in alginate-encapsulated formulations intended for controlling soil insects.

Starch pellets are produced through a process that encapsulates EPF spores as small, spherical starch particles, including maize, potato, or rice starch. Starch has been used to create microspheres that encapsulate *B. bassiana* spores, in addition to its application in starch pellets. Starch pellets enhance the retention and attachment of fungal spores on target insects, facilitating extended contact and a continuous supply of spores, which boosts the likelihood of successful infection. The starch-based microspheres shield fungal spores from environmental stressors like UV radiation and desiccation, improving their shelf life and effectiveness as bioinsecticides. Supplementary coatings can be useful to the pellets or microspheres to enhance the fortification and efficacy of alginate-based formulations in field conditions (Friuli et al. 2023). Bio-coatings can include biodegradable polymers, surfactant molecules, or bio-adhesive materials that enhance sticking ability to target insects, facilitate controlled release of propagules, or improve formulation compatibility for specific crops. Optimizing the manufacturing process for encapsulated-based formulations involves adjusting several processing parameters, including temperature, relative humidity, pH, stirring speed, drying circumstances, and the concentration of starch and asexual spores. These parameters affect size, shape, spore release kinetics, and protection of the pellets or microspheres as output. These microspheres serve as a protective obstacle, safeguarding the EPF spores and maintaining their sustainability until they encounter the target insect pests.

4.2.6 Biomaterials-Based Formulations

Biomaterials like oils, gels, and polymers are being used frequently to safeguard EPF asexual spores from the harmful effects of the environment. These materials create coatings to protect the asexual spores from harmful UV radiation and desiccation, which can reduce their sustainability and effectiveness as biocontrol agents. Formulations based on oil or emulsions (mineral or vegetable) serve to shield spores from UV exposure and simultaneously boost their ability to adhere to insect cuticles, thereby increasing the likelihood of successful infection. Oil formulations enhance fungal adhesion to hydrophobic insect cuticles, particularly in pests with waxy coatings. Studies have shown superior efficacy of oil-based EPF formulations compared to aqueous suspensions. Oil formulations consist of antioxidants, mineral oil or synthetic oil, fungal propagules like spores, dispersants, detergents, thickening agents for paste formulations, anti-saprophytic agents, and non-ionic surfactants.

Oil-based formulations significantly enhance the temperature tolerance of EPF (Hernádez-Velázquez et al. 2000). Incorporating vegetable oils into the conidia of EPF has demonstrated an improvement in protection against detrimental UV radiation. These emphasize the important advancements in abiotic stress tolerance, deterrence of spore desiccation, and the improved shelf life achieved through fluid formulations and cryopreservation (Saminathan et al. 2025). In the study by Hernádez-Velázquez et al. (2000), it was found that citroline oil-formulated *M. anisopliae* significantly reduced the median lethal time (MLT) in Central American locusts (*Schistocerca piceifrons*). Ravi Sankar and Padmaja (2014) reported that *B. bassiana* in sunflower, olive, castor, and coconut oils induced higher mortality in *Spodoptera litura* larvae.

Hydrogels are widely used in the moisture retention process essential for the germination of fungal spores on the host insect surface or plant surface (Saminathan et al. 2025). Additionally, the use of encapsulation methods with biocompatible polymers facilitates the controlled spore release, thereby enhancing their obtainability and effectiveness in field conditions (Saberi Riseh et al. 2021). Methods of microencapsulation enhance the viability of spores by offering protection against environmental stressors and facilitating a sustained release, thereby increasing the reliability of EPF in agricultural and urban pest management methods. Biomaterial-enhanced formulations are compatible within IPM frameworks, providing an eco-friendly substitute for chemical pesticides while ensuring minimal ecological impact. The advancements in biomaterial-enhanced fungal formulations demonstrate their practicality for large-scale insect pest control, with field applications revealing extended perseverance and improved efficacy across diverse ecological conditions (Friuli et al. 2023). Formulations of alginate gel were created by combining spores of *M. anisopliae* with a sodium alginate aqueous solution, resulting in the formation of a gel formulation. The gel was consequently converted into small micro-droplets and developed into crosslinks in a solution of calcium chloride, resulting in the formation of spherical beads. The beads were subsequently dried under aseptic cabinet conditions for 32–42 h at a temperature of 28 ± 2 °C. A novel formulation utilizing alginate and *B. bassiana* spores has been created to target the blood-sucking insect *Triatoma infestans*, a vector for Chagas disease in Latin America (Friuli et al. 2023). Also, starch gels developed containing EPF spores are formed into small, spherical pellets through methods like extrusion, drop-wise gelatization, or pelletization techniques.

The starch or alginate is largely preferred by factors like availability, cost, and the specific properties required in the gel preparation. The selected starch is typically mixed with water or a suitable solvent to create a starch gel. The gel is produced through the application of heat to a starch-water mixture under controlled conditions, which facilitates the swelling of starch granules as they take up water, culminating in a viscous, gel-like consistency. Following the development of the starch gel formulation, it is subsequently cooled to a designated temperature range suitable for the incorporation of the EPF spores. The spores, whether in suspension or as a formulated product, are integrated into the starch gel and mixed thoroughly to ensure uniform distribution. The starch pellets that have been formed are subjected to a drying process aimed at removing excess moisture, thereby ensuring their stabilization. Various drying techniques are employed, including air-drying, freeze-drying, and oven-drying, each selected based on the specific formulation requirements and the necessity to maintain the viability of fungal spores. Extrusion entails the process of forcing the gel through a nozzle or die to obtain the specified pellet size, while drop-wise gelation involves the careful deposition of the gel into a cross-linking solution to create solid spherical pellets (Friuli et al. 2023).

The prevailing methodology in pest management predominantly employs sprayable synthetic chemicals aimed at addressing the adult, egg, and larval phases of insect pests. However, these substances may adversely affect the ecosystem and play a role in fostering resistance among thrips and various other pests. An innovative strategy for environmental management that emphasizes the soil-dwelling phase of thrips, employing a distinctive mechanism of action through biocontrol agents (Li et al. 2021). A comprehensive in vitro bioassay was conducted utilizing seven isolates of *B. bassiana*, revealing that two particular isolates, JEF-350 and JEF-341, exhibited a significant degree of virulence against melon thrips. The evaluation of conidial production and thermotolerance was conducted for JEF-341 and JEF-350 in comparison to the commercial isolate *B. bassiana* ERL836. *B. bassiana* JEF-350 showed similar conidial production levels to ERL836 and displayed significantly higher thermotolerance than both JEF-341 and ERL836. The pathogenicity of JEF-350 was assessed under controlled pot experiments as well as in natural field settings.

Millet-based fungal granules containing JEF-350 were applied to the soil surface to effectively target the pupal stage of melon thrips. In the conducted pot and field trials, JEF-350 granules demonstrated a significant decrease in thrips population, exhibiting greater efficacy compared to both JEF-341 and ERL836 granules. The application of JEF-350 should be conducted in conjunction with chemical fungicides to effectively manage the pupal stage of melon thrips present in the soil (Li et al. 2021). A study by Wenzel Rodrigues et al. (2017) assessed various biopolymers for their compatibility with both *B. bassiana* and *M. anisopliae*. Notably, maltodextrin at a concentration of 1.5% exhibited moderate toxicity toward *B. bassiana*. Dimethyl sulfoxide (DMSO) exhibited moderate toxicity levels for *B. bassiana* at a concentration of 3.0%, as well as for *M. anisopliae* at concentrations of 2.0% and 3.0%. The formulations exhibited stability throughout the 12-month evaluation period. But the pure conidia of the fungus showed a viability of 46% by the 6th month at a temperature of 26.5 °C (Wenzel Rodrigues et al. 2017).

4.3 Nano-Based Formulations

Nano-based formulations of EPF represent a cutting-edge advancement in the sustainable management of emerging and invasive insect pests of crop plants worldwide. Globally, India leads in the production of nano-based agricultural products, followed by the UK and Japan. Recent advancements in EPF bioformulation have focused on the development of novel delivery systems that improve field performance, enable targeted release, and enhance environmental resilience. The integration of nanotechnology with EPF has opened new frontiers in sustainable IPM practices, offering precise, eco-friendly solutions against invasive and resistant insect pests. Among these innovations, EPF-synthesized nanoparticles have gained prominence as a biologically derived, environmentally benign alternative to chemical pesticides (Bihal et al. 2023). Nanoparticles developed through various synthesis methods, particularly those mediated by fungi, have emerged as potent biopesticides (Mondal 2024). Advances in nanobiotechnology have highlighted hypocrealean fungi as efficient "nano-factories," capable of producing nanoparticles with controlled size and morphology at an industrial scale (Adeleke et al. 2024). By encapsulating fungal conidia or metabolites within biodegradable nanoparticles (e.g., chitosan, silica, or polymeric carriers), these formulations enhance UV protection, moisture retention, and targeted delivery, significantly improving field persistence and efficacy (Bihal et al. 2023).

Nanoparticles possess distinct mechanisms toward toxicity in insects; in particular graphene oxide (GO) and silver nanoparticles (AgNPs) composite (GO-AgNPs) exert multiple toxic effects on insect pests by disrupting physiological processes. These include impairing antioxidant and detoxification enzymes, disturbing enzyme activity, inducing oxidative stress, causing cellular damage, and causing structural damage to gut cells (Alian et al. 2025). Documented effects also encompass morphological deformities such as setae loss, cuticular dyspigmentation, and necrosis, as well as molecular-level disruptions like DNA binding, cell lysis, and interference with respiratory enzymes, leading to reactive oxygen species (ROS) accumulation and apoptosis (Bihal et al. 2023).

The utilization of myco-synthesized silver nanoparticles (AgNPs) offers a novel and eco-friendly approach for the control of pest populations, particularly through their larvicidal properties. *B. bassiana*, *M. anisopliae*, *I. fumosorosea*, and *Trichoderma harzianum* represent intriguing EPF recognized for their exceptional capacity to produce silver nanoparticles. The process is facilitated by extracellular synthesis, employing the aqueous extracts derived from these fungi. The silver nanoparticles synthesized from EPF extracts demonstrated a diverse size distribution, ranging from 10 to 200 nm, and exhibited a variety of morphologies, including spherical, triangular, hexagonal, and tubular forms. Silver nanoparticles obtained from *M. anisopliae* culture extracts have the potential to act as effective biocontrol agents against the malaria vector *Anopheles culicifacies* mosquitoes. Median lethal concentration (LC_{50}) of the analyzed nanoparticles ranged from 32 to 51 mg L^{-1} across various larval stages, with a noted concentration of 60 mg L^{-1} for pupae. Nonetheless, the reduced concentrations of AgNPs inhibited the development of both larvae and pupae (Amerasan et al. 2016). Titanium nanoparticles (TiNPs) were produced through the utilization of *M. anisopliae*, in both its non-irradiated state and following exposure to irradiation at a dose of 0.4 KGy. The nanoparticles produced were thoroughly characterized using a range of techniques, such as UV-Vis spectrophotometry, transmission electron microscopy (TEM), and energy-dispersive spectrometry (EDX). The investigation further examined the interactions between unirradiated and irradiated *M. anisopliae* and larval cuticle, employing scanning electron microscopy (SEM) for comprehensive observation.

The insecticidal efficacy of both unirradiated and irradiated *M. anisopliae*, in conjunction with and without their mycosynthesized TiNPs, was evaluated against *Galleria mellonella* larvae. The highest larvicidal efficacy was noted when irradiated *M. anisopliae* was combined with its mycosynthesized TiNPs. The observed improvement in effectiveness was associated with elevated fungal enzymatic activities, an increased concentration of TiNPs, and a reduction in the size of the TiNPs. The conducted studies demonstrated a collaborative larvicidal effect between irradiated *M. anisopliae* cells and their mycosynthesized TiNPs, yielding a synergistic factor (SF) of 1.6 for LC50 and 4.2 for LC90. This investigation elucidates the preliminary results regarding the synergistic effects of irradiated *M. anisopliae* and their mycosynthesized TiNPs. This combination may represent a viable approach to reduce the likelihood of insects acquiring resistance to EPF (Yosri et al. 2018).

A study demonstrated the first successful synthesis of ZnO nanoparticles using *B. bassiana* exhibited significant efficacy against the *Trialeurodes vaporariorum* (greenhouse whitefly), highlighting their potential in IPM (Khooshe-Bast et al. 2016). This research investigates the effects of two nano-formulated EPFs, *P. fumosorosea* and *Metarhizium flavoviride*, assessed in laboratory, greenhouse, and field environments targeting the potato tuber moth, *Phthorimaea operculella*. The findings revealed that the LC50 values for the nano-formulated *P. fumosorosea* and *M. flavoviride* were $10^3 \times 10^4$ and 89×10^4 spores/ml, respectively. In semi-field conditions, the observed LC50 values were 50×10^4 and 47×10^4 spores/ml. The influence of nano-formulated EPF on the life cycle of *P. operculella* demonstrates a significant decrease in the egg-laying rate per female. The count decreased to 23 ± 8.4 and 366 ± 8.7 eggs per female following treatment with nano-*P. fumosorosea* and nano-*M. flavoviride*, respectively, in comparison to 366 ± 8.7 eggs per female in the control group. The egg hatching rates decreased significantly to 2% and 1% after treatment with *P. fumosorosea* and *M. flavoviride*, compared to the 100% hatching rate recorded in the control (Sabbour 2015).

To maximize efficacy, nanopesticides are engineered to improve solubility, enable controlled release, and prevent degradation (Vishnu et al. 2024). These formulations are classified based on their carrier systems, including nanoemulsions (nano-suspensions/nano-dispersions) and nanoencapsulation (hydrophilic or hydrophobic) techniques (nanospheres/nanogels), which facilitate better adhesion to insect cuticles, promoting faster infection and mortality in *S. frugiperda* and *Bemisia tabaci* (Vishnu et al. 2024). Additionally, nano-based carriers enable controlled release of fungal propagules, reducing application frequency while maintaining lethal concentrations. Typically, oil-in-water (O/W) systems, nanoemulsions disperse active ingredients as nanoscale droplets, stabilized by surfactants at the oil-water interface (Shin et al. 2023).

4.4 Smart Encapsulation and Controlled Release

Recent innovations in formulation technology have significantly enhanced the environmental resilience and field efficacy of EPF through smart encapsulation and controlled-release systems. Solid encapsulated formulations, including tablets and pellets, optimize ease of application and persistence. Microencapsulation techniques employ biodegradable polymers (e.g., alginate, chitosan) to create protective matrices around fungal spores, shielding them from UV radiation, desiccation, and temperature extremes while enabling gradual liberation in response to environmental triggers (Quesada-Moraga et al. 2024). Microencapsulation, a process that entraps fungal spores within biodegradable polymeric matrices such as gelatin, starch, or cellulose, provides robust protection against abiotic stressors while enabling sustained spore release. This research presents a cellulose nanomaterial (CNM)-stabilized pickering emulsion aimed at encapsulating *B. bassiana* spores (Shin et al. 2023). *B. bassiana* is widely used as an entomopathogenic fungus for controlling agricultural insect pests; however, it struggles to withstand sunlight and UV exposure when applied to crops in open field conditions. The composition of pickering emulsions for encapsulating *B. bassiana* conidia was formulated utilizing varying ratios of cellulose nanocrystals and cellulose nanofibrils. The study examined phase stability, emulsion size distribution, colloidal stability, and rheological properties. The encapsulated conidia maintained viability for germination and growth, exhibiting notable insecticidal activity despite exposure to UV radiation. This research demonstrated that CNM-stabilized pickering emulsions are capable of effectively encapsulating microbial cells, thereby aiding in the preservation of their viability under challenging environmental conditions (Shin et al. 2023).

Recent advancements in the development of alginate-based microcapsules have fulfilled essential needs for bacterial encapsulation, such as being biocompatible, biodegradable, and capable of supporting long-term survival and functionality. The study developed *B. bassiana* microcapsules using sodium humate coatings by spray-drying, improving UV and desiccation resistance (Aziz Qureshi et al. 2015). Similarly, capsule suspensions (CS), where bioagents are encased in polymeric shells and suspended in aqueous media, offer extended shelf life and gradual spore liberation, mitigating rapid environmental degradation. Hydrogel-based formulations, such as alginate-encapsulated *M. anisopliae*, further enhance solar radiation and thermal tolerance (Saberi Riseh et al. 2021). The study used biomass that includes mycelium (Stephan et al. 2021) or a combination of mycelium and submerged spores from the EPF. *B. bassiana*, *Cordyceps fumosorosea*, and *M. brunneum* were cultivated in liquid broth. The biomass was then separated, and concentrations were adjusted. Then, the EPF biomass was exposed to thermotolerance and adjustments to fluid-bed coating. Additionally, they explored the process of granules for colonization and sporulation. The colonization of granules was influenced by both the concentration of biomass and the specific strain used. In the case of *C. fumosorosea* and *B. bassiana*, using a concentration of 0.003% dry weight led to almost complete granule colonization, while *M. brunneum* at a concentration of 0.7% dry weight achieved only about 50% colonization. The moisture content had a significant impact on the conidiation observed on granules in sterile soil. Due to the unsatisfactory granule colonization of *M. brunneum*, they first pre-coated the nutrients and then applied a layer of biomass, submerged spores, or conidia. The addition of malt extract positively influenced the colonization of granules, enhancing both biomass and submerged spores. They found (Stephan et al. 2021) that the process of fluid-bed coating applied to fungal biomass proves to be an effective method for developing a granular formulation.

Innovations like capsule suspension further optimize shelf life and application efficiency by suspending microencapsulated spores in aqueous media for on-demand dilution. These technologies mitigate abiotic stress-based degradation and enhance host-specific delivery, reducing nontarget effects and application frequency. By integrating responsive materials (e.g., temperature- or pH-sensitive polymers), next-generation EPF formulations promise precision biocontrol with extended residual activity, bridging the gap between laboratory efficacy and field performance (Harish et al. 2023). These advanced delivery systems not only prolong fungal viability under field conditions but also improve precision in pest targeting, underscoring their potential in sustainable IPM strategies. These formulations, such as pellets, granules, and gels, provide numerous benefits compared to conventional chemical insecticides, including precise targeting, environmental safety, and sustainability. Biopolymer-based formulations are environmentally friendly and biodegradable, providing a sustainable option compared to chemical pesticides (Saberi Riseh et al. 2021; Friuli et al. 2023). To ensure the practicality and cost-effectiveness of these formulations, it is essential to address challenges such as production costs, stability, and shelf life. Various approaches could lead to the achievement of these goals.

4.5 Stimuli-responsive Formulations

Stimuli-responsive formulations are a novel advancement in EPF delivery systems, enhancing biocontrol efficacy through environmentally triggered spore release (Chaud et al. 2021). These formulations employ smart encapsulation technologies that utilize bioengineered coatings or encapsulation matrices that release fungal spores to specific abiotic or biotic cues, such as temperature fluctuations, humidity levels, pH changes, or host-derived chemical cues, ensuring targeted and optimized fungal activity. By protecting spores from detrimental environmental stressors (e.g., UV radiation, desiccation, or extreme temperatures) during storage and application (Quesada-Moraga et al. 2024), these systems significantly improve spore viability and field persistence. Mostly, temperature-responsive coatings can delay spore liberation until optimal thermal conditions for insect activity are assured; also, humidity-sensitive polymers release conidia only under favorable moisture levels, preventing premature inactivation. Host-specific formulations further refine precision by activating in response to insect cuticular compounds or physiological pH shifts, minimizing off-target effects. Such approaches extend the residual activity of EPF and reduce application frequency and environmental contamination. Examples include pH-responsive hydrogels for soil-dwelling pests and moisture-triggered microcapsules as foliar applications for sap-feeders. By integrating these novel innovations, stimuli-responsive formulations bridge the gap between laboratory efficacy and field performance, offering a sustainable, precision-based solution for IPM that aligns with ecological and agricultural demands. This technology holds potential for challenging environments where conventional biopesticides fail, offering the pathway for next-generation, climate-resilient EPF products.

EPF formulations have evolved significantly, offering diverse application methods tailored to different agroecosystems and pest behaviors. Still, traditional formulations like wettable powders and oil dispersions remain widely used, and emerging technologies such as nanoencapsulation and bait systems are primitive for improving precision and sustainability in pest management. The integration of EPF with metallic nanoparticles (e.g., silver or copper) can synergistically enhance antifungal and insecticidal properties, broadening their spectrum of activity (Crisan et al. 2021). Despite challenges in scalability, cost, and regulatory approval, nano-formulated EPFs offer promising solutions against pesticide-resistant pests and reduce nontarget effects.

5 Genetic Engineering and Molecular Innovations

5.1 Genome Editing

Genetic engineering has the potential to enhance the ability of EPF to effectively attack their hosts through various mechanisms (e.g., CRISPR/Cas). Recent biotechnological advances have revolutionized the development of EPF as enhanced biocontrol agents through targeted genetic engineering approaches (Parthasarathy 2023). Key research focuses include identifying and manipulating genes governing critical infection processes such as conidial adhesion, cuticle penetration, thermotolerance, and toxin production to create optimized fungal strains with elevated virulence and environmental resilience. In this case, by enhancing virulence through the introduction of specific genes responsible for toxins, enzymes, or other factors associated with virulence, researchers may develop engineered strains that exhibit increased lethality toward targeted pests. The genetically modified chitinases (*Bbchit1* and *Bbchit2*) play a vital role as a virulence factor in *B. bassiana*, particularly in its interactions with various lepidopteran insects. The virulence of mycotoxins can be enhanced through several approaches, including (i) modifying the genes of fungal strains and (ii) modifying insect proteins to increase the virulence of mycotoxins (Karthi et al. 2024).

Recombinant DNA technology has enabled the progress of EIPF capable of producing insecticidal proteins while colonizing host plants (Fadiji and Babalola 2020), while gene transfer techniques facilitate the production of plant-protective secondary metabolites. Strategic genetic modifications have yielded breakthroughs in fungal efficacy, including engineered strains producing cuticle-degrading enzymes to accelerate host penetration and transgenic *B. bassiana* (BbHV8) with overexpressed class I hydrophobin (*hyd1*) genes demonstrating superior virulence against *S. litura* (Wang et al. 2013). *P hyd1-t1* serves as an important resource for boosting the expression of beneficial proteins like vip3Aa1 in fungal conidia, which act as the active components of mycoinsecticides. Modern genomics tools, including next-generation sequencing (Illumina, 454 pyrosequencing) and metagenomic analysis, are further elucidating plant-fungal interactions and microbiome dynamics (Bamisile et al. 2021), enabling precision editing of fungal traits. These innovations collectively enhance EPF biocontrol precision, environmental adaptability, and scalability, positioning genetically optimized fungi as sustainable, next-generation alternatives to chemical pesticides in IPM systems.

5.2 Enhancing Virulence, Environmental Tolerance, and Host Range

Recent advances in genetic engineering and RNA interference (RNAi) have enabled the development of recombinant EPF with enhanced stress tolerance, virulence, and host specificity. By integrating insecticidal genes such as scorpion neurotoxins (Androctonus australis) into fungal genomes, researchers have significantly increased pathogenicity; for example, *M. anisopliae* engineered to express this toxin exhibited a ninefold increase in virulence against insect pests without compromising host specificity. However, the efficacy of EPF in field applications is often constrained by environmental stressors, including extreme temperatures, UV radiation, and fluctuating humidity, which impair spore viability and infectivity (Quesada-Moraga et al. 2024). To address these limitations, stress-tolerant fungal strains have been developed through genetic modifications, such as the incorporation of archaeal photoreactivation systems and pigment synthesis pathways, enhancing UV resistance. Additionally, naturally resilient isolates like *B. bassiana* PKDE exhibit exceptional tolerance to high temperatures (40 °C), osmotic stress, and oxidative conditions due to elevated trehalose and antioxidant enzyme levels (Nimisha et al. 2024).

Further research into metabolic pathways, such as mannitol biosynthesis (mediated by *MPD* and *MTD* genes) and heat shock proteins (*Hsp70a-f*), has provided insights into multi-stress resistance mechanisms in *B. bassiana* (Wang et al. 2020). The stress-induced auxiliary chaperone protein *Sti1*, analogous to the mammalian protein *Hop*, has demonstrated notable and varied roles in the adaptation of different fungi to their environments and their capacity to induce infections (Liu et al. 2025). The precise roles of *Sti1* in filamentous entomopathogens are still not well understood. This investigation examines the functions of *Sti1* in *B. bassiana*, a prominent entomopathogen frequently utilized as a fungal pesticide for the management of insect populations. The disruption of *Sti1* led to a notable 68.88% reduction in colony area at 33 °C and a 28.88% decrease in the median lethal time (LT50) of asexual conidia at 45 °C. Nevertheless, it exhibited comparable hyphal growth and conidial germination relative to the control strains, which comprised the wild-type strain and complementation mutants, under standard conditions. The removal of sti1 resulted in a decrease in conidiophores, subsequently leading to a reduction of 33.95–43.49% in conidiation after an incubation period of 4–5 days. Additionally, the *sti1*-deficient mutants exhibited compromised cell wall integrity and increased susceptibility to various stressors, including carbendazim, hydrogen peroxide, menadione, sodium chloride, and UV-B radiation. The findings corresponded with a reduction in the levels of glycerol, trehalose, mannitol, superoxide dismutases (SOD), and catalases (Litwin et al. 2020). The results suggest that *Sti1* is important in the asexual conidiation, adaptation to harsh environments and pathogenicity of *B. bassiana* (Liu et al. 2025). These innovations underscore the potential of genetically optimized EPF to overcome environmental challenges while maintaining high virulence, paving the way for more reliable and sustainable integrated pest management solutions.

5.3 Molecular Markers for Strain Tracking and Efficacy Monitoring

The development and application of molecular markers have revolutionized strain-specific tracking and performance assessment of EPF in both research and field applications. Advanced molecular techniques, including SSR (simple sequence repeat), ISSR (inter-simple sequence repeat), and SNP (single nucleotide polymorphism) markers, enable precise differentiation of closely related EPF strains, facilitating quality control during mass-production and post-application monitoring. Multilocus sequence typing (MLST) of housekeeping genes (e.g., β-tubulin, Bloc, TEF-1α, RPB1/RPB2 nrSSU, nrLSU, TEF) provides reliable phylogenetic discrimination of *Beauveria* and *Metarhizium* species complexes, while mitochondrial markers offer insights into population dynamics and environmental persistence (Bidochka and Hajek 1998; Karthi et al. 2024; Cai and Dimopoulos 2025). Quantitative PCR (qPCR) assays using strain-specific primers allow for sensitive detection and quantification of applied fungi in complex environmental samples, enabling researchers to correlate fungal biomass with biocontrol efficacy. Recent advances in whole-generation sequencing by next-generation sequencing have identified virulence-associated molecular markers, such as protease genes (Pr1) and toxin biosynthesis clusters, which serve as predictive indicators of strain pathogenicity (Litwin et al. 2020; Sharma and Sharma 2021).

Fluorescent protein tagging combined with confocal microscopy has further enhanced in situ tracking of fungal colonization patterns in both insect hosts and plant tissues. These molecular tools are proving invaluable for monitoring field persistence, assessing nontarget effects, and optimizing application strategies. Emerging technologies like CRISPR-based barcoding and metagenomic analysis promise even greater resolution for studying fungal establishment and ecological interactions in diverse environments (Saha et al. 2025). The integration of molecular marker data with GIS mapping and machine learning algorithms will establish precision biocontrol systems, where strain performance can be predicted based on environmental parameters and pest population genetics. This molecular toolkit strengthens quality assurance in commercial EPF products and also provides critical insights for the selection and improvement of elite strains through marker-assisted breeding programs.

6 Environmental and Ecological Applications

6.1 Soil Bioremediation

Recent research highlights the emerging potential of EPF as agents of soil bioremediation, extending their role beyond biological pest control. Certain EPF species, particularly from the *Beauveria* and *Metarhizium* genera, demonstrate remarkable capabilities in degrading environmental pollutants, including pesticides, heavy metals, and hydrocarbons. These EPF produce extracellular enzymes such as laccases, peroxidases, and esterases that break down complex organic compounds, while their filamentous growth enhances soil penetration and contaminant accessibility (Ahsan et al. 2024; Saha et al. 2025). Studies show *B. bassiana* can metabolize organophosphate pesticides through enzymatic detoxification, while *Metarhizium* spp. exhibit heavy metal tolerance and biosorption properties, immobilizing toxic elements in contaminated soils. Additionally, EPF form symbiotic relationships with plants in polluted environments, enhancing phytoremediation efficiency through improved nutrient uptake and stress tolerance. Their rhizosphere competence further facilitates the establishment of beneficial microbial consortia that synergistically degrade pollutants. Field trials demonstrate the ability of EPF to persist in contaminated soils and maintain bioremediation activity, suggesting their potential for large-scale soil restoration. However, optimizing fungal strains, formulations, and application methods remains critical to maximize remediation efficiency across diverse soil types and pollution scenarios. As sustainable alternatives to chemical remediation, EPF offer a dual benefit, simultaneously rehabilitating degraded soils and suppressing soil-borne pests, positioning them as innovative tools in ecological restoration and integrated crop and land management strategies. Future research should explore genetic engineering approaches to enhance their degradative capacities and environmental adaptability (Sharma et al. 2023).

EPF can carry out biotransformation, resulting in the production of derivatives that exhibit enhanced activity. These processes are advantageous, as they enhance the absorbability of products. A variety of filamentous fungi are involved in steroid transformations, and EPF demonstrates notable efficiency in these processes as well. Recent studies indicate that *I. farinosa* strains are capable of effectively transforming dehydroepiandrosterone (DHEA) into 7a- and 7b-hydroxy derivatives (Kozłowska et al. 2019). Among the strains studied, *P. farinosa* demonstrated a remarkable ability to transform androstenediol, androstenedione, adrenosterone, 17a-methyltestosterone, 17b-hydroxyandrost-1,4,6-triene-3-one, and progesterone into mono- or dihydroxy derivatives, achieving both high yield and stereoselectivity (Kozłowska et al. 2019). An independent study validated the biotransformation of estrogens (E1, E2, and EE2) by *P. fumosorosea*. *B. bassiana* is also utilized in the pharmaceutical sector due to its ability to produce high yields of valuable 11a-hydroxy intermediates of steroids (Saminathan et al. 2025). EPF can transform steroids on a significant scale, presenting potential applications for industrial use in the future. EPF appear to hold significant promise in these processes; however, at present, *Isaria* and *Beauveria* are the focus of frequent study, while the potential of *Metarhizium* remains largely unexplored in biotransformation (Litwin et al. 2020).

6.2 Soil Health Restoration

EPF demonstrate a multifunctionality as biocontrol agents, soil biostimulants, and plant growth promoters through their endophytic or rhizospheric associations (Harish et al. 2019). Much research reveals that EPF, such as *B. bassiana* and *Metarhizium* spp., enhance plant growth by phytohormone production, including auxins (e.g., indole acetic acid), gibberellins, and cytokinins, which regulate critical physiological processes (Liao et al. 2017). These hormones stimulate root architecture, nutrient uptake, and photosynthetic efficiency, leading to improved plant vigor and resilience against pathogens, evidenced by *B. bassiana*-inoculated plants exhibiting reduced disease severity from *Zucchini yellow mosaic virus* infection. Furthermore, EPF exhibit rhizosphere competence, with *M. anisopliae* conidia persisting longer in root-associated soils than in bulk soil (Bruck 2005), suggesting a synergistic role in soil health. Notably, *Metarhizium robertsii* promotes *Arabidopsis* growth by modulating IAA-dependent pathways, rescuing root defects in IAA-deficient mutants, and activating auxin-responsive genes (Liao et al. 2017). Such traits, coupled with their ability to enhance nutrient cycling and stress tolerance, position EPF as multifunctional bioagents for sustainable agriculture, bridging pest management and soil-plant microbiome optimization.

6.3 Potential Roles in Climate Change Mitigation

EPF may contribute to climate change mitigation through multiple ecological mechanisms, particularly by enhanced carbon sequestration in terrestrial ecosystems. As integral components of soil microbiomes (Qadri et al. 2020), EPF influence carbon cycling through their saprophytic activity, converting insect-derived chitin into stable soil organic matter. Species such as *B. bassiana* and *M. robertsii* form endophytic relationships with plants, stimulating root growth and exudation, which promotes soil carbon storage through increased rhizodeposition and microbial biomass. Their symbiotic associations with plants also enhance photosynthetic efficiency and biomass production, further amplifying carbon capture. Additionally, EPF-mediated decomposition of insect cadavers generates fungal biomass rich in melanized hyphae, which are resistant to degradation and contribute to long-term carbon stabilization in soils. Field studies suggest that EPF inoculation can increase soil organic carbon content by 10–15% in agroecosystems, while reducing reliance on synthetic pesticides lowers associated greenhouse gas emissions from chemical production and application. Some EPF strains demonstrate temperature and drought tolerance, potentially maintaining carbon sequestration functions under climate stress (Quesada-Moraga et al. 2024). However, the scalability of these effects requires further quantification across ecosystems. Additional research should explore genetic traits influencing carbon allocation in EPF, develop formulations to optimize their persistence in carbon-poor soils, and assess their integration with regenerative agricultural practices. These multifunctional organisms represent a promising biological potential for climate-smart land management strategies that must be explored further.

7 Emerging Trends in Applications

7.1 Integrated Pest Management (IPM)

EPF exhibit significant potential for integration with other biocontrol agents, including predators and parasitoids, enhancing their efficacy in IPM systems. Studies have demonstrated synergistic interactions between EPF and natural enemies, such as the combined use of *B. bassiana* and *M. brunneum* with the parasitoid *Aphidius colemani* for controlling *Myzus persicae* in sweet pepper (Jaber and Araj 2018). Similarly, *L. lecanii* has been successfully paired with aphid alarm pheromones and reduced insecticide doses to improve aphid suppression (Subramaniam et al. 2021). Further research highlights the compatibility of *B. bassiana* and *Hirsutella lixii* with parasitoids like *Diglyphus isaea* and *Phaedrotoma scabriventris* for managing *Liriomyza huidobrensis* (Akutse et al. 2014). Additionally, semiochemical-mediated fungal dissemination strategies, such as using *Z. radicans* conidia in attractant-based inoculation devices (Georgiev et al. 2013) and the integration of EPF with fungicides for postharvest disease control, highlight the adaptability of EPF in diverse pest management frameworks.

The compatibility of EPF with chemical and botanical pesticides varies significantly, influencing their applicability in IPM. While some insecticides, such as chlorantraniliprole, spinosad, and imidacloprid, support fungal growth, others, like chlorpyriphos insecticide and carbendazim fungicide, exhibit strong inhibitory effects. Similarly, *B. bassiana* isolates demonstrate high compatibility with certain insecticides (e.g., spinosad) but are suppressed by fungicides like hexaconazole (Nimisha et al. 2024). Botanical extracts, such as neem and pongamia oils, show strain-dependent interactions, with some enhancing EPF efficacy while others impair spore germination. *Cordyceps javanica* is a significant EPF that serves as a mycoinsecticide, effectively managing pests like aphids and whiteflies. Despite its significance, there has been limited research exploring how *C. javanica* affects aphids. They conducted a virulence assay to explore *C. javanica*, known for its strong virulence against *Acyrthosiphon pisum* (Wang et al. 2023). Notably, combinations of *C. javanica* with reduced-dose insecticides (e.g., thiamethoxam, clothianidin) or botanicals significantly improve pest mortality compared to standalone treatments.

The combined application of EPF, including *B. bassiana*, alongside various chemical pesticides has been explored. It has been proposed that embracing this combination could enhance resistance management strategies and mitigate ecosystem pollution resulting from the overuse of chemical insecticides. Over the last several decades, the integration of biological control agents using an autodissemination method has demonstrated notable success. This approach has demonstrated effectiveness for numerous strains, particularly when employing EPF alongside semiochemicals and various natural insect adversaries. These findings emphasize the need for tailored integration strategies, balancing fungal viability with chemical or botanical synergies to optimize pest control outcomes in sustainable agriculture. EPF presents a multifaceted and eco-friendly approach to

Table 18.2 Importance of EPF in sustainable pest management

Entomopathogenic fungus	Target pest	Reference
Aschersonia aleyrodis	Citrus whitefly, Glasshouse whitefly, and Scales	Ingle et al. (2022)
Beauveria bassiana	Apple leaf miner, Bihar hairy caterpillar, Boll weevil, Brinjal aphid, Brown plant hopper, Chinch bug, Citrus weevil, Colarado potato beetle, Coffee berry borer, Corn root worm, Cotton bollworms, Cotton leaf hopper, Guava whitefly, Japanese beetle, Mango mealy bug, Rice leaf folder, Soybean pod borer, Tea mosquito bug, Wheat aphid, and Yellowstem borer	Tahira et al. (2014) and Ambethgar (2018)
Entomophaga maimaiga	Gypsy moth	Tahira et al. (2014)
Entomophthora muscae	Flies	Tahira et al. (2014)
Hirsutella thompsonii	Citrus rust mite, Coconut eriophyid mite, Red spider mite, and Yellow mite	Ambethgar (2018) and Ambethgar et al. (2019)
Isaria fumosorosea	Rugose spiraling whitefly	Sumalatha et al. (2020)
Lecanicillium lecanii	Aphids, Green leafhopper, Potato leafhopper, Scales, and Whiteflies	Harper and Huang (1986)
Metarhizium anisopliae	Fall armyworm, Grasshopper, Locust, Rhinocerous beetle, Spittle bug, Sugarcane hopper, and Termite	Bischoff et al. (2009) and Quesada-Moraga et al. (2024)
M. brunneum	Western thrips	Ain et al. (2021)
M. rileyi (N. rileyi)	Cabbage looper, Castor semi-looper, Cotton bollworm, and Tobacco cutworm	Grewal et al. (2021)
Nerodia floridana	Mites	Tahira et al. (2014)
Neozygites fresenii	Aphids	Tahira et al. (2014)
Nosema locustae	Locusts, grasshoppers, and crickets	Bidochka and Hajek (1998)
Pandora delphacis	Brown plant hopper and green leaf hopper	Zhou et al. (2014)
P. neoaphidis	Aphids	Tahira et al. (2014)
Zoophthora radicans Entomophaga maimaiga	Aphids, and Gipsy moths	Georgiev et al. (2013)

pest management, demonstrating efficacy against a diverse array of pests such as aphids, whiteflies, hemipteran bugs, thrips, beetles, and lepidopteran caterpillars (Table 18.2).

The efficacy of EPF in soil ecosystems is significantly influenced by complex microbial interactions. Soil microbiota produce antimicrobial metabolites that can suppress EPF activity and have been reported to reduce *B. bassiana* virulence against Colorado potato beetles under high soil fungistasis. However, hypocrealean entomopathogens have evolved competitive strategies, producing secondary metabolites during host infection that inhibit opportunistic microorganisms during the saprotrophic phase (Litwin et al. 2020; Sharma et al. 2023). This ecological interplay suggests potential for synergistic microbial consortia development, where compatible microorganisms could enhance EPF persistence and activity while mitigating soil fungistasis. Strategic combinations of EPF with plant growth-promoting rhizobacteria or mycorrhizal fungi may create mutually beneficial relationships, improving both pest control and plant health. Future research should focus on identifying optimal microbial partnerships that maintain EPF virulence while leveraging the ecological services of soil microbiomes, offering a novel approach to sustainable pest management in agroecosystems. Their field applications have evolved significantly, with traditional methods such as deploying infected insect cadavers, either fresh or stored, and releasing laboratory-infected insects proving effective in suppressing pest populations.

The success of EPF as biocontrol agents depends on optimized application techniques that ensure uniform spore distribution, sufficient pest contact, and favorable environmental conditions for fungal germination and persistence. Innovations in field application methods include foliar spraying, soil incorporation with carrier materials, dry dust formulations, and strategic cadaver placement. Four primary EPF application strategies have emerged as particularly effective in IPM: seed dressing, where spores of *B. bassiana* and *M. robertsii* reduce aphid infestations in crops like wheat and beans (Ambethgar 2018); plant/root dipping, which protects transplantable crops such as cereals, millets, and vegetables during early growth stages; soil drenching, introducing fungi like *Verticillium lecanii* and *M. anisopliae* into the rhizosphere to combat soil-dwelling pests; and foliar spraying, enhanced by innovations such as combining *B. bassiana* with heat-resistant chitinase to increase pathogenicity. Commercially available mycopesticides, formulated with live fungal propagules, employ either inundative or inoculative strategies to suppress pest populations, demonstrating the expanding role of EPFs in sustainable agriculture (Sabbour 2015; Friuli et al. 2023; Ahsan et al. 2024).

7.2 Commercial Applications

As research continues to refine application techniques and formulations, EPFs are poised to play an increasingly critical role in eco-friendly crop protection, aligning with global trends toward reduced chemical pesticide reliance and enhanced ecological resilience (Mantzoukas and Eliopoulos 2020; Mantzoukas et al. 2022). Mycoinsecticides are integral to the biopesticide market. Their proportion of the worldwide crop protection market is currently at 5%, with an annual growth rate of 10% observed. Their market share in the global biopesticide sector is approximately 35%, placing them in the second position, trailing only bacterial pesticides. The production of fungal insecticides is predominantly concentrated in areas including Asia, America, Africa, and Europe. China has approved over 30 mycoinsecticide

Table 18.3 Commercially available entomopathogenic fungal products

EPF	Commercial formulation	Target pests
Beauveria bassiana	Mycotrol, BotaniGard, Velifer, Mycoguard Bb-L	Aphids, Whiteflies, Thrips etc.
Entomophthora muscae	Eco-PM	House flies
Isaria fumosorosea	PFR-97, Shatpada Rugose White Fly Kill	Whiteflies, Rugose white fly, Aphids, and Thrips
Lecanicillium lecanii	Naturalis-L, Verelac	Aphids, Whiteflies, and Thrips
Metarhizium anisopliae	Met52 or Kalichakra	Beetles and Caterpillars
Paecilomyces. fumosoroseus	NoFly, PreFeRal	Aphids and Whiteflies
Paecilomyces lilacinus	Arka Krishi Kawach	Root Knot Nematode

products, predominantly employing *B. bassiana* to manage notable lepidopteran pests, such as the pine moth and diamondback moth.

A diverse range of mycoinsecticides has demonstrated efficacy in managing various lepidopteran pest populations. Table 18.3 presents a compilation of commercially available products, including Mycotrol WP, which is derived from *B. bassiana* and produced by Emerald BioAgriculture Corp. (USA). This product exhibits notable pesticide effectiveness against aphids and whiteflies. Myco-Jaal, derived from *B. bassiana* and manufactured by Pest Control India Pvt. Ltd. (India), demonstrates notable insecticidal efficacy against the diamondback moth. Furthermore, Numoraea 50, which is obtained from *N. rileyi* by Ago Biocontrol in Colombia, exhibits significant insecticidal effectiveness against various lepidopteran pests. TRI-SIN is a commercial formulation that integrates *B. bassiana*, *M. anisopliae*, and *P. fumosoroseus*, created by Agrobiologicos del Noroeste S.A. de C.V. (Agrobionsa, Mexico), and exhibits notable pesticidal efficacy against psyllids. Entomopathogens are crucial in the global management of agricultural pests. Despite their significance, many mycoinsecticides face challenges in commercialization because of stringent regulatory standards. The discontinuation of specific products, such as Boverin, from the former USSR has transpired due to this scenario (Seethapathy 2024).

7.3 Veterinary and Medical Entomology

EPFs are increasingly recognized for their potential in the biological control of disease vectors such as mosquitoes, ticks, and flies. This approach presents an environmentally sustainable alternative to traditional chemical pesticides. Over the past 40 years, research has identified more than 70 new secondary metabolites originating from *Isaria* sp. (Berestetskiy and Hu 2021). Specifically, beauvericin, a non-ribosomal peptide metabolite that is a cyclic hexadepsipeptide mycotoxin derived from a strain of *P. fumosorosea*, has demonstrated antibacterial, anti-feedent, anti-larval, antiviral, insecticidal, and cytotoxic properties, making it a promising candidate for the development of new pesticides (Sharma et al. 2023; Saha et al. 2025). Scientists (Liao et al. 2017) engineered *Metarhizium* strains to shorten mosquito lifespans, impairing *Plasmodium* transmission and reducing disease spread without inducing resistance. Beyond genetic approaches, EPF has demonstrated efficacy in nanoparticle-based applications. *T. harzianum*-synthesized silver nanoparticles achieved up to 100% mortality in larvae and pupae (Saminathan et al. 2025). Additionally, innovative formulations like Merril 10 SP, a mineral oil-based *M. anisopliae* product, effectively suppressed cattle tick (*Rhipicephalus microplus*) populations (Karthi et al. 2024).

These developments highlight the expanding role of EPF in IPM, combining precision, sustainability, and adaptability for next-generation vector control strategies. Moreover, fumosorinone, a terpene compound sourced from the same fungal species, exhibits the potential to function as a classic non-competitive inhibitor of protein tyrosine phosphatase 1B (PTP1B). This indicates that the compound could be significant in medical applications concerning the medical treatment of type II diabetes and related metabolic disorders. Furthermore, there are many documented instances concerning the isolation and characterization of a range of biochemicals with medicinal importance obtained from diverse fungal species, such as *B. bassiana*, *B. brongniartii*, and *M. anisopliae* (Bamisile et al. 2021).

7.4 Industrial and Biotechnological Sector

EPF represents an untapped reservoir of bioactive compounds with significant industrial and biotechnological applications. These fungi produce a diverse array of secondary metabolites, including enzymes (chitinases, proteases, lipases), antimicrobial agents (beauvericin, oosporein), and insecticidal compounds (destruxins, bassianolide), which have potential uses across multiple sectors (Litwin et al. 2020). Cyclic depsipeptides like destruxins (Dtxs) show a wide range of biological effects, such as fighting infections, combating tumors, tackling fungal infections, reducing inflammation, and suppressing the immune response (Karthi et al. 2024). These diverse actions stem from their distinctive mechanisms and adaptable molecular structures. In agriculture, EPF-derived enzymes offer eco-friendly alternatives for pest control and soil bioremediation, while their antimicrobial metabolites show promise as novel antibiotics in pharmaceutical development. The pharmaceutical industry is particularly interested in EPF-produced compounds for their anticancer, immunosuppressive, and antiviral properties, with several molecules currently under preclinical investigation. Additionally, EPF synthesize unique bioactive compounds that could revolutionize biocatalysis and biomaterial production, such as fungal chitosan for wound healing and drug delivery systems. Recent advances in genomics and metabolomics have accelerated the discovery of novel EPF metabolites, enabling targeted exploration of their biosynthetic pathways for synthetic biology applications. The cosmetics industry is also exploring fungal melanins and antioxidants for UV protection and anti-aging formulations. However, challenges remain in scaling up production, optimizing extraction methods, and ensuring compound stability for commercial use. Future research should focus on strain improvement through genetic engineering, the development of cost-effective fermentation processes, and the exploration of extremophilic EPF strains for novel compound discovery. As sustainable production platforms, EPFs offer a compelling alternative to chemical synthesis, aligning with growing demand for bio-based products in the circular economy. Their dual role as biopesticide producers and biochemical factories positions EPF as multifaceted biological resources for next-generation industrial biotechnology (Saha et al. 2025).

8 Challenges and Limitations in Application

The potential for unintended residual impacts on predators, parasitoids, pollinators, and various nontarget organisms has been noted as well. This phenomenon has been investigated in various insect-pathogenic fungal species that exhibit a broad spectrum and extensive host range, such as *B. bassiana* and *B. brongniartii*. The use of nanopesticides poses unexpected risks, particularly in terms of their potential ecotoxicological impacts, such as toxicity to nontarget organisms, bioaccumulation, and interactions with other environmental pollutants, all of which could lead to significant economic consequences (Crisan et al. 2021). Standard ecotoxicological tests are inadequate for assessing these risks because the physicochemical properties of nanopesticides, their release dynamics, and their interactions with plants and other environmental factors differ significantly from those of conventional pesticides (Chaud et al. 2021). Temperature serves as the primary constraint on the infectivity potential of EPF, while UV-B, followed by humidity and temperature, are the principal factors leading to the depletion of propagules.

While mass-production and formulation methods offer minimal protection for propagules against environmental inactivation, they fall short of addressing the problem of virulence loss in entomopathogenic fungal strains (Seethapathy 2024). To ensure high virulence, the most effective approach is to select fungal strains that are environmentally competent, preferably from the same geographical origin or habitat, or from regions with similar climatic conditions to the agroecosystem in which the fungus will be applied. The advancement of microbial biocontrol solutions is becoming more accessible; however, it has faced numerous instances of insufficient environmental effectiveness. The creation of models that account for temperature variations to identify geographically suitable regions for the application of specific fungal strains, based on their thermal biology, could enhance the environmental effectiveness of EPF (Quesada-Moraga et al. 2022).

9 Future Prospects and Research Directions

Recent research has revealed that EPF exhibit multifaceted biological roles extending far beyond insect pest management. EIPF strains of *B. bassiana* and *Lecanicillium* spp. demonstrate significant antagonistic activity against phytopathogens through multiple mechanisms, including antibiosis, competitive exclusion, and induction of plant systemic resistance. Notably, *Lecanicillium lilacinum* suppresses cucumber powdery mildew, while *M. robertsii* protects beans against root rot (Kumar et al. 2025). EPF are also prolific producers of bioactive secondary metabolites with diverse applications. Mycotoxins and organic compounds derived from these fungi serve as valuable resources for antibiotic production, including medically important drugs like penicillin. Species such as *Cordyceps* have been utilized in traditional medicine for their immunomodulatory and anticancer properties (Wang et al. 2023). Furthermore, EPF contributes to industrial biotechnology through the production of enzymes, antibodies, and nutritional supplements. These emerging applications highlight EPF's expanding role in agriculture, medicine, and biotechnology, positioning them as versatile tools for sustainable development across multiple sectors. Their ability to simultaneously address crop protection, pharmaceutical needs, and industrial processes underscores their exceptional potential in an era increasingly focused on eco-friendly solutions.

From the early days of the biotechnology transformation, scientific inquiry has concentrated on the genetic modification of fungal EPF, primarily to enhance plant yields and bolster their defense mechanisms. The advent of gene modification techniques in EPF and advancements in RNAi technology have led to research focused on developing recombinant EPF strain *I. fumosorosea* to exhibit increased virulence against the TLR7 gene in *B. tabaci* (Chen et al. 2015). Genetic engineering presents promising strategies to enhance fungal virulence or to bolster fungal resilience in response to diverse stressors. The utilization of recombinant EIPF genes as biological control agents has increasingly garnered attention and importance over time. Recombinant EIPF produce anti-pest proteins that play a significant role in the control of insect pests, demonstrating a remarkable ability to colonize host plants effectively (Qadri et al. 2020). It is crucial to emphasize that recombinant EIPF demonstrating heightened virulence against insects may endanger pollinators, natural enemies, and populations of beneficial insects. Understanding the chemical pathways pertinent to biotechnological applications necessitates a focus on the transfer of genes from associated EPF fungi to the genomes of their hosts, which has become a fundamental principle to produce secondary metabolites. This process contributes to clarifying the multiple sources of chemical defenses throughout the evolutionary history of various plant species.

The growing concerns over chemical pesticide toxicity have accelerated interest in nano-mycopesticides as sustainable alternatives for pest management. While EPF presents an ecologically sound biocontrol strategy, their commercial success hinges on addressing key challenges: technical efficacy must be coupled with practical applicability, economic viability, and environmental safety. Achieving these demands requires multidisciplinary collaboration among researchers, industries, and policymakers, particularly for non-commercial strategies requiring long-term governmental support. A critical bottleneck lies in isolating and characterizing EIPF, many of which remain unculturable, complicating community structure analysis (Fadiji and Babalola 2020). Future efforts must prioritize optimizing nanoparticle biosynthesis for cost-effective nanopesticide production while rigorously assessing their ecological persistence and nontarget impacts. Additionally, advancing EPF deployment requires isolating more virulent strains, refining mass-production techniques, and innovating application methods. Understanding EPF-plant interactions, especially in staple crops like maize, will be pivotal for integrating these biocontrol agents into resilient agricultural systems. By addressing these gaps, EPF-based solutions can transition from niche applications to scalable, sustainable pest management tools, aligning with global agricultural and environmental goals. Biopesticides, including EPF, present significant potential for contributing to various United Nations Sustainable Development Goals (SDGs), particularly concerning SDG 2 (Zero Hunger) and SDG 1 (No Poverty).

The reviewed literature highlights the important role of EPF biopesticides in promoting sustainable agriculture and integrated pest management strategies. It emphasizes their effectiveness in maintaining crop yields, improving soil health, and minimizing adverse environmental impacts (Cai and Dimopoulos 2025).

10 Conclusion

The biopesticide sector is advancing rapidly due to expanding global demand, with significant research currently expected to drive innovation and expand the applications of EPF. Biopesticides presently account for only 5% of the total crop protection market. Nevertheless, there is an increasing demand for these products, evidenced by a projected global market value of USD 15 billion by 2030, accompanied by a compound annual growth rate (CAGR) of 25%. The ongoing demand has the potential to enhance the efficacy of benefits, such as waste recycling for biopesticide production and the rehabilitation of polluted ecosystems, thus aiding in the achievement of sustainable consumption and production goals. Future research in EPF should focus on optimizing biocompatible nanomaterials, field stability, and large-scale production to facilitate commercialization. As nanotechnology converges with microbial biocontrol, nano-EPF formulations are poised to revolutionize pest management, offering precision, sustainability, and resilience in global agriculture. Future research should focus on optimizing large-scale production, enhancing formulation stability under field conditions by nanoencapsulation methods, and integrating EPF with other biocontrol agents for synergistic effects. As agricultural systems increasingly shift toward sustainable practices, advanced EPF formulations will subsequently support managing invasive and notorious insect pests worldwide.

Acknowledgments The corresponding author is grateful for the award of the DBT-Biotechnology Industry Research Assistance Council (BIRAC), New Delhi, Fellowship (E28XZ) for the Development of Bioformulation of Mycoparasitic Fungi by the University Innovation Cluster (UIC) of Tamil Nadu Agricultural University, which is highly appreciated.

References

Adeleke BS, Olowe OM, Ayilara MS, Fasusi OA, Omotayo OP et al (2024) Biosynthesis of nanoparticles using microorganisms: a focus on endophytic fungi. Heliyon 10:e39636. https://doi.org/10.1016/j.heliyon.2024.e39636

Ahsan SM, Injamum-Ul-Hoque M, Das AK, Rahman MM, Mollah MM et al (2024) Plant-EPF interaction: recent progress and future prospects on endophytism-mediated growth promotion and biocontrol. Plants 13. https://doi.org/10.3390/plants13101420

Ain Q, Mohsin AU, Naeem M, Shabbir G (2021) Effect of EPF, *Beauveria bassiana* and *Metarhizium anisopliae*, on *Thrips tabaci* Lindeman (Thysanoptera: Thripidae) populations in different onion cultivars. Egypt J Biol Pest Control 31. https://doi.org/10.1186/s41938-021-00445-y

Akutse KS, Fiaboe KK, Van den Berg J, Ekesi S, Maniania NK (2014) Effects of endophyte colonization of *Vicia faba* (Fabaceae) plants on the life-history of leafminer parasitoids *Phaedrotoma scabriventris* (hymenoptera: braconidae) and *Diglyphus isaea* (hymenoptera: eulophidae). PLoS One 9(10):e109965. https://doi.org/10.1371/journal.pone.0109965

Alian RS, Flasz B, Kędziorski A, Rost-Roszkowska M, Rozpędek K et al (2025) Concentration-dependent disturbances of digestive functions in house cricket (Insecta: Orthoptera) exposed to GO-AgNP composite. Sci Rep 15:12699. https://doi.org/10.1038/s41598-025-97589-w

Ambethgar V (2018) Strategic approaches for applications of EPF to counter insecticide resistance in agriculturally important insect pests. In: Gehlot P Singh, J. (eds) Fungi and their role in sustainable development: current perspectives. 221−254 Springer, Singapore. doi:https://doi.org/10.1007/978-981-13-0393-7_13In

Ambethgar V, Kollam M, Chinnadurai C, Ramsubhag A, Jayaraman J (2019) Ecology of emerging vector-borne plant viruses and approaches for integrated management in vegetable production systems. Trop Agric 95:81–94

Amerasan D, Nataraj T, Murugan K, Panneerselvam C, Madhiyazhagan P et al (2016) Myco-synthesis of silver nanoparticles using *Metarhizium anisopliae* against the rural malaria vector *Anopheles culicifacies* Giles (Diptera: Culicidae). J Pest Sci 89:249–256. https://doi.org/10.1007/s10340-015-0675-x

Aziz Qureshi A, Vineela V, Vimala Devi PS (2015) Sodium humate as a promising coating material for microencapsulation of *Beauveria bassiana* conidia through spray drying. Drying Technol 33:162–168. https://doi.org/10.1080/07373937.2014.938814

Bamisile BS, Siddiqui JA, Akutse KS, Ramos Aguila LC, Xu Y (2021) General limitations to endophytic EPF use as plant growth promoters, pests and pathogens biocontrol agents. Plants 10:1–23. https://doi.org/10.3390/plants10102119

Berestetskiy A, Hu Q (2021) The chemical ecology approach to reveal fungal metabolites for arthropod pest management. Microorganisms 9:1379. https://doi.org/10.3390/microorganisms9071379

Bextine BR, Thorvilson HG (2002) Field applications of bait-formulated *Beauveria bassiana* alginate pellets for biological control of the red imported fire ant (Hymenoptera: Formicidae). Environ Entomol 31:746–752. https://doi.org/10.1603/0046-225X-31.4.746

Bidochka MJ, Hajek AE (1998) A nonpermissive entomophthoralean fungal infection increases activation of insect prophenoloxidase. J Invertebr Pathol 72:231–238. https://doi.org/10.1006/jipa.1998.4782

Bihal R, Al-Khayri JM, Banu AN, Kudesia N, Ahmed FK et al (2023) EPF: an eco-friendly synthesis of sustainable nanoparticles and their nanopesticide properties. Microorganisms 11:1617. https://doi.org/10.3390/microorganisms11061617

Bischoff JF, Rehner SA, Humber RA (2009) A multilocus phylogeny of the *Metarhizium anisopliae* lineage. Mycologia 101:512–530. https://doi.org/10.3852/07-202

Bruck DJ (2005) Ecology of *Metarhizium anisopliae* in soilless potting media and the rhizosphere: implications for pest management. Biol Control 32:155–163. https://doi.org/10.1016/j.biocontrol.2004.09.003

Cai P, Dimopoulos G (2025) Microbial biopesticides: a one health perspective on benefits and risks. One Health 20:100962. https://doi.org/10.1016/j.onehlt.2024.100962

Chaud M, Souto EB, Zielinska A, Severino P, Batain F et al (2021) Nanopesticides in agriculture: benefits and challenge in agricultural productivity, toxicological risks to human health and environment. Toxics 9:131. https://doi.org/10.3390/toxics9060131

Chavan B, Kadam JR, Saindane Y (2008) Bioefficacy of liquid formulation of *Verticillium lecanii* against red spider mite (*Tetranychus cinnabarinus*). Int J Plant Prot 1(2):48–51

Chen X, Li L, Hu Q, Zhang B, Wu W et al (2015) Expression of dsRNA in recombinant *Isaria fumosorosea* strain targets the TLR7 gene in *Bemisia tabaci*. BMC Biotechnol 15:1–8. https://doi.org/10.1186/s12896-015-0170-8

Crisan CM, Mocan T, Manolea M, Lasca LI, Tăbăran FA et al (2021) Review on silver nanoparticles as a novel class of antibacterial solutions. Appl Sci 11:1–18. https://doi.org/10.3390/app11031120

Davari A, Parker BL, Sullivan CF, Ghalehgolabbehbahani A, Skinner M (2021) Biological control of Western flower thrips, *Frankliniella occidentalis* using a self-sustaining granular fungal treatment. Bull Entomol Res 111:688–693. https://doi.org/10.1017/S0007485321000365

Dutta D, Puzari KC, Gogoi R, Dutta P (2014) Endophytes: exploitation as a tool in plant protection. Braz Arch Biol Technol 57:621–629. https://doi.org/10.1590/S1516-8913201402043

Fadiji AE, Babalola OO (2020) Exploring the potentialities of beneficial endophytes for improved plant growth. Saudi J Biol Sci 27:3622–3633. https://doi.org/10.1016/j.sjbs.2020.08.002

Friuli M, Pellegrino R, Lamanna L, Nitti P, Madaghiele M et al (2023) Materials engineering to help pest control: a narrative overview of biopolymer-based EPF formulations. J Fungi 9. https://doi.org/10.3390/jof9090918

Georgiev G, Mirchev P, Rossnev B, Petkov P, Georgieva M et al (2013) Potential of *Entomophaga maimaiga* Humber, Shimazu and Soper (Entomophthorales) for suppressing *Lymantria dispar* (Linnaeus) outbreaks in Bulgaria. Proceed Bulg Acad Sci 66. https://doi.org/10.7546/CR-2013-66-7-13101331-14

Gopal GS, Venkateshalu B, Nadaf AM, Guru PN, Pattepur S (2021) Management of the grape mealy bug, *Maconellicoccus hirsutus* (Green), using EPF and botanical oils: a laboratory study. Egypt J Biol Pest Control 31:100. https://doi.org/10.1186/s41938-021-00444-z

Grewal GK, Joshi N, Suneja Y (2021) Pathogenicity of *Metarhizium rileyi* (Farlow) Kepler, S.A. Rehner and Humber isolates against *Spodoptera litura* (Fabricius) and their extracellular enzymatic activities. Egypt J Biol Pest Control 31:59. https://doi.org/10.1186/s41938-021-00407-4

Harish S, Parthasarathy S, Durgadevi D, Anandhi K, Raguchander T (2019) Plant growth-promoting Rhizobacteria: harnessing its potential for sustainable plant disease management. In: Kumar a, Meena V (eds) Plant growth promoting rhizobacteria for agricultural sustainability. Springer, Singapore, pp 151–187. https://doi.org/10.1007/978-981-13-7553-8_8

Harish S, Parthasarathy S, Prabhukarthikeyan SR, Mahanty A, Teli B et al (2022) Omics: a potential tool to delineate the mechanism of biocontrol agents against plant pathogens. In: Sahu J, Vaishnav A, Singh HB (eds) Plant-microbe interactions. CRC Press, Boca Raton, pp 79–100

Harish S, Sendhilvel V, Rajendran L, Parthasarathy S, Raguchander T (2023) Plant associated endophytes as potential agents for the protection of crops from phytopathogens. In: Bastas KK, Kumar A, Sivakumar U (eds) Microbial biocontrol: molecular perspective in plant disease management. Springer Nature Singapore, Singapore, pp 89–115

Harper AM, Huang HC (1986) Evaluation of the Entomophagous fungus *Verticillium lecanii* (Moniliales: Moniliaceae) as a control agent for insects. Environ Entomol 15:281–284. https://doi.org/10.1093/ee/15.2.281

Hernádez-Velázquez VM, Berlanga-Padilla AM, Barrientos-Lozano L (2000) Vegetable and mineral oil formulations of *Metarhizium anisopliae* var. *acridum* to control the central American Locust (*Schistocerca piceifrons piceifrons* Walker) (Orthoptera: Acrididae). J Orthoptera Res 9: 223–227. https://doi.org/10.2307/3503650

Immediato D, Iatta R, Camarda A, Giangaspero A, Capelli G et al (2017) Storage of *Beauveria bassiana* conidia suspension: a study exploring the potential effects on conidial viability and virulence against *Dermanyssus gallinae* De Geer, 1778 Acari: Dermanyssidae. Ann Biol Sci 05(02). https://doi.org/10.21767/2348-1927.1000111

Ingle YV, Bhosale DN, Karande VD, Bramhankar SB, Mane SS et al (2022) Identification, pathogenesis and compatibility of *Aschersonia aleyrodis* (Webber) with selected fungicides and insecticides. Int J Trop Insect Sci 42:2093–2101. https://doi.org/10.1007/s42690-022-00730-9

Jaber LR, Araj SE (2018) Interactions among endophytic fungal entomopathogens (Ascomycota: Hypocreales), the green peach aphid *Myzus persicae* Sulzer (Homoptera: Aphididae), and the aphid endoparasitoid *Aphidius colemani* Viereck (Hymenoptera: Braconidae). Biol Control 116: 53–61. https://doi.org/10.1016/j.biocontrol.2017.04.005

Karthi S, Vasantha-Srinivasan P, Senthil-Nathan S, Han YS, Shivakumar MS et al (2024) EPF promising biocontrol agents for managing lepidopteran pests: review of current knowledge. Biocatal Agric Biotechnol 58:103146. https://doi.org/10.1016/j.bcab.2024.103146

Khooshe-Bast Z, Sahebzadeh N, Ghaffari-Moghadham M, Mirsheker A (2016) Insecticidal effects of zinc oxide nanoparticles and *Beauveria bassiana* TS11 on *Trialeurodes vaporariorum* (Westwood, 1856) (Hemiptera: Aleyrodidae). Acta Agric Slov 107:299–309. https://doi.org/10.14720/aas.2016.107.2.04

Kozłowska E, Dymarska M, Kostrzewa-Susłow E, Janeczko T (2019) Cascade biotransformation of estrogens by *Isaria fumosorosea* KCh J2. Sci Rep 9:1–8. https://doi.org/10.1038/s41598-019-47225-1

Kumar D, Roy S, Babu A, Pandey AK (2025) Harnessing fungal bioagents rich in volatile metabolites for sustainable crop protection: a critical review. J Basic Microbiol 65. https://doi.org/10.1002/jobm.70003

Lee SJ, Yu JS, Nai YS, Parker BL, Skinner M et al (2015) *Beauveria bassiana* sensu lato granules for management of brown planthopper, *Nilaparvata lugens* in rice. BioControl 60:263–270. https://doi.org/10.1007/s10526-014-9632-5

Li D, Park SE, Lee MR, Kim JC, Lee SJ et al (2021) Soil application of *Beauveria bassiana* JEF-350 granules to control melon thrips, *Thrips palmi* Karny (Thysanoptera: Thripidae). J Asia Pac Entomol 24:636–644. https://doi.org/10.1016/j.aspen.2021.05.010

Liao X, Lovett B, Fang W, St Leger RJ (2017) *Metarhizium robertsii* produces indole-3-acetic acid, which promotes root growth in *Arabidopsis* and enhances virulence to insects. Microbiol (United Kingdom) 163:980–991. https://doi.org/10.1099/mic.0.000494

Litwin A, Nowak M, Różalska S (2020) EPF: unconventional applications. Rev Environ Sci Biotechnol 19:23–42. https://doi.org/10.1007/s11157-020-09525-1

Liu X, Li R, Liu Y, Feng M, Zhou G et al (2025) Roles of Sti1, a homolog to mammalian Hop, in conidiation, environmental adaptation, and virulence of *Beauveria bassiana*. Fungal Biol 129: 101596. https://doi.org/10.1016/j.funbio.2025.101596

Lohse R, Jakobs-Schönwandt D, Vidal S, Patel AV (2015) Evaluation of new fermentation and formulation strategies for a high endophytic establishment of *Beauveria bassiana* in oilseed rape plants. Biol Control 88:26–36. https://doi.org/10.1016/j.biocontrol.2015.05.002

Maina U, Zakaria D, Galadima I, Gambo F (2018) A review on the use of EPF in the management of insect pests of field crops. J Entomol Zool Stud 6:27–32

Mantzoukas S, Eliopoulos PA (2020) Endophytic EPF: a valuable biological control tool against plant pests. Appl Sci 10. https://doi.org/10.3390/app10010360

Mantzoukas S, Kitsiou F, Natsiopoulos D, Eliopoulos PA (2022) EPF: interactions and applications. Encyclopedia 2:646–656. https://doi.org/10.3390/encyclopedia2020044

Mascarin GM, Golo PS, de Souza Ribeiro-Silva C, Muniz ER, de Oliveira Franco A et al (2024) Advances in submerged liquid fermentation and formulation of EPF. Appl Microbiol Biotechnol 108:451. https://doi.org/10.1007/s00253-024-13287-z

Mondal S (2024) Potential of fungal nanobiopesticides on entomopathogens. In: Deshmukh SK, Sridhar KR (eds) EPF. Springer, Singapore, pp 507–522

Mora MA, Castilho AM, Fraga ME (2017) Classification and infection mechanism of entomopathogenic fungi. Arq Inst Biol 84:e0552015

Nimisha T, Deepthy KB, Bhaskar H, Chellappan M (2024) Biochemical basis of abiotic stress tolerance in native isolates of *Beauveria Bassiana* (Balsamo) Vuillemin from Kerala. Indian J Entomol 86:867–874. https://doi.org/10.55446/IJE.2023.1288

Parthasarathy S (2023) Biological control of plant diseases: principles and practices, 1st edn. Kalyani Publishers, New Delhi

Parthasarathy S, Lakshmidevi P, Satya VK, Gopalakrishnan C (2024) Plant pathology and disease management principles and practices. CRC Press, London, p 456

Pelizza SA, Mariottini Y, Russo LM, Vianna FM, Scorsetti AC et al (2019) Application of *Beauveria bassiana* using different baits for the control of grasshopper pest Dichroplus maculipennis under field cage conditions. J King Saud Univ Sci 31:1511–1515. https://doi.org/10.1016/j.jksus.2019.02.014

Qadri M, Short S, Gast K, Hernández J, Wong AC (2020) Microbiome innovation in agriculture: development of microbial based tools for insect pest management. Front Sustain Food Syst 4:1–20. https://doi.org/10.3389/fsufs.2020.547751

Quesada-Moraga E, Garrido-Jurado I, Yousef-Yousef M, González-Mas N (2022) Multitrophic interactions of EPF in BioControl. BioControl 67:457–472. https://doi.org/10.1007/s10526-022-10163-5

Quesada-Moraga E, González-Mas N, Yousef-Yousef M, Garrido-Jurado I, Fernández-Bravo M (2024) Key role of environmental competence in successful use of EPF in microbial pest control. J Pest Sci 97:1–15. https://doi.org/10.1007/s10340-023-01622-8

Ravi Sankar U, Padmaja V (2014) Preparation and use of oil formulations of *Beauveria bassiana* and *Metarhizium anisopliae* against *Spodoptera litura* larvae. Afr J Microbiol Res 8:1638–1644. https://doi.org/10.5897/AJMR2013.6593

Sabbour M (2015) The toxicity effect of nano fungi *Isaria fumosorosea* and *Metarhizium flavoviride* against the Potato tuber moth, *Phthorimaea operculella*. Am J Biol Life Sci 3: 155–160

Saberi Riseh R, Skorik YA, Thakur VK, Moradi Pour M, Tamanadar E et al (2021) Encapsulation of plant biocontrol bacteria with alginate as a main polymer material. Int J Mol Sci 22:11165. https://doi.org/10.3390/ijms222011165

Saha N, Sharma A, Bora P (2025) Expanding the functional landscape of microbial entomopathogens in agriculture beyond pest management. Folia Microbiol (Praha):343–357. https://doi.org/10.1007/s12223-025-01251-x

Saminathan N, Subramanian J, Sankaran Pagalahalli S, Theerthagiri A, Mariappan P (2025) EPF: translating research into field applications for crop protection. Arthropod Plant Interact 19:1–27. https://doi.org/10.1007/s11829-024-10110-4

Sarma BD, Puzari KC, Dutta P, Pandey AK (2023) An alginate-based encapsulation enhances shelf life and bioactivity of the entomopathogenic fungus, *Metarhizium anisopliae*. Egypt J Biol Pest Control 33:69. https://doi.org/10.1186/s41938-023-00714-y

Savary S, Willocquet L, Pethybridge SJ, Esker P, McRoberts N et al (2019) The global burden of pathogens and pests on major food crops. Nat Ecol Evol 3:430–439. https://doi.org/10.1038/s41559-018-0793-y

Seethapathy P (2024) Mass production of *Beauveria bassiana*. In: Deshmukh SK, Sridhar KR (eds) The role of EPF in agriculture. CRC Press, Boca Raton, pp 298–320

Seethapathy P, Gurudevan T, Subramanian KS, Kuppusamy P (2016) Bacterial antagonists and hexanal-induced systemic resistance of mango fruits against *Lasiodiplodia theobromae* causing stem-end rot. J Plant Interact 11:158–166. https://doi.org/10.1080/17429145.2016.1252068

Seethapathy P, Sankarasubramanian H, Lingan R, Thiruvengadam R (2023) *Chaetomium* sp.: an insight into its antagonistic mechanisms, mass multiplication, and production cost analysis. In: Amaresan N, Dharumadurai D, Babalola OO (eds) Agricultural microbiology based entrepreneurship making money from microbes, Microorganisms for sustainability, vol 39. Springer, Singapore, pp 267–288. https://doi.org/10.1007/978-981-19-5747-5_17

Seethapathy P, Venugopal L, Binil M, Soman M (2025) Anguina. In: Amaresan N, Kumar K (eds) Compendium of phytopathogenic microbes in agro-ecology. Springer, Cham, pp 297–324. https://doi.org/10.1007/978-3-031-81999-5_16

Shanmugam V, Seethapathy P (2017) Isolation and characterization of white muscardine fungi *Beauveria bassiana* (Bals.) Vuill.-A causative of mulberry silkworm. J Entomol Zool Stud 5: 512–515

Sharma R, Sharma P (2021) Fungal entomopathogens: a systematic review. Egypt J Biol Pest Control 31. https://doi.org/10.1186/s41938-021-00404-7

Sharma A, Sharma S, Yadav PK (2023) EPF and their relevance in sustainable agriculture: a review. Cogent Food Agric 9:1–21. https://doi.org/10.1080/23311932.2023.2180857

Shin J, Shin S, Choi J, Jeong W, Kim JE et al (2023) Encapsulation of the biocontrol agent *Beauveria bassiana* in cellulose nanomaterial-stabilized Pickering emulsion for foliar application. ACS Agric Sci Technol 3:894–905. https://doi.org/10.1021/acsagscitech.3c00184

Sivasabari K, Parthasarathy S, Chandran D, Sankaralingam S, Ajaykumar R (2023) Vermicompost and vermi-leachate in pest and disease management. In: Mupambwa HA, Horn LN, Mnkeni PNS (eds) vermicomposting for sustainable food systems in Africa, sustainability sciences in Asia and Africa. Springer, Singapore, pp 279–304. https://doi.org/10.1007/978-981-19-8080-0_17

Skinner M, Parker BL, Kim JS (2014) Role of EPF in integrated pest management. In: Abrol DP (ed) Integrated pest management, current concepts and ecological perspective. Elsevier, pp 169–191

Stephan D, Bernhardt T, Buranjadze M, Seib C, Schäfer J et al (2021) Development of a fluid-bed coating process for soil-granule-based formulations of *Metarhizium brunneum*, *Cordyceps fumosorosea* or *Beauveria bassiana*. J Appl Microbiol 131:307–320. https://doi.org/10.1111/jam.14826

Subramaniam MSR, Babu A, Deka B (2021) *Lecanicillium lecanii* (Zimmermann) Zare & Gams, as an efficient biocontrol agent of tea thrips, *Scirtothrips bispinosus* Bagnall (Thysanoptera: Thripidae). Egypt J Biol Pest Control 31:38. https://doi.org/10.1186/s41938-021-00380-y

Sumalatha BV, Selvaraj K, Poornesha B, Ramanujam B (2020) Pathogenicity of entomopathogenic fungus *Isaria fumosorosea* on rugose spiralling whitefly *Aleurodicus rugioperculatus* and its effect on parasitoid *Encarsia guadeloupae*. Biocontrol Sci Tech 30:1150–1161. https://doi.org/10.1080/09583157.2020.1797632

Tahira HG, Saeed S, Khan FZA (2014) EPF as effective insect pest management tactic: a review EPF as effective insect pest management tactic : a review. Appl Sci Bus Econ 1:10–18

Vishnu M, Kannan M, Soundararajan RP, Suganthi A, Subramanian A et al (2024) Nano-bioformulations: emerging trends and potential applications in next generation crop protection. Environ Sci Nano 11:2831–2860. https://doi.org/10.1039/D4EN00263F

Wang ZL, Ying SH, Feng MG (2013) Recognition of a core fragment of *Beauveria bassiana* hydrophobin gene promoter (P hyd1) and its special use in improving fungal biocontrol potential. Microb Biotechnol 6(1):27–35. https://doi.org/10.1111/j.1751-7915.2012.00351.x

Wang J, Chen J, Hu Y, Ying SH, Feng MG (2020) Roles of six Hsp70 genes in virulence, cell wall integrity, antioxidant activity and multiple stress tolerance of *Beauveria bassiana*. Fungal Genet Biol 144:103437. https://doi.org/10.1016/j.fgb.2020.103437

Wang D, Xing PX, Diao HL, Zhou WW, Li XW et al (2023) Pathogenicity characteristics of the entomopathogenic fungus *Cordyceps javanica* IJ-tg19 to *Acyrthosiphon pisum*. BioControl 68: 447–458. https://doi.org/10.1007/s10526-023-10203-8

Wenzel Rodrigues IM, Batista Filho A, Giordano IB, Denadae BE, Fernandes JB et al (2017) Compatibility of polymers to fungi *Beauveria bassiana* and *Metarhizium anisopliae* and their formulated products stability. Acta Sci Agron 39:457. https://doi.org/10.4025/actasciagron.v39i4.32903

Yosri M, Abdel-Aziz MM, Sayed RM (2018) Larvicidal potential of irradiated myco-insecticide from *Metarhizium anisopliae* and larvicidal synergistic effect with its mycosynthesized titanium nanoparticles (TiNPs). J Radiat Res Appl Sci 11:328–334. https://doi.org/10.1016/j.jrras.2018.06.001

Zhou X, Wang DW, Guo K, Mao SF (2014) Germination and sporulation of Pandora delphacis (Entomophthoromycota: Entomophthorales) on the rice pest *Nilaparvata lugens*: SEM observation. Mycosystema 33(4):819–826. https://doi.org/10.1111/jen.12476

Entomopathogenic Fungi and Their Nano-formulations: Bridging Nature and Technology

19

Indrani Sarkar, Maunata Ghorui, Shouvik Chowdhury, and Sashidhar Burla

Abstract

Entomopathogenic fungi (EPF), such as *Beauveria bassiana* and *Metarhizium anisopliae*, are recognized as potent biological control agents due to their natural ability to infect and kill insect pests across various life stages. However, conventional EPF applications face several limitations, including environmental sensitivity, short shelf life, and slower action compared to chemical pesticides. The integration of nanotechnology presents a novel strategy to overcome these challenges by enhancing the efficacy, stability, and delivery of fungal biopesticides. Nano-formulations, including metal-based nanoparticles (e.g., silver, copper, zinc oxide, and selenium) and encapsulated fungal conidia, offer precision targeting, improved adhesion and penetration into insect cuticles, and protection from environmental degradation. These innovations not only increase pest mortality but also reduce non-target effects and environmental contamination. This review explores the taxonomy, infection mechanisms, and limitations of EPF, alongside recent advancements in nano-enabled delivery systems. It also highlights their synergistic potential with plant extracts and biocompatible carriers. It concludes by identifying future research needs, including omics-level insights, eco-safety evaluations, scalable formulation techniques, and integration into precision agriculture systems, emphasizing the transformative potential of EPF-nanotechnology fusion in sustainable pest management.

I. Sarkar
Amity Institute of Biotechnology, Amity University Noida, Noida, India

M. Ghorui · S. Chowdhury
Symbiotic Sciences Pvt. Ltd., Gurgaon, India

S. Burla (✉)
ATGC Biotech Pvt. Ltd., Hyderabad, India

S. K. Deshmukh et al. (eds.), *Frontiers in Entomopathogenic Fungi*,
https://doi.org/10.1007/978-981-95-6529-0_19

Keywords

Green synthesis · Integrated pest management · Nanocarriers · Nanoparticles · Nanotechnology

1 Introduction

Entomopathogenic fungi (EPF) are heterotrophic, filamentous, saprophytic eukaryotes that play a key role in regulating insect populations (Santos et al. 2022). These fungi are widespread across terrestrial habitats, with tropical forests hosting the greatest species diversity. Some EPF species are even adapted to extreme climates, including Arctic tundra regions (Sharma et al. 2023). The field of entomopathogenic mycology traces back to 1835, when Agostino Bassi discovered *Beauveria bassiana* as the cause of the fatal silkworm disease white muscardine. This fungus, now classified under Cordycipitaceae, Hypocreales, marked the beginning of scientific exploration into fungal biocontrol (Mantzoukas et al. 2022). Their ecological importance extends beyond pest control, contributing significantly to organic matter decomposition and nutrient cycling (Santos et al. 2022).

EPF fall into six principal groups (Fig. 19.1): Ascomycota (≈476 species), the primary saprophytic, insect-lethal group used in biocontrol; Entomophthoromycota (≈474), made up of obligate insect pathogens; Microsporidia (≈339), which live as intracellular parasites; Basidiomycota (≈238) and Chytridiomycota (≈238), both only occasionally insect-pathogenic; and the fungus-like Oomycetes (≈12), whose scant species play only a minor role in pest management. Together, these groups highlight the taxonomic diversity and ecological potential of EPF, with Ascomycota and Entomophthoromycota standing out as the most promising candidates for sustainable pest control solutions.

Herbivorous insects are responsible for approximately 18% of global crop losses, underscoring the essential role of EPF in modern integrated pest management (IPM)

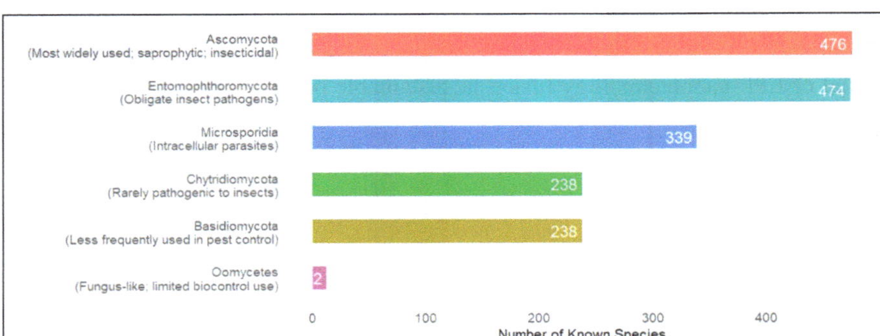

Fig. 19.1 Classification of entomopathogenic fungi and their biocontrol characteristics. (Adapted from Sharma et al. 2023)

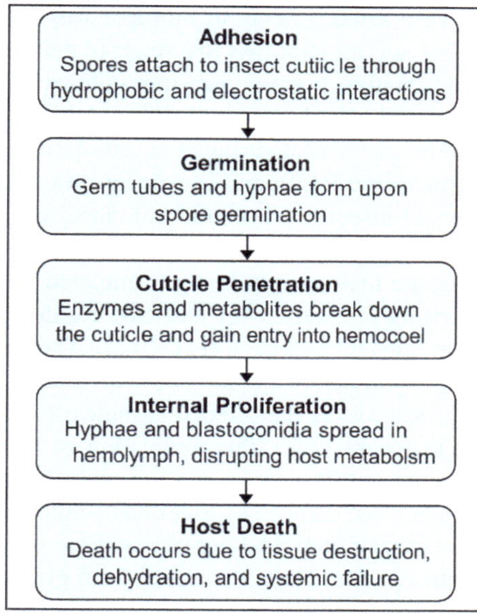

Fig. 19.2 Infection process of entomopathogenic fungi

strategies (Sharma et al. 2023). EPF offer a natural and effective means of pest suppression by infecting and killing insects from various orders (Álvarez et al. 2017). Among the most studied and widely utilized species are *M. anisopliae*, *B. bassiana*, and *Isaria fumosorosea*, which have demonstrated significant bioinsecticidal capabilities and are frequently used in both research and practical pest control applications (Santos et al. 2022). These fungi exhibit broad-spectrum activity, effectively targeting pests across all developmental stages—from eggs to adults.

Unlike chemical pesticides, which often depend on the feeding behavior of pests, EPF initiate infection externally. They adhere to the insect's cuticle and penetrate it using a combination of adhesins, hydrolytic enzymes, and secondary metabolites, making them effective even against non-feeding stages like eggs and pupae (Álvarez et al. 2017; Litwin et al. 2020). Once attached, EPF follow a well-defined infection process involving multiple stages (Fig. 19.2). First, spores adhere to the insect cuticle through hydrophobic and electrostatic interactions. This is followed by germination, where the spores produce germ tubes and hyphae. In the cuticle penetration phase, the fungi release enzymes and toxic metabolites to break down the host's protective barriers and gain entry into the hemocoel. Once inside the host, internal proliferation occurs, during which fungal hyphae and blastoconidia spread through the hemolymph, interfering with metabolic processes and weakening the host (Santos et al. 2022). This systemic infection ultimately causes host death through tissue destruction, dehydration, and physiological failure. In the final stage, sporulation, new spores emerge from the cadaver, ensuring fungal propagation and infection of new hosts (Ghorui et al. 2024). Post-infection, hyphae absorb nutrients from the host and lead to hardening of the insect's body.

EPF affect insects in multiple ways; they damage the insect's cuticle, causing water and mineral loss that leads to dehydration. These fungi also interfere with the reproductive system, causing failures such as defects in gonadotropin-releasing hormone, which reduce fertility. Additionally, they induce physiological stress by promoting oxidative damage, increasing the activity of detoxifying enzymes like cytochrome P450, and impairing protein synthesis. Together, these effects disrupt normal insect function and ultimately contribute to insect mortality.

Due to their natural origin, effectiveness, and minimal environmental impact, EPF are increasingly being formulated into bioinsecticides, particularly in response to rising concerns over pesticide residues, environmental degradation, and insecticide resistance (Santos et al. 2022). Their ecological function as natural regulators of insect populations, combined with their versatility and well-characterized metabolic profiles, makes them indispensable tools in sustainable agriculture.

In recent years, the application of EPF in biological pest control has received growing interest as a safer and more sustainable alternative to chemical pesticides, which often carry risks to human health and to the environment. EPF possess broad-spectrum insecticidal activity and are capable of targeting pests at all life stages—from eggs to adults—making them effective tools in integrated pest management. EPF like *B. bassiana*, *M. anisopliae*, and *L. lecanii* are widely deployed in open fields and greenhouses (Donzelli et al. 2010).

2 Challenges in Conventional Application

Despite their promise, the practical deployment of EPF in agriculture faces several challenges (Fig. 19.3). Limitations include relatively high production costs, slower pest control action compared to chemical alternatives, short shelf life, low environmental persistence, and reduced efficacy under suboptimal conditions (Santos et al. 2022; Bihal et al. 2023). EPF are particularly vulnerable to environmental stressors such as UV radiation, high temperatures, and low humidity. For example, extended exposure to UV-B radiation can significantly diminish spore viability, thereby compromising their pest-infecting potential (de Souza et al. 2022). Souza et al. (2014) investigated the heat tolerance of different EPF species to evaluate their sensitivity to environmental stress. Among them, *M. anisopliae* exhibited greater susceptibility to elevated temperatures, resulting in reduced spore viability. Similarly, Castrillo et al. (2010) examined the environmental persistence of *B. bassiana* on plant surfaces by monitoring the recovery of colony-forming units (CFUs) from both leaf and bark surfaces after application.

The findings revealed a marked reduction in fungal viability within just a week, with the decline being more pronounced on leaves than on bark (de Souza et al. 2022). Additional studies have shown that conidia of *B. bassiana* and *M. anisopliae* are highly vulnerable to UV radiation. Sunlight, particularly in the UV range of 290–400 nm, significantly compromises the longevity of fungal spores deposited on foliage (Acheampong et al. 2020). High ambient temperatures above 35 °C further reduce the durability and viability of spores, while cooler temperatures appear to

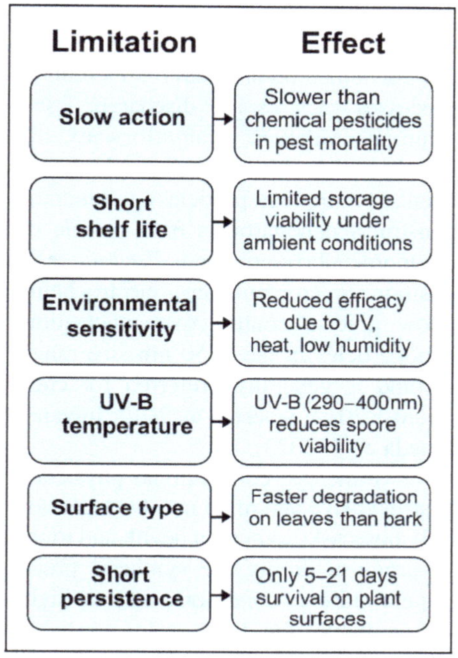

Fig. 19.3 Limitations and environmental constraints of entomopathogenic fungi

have a lesser impact (de Souza et al. 2022). Notably, *B. bassiana* has been reported to persist on plant surfaces for about 5 to 21 days, depending on external conditions (Islam et al. 2021; de Souza et al. 2022). To achieve effective pest suppression, large quantities and thorough application are often required.

3 Overview of Nanotechnology

The term "*nano*" originating from the Greek word meaning "dwarf," refers to extremely small dimensions, typically ranging from 10^{-9} to 10^{-7} m (Tyagi et al. 2019). Nanoparticles have recently garnered considerable attention due to their versatile applications across diverse fields. These include their roles as antimicrobial agents and tools in bioremediation and biosensing, as well as in cancer therapies and targeted drug delivery. Their appeal lies in their unique physicochemical attributes, such as ultra-small size, high surface area-to-volume ratio, enhanced catalytic potential, and varied morphologies (Tyagi et al. 2019).

In the agricultural sector, nanoparticles offer significant promise, especially in enhancing crop yield and promoting eco-friendly practices. They improve fertilizer efficiency by allowing for targeted and controlled nutrient release and aid in the accelerated degradation of pesticide residues in soil and water, thereby reducing environmental contamination (Shatalova et al. 2022).

Nanoparticles can be synthesized using physical or chemical techniques, broadly classified into two main strategies: the top-down and bottom-up approaches. The top-down method involves fragmenting bulk materials into nanoscale particles using processes such as arc discharge, laser ablation, evaporation-condensation, and grinding. Although straightforward, this approach often results in a wide size distribution (~100 nm), lacks uniformity in shape, and typically requires stabilizing agents to prevent particle agglomeration. Conversely, the bottom-up approach constructs nanoparticles from atomic or molecular precursors like metal salts or organometallic compounds. Techniques such as chemical reduction, thermal decomposition, microemulsions, electrochemical synthesis, or photoinduced methods allow precise control over nucleation and growth, often producing uniform nanoparticles in the 1–50 nm size range. Due to these advantages, the bottom-up method is generally preferred for chemical nanoparticle synthesis (Prabhu and Poulose 2012; Sweet et al. 2012; Iravani et al. 2014; Kumar and Dixit 2018; Beltrán Pineda et al. 2023).

Nevertheless, conventional physical and chemical synthesis techniques can be cost-intensive and often involve hazardous or toxic chemicals. These substances not only pose risks to human health and to the environment but may also lead to harmful by-products during the synthesis process. Furthermore, physical methods often require extreme conditions such as high temperatures and pressures, making them less suitable for large-scale production. In contrast, biological synthesis of nanoparticles has emerged as a safer, more sustainable alternative. Particularly in the production of silver nanoparticles, biological methods offer economic and ecological advantages by eliminating the use of harmful reagents and reducing toxicity risks (Tyagi et al. 2019; Soleimani et al. 2022).

4 Nano-formulations of Entomopathogenic Fungi and Their Impacts

Recent advances in nanotechnology have provided promising solutions to overcome the environmental challenges faced by EPF formulations (Deshmukh et al. 2023) (Fig. 19.4). Integrating nanoparticles into EPF preparations has emerged as an innovative approach in pest control, offering protection to fungal spores against stressors such as UV radiation, desiccation, and heat, while also extending their shelf life (Álvarez et al. 2017; Bihal et al. 2023). The efficacy of these nanoparticles largely depends on their physicochemical characteristics—namely size, shape, and chemical composition, which significantly influence their insecticidal activity and overall performance (Hayles et al. 2017). Nanotechnology also facilitates the creation of advanced, eco-friendly pesticide systems by enabling controlled-release, targeted delivery, and a reduced environmental footprint (Álvarez et al. 2017; Bihal et al. 2023). EPF-based nano-formulations can be used alone or in combination with plant-derived bioactive compounds to produce synergistic effects in pest management.

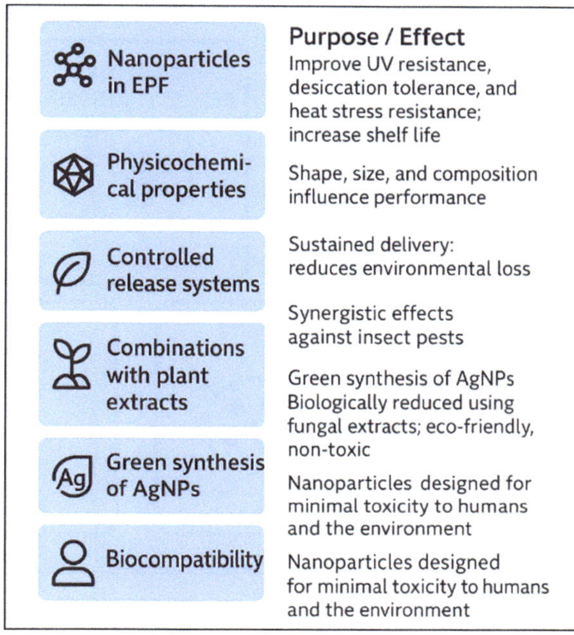

Fig. 19.4 Nano-strategies in pest management using entomopathogenic fungus formulations

A key advantage of these nanosystems is their ability to provide sustained and localized release of active agents, including nutrients and biocontrol substances, ensuring continuous exposure at the infection site (Fig. 19.5). This leads to enhanced plant uptake, longer-lasting pest suppression, improved pesticide and fertilizer efficiency, minimized resource waste, and reduced environmental contamination, ultimately boosting agricultural productivity (Wang et al. 2025). Compared to conventional chemical pesticides—which are often associated with toxic residues, groundwater contamination, bioaccumulation, and unintended harm to non-target organisms, including beneficial insects, animals, and humans. Nanopesticides pose significantly lower risks to human health and the environment (Wang et al. 2025). This is particularly true for nanoparticles made from biocompatible and biodegradable materials designed for precision and efficiency (Wang et al. 2025).

Fungi-synthesized silver or selenium nanoparticles are coated naturally with stabilizing protein-organic layers that prevent aggregation, enhance stability, and maintain bioactivity (Bihal et al. 2023). These biologically produced nanoparticles are non-toxic to humans and possess inherent insecticidal properties (Tyagi et al. 2019). The green synthesis of silver nanoparticles (AgNPs) using EPF involves the biological reduction of silver salts by fungal extracts, offering an environmentally friendly and cost-effective alternative to chemical methods without requiring harmful chemicals or high energy inputs.

Several studies have documented the insecticidal efficacy of these biologically synthesized nanoparticles. For instance, AgNPs produced by *Isaria fumosorosea* showed strong activity against mosquito species such as *Aedes aegypti* and *Culex*

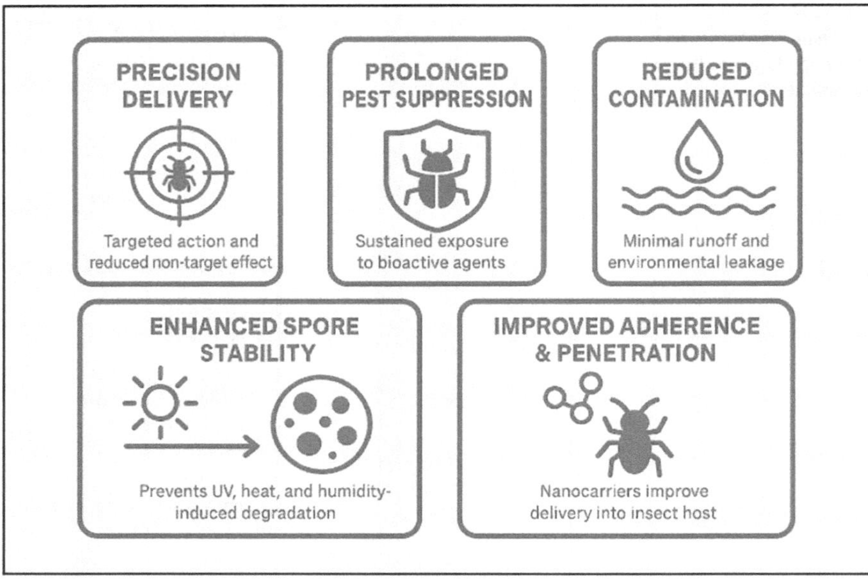

Fig. 19.5 Benefits of nano-formulated entomopathogenic fungus in pest management

quinquefasciatus (Banu and Balasubramanian 2014). Likewise, AgNPs derived from *B. bassiana* and *M. anisopliae* were effective against mustard aphids (*Lipaphis erysimi*) and *Culex pipiens* (Santos et al. 2022).

Moreover, nanocarriers improve the adhesion, penetration, and delivery efficiency of bioactive compounds. Biodegradable polymer-based nanocarriers not only shield fungal spores from environmental degradation but also enhance their interaction with insect pests, increasing the effectiveness of the biocontrol agents (Wang et al. 2025).

Nano-formulations stem from a diverse array of sources. EPF bind with inorganic nanoparticles, which are metal-based formulations like silver (Ag), zinc oxide (ZnO), and copper (Cu) nanoparticles, and impacts the insects (Table 19.1).

4.1 Silver Nanoparticles

Silver nanoparticles (AgNPs) produced via EPF are known for their rapid biomass and extracellular protein production under lab settings, which facilitates efficient nanoparticle biosynthesis (Tyagi et al. 2019). In a study by Tyagi et al. (2019), AgNPs were synthesized from the EPF isolates of *M. anisopliae* (TT1) and three strains of *Beauveria bassiana* (JS1, JS2, KA75). Optimal nanoparticle formation occurred at neutral pH and 60 °C, with the study focusing on the antimicrobial efficacy of AgNPs and their combinations with antibiotics. Notably, this work was among the first to explore the synergistic antimicrobial effects of AgNPs and

Table 19.1 Impact of various entomopathogenic-based nanoparticles on targeted pests

EPF used	Metal used in nano-formulation	Target pest	Result (LD_{50}/LC_{50}/ mortality %)	Reference
Beauveria bassiana	Nanoform (conidia)	*Sitophilus oryzae*	LC_{50}—4.5* 10^5 conidia/ml	Sabbour (2015)
Beauveria bassiana	Silver (AgNPs)	*Lipaphis erysimi* (mustard aphid)	LC_{50}—0.05 * 10^7 spores/ml at 120 h	Kamil et al. (2017)
Beauveria bassiana	Silver (AgNPs)	*Rhynchophorus ferrugineus*	Mortality—84% (eggs), 60% (larvae), 70% (adults)	Abdel-Raheem et al. (2020)
Fusarium solani	Selenium (SeNPs)	*Callosobruchus chinensis* (stored bean pest)	Mortality—23% (eggs) Survival—18% (egg-to-adult)	Helmy et al. (2023)
Isaria fumosorosea	Silver (AgNPs)	*Plutella xylostella* (Cabbage moth)	LC_{50}—0.691 mg/ml	Santos et al. (2022)
Metarhizium anisopliae	Nanoform (conidia)	*Sitophilus oryzae*	LC_{50}—5.7 * 10^5 conidia/ml	Sabbour (2015)
Metarhizium anisopliae	Silver (AgNPs)	*Rhynchophorus ferrugineus* (Red palm weevil)	Mortality—90% (eggs), 95% (larvae), 77% (adults)	Abdel-Raheem et al. (2020)
Metarhizium anisopliae	Silver (AgNPs)	*Aedes aegypti*	Mortality—100% at 3.2 and 1.8 μL/cm^2, 60% at 0.8 μL/cm^2	Mehmood et al. (2024)
Metarhizium robertsii	Copper (CuNPs)	*Aedes aegypti*	LC_{50}—7.796 μg/ml (larvae)	Vivekanandhan et al. (2024)
Metarhizium robertsii	Copper (CuNPs)	*Aedes albopictus*	LC_{50} at 48 h—74.873 mg/L (larvae), 76.101 mg/L (pupae), 136.645 mg/L (adults)	Vivekanandhan et al. (2024)
Metarhizium robertsii	Copper (CuNPs)	*Culex quinquefasciatus*	LC_{50}—3.136 μg/ml (larvae)	Vivekanandhan et al. (2024)
Metarhizium anisopliae	Zinc oxide (ZnO)	*Spodoptera frugiperda* (fall armyworm)	Mortality—100 μg/ml after 48 h (larvae)	Kumaravel et al. (2021)
Trichoderma culture filtrate	Selenium (SeNPs)	*Spodoptera litura* (larvae)	LC_{50}—39.739 ppm	Arunthirumeni et al. (2022)
Verticillium lecanii	Silver (AgNPs)	*Rhynchophorus ferrugineus*	Mortality—58% (eggs), 60% (larvae), 48% (adults)	Abdel-Raheem et al. (2020)

antibiotics against *Escherichia coli*, *Staphylococcus aureus*, and *Pseudomonas aeruginosa*. In another study, Santos et al. (2022) used aqueous fungal extracts of *B. bassiana*, *M. anisopliae*, and *Isaria fumosorosea* to biosynthesize AgNPs and evaluate their insecticidal action against the insect pest *Plutella xylostella*. The synthesized particles, characterized via UV-Vis and DLS analysis, demonstrated strong larvicidal activity, with a 78% cumulative mortality rate by the end of the larval stage and an LC_{50} of 0.691 mg/mL after 72 h.

I. fumosorosea showed the highest nanoparticle yield, underscoring its potential in green pest management. *Fusarium oxysporum*, another EPF, was used to generate AgNPs under variable pH conditions. The nanoparticles, analyzed through TEM and SEM, showed that fungal proteins played a crucial role in reducing silver ions and enhancing biological activity. AgNPs produced at pH 4 and 5 induced 100% mortality of *Rhipicephalus microplus* ticks within 24 h (Sumera et al. 2024). Similarly, *Metarhizium niley*-derived AgNPs were studied by Jamunarani et al. (2021), with a visible color shift indicating successful synthesis. These AgNPs effectively killed *Spodoptera litura* larvae, achieving up to 80% mortality in second instars at 1.75 mM and 73.75–78.75% in pot trials. In another experiment, mycogenic AgNPs synthesized from *Fusarium pallidoroseum* biomass were tested on third-instar *Holotrichia* grubs, serious pests in sugarcane cultivation. The larvae showed visible physiological changes and complete mortality within 24 h, with an LD_{50} of 92.68 ppm (Shukla et al. 2022). Kteo et al. (2025) explored the use of AgNPs, conidial suspensions, and fungal filtrates derived from *M. anisopliae* to combat *Tribolium castaneum* and *Trogoderma granarium*. After 8 weeks, AgNP-treated bags had significantly fewer insects (26.7–66 per bag) compared to untreated ones (366.7–586 per bag), validating the efficacy of AgNPs in stored grain protection.

4.2 Zinc Oxide Nanoparticles

Green synthesis of zinc oxide nanoparticles (ZnO NPs) is gaining traction as an environmentally responsible alternative to conventional methods. These ZnNPs are increasingly applied in agriculture for their antibacterial and insecticidal properties (Bihal et al. 2023). In a study by Kumaravel et al. (2021), *M. anisopliae* was used to biosynthesize bimetallic ZnO nanoparticles (BMNPs) that were evaluated against *Spodoptera frugiperda*, a key pest affecting crops like maize, cotton, and rice. The BMNPs showed superior larvicidal activity, with the highest mortality observed at 100 μg/mL after 48 h. These particles also demonstrated strong antifeedant and pupicidal properties, highlighting their multifaceted insecticidal potential.

4.3 Selenium Nanoparticles

Selenium nanoparticles (SeNPs) are known for their diverse biological properties, including antimicrobial, antioxidant, anticancer, and plant growth-promoting effects, making them highly suitable for agricultural and medicinal use (Helmy et al. 2023). Helmy et al. synthesized SeNPs and other metal nanoparticles from *Fusarium solani* extracts. Among these, SeNPs were particularly effective against stored grain pests like *Callosobruchus chinensis*, reducing egg hatching by 23% and shortening developmental time by 18%. In a separate study, Arunthirumeni et al. (2022) synthesized SeNPs using *Trichoderma* fungi and tested their efficacy against *Spodoptera litura*. Significant larvicidal effects were observed at 48 h, with LC_{90}

and LC_{50} values of 142.839 ppm and 39.739 ppm, respectively. Maximum antifeedant activity was recorded at 100 μg/mL. These results highlight the potential of SeNPs as biopesticides in sustainable pest control.

4.4 Copper Nanoparticles

Copper nanoparticles (CuNPs) disrupt insect metabolism by targeting oxidative stress enzymes, leading to increased mortality. These nanoparticles can penetrate cell membranes and bind with biomolecules, resulting in cell death (Vivekanandhan et al. 2024). *Metarhizium robertsii* was used to biosynthesize CuNPs with strong mosquitocidal activity. Vivekanandhan et al. (2024) demonstrated their effectiveness against *Aedes albopictus*, achieving 97.33% larval mortality and 74.66% adult mortality within 48 h, with LC_{50} values of 74.873 mg/L (larvae) and 136.645 mg/L (adults). These CuNPs were also tested on other vector species like *Anopheles stephensi*, *Aedes aegypti*, and *Culex quinquefasciatus*, achieving significant larval mortality at just 25 μg/mL. LC_{50} and LC_{90} values ranged from 3.136 to 65.144 μg/mL across species. Notably, the biogenic CuNPs exhibited specificity with minimal effects on non-target organisms, suggesting their suitability for integrated vector management (Vivekanandhan et al. 2024).

5 Mechanisms of Action of Formulations

Nanoparticles (NPs), typically 1–100 nm in size, exhibit unique physicochemical properties, such as high surface area, optical activity, and chemical reactivity, which contribute to their insecticidal potential (Sahayaraj et al. 2016). However, detailed insights into their mechanisms of entry and action in insects remain limited. The different modes of nanotoxicological effects of nanoparticles on insects are illustrated in Fig. 19.6. Entry into insect systems is size-dependent: smaller nanoparticles are internalized via endocytosis, ~500 nm particles through phagocytosis, and aggregates by micropinocytosis (Foldbjerg et al. 2015). They can penetrate insect bodies through contact, inhalation, or ingestion, entering via spiracles, mouthparts, anal prologs, abdominal prolegs, and setae (Shahzad and Manzoor 2021; Lade and Gogle 2019). Once inside, nanoparticles disrupt insect physiology by dehydrating the cuticle, such as alumina NPs that strip wax layers via triboelectric forces (Stadler et al. 2017), or by inducing structural changes, as observed with *Pimpinella anisum* nanoemulsions that cause epidermal necrosis in *Tribolium castaneum* (Hashem et al. 2018).

Silver nanoparticles (AgNPs) interfere with melanin synthesis, downregulate copper-dependent enzymes, and sequester copper ions, leading to reduced flight ability in *Drosophila melanogaster* (Armstrong et al. 2013). Carbon-silver nanohybrids bind to phosphorus/sulfur in DNA and proteins, causing gut and cuticle disruption in *Culex quinquefasciatus* and *Anopheles stephensi* (Sultana et al. 2018). In *Chironomus riparius*, nanosilver exposure alters gene expression, downregulating

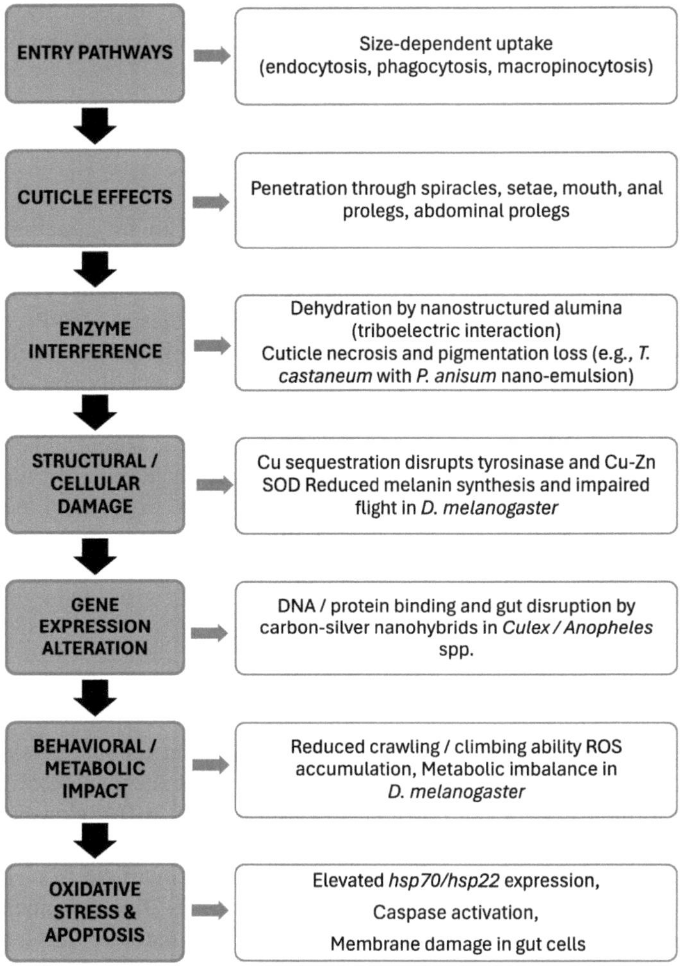

Fig. 19.6 Diagrammatic representation of the nanotoxicological effects of nanoparticles on insects. The flowchart illustrates key impact categories, including entry pathways, cuticle disruption, enzymatic interference, structural damage, gene expression changes, behavioral alterations, and oxidative stress leading to apoptosis

ribosomal genes like CrL15 and upregulating stress and reproductive genes (CrGnRH1, CrBR2.2), suggesting reproductive failure and defensive responses (Nair and Choi 2011). AgNP ingestion in early larval stages leads to impaired motor functions, metabolic alterations, and increased reactive oxygen species (ROS) levels (Raj et al. 2017). Additionally, AgNPs induce expression of hsp70 and hsp22 in midgut cells, promoting caspase activation, membrane instability, and apoptosis (Pandey et al. 2013).

6 Future Perspectives and Research Needs

The integration of EPF with nanotechnology represents a transformative frontier in biological pest control. While EPF, such as *B. bassiana* and *M. anisopliae*, have long been valued for their eco-friendly pest-killing abilities, their real-world application is often limited by environmental sensitivity, slow action, and formulation challenges. Nano-formulations offer a powerful solution to these limitations, opening new avenues for precision pest management.

Enhanced Stability and Delivery Systems One of the most significant contributions of nanotechnology is the protection of fungal spores from degradation caused by UV radiation, desiccation, and temperature extremes. Nano-encapsulation and controlled-release systems can ensure that fungal propagules remain viable and effective over longer periods, even under field conditions. This advancement could lead to more reliable and shelf-stable biopesticide formulations.

Targeted Action and Uptake Mechanisms Nanocarriers, including lipid-based nanoparticles, polymers, and nanoemulsions, can be designed to enhance adhesion to insect cuticles and facilitate penetration through natural entry points like spiracles or intersegmental membranes. This not only increases efficacy but can also reduce the required dosage, making applications more economical and environmentally sound.

Synergistic Combinations with Other Bioagents or Actives The future may see nano-formulated fungi combined with essential oils, plant extracts, or other biocontrol agents, enhancing their collective impact. For instance, encapsulating EPF with nanoemulsions of insecticidal phytochemicals may produce a dual-action effect—attacking the pest both physiologically and chemically while still aligning with organic farming principles.

Eco-safety and Non-target Considerations As nano-enabled fungal biopesticides become more prevalent, ecotoxicological studies will be critical. Understanding their fate in soil, water, and within non-target organisms (e.g., pollinators or beneficial arthropods) will ensure they remain a safe alternative to synthetic pesticides.

Molecular and Omics-Level Insights Emerging research in genomics, transcriptomics, and proteomics can help to decode how fungal pathogens interact with insect hosts at a molecular level, particularly under the influence of nano-formulations. These insights may allow researchers to engineer fungal strains or nanocarriers for even greater specificity and virulence.

Smart Delivery Systems (AI, Drones, IoT Integration) The future of EPF nano-formulation lies in digital agriculture and precision pest management. Smart technologies like AI-driven decision support systems, Internet of Things (IoT) sensors, and drone-based application platforms can facilitate real-time monitoring of pest pressure and optimize fungal bioinsecticide deployment. Drones equipped with image recognition and geospatial mapping software can apply nano-formulated EPF with pinpoint accuracy, reducing off-target effects and labor costs.

Scalable Formulation and Field Deployment Despite promising lab results, many nano-biopesticides face hurdles in scaling up production and meeting regulatory standards. Future research should focus on cost-effective, field-ready formulation techniques that preserve the fungi's viability while meeting the shelf life and usability expectations of farmers and agribusinesses.

Digital Agriculture Integration In the long term, nano-formulated EPF products may be integrated into precision agriculture systems, including drones for targeted application, sensors to monitor pest pressure, and AI-driven decision support systems. This will help optimize timing, reduce inputs, and improve overall pest control efficiency.

By fusing the biological prowess of EPF with the precision and efficiency of nanotechnology, we can move toward a future of resilient, sustainable, and eco-conscious pest management. Continued interdisciplinary research, smart regulatory frameworks, and inclusive farmer adoption strategies will be key to realizing this vision.

7 Conclusion

The integration of EPF with nanotechnology offers a promising and sustainable approach to pest management. Nano-formulations enhance the stability, effectiveness, and environmental persistence of EPF by protecting fungal spores from UV radiation, desiccation, and temperature extremes while enabling controlled and targeted delivery. Biologically synthesized nanoparticles, such as silver, zinc oxide, selenium, and copper, have demonstrated strong insecticidal activity and show potential for use in integrated pest management programs. However, their widespread application raises important toxicological concerns. Potential risks include unintended effects on non-target organisms such as pollinators and soil microbes, bioaccumulation in ecosystems, and possible human exposure during handling or through residues in food. While fungal-based nanopesticides are generally considered safer than synthetic chemicals, thorough ecotoxicological and human health assessments are essential. Furthermore, regulatory frameworks must adapt to address the unique properties of these nano-biopesticides. Future research should focus on ensuring safety, environmental compatibility, and scalability, paving the way for their responsible use in sustainable agriculture. Integrating nano-formulated EPF with digital tools in precision agriculture—such as drones, sensors, and AI—can further optimize pest control outcomes. The combination of nature-based biocontrol and technological innovation may contribute more to resilient and environmentally conscious agricultural ecosystems.

References

Abdel-raheem MA, Reyad N, Alghamdi H (2020) Virulence of nano–particle preparation of entomopathogenic fungi and entomopathogenic bacteria against red palm weevil *Rhynchophorus ferrugineus* (Olivier) (Coleoptera: Curculionidae). Rom Biotechnol Lett 25(1):1151–1159. https://doi.org/10.25083/rbl/25.1/1151.1159

Acheampong MA, Hill MP, Moore SD, Coombes CA (2020) UV sensitivity of Beauveria bassiana and *Metarhizium anisopliae* isolates under investigation as potential biological control agents in South African citrus orchards. Fungal Biol 124(5):304–310. https://doi.org/10.1016/j.funbio.2019.08.009

Álvarez SP, Tapia MA, Pérez KI, Guerrero AM (2017) Agriculture applications of entomopathogenic fungi using nanotechnology. In: Prasad R (ed) Fungal nanotechnology: applications in agriculture, industry, and medicine. Springer, Cham, pp 35–53. https://doi.org/10.1007/978-3-319-68424-6_2

Armstrong N, Ramamoorthy M, Lyon D, Jones K, Duttaroy A (2013) Mechanism of silver nanoparticles action on insect pigmentation reveals intervention of copper homeostasis. PLoS One 8(1):e53186. https://doi.org/10.1371/journal.pone.0053186

Arunthirumeni M, Veerammal V, Shivakumar MS (2022) Biocontrol efficacy of mycosynthesized selenium nanoparticle using *Trichoderma* sp. on insect pest *Spodoptera litura*. J Clust Sci 33: 1645–1653. https://doi.org/10.1007/s10876-021-02095-4

Banu AN, Balasubramanian C (2014) Optimization and synthesis of silver nanoparticles using *Isaria fumosorosea* against human vector mosquitoes. Parasitol Res 113(10):3843–3851. https://doi.org/10.1007/s00436-014-4052-0

Beltrán Pineda ME, Lizarazo Forero LM, Sierra YC (2023) Mycosynthesis of silver nanoparticles: a review. Biometals 36(4):745–776. https://doi.org/10.1007/s10534-022-00479-1

Bihal R, Al-Khayri JM, Banu AN, Kudesia N, Ahmed FK, Sarkar R et al (2023) Entomopathogenic fungi: an eco-friendly synthesis of sustainable nanoparticles and their nanopesticide properties. Microorganisms 11(6):1617. https://doi.org/10.3390/microorganisms11061617

Castrillo LA, Griggs MH, Liu H, Bauer LS, Vandenberg JD (2010) Assessing deposition and persistence of *Beauveria bassiana* GHA (Ascomycota: Hypocreales) applied for control of the emerald ash borer, Agrilus planipennis (Coleoptera: Buprestidae), in a commercial tree nursery. Biol Control 54(1):61–67. https://doi.org/10.1016/j.biocontrol.2010.03.005

de Souza TD, Fernandes FO, Sanches AC, do Nascimento J, Pinto AA et al (2022) Relation between *Helicoverpa armigera* (Hubner) (Lepidoptera/Noctuidae) mortality and entomopathogenic fungi persistence in soybean leaflets. Egypt J Biol Pest Control 32(1):10. https://doi.org/10.1186/s41938-022-00508-8

Deshmukh SK, Kochar M, Kaur P, Singh PP (2023) Nanotechnology in agriculture and environmental science. CRC Press, p 332

Donzelli BGG, Krasnoff SB, Churchill ACL, Vandenberg JD, Gibson DM (2010) Identification of a hybrid PKS–NRPS required for the biosynthesis of NG-391 in *Metarhizium robertsii*. Curr Genet 56(2):151–162. https://doi.org/10.1007/s00294-010-0288-0

Foldbjerg R, Jiang X, Miclăuş T, Chen C, Autrup H et al (2015) Silver nanoparticles–wolves in sheep's clothing? Toxicol Res 4(3):563–575. https://doi.org/10.1039/c4tx00110a

Ghorui M, Chowdhury S, Burla S (2024) The science behind entomopathogenic fungi: mechanisms and applications. In: Deshmukh SK, Sridhar KR (eds) Entomopathogenic fungi: prospects and challenges. Springer, Singapore, pp 3–35. https://doi.org/10.1007/978-981-97-5991-0_1

Hashem AS, Awadalla SS, Zayed GM, Maggi F, Benelli G (2018) *Pimpinella anisum* essential oil nanoemulsions against *Tribolium castaneum*—insecticidal activity and mode of action. Environ Sci Pollut Res 25(19):18802–18812. https://doi.org/10.1007/s11356-018-2068-1

Hayles J, Johnson L, Worthley C, Losic D (2017) Nanopesticides: a review of current research and perspectives. In: New pesticides and soil sensors. Elsevier, pp 193–225. https://doi.org/10.1016/B978-0-12-804299-1.00006-0

Helmy EA, San PP, Zhang YZ, Adarkwah C, Tuda M (2023) Entomotoxic efficacy of fungus-synthesized nanoparticles against immature stages of stored bean pests. Sci Rep 13(1):8508. https://doi.org/10.1038/s41598-023-35697-1

Iravani S, Korbekandi H, Mirmohammadi SV, Zolfaghari B (2014) Synthesis of silver nanoparticles: chemical, physical and biological methods. Res Pharm Sci 9(6):385–406. http://journals.lww.com/rips

Islam W, Adnan M, Shabbir A, Naveed H, Abubakar YS et al (2021) Insect-fungal-interactions: a detailed review on entomopathogenic fungi pathogenicity to combat insect pests. Microb Pathog 159:105122. https://doi.org/10.1016/j.micpath.2021.105122

Jamunarani GS, Ramanagoljda SH, Satyanarayana C, Mulla SR, Vinaykumar MM et al (2021) Efficacy of biosynthesised silver nanoparticles from entomopathogenic fungi *Metarhizium* Rileyi against *Spodoptera litura* on cabbage. Indian J Entomol 83(4):558–563. https://doi.org/10.5958/0974-8172.2020.00231.X

Kamil D, Prameeladevi T, Ganesh S, Prabhakaran N, Nareshkumar R et al (2017) Green synthesis of silver nanoparticles by entomopathogenic fungus *Beauveria bassiana* and their bioefficacy against mustard aphid (*Lipaphis erysimi* Kalt.). Indian J Exp Biol 55:555–561

Kteo BQ, Mohammed AA, Al-Saffar HH (2025) Efficacy of different conidial concentrations, fungal filtrations and silver nanoparticles mycosynthesis using two isolates of *Metarhizium anisoplae* against two stored-grain insects. Kufa J Agric Sci 17(1):61–72. https://doi.org/10.36077/kjas/2025/v17i1.11900

Kumar A, Dixit CK (2018) Methods for characterization of nanoparticles. In: Advances in nanomedicine for the delivery of therapeutic nucleic acids. Woodhead Publishing, pp 43–58. https://doi.org/10.1016/B978-0-08-100557-6.00003-1

Kumaravel J, Lalitha K, Arunthirumeni M, Shivakumar MS (2021) Mycosynthesis of bimetallic zinc oxide and titanium dioxide nanoparticles for control of *Spodoptera frugiperda*. Pestic Biochem Physiol 178:104910. https://doi.org/10.1016/j.pestbp.2021.104910

Lade BD, Gogle DP (2019) Nano-biopesticides: synthesis and applications in plant safety. In: Abd-Elsalam K, Prasad R (eds) Nanobiotechnology applications in plant protection, Nanotechnology in the life sciences, vol 2. Springer, Cham, pp 169–189. https://doi.org/10.1007/978-3-030-13296-5_9

Litwin A, Nowak M, Różalska S (2020) Entomopathogenic fungi: unconventional applications. Rev Environ Sci Biotechnol 19(1):23–42. https://doi.org/10.1007/s11157-020-09525-1

Mantzoukas S, Kitsiou F, Natsiopoulos D, Eliopoulos PA (2022) Entomopathogenic fungi: interactions and applications. Encyclopedia 2(2):646–656. https://doi.org/10.3390/encyclopedia2020044

Mehmood N, Hassan A, Zhong X, Zhu Y, Ouyang G et al (2024) Entomopathogenic fungi-based silver nanoparticles: a potential substitute of synthetic insecticides to counter behavioral and physiological immunity in *Aedes aegypti* mosquito (Diptera: Culicidae). Environ Sci Pollut Res 31(21):30793–30805. https://doi.org/10.1007/s11356-024-33292-z

Nair PM, Choi J (2011) Identification, characterization and expression profiles of *Chironomus riparius* glutathione S-transferase (GST) genes in response to cadmium and silver nanoparticles exposure. Aquat Toxicol 101(3–4):550–560. https://doi.org/10.1016/j.aquatox.2010.12.006

Pandey A, Chandra S, Chauhan LK, Narayan G, Chowdhuri DK (2013) Cellular internalization and stress response of ingested amorphous silica nanoparticles in the midgut of *Drosophila melanogaster*. Biochimica et Biophysica Acta (BBA) 1830(1):2256–2266. https://doi.org/10.1016/j.bbagen.2012.10.001

Prabhu S, Poulose EK (2012) Silver nanoparticles: mechanism of antimicrobial action, synthesis, medical applications, and toxicity effects. Int Nano Lett 2(1):32. https://doi.org/10.1186/2228-5326-2-32

Raj A, Shah P, Agrawal N (2017) Sedentary behavior and altered metabolic activity by AgNPs ingestion in *Drosophila melanogaster*. Sci Rep 7(1):15617. https://doi.org/10.1038/s41598-017-15645-6

Sabbour M (2015) A novel pathogenicity of nano-*Beauveria bassiana* and *Metarhizium anisopliae* against *Sitophilus oryzae* (l.) (coleoptera: curculiondae) under laboratory and store conditions. Int J Eng Res 6(12):938–945

Sahayaraj K, Madasamy M, Radhika SA (2016) Insecticidal activity of bio-silver and gold nanoparticles against *Pericallia ricini* fab. (lepidaptera: archidae). J Biopest 9(1):63–72. https://doi.org/10.57182/jbiopestic.9.1.63-72

Santos TS, de Souza Varize C, Sanchez-Lopez E, Jain SA, Souto EB et al (2022) Entomopathogenic fungi-mediated AgNPs: synthesis and insecticidal effect against *Plutella xylostella* (Lepidoptera: Plutellidae). Materials 15(21):7596. https://doi.org/10.3390/ma15217596

Shahzad K, Manzoor F (2021) Nanoformulations and their mode of action in insects: a review of biological interactions. Drug Chem Toxicol 44(1):1–11. https://doi.org/10.1080/01480545.2018.1525393

Sharma A, Sharma S, Yadav PK (2023) Entomopathogenic fungi and their relevance in sustainable agriculture: a review. Cogent Food Agric 9(1):2180857. https://doi.org/10.1080/23311932.2023.2180857

Shatalova EI, Grizanova EV, Dubovskiy IM (2022) The effect of silicon dioxide nanoparticles combined with entomopathogenic bacteria or fungus on the survival of Colorado potato beetle and cabbage beetles. Nanomaterials 12(9):1558. https://doi.org/10.3390/nano12091558

Shukla G, Gaurav SS, Singh A, Rani P (2022) Synthesis of mycogenic silver nanoparticles by *Fusarium pallidoroseum* and evaluation of its larvicidal effect against white grubs (*Holotrichia* sp.). Mater Today Proc 49(8):3517–3527. https://doi.org/10.1016/j.matpr.2021.07.238

Soleimani P, Mehrvar A, Michaud JP, Vaez N (2022) Optimization of silver nanoparticle biosynthesis by entomopathogenic fungi and assays of their antimicrobial and antifungal properties. J Invertebr Pathol 190:107749. https://doi.org/10.1016/j.jip.2022.107749

Souza RK, Azevedo RF, Lobo AO, Rangel DE (2014) Conidial water affinity is an important characteristic for thermotolerance in entomopathogenic fungi. Biocontrol Sci Tech 24(4): 448–461. https://doi.org/10.1080/09583157.2013.871223

Stadler T, López García GP, Gitto JG, Buteler M (2017) Nanostructured alumina: biocidal properties and mechanism of action of a novel insecticide powder. Bull Insectol 70(1):17–25. https://www.lihoplus.com.ar/liho/wp-content/uploads/2018/06/30.-Stadleretal.pdf

Sultana N, Raul PK, Goswami D, Das B, Gogoi HK et al (2018) Nanoweapon: control of mosquito breeding using carbon-dot-silver nanohybrid as a biolarvicide. Environ Chem Lett 16(3): 1017–1023. https://doi.org/10.1007/s10311-018-0712-0

Sumera NS, Iqbal SS, Khan ST (2024) *Fusarium oxysporum*; its enhanced entomopathogenic activity with acidic silver nanoparticles against *Rhipicephalus microplus* ticks. Braz J Biol 84: e266741. https://doi.org/10.1590/1519-6984.266741

Sweet MJ, Chessher A, Singleton I (2012) Metal-based nanoparticles; size, function, and areas for advancement in applied microbiology. Adv Appl Microbiol 80:113–142. https://doi.org/10.1016/B978-0-12-394381-1.00005-2

Tyagi S, Tyagi PK, Gola D, Chauhan N, Bharti RK (2019) Extracellular synthesis of silver nanoparticles using entomopathogenic fungus: characterization and antibacterial potential. SN Appl Sci 1(12):1545. https://doi.org/10.1007/s42452-019-1593-y

Vivekanandhan P, Swathy K, Sarayut P, Krutmuang P (2024) Effects of copper nanoparticles synthesized from the entomopathogen *Metarhizium robertsii* against the dengue vector *Aedes albopictus* (Skuse, 1894). PLoS One 19(11):e0314279. https://doi.org/10.1371/journal.pone.0314279

Wang Y, Chang J, Wang R, Liu J, Hou R et al (2025) Application of an environmentally friendly nanopesticide translocation system based on *Metarhizium anisopliae*: using biodegradable chitosan nanopesticides to control *Spodoptera frugiperda*. Int J Biol Macromol 310:143261. https://doi.org/10.1016/j.ijbiomac.2025.143261

Commercialization of Entomopathogenic Fungi

20

Mansi Kothari and Sardul Singh Sandhu

Abstract

The agriculture sector is a source of food for the growing population. To meet food demand, farmers are applying numerous synthetic pesticides to protect their crops. Synthetic pesticides not only protect crops from pest populations but also pollute the ecosystem and harm beneficial microorganisms. To maintain an ecosystem balance without destroying beneficial microorganisms and enhancing crop productivity, there is a need to find an alternative prospect that suppresses pest populations without affecting the ecosystem. Biopesticides are the best alternative approach, exhibiting characteristics similar to those of synthetic pesticides. Biopesticides include living agents that help to suppress pest populations. Among different biological agents, fungi are an excellent substitute for all, as they infect the host pest, adjust according to their immunity, consume them as food, proliferate, and suppress their metabolic activity, ultimately causing them to die. To utilize EPF as a sustainable solution to enhance food productivity, it is necessary to formulate EPF-based biopesticides in mass quantities and commercialize them. EPF-based biopesticide commercialization provides many opportunities, such as eco-friendly pest management, a broad host range, and enhanced plant health that help in developing sustainable agriculture. To utilize these opportunities, it is necessary to adapt biopesticides in farming practices. Adaptation is not as easy as it looks; various factors require overcoming certain challenges, such as quality maintenance issues, limited storage life, insufficient knowledge, costly manufacturing hurdles, sensitivity to ecological stressors, and

M. Kothari
Fungal Biotechnology and Invertebrate Pathology Laboratory, Department of Biological Sciences, Rani Durgavati University, Jabalpur, Madhya Pradesh, India

S. S. Sandhu (✉)
Bio-Design Innovation Centre, Rani Durgavati University, Jabalpur, Madhya Pradesh, India

S. K. Deshmukh et al. (eds.), *Frontiers in Entomopathogenic Fungi*,
https://doi.org/10.1007/978-981-95-6529-0_20

a lack of suitable formulations. Challenges faced during commercialization can also be overcome through effective strategies when integrated with agricultural practices.

Keywords

Biopesticides · Challenges · Formulation · Opportunities · Policy guidelines · Regulation

1 Introduction

Agriculture is a major sector of the economy. The agriculture sector meets the food requirements by facing numerous challenges in feeding the global population. Pest destruction caused to plants is one of the major issues faced by the agricultural sector. Pests are responsible for ruining a huge number of crops. Pests consume plants and derive energy from them, posing one of the most alarming threats to the agricultural sector. The global population is continuously growing; therefore, the food production rate should also increase. To prevent crops from pests and enhance agricultural productivity, many chemical pesticides have been used since ancient times. These pesticides can control pest populations but also deteriorate soil fertility, harm the environment, and cause various fatal diseases in humans. Pesticides are chemical pollutants used to eliminate pests. Additionally, chemical pesticides also harm plant-associated beneficial microorganisms that either help control pest populations or enrich plant growth. The World Health Organization (WHO) reports that approximately 220,000 deaths occur annually due to pesticide pollution. It is essential to develop a sustainable technique that restricts pest pollution, enhances plant productivity, and safeguards the ecosystem (Avinash et al. 2024; Jhala et al. 2020; Singh et al. 2018).

The sustainable agriculture concept is based on the primary objectives of ecological health, financial profitability, economic viability, and social equity (Jeranyama et al. 2020). Sustainable agriculture is a comprehensive system designed to meet food requirements, enrich the ecosystem, and efficiently utilize all plant production-related resources, thereby sustaining farm viability for an extended period. Sustainable agriculture enriches human life by fulfilling their necessity without compromising the necessities of upcoming generations (Malo 2020). Sustainable agriculture is essential for humankind, as it prohibits the use of chemical substances, conserves raw resources, and promotes financial self-sufficiency (Singh et al. 2018). Biopesticides are the most sustainable technique that resembles the properties of pesticides and, therefore, can be utilized as a substitute. Biopesticides are living agents that control pest populations, are environmentally friendly, are long-term effective, and are biodegradable (Ayilara et al. 2023; Booth et al. 2022; Yadav et al. 2022).

To sustain agriculture by utilizing sustainable substitutes, such as biopesticides, it is essential to mass-produce and commercialize them, thereby reducing the market share of synthetic pesticides. Commercialization of EPF-based biopesticides offers many opportunities over synthetic pesticides. Still, it also comes with many challenges that can be overcome through several strategies and cooperative efforts exhibited by the public and private sectors. This article reviews EPF-based biopesticides, commercialization, commonly commercialized EPFs, opportunities for commercializing EPF-based biopesticides, challenges faced in adapting biopesticides, and their future outlook.

2 Biopesticides

According to the United States Environmental Protection Agency (USEPA), biopesticides are substances derived from biological agents, including microorganisms, plants, and animals. Biological agents that control pest populations are known as biocontrol or biological control. The National Academy of Science, situated in the United States, describes the term biological control as the utilization of living organisms to decrease the consequences caused by non-beneficial microorganisms and promote the growth of beneficial microorganisms. Microorganisms that can suppress the growth of pathogens are known as biological control agents (BCAs) (Bajsa et al. 2023). The biological control process does not require any human intervention; however, to enhance the pest suppression rate, humans can take appropriate action (De Sousa et al. 2025). Biocontrol agents are sustainable, target-specific organisms that can destroy pests.

Biopesticides enhance plant viability, optimize production, and therefore benefit farmers. The biodegradable nature of biopesticides prevents environmental pollution. Food safety is essential for the world's rising population, and it can be ensured by enhancing soil fertility. Pest infestations injure beneficial microbial populations, thereby reducing soil richness. Biopesticides suppress pest populations and enhance soil fertility. These opportunities exhibited by biopesticides will expand commercialization with time. Biopesticide formulations should be improved to enhance their yield, and their cost should be minimized; only then can farmers replace the use of chemical pesticides with biopesticides. Once replaced, it can bestow health and shield food.

Biopesticides are an ideal component of integrated pest management (IPM) programs (Vaghasiya et al. 2021) that aim to reduce the dependency on chemical pesticides and control pests using biological means. Biopesticides integrated with IPM offer several benefits, such as multiple mechanisms of action, being eco-friendly, and being highly specific toward targeted organisms and pest populations (Mawcha et al. 2025). In India, the utility of biopesticides comprising IPM has risen, and a minimum of 15 biological control agents have been formulated as biopesticides in 2017, and the number of registered formulated biopesticides was 970. The biopesticides market in India faces various challenges, including restricted resources for mass production, limited acceptance, and trade of unregistered products (Kumar et al. 2019).

Table 20.1 Major classification of biopesticides

Active substance/ mechanism of action	Effect	Result
Based on the active substance	Plant-incorporated protectants	Pesticidal compounds generated by plants after the incorporation of genetic material
	Microbial pesticides	Fungi, yeast, algae, and bacteria
	Biochemical pesticides	Semiochemicals, insect growth regulators, inorganic substances, and herbal extracts
Based on the mechanism of action	Repellency	Pest inhibition by crop-harming
	Predation	Pest control through killing and consumption as food
	Microbial action	Microorganisms infect the pest and ultimately cause death
	Metabolic disruption	Obstruction of metabolic processes

According to Akutse et al. (2020), USEPA classified biopesticides according to active substance into three major categories: microbial, plant-incorporated protectants, and biochemical pesticides. Microorganisms are the active ingredient of microbial pesticides, such as fungi, yeast, bacteria, and algae (Akutse et al. 2020). Microorganisms that exert a detrimental effect on the pest population are usually used as microbial-derived pesticides (Nanda et al. 2021). Natural compounds such as semiochemicals, insect growth regulators, inorganic substances, and herbal extracts are active ingredients of biochemical pesticides that control pest populations and exhibit a harmless mode of action. Pesticidal compounds generated by plants after incorporation of genetic material act as active ingredients of plant-incorporated protectants (Table 20.1).

According to Mawcha et al. (2025), biopesticides suppress pest populations by applying various mechanisms of action, which are categorized into the following types: Inhibiting pests from harming the crops (repellency), the pest population gets inhibited either by being killed and consumed by pests as food (predation), pests get infected by microorganisms (microbial action), and metabolic processes get obstructed (metabolic disruption) (Table 20.1). Biopesticides play an essential role in the IPM system, participating in sustainable farming by decreasing the ecological impact and reducing the resistance capability against any pests.

2.1 Entomopathogenic Fungi

Entomopathogens are infectious agents, such as fungi, that infect living pests and survive on them, ultimately causing mortality. This mode of action is known as microbial control. EPF plays a vital role in lowering the number of pests existing in nature (Misra et al. 2014) and is beneficial in integrated pest management. EPF is a eukaryotic, heterotrophic, evolutionarily diverse group of fungi that multiply either asexually, sexually, or both. Fungi are multicelled or unicelled organisms whose cell membranes, composed of chitin, are mostly motionless. In insects, approximately 80% of infections are caused by fungal agents. Millions of fungal species exist on Earth, but till now, only 100 genera that consist of 750–1000 fungal entomopathogens were known (Mora et al. 2017). In Myanmar, *Paleoophiocordyceps coccophagus*, a parasitic fungal fossil, was found, indicating that EPF existed from the Cretaceous period. EPFs are classified into various taxonomic categories, including Ascomycota, Oomycetes, Basidiomycota, Chytridiomycota, Entomophthoromycota, and Microsporidia (Vivekanandhan et al. 2024b).

Fungal pathogenicity was first recognized by Louis Pasteur (Kothari and Sandhu 2025). Agostino Bassi unearthed the first EPF in 1835 and termed it corpuscles. Paul Vuillemin in 1912 reclassified fungi and named them *Beauveria bassiana*. In insects, it causes white muscardine infection. In 1879, the cockchafer, a pest, was found infected with a fungal agent known as *Entomophthora anisopliae* that causes green muscardine infection. "Green muscardine" was named due to the germination of green-colored conidial cells. Later, it was found that *Cleonus punctiventis* was infected by the same fungal agent, which was renamed by Ilya Ilyich Metchnikoff and Mikhail Stepanovich Sorokin as *Metarhizium anisopliae*. After these two discoveries in the late nineteenth century, Lewis Pasteur (the father of microbiology) experimented to determine the potential of fungi as a biological control agent. Later, chemical insecticides were introduced into the trade, causing hindrance in the establishment of fungi as a biological control agent (Kelwatkar et al. 2018; Mantzoukas et al. 2022).

Fungal biopesticides are potent biological control agents and are safe for the ecosystem as well as for humans. Fungi come into contact with insect pests and adjust to their immunity, then initiate proliferation, and the pests become unable to resist infection. Fungi can proliferate rapidly, and therefore, fungi are an ideal organism to be used as a biopesticide (Chiew et al. 2022).

EPF has a wide range of actions and can infect different types of pest species. EPF is placed under 6 phyla that are subdivided into 12 classes within 4 groups, including Hyphomycetes, Laboulbeniales, Zygomycetes, and Pyrenomycetes (Opender Koul 2012). More awareness about organic insecticides and the growing demand for chemical-free food has enhanced the utilization of EPF in the agriculture sector. EPF is an essential tool in integrated pest management (De Sousa et al. 2025). All around the world, the trade of microbial biopesticides was about US$3.3 billion, of which 9% of the trade contribution was of EPF, making it the 2nd most traded microbial agent. Fungal biopesticide sales were high because of their sustainable nature, which can control various types of pest populations. EPF enhances sustainable farming and reduces chemical pollution. EPF suppresses detrimental pests and promotes the biological diversity of the soil, thereby helping in maintaining ecosystem balance (Vivekanandhan et al. 2024b). Globally, over 170 EPF-based biopesticides were formulated, out of which 75% of biopesticides are applied or registered for their application in the agriculture sector. The yearly growth of EPF-based biopesticide trade is 15% (Clifton et al. 2020).

EPF gives rise to spores that attach to the host pest's cuticle. Spores initiate germination and form an appressorium. Appressorium develops a mechanical force on the pest cuticle and liberates enzymes such as lipase, chitinase, and certain types of toxins, which help in the rupture of the pest cuticle. The hypha tube penetrates toward the haemolymph, where massive sporulation takes place, resulting in the development of infection within the host body. Fungal metabolites cause paralysis and repress the pest's immune system. The infectious period lasts for about 14 days. The fungus sucks up fluid from the host insect, due to which the host's body becomes rigid. Fungi emerge from the cadaver's natural pores, producing more spores to extend their lifespan, thereby controlling the pest population (Gull et al. 2023; Pradhan et al. 2024). Pest insects are directly treated with fungal biopesticides and mixed with soil; plants are treated with biopesticides. Biopesticide application is based on the fungal strain's mechanism of action and the occurrence of the type of host pest in an agricultural field (Goldstein 2024).

2.2 Entomopathogenic Fungal-Based Biopesticides

The term commercialization here refers to strategies designed to gain profit from a newly developed product (Aslani et al. 2015). Commercialization is the concluding step of the development procedure (Ulrich et al. 1995). A novel biopesticidal product is transferred from industry (origin site) to the market, where it can be commercialized. If the product generates economic profit, then commercialization is successful (Kim et al. 2019; Schilling 2008). To commercially produce EPF-based biopesticides, first select a virulent fungal isolate, then mass-produce and formulate it (Sharma et al. 2023). Formulation and commercialization of EPF-based biopesticide by-product includes several steps: biodiscovery, identification, effective strain selection, enhancing mass production, developing quality assurance, field effectiveness, biosecurity evaluation, registration, and commercialization (Srinivasan et al. 2019).

2.2.1 Biodiscovery

Initially, different sites have to be explored where the pest of concern can be found. EMF can be detected by observing fungal infections on insects. Early infection exhibits necrotic tissue, whereas later infection symptoms exhibit unhealthy tissues. Sometimes infections also exhibit a fluid-filled body, or the body shape gets distorted. Fungal infection emerges from the body through natural openings. EMF can also be segregated from the environmental samples.

2.2.2 Virulence Trial

The host insect is exposed to fungal spores. The insect is dipped into the fungal spore suspension, or fungal spores are directly sprayed on the host insect. In another technique, the plant leaf is dipped in a fungal spore suspension, and the insect is left to feed on it for about a week at a constant temperature. Then the infection rate can be determined. Using this technique, virulent strains can be selected for pest suppression in field applications.

2.2.3 Effective Strain Selection

Fungal characteristics vary from species to species. Fungal strains should be selected very specifically; they should not only show virulence, but it should also possess the best traits in different parameters such as temperature, medium, time, and moisture content that will improve complete performance. Reinoculation after a fixed time using a host pest helps in maintaining virulence (Parker et al. 2000).

2.2.4 Mass Production

Large-scale production of fungi (biocontrol agents) can be achieved using the fermentation technique. Submerged fermentation supports fungi in developing mycelia, which are later desiccated and distributed in the agricultural field. Fungus to be commercialized should be grown in a cheap, easily available medium containing vital nutrients. Fungal growth rate is an essential parameter. The fermentation process should be short to avoid contamination (Auld 1992). Fungi are living creatures; therefore, they require food to develop and proliferate. Fungi consume substrate as a food on which they survive. All substrates are not equally beneficial for every fungus. Optimum fungal growth can be obtained by growing them on a suitable, cheap medium that promotes a fast proliferation rate without depriving virulence. Major challenges faced during fungal mass production include an insufficient amount of good substratum. Fungal biopesticide mass production is essential for pest control, yet no new method has been explored in farming systems. Large-scale development of EPFs and their virulence assay is an essential parameter to be detected before applying them in the field as biopesticides (Kelwatkar et al. 2018; Prasad and Pal 2014).

2.2.5 Laboratory Trial

The mass-produced EPF strain has to be tested in a small quantity against the host pest. EPF strain effectiveness is determined after a certain time interval. Effectiveness is determined by the insect mortality rate, stopping feeding, and enhancing yield by reducing crop damage. Effectiveness also depends upon environmental conditions and the technique applied. Biopesticide formulation time and quantity depend on the trial run. Biopesticides that pass all parameters of the trial run can only be mass-produced. Without a trial run, no fungal biopesticides can be transferred to farmers. The trial run failed; fungal biopesticides have to be destroyed (Parker et al. 2000).

2.2.6 Formulation

The virulence of a fungal biopesticide depends on the metabolic activity of a fungus. Formulation of a fungal agent is required so that it can be efficiently used and preserved. Biopesticides are susceptible to environmental conditions and therefore require protective formulation to conserve their potential (Mawcha et al. 2025). Formulation and carriage are complicated steps. Spore formulation is a challenging step, as it should improve stability and adherence capability with the pest cuticle in environmental conditions. Large-scale production of conidia for commercialization is expensive and technically complex. Few strains of EPF can harm beneficial pests, and this is a point of worry as it can disturb the ecosystem balance (Ruiu 2018).

Formulation of a fungal-based biopesticide product requires a fungal agent (active ingredient) combined with different functional and inactive substances. This generated biopesticidal product is suitable for application over the pest population using in-use equipment. The fungal agent is integrated with support material and different preservatives to enhance the efficiency and longevity of biopesticides. Improving product formulation considerably enhances the consistency of field performance. The most essential step in biopesticide development is formulation. Easily usable formulation requires several essential measures, including protecting and exhibiting the fungal pesticidal characteristic, increasing storage life minimally up to half a year or ideally for 2 years under environmental conditions, and facilitating the use of existing tools (Mawcha et al. 2025).

EPF has the capability of controlling the pest population. To fulfill the requirements of the agriculture sector, it is essential to mass-produce the EPF. Commercialization of EPF is done in various formulations (De Jesus Seabra et al. 2024). Commercially biopesticide products are sub-classified into two physical conditions, that is, liquid-based formulation (emulsion, suspension concentrate, and oil dispersion) and solid-based formulation (dusts, powder, granules, and wettable powders). An emulsion is a liquid globule that is dispersed (0.1 to 10 μm) in a non-miscible liquid. It is important to select a proper emulsifier to achieve stability. At the external phase of the emulsion, oil exists, which gets lost during evaporation, and pesticide drift is insignificant regarding water-in-oil emulsion. In suspension, the concentrated pulverized component (1 − 10 μm) is dispersed in a liquid state. During the pulverizing procedure, non-reactive materials are adsorbed onto the molecule surface, thus blocking tiny molecules from accumulating. A big surface allows faster access to the active ingredients of plant substances. Bio-effectiveness of small substances is higher. Oil dispersion is an active solid substance that is dispersed in an anhydrous liquid. Dusts are composed of active substances along with pulverized solids (50 − 100 μm). Dust can be directly applied to the target pest either by machine or by hand. Formulated dust improves the adsorption of non-reactive substances, such as UV protectants and anti-sticking agents. Powder consists of active substances that are mixed with a carrier and supporters that promote adhesion of the by-product with the seed. Granules are composed of pulverized plant material, polymers, silica, starch, kaolin, and attapulgite. Initially, paste is prepared by mixing liquid with powder in an appropriate amount, which is then squeezed out and dehydrated to generate the active substance. Generally, granules are applied to the soil to control weeds, soil pests, etc. *M. anisopliae* is a fungal agent that requires soil humidity to release its active substance. Wettable powder formulation incorporates an active ingredient with dispersion, wetting, surfactants, and inactive filler. The mixture is then ground up to a particle dimension of about 5 μm (Fenibo et al. 2022). Ultra-low volume liquid formulation composed of a dense concentration of active substances that is greatly soluble in a sustainable liquid. The formulation is not dissolved in liquid before application (Soyel et al. 2022).

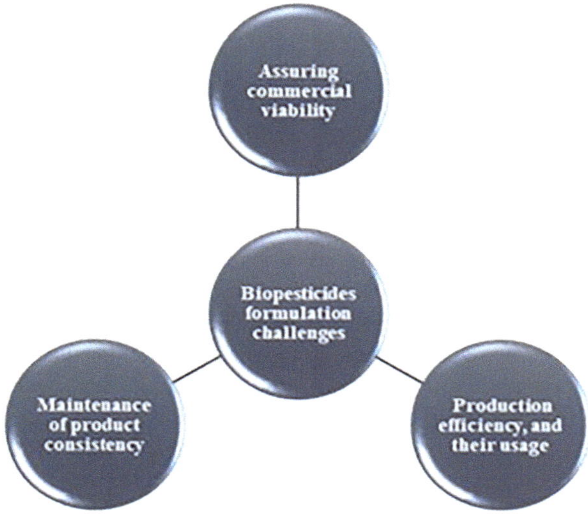

Fig. 20.1 Challenges faced at the time of biopesticide formulation

The purpose of formulation includes enhancing the product consistency, biological activity, transport, and incorporation into pest control. A promising formulation places consumer benefits in preference, harmonious work with tools and their operations, and potency at an efficient rate. Various challenges are faced in formulating microbial pesticides, such as assuring commercial viability, production efficiency, and their usage; maintenance of product consistency during stockpiling; and transferring (Fig. 20.1). Various factors that restrict commercialization, such as quality maintenance problems, storage life, limited knowledge, costly hurdles in manufacturing, sensitivity to ecological stressors, and an absence of appropriate formulation (Mawcha et al. 2025; Samada and Tambunan 2020).

2.2.7 Delivery Method

In plants, EPF is inoculated using different methods such as foliar application, stem injection, seed dressing, and soil sprinkling. Foliar application is an effective mechanism for controlling pests in the phyllosphere region. EPF penetrates inside the plant tissue, nourishes itself, and controls pests. *B. bassiana* and *M. brunneum* are EPF that colonize the phyllosphere of plants, including melons, tomatoes, alfalfa, and sweet peppers, temporarily. Commercially formulated EPF storage life is 3–6 months, respectively. Formulated biopesticides contain 10^9 to 10^{10} fungal spores per gram. Application dose ranges correspond to the formulation, intensity of infection, pest type, and ecological conditions. Fungal biopesticides are provided with instructions for application (Sharma et al. 2023).

2.2.8 Efficiency Trial, Registration, and Commercialization

Biopesticides were introduced to protect plants, and the next generation of biopesticides is being developed to control pests (Kesho 2020). Evolving EPFs as a potential biopesticide, which has been approved for application in agricultural fields, can be a lengthy and costly process. Therefore, ample research on assaying different EP fungal strains that exhibit virulence against insect pests and can be formulated as biopesticides is ongoing. Once EPF-based biopesticides have been commercially developed, they should be tested in an agricultural field without any delay (Clifton et al. 2020). Various steps have been involved in the development of biopesticides. Biopesticide commercialization requires large-scale production of a fungal agent and protection for storage, and living agents are utilized for formulation to enhance their viability (Kesho 2020). In India, the commercialization of EPF biopesticides has constantly enhanced. Many companies in India are formulating and commercializing EPF-based biopesticides, which are making it easier for farmers to access, thereby promoting sustainable agricultural practices. Multiplex and T. Stanes are companies formulating fungal biopesticides that help to protect crops from pests (Pavan et al. 2024). In 30 nations, it has been estimated that more than 200 biopesticides are accessible that have been approved by the Organization for Economic Co-operation and Development (OECD). Due to the enhancement in technology, registration of biopesticides all across the world is constantly increasing (Kesho 2020). The USEPA has approved over 100 fungi-derived biocontrol by-products over the past 30 years (Parveen and Rashtrapal 2024). Over 14 agencies are mass-producing and commercializing *B. brongniartii* and *B. bassiana*, whereas over 10 agencies are mass-producing and commercializing *M. anisopliae* globally (Elbashir et al. 2014).

According to Ruiu (2018), various EPF-based pesticidal products have distinct types of fungal strains. The representative operating names are expressed on the related organization websites, from which the source should be created. Biopesticide awareness, health hazard knowledge, and ecosystem imbalance due to synthetic biopesticides have resulted in the growth of the biopesticides market during the preceding years. Mainly, commercially available EPF-based pesticide products are derived from conidial suspension.

Beauveria bassiana is a highly utilized fungal biopesticide and the first pest control agent in the late nineteenth century. *B. brongniartii* and *B. bassiana* are fungal agents utilized in various commercially formulated biopesticides, exhibiting diverse levels of pathogenicity against various pests.

Metarhizium anisopliae is able to infect a wide variety of pests. *M. anisopliae* produces varied types of pesticidal toxins and virulence determinants. Globally, several fungal agents such as *Lecanicillium* spp., *Paecilomyces* spp., *Verticillium lecanii.*, *Isaria* spp., *Hirsutella* spp. are commercially utilized to control pests.

According to Parveen and Rashtrapal (2024), the USEPA has approved over 100 fungal agents and bio-based microbial control products since 1995. Mycotrol is a commercially accessible product incorporating *Beauveria* to eradicate grasshoppers, whiteflies, aphids, and thrips. Green Muscle, incorporating *M. anisopliae*, is a product used to eradicate locusts and grasshoppers. These formulated products can be contagious for a long period when preserved at 25–30 °C. Table 20.2 shows commercially accessible EPF-based products and the targeted pests.

Table 20.2 Commercialized entomopathogenic fungal-based biopesticides

Entomopathogenic fungi	Target pest	Commercial product name	Reference
Beauveria bassiana	Sucking insects	Naturalis™, Tray Bioscience	Kumar et al. (2020)
	Whiteflies/Aphids/Thrips	Mycontrol-WP/Mycotech Corp	
	Corn borer	Ostrinil/Natural plant protection	
	Varied types of insects and mites	Bio-power, Biorin/Kargar, Botanigard, Daman, Nagestra, Beauvitech-WP, Bb-Protec, and Racer, Mycotrol	Ruiu (2018)
	Coleoptera (Curculionidae, Scarabaeidae), Lepidoptera (Castniidae)	Bauveril	
	Coleoptera (Chrysomelidae)	Boverol	
	Colorado potato beetle; Leptinotarsa decemlineata, and the codling moth, Cydia pomonella	Boverin	
	Coleoptera (Curculionidae)	Conidia	
	Lepidoptera (Noctuidae)	Proecol	
	Coleoptera (Curculionidae, Scarabaeidae), Lepidoptera (Castniidae, Pieridae), Hemiptera (Aleyrodidae), Thysanoptera (Thripidae)	Trichobass-L and Trichobass-P	
Beauveria brongniartii	Helicoverpa armigera, Berry borer, Root grubs	Bas-Eco-	
	European cockchafer and other scarab beetle species	Beauveria Schweizer	
	Coleoptera (Scarabaeidae)	Betel	
	Coleoptera (Cerambycidae)	Biolisa-Kamikiri	
	European cockchafer and other scarab beetle species	Engerlingspilz	
Isaria farinosa	Apple moth, Siberian pine caterpillar, and larch caterpillar	Paecilomin,	
Metarhizium anisopliae	Pacer-beetles and caterpillar pests; grasshoppers, termites	Biomet/Ankush, Bio-magic, Devastra, Kalichakra, Novacrid and Met52/ BIO1020 granular	
	Termites	Bio-path™/ EcoScience/ USA	Kumar et al. (2020)
	Locusts, Grasshoppers, Red-headed, cockchafer, Pasture/Turf	Biogreen/Biocare Technology Pvt. Ltd.	

(continued)

Table 20.2 (continued)

Entomopathogenic fungi	Target pest	Commercial product name	Reference
	Tetranynchus urticae	Achieve	
	Isoptera (Kalotermitidae, Rhinotermitidae, Termopsidae)	BioBlast	
	Sugarcane pest: grayback canegrub	Bio-Cane	
	Hemiptera (Aleyrodidae, Aphididae)	Bio-Catch-M	
	Mealbugs	Campaign	
	Coleoptera (Scarabaeidae), Hemiptera (Cercopidae, Aphididae)	Cobican	
	Coleoptera (Scarabaeidae, Curculionidae, Nitidulidae)	Gran Met-P	
	Orthoptera (Acrididae)	Green guard	
	Locust and grasshopper	Green muscle	
	Turticae	Mazao achieve	
	Mealybugs	Mazao campaign	
	Aphids	Mazao supreme	
	Sugarcane spittlebug; Mahanarva postica	Metaquino	
	Coleoptera (Scarabaeidae)	Metarhizium Schweizer	
	Thrips	Metathripol	
	Isoptera	Pacer	
	Mealybugs	Real metarhizium SC	
	Coleoptera (Curculionidae, Scarabaeidae), Diptera (Ephydridae, Mycetophilidae, Sciaridae, Tipulidae), Thysanoptera (Thripidae)	Taenure granular Bioinsecticide	
	Acari (Ixodidae), Coleoptera (Scarabaeidae)	Tick-Ex	
	Coleopterans; Elateridae, Curculionidae	TAE-001 technical Bioinsecticide	
Hirsutella thompsonii	Spider mites	No-Mite	Ruiu (2018)
Isaria fumosorosea	Whitefly, aphids, thrips, psyllids, mealybugs, and fungus gnats	Nofly	
Paecilomyces lilacinus-	Plant pathogenic nematodes	Bio-Nematon, MeloCon, Mytech-WP, and Paecilo	
Paecilomyces fumosoroseus	Insects, Mites, Nematodes, Thrips	Bioact WG, No-Fly-WP, and Paecilomite	

(continued)

Table 20.2 (continued)

Entomopathogenic fungi	Target pest	Commercial product name	Reference
Lecanicillium lecanii	Aphids, leafminers, mealybugs, scale insects, thrips, whiteflies	Lecatech-WP and Varunastra	
Myrothecium verrucaria-	Nematodes	DiTera	
Metarhizium brunneum-	*Agriotes* spp.	Attracap-	Ruiu (2018)
Metarhizium flavoviride var. flavoviride	Grasshoppers and locusts	Biogreen	Skinner et al. (2014)
Trichoderma viride	Effective against rot disease	EcosomTV, Tricon, and Trieco	Rajamani and Negi (2020)
Trichoderma harzianum	Effective against a variety of soils pathogens	Rootshield, BioTrek, and Supresivit	Rajamani and Negi (2020)
Verticillium lecanii,	Aphids, Whiteflies, and Thrips,	Mycotal/Koppert	Sandhu et al. (2012)
	Mealybugs and sucking insects	Bio-Catch, Mealikil, and Bioline/Verti-Star	Ruiu (2018)

A total of 410 biopesticide manufacturing units were established in India, out of which 280 units are governed by the government, and the remaining 130 units are under the private sector. The government holds central IPM centers, or CIPMCs (26 units), ICAR (31 units), DBT (22 units), and different units such as biological control laboratories. Microbial research facilities were established by the state Ministry of Agriculture in different states, including Karnataka, Andhra Pradesh, Uttar Pradesh, Tamil Nadu, Kerala, and Gujarat. Biopesticides are also manufactured by ICAR institutes and agricultural institutions. Only 12 biopesticides were registered in India, as referred to in the 1968 Insecticides Act, including three varieties of *Bacillus thuringiensis*, including *B. thuringiensis* var. *kurstaki*, *B. thuringiensis* var. *israelensis*, and *B. thuringiensis* var. *galleriae*. Two fungal species belong to *Trichoderma*, that is, *T. harzianum* and *T. viride*. Other registered biopesticides were *Pseudomonas fluorescens*, *Beauveria bassiana*, *Cymbopogon*, *Bacillus sphaericus*, NPV of *Spodoptera litura*, and NPV of Helicoverpa armigera and neem-based pesticides, respectively (Chakraborty et al. 2023).

Commercialization of biopesticides in India is about 4.2% of global commercialized pesticides. Agriculture is the major source of income, as 58% of the population depends on it (Chaitra and Harish 2024). Biopesticides are utilized by 8% of Indian cultivators (Parvathy et al. 2024).

2.3 Commercially Utilized Fungal Agents as Biopesticides

Microorganisms are an essential component of the environment, as they help in establishing ecological farming. Microorganisms provide many ecology-related advantages, including ecological restoration, improving the persistence of plant associations, recovering vanished plant species, and modifying approaches for biodiversity preservation (Singh et al. 2018). Fungal biopesticides are an eco-friendly, excellent pest control agent. They exhibit desirable character as biological control agents. Various fungal agents are formulated, mass-produced, and commercialized as mycoinsecticides. EPF potential is exploited to employ IPM strategies. Mass cultivation methods and environmental approaches are continuously evolving to enhance their efficiency and secure continuous integration with eco-farming practices. Fungal biopesticides, merged with traditional pesticides, elevate the pest mortality rate (Mawcha 2025). In the agriculture sector, the most regularly utilized fungal agents as biopesticides in India include *Verticillium lecanii*, *Trichoderma* spp., *Metarhizium anisopliae*, *Nomuraea* sp., *Paecilomyces* sp., *Isaria fumosorosea*, *Hirsutella thompsonii*, and *Beauveria bassiana*.

Trichoderma spp. is a fungal agent that acts as a competitor for insect pests. Fungal agents are responsible for causing several diseases in plants. For survival, *Trichoderma* species compete with infectious fungi. Inhibit pathogen growth by parasitizing and releasing enzymes on them. *Trichoderma*-formulated biopesticides are applied to various types of crops, such as pulses and cereals. These biopesticides are generally implemented on the soil. *Trichoderma* supports root growth and enhances nutrient supply, thereby enhancing plant development (Pavan et al. 2024).

Beauveria bassiana is applied commercially to control a broad range of pest populations, including beetles, whiteflies, caterpillars, and aphids. Fungus causes disease in the host pest and multiplies within the host pest, releasing a toxin known as beauvericin, ultimately causing mortality (Shree et al. 2024). Mortality rate is temperature-sensitive, such as *Beauveria* spores causing mortality in aphids at 720 °F with a time duration of 3–5 days (Arora et al. 2017). Botanigard® is a commercially formulated biopesticide containing *B. bassiana* as an active fungal agent (Shree et al. 2024). *B. bassiana* is formulated in India and is applied to the soil so that the plant can uptake the microbial chemicals or biopesticides, which are directly sprayed on plants. *B. bassiana* is an essential component in a sustainable pest management program, as this fungal agent does not cause infection in non-targeted pests; it only infects specific pests (Pavan et al. 2024).

M. anisopliae is an EPF capable of infecting numerous types of pests, such as locusts, termites, and beetles. *M. anisopliae* causes disease in pests, multiplying within them, and produces isarolides A, B, C, and D, resulting in mortality. In India, *M. anisopliae*-based formulated biopesticides are utilized to control pests in agricultural fields, where these biopesticides are directly sprayed on plants or soil is supplemented with biopesticides. Met52® is a commercially formulated biopesticide containing *M. anisopliae* as an active fungal agent (Shree et al. 2024).

V. lecanii is an EPF that specifically harms soft-bodied pest populations, including scales, aphids, and whiteflies. Like other EPF, *V. lecanii* infects pests, proliferates, and causes mortality. *V. lecanii* virulence is enhanced when applied under glasshouse conditions, as in a closed chamber moisture content is high, which promotes fungal growth rate. *V. lecanii* can act as an essential tool in a sustainable pest management program, as it does not infect the non-target pest population, and it selectively infects pests. In India, crops such as tea, cotton, and vegetables are treated with biopesticides (Pavan et al. 2024). Mycotal® is a commercially formulated biopesticide containing *V. lecanii* as an active fungal agent (Shree et al. 2024).

Nomuraea rileyi is a dimorphic EPF belonging to the hyphomycete group, capable of causing mortality in diverse types of pests. *N. rileyi* is a host-specific and sustainable fungal agent utilized as a biopesticide. *N. rileyi* can infect various types of pests, such as *Heliothis zea*, *Bombyx mori*, *Anticarsia gemmatalis*, *Plathypena scabra*, *Trichoplusia ni*, and *Pseudoplusia*.

Paecilomyces fumosoroseus is the most vital enemy of whiteflies all across the world, which causes a disease known as "Yellow Muscardine." *P. fumosoroseus* causes an epidemic in *Trialeurodes* spp. and *Bemisia* spp. in the glasshouse and farmland. *P. fumosoroseus* grows on the surface of a leaf in the presence of moisture, and this is a quality that promotes proliferation among whiteflies. According to Kim et al. (2002), *P. fumosoroseus* is a fungal agent that infects whiteflies and develops mycelia all over the body and then adheres to the whiteflies' undersurface and the plant leaves (Rai et al. 2014). *Paecilomyces* spp. produces beauvericin, which causes mortality.

I. fumosorosea infects pest populations, including aphids, thrips, whiteflies, and diverse types of caterpillars. *I. fumosorosea* also targets soft-bodied pest populations. It specifically harms whiteflies that feed on vegetables and cotton crops. PFR97® is a commercially formulated biopesticide containing *I. fumosorosea* as an active fungal agent. *Isaria* spp. produces isarin, which causes mortality in pests.

The *H. thompsonii* is an EPF that harms the pest population. Specifically target arthropod mites, whereas less harmful arthropods such as honey bees remain uninfected. Mostly implemented on citrus farms to enhance citrus yield. Thereby, it can be used as an essential fungal agent in a sustainable pest management program (Shree et al. 2024).

2.4 Challenges of Biopesticide Commercialization

2.4.1 Biopesticides Registration—Regulatory Hurdles

Various types of microbial (fungi, bacteria, and viruses) biopesticide products have been designed nowadays and used for the management of pest populations. From the time of biopesticides' origin, a huge number of products have been registered and utilized in the agricultural market (Jhala et al. 2020). Biopesticide registration is growing globally due to their efficiency toward pests and their ecologically safe nature. By the year 2010, the commercialization of biopesticides was expected to surpass $1 billion (Boyetchko and Svircev 2011). Around the world, biopesticide sales were estimated to be 1.3 billion in 2011, 3 billion in 2013, and 3.2 billion in

2017, accounting for the world pesticide market. By 2023, it is expected that the market will expand to $4.5 billion. By 2040–2050, it is expected that the chemical pesticide and biopesticide marketing will be equal. The expansion of the biopesticide trade is expected to continue (Kumar et al. 2019; Akutse et al. 2020; Shukla et al. 2019). According to Wend et al. (2024), the slow adaptation of biopesticides is due to the expanded, costly, and two-phase process for biopesticide approval was detected.

Registration is the largest obstacle for biopesticide development. Mostly, production is cheap compared to the registration process. Biopesticide transportation is tougher than that of synthetic pesticides. The government can approve novel biomolecules, but the process is long and complicated. A new fungal biopesticide can be registered by submitting specific data to the environmental regulator. Before registration, the safety and potency parameters should be passed; this process can retard the launch of novel EPF-based biopesticides into the market. The government can establish rules worldwide through workshops, events, and meetings to enhance the level of bioproducts. Registration rules for biopesticide utilization and transportation vary among countries. Rules can impose uniform laws that should be accepted worldwide, and then there will be the same strategy regarding biopesticide applications (Fenibo et al. 2022; Goldstein 2024).

For biopesticides, product registration requires certain criteria, including that the fungal strain should have the potential to infect. Fungal biopesticides should not harm farmers, food consumers, and the ecosystem. Beneficial microorganisms should not be harmed. Biopesticides are decomposable; therefore, no noxious components reside in the ecosystem, and the food chain is not affected. EPF spores do not resist elevated temperatures and cannot last on vegetative leaves for long. Diseased cadavers sporulate under favorable microclimates. EPF biopesticides have various modes of pathogenesis. Many EPF-based biopesticides have been commercialized worldwide (Kesho 2020).

Policy Guideline for Biopesticides

It is crucial to apply certain policy guidelines to minimize the application of synthetic pesticides and enhance biopesticide manufacturing and perform more investigations in the field.

- Enhance sustainable farming by developing pest-protected and disease-free crops, prudent intercropping, and decreasing the practice of growing various crops on the same agricultural field.
- Enhance the IPM training intensity. Training effectiveness and the number of cultivators should be the point of focus. Instructor cultivators should provide regular instruction.
- The government bodies promote pest management strategies, and these bodies have a high influence on the University of Agriculture; therefore, more attention should be focused on fungal biopesticides.
- To increase the application of fungal biopesticides, more investigation and testing on region-specific and plant-specific formulations are required.

- Regular investment in skills for the exploration, evolution, and execution of fungal biopesticide development in commercial products is required to promote the registration of a large number of biopesticides. Government and university scientific institutes are sectors where basic fundamental research is conducted. Private and public organizations are required to instruct agriculturalists, dealers, and the general public on the applications and values of biopesticides.
- Charitable bodies should step forward to investigate new and inventive formulations of biopesticides. Negligible impact was shown by the government bodies on promoting IPM and biopesticide applications. Lack of active campaigning by the Department of Agriculture, resulting in no growth in marketing, has been enhanced due to retailers not maintaining a stock and trade of biopesticides (Yadav et al. 2022).

2.4.2 Competition with Chemical Pesticides

In the agriculture sector, EPF is a sustainable substitute for synthetic pesticides. When certain parameters of EPF are compared with synthetic pesticides, a noticeable difference is shown. Synthetic pesticides show a quick response against pests, but they harm the ecosystem, hindering growth and killing beneficial microorganisms. In comparison, EPF acts slowly, and the pest population is controlled for a long period with hardly any environmental aftereffects. Synthetic pesticides affect pest populations within a short period. However, pesticides have to be frequently applied in the field to restrict the pest population. With time, diminishing marginal productivity occurs because strong doses or more novel chemical formulations are required. Conversely, EPF acts slowly and takes time to cause pest mortality, and this is one of the biggest drawbacks, as at the time of pest attack, instant action is essential. EPF can persist in the ecosystem, and therefore, for a long period, pests can be infected and do not require periodic applications. Primarily, synthetic pesticides are cheap, easily available, and standard operating procedure. However, their application causes ecosystem damage, biodiversity loss, and pest tolerance development, creating uneconomical conditions for sustainability. Reapplication of synthetic pesticides makes them expensive. EPF mass production and formulation are more costly. EPF does not require supplementary pest control inputs and encourages a sustainable environment, enhancing the pathogen population that suppresses pests. Synthetic pesticides are considerably more impactful than EPF-based biopesticides. A small volume of synthetic pesticides can control a large population of pests; therefore, farmers prefer using synthetic pesticides as compared to biopesticides. Lack of synthetic pesticides can increase crop loss all across the world. Expensive biopesticides and their absence in the market discourage farmers from utilizing them (Vivekanandhan et al. 2024a,; Fenibo et al. 2022).

2.4.3 Ecological Conditions

Biopesticide's efficiency is influenced by several ecological factors, including exposure to UV, humidity, and temperature. These particular ecological conditions restrict the efficiency of EPF to infect pests. EPF requires high humidity conditions to invade the pest's cuticle. Low moisture level and high temperature hinder the efficiency. Biopesticide formulation manufacturing requires careful evaluation based on these factors (Devi et al. 2020; Mawcha et al. 2025; Samada and Tambunan 2020).

An immense number of EPF species populate the soil. EPF efficiency and viability in an agricultural field depend on ecological factors. Temperature influences the sprouting rate, maturation, effectiveness, and shelf life of fungal spores. Another ecological factor that affects the viability of fungal EP is humidity. Fungal spores require high humidity to obtain optimum growth. Another essential parameter includes ultraviolet rays that influence the viability of fungal conidia. UV-A (longer wavelength radiation) and UV-B (shorter wavelength radiation) influence the viability of fungal spores and decrease their effectiveness. Plant type is also a major factor, as it influences the life span of fungal conidia. Pests that reside in the soil, when they come in contact with fungal conidia under favorable ecological conditions, germinate and complete their infectious process. Freshly formed conidia released from cadavers remigrate into the soil, where spores remain retained until a new disease cycle begins (Bamisile et al. 2021).

2.4.4 Storage Life and Low Quality Hinder Commercialization

Low quality is a major issue that restricts biopesticide commercialization. Various EPF-based biopesticide products are commercialized that contain poor-quality fungal agents and are contaminated with other microbial cultures. Low performance in the agriculture field is due to the short storage life of biopesticide products and decreased demand by farmers. Improper storage management by the manufacturer, trader, and farmers reduced biopesticides' viability, creating a challenge for crop cultivators. The storage life of fungal biopesticides relies on formulation, transportation, casing material, and the mechanism of transportation. Humidity must be very low to enhance the storage life of fungal biopesticides. Therefore, the fungal agent remains inactive, protected from ecological stress, and hard to contaminate. To maintain product quality and increase storage life, it is necessary to formulate the product in an appropriate manner (Goldstein 2024; Fenibo et al. 2022).

2.4.5 Slow Action

Synthetic pesticides suppress the pest population quickly compared with fungal biopesticides. Synthetic pesticides cause pest mortality in less than a few hours, whereas EPF requires a few days or weeks to cause pest mortality, and at this time, pests continue to harm crops. Due to this lag period caused by EPF restricting farmers, they opt for a fast-action solution (Devi et al. 2020; Ruiu 2018).

2.4.6 Unawareness Among Farmers

Farmers are susceptible to the toxic chemicals exhibited by pesticides that help in controlling the pest population. These toxic chemicals cause serious health problems in farmers. Technical constraints and the absence of shielding equipment increase the chance of exposure to pesticides, and it is a major challenge faced by farmers at the time of field applications. To be aware that farmers need to handle the pesticides requires technical education and personal shielding equipment. Less awareness about pesticides among farmers increases the risk of health hazards, whereas providing knowledge to farmers can lower the risk of health-related diseases (Ahmadipour and Nakhei 2024). Lack of adaptation of biopesticides among farmers is due to less awareness, which can be enhanced by providing education and different approaches to getting information. Awareness can enhance biopesticide applications (Deshmukh and Sharma 2023). Farmers require environmental interaction, biological interaction between pest and pathogen, and special implementation strategies for optimum success (Krishnamoorthi et al. 2024).

2.4.7 Limited Knowledge of Plant-Fungal Interactions

EPF can colonize the tissues of plants asymptomatically, enhance the growth of plants, and guard the plant against biological stress as well as ecological stress. A fungal persistence time inside plant organs or tissues varies among different strains. EF can also colonize and persist in the rhizosphere soil zone where root exudates ooze out. Several types of microorganisms survive in this zone. EPF germinated from the soil has a rhizosphere as a favorable habitat. EF can exhibit pest suppression activity and can enhance plant development. EPF controls the pest population by causing infection and producing toxins. Alternative methods involve the production of metabolites in plants or activating the defense system of the plant (Quesada-Moraga et al. 2022; Mantzoukas et al. 2022). EPF plays the role of an antagonist against the pest population. Antagonism is a mechanism in which the pest disease development ability gets suppressed by producing different types of metabolites, including antibiotics, enzymes, and VOC (volatile organic compounds). Parasitism, systemic acquired resistance, competition, reduced virulence, and promotion of plant development. EPF such as *B. bassiana* and *Lecanicillium* sp. exhibit antagonism mechanisms (Shahid et al. 2012).

2.4.8 Particularity and Limited Range

Most of the microbial agents target particular species of pests or development stages. This characteristic of attacking a particular species is beneficial for storing useful microorganisms but restricts effectiveness in controlling diverse pest populations.

2.4.9 Potential of Disruption and Unintended Effect

Introducing a microbial agent into new ecological conditions can give rise to unforeseen outcomes, such as non-beneficial pest species being affected, inadvertent disruption of the natural ecosystem, or exotic species establishment.

2.4.10 Price and Expense

Manufacturing of biological pesticides is expensive, which limits their utilization by farmers in limited growing countries, and it is one of the chief hurdles for biopesticides' commercialization (Goldstein 2024). Synthetic pesticides are long-lasting; therefore, they require only a single application. Microbial pest control requires detailed analysis, formulation, and application in the agricultural field with an appropriate approach. Continuous observation and control need constant expense.

2.4.11 Incorporation with Pest Control Approaches

Pest management approaches are highly potent when incorporated with wide IPM approaches that integrate various approaches. However, the incorporation of pest management with synthetic pesticides needs strategic planning to evade conflict and maximize outcomes.

2.4.12 Modification Plans

A pest exposed to microbial control agents should be regularly observed. Progress on biotic stress should be monitored with modified plans and the management of pests.

2.4.13 Social View and Adoption

Social view of microbial control strategies, specifically those including bioengineered organisms, may affect the approval and adoption. Ecological impact, protection, and ethical factors are important for wide social adoption (Krishnamoorthi et al. 2024).

2.5 Opportunities for Commercialization

According to Krishnamoorthi et al. (2024), utilizing EPF-based biopesticides poses several opportunities, such as crop resilience and enhancing sustainable farming (Fig. 20.2).

2.5.1 Host Specificity

EPF is applied to control detrimental insects and does not harm non-detrimental pests and useful pest populations. Biopesticides do not harm mammals. EPF endures for a long period after implementation and is used for the development of new research (Güner et al. 2023).

2.5.2 Highly Efficient

EPF is highly efficient against the pest population and therefore highly utilized for biodynamic farming. Many EPFs develop scant growth on the exterior surface of the pest body, due to which the C/N ratio of the soil does not get disturbed.

2.5.3 Eco-friendly Pest Management

Humans are secure from EPF-based biopesticides, which are non-toxic to unintended pests and are ecologically safe as no harmful residue remains; therefore, contamination gets decreased and ecological health gets improved (Krishnamoorthi et al. 2024).

2.5.4 Integration of Pest Management and Sustainability

Several types of EPF can be merged with chemical pesticides to generate a synergism. EPF targets specific pests; therefore, in the IPM system, it has tremendous use. Efficient in small amounts and quickly degrades. Fungi are naturally occurring

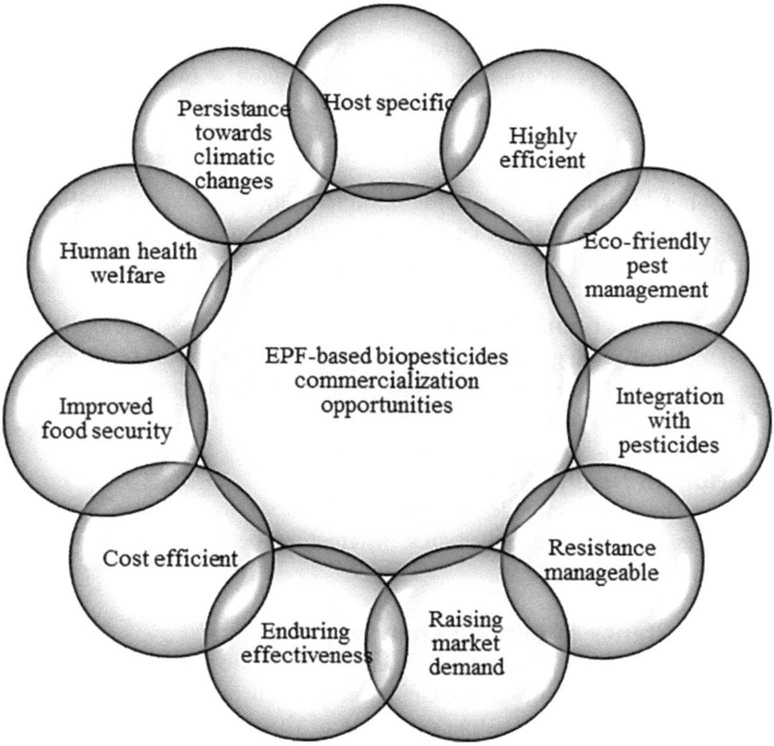

Fig. 20.2 Opportunities for commercializing EPF-based biopesticides

organisms (Misra 2014) and a sustainable resource. Fungi merged with IPM enhance sustainability by reducing synthetic substitutes and encouraging biodiversity (Phatale 2020).

2.5.5 Ecological Persistence

Plant growth does not get affected by endophytic EPF. In favorable ecological conditions, EPF can self-sustain and be inexpensive in the long term. EPF exhibits a unique mechanism of action that helps in decreasing the dependency of farmers on synthetic pesticides (Sharma et al. 2023).

2.5.6 Raising Market Demand

Fungal biopesticide demand is continuously increasing due to its ecologically safe and effective nature. Fungal agents are microorganisms that secrete metabolites, resulting in pest mortality. The global biopesticides commercialization in 2009 was $1.6 billion. In 2014, it was expected that the market would rise to $3.3 billion. The Lux Research agency estimated the trade of biopesticides in 2023 to be about $4.5 billion (Marrone 2014).

2.5.7 Enduring Effectiveness

Microbial control is an endurable solution. Beneficial fungal agents can last and keep going to control pest populations over several cropping seasons without a constant supply.

2.5.8 Cost Efficient

Low-cost labor for fungal biopesticide implementation, decreased pesticide implementation, and reduced requirement for pest control inputs promote a cost-efficient way of controlling pests.

2.5.9 Improved Food Security

Microbial control strategies can promote food yield without synthetic pesticide remains. The user chooses the yield that grows with hardly any chemical involvement, standardized with food security.

2.5.10 Human Health Welfare

By reducing chemical pesticide utilization to yield for workers and users, there are fewer chances of getting health issues, whereas biopesticides enhance health and provide a safe operating place.

2.5.11 Persistence of Climate Change

Microbial control promotes agricultural persistence in the climatic crisis. Pathogenic and beneficial fungal agents can adapt to ecological changes. Crops can tolerate different ecological actors (Krishnamoorthi et al. 2024).

2.6 Future Opportunities

Commercialization faces many challenges to overcome; it is necessary to integrate modern techniques, such as introducing genetic modification and molecular manufacturing with biopesticides, to develop sustainable farming with enhanced performance. EPF-based biopesticides are a safer substitute for synthetic chemicals; therefore, more exploration is needed to enhance effectiveness, field relevance, and stability. Future exploration must include these challenges to develop more potent and cheaper fungal biopesticides that can resist ecological conditions and enhance their performance. Chemical constituents of EPF can help in gaining knowledge about the active substance that is responsible for pesticidal activity (Vivekanandhan et al. 2024a).

2.6.1 Molecular Advancement

Researchers are investigating techniques that help in improving EPF virulence through genetic modification. Genetic modification should be done to enhance resistance mitigation strategies, ecological assessment, incorporation into sustainable farming, and synergistic relation with other pest management methods. Innovative formulations, delivery methods, fungal agents, general awareness among the public, and acceptance are necessary to enhance fungal biopesticide formulation and its commercialization. EPF-based biopesticide development by merging RNAi and CRISPR/Cas9 can enhance effectiveness, ecological safety, and specificity (Mawcha et al. 2025). RNA interference is a technique that represses or silences the expression of genes when introduced into a living agent. The CRISPR/Cas9 technique can produce pest-resistant plants by causing alterations in plant DNA. These techniques provide chances for sustainable farming (Verma et al. 2023).

2.6.2 Enhanced Formulation and Supplementary Pest Management Techniques

Formulations should be developed that enhance storage life and improve fungal conidial application and stability to control pest populations. Incorporating EPF with supplementary pest management techniques, including microbial pesticides and pheromones, can increase the efficiency of formulated biopesticides (Devi et al. 2020).

2.6.3 Novel Formulations

Nanobiopesticides (NBPs) are living agents extracted from chemical compounds with nanoparticles designed to be utilized as biopesticides. NBPs are designed using nanotechnology to enhance potency and decrease the utilization of synthetic pesticides in the ecosystem. NBPs' potency is increased by utilizing macromolecules, active micelles, metallic oxide, etc. NBPs are formulated in a small size that is immensely capable of pest population management with minimum pesticide residue and is ecologically safe (Lade et al. 2019; Adegbola et al. 2024). Nanomaterials can go through the biomembrane, attack organs, and access the bloodstream; as a result, genetic material gets damaged and chemical processes get altered (Yousef et al. 2023).

According to Sarmah et al. (2023), when pests are exposed to nanoparticles such as graphene oxide and silver, they exhibit significant effects. Enzymes get detoxified, and antioxidant substances are released, leading to cellular damage in pests. Acetylcholinesterase (enzyme) activity gets reduced by the addition of silver nanoparticles. The growth and reproduction process is disrupted in the presence of gold nanoparticles by the inhibitor trypsin. Nucleic acids and proteins contain phosphorus and sulfur that get linked with nanoparticles, resulting in a decrease in the permeability of membranes, enzymes, and organelles, which ultimately leads to cell death. Pest genes can also be elevated or suppressed in the presence of silver nanoparticles, hindering the synthesis of protein and release of gonadotropin, resulting in growth disability and an infertile reproductive system. Pests come in contact with aluminum oxide and silicon dioxide nanoparticles, resulting in dehydration.

3 Future Approaches

Nature-gifted resources are continuously being exploited by humankind for their welfare, and adverse weather conditions and chemical pesticides in agricultural fields are the main causes of crop destruction. Pest control is a dynamic method.

Pest control techniques should be cheap, preventive, and non-hazardous to the ecosystem. The government should put certain strict safety rules on synthetic pesticides, which can lower their commercialization and thereby enhance the biopesticides market. The majority of farmers should choose biopesticides; then only synthetic pesticide application will be reduced, and the ecosystem will be secure from toxic chemical components. Biopesticides have been utilized for decades to protect crops from pests, but compared to chemical inputs, biopesticide demand is a long way off. To enhance biopesticides' commercialization public and private sectors have to collaborate with farmers and work on all the challenges faced by them; then only trust among farmers can be developed (Rani et al. 2021). To enhance biopesticide adoption, the following issues should be considered.

- Massive marketing is obligatory to publicize the skill of bio-IPM approaches and sustainable farming.
- Develop appropriate quality assurance and authorization for biopesticides that are commercially accessible.
- The private and public sectors should cooperatively conduct a public education campaign among manufacturers and customers on the merits of biopesticides for ecosystem and human health.
- Develop knowledge among vendors and cultivators about the appropriate application method of fungal biopesticides and their secure storage.
- Involved in coordinating biopesticides investigation at the international as well as regional level.
- Bio-IPM research should be conducted for various types of crops.
- Develop trans-country training research programs, agricultural advisors, private enterprise, quality assurance, and biopesticide application.
- Increase government research strategies to evolve and adopt environmentally safe fungal biopesticides.
- Advise governments on how schemes and the private domain promote and boost biopesticide applications and influence farming sectors (Hossain and Pandit 2024).

4 Conclusion

Globally, farmers are utilizing synthetic pesticides for pest population management. Synthetic pesticides lower the pest population but pollute the ecosystem. Therefore, a sustainable approach should be developed that suppresses the pest population without harming the ecosystem. Incorporating farming with biopesticides represents an encouraging transition that is a sustainable and eco-friendly pest control. Biopesticides such as entomopathogenic fungi are an ecologically safer alternative to synthetic pesticides. Fungal biopesticides have a distinct benefit over other microbial agents and techniques. The major benefit of utilizing EPF-based biopesticides is that they are ecologically safe and host-specific. EPF can cause epizootics, as it can persist in soil and transmit efficiently. The development and registration cost of the EPF-based biopesticides is less than synthetic pesticides. EPF-based biopesticides can regulate the pest population for a long period. Biopesticides possess various benefits, but their commercialization also exhibits many challenges, such as unawareness among the cultivators. EPF exhibits slow action and low storage life and is influenced by ecological conditions such as temperatures, humidity, and UV exposure. However, more research is required to investigate more virulent fungal strains and formulate them to apply as effective biopesticides. More research has to be done to reduce limitations exhibited by EPF-based biopesticides. Government and private sectors should come forward to encourage and train farmers about biopesticide applications; therefore, commercialization can be supported.

Acknowledgments The authors wish to express gratitude to the Fungal Biotechnology and Invertebrate Pathology Laboratory, Department of Biological Sciences, Rani Durgavati University, Jabalpur (Madhya Pradesh, India), for providing their support and guidance throughout my work.

References

Adegbola RQ, Atanda SA, Jimoh MO, Okparavero NF, Aremu MB et al (2024) Nanobiopesticides in post-harvest management of insect pests of crops: present status, challenges and prospects review. FUDMA J Sci 8(4):40–54. https://doi.org/10.33003/fjs-2024-0804-2557

Ahmadipour H, Nakhei Z (2024) The effect of education on safe use of pesticides based on the health belief model. BMC Res Notes 17(1):134. https://doi.org/10.1186/s13104-024-06797-6

Akutse KS, Subramanian S, Maniania NK, Dubois T, Ekesi S (2020) Biopesticide research and product development in Africa for sustainable agriculture and food security—experiences from the international centre of insect physiology and ecology (*icipe*). Front Sustain Food Syst 4: 563016. https://doi.org/10.3389/fsufs.2020.563016

Arora A, Kaur P, Kumar M, Saini V (2017) Production of biopesticides, namely *Trichoderma viride* and *Beauveria bassiana*. Ind J Sci Technol 10:1–7. https://doi.org/10.17485/ijst/2017/v10i26/110796

Aslani A, Eftekhari H, Hamidi M, Nabavi B (2015) Commercialization methods of a new product/service in ICT industry: case of a science & technology park. Organ 48(2):131–138

Auld BA (1992) Development and commercialization of biocontrol agents. In: Proc. 1st Int. Weed Control Congr, vol. 1:269–272

Avinash GP, Namasivayam SKR, Bharani RA, Samrat K, Tamilselvi M (2024) Eco-friendly formulation preparation of fungal biopesticide *Metarhizium anisopliae* and its impact on the viability, insecticidal activity and pesticidal compatibility. J Environ Biol 45(6):686–694. https://doi.org/10.22438/jeb/45/6/MRN-5476

Ayilara MS, Adeleke BS, Akinola SA, Fayose CA, Adeyemi UT et al (2023) Biopesticides as a promising alternative to synthetic pesticides: a case for microbial pesticides, phytopesticides, and nanobiopesticides. Front Microbiol 14:1040901. https://doi.org/10.3389/fmicb.2023.1040901

Bajsa N, Fabiano E, Rivas-Franco F (2023) Biological control of phytopathogens and insect pests in agriculture: an overview of 25 years of research in Uruguay. Environ Sustain 6(2):121–133. https://doi.org/10.1007/s42398-023-00275-8

Bamisile BS, Siddiqui JA, Akutse KS, Ramos Aguila LC, Xu Y (2021) General limitations to endophytic entomopathogenic fungi use as plant growth promoters, pests and pathogens biocontrol agents. Plants 10:2119. https://doi.org/10.3390/plants10102119

Booth J, Schenk PM, Mirzaee H (2022) Microbial biopesticides against bacterial, fungal and oomycete pathogens of tomato, cabbage and chickpea. Appl Microbiol 2(1):288–301. https://doi.org/10.3390/applmicrobiol2010021

Boyetchko SM, Svircev AM (2011) Canadian biopesticides and bioherbicides. In: Proceedings of the soils and crops workshop. University of Saskatchewan, Saskatoon, p 13

Chaitra G, Harish D (2024) A study on the perception of farmers about biopesticides in Karnataka. J Farm Sci 37(01):80–82. https://doi.org/10.61475/JFS.2024.v37i1.18

Chakraborty N, Mitra R, Pal S, Ganguly R, Acharya K et al (2023) Biopesticide consumption in India: insights into the current trends. Agriculture 13(3):557. https://doi.org/10.3390/agriculture13030557

Chiew BFJ, Ong GH, Wong RR, Wong KK, Loh KE (2022) Safeness and effectiveness of entomopathogenic fungi for use as bioinsecticide: a mini review. J Biol Control 36(1):1–6. https://doi.org/10.18311/jbc/2022/30415

Clifton EH, Jaronski ST, Hajek AE (2020) Virulence of commercialized fungal entomopathogens against Asian longhorned beetle (Coleoptera: Cerambycidae). J Insect Sci 20(2):1. https://doi.org/10.1093/jisesa/ieaa006

De Jesus Seabra MG, Santos TS, de Souza Varize C, Souto EB, Severino P et al (2024) New *Beauveria bassiana* aerial conidia-based bioinsecticide obtained by spray-dried microencapsulation of the entomopathogenic fungi in biopolymers for crop protection. Egypt J Biol Pest Control 34:58. https://doi.org/10.1186/s41938-024-00825-0

De Sousa TK, da Silva AT, de Freitas Soares FE (2025) Fungi-based bioproducts: a review in the context of one health. Pathogens 14(5):463. https://doi.org/10.3390/pathogens14050463

Deshmukh R, Sharma B (2023) Empirical study on usage of biopesticides by farm growers in Kolhapur region. Seybold Rep 18(5):581–593. https://doi.org/10.17605/OSF.IO/WUZ4C

Devi M, Shekhawat SS, Ishika (2020) Use of entomopathogenic fungi in pest control. Just Agric 5(3):67–71

Elbashir MI, Bishwajeet P, Shankarganesh K, Gautam RD, Sharma P (2014) Pathogenicity of Indian isolates of entomopathogenic fungi against important insect pests and natural enemies. Ind J Entomol 76:37–43

Fenibo EO, Ijoma GN, Matambo T (2022) Biopesticides in sustainable agriculture: current status and future prospects. In: Mandal SD, Ramkumar G, Karthi S, Jin F (eds) New and future development in biopesticide research: biotechnological exploration. Springer, Singapore, pp 1–53. https://doi.org/10.1007/978-981-16-3989-0_1

Goldstein J (2024) Advancements in fungal biopesticides: innovative solutions for sustainable crop protection. Virol Myco 13(4):1000305. https://doi.org/10.35248/2161-0517.24.13.305

Gull S, Tariq A, Khanday AL, Meethal SP, Rashid G (2023) Pathogenicity of the entomopathogenic fungi against *Myllocerus fotedari* Ahmad, 1974 (Coleoptera: Curculionidae) under laboratory conditions in India. J For Sci 69(7):277–286. https://doi.org/10.17221/10/2023-JFS

Güner P, Aşkun T, Er A (2023) Entomopathogenic fungi and their potential role in the sustainable biological control of storage pests. Commagene J of Biol 7(1):90–97. https://doi.org/10.31594/commagene.1284354

Hossain ST, Pandit T (2024) Applications and commercialization of biopesticides in Bangladesh. FFTC J Agric Policy 7:22–31

Jeranyama P, Shrestha A, Neupane N (2020) Sustainable food systems: diversity, scope and challenges. In: Rusinamhodzi L (ed) The role of ecosystem services in sustainable food systems. Academic, pp 1–16. https://doi.org/10.1016/B978-0-12-816436-5.00001-9

Jhala J, Baloda AS, Rajput VS (2020) Role of bio-pesticides in recent trends of insect pest management: a review. J Pharmacogn Phytochem 9(1):2237–2240

Kelwatkar NM, Kankale MD, Das SB (2018) Identification and evaluation of economic substrate for mass multiplication of *Metarhizium anisopliae*. J Entomol Zool Stud 6(3):622–627

Kesho A (2020) Microbial bio-pesticides and their use in integrated pest management. Chem Biomol Eng 5(1):26–34. https://doi.org/10.11648/j.cbe.20200501.15

Kim JJ, Lee MH, Yoon CS, Kim HS, Yoo JK et al (2002) Control of cotton aphid and greenhouse whitefly with a fungal pathogen. J Natl Inst Agric Sci Technol 2002:7–14

Kim M, Park H, Sawng YW, Park SY (2019) Bridging the gap in the technology commercialization process: using a three-stage technology–product–market model. Sustainability 11(22):6267. https://doi.org/10.3390/su11226267

Kothari M, Sandhu SS (2025) Entomopathogenic perspectives of *Nomuraea rileyi*. In: Deshmukh SK, Sridhar KR (eds) The role of entomopathogenic fungi in agriculture. CRC Press, pp 140–161

Krishnamoorthi A, Patel PK, Kumar S, Khambhu CV, Kumar V et al (2024) The future of farming with advances in biological control techniques for crop health. Microbiol Res J Int 34(7): 93–112. https://doi.org/10.9734/mrji/2024/v34i71461

Kumar KK, Sridhar J, Murali-Baskaran RK, Senthil-Nathan S, Kaushal P et al (2019) Microbial biopesticides for insect pest management in India: current status and future prospects. J Invertebr Pathol 165:74–81. https://doi.org/10.1016/j.jip.2018.10.008

Kumar S, Kumar S, Bhandari D, Gautam MP (2020) Entomopathogens, pathological symptoms and their role in present scenario of agriculture: a review. Int J Curr Microbiol App Sci 9(12): 2110–2124

Lade BD, Gogle DP, Lade DB, Moon GM, Nandeshwar SB et al (2019) Nanobiopesticide formulations: application strategies today and future perspectives. In: Koul O (ed) Nanobiopesticides today and future perspectives. Academic, pp 179–206. https://doi.org/10.1016/B978-0-12-815829-6.00007-3

Mora MAE, Castilho AMC, Fraga ME (2017) Classification and infection mechanism of entomopathogenic fungi. Arq Inst Biol 84:e0552015. https://doi.org/10.1590/1808-1657000552015

Malo M (2020) Sustainable agriculture: need of the hour. Agric Food E Newsl 2:298–300

Mantzoukas S, Kitsiou F, Natsiopoulos D, Eliopoulos PA (2022) Entomopathogenic fungi: interactions and applications. Encyclopedia 2(2):646–656. https://doi.org/10.3390/encyclopedia2020044

Marrone PG (2014) The market and potential for biopesticides. In: Biopesticides: state of the art and future opportunities. American Chemical Society, pp 245–258. https://doi.org/10.1021/bk-2014-1172.ch016

Mawcha KT, Malinga L, Muir D, Ge J, Ndolo D (2025) Recent advances in biopesticide research and development with a focus on microbials. F1000Res 13:1071. https://doi.org/10.12688/f1000research.154392.5

Misra HP (2014) Role of botanicals, biopesticides, and bioagents in integrated pest management

Nanda S, Mondal T, Yadav SK, Kumar G (2021) Biopesticides and their encapsulation techniques: current updates and future prospective. Int J Theor Appl Sci 13(1):64–67

Opender Koul OK (2012) Microbial biopesticides: opportunities and challenges. Cabi Rev 2011:1–26. https://doi.org/10.1079/PAVSNNR2011605

Parker BL, Skinner M, Brownbridge M, Mustapha El-Bouhssini MEB (2000) Control of insect pests with entomopathogenic fungi. Arab J Plant Prot 18(2):133–138

Parvathy U, Redhu NS, Taunk J, Yashveer S, Panwar A (2024) Biopesticides in agriculture. GreenariA 02(11):106–109

Parveen SS, Rashtrapal PS (2024) Integrated pest management strategies using endophytic entomopathogenic fungi. Plant Sci Today 11(1):568–574. https://doi.org/10.14719/pst.2740

Pavan JS, Vireesha P, Chaitanya M, Raja GP, Komal M (2024) Biopesticides in Indian agriculture: advances, challenges, and pathways to sustainable crop protection. Int J Adv Biochem Res SP-8 (8):936–943. https://doi.org/10.33545/26174693.2024.v8.i8Sm.1966

Phatale PN (2020) The bright side of bio-pesticides generating business opportunities. IJIIRD 04(02):207–212

Pradhan PP, Sahoo S, Nayak U, Sarangi S (2024) Scope of Entomopathogenic fungi in the mustard ecosystem: a review. Agric Sci Dig:1–8. https://doi.org/10.18805/ag.D-6069

Prasad CS, Pal R (2014) Mass production and economics of entomopathogenic fungus, Beauveria bassiana, *Metarhizium anisopliae* and *Verticillium lecanii* on agricultural and industrial waste. Sch J Agric Vet Sci 1(1):28–32. https://doi.org/10.36347/sjavs.2014.v01i01.005
Quesada-Moraga E, Garrido-Jurado I, Yousef-Yousef M, González-Mas N (2022) Multitrophic interactions of entomopathogenic fungi in biocontrol. BioControl 67(5):457–472. https://doi.org/10.1007/s10526-022-10163-5
Rai D, Updhyay V, Mehra P, Rana M, Pandey AK (2014) Potential of entomopathogenic fungi as biopesticides. Indian J Sci Res Technol 2(5):7–13
Rajamani M, Negi A (2020) Biopesticides for pest management. In: Venkatramanan V, Shah S, Prasad R (eds) Sustainable bioeconomy: pathways to sustainable development goals. Springer, Singapore, pp 239–266. https://doi.org/10.1007/978-981-15-7321-7_11
Rani AT, Kammar V, Keerthi MC, Rani V, Majumder S et al (2021) Biopesticides: an alternative to synthetic insecticides. In: Bhatt P, Gangola S, Udayanga D, Kumar G (eds) Microbial technology for sustainable environment. Springer, Singapore. https://doi.org/10.1007/978-981-16-3840-4_23
Ruiu L (2018) Microbial biopesticides in agroecosystems. Agronomy 8(11):235. https://doi.org/10.3390/agronomy8110235
Samada LH, Tambunan US (2020) Biopesticides as promising alternatives to chemical pesticides: a review of their current and future status. Online J Biol Sci 20(2):66–76. https://doi.org/10.3844/ojbsci.2020.66.76
Sandhu SS, Sharma AK, Beniwal V, Goel G, Batra P et al (2012) Myco-biocontrol of insect pests: factors involved, mechanism, and regulation. J Pathog 2012(1):126819. https://doi.org/10.1155/2012/126819
Sarmah K, Borah R, Boruah S, Sarmah D (2023) Nanopesticides: revolutionizing pest management with nanotechnology. Biol Forum-Int J 15(7):210–218
Schilling MA (2008) Strategic management of technological innovation, vol 2. McGraw-Hill/Irwin, New York
Shahid AA, Rao QA, Bakhsh A, Husnain T (2012) Entomopathogenic fungi as biological controllers: new insights into their virulence and pathogenicity. Arch Biol Sci 64(1):21–42. https://doi.org/10.2298/ABS1201021S
Sharma A, Sharma S, Yadav PK (2023) Entomopathogenic fungi and their relevance in sustainable agriculture: a review. Cogent Food Agric 9(1):2180857. https://doi.org/10.1080/23311932.2023.2180857
Shree D, Jarpla M, Bandhavi HL, Manjula (2024) Entomopathogenic fungi: their role in controlling insect pests and disease suppression. Int J Adv Biochem Res 8(8):1435–1440. https://doi.org/10.33545/26174693.2024.v8.i8r.2033
Shukla N, Singh EANA, Kabadwa BC, Sharma R, Kumar J (2019) Present status and future prospects of bio-agents in agriculture. Int J Curr Microbiol App Sci 8(4):2138–2153
Singh RP, Singh PK, Gupta R, Singh RL (2018) Biotechnological tools to enhance sustainable production. In: Singh RL, Mondal S (eds) Biotechnology for sustainable agriculture, emerging approaches and strategies. Academic, pp 19–66. https://doi.org/10.1016/B978-0-12-812160-3.00002-7
Skinner M, Parker BL, Kim JS (2014) Role of entomopathogenic fungi in integrated pest management. In: Abrol DP (ed) Integrated pest management, current concepts and ecological perspective. Academic, pp 169–191. https://doi.org/10.1016/B978-0-12-398529-3.00011-7
Soyel SA, Ruidas S, Roy P, Mondal S, Bhattacharyya S et al (2022) Biopesticides as eco-friendly substitutes to synthetic pesticides: an insight of present status and future prospects with improved bio-effectiveness, self-lives, and climate resilience. IJESP 2(2):1–12. https://doi.org/10.35745/ijesp2022v02.02.0001
Srinivasan R, Sevgan S, Ekesi S, Tamò M (2019) Biopesticide-based sustainable pest management for safer production of vegetable legumes and brassicas in Asia and Africa. Pest Manag Sci 75(9):2446–2454. https://doi.org/10.1002/ps.5480

Ulrich KT, Eppinger SD, Yang MC (1995) Product design and development, vol 384. McGraw-Hill, New York

Vaghasiya PD, Chouhan BP, Nair A (2021) Evaluation of various parameters in mass multiplication of Beauveria bassiana in modified method. Int J Environ Agric Res 7:29–34

Verma NS, Kuldeep DK, Chouhan M, Prajapati R, Singh SK (2023) A review on eco-friendly pesticides and their rising importance in sustainable plant protection practices. Int J Plant Soil Sci 35(22):200–214. https://doi.org/10.9734/ijpss/2023/v35i224126

Vivekanandhan P, Alford L, Krutmuang P (2024a) Editorial: role of entomopathogenic fungi in sustainable agriculture. Front Microbiol 15:1504175. https://doi.org/10.3389/fmicb.2024.1504175

Vivekanandhan P, Swathy K, Sarayut P, Patcharin K (2024b) Classification, biology and entomopathogenic fungi-based management and their mode of action against *Drosophila* species (Diptera: Drosophilidae): a review. Front Microbiol 15:1443651. https://doi.org/10.3389/fmicb.2024.1443651

Wend K, Zorrilla L, Freimoser FM, Gallet A (2024) Microbial pesticides–challenges and future perspectives for testing and safety assessment with respect to human health. Environ Health 23(1):49. https://doi.org/10.1186/s12940-024-01090-2

Yadav R, Singh S, Sing AN (2022) Biopesticides: current status and future prospects. Proc Int Acad Ecol Environ Sci 12(3):211–233

Yousef HA, Fahmy HM, Arafa FN, Abd Allah MY, Tawfik YM et al (2023) Nanotechnology in pest management: advantages, applications, and challenges. Int J Trop Insect Sci 43:1387–1399. https://doi.org/10.1007/s42690-023-01053-z

MIX
Papier aus verantwortungsvollen Quellen
Paper from responsible sources
FSC® C105338

If you have any concerns about our products,
you can contact us on
ProductSafety@springernature.com

In case Publisher is established outside the EU,
the EU authorized representative is:
Springer Nature Customer Service Center GmbH
Europaplatz 3, 69115 Heidelberg, Germany

Printed by Libri Plureos GmbH
in Hamburg, Germany